Mathematical
Surveys
and
Monographs

Volume 216

An Introduction to the Theory of Higher-Dimensional Quasiconformal Mappings

Frederick W. Gehring
Gaven J. Martin
Bruce P. Palka

American Mathematical Society
Providence, Rhode Island

2010 *Mathematics Subject Classification.* Primary 30C65, 30C62.

Library of Congress Cataloging-in-Publication Data

Names: Gehring, Frederick W. | Martin, Gaven J. | Palka, Bruce P.
Title: An introduction to the theory of higher-dimensional quasiconformal mappings / Frederick W. Gehring, Gaven J. Martin, Bruce P. Palka.
Description: Providence, Rhode Island : American Mathematical Society, [2017] | Series: Mathematical surveys and monographs ; volume 216 | Includes bibliographical references and index.
Identifiers: LCCN 2016029235 | ISBN 9780821843604 (alk. paper)
Subjects: LCSH: Quasiconformal mappings. | Conformal mapping. | Mappings (Mathematics) | AMS: Functions of a complex variable – Geometric function theory – Quasiconformal mappings in $\mathbf{R}^n$. msc | Functions of a complex variable – Geometric function theory – Quasiconformal mappings in the plane. msc
Classification: LCC QA360 .G437 2016 | DDC 515/.93–dc23 LC record available at https://lccn.loc.gov/2016029235

---

10 9 8 7 6 5 4 3 2 1 22 21 20 19 18 17

To the memory of Fred Gehring, advisor and friend,
and to our partners Lois Gehring, Dianne Brunton, and Mary Ann Palka

Contents

Preface

This book presents a fairly comprehensive account of the modern theory of quasiconformal mappings in Euclidean n-space for $n \geq 2$, starting from the elementary theory of conformal mappings and building towards the more general aspects by carefully developing the necessary analytic and geometric tools. This book is primarily aimed at graduate students and researchers who seek to understand quasiconformal mappings, particularly in three or more dimensions, perhaps after having seen applications of the two-dimensional theory in Teichmüller spaces of Riemann surfaces, or in conformal dynamical systems and elsewhere. However, as we carefully develop most of the necessary analytic theory only a basic background course in multi-dimensional real analysis is assumed.

The theory of quasiconformal mappings seeks to generalise the remarkable geometric and analytic theory of conformal mappings in the plane to higher dimensions. This is since Liouville's rigidity theorem implies an extreme paucity—a finite dimensional family—of conformal mappings defined on domains $\Omega \subset \mathbb{R}^n$, $n \geq 3$. Of course in two dimensions the conformal mappings of a domain form an infinite-dimensional family and one has the Riemann mapping theorem. The reasons for seeking this generalisation are manifold with wide application. For instance in the theory of partial differential equations, quasiconformal mappings preserve the ellipticity of second order equations of divergence type—those with the widest application in physics—so the solution to mapping problems enables the transfer of equations from one domain to another, potentially nicer, domain where a solution might be found.

In higher dimensions few manifolds admit a conformal structure, yet D. Sullivan has shown that every topological manifold admits a quasiconformal structure, that is, a covering with quasiconformal local coordinate charts. This presents the opportunity to compute analytic invariants on a topological manifold or to compute topological invariants analytically—for instance in the work of A. Connes, D. Sullivan, and N. Teleman. Unfortunately we will only touch on these deep applications in this work. Nevertheless the reader will find—for the first time in book form—a solid foundation to explore these remarkable results and applications.

We approach the theory of quasiconformal mappings from the geometric point of view, using conformal invariants such as the moduli of curve families and capacities. These ideas are of independent interest and again of wide utility in many areas of mathematics, and so we give a fairly thorough account of them.

We begin by developing the basics of the theory—including the study of conformal mappings in space, elementary aspects of higher-dimensional hyperbolic geometry and its isometries, along with the associated matrix groups. This leads quickly to the celebrated rigidity theorem of Liouville for smooth mappings, the proof for which follows an argument of Nevanlinna. To get Liouville's theorem in complete generality, more theory—in particular the theory of conformal modulus—is developed. The geometric aspects of the theory of quasiconformal mappings rely to a great deal on understanding and estimating these conformal invariants. Indeed the very definition of a quasiconformal mapping here is via the distortion of moduli by a multiplicative factor.

We then consider deeper properties of conformal modulus such as symmetrisation, continuity, the structure of sets of capacity zero and the existence and uniqueness of extremal functions. These give us powerful tools to study quasiconformal mappings which enable us to not only establish analytic properties, but also to develop the compactness and normal family properties of sequences of quasiconformal mappings.

We then turn our attention to the mapping problem in its various forms, basically seeking a higher-dimensional version of the Riemann mapping theorem for the class of quasiconformal mappings. We present the classical geometric obstructions to existence and then turn to positive results. We give a proof for the Schoenflies theorem in the quasiconformal category and subsequently give a fairly complete proof of Väisälä's mapping theorem for cylindrical domains, perhaps the best result to date answering this question.

We then present the sophisticated and important work of Tukia-Väisälä developed using Sullivan's machinery. In particular we give a proof for their solution of the lifting problem. Many of the last chapters of this book—part of a central theme in the area to develop quasiconformal versions of classical theorems in geometric topology—have never previously appeared in book form. Indeed many aspects of the approach to the theory given here are novel among recent monographs on the subject, as these primarily focus on the analytic approach through the associated nonlinear partial differential equations and differential inequalities.

We close with a presentation of the Mostow rigidity theory, one of the most compelling and important applications of the higher-dimensional theory of quasiconformal mappings. We take a fairly roundabout approach here so as to be able to clearly exhibit the remarkable interaction between quasiconformal theory, hyperbolic geometry, and modern aspects of geometric group theory. In particular we give a fairly comprehensive discussion of quasi-isometries and isomorphisms of hyperbolic groups.

During the long gestation of this book the first-named author Fred Gehring passed away. He was of course a major figure in the area, and much of the important work presented in this book is due to him and his coauthors. He is sadly missed.

It is our pleasure to acknowledge the wide-ranging support we have had from a number of places that has made this book possible. We have all been partly supported by the Academy of Finland, the Marsden Fund of New Zealand, and the National Science Foundation of the United States at one time or another. Also the Aalto Science Institute deserves thanks for providing the time and support needed to finally complete this project.

We would also like to thank the team at the American Mathematical Society (in particular Ina Mette who tirelessly pressed us to complete) who skillfully guided us through the production process and whose considerable efforts improved this book.

Gaven Martin and Bruce Palka

Auckland and Washington, 2015.

CHAPTER 1

Introduction

The geometric theory of functions in higher dimensions is largely concerned with generalisations to $\mathbb{R}^n$, $n \geq 3$, of various aspects of the theory of analytic functions and conformal mappings in two-dimensions - particularly the geometric and function theoretic properties. It also seeks to identify the often stark contrasts between planar theory and what happens in higher dimensions. In the latter case the most well known of these is the rigidity theorem of J. Liouville, dating back to 1850 [**100**], but there are others such as V.A. Zorich's local to global injectivity theorems from the 1960s [**171, 172**] and their various generalisations which can be found in the foundational work of O. Martio, S. Rickman, and J. Väisälä, [**109, 110, 111**]. Because of the rigidity theorem, established in its most useful form in the 1970s independently and by different methods by F.W. Gehring [**46**] and Yu. G. Reshetnyak [**140, 141**], the class of conformal mappings in dimension $n \geq 3$ is quite meagre, and, as we will see in Section 2.6, it is a finite-dimensional Lie group. The Liouville theorem implies that in dimension $n \geq 3$ any domain conformally equivalent to the unit ball is a round ball or half-space. In comparison, the Riemann mapping theorem (first stated by Riemann for smoothly bounded domains in 1851 in his PhD thesis) shows that in two dimensions conformal mappings are quite plentiful and certainly form an infinite-dimensional family.

Researchers have a wide variety of motivations to overcome this lack of flexibility in the higher-dimensional geometric theory of mappings with the steadfast desire to retain as much as possible of the powerful theory of conformal mappings and their interaction and application with other areas of mathematics, particularly geometry, topology and analysis.

Infinitesimally conformal mappings have no distortion, for pointwise the derivative of a conformal mapping is a scalar multiple of an orthogonal transformation. This is expressed by the Cauchy–Riemann equations, and it is remarkable how little regularity is needed for the solutions to these equations to recoup the full beauty of the theory of analytic functions. The category of maps that one usually considers in the higher-dimensional theory are the *quasiconformal* mappings—introduced by H. Grötzsch in 1928 [**66**] and named by L.V. Ahlfors in 1935 in his seminal work for which he was awarded the Fields Medal [**2, 4**]. Both Ahlfors and Grötzsch were only concerned with the two-dimensional theory at that time, however their geometric ideas and techniques had clear generalities to higher dimensions. It was not until the 1950s that the higher-dimensional theory of quasiconformal mappings was initiated in earnest - perhaps Ahlfors' 1953 paper [**5**] marks this point. About five years later D.E. Callender [**27**] followed R. Finn and J. Serrin [**38**] in establishing Hölder continuity estimates, and therefore equicontinuity, for higher-dimensional quasiconformal mappings. These early beginnings were taken up in earnest and the modern theory was born. Its connections with geometry and nonlinear partial differential equations were first explored in the work of Yu G. Reshetnyak (USSR), F.W. Gehring (USA), and J. Väisälä (Finland) in the early 60s, probably under the urging and watchful eye of Ahlfors, a Finn at Harvard where Gehring had been a postdoc. We note that one of the most famous applications of the theory of higher-dimensional quasiconformal mappings, Mostow's rigidity theorem from 1967 [**125**], came quite soon after the basic foundations were laid, and five years after Gehring's proof of the Liouville theorem without any a priori assumptions—a key ingredient of Mostow's proof.

It should be noted that quasiconformal mappings solve natural partial differential equations closely, analogous to the familiar Cauchy–Riemann and Beltrami equations of the plane. The primary difference is that in higher dimensions these equations necessarily become nonlinear and overdetermined—and from this stems a certain amount of rigidity. Other desirable properties for a theory of the geometry of mappings are that they should preserve the natural Sobolev spaces which arise in consideration of the function theory on subdomains of $\mathbb{R}^n$, or more generally on n-dimensional manifolds. It is therefore important to prove—in any monograph such as this—that quasiconformal mappings do have these properties. This is not an altogether trivial endeavour, and some quite sophisticated analysis is needed. A good part of this book will be taken up establishing these basics.

Quasiconformal mappings provide a class of deformations which lie *between* homeomorphisms and diffeomorphisms, but which enjoy compactness properties that neither of these families do. The compactness properties of families of quasiconformal mappings make them ideal tools for solving various problems in n-dimensional analysis, for instance in the study of energy minimising deformations of elastic bodies and the related extremals for variational integrals. Here we give a thorough account of these compactness theorems for the reason that so much of the potential applications of the theory of higher-dimensional quasiconformal mappings in nonlinear elasticity and materials science has yet to be fully explored. In fact it is fair to say that these connections and applications are still in their infancy. However many of these problems are not simple at all and researchers are only coming to grips with them in higher dimensions; see for instance the recent work of T. Iwaniec and J. Onninen, [**86, 87**] and the references therein.

There are several good surveys of the mathematical theory of quasiconformal mappings. For instance [**13**] gives a list of the invited talks at the International Congress of Mathematics on the subject in its quite extensive bibliography, while F.W. Gehring's paper [**48**] is an outstanding survey of the area up to 2005 and the Handbook and Bibliography [**93**] of R. Kühnau is in general a valuable reference for anyone working in the area of geometric function theory.

There are still further generalisations and applications of these ideas in the recent field of geometry and analysis on metric spaces. The connections among this theory, the nonlinear potential theory, and the higher-dimensional theory of quasiconformal mappings was pioneered by J. Heinonen and P. Koskela [**76**] and is a very active area of research today. However to fully appreciate these generalisations and why they are made, a good understanding of the classical theory is necessary and the aim of this book is to provide it.

Below is a sample of the successful and diverse applications that the modern higher-dimensional theory of quasiconformal mappings touches upon—in no particular order:

- the theory of higher-dimensional conformal invariants and moduli;
- compactness, equicontinuity, and local to global distortion estimates;
- the Liouville theorem and other stability and rigidity phenomena;
- improved regularity and higher integrabilty;
- Mostow rigidity—uniqueness of hyperbolic structures ($n \geq 3$);
- Sullivan's uniformisation theorem—the existence of quasiconformal structures on topological n-manifolds ($n \neq 4$);
- Rickman's versions of the Picard theorem and Nevanlinna theory;
- applications in nonlinear partial differential equations;
- the nonlinear potential theory, A-harmonic functions, and nonlinear elasticity;
- Tukia-Väisälä's "quasiconformal geometric topology";
- quasiconformal group actions, geometric group theory, and hyperbolic groups;
- Donaldson and Sullivan's "quasiconformal Yang-Mills theory";
- Painlevé type theorems and the structure of singularities;
- quasiconformal maps in metric spaces with controlled geometry;
- analysis and geometric measure theory in metric spaces.

Mindful of the readership of a book such as this, we will not strive to present the most recent and complex results. However, after familiarisation with the many techniques developed here the reader should be in a good position to approach this more advanced literature.

Finally, so as not to forget the past, we cannot fail to mention the classic texts in the area and some of the more recent and relevant books. For the two-dimensional theory—really a part of modern complex analysis—we have Ahlfors, *Lectures on quasiconformal mappings*, 1966, [**3**] as well as O. Lehto and K. Virtanen, *Quasiconformal mappings in the plane*, 1973, [**97**] and the recent monograph of K. Astala, T. Iwaniec, and G. Martin, [**15**].

In higher dimensions there is Väisälä's *Lectures on n-dimensional quasiconformal mappings*, 1971, [**160**], from which many of us learnt the basics of higher-dimensional theory. Further aspects of the geometric and conformal invariants' point of view are M. Vuorinen's 1988 lecture notes [**165**] and his book with G. Anderson and M. Vamanamurthy, [**12**], from 1997. In those books there is a wealth of technical details and calculations concerning the special functions—for instance the distortion function which we will come to define—that are necessary if one is to understand the best constant or the precise asymptotics of general theorems. From the analytic point of view, viewing quasiconformal mappings as solutions to nonlinear partial differential equations with higher regularity properties, there is the recent research monograph of T. Iwaniec and G. Martin [**84**] from 2001.

CHAPTER 2

Topology and Analysis

In this chapter we collect some very basic results from topology, linear algebra, and analysis which will be well known to most. This allows us to set up fairly standard notations which will be used throughout the text.

2.1. Euclidean n-space

The symbol $\mathbb{R}$ designates the field of real numbers. For $n \geq 1$ the set of ordered n-tuples $x = (x_1, x_2, \ldots, x_n)$ of real numbers is denoted by $\mathbb{R}^n$ (naturally, $\mathbb{R}^1 = \mathbb{R}$). With the operations of addition and scalar multiplication defined in a coordinatewise fashion, $\mathbb{R}^n$ becomes an n-dimensional vector space over $\mathbb{R}$; the vectors

$$e_1 = (1, 0, \ldots, 0)\ ,\ e_2 = (0, 1, 0, \ldots, 0)\ ,\ \ldots\ ,\ e_n = (0, \ldots, 0, 1)$$

furnish a basis for $\mathbb{R}^n$ over $\mathbb{R}$, its *standard basis*. For x and y in $\mathbb{R}^n$, the Euclidean inner product $\langle x, y \rangle$ and the Euclidean norm $|x|$ are defined by the formulae

$$\langle x, y \rangle = x_1 y_1 + x_2 y_2 + \cdots + x_n y_n, \qquad |x| = \sqrt{\langle x, x \rangle} = \sqrt{x_1^2 + x_2^2 + \cdots + x_n^2}\,.$$

We use $\theta(x, y)$ to indicate the Euclidean measurement in $[0, \pi]$ of the angle between nonzero vectors x and y in $\mathbb{R}^n$:

$$\theta(x, y) = \arccos\left(\frac{\langle x, y \rangle}{|x|\,|y|}\right)\,.$$

The vector space $\mathbb{R}^n$, outfitted with the extra geometric structure that stems from the Euclidean inner product, is called a *Euclidean n-space*. We employ the following notation for Euclidean balls and spheres: for x in $\mathbb{R}^n$ and $0 < r < \infty$,

$$\begin{aligned} B^n(x, r) &= \{y \in \mathbb{R}^n : |y - x| < r\}, \\ \overline{B}^n(x, r) &= \{y \in \mathbb{R}^n : |y - x| \leq r\}, \\ \mathbb{S}^{n-1}(x, r) &= \{y \in \mathbb{R}^n : |y - x| = r\}. \end{aligned}$$

We also make use of the abbreviations $B^n(r) = B^n(0, r)$, $\overline{B}^n(r) = \overline{B}^n(0, r)$, $\mathbb{S}^{n-1}(r) = \mathbb{S}^{n-1}(0, r)$, $B^n = B^n(1)$, $\overline{B}^n = \overline{B}^n(1)$, $\mathbb{S}^{n-1} = \mathbb{S}^{n-1}(1)$.

In case $m < n$ we shall routinely treat $\mathbb{R}^m$ as a subset of $\mathbb{R}^n$ by identifying the former with a specific linear subspace of the latter, that being $\{x \in \mathbb{R}^n : x_i = 0$ for $i \geq m + 1\}$.

The formula $\operatorname{dist}(x, y) = |x - y|$ defines a distance in $\mathbb{R}^n$. Associated with d is a metric topology, the *standard topology* of $\mathbb{R}^n$: the collection of balls $B^n(x, r)$ $(0 < r < \infty)$ constitutes a base for the neighbourhoods of any point x of $\mathbb{R}^n$. If A and B are nonempty subsets of $\mathbb{R}^n$, then $\operatorname{diam}(A)$ and $\operatorname{dist}(A, B)$ denote the

Euclidean diameter of A and the Euclidean distance between A and B, respectively; thus

$$\operatorname{diam}(A) = \sup\{|x-y| : x, y \in A\}, \qquad \operatorname{dist}(A, B) = \inf\{|x-y| : x \in A\ ,\ y \in B\}.$$

We shorten $\operatorname{dist}(\{x\}, B)$ to the more compact $\operatorname{dist}(x, B)$.

2.2. Möbius n-space

Throughout much of this book we shall want to work not only in the Euclidean space $\mathbb{R}^n$, but also in its one-point compactification, the *Möbius space* $\hat{\mathbb{R}}^n = \mathbb{R}^n \cup \{\infty\}$.

To be precise, $\hat{\mathbb{R}}^n$ is the topological space created by adjoining to $\mathbb{R}^n$ an ideal point ∞ and taking as a base for the neighbourhoods of any point x of $\hat{\mathbb{R}}^n$ the balls $B^n(x, r)$, $0 < r < \infty$. Here

$$B^n(\infty, r) = \{x \in \mathbb{R}^n : |x| > r^{-1}\} \cup \{\infty\}.$$

Endowed with this topology, the set $\hat{\mathbb{R}}^n$ becomes a compact Hausdorff space that contains $\mathbb{R}^n$ as a dense, open subset. As for Euclidean spaces, we regard $\hat{\mathbb{R}}^m$ as a subset of $\hat{\mathbb{R}}^n$ whenever $m < n$.

For a subset A of $\hat{\mathbb{R}}^n$, we use ∂A, $\bar{A}$, and $\operatorname{int}(A)$ to denote the boundary, closure, and interior of A in the topology of $\hat{\mathbb{R}}^n$. (This statement applies even in the event that A lies in $\mathbb{R}^n$. In fact, barring an explicit indication to the contrary, the reader should interpret any remark in this book touching upon topology in n-space, be it $\mathbb{R}^n$ or $\hat{\mathbb{R}}^n$, as referring to the topology of $\hat{\mathbb{R}}^n$.)

We most often write A^c for $\hat{\mathbb{R}}^n \setminus A$, the complement of A.

There is a well-known homeomorphism $\pi : \hat{\mathbb{R}}^n \to \mathbb{S}^n$, where the sphere $\mathbb{S}^n$ has its usual (= Euclidean metric) topology. The function π in question is the *stereographic projection* of $\hat{\mathbb{R}}^n$ onto $\mathbb{S}^n$, defined as follows: $\pi(\infty) = e_{n+1}$, and for x in $\mathbb{R}^n$ (viewed as a subset of $\mathbb{R}^{n+1}$), $\pi(x)$ is the point of intersection with $\mathbb{S}^n$ of the Euclidean ray that issues from e_{n+1} and passes through x.

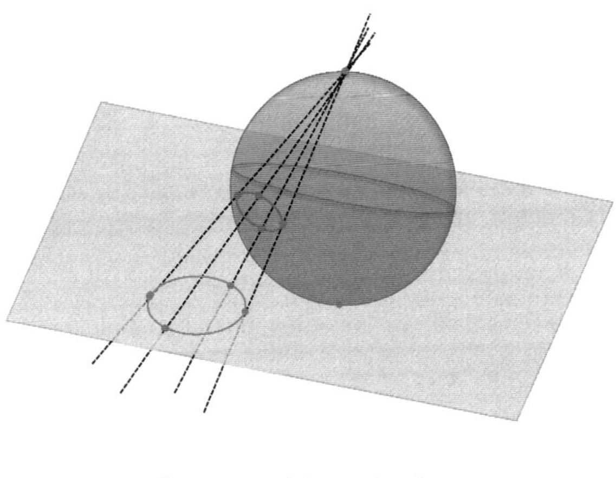

Stereographic projection.

Using elementary analytic geometry, one readily derives the formula for $\pi(x)$ as

$$\pi(x) = \left(\frac{2x_1}{|x|^2+1}, \ldots, \frac{2x_n}{|x|^2+1}, \frac{|x|^2-1}{|x|^2+1} \right) . \tag{2.1}$$

One then verifies without difficulty that $\pi^{-1} : \mathbb{S}^n \to \hat{\mathbb{R}}^n$ is given by

$$\pi^{-1}(y) = \left(\frac{y_1}{1-y_{n+1}}, \frac{y_2}{1-y_{n+1}}, \ldots, \frac{y_n}{1-y_{n+1}} \right) \tag{2.2}$$

for $y \neq e_{n+1}$, while $\pi^{-1}(e_{n+1}) = \infty$. In particular, (2.1) and (2.2) imply that $\pi(x) \to e_{n+1}$ as $x \to \infty$ and that $\pi^{-1}(y) \to \infty$ as $y \to e_{n+1}$, showing that π is continuous at ∞ and π^{-1} is continuous at e_{n+1}. As (2.1) and (2.2) also make evident the continuity of π and π^{-1} at all remaining points of their respective domains, we conclude that π is indeed a homeomorphism of $\hat{\mathbb{R}}^n$ onto $\mathbb{S}^n$.

Stereographic projection provides a natural mechanism for introducing a metric structure into $\hat{\mathbb{R}}^n$; namely, we can define a distance function q by setting

$$q(x,y) = |\pi(x) - \pi(y)|$$

for points x and y of $\hat{\mathbb{R}}^n$. The distance q is known as the *chordal metric* on $\hat{\mathbb{R}}^n$. (It is not to be mistaken for the *spherical metric* $\tilde{q}$ on $\hat{\mathbb{R}}^n$, which has $\tilde{q}(x,y)$ equal to the distance between $\pi(x)$ and $\pi(y)$ in the standard intrinsic Riemannian geometry of $\mathbb{S}^n$, as opposed to the distance between these points in the ambient space $\mathbb{R}^{n+1}$.) With the aid of (2.1) it is a simple matter to check that

$$q(x,y) = \frac{2|x-y|}{(1+|x|^2)^{1/2}(1+|y|^2)^{1/2}}, \qquad \text{if } x, y \in \mathbb{R}^n , \tag{2.3}$$

$$q(x,y) = \frac{2}{(1+|x|^2)^{1/2}}, \qquad \text{if } x \in \mathbb{R}^n,\ y = \infty . \tag{2.4}$$

From the fact that π is a homeomorphism from $\hat{\mathbb{R}}^n$ onto $\mathbb{S}^n$ and at the same time an isometry between the metric space $(\hat{\mathbb{R}}^n, q)$ and $\mathbb{S}^n$ with its Euclidean metric, we infer that the topology of $\hat{\mathbb{R}}^n$ is the same as the metric topology associated with q.

Let A and B be nonempty subsets of $\hat{\mathbb{R}}^n$. We write $q(A)$ for the chordal diameter of A and $q(A,B)$ for the chordal distance between A and B. Thus

$$q(A) = \sup\{q(x,y) : x, y \in A\}, \qquad q(A,B) = \inf\{q(x,y) : x \in A\ ,\ y \in B\} .$$

We likewise abbreviate $q(\{x\}, B)$ with $q(x,B)$. Finally, we adopt the convention that $\operatorname{dist}(x,\infty) = \infty$ for all x in $\mathbb{R}^n$.

2.3. Recollections from linear algebra

Let $T : \mathbb{R}^n \to \mathbb{R}^m$ be a linear transformation. We define the *maximal stretching* $L(T)$ and *minimal stretching* $\ell(T)$ of T by

$$L(T) = \max_{|x|=1} |T(x)|, \qquad \ell(T) = \min_{|x|=1} |T(x)| .$$

The quantity $L(T)$ frequently goes by a different name, the *operator norm* of T, under the alternate notation $\|T\|$. For the composition ST of linear transformations $T : \mathbb{R}^n \to \mathbb{R}^m$ and $S : \mathbb{R}^m \to \mathbb{R}^p$, it is true that

$$L(ST) \le L(S)L(T), \qquad \ell(ST) \ge \ell(S)\ell(T) . \tag{2.5}$$

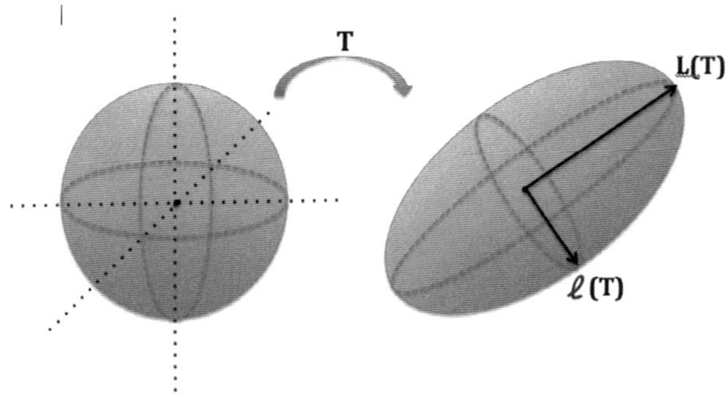

A linear transformation $T : \mathbb{R}^n \to \mathbb{R}^n$ is nonsingular if and only if $\ell(T) > 0$, in which event

$$L(T^{-1}) = \ell(T)^{-1}, \qquad \ell(T^{-1}) = L(T)^{-1}. \tag{2.6}$$

Assertions (2.5) and (2.6) are straightforward to confirm. Recall that the nonsingular linear transformations of $\mathbb{R}^n$ form a group under composition, called the *general linear group* $\mathrm{GL}(n)$.

A linear transformation $T : \mathbb{R}^n \to \mathbb{R}^n$ is called an *orthogonal transformation* if $|T(x)| = |x|$ for every x in $\mathbb{R}^n$ or, equivalently, if $\langle T(x), T(y)\rangle = \langle x, y\rangle$ for all x and y in $\mathbb{R}^n$.

The orthogonal transformations of $\mathbb{R}^n$ constitute a subgroup of $\mathrm{GL}(n)$, the *orthogonal group* $\mathrm{O}(n)$. An element T of $\mathrm{GL}(n)$ belongs to $\mathrm{O}(n)$ if and only if $T^{-1} = T^*$, where T^* denotes the *adjoint* of T, the unique linear transformation $T^* : \mathbb{R}^n \to \mathbb{R}^n$ that satisfies

$$\langle T(x), y\rangle = \langle x, T^*(y)\rangle$$

for all x and y in $\mathbb{R}^n$.

If U is an element of $\mathrm{O}(n)$—we typically represent orthogonal transformations by the letters U and V—then $L(U) = \ell(U) = 1$ (this property clearly characterizes the members of $\mathrm{O}(n)$), while the determinant $\det(U)$ of U is either 1 or -1.

If an orthogonal transformation U has $\det(U) = 1$, we call U a *rotation* of $\mathbb{R}^n$. The *special orthogonal group* $\mathrm{SO}(n)$ is the subgroup of $\mathrm{O}(n)$ consisting of all such rotations. It is not hard to see that

$$L(VTU) = L(T), \qquad \ell(VTU) = \ell(T) \tag{2.7}$$

for any linear transformation $T : \mathbb{R}^n \to \mathbb{R}^m$ whenever $U \in \mathrm{O}(n)$ and $V \in \mathrm{O}(m)$.

By an *affine transformation* of $\mathbb{R}^n$ we understand a function $f : \mathbb{R}^n \to \mathbb{R}^n$ of the type $f = T + b$ (our shorthand for $f(x) = Tx + b$), where T belongs to $\mathrm{GL}(n)$ and b to $\mathbb{R}^n$.

The affine transformations of $\mathbb{R}^n$ also form a group under composition, the *affine group* $\mathrm{A}(n)$. It contains as a subgroup the *Euclidean group* $\boldsymbol{E}(n)$, the group of Euclidean isometries of $\mathbb{R}^n$: every $f \in \boldsymbol{E}(n)$ has the structure $f = U + b$ with U in $\mathrm{O}(n)$ and b in $\mathbb{R}^n$. To verify this, fix f in $\boldsymbol{E}(n)$ and set $b = f(0)$. Then $U = f - b$

is a member of $\boldsymbol{E}(n)$, $U(0) = 0$, and

$$\begin{aligned} \langle U(x), U(y)\rangle &= \frac{|U(x)|^2 + |U(y)|^2 - |U(x) - U(y)|^2}{2} \\ &= \frac{|x|^2 + |y|^2 - |x - y|^2}{2} = \langle x, y\rangle \end{aligned}$$

for all x and y in $\mathbb{R}^n$. Therefore the vectors $U(e_1), U(e_2), \ldots, U(e_n)$ constitute an orthonormal basis for $\mathbb{R}^n$. It follows that

$$U(x) = \sum_{i=1}^{n} \langle U(x), U(e_i)\rangle U(e_i) = \sum_{i=1}^{n} \langle x, e_i\rangle U(e_i) = \sum_{i=1}^{n} x_i U(e_i)$$

for every x in $\mathbb{R}^n$, which shows that U is a linear transformation. This gives $U \in \mathrm{O}(n)$ and thus gives f the indicated structure. Of course, every function f of this form is an element of $\boldsymbol{E}(n)$.

When a linear transformation $S : \mathbb{R}^n \to \mathbb{R}^n$ enjoys the property that $S^* = S$, we say that S is *symmetric* (or *self-adjoint*).

For example, if $T : \mathbb{R}^n \to \mathbb{R}^n$ is an arbitrary linear transformation, then both T^*T and TT^* are symmetric. One of the linchpins of Euclidean linear algebra is:

THEOREM 2.3.1. *If S is a symmetric linear transformation of $\mathbb{R}^n$, then there exists $U \in \mathrm{O}(n)$ such that $D = U^{-1}SU$ has the form*

$$D(x) = (\lambda_1 x_1, \lambda_2 x_2, \ldots, \lambda_n x_n).$$

In particular, the eigenvalues of S (that is, the numbers $\lambda_1, \lambda_2, \ldots, \lambda_n$) are real.

Concerning the transformation D in Theorem 2.3.1—transformations of this general type are called *diagonal transformations*—we observe that

$$\min_{1\le i\le n} |\lambda_i| \le |D(x)| = \sqrt{\lambda_1^2 x_1^2 + \lambda_2^2 x_2^2 + \cdots + \lambda_n^2 x_n^2} \le \max_{1\le i\le n} |\lambda_i|$$

whenever $|x| = 1$, so $L(D) = \max_{1\le i\le n} |\lambda_i|$ and $\ell(D) = \min_{1\le i\le n} |\lambda_i|$. Of course, $\det(D) = \lambda_1\lambda_2\cdots\lambda_n$. Through the proper choice of U, one can always arrange in Theorem 2.3.1 that $\lambda_1 \ge \lambda_2 \ge \cdots \ge \lambda_n$.

To say that a linear transformation $T : \mathbb{R}^n \to \mathbb{R}^n$ is *positive definite* (respectively, *positive semidefinite*) means that $\langle x, T(x)\rangle > 0$ (respectively, $\langle x, T(x)\rangle \ge 0$) for every nonzero vector x in $\mathbb{R}^n$. For instance, if $T : \mathbb{R}^n \to \mathbb{R}^n$ is an arbitrary linear transformation, then T^*T and TT^* are positive semidefinite. If, in addition, T is nonsingular, then T^*T and TT^* are actually positive definite. Any real eigenvalue of a positive definite (respectively, positive semidefinite) linear transformation is positive (respectively, nonnegative).

Suppose now that a linear transformation $T : \mathbb{R}^n \to \mathbb{R}^n$ is both symmetric and positive semidefinite. On the basis of Theorem 2.3.1 and the subsequent comments, we can list the eigenvalues of T in the manner $\lambda_1 \ge \lambda_2 \ge \cdots \ge \lambda_n \ge 0$ and assert the existence of U in $\mathrm{O}(n)$ for which $U^{-1}TU = D$, where $D(x) = (\lambda_1 x_1, \lambda_2 x_2, \ldots, \lambda_n x_n)$. Appealing to (2.7), we conclude that

$$L(T) = L(U^{-1}TU) = L(D) = \lambda_1 .$$

Similarly, $\ell(T) = \lambda_n$ and $\det(T) = \lambda_1\lambda_2\cdots\lambda_n$.

Another basic result that we will soon need to use describes the analogue of polar coordinates for linear transformations.

THEOREM 2.3.2. *Any linear transformation $T : \mathbb{R}^n \to \mathbb{R}^n$ can be factored as $T = PU$, where U belongs to $O(n)$ and P is both symmetric and positive semidefinite.*

The linear transformation P in Theorem 2.3.2 is uniquely determined by T: denoting by I the identity transformation of $\mathbb{R}^n$, we compute

$$TT^* = (PU)(PU)^* = PUU^*P^* = PIP = P^2 ,$$

and learn that P is the (known to be unique) symmetric, positive semidefinite square root of TT^*.

This observation entitles us to list the eigenvalues of P as $\lambda_1^{1/2}, \lambda_2^{1/2}, \ldots, \lambda_n^{1/2}$, where $\lambda_1 \geq \lambda_2 \geq \cdots \geq \lambda_n \geq 0$ are the eigenvalues of TT^* (which, by the way, are the same as the eigenvalues of T^*T). Theorem 2.3.2 gives us a firm grip on the geometry of the transformation T by telling us how T transforms the unit sphere $\mathbb{S}^{n-1}$ and, hence, by reason of its linearity, how T transforms an arbitrary sphere or ball in $\mathbb{R}^n$.

According to Theorem 2.3.1, there is an orthogonal transformation V such that $V^{-1}PV = D$, with $y = D(x)$ given as follows:

$$y_1 = \lambda_1^{1/2} x_1 \, , \, y_2 = \lambda_2^{1/2} x_2, \ldots, y_n = \lambda_n^{1/2} x_n .$$

Assume initially that T is nonsingular. Then all the eigenvalues of TT^* are positive, which makes it apparent that D maps the sphere $\mathbb{S}^{n-1}$ bijectively to the ellipsoid E whose equation is $(y_1^2/\lambda_1) + (y_2^2/\lambda_2) + \cdots + (y_n^2/\lambda_n) = 1$. Because $U(\mathbb{S}^{n-1}) = V(\mathbb{S}^{n-1}) = \mathbb{S}^{n-1}$, we discover that

$$T(\mathbb{S}^{n-1}) = PU(\mathbb{S}^{n-1}) = P(\mathbb{S}^{n-1}) = PV(\mathbb{S}^{n-1}) = VD(\mathbb{S}^{n-1}) = V(E) .$$

Therefore $T(\mathbb{S}^{n-1})$, a set obtained from E via a rigid motion of $\mathbb{R}^n$, is itself an ellipsoid, one whose longest semiaxis has length $\sqrt{\lambda_1} = L(P) = L(PU) = L(T)$ and whose shortest semiaxis has length $\ell(T)$.

Should the transformation T be singular, the image of $\mathbb{S}^{n-1}$ under T would be a lower-dimensional ellipsoid provided we include as degenerate cases under this heading line segments and points.

A further consequence of Theorem 2.3.2 is the useful pair of inequalities

$$|\det (T)| \leq L(T)^n, \qquad \ell(T)^n \leq |\det (T)| \, , \tag{2.8}$$

valid for any linear transformation $T : \mathbb{R}^n \to \mathbb{R}^n$. They follow from the observation that

$$L(T) = L(P) = \lambda_1^{1/2} \, , \quad \ell(T) = \ell(P) = \lambda_n^{1/2}$$

and that

$$|\det (T)| = \det (P) = (\lambda_1 \lambda_2 \cdots \lambda_n)^{1/2} \, ,$$

in which $\lambda_1, \lambda_2, \ldots, \lambda_n$ are once again the eigenvalues of TT^* listed in order of diminishing size.

Together, Theorems 2.3.1 and 2.3.2 provide us with the following useful bit of information.

THEOREM 2.3.3. *Let $T : \mathbb{R}^n \to \mathbb{R}^n$ be a linear transformation, and list the eigenvalues $\lambda_1, \lambda_2, \ldots, \lambda_n$ of TT^* in nonincreasing order. Then there exist orthogonal transformations $U, V \in O(n)$ such that $VTU = D$, where*

$$D(x) = (\lambda_1^{1/2} x_1, \lambda_2^{1/2} x_2, \ldots, \lambda_n^{1/2} x_n).$$

2.4. Dilatation and distortion of linear maps

The *dilatation* $H(T)$, *inner dilatation* $H_I(T)$, and *outer dilatation* $H_O(T)$ of a nonsingular linear transformation $T : \mathbb{R}^n \to \mathbb{R}^n$ are the quantities defined by

$$H(T) = \frac{L(T)}{\ell(T)}, \qquad H_I(T) = \frac{|\det(T)|}{\ell(T)^n}, \qquad H_O(T) = \frac{L(T)^n}{|\det(T)|}. \tag{2.9}$$

(For singular T it is customary to set $H(T) = H_I(T) = H_O(T) = \infty$.)

In geometric terms, $H(T)$ measures the eccentricity of the ellipsoid $T(\mathbb{S}^{n-1})$, while $H_I(T)$ and $H_O(T)$ relate the volume of $T(B^n)$ to the volumes of the balls centered at the origin that are, respectively, inscribed in and circumscribed about $T(\mathbb{S}^{n-1})$.

We remark that

$$H(T^{-1}) = H(T), \qquad H_I(T^{-1}) = H_O(T), \qquad H_O(T^{-1}) = H_I(T) \tag{2.10}$$

and that

$$H(VTU) = H(T), \qquad H_I(VTU) = H_I(T), \qquad H_O(VTU) = H_O(T) \tag{2.11}$$

if U and V are members of $\mathrm{O}(n)$. Owing to (2.11) and Theorem 2.3.3, it involves little effort to demonstrate that

$$1 \le H_O(T) \le H_I(T)^{n-1}, \qquad 1 \le H_I(T) \le H_O(T)^{n-1} \tag{2.12}$$

and also that

$$\begin{aligned} 1 &\le \min\{H_I(T), H_O(T)\} \le H(T)^{n/2} \\ &\le \max\{H_I(T), H_O(T)\} \le H(T)^{n-1}. \end{aligned} \tag{2.13}$$

Finally, we note that as a consequence of (2.13) $H(T) = H_I(T) = H_O(T)$ when $n = 2$.

2.5. Partial derivatives

Given a subset A of $\mathbb{R}^n$ and a function $f : A \to \mathbb{R}^m$, we use $\partial_i f$ to symbolize the partial derivative of f with respect to the ith-coordinate variable: $\partial_i f$ is the function defined by

$$\partial_i f(x) = \lim_{t\to 0} \frac{f(x + te_i) - f(x)}{t},$$

at any point x in the interior of A for which a limit exists in $\mathbb{R}^m$.

Writing $f = (f_1, f_2, \ldots, f_m)$, it is clear that $\partial_i f(x)$ exists if and only if $\partial_i f_j(x)$ exists for $j = 1, 2, \ldots, m$, in which case for such $x \in A$

$$\partial_i f(x) = (\partial_i f_1(x), \partial_i f_2(x), \ldots, \partial_i f_m(x)) \in \mathbb{R}^m.$$

Let U be a nonempty open subset of $\mathbb{R}^n$. We define the function spaces $C^k(U, \mathbb{R}^m)$ for $k = 0, 1, 2, \ldots$ as usual: $C^0(U, \mathbb{R}^m)$ is the class of all continuous functions $f : U \to \mathbb{R}^m$; for $k \ge 1$, $C^k(U, \mathbb{R}^m)$ consists of the functions f in $C^0(U, \mathbb{R}^m)$ such that all partial derivatives $\partial_{i_1, i_2, \ldots, i_\ell} f = \partial_{i_\ell} \partial_{i_{\ell-1}} \cdots \partial_{i_1} f$ with $1 \le \ell \le k$ and $1 \le i_1, i_2, \ldots, i_\ell \le n$ are defined and continuous in U.

The class $C^\infty(U, \mathbb{R}^m)$ is the intersection of the classes $C^k(U, \mathbb{R}^m)$, $k = 0, 1, 2, \ldots$,

$$C^\infty(U, \mathbb{R}^m) = \bigcap_{k=0}^{\infty} C^k(U, \mathbb{R}^m).$$

For $0 \leq k \leq \infty$ we use $C_0^k(U, \mathbb{R}^m)$ to denote the subclass of $C^k(U, \mathbb{R}^m)$ comprising those functions f for which the support of f, $\mathrm{supp}(f)$, is a compact subset of U.

(Assuming that a function f belongs to $C^0(U, \mathbb{R}^m)$, we employ $\mathrm{supp}(f)$ to designate its *support*, meaning the closure relative to U of $\{x \in U : f(x) \neq 0\}$.)

We abbreviate $C^k(U, \mathbb{R})$ to $C^k(U)$ and $C_0^k(U, \mathbb{R})$ to $C_0^k(U)$.

If a subset A of $\mathbb{R}^n$ contains U, we refer to any function $f : A \to \mathbb{R}^m$ whose restriction to U is in $C^k(U, \mathbb{R}^m)$ as a *C^k-function in U* or a *function of class C^k in U*.

The situation in two dimensions has an extra feature that warrants special commentary. We can identify $\mathbb{R}^2$ with the complex plane $\mathbb{C}$ via the correspondence $(x, y) \mapsto z = x + iy$ and then look at functions $f : A \to \mathbb{R}^2$, where A is a subset of $\mathbb{R}^2$, as complex-valued functions of the complex variable z. From this point of view the differential operators

$$\partial = \frac{1}{2}(\partial_1 - i\partial_2) = \frac{1}{2}\Big(\frac{\partial}{\partial x} - i\frac{\partial}{\partial y}\Big) \quad \text{and} \tag{2.14}$$

$$\overline{\partial} = \frac{1}{2}(\partial_1 + i\partial_2) = \frac{1}{2}\Big(\frac{\partial}{\partial x} + i\frac{\partial}{\partial y}\Big) \tag{2.15}$$

acquire special significance. For instance, $\overline{\partial} f(z) = 0$ is equivalent to the statement that the real and imaginary parts of f satisfy the Cauchy-Riemann relations at z. Hence these operators play a very important role in modern complex analysis and the planar theory of quasiconformal mappings, [**15**].

2.6. Differentiability

A function $f : A \to \mathbb{R}^m$, where A is a subset of $\mathbb{R}^n$, is *differentiable at a point* x if x is an interior point of A and if there is a linear transformation $T : \mathbb{R}^n \to \mathbb{R}^m$ with the property that for all y in A,

$$f(y) = f(x) + T(y - x) + |y - x|\, \varepsilon(y)\,, \tag{2.16}$$

where $\lim_{y \to x} \varepsilon(y) = 0$.

Given that such a transformation T exists, it is uniquely determined by f, and (2.16) plainly demands that

$$T(h) = \lim_{t \to 0} \frac{f(x + th) - f(x)}{t} \tag{2.17}$$

for every h in $\mathbb{R}^n$.

We call T the (*Fréchet*) *derivative of f at x* (or the *differential of f at x*) and denote it by $f'(x)$. (Except in circumstances where doing so might cause confusion, we write $f'(x)h$, rather than the more awkward $f'(x)(h)$, to indicate the value of $f'(x)$ at h.)

To avoid confusion with the complex derivative in two dimensions,

$$f'(z) = \lim_{\mathbb{C} \ni h \to 0} \frac{f(z + h) - f(z)}{h}\,,$$

the alternate notation $Df(x)$ for this derivative also finds itself in common use.

It will be seen shortly that in the case $m = 1$, that is, for real-valued functions, we can identify $f'(x)$ with the *gradient of f at x*, that is, with the vector

$$\nabla f(x) = (\partial_1 f(x), \partial_2 f(x), \ldots, \partial_n f(x)),$$

giving us yet a third notation for the derivative. Indeed, $f'(x)h = \langle \nabla f(x), h \rangle$ when $f : A \to \mathbb{R}$ is differentiable at x.

If we express a function f as $f = (f_1, f_2, \ldots, f_m)$, then by considering the individual components of (2.16) we become aware that f is differentiable at x if and only if each of the real-valued functions $f_1, f_2, \ldots, f_m$ has that property. We quickly deduce from (2.16) that differentiability on the part of f at a point x entails the continuity of f at x.

Further, the existence of $f'(x)$ ensures the existence of the partial derivatives $\partial_1 f(x), \partial_2 f(x), \ldots, \partial_n f(x)$. In fact, (2.17) reveals that $\partial_i f(x) = f'(x)e_i$, so the matrix of $f'(x)$ with respect to the standard bases for $\mathbb{R}^n$ and $\mathbb{R}^m$ is

$$\begin{bmatrix} \partial_1 f_1(x) & \partial_2 f_1(x) & \cdots & \partial_n f_1(x) \\ \partial_1 f_2(x) & \partial_2 f_2(x) & \cdots & \partial_n f_2(x) \\ \vdots & \vdots & \ddots & \vdots \\ \partial_1 f_m(x) & \partial_2 f_m(x) & \cdots & \partial_n f_m(x) \end{bmatrix}, \tag{2.18}$$

where $f_1, f_2, \ldots, f_m$ are the component functions of f.

Here there is in fact a mild abuse of notation: we use $f'(x)$ to symbolize the matrix in (2.18) as well as the derivative of f at x. The expression $f'(x)h$ for the value of $f'(x)$ at h is then correctly interpreted as denoting a matrix multiplication, provided we first agree to transpose h into a column-vector, then perform the multiplication, and finally convert the product back to row-vector form.

The fact that $\partial_1 f(x), \partial_2 f(x), \ldots, \partial_n f(x)$ all exist does not, in general, imply the differentiability of f at x (although it does permit one to write the matrix in (2.18), which in the absence of definite information regarding differentiability is called the *formal derivative matrix of f at x*).

If, however, the functions $\partial_1 f, \partial_2 f, \ldots, \partial_n f$ are all defined in some neighbourhood of x and each is continuous at x, then f is differentiable at x.

In particular, any function from the class $C^1(U, \mathbb{R}^m)$, with U an open set in $\mathbb{R}^n$, is differentiable at every point of U. For example, a function $f : \mathbb{R}^n \to \mathbb{R}^m$ of the type $f = T + b$, where $T : \mathbb{R}^n \to \mathbb{R}^m$ is a linear transformation and b is an element of $\mathbb{R}^m$, is differentiable and has $f'(x) = T$ at each point x of $\mathbb{R}^n$.

Let A be a subset of $\mathbb{R}^2$, and let $f : A \to \mathbb{R}^2$ be differentiable at a point z. Then we can consider $T = Df(z)$ to be a linear transformation of the complex plane $\mathbb{C}$ viewed as a two-dimensional vector space over $\mathbb{R}$. In this context, it would obviously be desirable to describe the action of T in complex terms. Fortunately, this is readily accomplished: with A regarded as a subset of $\mathbb{C}$ and f as a complex-valued function on A, the linear T takes the form

$$T(h) = \partial f(z)h + \bar{\partial} f(z)\bar{h} \tag{2.19}$$

for $h \in \mathbb{C}$.

Here the indicated algebraic operations are just complex addition and multiplication. A consequence of (2.19) is the following: in order for T to be a linear transformation of $\mathbb{C}$ as a one-dimensional complex vector space, it is necessary and sufficient that $\bar{\partial} f(z) = 0$; thus the real and imaginary parts of f obey the Cauchy-Riemann equations at z. Since f is assumed to be differentiable at z, this happens precisely when f has a complex derivative at z.

2.7. Maximal and minimal stretchings

Let A be a subset of $\mathbb{R}^n$ and x be an interior point of A. We define the *maximal stretching* $L_f(x)$ and *minimal stretching* $\ell_f(x)$ at x of a function $f : A \to \mathbb{R}^m$ by

$$L_f(x) = \limsup_{h\to 0} \frac{|f(x+h) - f(x)|}{|h|}, \tag{2.20}$$

$$\ell_f(x) = \liminf_{h\to 0} \frac{|f(x+h) - f(x)|}{|h|}. \tag{2.21}$$

Should f be differentiable at the point x, we would have

$$L_f(x) = L\big[f'(x)\big], \qquad \ell_f(x) = \ell\big[f'(x)\big]. \tag{2.22}$$

To verify the first relation in (2.22), we note that (2.16) yields the estimate

$$|f(x+h) - f(x)| = \big|f'(x)h + |h|\varepsilon(h)\big| \le |h|\, L\big[f'(x)\big] + |h| \cdot |\varepsilon(h)|\,,$$

whenever h is sufficiently close to 0 that $x + h$ belongs to A. Hence.

$$L_f(x) \le \limsup_{h\to 0} \Big\{ L\big[f'(x)\big] + |\varepsilon(h)| \Big\} = L\big[f'(x)\big]\,.$$

To establish the reverse inequality, we select h in $\mathbb{S}^{n-1}$ for which $L[f'(x)] = |f'(x)h|$ and infer from (2.17) that

$$L_f(x) \ge \lim_{t\to 0^+} \frac{|f(x+th) - f(x)|}{t} = |f'(x)h| = L\big[f'(x)\big]\,.$$

The assertion concerning $\ell_f(x)$ is handled similarly. We remark in passing that, for a function $f : A \to \mathbb{C}$ with A contained in $\mathbb{C}$,

$$L_f(z) = |\partial f(z)| + |\overline{\partial} f(z)|, \qquad \ell_f(z) = \big|\, |\partial f(z)| - |\overline{\partial} f(z)|\, \big| \tag{2.23}$$

at any point z where f is differentiable. In these expressions $|\cdot|$ denotes the modulus or absolute value of a complex number. The relations in (2.10) are obtained in a straightforward fashion from (2.22) and (2.19).

2.8. Diffeomorphisms

An important fact about derivatives is the validity of the "chain rule":

THEOREM 2.8.1. *Let A be a subset of $\mathbb{R}^n$ and B be a subset of $\mathbb{R}^m$. If $f : A \to \mathbb{R}^m$ is differentiable at a point x and $g : B \to \mathbb{R}^p$ is differentiable at $y = f(x)$, then the composite function $g \circ f : A \cap f^{-1}(B) \to \mathbb{R}^p$ is differentiable at x, where its derivative (derivative matrix) satisfies the relation $(g \circ f)'(x) = g'(y)f'(x)$.*

Suppose next that f belongs to $C^1(U, \mathbb{R}^m)$ and g to $C^1(V, \mathbb{R}^p)$, where U is an open set in $\mathbb{R}^n$, V is an open set in $\mathbb{R}^m$, and $f(U)$ is contained in V. Theorem 2.8.1 informs us that $g \circ f$ is differentiable everywhere in U. It also gives a formula for each first order partial derivative of $g \circ f$:

$$\partial_i(g_j \circ f) = \sum_{k=1}^{m} (\partial_k g_j \circ f) \cdot \partial_i f_k$$

for $1 \le i \le n$ and $1 \le j \le p$. Here $f = (f_1, f_2, \ldots, f_m)$ and $g = (g_1, g_2, \ldots, g_p)$. On the basis of this formula, we are able to conclude that $g \circ f$ is a function of class $C^1(U, \mathbb{R}^p)$. More generally, an induction argument proves that the composition of C^k-functions $1 \le k \le \infty$ is again a C^k-function.

Inverse functions will occasionally be a matter of some interest in this book. We record two results that will help us to deal with them. They can be found in S. Lang's book on the fundamentals of differential geometry, [**95**].

THEOREM 2.8.2. *Let A be a subset of $\mathbb{R}^n$, and let $f : A \to \mathbb{R}^n$ be an injective function. Suppose that f is continuous in some neighbourhood of a point x, that f is differentiable at x itself, and that $f'(x)$ is nonsingular. Then $y = f(x)$ is an interior point of $f(A)$, the function $g = f^{-1} : f(A) \to A$ is differentiable at y, and $g'(y) = f'(x)^{-1}$.*

Consider an open set U in $\mathbb{R}^n$ and an injective function f belonging to $C^k(U, \mathbb{R}^n)$ for some $k \geq 1$. In case $k \geq 1$ assume additionally that $J_f(x) \neq 0$ for every x in U, where J_f denotes the *Jacobian determinant* of f, i.e., $J_f(x) = \det [f'(x)]$. Then $V = f(U)$ is an open set, and f is a homeomorphism of U onto V. (The truth of these two assertions is certified by the "invariance of domain" theorem from topology, [**61**].)

When $k \geq 1$, Theorem 2.8.2 shows that $g = f^{-1}$ is differentiable at each point y of V, with $g'(y) = f'[g(y)]^{-1}$. Cramer's rule leads to an explicit representation for the partial derivative $\partial_i g_j(y)$ $(1 \leq i, j \leq n)$ as the quotient of a polynomial in various components of $\partial_1 f[g(y)], \partial_2 f[g(y)], \ldots, \partial_n f[g(y)]$, and the quantity $J_f[g(y)]$. From this representation it can be inferred that g is in the class $C^k(V, \mathbb{R}^n)$, making f a C^k-diffeomorphism.

For k a positive integer or ∞, a *C^k-diffeomorphism* f between open sets U and V in $\mathbb{R}^n$ is a bijection $f : U \to V$ such that both f and f^{-1} are of class C^k. Unless otherwise stated, "diffeomorphism" will mean a C^1-diffeomorphism. If $k \geq 1$ and the set U is connected, then either $J_f(x) > 0$ for every x in U or $J_f(x) < 0$ for all such x. In the former case we call f a *sense-preserving* diffeomorphism; in the latter case we say that f is a *sense-reversing* diffeomorphism.

We shall not elaborate on the precise significance of both of the terms "sense-preserving" and "sense-reversing", because to do so would entail a major foray into topology. These notions are discussed in most introductory texts to algebraic topology, for instance L. Greenberg's book [**61**].

We close this section by quoting a criterion that enables us to determine whether a function is, at least on a local level, a diffeomorphism.

THEOREM 2.8.3. *Let U be an open set in $\mathbb{R}^n$, and let f be a member of the class $C^k(U, \mathbb{R}^n)$ with $1 \leq k \leq \infty$. If x is a point of U for which $J_f(x) \neq 0$, then there exists an $r > 0$ such that the restriction of f to the ball $B = B^n(x, r)$ is a C^k-diffeomorphism of B onto $f(B)$.*

CHAPTER 3

Conformal Mappings in Euclidean Space

Rigidity will prevent a wide and flexible class of conformal transformations in dimensions $n \geq 3$, and as we have said, this is a prime motivation for the general theory of quasiconformal mappings.

However the theory of conformal geometry in higher dimensions is interesting and important and connects with many other fields of mathematics. It is the purpose of this chapter to collect together many of these ideas and connections, not only for their interest but also to motivate and describe potential tools and results for quasiconformal mappings. These things include higher-dimensional cross ratios, the extension theorem of Poincaré, hyperbolic geometry, and the compactness and normal families properties of conformal transformations.

The contents of a significant part of the latter portions of this chapter are a synthesis of material found in Lars Ahlfors' Ordway Lectures [**9**] or in Alan Beardon's book *The Geometry of Discrete Groups*, [**16**]. Each of these we warmly recommend to the reader if they wish to discover more of the subject for themselves.

3.1. Linear conformal transformations

Let $n \geq 2$. A linear transformation $T : \mathbb{R}^n \to \mathbb{R}^n$ is *conformal* if T is nonsingular and preserves Euclidean angles, in the sense that

$$\theta[T(x), T(y)] = \theta(x, y)$$

for all nonzero vectors x and y in $\mathbb{R}^n$. It follows from statement (ii) in Theorem 3.1.1 that conformal linear transformations are actually "shape preserving", which is more faithful than "angle- preserving" to the literal meaning of "conformal". The foregoing definition permits a conformal linear transformation T to be either sense-preserving, the situation when $\det(T) > 0$, or sense-reversing, which occurs when $\det(T) < 0$.

The experienced reader will realize that this is at variance with the standard usage of "conformal" in classical complex analysis. In that setting, the term is traditionally reserved for transformations, linear or otherwise, that preserve angles not merely in size but also in orientation, while the expression "anti-conformal" is used when they are sense-reversing.

We note that the conformal linear transformations of $\mathbb{R}^n$ constitute a subgroup of $\mathrm{GL}(n)$, as do the sense-preserving conformal linear transformations. If a conformal linear transformation $T : \mathbb{R}^n \to \mathbb{R}^n$ leaves $\mathbb{R}^m$ $(2 \leq m < n)$ invariant, then the restriction of T to $\mathbb{R}^m$ is a conformal linear transformation of the lower-dimensional space.

It will prove convenient to have several ways of characterizing conformal linear transformations, and we now set about trying to determine these criteria.

THEOREM 3.1.1. *Let $n \geq 2$. The following statements concerning a linear transformation $T \in \mathrm{GL}(n)$ are equivalent:*

(i) *T is conformal,*
(ii) *$T = \lambda U$, where λ is a positive number and U belongs to $\mathrm{O}(n)$,*
(iii) *$H(T) = 1$,*
(iv) *$\|T\|^n = |\det(T)|$,*
(v) *$T^*T = |\det(T)|^{2/n} I$,*

where I denotes the identity matrix.

PROOF. The implications (ii) $\Rightarrow$ (i), (ii) $\Rightarrow$ (iii), (ii) $\Rightarrow$ (iv), and (ii) $\Rightarrow$ (v) are clear. To demonstrate that (i) $\Rightarrow$ (ii), let $v_i = T(e_i)$ for $i = 1, 2, \ldots, n$ and set $\lambda = |v_1| > 0$. Because T is conformal, the vectors $v_1, v_2, \ldots, v_n$ are pairwise orthogonal. For $i \geq 2$, let Δ_i be the isosceles triangle in $\mathbb{R}^n$ with vertices 0, e_1, and e_i. Then T transforms Δ_i to a triangle Δ_i' with vertices $0, v_1$, and v_i. Conformality dictates that the angle of Δ_i' at any of its vertices be equal to the angle of Δ_i at the corresponding vertex. As a consequence, Δ_i' is also isosceles. It follows that $|v_i| = \lambda$ for $1 \leq i \leq n$. Thus $U = \lambda^{-1}T$ transforms $e_1, e_2, \ldots, e_n$ to another orthonormal basis for $\mathbb{R}^n$, which forces U into $\mathrm{O}(n)$ and gives T the stated form.

We next show that (iii) $\Rightarrow$ (ii). Writing $\lambda = L(T) = \ell(T) > 0$, we conclude that $|T(x)| = \lambda$ for every x in $\mathbb{S}^{n-1}$, so $|T(x)| = \lambda|x|$ for every x in $\mathbb{R}^n$. Accordingly, $U = \lambda^{-1}T$ has $|U(x)| = |x|$ for all x, i.e., U belongs to $\mathrm{O}(n)$. Again T has the required structure.

Assuming (iv), we let $\lambda_1 \geq \lambda_2 \geq \cdots \geq \lambda_n > 0$ denote the eigenvalues of TT^*. The observation used to verify 2.8 leads to

$$\lambda_1^n = L(T)^{2n} = \|T\|^{2n} = |\det(T)|^2 = |\det(TT^*)| = \lambda_1 \lambda_2 \cdots \lambda_n \,,$$

which implies that $\lambda_1 = \lambda_2 = \cdots = \lambda_n$. In particular, $L(T) = \sqrt{\lambda_1} = \sqrt{\lambda_n} = \ell(T)$. Therefore $H(T) = 1$, so (iv) $\Rightarrow$ (iii).

Finally, given that (v) holds, we compute for any unit vector x in $\mathbb{R}^n$:

$$|T(x)|^2 = \langle T(x), T(x)\rangle = \langle T^*T(x), x\rangle = |\det(T)|^{2/n}\langle x, x\rangle = |\det(T)|^{2/n} \,.$$

From this we easily conclude that $\|T\|^n = |\det(T)|$. In other words, (v) $\Rightarrow$ (iv). This completes the proof of the theorem. □

A straightforward consequence of Theorem 3.1.1 is that

$$\|ST\| = \|S\| \cdot \|T\| \tag{3.1}$$

whenever S and T are conformal linear transformations of $\mathbb{R}^n$.

Let D be a domain in $\mathbb{R}^n$ with $n \geq 2$—by a *domain* in a topological space X we understand a nonempty, open, connected subset of X—and let $f : D \to \mathbb{R}^n$ be a continuous injection (hence, a homeomorphism of D onto the domain $D' = f(D)$). We say that f is *conformal at a point* x of D under the condition that f is differentiable at x and its derivative $f'(x)$ is a conformal linear transformation.

Couched in geometric terms, this definition requires that f be infinitesimally angle-preserving (or, more generally, shape-preserving) at x. It demands, for instance, that f transform any pair of curves (or hypersurfaces) in D that intersect at x in an angle θ to curves (or hypersurfaces) whose angle of intersection at the point $f(x)$ is likewise θ.

We say that f is a *conformal mapping* of D onto D' provided that f is conformal at each point of D.

As a homeomorphism between domains in $\mathbb{R}^n$, such a mapping f must be either sense-preserving or sense-reversing. This fact finds its analytic expression in the statement that either the Jacobian determinant $J_f(x) > 0$ for all x in D, or $J_f(x) < 0$ for all x in D.

The preceding would be an elementary observation were f a diffeomorphism, but is not quite so obvious in the situation at hand, where no continuity conditions are imposed upon f'.

If f maps D conformally onto D' and g maps D' conformally onto D'', then $g \circ f$ is a conformal mapping of D onto D''. This follows from the chain rule and the fact that the conformal linear transformations of $\mathbb{R}^n$ form a group under composition. If f maps D conformally onto D', then Theorem 2.8.2 certifies that f^{-1} maps D' conformally onto D. If D is a domain in $\mathbb{R}^n$ with the property that $G = D \cap \mathbb{R}^m$ is a domain in $\mathbb{R}^m$ $(2 \leq m < n)$ and if f, a conformal mapping of D onto some domain in $\mathbb{R}^n$, has the feature that $f(G)$ is a subset of $\mathbb{R}^m$, then $g = f|G$ is a conformal mapping of G onto $f(G)$. In fact, we deduce from (2.17) that for each x in G the linear transformation $f'(x)$ has $\mathbb{R}^m$ as an invariant subspace and infer that $g'(x) = f'(x)|\mathbb{R}^m$ is a conformal linear transformation of $\mathbb{R}^m$.

Consider a sense-preserving conformal mapping $f : D \to D'$ between domains D and D' in $\mathbb{R}^2$, which we now think of as the complex plane. A simple calculation confirms that, when viewed as a complex-valued function of a complex variable, any differentiable function $g : D \to \mathbb{C}$ has

$$J_g(z) = |\partial g(z)|^2 - |\overline{\partial} g(z)|^2 , \tag{3.2}$$

which gives us a particularly nice formula for the Jacobian determinant.

In the present case we have $J_f > 0$ in D, so the conformality of f in combination with Theorem 3.1.1(iv), (2.22), (2.23), and (3.2) provides the information that

$$\left(|\partial f(z)| + |\overline{\partial} f(z)|\right)^2 = L_f(z)^2 = J_f(z) = |\partial f(z)|^2 - |\overline{\partial} f(z)|^2 \tag{3.3}$$

for every z in D. Thanks to the positivity of J_f, we can be certain that $|\partial f(z)| + |\overline{\partial} f(z)| > 0$ throughout D. Therefore, (3.3) reduces to

$$|\partial f(z)| + |\overline{\partial} f(z)| = |\partial f(z)| - |\overline{\partial} f(z)|$$

for each z in D.

The last equation demands that $\overline{\partial} f$ vanish identically in D, hence it forces f to be analytic there. The consequence of this discussion is that if $f : D \to \mathbb{C}$ is a sense-preserving conformal mapping of a plane domain D onto $D' = f(D)$, then f is an injective analytic function in D. The converse of that statement is also true—and not hard to prove. Similar reasoning reveals that the sense-reversing conformal mappings in the plane are the functions of the type $\bar{f}$, where $f : D \to \mathbb{C}$ is an injective analytic function.

The family of conformal mappings from $\mathbb{R}^n$ into itself plainly includes all mappings of the type $f = \lambda U + b$, in which λ is a positive number, U belongs to $\mathrm{O}(n)$, and b is an element of $\mathbb{R}^n$. The class of such mappings, which we refer to as *similarity transformations*, is a subgroup of the affine group $\mathrm{A}(n)$. We call it the (*general*) *similarity group* of $\mathbb{R}^n$ and denote it by $\mathrm{GS}(n)$. Again, we point out the divergence of this terminology from more common usage, which would exclude a sense-reversing mapping from inclusion under the "similarity transformation" heading. The sense-preserving similarities form a subgroup $\mathrm{GS}^+(n)$ of $\mathrm{GS}(n)$. Certain members of $\mathrm{GS}^+(n)$ bear distinguishing titles: a mapping of the type $f(x) = x + b$, with $b \neq 0$, is called a *translation*; a *dilation* (or *homothety*) is a transformation of the form $f(x) = \lambda x$ with $\lambda > 0$ but $\lambda \neq 1$.

We make the following relevant remark. Let f be an affine transformation of $\mathbb{R}^n$, say $f = T + b$ with $T \in \mathrm{GL}(n)$ and $b \in \mathbb{R}^n$. Since

$$|f(x)| \geq |T(x)| - |b| \geq \ell(T)|x| - |b| \to \infty$$

as $|x| \to \infty$, we see that by defining $f(\infty) = \infty$ we may extend f to a homeomorphism of $\hat{\mathbb{R}}^n$ onto itself. By extending each of its members in this way, we are free to regard the group $\mathrm{A}(n)$ as a group of homeomorphisms of $\hat{\mathbb{R}}^n$, which we shall tacitly do from now on. We note especially that, under this convention, any similarity transformation of $\mathbb{R}^n$ is a homeomorphism of $\hat{\mathbb{R}}^n$.

3.2. Reflections

Let P be a hyperplane in $\mathbb{R}^n$; that is, P is an $(n-1)$-dimensional affine subspace of $\mathbb{R}^n$. We can describe P by means of a linear equation, say

$$P : \langle \nu, x \rangle = d,$$

where ν is a unit vector in $\mathbb{R}^n$ and $d \geq 0$. As a matter of fact, subject to the stated normalizations, ν and d are uniquely determined by P except when $d = 0$, in which case there are two choices for ν, one the negative of the other.

The function $R : \hat{\mathbb{R}}^n \to \hat{\mathbb{R}}^n$ defined by

$$R(x) = x - 2\big(\langle \nu, x \rangle - d\big)\nu \tag{3.4}$$

for x in $\mathbb{R}^n$ and $R(\infty) = \infty$ is called the *reflection in* P. This term is geometrically motivated: R fixes every point of P and moves any point x of $\mathbb{R}^n \setminus P$ to its mirror image relative to P, which means to the point $R(x)$ such that P is the perpendicular bisector of the line segment with endpoints x and $R(x)$. In particular, R interchanges the two half-spaces into which P partitions $\mathbb{R}^n$.

We deduce from (3.4) by an elementary calculation that R is a Euclidean isometry. This makes R an affine transformation, one that is conformal when $n \geq 2$. Furthermore, if $n \geq 2$ and if the vectors $v_1, v_2, \ldots, v_{n-1}$ form a basis for the linear subspace P_0 of $\mathbb{R}^n$ defined by $P_0 : \langle \nu, x \rangle = 0$, then we see from (2.17) that the matrix of $T = R'(x)$ (in this instance $R'(x)$ is the same for all points x of $\mathbb{R}^n$) with respect to the basis $\nu, v_1, v_2, \ldots, v_{n-1}$ for $\mathbb{R}^n$ is

$$\begin{bmatrix} -1 & 0 & \cdots & 0 \\ 0 & 1 & \cdots & 0 \\ \vdots & \vdots & & \vdots \\ 0 & 0 & \cdots & 1 \end{bmatrix}.$$

It follows that $\det(T) = -1$, hence that R is sense-reversing. As a member of $\mathrm{A}(n)$, R is a homeomorphism of $\hat{\mathbb{R}}^n$ onto itself. On top of this, R is an involution, meaning that $R^{-1} = R$, or $R \circ R = I$, the identity transformation.

Next, let $S = \mathbb{S}^{n-1}(x_0, r)$ be a Euclidean sphere in $\mathbb{R}^n$. The *reflection in S* (also known as the *inversion in S*) is the function $R : \hat{\mathbb{R}}^n \to \hat{\mathbb{R}}^n$ defined by the rule

$$R(x) = x_0 + \frac{r^2(x - x_0)}{|x - x_0|^2} \tag{3.5}$$

for x in $\mathbb{R}^n \setminus \{x_0\}$, while $R(x_0) = \infty$ and $R(\infty) = x_0$.

Formula (3.5) again has a simple geometric interpretation: R transforms a point x of $\mathbb{R}^n \setminus \{x_0\}$ to the unique point $R(x)$ that lies on the Euclidean ray issuing from x_0 and passing through x and that satisfies the condition

$$|R(x) - x_0| \cdot |x - x_0| = r^2.$$

Just like the reflection in a hyperplane, the reflection in the sphere $S = \mathbb{S}(x_0, r)$ is an involutive homeomorphism of $\hat{\mathbb{R}}^n$. It leaves fixed every point of S and interchanges the two components of S^c, $B^n(x_0, r)$, and $\hat{\mathbb{R}}^n \setminus \bar{B}^n(x_0, r)$. This function is also a C^∞-diffeomorphism of $\mathbb{R}^n \setminus \{x_0\}$ onto itself, as (3.5) makes readily apparent.

We shall examine R_0, the reflection in the unit sphere $\mathbb{S}^{n-1}$, more closely. In this situation we have $R_0(x) = x/|x|^2$ for x different from 0 and ∞. As a function of class C^∞ in $\mathbb{R}^n \setminus \{0\}$, R_0 is differentiable at every point x of this set, and its derivative is not difficult to compute: the derivative matrix of R_0 at x is

$$R_0'(x) = |x|^{-2}[I - 2Q(x)], \tag{3.6}$$

where $Q(x)$ is the matrix whose $(i,j)^{th}$-entry is

$$(R_0'(x))_{i,j} = x_i x_j / |x|^2.$$

Now $Q = Q(x)$ is the matrix of a symmetric linear transformation, also labeled Q. It is an easy exercise to see that Q satisfies

$$Q^2 = Q.$$

As a consequence, we compute

$$(I - 2Q)(I - 2Q)^* = (I - 2Q)^2 = I - 4Q + 4Q^2 = I - 4Q + 4Q = I$$

and therefore discover that $|x|^2 R_0'(x) = I - 2Q(x)$ belongs to the orthogonal group $\mathrm{O}(n)$. This has the immediate consequence that

$$\det(R_0'(x)) = -|x|^{-2n}. \tag{3.7}$$

Assume now that $n \geq 2$. On the basis of Theorem 3.1.1(ii) we can assert that, for each x in $\mathbb{R}^n \setminus \{0\}$, $R_0'(x)$ is a conformal linear transformation of $\mathbb{R}^n$. In other words, R_0 is a conformal self-mapping of $\mathbb{R}^n \setminus \{0\}$. Moreover,

$$\|R_0'(x)\| = \frac{1}{|x|^2}. \tag{3.8}$$

Since the matrix corresponding to $R_0'(e_1)$, which once again is

$$\begin{bmatrix} -1 & 0 & \cdots & 0 \\ 0 & 1 & \cdots & 0 \\ \vdots & \vdots & & \vdots \\ 0 & 0 & \cdots & 1 \end{bmatrix},$$

has determinant -1, we conclude that R_0 is sense-reversing.

The reflection R in an arbitrary sphere $\mathbb{S}^{n-1}(x_0, r)$ is expressible as the composition $R = f^{-1} \circ g \circ R_0 \circ f$, where $f(x) = x - x_0$ and $g(x) = r^2 x$. Because $f'(x) = I$ and $g'(x) = r^2 I$ for all x in $\mathbb{R}^n$, the chain rule yields

$$R'(x) = r^2 R_0'(x - x_0) \tag{3.9}$$

for every x in $\mathbb{R}^n \setminus \{x_0\}$. In case $n \geq 2$, we conclude that R is a sense-reversing conformal mapping of $\mathbb{R}^n \setminus \{x_0\}$ onto itself and, in view of (3.8), that

$$\|R'(x)\| = \frac{r^2}{|x - x_0|^2} \tag{3.10}$$

for x other than x_0 or ∞. We can take the determinant of both sides of (3.9) to obtain the following lemma.

LEMMA 3.2.1. *Let R be the reflection in an arbitrary sphere $\mathbb{S}^{n-1}(x_0, r)$. Then the Jacobian determinant of the mapping R is*

$$J_R(x) = -\frac{r^{2n}}{|x - x_0|^{2n}}. \tag{3.11}$$

One special case covered by the preceding discussion is worth highlighting. Suppose that $R : \hat{\mathbb{R}}^{n+1} \to \hat{\mathbb{R}}^{n+1}$ is the inversion in the sphere $\mathbb{S}^n(e_{n+1}, \sqrt{2}\,)$. For x in $\mathbb{R}^n$ we have $|x - e_{n+1}|^2 = |x|^2 + 1$, so for such x we obtain

$$R(x) = e_{n+1} + \frac{2(x - e_{n+1})}{|x|^2 + 1} = \left(\frac{2x_1}{|x|^2 + 1}, \dots, \frac{2x_n}{|x|^2 + 1}, \frac{|x|^2 - 1}{|x|^2 + 1}\right) = \pi(x)\ ,$$

where $\pi : \hat{\mathbb{R}}^n \to \mathbb{S}^n$ is the stereographic projection. Therefore, π is the restriction to $\hat{\mathbb{R}}^n$ of a reflection operating in $\hat{\mathbb{R}}^{n+1}$!

This explains the well-known conformality of π as a mapping from $\mathbb{R}^n$ to $\mathbb{S}^n$, the latter endowed with the Riemannian structure it acquires as a smooth submanifold of $\mathbb{R}^{n+1}$.

Several consequences of this observation will be of value later on. One is that the function $f = R \circ S \big| \mathbb{H}^{n+1}$, in which

$$\mathbb{H}^{n+1} = \{x \in \mathbb{R}^{n+1} : x_{n+1} > 0\}$$

is the open "upper" half-space in $\mathbb{R}^{n+1}$ and S is reflection in $\mathbb{R}^n$ regarded as a hyperplane in $\mathbb{R}^{n+1}$, provides a sense-preserving conformal mapping of $\mathbb{H}^{n+1}$ onto B^{n+1}.

This is simply because R maps the domain $D = \{x \in \mathbb{R}^{n+1} : x_{n+1} < 0\}$ in a sense-reversing, conformal fashion into $\mathbb{R}^{n+1}$; being a homeomorphism of $\hat{\mathbb{R}}^{n+1}$ with $R(\partial D) = \mathbb{S}^n$, R maps D to one of the two components of $\hat{\mathbb{R}}^{n+1} \setminus \mathbb{S}^n$; as $R(-e_{n+1}) = 0$, said component has to be B^{n+1}; the restriction of S to $\mathbb{H}^{n+1}$ is a sense-reversing conformal mapping of $\mathbb{H}^{n+1}$ onto D; thus, f has all the stated properties.

Another is that $\pi : \mathbb{R}^n \to \mathbb{R}^{n+1}$ has

$$\|\pi'(x)\| = \frac{2}{1 + |x|^2}\ , \tag{3.12}$$

which is true because $\pi'(x) = R'(x)|\mathbb{R}^n$ once we recall (3.10).

In order to consolidate the notion of reflection in a sphere with that of reflection in a hyperplane, we introduce some convenient terminology. We call a subset Σ of

$\hat{\mathbb{R}}^n$ a *chordal sphere* if Σ is either a Euclidean sphere in $\mathbb{R}^n$ or a set of the type $\Sigma = P \cup \{\infty\}$, where P is a hyperplane in $\mathbb{R}^n$.

We will soon see that all the chordal spheres in $\hat{\mathbb{R}}^n$ are precisely the spheres of radius $\sqrt{2}$ or less for the chordal metric on $\hat{\mathbb{R}}^n$.

Every chordal sphere Σ in $\hat{\mathbb{R}}^n$ admits descriptions of the type

$$\Sigma : a|x|^2 - 2\langle b, x\rangle + c = 0$$

with $a, c \in \mathbb{R}$, $b \in \mathbb{R}^n$, and $ac < |b|^2$, provided we establish the convention that ∞ is to be counted as a solution of such an equation if and only if $a = 0$. With the notation interpreted in the obvious way, the vector (a, b, c) in $\mathbb{R}^{n+2}$ is known as a *coefficient vector* for Σ. Any two coefficient vectors for Σ are nonzero scalar multiples of each another. Conversely, if the stated convention is respected, the solution locus in $\hat{\mathbb{R}}^n$ of an equation fitting the above description is a chordal sphere. With each chordal sphere Σ in $\hat{\mathbb{R}}^n$ is associated the reflection in Σ, a mapping we denote by R_Σ.

3.3. The Möbius group

The *(general) Möbius group* $\text{Möb}(n)$ is the group of homeomorphisms of $\hat{\mathbb{R}}^n$ generated by the reflections R_Σ, where Σ ranges over all chordal spheres in $\hat{\mathbb{R}}^n$. Its members are known as the *Möbius transformations* of $\hat{\mathbb{R}}^n$. The *special Möbius group* $\text{Möb}^+(n)$ is the subgroup of $\text{Möb}(n)$ consisting of those Möbius transformations that can be obtained by composing an even number of reflections.

Since any reflection R is a sense-reversing homeomorphism of $\hat{\mathbb{R}}^n$, the mappings in $\text{Möb}^+(n)$ are in fact the sense-preserving members of $\text{Möb}(n)$. From the fact that each reflection R is a diffeomorphism of the set $\hat{\mathbb{R}}^n \setminus \{\infty, R^{-1}(\infty)\}$ onto itself, we make the following inference: for any Möbius transformation f there is a finite subset E of $\hat{\mathbb{R}}^n$ such that the restriction of f to the set $D = \hat{\mathbb{R}}^n \setminus E$ is a diffeomorphism—a conformal diffeomorphism if $n \geq 2$—of D onto $f(D)$.

We shall discover later that we can always take $E = \{\infty, f^{-1}(\infty)\}$, but this is not yet entirely obvious from our definition of a Möbius transformation.

If $\pi : \hat{\mathbb{R}}^n \to \mathbb{S}^n$ is a stereographic projection, then

$$\text{Con}(n) = \pi \circ \text{Möb}(n) \circ \pi^{-1}$$

is an important and much studied group of homeomorphisms of $\mathbb{S}^n$, the *conformal group* of $\mathbb{S}^n$.

The Möbius group—indeed, the special Möbius group—contains all dilations and translations of $\mathbb{R}^n$: if $f(x) = \lambda x$ with $\lambda > 0$, we have $f = R \circ R_0$, where R_0 is the inversion in $\mathbb{S}^{n-1}$ and R is the inversion in $\mathbb{S}^{n-1}(\sqrt{\lambda})$; if $f(x) = x + b$ with $b \neq 0$, we can express f in the form $f = R_2 \circ R_1$, where R_1 is the reflection in the hyperplane $P_1 : \langle b, x\rangle = 0$ and R_2 is the reflection in $P_2 : \langle b, x\rangle = |b|^2/2$. The former assertion follows directly from (3.5), while the latter is established by a straightforward calculation, once it is observed that

$$R_1(x) = x - 2\langle \nu, x\rangle \nu, \qquad R_2(x) = x - 2\left(\langle \nu, x\rangle - \frac{|b|}{2}\right)\nu$$

with $\nu = b/|b|$. Less apparent is the fact that every orthogonal linear transformation of $\mathbb{R}^n$ belongs to $\text{Möb}(n)$.

LEMMA 3.3.1. *If U is a member of the orthogonal group $\mathrm{O}(n)$, then U can be represented as a composition of n or fewer reflections in hyperplanes that pass through the origin of $\mathbb{R}^n$.*

PROOF. If U is the identity, then $U = R \circ R$ for any such reflection. We can therefore assume $U \neq I$. We construct in a step-by-step fashion transformations $V_1, V_2, \ldots, V_n$ in $\mathrm{O}(n)$, each of them either the identity transformation I or a reflection in a hyperplane through the origin, so that for $m = 1, 2, \ldots, n$ the orthogonal linear transformation $U_m = V_m V_{m-1} \cdots V_1 U$ leaves fixed all of the vectors $e_1, e_2, \ldots, e_m$. If we can accomplish this, then U_n will fix $e_1, e_2, \ldots, e_n$, compelling the conclusion that $U_n = I$—hence, that $U = V_1 V_2 \cdots V_n$ (remember: $V_j^{-1} = V_j$) is a composition of n or fewer reflections in hyperplanes that contain the origin. To start the construction, write $b_1 = U(e_1) - e_1$. If $b_1 = 0$, take $V_1 = I$; if $b_1 \neq 0$, let V_1 be the reflection in the hyperplane $P_1 : \langle b_1, x\rangle = 0$. In the latter instance, we note that the vector $U(e_1) + e_1$ lies in P_1 for

$$\big\langle b_1, U(e_1) + e_1\big\rangle = \big\langle U(e_1) - e_1, U(e_1) + e_1\big\rangle = |U(e_1)|^2 - |e_1|^2 = 1 - 1 = 0\,.$$

In either case, the transformation $U_1 = V_1 U$ fixes e_1. This is trivial if $b_1 = 0$, since then $U_1(e_1) = U(e_1) = e_1$; if $b_1 \neq 0$, we compute

$$\begin{aligned} U_1(e_1) &= V_1\big(U(e_1)\big) = V_1\left(\frac{U(e_1) + e_1 + b_1}{2}\right) \\ &= \frac{1}{2}V_1\big(U(e_1) + e_1\big) + \frac{1}{2}V_1(b_1) = \frac{1}{2}\big(U(e_1) + e_1\big) - \frac{1}{2}b_1 = e_1\,. \end{aligned}$$

Assuming that $m < n$ and that we have been successful in producing transformations $V_1, V_2, \ldots, V_m$ of the type indicated such that U_m fixes e_j when $1 \leq j \leq m$, we proceed to construct V_{m+1}. Let $b_{m+1} = U_m(e_{m+1}) - e_{m+1}$. Mimicking what we did to begin the construction, we set $V_{m+1} = I$ if $b_{m+1} = 0$ and let V_{m+1} be the reflection in $P_{m+1} : \langle b_{m+1}, x\rangle = 0$ if $b_{m+1} \neq 0$. Exactly as before, we see that $U_{m+1} = V_{m+1} U_m$ fixes the vector e_{m+1}. It also fixes e_j for $j = 1, 2, \ldots, m$. Again this is clear when $b_{m+1} = 0$ (then $U_{m+1} = U_m$), while for $b_{m+1} \neq 0$ it is a consequence of the fact that U_m fixes e_j and that

$$\begin{aligned} \langle b_{m+1}, e_j\rangle &= \big\langle U_m(e_{m+1}) - e_{m+1}, e_j\big\rangle = \big\langle U_m(e_{m+1}), e_j\big\rangle \\ &= \big\langle U_m(e_{m+1}), U_m(e_j)\big\rangle - \langle e_{m+1}, e_j\rangle = 0\;, \end{aligned}$$

which places e_j in the fixed point locus of V_{m+1}. The procedure can thus be continued until the desired end result is achieved. □

We have shown that $\mathrm{Möb}(n)$ includes $\mathrm{GS}(n)$, the group of similarity transformations of $\mathbb{R}^n$. In fact, it is not difficult to see that $\mathrm{Möb}(n)$ is generated by the dilations, translations, and orthogonal linear transformations of $\mathbb{R}^n$, together with the inversion in $\mathbb{S}^{n-1}$. We also note that both $\mathrm{Möb}(n)$ and $\mathrm{Möb}^+(n)$ act transitively on $\hat{\mathbb{R}}^n$. This means that given $x, y \in \hat{\mathbb{R}}^n$ there is $f \in \mathrm{Möb}^+(n)$ such that $f(x) = y$.

It is a simple exercise to show that $\mathrm{Möb}(1)$ consists of all functions $f : \hat{\mathbb{R}} \to \hat{\mathbb{R}}$ that admit representations of the type $f(x) = (ax + b)/(cx + d)$, where a, b, c, and d are real numbers with $ad - bc \neq 0$. Those f for which $ad - bc > 0$ constitute the group $\mathrm{Möb}^+(1)$. In the classical case $n = 2$, members of the Möbius group also assume familiar forms.

THEOREM 3.3.2. *The group* Möb(2) *consists of the functions* $f : \hat{\mathbb{R}}^2 \to \hat{\mathbb{R}}^2$ *that can be represented with the aid of complex notation in either the form*

$$f(z) = \frac{az + b}{cz + d} \tag{3.13}$$

or the form

$$f(z) = \frac{a\bar{z} + b}{c\bar{z} + d} , \tag{3.14}$$

where $a, b, c,$ *and* d *are complex numbers and* $ad - bc \neq 0$. *The group* $\text{Möb}^+(2)$ *is made up of functions of the first type.*

PROOF. In order for the above formulas to define functions from $\hat{\mathbb{R}}^2 = \mathbb{C}$ to itself, they must be properly interpreted at certain troublesome points: if $c = 0$, $f(\infty) = \infty$; if $c \neq 0$, $f(\infty) = a/c$, $f(-d/c) = \infty$ in (3.13), and $f(-\bar{d}/\bar{c}) = \infty$ in (3.14). Modulo these conventions, a function given by (3.13) or (3.14) is a homeomorphism of $\mathbb{C}$ onto itself.

Under composition the collection of such functions forms a group, call it G. The functions of character (3.13) constitute a subgroup H of G.

Any line L in $\mathbb{C}$ admits complex equations of the kind $L : Az + \bar{A}\bar{z} + B = 0$, where $A \neq 0$ and B is real. The reflection R in L is then expressible as

$$R(z) = -(\bar{A}/A)\bar{z} - (B/A) ,$$

which has the form (3.14). The inversion R in $S^1(z_0, r)$ can be written as

$$R(z) = z_0 + \frac{r^2}{\bar{z} - \bar{z}_0} = \frac{z_0\bar{z} + r^2 - |z_0|^2}{\bar{z} - \bar{z}_0} ,$$

which is once more of type (3.14). Since the generators of Möb(2) lie in G, Möb(2) is definitely a subgroup of G. Furthermore, the composition of an even number of reflections clearly takes the form (3.13), making $\text{Möb}^+(2)$ a subgroup of H.

In complex notation the members of $\text{O}(2) \setminus \{identity\}$ have the appearance $f(z) = e^{i\theta}z$ or $f(z) = e^{i\theta}\bar{z}$, with θ real and not an integral multiple of 2π.

The function $f(z) = e^{i\theta}z$ is not itself a reflection, so in view of Lemma 3.3.1 it must be a composition of two reflections; that is, f belongs to $\text{Möb}^+(2)$.

It follows that $\text{Möb}^+(2)$ includes all functions having the structure $f(z) = az+b$ with $a \neq 0$. Indeed, we have $f = k \circ h \circ g$, where $g(z) = e^{i\theta}z$ for $\theta = \text{Arg}\, a$, $h(z) = |a|z$ is a dilation, and $k(z) = z + b$ is a translation. The group $\text{Möb}^+(2)$ also contains the function $f(z) = z^{-1}$ for $f = R_1 \circ R_0$, in which R_0 is reflection in S^1 and R_1 is reflection in the real axis. When $c \neq 0$, formula (3.13) can be rewritten

$$f(z) = \frac{a}{c} + \frac{bc - ad}{c^2} \, \frac{1}{z + (d/c)} ,$$

which implies that any such f belongs to $\text{Möb}^+(2)$. We have thus shown that H is a subgroup of $\text{Möb}^+(2)$. As a result, $\text{Möb}^+(2) = H$. Finally, since G is generated by H and the reflection R_1, itself a member of Möb(2), we conclude that $\text{Möb}(2) = G$. □

It is customary to normalize representations (3.13) and (3.14) by imposing the requirement that the "determinant" $ad - bc = 1$.

Such normalization is possible because the functions defined via these formulas are not affected when a, b, c, and d are replaced by $\lambda a, \lambda b, \lambda c$, and λd, where λ is any nonzero complex number.

One of the most important geometric features of Möbius transformations is the subject of the next theorem.

THEOREM 3.3.3. *If f is a Möbius transformation of $\hat{\mathbb{R}}^n$ and Σ is a chordal sphere in $\hat{\mathbb{R}}^n$, then $f(\Sigma)$ is also a chordal sphere.*

PROOF. If functions f and g map all chordal spheres to chordal spheres, and so does their composition. That similarity transformations of $\mathbb{R}^n$ transform chordal spheres to chordal spheres is evident. Because $\text{Möb}(n)$ is generated by $\text{GS}(n)$ and R_0, the inversion in $\mathbb{S}^{n-1}$, the proof boils down to confirming the assertion of the theorem for $f = R_0$.

Fix an equation for the chordal sphere Σ, say

$$\Sigma : a|x|^2 - 2\langle b, x\rangle + c = 0.$$

Then for x different from 0 and ∞ the point $y = R_0(x) = x/|x|^2$ satisfies

$$c|y|^2 - 2\langle b, y\rangle + a = \frac{c}{|x|^2} - 2\left\langle b, \frac{x}{|x|^2}\right\rangle + a = \frac{c - 2\langle b, x\rangle + a|x|^2}{|x|^2},$$

so x lies on Σ precisely when y lies on the chordal sphere $\Sigma' : c|y|^2 - 2\langle b, y\rangle + a = 0$. Furthermore, ∞ belongs to Σ if and only if $a = 0$, which occurs if and only if $0 = R_0(\infty)$ is on Σ'. Similarly, 0 is a point of Σ if and only if $\infty = R_0(0)$ lies on Σ'. Therefore $R_0(\Sigma) = \Sigma'$ is a chordal sphere, as claimed. □

When $n \geq 2$ the property of mapping chordal spheres to chordal spheres actually characterizes Möbius transformations:

LEMMA 3.3.4. *If $n \geq 2$ and $f : \hat{\mathbb{R}}^n \to \hat{\mathbb{R}}^n$ is an injective function with the property that $f(\Sigma)$ is a chordal sphere for every chordal sphere Σ, then f is a Möbius transformation of $\hat{\mathbb{R}}^n$.*

Since we do not intend to make any use of this observation, we leave its proof as an amusing and instructive exercise for the reader.

Theorem 3.3.3 has a generalization covering lower-dimensional chordal spheres:

LEMMA 3.3.5. *If f is a Möbius transformation of $\hat{\mathbb{R}}^n$ and Γ is a chordal p-sphere in $\hat{\mathbb{R}}^n$, then $f(\Gamma)$ is a chordal p-sphere.*

By a chordal p-sphere in $\hat{\mathbb{R}}^n$, $1 \leq p \leq n-1$, we mean either a p-dimensional Euclidean sphere contained in some $(p+1)$-dimensional affine subspace of $\mathbb{R}^n$ or a set of the type $V \cup \{\infty\}$, where V is a p-dimensional affine subspace of $\mathbb{R}^n$.

Thus "chordal sphere" is short for "chordal $(n-1)$-sphere".

As was the situation in Theorem 3.3.3, difficulties in the proof of its generalized version surface only for $f = R_0$, the inversion in $\mathbb{S}^{n-1}$. They are easily handled by doing an induction on n, using Theorem 3.3.3 and the fact that $U \circ R_0 = R_0 \circ U$ for every U from $\text{O}(n)$. The details are again left to the reader.

Let Σ be a chordal sphere in $\hat{\mathbb{R}}^n$. We can represent Σ uniquely in the manner $\Sigma = \tilde{\Sigma} \cap \hat{\mathbb{R}}^n$, in which $\tilde{\Sigma}$ is a chordal sphere in $\hat{\mathbb{R}}^{n+1}$ that is orthogonal to $\mathbb{R}^n$.

If $\pi : \hat{\mathbb{R}}^n \to \mathbb{S}^n$ denotes stereographic projection, then we are able to profit from an earlier discussion by remarking that

$$\pi(\Sigma) = \pi(\tilde{\Sigma} \cap \hat{\mathbb{R}}^n) = R(\tilde{\Sigma} \cap \hat{\mathbb{R}}^n) = R(\tilde{\Sigma}) \cap \mathbb{S}^n ,$$

where R denotes the inversion in $\mathbb{S}^n(e_{n+1}, \sqrt{2}\,)$. As the nondegenerate intersection of $\mathbb{S}^n$ with either another Euclidean sphere or a hyperplane, the set $\pi(\Sigma)$ is a Euclidean sphere of dimension $n-1$ sitting in $\mathbb{S}^n$. A moment's thought reveals that there must exist a point p_0 of $\mathbb{S}^n$ (there may be two choices for p_0) and a number r in $(0, \sqrt{2}\,]$ such that $\pi(\Sigma) = \{p \in \mathbb{S}^n : |p - p_0| = r\}$. From this we infer that $\Sigma = \{x \in \hat{\mathbb{R}}^n : q(x, x_0) = r\}$, in which $x_0 = \pi^{-1}(p_0)$. We conclude that Σ really is a sphere in the chordal metric on $\hat{\mathbb{R}}^n$.

The group $\text{Möb}(n)$ acts transitively on the family of chordal spheres in $\hat{\mathbb{R}}^n$: *if Σ and Σ' are chordal spheres in $\hat{\mathbb{R}}^n$, then there exists a Möbius transformation f of $\hat{\mathbb{R}}^n$ with the property that $f(\Sigma) = \Sigma'$.* This readily verified observation plays a role in the proof of the ensuing theorem, which would be false were $n = 1$.

THEOREM 3.3.6. *Let Σ be a chordal sphere in $\hat{\mathbb{R}}^n$ with $n \geq 2$, and let f be a member of* $\text{Möb}(n)$ *that fixes every point of Σ. Then f is either the identity transformation of $\hat{\mathbb{R}}^n$ or the reflection in Σ.*

PROOF. We deal initially with the special case $\Sigma = \mathbb{R}^{n-1} \cup \{\infty\}$. Then, in particular, $f(\infty) = \infty$. Consider a sphere $\Sigma' = \mathbb{S}^{n-1}(x_0, r)$ with x_0 in $\mathbb{R}^{n-1}$. Since ∞ is not a point of Σ', the set $f(\Sigma')$ must also be a Euclidean sphere, one for which

$$f(\Sigma') \cap \mathbb{R}^{n-1} = f(\Sigma') \cap f(\mathbb{R}^{n-1}) = f(\Sigma' \cap \mathbb{R}^{n-1}) = \Sigma' \cap \mathbb{R}^{n-1} .$$

Moreover, because Σ' is orthogonal to $\mathbb{R}^{n-1}$, the conformality of f in the complement of some finite subset of $\hat{\mathbb{R}}^n$ requires that $f(\Sigma')$, too, be orthogonal to $\mathbb{R}^{n-1}$. This information is enough to pin down the image of Σ' under f; namely, $f(\Sigma') = \Sigma'$.

Let x be an element of $\mathbb{R}^n \setminus \mathbb{R}^{n-1}$. We have seen that for each x_0 in $\mathbb{R}^{n-1}$ the sphere $S^{n-1}(x_0, |x - x_0|)$ remains invariant under f. Therefore $y = f(x)$ satisfies the condition $|y - x_0| = |x - x_0|$ for every x_0 in $\mathbb{R}^{n-1}$. The choice $x_0 = 0$ exposes the fact that $|y| = |x|$. We deduce that

$$\langle y, x_0 \rangle = \frac{|y|^2 + |x_0|^2 - |y - x_0|^2}{2} = \frac{|x|^2 + |x_0|^2 - |x - x_0|^2}{2} = \langle x, x_0 \rangle$$

for all x_0 in $\mathbb{R}^{n-1}$. Taking $x_0 = e_i$ yields $y_i = \langle y, e_i \rangle = \langle x, e_i \rangle = x_i$ for $1 \leq i \leq n-1$. Because we also know that $|y| = |x|$, either $y_n = x_n$ or $y_n = -x_n$. Now either the homeomorphism f leaves each component of $\mathbb{R}^n \setminus \mathbb{R}^{n-1}$ invariant or it interchanges these components. It follows that either $f(x) = x$ for every x in $\hat{\mathbb{R}}^n$ or $f(x) = (x_1, x_2, \ldots, -x_n)$ for all such x. In the latter case f is the reflection in $\mathbb{R}^{n-1}$.

Suppose finally that Σ is an arbitrary chordal sphere in $\hat{\mathbb{R}}^n$. We choose g in $\text{Möb}(n)$ for which $g(\Sigma_0) = \Sigma$, where $\Sigma_0 = \mathbb{R}^{n-1} \cup \{\infty\}$. Write $R = R_\Sigma$, $S = R_{\Sigma_0}$. The function $g^{-1} \circ R \circ g$ belongs to $\text{Möb}(n)$, is not the identity mapping of $\hat{\mathbb{R}}^n$, yet fixes Σ_0 pointwise. By the first part of the proof, $g^{-1} \circ R \circ g = S$. The Möbius transformation $g^{-1} \circ f \circ g$ also fixes every point of Σ_0. Again appealing to the special case treated first, we conclude that either $g^{-1} \circ f \circ g = I$, making $f = I$, or $g^{-1} \circ f \circ g = S$, in which case $f = g \circ S \circ g^{-1} = R$. □

The proof of Theorem 3.3.6 demonstrates that the reflection R_Σ associated with a chordal sphere Σ in $\hat{\mathbb{R}}^n$ is conjugate in the group Möb(n) to R_{Σ_0}, where $\Sigma_0 = \mathbb{R}^{n-1} \cup \{\infty\}$. Theorem 3.3.6 has the further consequence that

$$R_{f(\Sigma)} \circ f = f \circ R_\Sigma \tag{3.15}$$

for every f in Möb(n). To verify this, simply observe that the Möbius transformation $f^{-1} \circ R_{f(\Sigma)} \circ f$ fixes each point of Σ but is not the identity transformation. Accordingly, $f^{-1} \circ R_{f(\Sigma)} \circ f = R_\Sigma$, which is a rephrasing of (3.15). Recalling that points x and y of Σ^c are *symmetric with respect to* Σ provided $R_\Sigma(x) = y$ (hence, also, $R_\Sigma(y) = x$), we derive directly from (3.15) a useful symmetry principle (the case $n = 1$ requires a separate proof, which we omit).

THEOREM 3.3.7. *If points x and y are symmetric with respect to a chordal sphere Σ in $\hat{\mathbb{R}}^n$ and if f belongs to* Möb(n), *then $f(x)$ and $f(y)$ are symmetric with respect to $f(\Sigma)$.*

Symmetry with respect to a chordal sphere admits the following characterisation.

COROLLARY 3.3.8. *Let Σ be a chordal sphere in $\hat{\mathbb{R}}^n$ with $n \geq 2$, and let x and y be points of Σ^c, the complement of Σ in $\hat{\mathbb{R}}^n$. Then x and y are symmetric with respect to Σ if and only if every chordal sphere that passes through both x and y intersects Σ orthogonally.*

PROOF. The assertions of the theorem are evident in the case of $\Sigma = \mathbb{R}^{n-1} \cup \{\infty\}$. Because of Theorem 3.3.7 and the conformality of Möbius transformations, both the symmetry property and the orthogonality property are Möbius invariant. This fact enables us to reduce the general situation to the special one, merely by selecting a transformation f in Möb(n) that maps a given chordal sphere Σ to $\mathbb{R}^{n-1} \cup \{\infty\}$. □

3.3.1. Cross ratios. If x, y, u, and v are distinct points of $\hat{\mathbb{R}}^n$, then the *chordal cross-ratio* $[x, y, u, v]$ is the quantity defined by

$$[x, y, u, v] = \frac{q(x,u) \cdot q(y,v)}{q(x,y) \cdot q(u,v)}. \tag{3.16}$$

When all four points are finite, equation (3.16) reduces to

$$[x, y, u, v] = \frac{|x-u| \cdot |y-v|}{|x-y| \cdot |u-v|}. \tag{3.17}$$

Euclidean expressions for cross-ratios in which one of the four points involved is ∞ are:

$$\begin{cases} [\infty, y, u, v] = \dfrac{|y-v|}{|u-v|}, & [x, \infty, u, v] = \dfrac{|x-u|}{|u-v|}, \\ [x, y, \infty, v] = \dfrac{|y-v|}{|x-y|}, & [x, y, u, \infty] = \dfrac{|x-u|}{|x-y|}. \end{cases} \tag{3.18}$$

Möbius transformations are characterized by their preservation of chordal cross-ratios or Euclidean cross-ratios.

THEOREM 3.3.9. *Suppose that $f : \hat{\mathbb{R}}^n \to \hat{\mathbb{R}}^n$ is an injective function. Then f belongs to* Möb(n) *if and only if*

$$[x, y, u, v] = \big[f(x), f(y), f(u), f(v)\big] \tag{3.19}$$

whenever x, y, u, and v are distinct points of $\hat{\mathbb{R}}^n$.

PROOF. Assume first that f is a member of $\text{Möb}(n)$. If f is a dilation or a translation or an orthogonal linear transformation, then $f(\infty) = \infty$ and there is a constant $\lambda > 0$ with the property that $|f(z) - f(w)| = \lambda|z - w|$ for all points z and w of $\mathbb{R}^n$. It follows almost immediately from (3.17) and (3.18) that (3.19) holds for such f.

Consider next the inversion R_0 in $\mathbb{S}^{n-1}$. An elementary computation leads to the relation

$$|R_0(z) - R_0(w)| = \frac{|z - w|}{|z| \cdot |w|} \tag{3.20}$$

for z and w in $\mathbb{R}^n \setminus \{0\}$. With (3.17), identity (3.20) confirms (3.19) for $f = R_0$, provided that none of the points in question is 0 or ∞.

The continuity of cross-ratios is obvious from definition (3.16). From this continuity, (3.19) remains valid for this particular transformation when 0 and ∞ are included. Since (3.19) holds for $g \circ f$ whenever it holds for both f and g and since $\text{Möb}(n)$ is generated by $\text{GS}(n)$ and R_0, (3.19) obtains an arbitrary Möbius transformation f.

In proving the converse we are free to assume that $f(\infty) = \infty$. Should $f(\infty) \neq \infty$, we would simply consider $h = g \circ f$, where g is a Möbius transformation that takes $f(\infty)$ to ∞. Then h obeys (3.19), h fixes ∞, and $f = g^{-1} \circ h$ belongs to $\text{Möb}(n)$ if h does.

Let x, y, u, and v be distinct points in $\mathbb{R}^n$. From (3.18) and the invariance assumption (3.19) we learn that

$$\frac{|y - v|}{|u - v|} = [\infty, y, u, v] = \big[\infty, f(y), f(u), f(v)\big] = \frac{|f(y) - f(v)|}{|f(u) - f(v)|}$$

and

$$\frac{|y - v|}{|x - y|} = [x, y, \infty, v] = \big[f(x), f(y), \infty, f(v)\big] = \frac{|f(y) - f(v)|}{|f(x) - f(y)|}.$$

These equations enable us to conclude that

$$\frac{|f(x) - f(y)|}{|x - y|} = \frac{|f(y) - f(v)|}{|y - v|} = \frac{|f(u) - f(v)|}{|u - v|},$$

the inference being that the ratio $|f(z) - f(w)|/|z - w|$ is the same for all pairs of distinct points z and w in $\mathbb{R}^n$. If we call the ratio in question r, then the function $r^{-1}f$ is a Euclidean isometry. This makes f a member of $\text{GS}(n)$, a subgroup of $\text{Möb}(n)$. □

Picking through the details of the preceding proof, we glean some additional information on the structure of Möbius transformations.

First, we draw attention to the analogue of (3.20) for the inversion $R = R_\Sigma$ in an arbitrary Euclidean sphere $\Sigma = \mathbb{S}^{n-1}(x_0, r)$:

$$|R(x) - R(y)| = \frac{r^2|x - y|}{|x - x_0| \cdot |y - x_0|} \tag{3.21}$$

for x and y in $\mathbb{R}^n \setminus \{x_0\}$. This relation is derived from (3.20) by using the fact that $R = g^{-1} \circ f \circ R_0 \circ g$ with $g(x) = x - x_0$ and $f(x) = r^2 x$.

Next we point out:

THEOREM 3.3.10. *Let f be a member of* Möb(n). *If $f(\infty) = \infty$, then f is a similarity transformation. If $f(\infty) \neq \infty$, then $\Sigma = \{x \in \mathbb{R}^n : \|f'(x)\| = 1\}$ is a Euclidean sphere centered at $x_0 = f^{-1}(\infty)$, and f can be written as the composition $f = g \circ R_\Sigma$, in which g is a Euclidean isometry.*

PROOF. That a Möbius transformation fixing ∞ necessarily belongs to GS(n) is knowledge that emerged in the last part of the proof of Theorem 3.3.9. Assuming now that $f(\infty) \neq \infty$, let $x_0 = f^{-1}(\infty)$ and let R be the inversion in the sphere $\mathbb{S}^{n-1}(x_0, 1)$. Then $h = f \circ R$ is a Möbius transformation that fixes ∞. Thus h is in GS(n), say $h = \lambda T$ with $\lambda > 0$ and T in the group $\boldsymbol{E}(n)$. In particular, $\|h'(x)\| = \lambda$ for every x in $\mathbb{R}^n$. Because $f = h \circ R^{-1} = h \circ R$, we conclude that f is everywhere differentiable in the domain $\mathbb{R}^n \setminus \{x_0\}$, where

$$\|f'(x)\| = \|h'[R(x)]\| \cdot \|R'(x)\| = \frac{\lambda}{|x - x_0|^2},$$

where we have recalled (3.1) and (3.10).

From this statement we extract the information that

$$\Sigma = \{x \in \mathbb{R}^n : \|f'(x)\| = 1\} = \mathbb{S}^{n-1}(x_0, \sqrt{\lambda}).$$

Moreover, (3.21) gives

$$|f(x) - f(y)| = \lambda |R(x) - R(y)| = \frac{\lambda |x - y|}{|x - x_0| \cdot |y - x_0|} = |x - y| \tag{3.22}$$

whenever x and y lie on Σ. The function $g = f \circ R_\Sigma$ is yet another similarity transformation, one whose restriction to Σ is shown by (3.22) to be a Euclidean isometry. We infer that g itself is a Euclidean isometry. And, of course, $f = g \circ R_\Sigma$. □

If f is a Möbius transformation that does not fix the point ∞, then the set $\Sigma = \{x \in \mathbb{R}^n : \|f'(x)\| = 1\}$ is called the *isometric sphere* of f. Theorem 3.3.10 ties up a tiny loose end left hanging at the beginning of this section.

We are now in a position to state unequivocally: *a Möbius transformation f of $\hat{\mathbb{R}}^n$ restricts to a C^∞ diffeomorphism of the domain $\mathbb{R}^n \setminus [f^{-1}(\infty)]$ onto $\mathbb{R}^n$. This diffeomorphism is conformal when $n \geq 2$.*

As a second corollary of Theorem 3.3.10 we obtain a criterion for deciding whether two potentially different Möbius transformations actually are the same.

COROLLARY 3.3.11. *Suppose that f and g are both Möbius transformations of $\hat{\mathbb{R}}^n$. If the set $E = \{x \in \hat{\mathbb{R}}^n : f(x) = g(x)\}$ does not lie on any chordal sphere, then $f \equiv g$.*

PROOF. We demonstrate that the function $h = g^{-1} \circ f$, which fixes E pointwise, must be the identity transformation of $\hat{\mathbb{R}}^n$. Through any set of $n+1$ or fewer points of $\hat{\mathbb{R}}^n$ there always passes at least one chordal sphere. Thus the set E here can contain no fewer than $n + 2$ points. By performing a preliminary conjugation we are therefore at liberty to assume that both 0 and ∞ belong to E.

The nonzero finite points of E must include a basis for $\mathbb{R}^n$. Otherwise the linear subspace of $\mathbb{R}^n$ spanned by such points would have dimension no greater than $n - 1$, so E would definitely be contained in some chordal sphere. Since $h(0) = 0$

and $h(\infty) = \infty$, we infer from Theorem 3.3.10 that h is a linear transformation of $\mathbb{R}^n$; since h also fixes the members of a basis for $\mathbb{R}^n$, $h = I$, the identity. Thus $f \equiv g$. □

It is a direct consequence of Theorem 3.3.9 that every isometry of the metric space $(\hat{\mathbb{R}}^n, q)$ is a Möbius transformation. The chordal isometry group of $\hat{\mathbb{R}}^n$ is therefore a subgroup of Möb(n). As a matter of fact, the chordal isometries of $\hat{\mathbb{R}}^n$ are quite easy to characterize.

THEOREM 3.3.12. *The chordal isometries of* $\hat{\mathbb{R}}^n$ *are the functions* $f = \pi^{-1} \circ U \circ \pi$, *where* $\pi : \hat{\mathbb{R}}^n \to \mathbb{S}^n$ *is a stereographic projection and* U *is an arbitrary member of* $\mathrm{O}(n+1)$. *In particular, the chordal isometry group acts transitively on* $\hat{\mathbb{R}}^n$.

PROOF. Every function f of the type described is obviously a chordal isometry. Because $\mathrm{O}(n+1)$ acts transitively on $\mathbb{S}^n$, the last assertion in the theorem then becomes clear as well. It remains to show that an arbitrary chordal isometry f has the structure $f = \pi^{-1} \circ U \circ \pi = g_U$ for some U in $\mathrm{O}(n+1)$. To demonstrate this, we first choose V from $\mathrm{O}(n+1)$ in such a way that $h = g_V \circ f$ maps ∞ to ∞. Then h is also a chordal isometry and, because

$$q[h(0), \infty] = q[h(0), h(\infty)] = q(0, \infty) = 2 \ ,$$

we find that $h(0) = 0$. We conclude with the help of Theorem 3.3.10 that h takes the form $h = \lambda T$, where $\lambda > 0$ and T belongs to $\mathrm{O}(n)$. Since $|h(e_1)| = \lambda$, we see that

$$\frac{\lambda}{\sqrt{1+\lambda^2}} = q\big[h(e_1), 0\big] = q(e_1, 0) = \frac{1}{\sqrt{2}} \ ,$$

which yields $\lambda = 1$ and places h in $\mathrm{O}(n)$. There is a unique transformation U in $\mathrm{O}(n+1)$ that satisfies $U(e_{n+1}) = e_{n+1}$ and $U|\mathbb{R}^n = h$. The transformation U maps the ray that emanates from e_{n+1} and passes through a point x of $\mathbb{R}^n$ to the analogous ray determined by e_{n+1} and $h(x)$. This ensures that $\pi \circ h = U \circ \pi$, so $h = g_U$. Finally, $f = g_V^{-1} \circ h = g_V^{-1} \circ g_U = g_{V^{-1}U}$, giving f the prescribed structure. □

In a number of upcoming discussions it will be important to know the exact nature of the Möbius transformations that leave invariant certain special domains in $\hat{\mathbb{R}}^n$. When D is a domain in $\hat{\mathbb{R}}^n$, we use the notation Möb(D) to represent $\{f \in \text{Möb}(n) : f(D) = D\}$, while Möb$^+(D)$ stands for Möb$(D) \cap$ Möb$^+(n)$. Both Möb(D) and Möb$^+(D)$ are subgroups of Möb(n).

We shall sometimes ignore the fact that the members of Möb(D) are global mappings of $\hat{\mathbb{R}}^n$ and think of them simply as homeomorphisms of D onto itself, blurring in the process the distinction between a transformation from Möb(D) and its restriction to D.

We now proceed to determine the Möbius transformations that belong to Möb$(\mathbb{H}^n)$ and Möb(B^n) for $n \geq 2$. In describing the first class we write GS$(\mathbb{H}^n)$ for Möb$(\mathbb{H}^n) \cap$ GS(n).

THEOREM 3.3.13. *Let* $n \geq 2$. *A similarity transformation* f *is in* GS$(\mathbb{H}^n)$ *if and only if* f *has the form*

$$f = \lambda U + b,$$

in which $\lambda > 0$, b *is an element of* $\mathbb{R}^{n-1} = \partial\mathbb{H}^n$, *and* U *is an orthogonal transformation in* $\mathrm{O}(n)$ *that fixes* e_n.

A Möbius transformation f *belongs to* $\mathrm{M\ddot{o}b}(\mathbb{H}^n)$ *if and only if* f *is in* $\mathrm{GS}(\mathbb{H}^n)$ *or has the structure*

$$f = g \circ R_\Sigma,$$

where g *is in* $\mathrm{GS}(\mathbb{H}^n)$ *and* Σ *is a Euclidean sphere that is orthogonal to* $\mathbb{R}^{n-1}$. *In the latter case one can take* Σ *to be the isometric sphere of* f, *in which event* g *is a Euclidean isometry.*

PROOF. Let f belong to $\mathrm{GS}(\mathbb{H}^n)$, say $f = \lambda U + b$ with $\lambda > 0$, b in $\mathbb{R}^n$, and U in $\mathrm{O}(n)$. Since $f(\mathbb{H}^n) = \mathbb{H}^n$, it follows that $f(\hat{\mathbb{R}}^{n-1}) = \hat{\mathbb{R}}^{n-1}$. Then $b = f(0)$ lies in $\mathbb{R}^{n-1}$, so $U(e_n) = \lambda^{-1}(f(e_n) - b)$ is a point of $\mathbb{H}^n$. On the other hand, $U(e_n)$ is a unit vector that is orthogonal to $U(\mathbb{R}^{n-1}) = f(\mathbb{R}^{n-1}) = \mathbb{R}^{n-1}$ and conformal mappings preserve angles. These conditions mandate that $U(e_n) = e_n$ and that f has the stated form.

Conversely, every similarity transformation of the type under consideration is clearly an element of $\mathrm{GS}(\mathbb{H}^n)$.

Next, an arbitrary transformation f in $\mathrm{M\ddot{o}b}(\mathbb{H}^n)$ either fixes ∞, in which case Theorem 3.3.10 places f in $\mathrm{GS}(\mathbb{H}^n)$, or has the structure $f = g \circ R_\Sigma$, where Σ is the isometric sphere of f and g is a Euclidean isometry. In the latter case, the center $x_0 = f^{-1}(\infty)$ of Σ must lie in $\mathbb{R}^{n-1}$, which lets us know that Σ is orthogonal to $\mathbb{R}^{n-1}$. The inversion R_Σ leaves invariant every ray emanating from x_0, a fact which reveals that R_Σ preserves $\mathbb{H}^n$. This causes R_Σ and, consequently, $g = f \circ R_\Sigma$ to be members of $\mathrm{M\ddot{o}b}(\mathbb{H}^n)$. Because g fixes ∞, it thus belongs to $\mathrm{GS}(\mathbb{H}^n)$. Lastly, a Euclidean sphere that meets $\mathbb{R}^{n-1}$ orthogonally has its center in $\mathbb{R}^{n-1}$, implying as earlier that the associated reflection R transforms $\mathbb{H}^n$ to itself, hence that R and $g \circ R$ for any g from $\mathrm{GS}(\mathbb{H}^n)$ are transformations in $\mathrm{M\ddot{o}b}(\mathbb{H}^n)$. □

The group $\mathrm{M\ddot{o}b}(\mathbb{H}^n)$ for $n \geq 2$ is generated by the dilations of $\mathbb{R}^n$, the translations of $\mathbb{R}^n$ in directions parallel to $\mathbb{R}^{n-1}$, the transformations from $\mathrm{O}(n)$ that fix the vector e_n, and the inversion R_0 in $\mathbb{S}^{n-1}$. The group $\mathrm{M\ddot{o}b}^+(\mathbb{H}^n)$ is generated by the sense-preserving members of $\mathrm{GS}(\mathbb{H}^n)$ together with the transformation $f_0 = g_0 \circ R_0$, with g_0 being the reflection in the hyperplane $P : x_1 = 0$. Notice that the group $\mathrm{GS}^+(\mathbb{H}^n)$—hence, each of the groups $\mathrm{M\ddot{o}b}(\mathbb{H}^n)$ and $\mathrm{M\ddot{o}b}^+(\mathbb{H}^n)$—acts transitively on $\mathbb{H}^n$. In the classical setting of the complex plane, the mappings in $\mathrm{M\ddot{o}b}(\mathbb{H}^2)$ take one of the normalized forms

$$f(z) = \frac{az + b}{cz + d},$$

with a, b, c, and d real and $ad - bc = 1$, or

$$f(z) = \frac{a\bar{z} + b}{c\bar{z} + d},$$

in which a, b, c, and d are purely imaginary and $ad - bc = 1$. The transformations of the first kind make up $\mathrm{M\ddot{o}b}^+(\mathbb{H}^2)$.

We note in passing that when $n \geq 2$ the quantity $|y - x|^2/(y_n x_n)$ is conserved under the action of $\mathrm{M\ddot{o}b}(\mathbb{H}^n)$, meaning that

$$\frac{|f(y) - f(x)|^2}{f_n(y) f_n(x)} = \frac{|y - x|^2}{y_n x_n} \tag{3.23}$$

whenever f is a member of $\text{Möb}(\mathbb{H}^n)$ and the points x and y belong to $\mathbb{H}^n$. That this holds for any function f in $\text{GS}(\mathbb{H}^n)$ is all but trivial: writing $f = \lambda U + b$ as described in Theorem 3.3.13, we see that $|f(y) - f(x)| = \lambda|y - x|$, $f_n(y) = \lambda y_n$, and $f_n(x) = \lambda x_n$ from which (3.23) falls out easily. If $f = R_\Sigma$ for a Euclidean sphere $\Sigma = \mathbb{S}^{n-1}(x_0, r)$ orthogonal to $\mathbb{R}^{n-1}$, then

$$f(x) = x_0 + \frac{r^2(x - x_0)}{|x - x_0|^2}. \tag{3.24}$$

Thus, remembering that x_0 lies in $\mathbb{R}^{n-1}$, we get

$$f_n(y) = \frac{r^2 y_n}{|y - x_0|^2}, \qquad f_n(x) = \frac{r^2 x_n}{|x - x_0|^2}. \tag{3.25}$$

Also, by (3.21),

$$|f(y) - f(x)|^2 = \frac{r^4|y - x|^2}{|y - x_0|^2|x - x_0|^2}.$$

These observations combine to give (3.23) for $f = R_\Sigma$. Because relation (3.23) is preserved under composition, we can certify its correctness for an arbitrary f in $\text{Möb}(\mathbb{H}^n)$ by an appeal to Theorem 3.3.13.

An extremely significant consequence of (3.23), in tandem with the conformality of Möbius transformations, is the identity

$$\|f'(x)\| = \lim_{y \to x} \frac{|f(y) - f(x)|}{|y - x|} = \frac{f_n(x)}{x_n} \tag{3.26}$$

which holds for all x in $\mathbb{H}^n$ once we recall (2.21), (2.22), and Theorem 3.1.1, provided f comes from the group $\text{Möb}(\mathbb{H}^n)$.

If Σ is a chordal sphere in $\hat{\mathbb{R}}^n$, then as remarked earlier there is a unique chordal sphere $\tilde{\Sigma}$ in $\hat{\mathbb{R}}^{n+1}$ that is orthogonal to $\mathbb{R}^n$ and has $\tilde{\Sigma} \cap \hat{\mathbb{R}}^n = \Sigma$. It is apparent from the formulas defining the reflections $R = R_\Sigma$ and $\tilde{R} = R_{\tilde{\Sigma}}$ that $R = \tilde{R}|\hat{\mathbb{R}}^n$. It is equally apparent that $\tilde{R}$ leaves $\mathbb{H}^{n+1}$ invariant. On the strength of these comments, we are able to conclude that each member f of $\text{Möb}(n)$ can be extended to a Möbius transformation $\tilde{f}$ belonging to $\text{Möb}(\mathbb{H}^{n+1})$, a transformation known as the *Poincaré extension* of f. There can be only one such extension. If $\tilde{f}_1$ were a second extension, then $g = \tilde{f}^{-1} \circ \tilde{f}_1$ would be a member of $\text{Möb}(n+1)$ that fixed $\hat{\mathbb{R}}^n$ pointwise, yet was neither the identity ($\tilde{f} \neq \tilde{f}_1$) nor the reflection in $\hat{\mathbb{R}}^n$ (g preserves $\mathbb{H}^{n+1}$) contrary to Theorem 3.3.6. By Theorem 3.3.13 each function in $\text{Möb}(\mathbb{H}^{n+1})$ arises as the Poincaré extension of a unique member of $\text{Möb}(n)$. The members of $\text{Möb}^+(\mathbb{H}^{n+1})$ are the mappings $\tilde{f}$ with f in $\text{Möb}^+(n)$.

One can directly compute a formula for the Poincaré extension of an element $f \in \text{Möb}^+(n)$ to an element of $\text{Möb}^+(\mathbb{H}^{n+1})$.

3.3.2. The Poincaré extension. Suppose that $f \in \text{Möb}^+(n-1)$. There are two cases to consider. If $f(\infty) = \infty$, then f is a similarity of the form $f : \lambda U + b$ with $U \in O(n-1)$ and $b \in \mathbb{R}^{n-1}$. The obvious thing to do is set

$$\tilde{U} = \begin{pmatrix} U & 0^t \\ 0 & 1 \end{pmatrix}, \qquad \tilde{b} = (b_1, b_2, \ldots, b_{n-1}, 0)$$

and set $\tilde{f} = \lambda\tilde{U} + \tilde{b}$. Here $0 \in \mathbb{R}^{n-1}$. Thus $(x,t) \mapsto (f(x), \lambda t)$, so to write the extension we need to identify only the scale factor λ which we must choose positive to preserve $\mathbb{H}^n$. Then $\lambda = |f(e_1) - f(0)|$, and if we define

$$\tilde{f}(x,t) = (f(x), |f(e_1) - f(0)|\, t),$$

it is easy to check that this is a Möbius transformation of $\mathbb{H}^n$ and $\tilde{f}|\mathbb{R}^{n-1} = f$.

Next, if $f(\infty) = y_0 \neq \infty$, we put

$$g(x) = y_0 + \frac{f(x) - y_0}{|f(x) - y_0|^2}$$

to see that g is a Möbius transformation for which $g(\infty) = \infty$. Then g has an extension $\tilde{g}$ as above, and once we note that $g = \Phi \circ f$, where Φ is inversion in the unit sphere about y_0 as per (3.24), then the extension we seek will be $\tilde{\Phi}^{-1} \circ \tilde{g}$. The necessary calculation is simplified by observing that all we need to do is calculate the height (that is, the nth coordinate) of $(\tilde{\Phi}^{-1} \circ \tilde{g})(x,t) = (\tilde{\Phi} \circ \tilde{g})(x,t)$, as we already know that this map is f on $\mathbb{R}^{n-1}$. We can therefore use (3.25) to obtain

$$\begin{aligned} s &= \tilde{g}(x,t)_n = |g(e_1) - g(0)|t \\ &= \left| \frac{f(e_1) - f(\infty)}{|f(e_1) - f(\infty)|^2} - \frac{f(0) - f(\infty)}{|f(0) - f(\infty)|^2} \right| t, \\ \tilde{\Psi}(\tilde{g}(x,t))_n &= \frac{s}{|\tilde{g}(x,t) - (y_0, 0)|^2} = \frac{s}{|g(x) - y_0|^2 + \tilde{g}(x,t)_n^2} \\ &= \frac{s}{|f(x) - f(\infty)|^{-2} + s^2}. \end{aligned}$$

This gives a simple procedure for writing the Poincaré extension.

LEMMA 3.3.14. *Let $f \in \text{Möb}^+(n-1)$ with $f(\infty) \neq \infty$ and set*

$$\alpha = \left| \frac{f(e_1) - f(\infty)}{|f(e_1) - f(\infty)|^2} - \frac{f(0) - f(\infty)}{|f(0) - f(\infty)|^2} \right|.$$

Then the Poincaré extension $\tilde{f} \in \text{Möb}^+(\mathbb{H}^n)$ is given by the formula

$$\tilde{f}(x,t) = \left(f(x), \frac{\alpha t}{|f(x) - f(\infty)|^{-2} + \alpha^2 t^2} \right). \tag{3.27}$$

Of course an alternative approach to the proof of this lemma is via the cross-ratio. Here we tacitly use the inclusion $\mathbb{R}^{n-1} \hookrightarrow \mathbb{R}^n$. Then we must have from Theorem 3.3.9 and (3.18),

$$\begin{aligned} [0, e_1, \infty, (x,t)] &= [\tilde{f}(0), \tilde{f}(e_1), \tilde{f}(\infty), \tilde{f}(x,t)] = [f(0), f(e_1), f(\infty), (f(x), \tilde{t})], \\ |x - e_1|^2 + t^2 &= \frac{|f(0) - f(\infty)|^2 |f(e_1) - (f(x), \tilde{t})|^2}{|f(0) - f(e_1)|^2 |f(\infty) - (f(x), \tilde{t})|^2} \\ &= \frac{|f(0) - f(\infty)|^2 \left(|f(e_1) - f(x)|^2 + \tilde{t}^2\right)}{|f(0) - f(e_1)|^2 \left(|f(\infty) - f(x)|^2 + \tilde{t}^2\right)}. \qquad (3.28) \end{aligned}$$

From this one can solve for a positive $\tilde{t}$, and the result will follow. The appearance of the variable x in the formula for $\tilde{t}$ obtained seems not to coincide with (3.27).

This is accounted for in the invariance of another cross-ratio which relates x and $f(x)$. For instance if $f(\infty) = \infty$, $f : \lambda U + b$, $\lambda > 0$, (3.28) becomes

$$\begin{aligned} |x - e_1|^2 + t^2 &= \frac{|f(e_1) - f(x)|^2 + \tilde{t}^2}{|f(0) - f(e_1)|^2} = \frac{\lambda^2 |x - e_1|^2 + \tilde{t}^2}{\lambda^2}, \\ \lambda^2 t^2 &= \tilde{t}^2, \end{aligned}$$

as before. Indeed there are even further approaches using the hyperbolic geometry of lines and their perpendiculars which the reader may care to explore.

3.3.3. The group Möb(B^n). Before attempting to characterize the transformations in the group Möb(B^n) for $n \geq 2$, we make some preparatory remarks. Recall that the function $\Phi = R \circ S$, where S is reflection in $\mathbb{R}^{n-1}$ and R is reflection in $\mathbb{S}^{n-1}(e_n, \sqrt{2}\,)$, defines a Möbius transformation that maps $\mathbb{H}^n$ to B^n. Let Σ be a chordal sphere that is orthogonal to $\mathbb{S}^{n-1}$. The chordal sphere $\Sigma' = \Phi^{-1}(\Sigma)$ is orthogonal to $\mathbb{R}^{n-1}$, and $R_\Sigma = \Phi \circ R_{\Sigma'} \circ \Phi^{-1}$. We already know that $R_{\Sigma'}$ preserves $\mathbb{H}^n$, allowing us to conclude: *the reflection R_Σ in any chordal sphere Σ that meets $\mathbb{S}^{n-1}$ orthogonally leaves B^n invariant.* A hyperplane P in $\mathbb{R}^n$ is orthogonal to $\mathbb{S}^{n-1}$ if and only if P passes through the origin.

The question then is to determine under what conditions does the Euclidean sphere Σ with center x_0 and radius r intersect $\mathbb{S}^{n-1}$ orthogonally? We maintain that this happens precisely when

$$|x_0|^2 = 1 + r^2 \,. \tag{3.29}$$

Given that $\Sigma \cap S^{n-1} \neq \emptyset$, the law of cosines determines the angle θ formed by radii of the two spheres at a point of intersection:

$$\cos\theta = \frac{1 + r^2 - |x_0|^2}{2r} \,.$$

To get $\theta = \pi/2$, (3.29) is necessary. Conversely, if (3.29) holds, then the center of Σ lies in $\mathbb{R}^n \setminus \bar{B}^n$. The nearest point of Σ to the origin has norm $|x_0| - r$, which is a number in $(0, 1)$ —a glance at (3.29) shows that $|x_0| > r$ and $(|x_0| - r)(|x_0| + r) = 1$— so Σ intersects B^n. Therefore Σ intersects $\mathbb{S}^{n-1}$, and the law of cosines affirms that the angle of intersection is a right angle. With these comments behind us, we may prove the following theorem.

THEOREM 3.3.15. *Let $n \geq 2$. A Möbius transformation f belongs to* Möb(B^n) *if and only if f is an orthogonal transformation in* O(n) *or has the structure*

$$f = g \circ R_\Sigma,$$

where $g \in$ O(n) and Σ is a Euclidean sphere that is orthogonal to $\mathbb{S}^{n-1}$.

PROOF. The transformations of the two types mentioned are definitely included in Möb(B^n). To establish the converse, consider an arbitrary member f of Möb(B^n). Out of necessity f transforms $\mathbb{S}^{n-1}$ to itself. If $f(0) = 0$, f also fixes ∞ (Theorem 3.3.7). Theorem 3.3.10 then implies that f is a linear transformation, and a linear transformation that leaves B^n invariant must be an orthogonal transformation. Assuming next that $x_0 = f^{-1}(0)$ is nonzero, let $r = (|x_0^*|^2 - 1)^{1/2}$ and $\Sigma = \mathbb{S}^{n-1}(x_0^*, r)$, where x_0^* is the point that is symmetric to x_0 with respect to $\mathbb{S}^{n-1}$. By reason of (3.29) the sphere Σ is orthogonal to $\mathbb{S}^{n-1}$, so R_Σ maps B^n to itself. Since $R_\Sigma(\infty) = x_0^*$, Theorem 3.3.7 guarantees that $R_\Sigma(0) = x_0$.

We infer that $g = f \circ R_\Sigma$ is a transformation from $\text{Möb}(B^n)$ and fixes the origin. The first part of our discussion implies that g belongs to $\text{O}(n)$ and $f = g \circ R_\Sigma$. $\square$

In light of Lemma 3.3.1, Theorem 3.3.15 shows that $\text{Möb}(B^n)$ is generated by the family of reflections R_Σ in which Σ is a chordal sphere orthogonal to $\mathbb{S}^{n-1}$. Further, an analogous statement is true for $\text{Möb}(\mathbb{H}^n)$: it is generated by the reflections R_Σ with Σ orthogonal to $\mathbb{R}^{n-1}$.

Of course it follows that the members of the group $\text{Möb}^+(B^n)$ are the transformations from $\text{SO}(n)$ and the mappings of the type $f = g \circ R_\Sigma$, where g is a sense-reversing orthogonal transformation and Σ is a Euclidean sphere that meets $\mathbb{S}^{n-1}$ at a right angle.

We draw particular attention to the transformations T_b in $\text{Möb}^+(B^n)$ defined for b in B^n as follows: $T_0 = I$, and for $b \neq 0$

$$T_b = g_b \circ R_b \ , \tag{3.30}$$

where R_b is the inversion in the Euclidean sphere of radius $r = (|b^*|^2 - 1)^{1/2}$ centered at b^* (notation as earlier), and g_b is the reflection in the hyperplane P: $\langle b, x\rangle = 0$.

The transformation T_b maps b to the origin. If $b \neq 0$, T_b fixes the points $b/|b|$ and $-b/|b|$. It thus leaves invariant the diameter of B^n through b. The presence of the transformations T_b in $\text{Möb}^+(B^n)$ shows directly that the action of this group on B^n is transitive. The mappings in $\text{Möb}(B^2)$ admit complex representations of the character

$$f(z) = \frac{az + b}{\bar{b}z + \bar{a}} \quad \text{or} \quad f(z) = \frac{a\bar{z} + b}{\bar{b}\bar{z} + \bar{a}} \ ,$$

in which a and b are complex numbers satisfying $|a|^2 - |b|^2 = 1$. Those of the first type are the mappings in $\text{Möb}^+(B^2)$.

The identities (3.23) and (3.26) have counterparts for the group $\text{Möb}(B^n)$ with $n \geq 2$:

$$\frac{|f(y) - f(x)|^2}{(1 - |f(y)|^2)(1 - |f(x)|^2)} = \frac{|y - x|^2}{(1 - |y|^2)(1 - |x|^2)} \tag{3.31}$$

whenever f belongs to $\text{Möb}(B^n)$ and the points x and y lie in B^n; if f is in $\text{Möb}(B^n)$, then

$$\|f'(x)\| = \frac{1 - |f(x)|^2}{1 - |x|^2} \tag{3.32}$$

for all x in B^n. The relation (3.31) is preserved under composition and holds trivially if f is in $\text{O}(n)$. To complete the verification of this fact, we must check that it holds if $f = R_\Sigma$ for a Euclidean sphere Σ perpendicular to $\mathbb{S}^{n-1}$. Let Σ have center x_0 and radius r, and write $R = R_\Sigma$. Then $|x_0|^2 = 1 + r^2$, $R(x_0^*) = 0$, and by (3.21)

$$\begin{aligned} |R(x)| &= |R(x) - R(x_0^*)| = \frac{r^2|x - x_0^*|}{|x - x_0| \cdot |x_0^* - x_0|} = \frac{r^2|x_0| \cdot |x - x_0^*|}{|x - x_0|(|x_0|^2 - 1)} \\ &= \frac{|x_0| \cdot |x - x_0^*|}{|x - x_0|} \end{aligned}$$

for every $x \in B^n$. Consequently,

$$\begin{aligned} 1 - |R(x)|^2 &= 1 - \frac{|x_0|^2 |x - x_0^*|^2}{|x - x_0|^2} \\ &= \frac{|x|^2 - 2\langle x, x_0\rangle + |x_0|^2 - |x_0|^2|x|^2 + 2\langle x, x_0\rangle - 1}{|x - x_0|^2}, \end{aligned}$$

which simplifies to

$$1 - |R(x)|^2 = \frac{r^2(1 - |x|^2)}{|x - x_0|^2}.$$

When coupled with (3.21), this yields (3.31) for $f = R_\Sigma$. Identity (3.32) is derived from (3.31) in the same manner that (3.26) was derived from (3.23):

$$\|f'(x)\| = \lim_{y \to x} \frac{|f(y) - f(x)|}{|y - x|} = \frac{1 - |f(x)|^2}{1 - |x|^2}.$$

Just as every f in Möb(n) has its Poincaré extension $\tilde{f}$ in Möb$(\mathbb{H}^{n+1})$, so each f in the conformal group Con(n) can be extended in a unique way to a Möbius transformation $\tilde{f}$ (also called the Poincaré extension of f) belonging to Möb(B^{n+1}). This follows from the fact that we know a transformation Φ in Möb$(n+1)$ that maps $\mathbb{H}^{n+1}$ to B^{n+1} and agrees with stereographic projection on $\hat{\mathbb{R}}^n$. Thus Möb$(B^{n+1}) = \Phi \circ$ Möb$(\mathbb{H}^{n+1}) \circ \Phi^{-1}$ and Con$(n) = \Phi \circ$ Möb$(n) \circ \Phi^{-1}$.

3.4. Hyperbolic geometry

In this section we derive some elementary aspects of hyperbolic geometry which are not only interesting in and of themselves, but which will prove useful for later applications. Further, the famous Mostow rigidity theorem we shall establish in Chapter 9 concerns itself with deformations of certain groups of isometries of hyperbolic space, and so we will need to know not only what these are, but will need to have some familiarity with hyperbolic geometry. There are many books which explore this subject in great depth, and we would recommend the monograph of J. G. Ratcliffe [**138**] at least as a starting point.

Here we start with a fairly general construction of metrics on domains in Euclidean space.

3.4.1. Conformally Euclidean metrics. Let D be a domain in $\mathbb{R}^n$. There is a standard procedure for modifying the Euclidean geometry of D in such a way that the interpretations of length, area, volume, and similar quantities undergo modification, while the measurement of angles is unaffected. These are the conformally Euclidean metrics.

We start by choosing from the class $C^\infty(D)$ a positive function ρ, termed in this context as a *metric density* in D.

Given a piecewise smooth path $\gamma : [a, b] \to D$ (γ is piecewise C^1, say), one defines its *ρ-length* $\ell_\rho(\gamma)$ by

(3.33) $$\ell_\rho(\gamma) = \int_\gamma \rho(x)\,|dx| = \int_a^b \rho(\gamma(t)) \cdot |\dot{\gamma}(t)|\,dt.$$

The *ρ-distance* $d_\rho(x, y)$ between points x and y of D is then defined by the rule

(3.34) $$d_\rho(x, y) = \inf_\gamma \ell_\rho(\gamma),$$

where the infimum is taken over the class of piecewise smooth paths γ in D with initial point x and endpoint y.

Again we point out that, at least in the locally Euclidean setting, we need only γ to be locally rectifiable so that (3.33) is defined (possibly equal to $+\infty$) and we could take the infimum over such paths. It is easily seen that these routes all lead to the same distance function for a given metric density ρ.

When $d_\rho : D \times D \to [0, \infty)$ is defined in this way, the pair (D, d_ρ) becomes a metric space. The only possible issue is with the triangle inequality, but this follows easily by joining paths in the obvious manner.

In fact, d_ρ is the distance function associated with the Riemannian metric in D whose fundamental form is

$$ds^2 = \rho^2(x)(dx_1^2 + dx_2^2 + \cdots + dx_n^2). \tag{3.35}$$

For instance, the choice $\rho \equiv 1$ gives rise to the *relative Euclidean metric* in D, which may differ radically from the restriction to D of the ordinary Euclidean metric unless the domain D is convex, since our infimum is taken only over paths lying in D, and the Euclidean line segment between two points of D may not.

In general, metrics of the type under discussion here are said to be *conformally Euclidean metrics* since if u and v are nonzero vectors based at a point x of D, then the angle between u and v in the Riemannian geometry of D corresponding to such a metric happens to be the same as the Euclidean angle between u and v. This follows from the form of the fundamental form at (3.35)—the inner product on the tangent space at x, where u and v lie, is a scalar multiple (in fact $\rho(x)$) times the Euclidean inner product. Thus the angles are identical.

The reader should note that in fact smoothness of the metric density is not a key feature in much of our discussion and the definitions of ℓ_ρ and d_ρ continue to make perfectly good sense if our insistence that the density ρ belong to $C^\infty(D)$ is relaxed to the demand that ρ be merely a positive continuous function in D. The pair (D, d_ρ) is still a metric space, albeit one possibly without the full array of geometric structure that would normally accompany a Riemannian metric in D. However many of these geometric structures—such as the existence of geodesics—remain with mild growth assumptions on ρ near ∂D.

The continuity of ρ allows the conclusion that

$$\lim_{y \to x} \frac{d_\rho(x, y)}{|x - y|} = \rho(x) \tag{3.36}$$

for every x in D. From (3.36) it is a short step to the observation that a sequence $\langle x_\nu \rangle$ in D converges to a point x of D in the Euclidean metric if and only if $d_\rho(x_\nu, x) \to 0$ as $\nu \to \infty$. In other words, the metric topology in D associated with d_ρ agrees with the topology induced in D by the standard topology of $\mathbb{R}^n$, independent of ρ.

Assume that D is a proper subdomain of $\mathbb{R}^n$. We give an example of the foregoing construction which will later play a very important role in the theory. We take as a metric density in D the function

$$\rho(x) = \frac{1}{\operatorname{dist}(x, \partial D)},$$

which is continuous but not usually differentiable in D. The metric d_ρ that arises from this choice of density is known as the *quasi-hyperbolic metric in* D and is usually denoted by k_D. We will next see the motivation for the terminology "quasi-hyperbolic". It can be shown that k_D is a complete metric—every quasi-hyperbolic Cauchy sequence in D converges to some point of D—a property definitely not shared by the Euclidean metric in D, be it the restricted metric or the relative one. Metrics such as this, and which reflect the geometry of the domain in question, are very useful tools for studies in conformal geometry. The quasi-hyperbolic metric was introduced in [**52**] where they further showed the existence of geodesics for this metric. It has since proven to be a useful tool.

3.4.2. Hyperbolic spaces. Let $n \geq 2$. The function

$$\rho(x) = \frac{1}{x_n} \tag{3.37}$$

is a metric density of class C^∞ in the upper half-space $\mathbb{H}^n$. With ρ is associated a conformally Euclidean metric d_ρ in $\mathbb{H}^n$, a metric known as the *hyperbolic metric* or *Poincaré metric*. The metric space $(\mathbb{H}^n, d_\rho)$ is called the *Poincaré half-space model for n-dimensional hyperbolic space* $\mathbb{H}^n$, the name deriving from the fact that the Riemannian geometry of $\mathbb{H}^n$ given by ρ is an n-dimensional non-Euclidean hyperbolic geometry and, for instance, the sum of the vertex angles of any geodesic triangle in this geometry is less than π.

In fact, to phrase things in the standard terminology of differential geometry, hyperbolic n-space $\mathbb{H}^n$ is the unique—up to isometry—complete, simply connected, n-dimensional Riemannian manifold having constant sectional curvature equal to -1. This abstract space has a number of concrete realizations, of which $\mathbb{H}^n$ endowed with the Poincaré metric is one such.

In this setting we switch to the notation $\ell_{\mathbb{H}}$ and $d_{\mathbb{H}}$ in place of the generic ℓ_ρ and d_ρ—we do not include the dimension $n \geq 2$, which will always be understood from the circumstances, to avoid clumsiness. The set $\hat{\mathbb{R}}^{n-1} = \partial\mathbb{H}^n$ also acquires a special name here; it is designated the *sphere at infinity* as one easily sees that the distance from a finite point x (an element of $\mathbb{H}^n$) to the boundary $\partial\mathbb{H}^n$ is infinite, by which we mean

$$\lim_{\mathbb{H}^n \ni y \to \partial\mathbb{H}^n} d_{\mathbb{H}}(x, y) = +\infty.$$

This fact is an easy consequence of the distance formula (3.39) established below.

Suppose f belongs to the group Möb$(\mathbb{H}^n)$. If $\gamma : [a, b] \to \mathbb{H}^n$ is a piecewise smooth path, then the path $\beta = f \circ \gamma$ is piecewise smooth as well and, because of the conformality of f, satisfies

$$|\dot{\beta}(t)| = |f'[\gamma(t)]\dot{\gamma}(t)| = \|f'[\gamma(t)]\| \cdot |\dot{\gamma}(t)|$$

whenever t is a point at which γ is differentiable. Recalling equation (3.26), we compute that

$$\begin{aligned} \ell_{\mathbb{H}}(\beta) &= \int_\beta \frac{|dx|}{x_n} = \int_a^b \frac{|\dot{\beta}(t)|\,dt}{\beta_n(t)} = \int_a^b \frac{\|f'[\gamma(t)]\| \cdot |\dot{\gamma}(t)|\,dt}{f_n[\gamma(t)]} \\ &= \int_a^b \frac{|\dot{\gamma}(t)|\,dt}{\gamma_n(t)} = \int_\gamma \frac{|dx|}{x_n} = \ell_{\mathbb{H}}(\gamma)\,. \end{aligned}$$

We thus recognize that the hyperbolic path length is invariant under f. This, in turn, makes certain that

$$d_{\mathbb{H}}\big[f(x), f(y)\big] = d_{\mathbb{H}}(x, y)$$

for all points x and y of $\mathbb{H}^n$ whenever f belongs to $\text{Möb}(\mathbb{H}^n)$. Stated differently, the mappings in $\text{Möb}(\mathbb{H}^n)$ are hyperbolic isometries. It is a fact that the group $\text{Möb}(\mathbb{H}^n)$ includes every hyperbolic isometry of $\mathbb{H}^n$, as we shall subsequently demonstrate.

Let r and s satisfy $0 < r < s < \infty$ and let $\gamma : [a, b] \to \mathbb{H}^n$ be a piecewise smooth path with $\gamma(a) = re_n$ and $\gamma(b) = se_n$. Then

$$\begin{aligned} \ell_{\mathbb{H}}(\gamma) &= \int_a^b \frac{|\dot\gamma(t)|\,dt}{\gamma_n(t)} \geq \int_a^b \frac{|\dot\gamma_n(t)|\,dt}{\gamma_n(t)} \geq \left| \int_a^b \frac{\dot\gamma_n(t)\,dt}{\gamma_n(t)} \right| \\ &= |\log \gamma_n(b) - \log \gamma_n(a)| = \log \frac{s}{r} \,. \end{aligned}$$

Moreover, $\ell_{\mathbb{H}}(\gamma) > \log(s/r)$ will definitely prevail unless two conditions are met. The first is that $\dot\gamma_1 = \dot\gamma_2 = \cdots = \dot\gamma_{n-1} = 0$ on $[a, b]$, which implies that $\gamma_1 = \gamma_2 = \cdots = \gamma_{n-1} = 0$ (remember that $\gamma_1(a) = \gamma_2(a) = \cdots = \gamma_{n-1}(a) = 0$). The second is that, once allowances are made for a finite number of points t at which $\dot\gamma_n(t)$ might fail to exist, $\dot\gamma_n$ is nonnegative on $[a, b]$, making γ_n a nondecreasing function on this interval. On the other hand, taking $\gamma(t) = te_n$ for $r \leq t \leq s$ we find that $\ell_{\mathbb{H}}(\gamma) = \log(s/r)$. The upshot of this discussion is that

$$d_{\mathbb{H}}(re_n, se_n) = \left| \log \frac{s}{r} \right| \tag{3.38}$$

for r and s in $(0, \infty)$ and that any piecewise smooth path γ between re_n and se_n with a chance of minimizing hyperbolic length (i.e., satisfying $\ell_{\mathbb{H}}(\gamma) = d_{\mathbb{H}}(re_n, se_n)$) must have the Euclidean line segment with endpoints re_n and se_n as its trajectory.

Given points x and y in $\mathbb{H}^n$, one can easily produce a transformation f in $\text{Möb}(\mathbb{H}^n)$ such that $f(x) = e_n$ and $f(y) = re_n$ with $r \geq 1$. In conjunction with (3.38), this comment leads to a nice formula for $d_{\mathbb{H}}(x, y)$:

$$\cosh d_{\mathbb{H}}(x, y) = 1 + \frac{|x - y|^2}{2x_n y_n} \,. \tag{3.39}$$

Indeed, (3.38) proves that (3.39) holds when e_n and re_n are substituted for x and y, respectively. However, both sides of (3.39) are invariant under $\text{Möb}(\mathbb{H}^n)$—don't forget (3.23)—so the validity of (3.39) persists for $x = f^{-1}(e_n)$ and $y = f^{-1}(re_n)$.

Let us employ (3.39), for example, to compute $d_{\mathbb{H}}(x, |x|e_n)$ for x in $\mathbb{H}^n$. We claim that

$$d_{\mathbb{H}}(x, |x|e_n) = \log\big[\cot(\varphi/2)\big] \,, \tag{3.40}$$

in which $\varphi = (\pi/2) - \theta(x, e_n)$. We assume that x is not a scalar multiple of e_n, for (3.40) holds trivially otherwise. Since both sides of (3.40) go unchanged when x is replaced by λx with $\lambda > 0$, we may further suppose that $|x| = 1$. Write $x = ae_n + bu$ with $a > 0$, $b > 0$, and u a unit vector in $\mathbb{R}^{n-1}$. Considering that $a^2 + b^2 = |x|^2 = 1$, we deduce from (3.39) that

$$\begin{aligned}\cosh d_{\mathbb{H}}(x, e_n) &= 1 + \frac{|(a-1)e_n + bu|^2}{2a} = 1 + \frac{(a-1)^2 + b^2}{2a} \\ &= \frac{1 + a^2 + b^2}{2a} = \frac{1}{a} = \csc\varphi\,.\end{aligned}$$

Solving for $d_{\mathbb{H}}(x, e_n)$ gives (3.40) when $|x| = 1$.

Formula (3.39) has a number of other noteworthy implications. One is that the hyperbolic sphere of radius r centered at a point x of $\mathbb{H}^n$, by which we mean the set

$$S_{\mathbb{H}}^{n-1}(x, r) = \{y \in H^n : d_{\mathbb{H}}(y, x) = r\},$$

is actually a Euclidean sphere. In fact, a brief calculation shows us that

$$S_{\mathbb{H}}^{n-1}(x, r) = \mathbb{S}^{n-1}(\tilde{x}, s)$$

with

$$\tilde{x} = (x_1, x_2, \ldots, x_{n-1}, x_n \cosh r) \quad \text{and} \quad s = x_n \sinh r.$$

Thus the families of closed hyperbolic balls in $\mathbb{H}^n$ and closed Euclidean balls in $\mathbb{H}^n$ coincide, although the hyperbolic center and hyperbolic radius of any such ball may differ from their Euclidean counterparts.

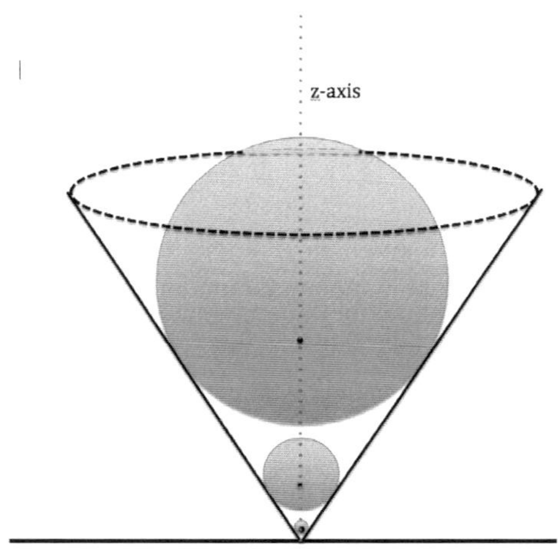

Hyperbolic balls of the same radius in $\mathbb{H}^3$. The cone consists of points a fixed distance from the z-axis.

A subset L of $\mathbb{H}^n$ is called a *hyperbolic line* (an $\mathbb{H}$-*line*, for short) if L can be written as $L = \Gamma \cap \mathbb{H}^n$, where Γ is a chordal circle in $\hat{\mathbb{R}}^n$ that is orthogonal to $\mathbb{R}^{n-1}$.

Through any pair of distinct points x and y of $\mathbb{H}^n$ there passes one and only one hyperbolic line L; the subarc of L with endpoints x and y is called the *hyperbolic segment* (or just the $\mathbb{H}$-*segment*) between x and y. A hyperbolic line will have two endpoints on the sphere at infinity; these are not, however, points of hyperbolic space.

It is an easy exercise to verify that the group Möb($\mathbb{H}^n$) acts transitively on the family of all hyperbolic lines in $\mathbb{H}^n$—although this is more easily seen in other models of hyperbolic space. Thus remarks in the preceding paragraphs quickly bring one to the realization that any (piecewise smooth) path from x to y whose hyperbolic length is minimal among all paths joining x to y must have the $\mathbb{H}$-segment between these points as its trajectory. For this reason we often speak of $\mathbb{H}$-segments as *hyperbolic geodesic segments* and of $\mathbb{H}$-lines as *hyperbolic geodesics*. The *standard* (or *hyperbolic arclength*) *parametrization* γ_0 of the $\mathbb{H}$-segment between x and y, described in the direction from x to y, is given as follows:

$$\gamma_0(t) = f^{-1}(e^t e_n), \qquad 0 \le t \le \log r,$$

where f is any member of $\text{Möb}(\mathbb{H}^n)$ that satisfies $f(x) = e_n$ and $f(y) = re_n$ with $r > 1$. Notice that a point z of $\mathbb{H}^n$ lies on this $\mathbb{H}$-segment if and only if $d_{\mathbb{H}}(x,z) + d_{\mathbb{H}}(z,y) = d_{\mathbb{H}}(x,y)$.

With regard to angles in hyperbolic geometry, we reiterate something that was pointed out earlier in reference to angles in arbitrary conformally Euclidean geometries; namely, that the hyperbolic angle determined at a point x of $\mathbb{H}^n$ by two hyperbolic rays issuing from x is just the Euclidean angle between these rays, which is to say the Euclidean angle between the vectors tangent to these rays at x.

Let $\bar{B}$ be a closed ball in $\mathbb{H}^n$. The set $\bar{B}$ is *hyperbolically convex* (abbreviated $\mathbb{H}$-*convex*), which means that for every pair of points x and y in B the entire $\mathbb{H}$-segment between x and y is contained in $\bar{B}$.

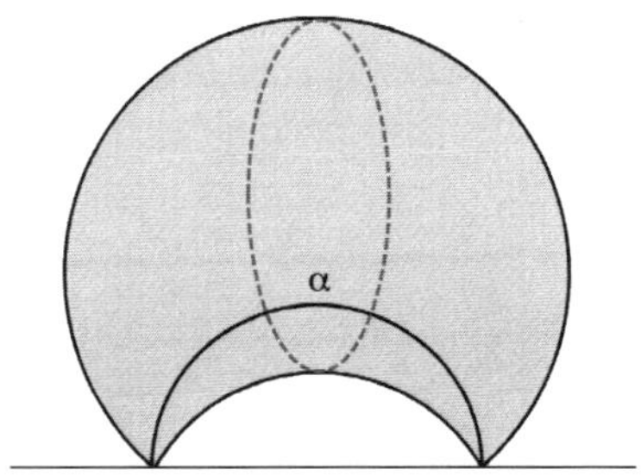

The collection of points a fixed distance from a hyperbolic line $\alpha \in \mathbb{H}^3$ is hyperbolically convex but not convex unless α is also a Euclidean line.

To recognize this it is enough to notice that the intersection of $\bar{B}$ with any hyperbolic line L, if nonempty, is an $\mathbb{H}$-segment or a single point. The truth of this observation is clear for $L = L_0$, the positive x_n-axis; its validity for arbitrary L is established by transforming L to L_0 with a member of $\text{Möb}(\mathbb{H}^n)$ and remembering that the mappings in this group preserve the classes of $\mathbb{H}$-lines, $\mathbb{H}$-segments, and closed balls in $\mathbb{H}^n$.

We may exploit the $\mathbb{H}$-convexity of $\bar{B}$ in deriving the bounds

$$a|x-y| \le d_{\mathbb{H}}(x,y) \le b|x-y| \tag{3.41}$$

for all x and y in $\bar{B}$, where

$$a = \min\{x_n^{-1} : x \in \bar{B}\}, \qquad b = \frac{\pi}{2}\max\{x_n^{-1} : x \in \bar{B}\}.$$

Consider a pair of points x and y of $\bar{B}$, which we assume to be distinct, and let γ_0 denote the standard parametrization of the $\mathbb{H}$-segment A between x and y.

From elementary geometry we obtain estimates for the Euclidean length $\ell(A)$ of A,

$$|x-y| \le \ell(A) \le \frac{\pi|x-y|}{2}.$$

Since A lies completely in the ball $\bar{B}$, we discover that

$$\begin{aligned} a|x-y| \le a\ell(A) &= a\int_{\gamma_0} |dx| \le \int_{\gamma_0} \frac{|dx|}{x_n} = d_{\mathbb{H}}(x,y) \le \frac{2b}{\pi}\int_{\gamma_0} |dx| \\ &= \frac{2b\ell(A)}{\pi} \le b|x-y|. \end{aligned}$$

The estimates in (3.41) are instrumental in the proof of an important fact.

THEOREM 3.4.1. *The metric space* $(\mathbb{H}^n, d_{\mathbb{H}})$ *is complete.*

PROOF. Let $\langle x_\nu \rangle$ be a hyperbolic Cauchy sequence. Then $r = \sup_\nu d_{\mathbb{H}}(x_\nu, x_1) < \infty$, so $\langle x_\nu \rangle$ is a sequence in the closed ball $\bar{B} = \{x \in \mathbb{H}^n : d_{\mathbb{H}}(x, x_1) \le r\}$. The first inequality in (3.41) implies that $\langle x_\nu \rangle$ is a Euclidean Cauchy sequence, hence that it converges in the Euclidean metric to some point x_0 of the closed set $\bar{B}$. The second half of (3.41) then informs us that $d_{\mathbb{H}}(x_\nu, x_0) \to 0$. Thus $\langle x_\nu \rangle$ converges to a finite point x_0 in the hyperbolic metric. $\square$

We make note of the fact that, for fixed x in $\mathbb{H}^n$, $d_{\mathbb{H}}(y,x) \to \infty$ as $y \to \partial\mathbb{H}^n$ (i.e., as $q(y, \partial\mathbb{H}^n) \to 0$). Otherwise there would be a sequence $\langle y_\nu \rangle$ in $\mathbb{H}^n$ such that $y_\nu \to \partial\mathbb{H}^n$, yet $r = \sup_\nu d_{\mathbb{H}}(y_\nu, x) < \infty$. But the closed ball $\bar{B} = \{y \in \mathbb{H}^n : d_{\mathbb{H}}(y,x) \le r\}$ is a compact set and is disjoint from $\partial\mathbb{H}^n$, implying that $\inf_\nu q(y_\nu, \partial\mathbb{H}^n) \ge q(\bar{B}, \partial\mathbb{H}^n) > 0$ and thereby contradicting the assumption that $q(y_\nu, \partial\mathbb{H}^n) \to 0$.

As indicated earlier, there are a number of alternative models for the hyperbolic space $\mathbb{H}^n$, one of which is the *Poincaré ball model.* It is provided by the unit ball B^n when it is equipped with the conformally Euclidean metric that corresponds to the density

$$\rho(x) = \frac{2}{1-|x|^2}. \tag{3.42}$$

We continue to write $\ell_{\mathbb{H}}$ and $d_{\mathbb{H}}$ for ρ-length and ρ-distance here, letting the context determine whether we are working in $\mathbb{H}^n$ or B^n.

As a matter of fact, most of what has been said about $(\mathbb{H}^n, d_{\mathbb{H}})$ can be transferred without effort to the setting of $(B^n, d_{\mathbb{H}})$ simply by using the mapping Φ that has aided us on several previous occasions: $\Phi = R \circ S$, where S is the reflection in $\mathbb{R}^{n-1}$ and R is the inversion in $\mathbb{S}^{n-1}(e_n, \sqrt{2})$. We know that Φ belongs to $\text{Möb}^+(n)$ and has $\Phi(\mathbb{H}^n) = B^n$. Taking heed of the fact that $R(-e_n) = 0$ and making reference to (3.21) and (3.10), we realize that for x different from e_n and ∞ it is true that

$$|R(x)|^2 = |R(x) - R(-e_n)|^2 = \frac{|x+e_n|^2}{|x-e_n|^2} = 1 + \frac{4x_n}{|x-e_n|^2}$$

and

$$\|R'(x)\| = \frac{2}{|x-e_n|^2}.$$

Accordingly, for any x in $\mathbb{H}^n$ we get

$$1 - |\Phi(x)|^2 = 1 - \big|R\big[S(x)\big]\big|^2 = \frac{4x_n}{|S(x) - e_n|^2} \tag{3.43}$$

and, by an appeal to the conformality of S in $\mathbb{H}^n$ and of R in $S(\mathbb{H}^n)$,

$$\|\Phi'(x)\| = \big\|R'\big[S(x)\big]\big\| \cdot \|S'(x)\| = \frac{2}{|S(x) - e_n|^2} \,. \tag{3.44}$$

From (3.43) and (3.44) it follows that

$$\frac{2\|\Phi'(x)\|}{1 - |\Phi(x)|^2} = \frac{1}{x_n} \tag{3.45}$$

whenever x lies in $\mathbb{H}^n$. If $\gamma : [a, b] \to \mathbb{H}^n$ is a piecewise smooth path and $\beta = \Phi \circ \gamma$, then (3.45) now yields

$$\begin{aligned} \ell_{\mathbb{H}}(\beta) &= \int_\beta \frac{2|dx|}{1 - |x|^2} = \int_a^b \frac{2|\dot\beta(t)|\,dt}{1 - |\beta(t)|^2} = \int_a^b \frac{2\|\Phi'[\gamma(t)]\| \cdot |\dot\gamma(t)|\,dt}{1 - |\Phi[\gamma(t)]|^2} \\ &= \int_a^b \frac{|\dot\gamma(t)|\,dt}{\gamma_n(t)} = \int_\gamma \frac{|dx|}{x_n} = \ell_{\mathbb{H}}(\gamma) \,, \end{aligned}$$

which implies that

$$d_{\mathbb{H}}\big(\Phi(x), \Phi(y)\big) = d_{\mathbb{H}}(x, y) \tag{3.46}$$

for all points x and y of $\mathbb{H}^n$. Equation (3.46) therefore expresses the fact that Φ is an isometry between the metric spaces $(\mathbb{H}^n, d_{\mathbb{H}})$ and $(B^n, d_{\mathbb{H}})$.

Once in possession of this knowledge we can painlessly translate statements about the Poincaré half-space to statements about the Poincaré ball, and vice versa. For instance the transformations in $\text{Möb}(B^n) = \Phi \circ \text{Möb}(\mathbb{H}^n) \circ \Phi^{-1}$ are hyperbolic isometries of B^n; the hyperbolic geodesics in B^n are the sets L of the form $L = \Gamma \cap B^n$, where Γ is a chordal circle in $\hat{\mathbb{R}}^n$ that meets $\mathbb{S}^{n-1}$ at right angles, and so forth.

It will later be important to know that

$$d_{\mathbb{H}}(0, x) = \log \frac{1 + |x|}{1 - |x|} \tag{3.47}$$

for every x in B^n. Since for any x in B^n there is an orthogonal linear transformation U with $U(x) = |x|e_n$, since the members of $\mathrm{O}(n)$ are simultaneously hyperbolic and Euclidean isometries of B^n, and since Φ maps the ray $\{re_n : r \geq 1\}$ to the radial segment of B^n from 0 to e_n, it suffices to prove (3.47) for $x = \Phi(re_n)$ with $r > 1$.

Formula (3.38) gives

$$d_{\mathbb{H}}(0, x) = \log r \,,$$

while (3.43) tells us that

$$1 - |x|^2 = \frac{4r}{(r + 1)^2} \,.$$

This equation can be solved for r to get

$$r = \frac{1 + |x|}{1 - |x|} \,,$$

from which (3.47) directly follows.

We next show that $\text{Möb}(B^n)$ and $\text{Möb}(\mathbb{H}^n)$ are the full isometry groups of the metric spaces $(B^n, d_{\mathbb{H}})$ and $(\mathbb{H}^n, d_{\mathbb{H}})$. We prepare the way with two lemmas. The first characterizes local Euclidean isometries.

LEMMA 3.4.2. *Let B be an open Euclidean ball in $\mathbb{R}^n$, and let $f : B \to \mathbb{R}^n$ be an isometric embedding, meaning that $|f(x) - f(y)| = |x - y|$ for all x and y in B. Then f is the restriction to B of a unique transformation g from the group $\boldsymbol{E}(n)$ of Euclidean isometries.*

PROOF. Suppose $B = B^n(x_0, r)$. We may assume that $x_0 = f(x_0) = 0$. (If not, consider in place of f the mapping f_0 defined in $B^n(r)$ by $f_0(x) = f(x+x_0) - f(x_0)$. If true for f_0, the lemma is also true for f.) For any x and y in B we have $\langle f(x), f(y)\rangle = \langle x, y\rangle$. Indeed,

$$\begin{aligned} \langle f(x), f(y)\rangle &= \frac{|f(x)|^2 + |f(y)|^2 - |f(x) - f(y)|^2}{2} = \frac{|x|^2 + |y|^2 - |x - y|^2}{2} \\ &= \langle x, y\rangle\,. \end{aligned}$$

Next, fixing a number s in $(0, r)$, we observe that the vectors $u_1 = s^{-1}f(se_1)$, $u_2 = s^{-1}f(se_2), \ldots, u_n = s^{-1}f(se_n)$ form an orthonormal basis for $\mathbb{R}^n$. Thus we discover that

$$f(x) = \sum_{i=1}^{n}\langle f(x), u_i\rangle u_i = s^{-1}\sum_{i=1}^{n}\langle f(x), f(se_i)\rangle u_i = s^{-1}\sum^{n}\langle x, se_i\rangle u_i = \sum_{i=1}^{n} x_i u_i$$

for every x in B. Now the function $g : \mathbb{R}^n \to \mathbb{R}^n$ defined by $g(x) = \sum_{i=1}^{n} x_i u_i$ is an orthogonal linear transformation, so $f = g|B$ with g a member of $\boldsymbol{E}(n)$. The uniqueness of g is a consequence of Corollary 3.3.11. □

The second lemma recasts Lemma 3.4.2 in its "infinitesimal" formulation.

LEMMA 3.4.3. *Let B be an open Euclidean ball in $\mathbb{R}^n$, and let $f : B \to \mathbb{R}^n$ be an injective mapping with the property that*

$$\lim_{y\to x}\frac{|f(y) - f(x)|}{|y - x|} = 1 \tag{3.48}$$

for every point x of B. Then f is an isometric embedding of B into $\mathbb{R}^n$—hence, is the restriction to B of a unique Euclidean isometry $g \in \boldsymbol{E}(n)$. In particular, $B' = f(B)$ is an open Euclidean ball. If both B and B' are centered at the origin, then $B = B'$ and g is an orthogonal linear transformation.

PROOF. As in the proof of Lemma 3.4.2, it suffices to consider the situation where $B = B^n(0, r)$ and $f(0) = 0$, in which case the claim is that $f(B) = B$ and that $f = g|B$, with g an orthogonal transformation of $\mathrm{O}(n)$. Condition (3.48) plainly implies that f is a continuous function. Thus f maps B homeomorphically onto a domain that contains the origin, a domain we label D. Also, it is an easy matter to check that (3.48) is true with f^{-1} in place of f at each point of D. The first step in the proof of the lemma is to demonstrate that $|f(y) - f(x)| \le |y - x|$ holds for all x and y in B. Fix x and y in B, with $x \ne y$, and $m > 1$. We proceed to verify that

$$|f(y) - f(x)| \le m|y - x|\,. \tag{3.49}$$

However, as $m > 1$ is arbitrary, this enables us to conclude that $|f(y) - f(x)| \le |y - x|$. Write $u = y - x$ and let $A = \{t \in [0, 1] : |f(x + tu) - f(x)| \le mt|u|\}$.

Then $0 \in A$, so this set is nonempty, $A \neq \emptyset$. By the continuity of f, $t_0 = \sup A$ belongs to A. Suppose that t_0 were smaller than 1. On the basis of 3.48 we would have

$$\lim_{t \to t_0^+} \frac{|f(x+tu) - f(x+t_0u)|}{(t-t_0)|u|} = 1 < m \,,$$

permitting us to pick t in the interval $(t_0, 1]$ for which

$$|f(x+tu) - f(x+t_0u)| < m(t-t_0)|u| \,.$$

The triangle equality would then give

$$|f(x+tu) - f(x)| \leq mt|u|$$

and would thus place t in A, in contradiction with the definition of t_0. To avoid a contradiction we must have $t_0 = 1$. Therefore $|f(y) - f(x)| \leq m|y-x|$, which proves (3.49). We can therefore be certain that $|f(y) - f(x)| \leq |y-x|$ is obtained for all x and y in B. One implication of this inequality is that B contains the domain D.

Now let $B' = B^n(0,d)$, where $d = d(0, \partial D)$. We can apply the foregoing argument to the function f^{-1} in the ball B'. It leads to the conclusion that $|f^{-1}(y) - f^{-1}(x)| \leq |y-x|$ whenever x and y lie in B'. On the other hand, the statement

$$|y-x| = \left|f\left[f^{-1}(y)\right] - f\left[f^{-1}(x)\right]\right| \leq |f^{-1}(y) - f^{-1}(x)|$$

is also valid for such x and y, showing that the restriction of f^{-1} to B' is an isometric imbedding. Lemma 3.4.2 tells us that f^{-1} coincides in B' with a unique mapping h from $\boldsymbol{E}(n)$. Indeed, since $h(0) = f^{-1}(0) = 0$, h is an orthogonal linear transformation. If y is a point of ∂D with $|y| = d$, then $f^{-1}(x)$ tends to ∂B as $x \to y$ through B', whence

$$d = \lim_{x \to y} |x| = \lim_{x \to y} |f^{-1}(x)| = r \,.$$

From this we infer that $B' = B = D$ and that f is the restriction to B of $g = h^{-1}$, as maintained. □

We are now in a position to establish the result we have really been after.

THEOREM 3.4.4. *The isometry group of* $(B^n, d_{\mathbb{H}})$ *is* $\text{Möb}(B^n)$*; the isometry group of* $(\mathbb{H}^n, d_{\mathbb{H}})$ *is* $\text{Möb}(\mathbb{H}^n)$.

PROOF. We have already shown that the members of $\text{Möb}(B^n)$ and $\text{Möb}(\mathbb{H}^n)$ are hyperbolic isometries. Let $f : B^n \to B^n$ be an arbitrary isometry of $(B^n, d_{\mathbb{H}})$. We maintain that $f = g|B^n$ for some g in $\text{Möb}(B^n)$. In proving this we are free to assume that $f(0) = 0$. (Otherwise choose h in $\text{Möb}(B^n)$ that sends $f(0)$ to 0 and consider $h \circ f$ instead of f.) Consider a point x of B^n. Then $d_{\mathbb{H}}[0, f(x)] = d_{\mathbb{H}}(0,x)$, so (3.47) implies that $|f(x)| = |x|$. Recalling (3.36), we find that

$$\begin{aligned} \lim_{y \to x} \frac{|f(y) - f(x)|}{|y-x|} &= \lim_{y \to x} \frac{|f(y) - f(x)|/d_{\mathbb{H}}[f(y), f(x)]}{|y-x|/d_{\mathbb{H}}(y,x)} \\ &= \frac{(1-|f(x)|^2)/2}{(1-|x|^2)/2} = \frac{1-|x|^2}{1-|x|^2} = 1 \end{aligned}$$

for every x in B^n. Lemma 3.4.3 informs us that $f = g|B^n$ for some g in $\text{O}(n)$. Finally, if f is a general hyperbolic isometry of $\mathbb{H}^n$, then $\Phi \circ f \circ \Phi^{-1}$ is a hyperbolic isometry of B^n. It follows that there is a transformation g in $\text{Möb}(B^n)$ such that $\Phi \circ f \circ \Phi^{-1} = g|B^n$. This makes f the restriction to $\mathbb{H}^n$ of $\Phi^{-1} \circ g \circ \Phi$, which belongs to $\text{Möb}(\mathbb{H}^n)$. □

3.5. Classification of hyperbolic isometries

Let $n \geq 2$ and let $f \in \text{Möb}(B^n)$ be a Möbius transformation different from the identity. The Brouwer fixed point theorem implies the existence of at least one fixed point for f in $\bar{B}^n$, though f need not fix any point of the open ball B^n.

3.5.1. Elliptic transformations. If f does have a fixed point in B^n, then we say that f is an *elliptic transformation*. In this case we can select an arbitrary point x_0 of B^n fixed by f and choose a transformation g in $\text{Möb}(B^n)$ with $g(0) = x_0$. The mapping $f_0 = g^{-1} \circ f \circ g$ belongs to $\text{Möb}(B^n)$ and fixes the origin, so f_0 is in $\text{O}(n)$. The set $\text{fix}(f_0)$ defined by

$$\text{fix}(f_0) = \{x \in \hat{\mathbb{R}}^n : f_0(x) = x\}$$

is then of the form $T \cup \{\infty\}$, where T is a linear subspace of $\mathbb{R}^n$. Moreover, $\text{fix}(f) = g[\text{fix}(f_0)]$. On the strength of these comments we can assert the following about f, should it be elliptic: f is conjugate in $\text{Möb}(B^n)$ to an orthogonal transformation; $\text{fix}(f)$ is either a two-point set or a chordal p-sphere $(1 \leq p \leq n-1)$ that is orthogonal to $\mathbb{S}^{n-1}$; f leaves invariant any hyperbolic sphere or ball in B^n whose hyperbolic center is fixed by f.

3.5.2. Loxodromic and parabolic transformations. Looking next at a transformation f from $\text{Möb}(B^n)$ that is free of fixed points in B^n, we remark that $\text{fix}(f)$ must be a subset of $\mathbb{S}^{n-1}$. If f were to fix a point x of $\hat{\mathbb{R}}^n \setminus \bar{B}^n$, then by Theorem 3.3.7 it would also fix x^*, the point symmetric with x relative to $\mathbb{S}^{n-1}$, which is a point of B^n. Moreover, we shall see in a moment that $\text{fix}(f)$ can contain no more than two elements. A transformation from $\text{Möb}(B^n)$ with exactly two fixed points on $\mathbb{S}^{n-1}$ and none in B^n is termed *loxodromic*; a member of $\text{Möb}(B^n)$ whose one and only fixed point lies on $\mathbb{S}^{n-1}$ is said to be *parabolic*.

Our proposed classification of the nonidentity members of $\text{Möb}(B^n)$ with $n \geq 2$ into elliptic, loxodromic, and parabolic transformations has an obvious parallel in $\text{Möb}(\mathbb{H}^n)$. The following fact makes certain that in both cases the classification is exhaustive: if f in $\text{Möb}(\mathbb{H}^n)$ fixes three points of ∂H^n, then f must have a fixed point in $\mathbb{H}^n$. Indeed, by passing from f to one of its conjugates in $\text{Möb}(\mathbb{H}^n)$ we may suppose that f fixes 0, e_1, and ∞.

According to Theorem 3.3.13, we can write $f = \lambda U + b$, where $\lambda > 0$, b is in $\mathbb{R}^{n-1}$, and U is a member of $\text{O}(n)$ that fixes e_n. Then $b = f(0) = 0$, whence $\lambda = |\lambda U(e_1)| = |f(e_1)| = |e_1| = 1$. We conclude that $f(e_n) = U(e_n) = e_n$, thereby exhibiting a fixed point for f in $\mathbb{H}^n$. Notice that in both B^n and $\mathbb{H}^n$ the classification of hyperbolic isometries is conjugacy invariant: if f and g are members of $\text{Möb}(B^n)$ (respectively, $\text{Möb}(\mathbb{H}^n)$) and if f is not the identity transformation, then f and $g^{-1} \circ f \circ g$ fall into the same category in the classification.

Elliptic transformations are easiest to picture in the ball model of hyperbolic space, where we have already observed that they are conjugate to orthogonal linear transformations. The structures, up to conjugacy, of parabolic and loxodromic transformations become more transparent in the Poincaré half-space. To set the stage for an analysis of these transformations, we make a remark concerning the nonexistence of finite fixed points for certain Euclidean isometries.

Let f in $\boldsymbol{E}(n)$ be of the form $f = U + b$, with b a nonzero vector in $\mathbb{R}^n$ and U in $\mathrm{O}(n)$. We claim that the fixed point set $\mathrm{fix}(f) = \{\infty\}$ if and only if b does not belong to $T^\perp$, where $T = \mathrm{fix}(U)$ and $T^\perp$ indicates the orthogonal complement of the linear space T. Note that under either condition, $T \neq \{0\}$. This fact is obvious if b is not in $T^\perp$, for then it is immediate that $T^\perp \neq \mathbb{R}^n$ and hence that $T \neq \{0\}$. If $T = \{0\}$, then 1 is not an eigenvalue of U, the linear transformation $U - I$ is nonsingular, and the equation $U(x) - x = -b$, which is equivalent to $f(x) = x$, has a finite solution.

Consequently, $\mathrm{fix}(f) \neq \{\infty\}$ when $T = \{0\}$. In substantiating this claim we make use of the direct sum decomposition $\mathbb{R}^n = T \oplus T^\perp$. Thus each vector x in $\mathbb{R}^n$ has a unique representation $x = x' + x''$ with x' in T and x'' in $T^\perp$. In particular, $b = b' + b''$. As $U(x') = x'$, for x in $\mathbb{R}^n$ we can express $f(x)$ as

$$\begin{aligned} f(x) &= f(x' + x'') = U(x') + U(x'') + b' + b'' \\ &= (x' + b') + \big[U(x'') + b''\big]. \end{aligned} \tag{3.50}$$

Because U is an orthogonal transformation, $T^\perp$ is invariant under U; hence $U(x'') \in T^\perp$ and also $U(x'') + b'' \in T^\perp$. We infer from (3.50) and the uniqueness of the orthogonal decomposition that $f(x) = x$ if and only if $x' + b' = x'$ and $U(x'') + b'' = x''$. By construction, 1 is not an eigenvalue of $U\big|T^\perp$, so the equation $U(x'') + b'' = x''$ definitely has a solution x'' in $T^\perp$. In order for f to be free of fixed points in $\mathbb{R}^n$ it is therefore both necessary and sufficient that $b' \neq 0$, which is equivalent to the requirement that b not be an element of $T^\perp$.

Formula (3.50) suggests another interesting observation about a Euclidean isometry f with $\mathrm{fix}(f) = \{\infty\}$. Writing $f^0 = I$, $f^k = f \circ f \circ \cdots \circ f$ (k factors) for $k = 1, 2, \ldots$ and $f^{-k} = (f^{-1})^k$ for $k = 1, 2, \ldots$, we learn from (3.50) that $f^k(x) = x' + kb' + z_k$, where z_k is a vector from $T^\perp$. Since $b' \neq 0$, we infer that

$$|f^k(x)| = \left(|x' + kb'|^2 + |z_k|^2\right)^{1/2} \geq |x' + kb'| \geq k|b'| - |x'| \to \infty$$

as $k \to \infty$. That is,

$$f^k(x) \to \infty \qquad \text{as } k \to \infty \text{ for every } x \text{ in } \hat{\mathbb{R}}^n.$$

Because f^{-1} has the same structure as f, we see also that $f^k(x) \to \infty$ as $k \to -\infty$ for all x in $\hat{\mathbb{R}}^n$.

Consider now a parabolic transformation f from $\mathrm{Möb}(\mathbb{H}^n)$ whose sole fixed point is ∞. We know from Theorem 3.3.13 that $f = \lambda U + b$ for some $\lambda > 0$, b in $\mathbb{R}^{n-1}$, and U in $\mathrm{O}(n)$ fixing e_n. We must have $b = f(0) \neq 0$. Also, $\lambda = 1$ is required, for $\lambda \neq 1$ would mean that λ^{-1} is certainly not an eigenvalue of U and hence would make $U - \lambda^{-1} I$ an invertible linear transformation. As a result, the equation $\lambda U(x) + b = x$ would have a solution in $\mathbb{R}^n$, a situation incompatible with our assumptions about f having the unique fixed point $\{\infty\}$. We have thus shown that f has the form $f = U + b$, where U is a transformation in $\mathrm{O}(n)$ that fixes e_n and b is a vector in $\mathbb{R}^{n-1}$ that, by our preliminary remarks, does not lie in $\mathrm{fix}(U)^\perp$. Of course, every transformation of the type just described is a parabolic transformation in $\mathrm{Möb}(\mathbb{H}^n)$ that fixes ∞, while an arbitrary parabolic isometry of $(\mathbb{H}^n, d_{\mathbb{H}})$ is conjugate in $\mathrm{Möb}(\mathbb{H}^n)$ to such a transformation.

These facts bring to light two key features of a general parabolic transformation f from $\text{Möb}(\mathbb{H}^n)$, say one with $\text{fix}(f) = \{p\}$. First, f leaves invariant each horosphere to ∂H^n at p. Here, by a *horosphere* to a chordal sphere Σ at a point p we mean a chordal sphere Σ' such that $\Sigma \cap \Sigma' = \{p\}$. Second, $f^k(x) \to p$ as $|k| \to \infty$ for every x in $\hat{\mathbb{R}}^n$. Both of these facts are easily seen when $p = \infty$—the horospheres are the hyperplanes $\{x \in \mathbb{H}^n : x_n = c > 0\}$; the general case then follows by means of conjugation. The obvious analogues of these two properties are enjoyed by parabolic transformations belonging to $\text{Möb}(B^n)$.

Suppose next that f is a loxodromic transformation in $\text{Möb}(\mathbb{H}^n)$ for which $\text{fix}(f) = \{0, \infty\}$. It follows from Theorem 3.3.13 that f can be represented in the manner $f = \lambda U$, where $\lambda > 0$, $\lambda \neq 1$, and U is an orthogonal transformation that fixes e_n. If x lies on the hyperbolic line L with endpoints 0 and ∞, we have $f(x) = \lambda x$, so L is invariant under f and the geometric effect of f on L is to translate each point a hyperbolic distance $|\log \lambda|$ along L, in the direction toward the origin if $\lambda < 1$ and away from it if $\lambda > 1$. Furthermore, for the iterate f^k we get $|f^k(x)| = \lambda^k |x|$ for all x in $\mathbb{R}^n$, which yields the information that

$$f^k(x) \to 0 \text{ and } f^{-k}(x) \to \infty \quad \text{as } k \to \infty \text{ for every } x \in \hat{\mathbb{R}}^n \setminus \{0, \infty\}$$

when $\lambda < 1$, whereas

$$f^k(x) \to \infty \text{ and } f^{-k}(x) \to 0 \quad \text{as } k \to \infty \text{ for every } x \in \hat{\mathbb{R}}^n \setminus \{0, \infty\}$$

when $\lambda > 1$.

Then through the process of conjugation we draw the following conclusions about a general loxodromic transformation f in $\text{Möb}(\mathbb{H}^n)$ or $\text{Möb}(B^n)$, assuming that $\text{fix}(f) = \{p, q\}$: f leaves invariant the hyperbolic line L whose terminal points are p and q and L is called the *axis* of f; on L the transformation f acts as a hyperbolic translation; the labeling of the fixed points of f can be done so that p is its *attracting fixed point* and q its *repelling fixed point* (i.e., $f^k(x) \to p$ and $f^{-k}(x) \to q$ as $k \to \infty$ for every x in $\hat{\mathbb{R}}^n \setminus \{p, q\}$).

We remark that a transformation f from the Möbius group $\text{Möb}(n)$ or from the conformal group $\text{Con}(n)$ is classified as parabolic, loxodromic, or elliptic under the condition that its Poincaré extension $\tilde{f}$, which is a member of $\text{Möb}(\mathbb{H}^{n+1})$ when f belongs to $\text{Möb}(n)$ and of $\text{Möb}(B^{n+1})$ when f is in $\text{Con}(n)$, is so classified. We can summarise the above discussions in a theorem

THEOREM 3.5.1. *Assume that $f \in \text{Möb}(n)$ is not the identity mapping of $\hat{\mathbb{R}}^n$. Then*

- *f is parabolic if it has a unique fixed point;*
- *f is loxodromic if its Poincaré extension has exactly two fixed points;*

in all other cases—in particular, when f has no fixed points—f is elliptic.

Of course a similar result is true for $f \in \text{Con}(n)$.

3.6. The distortion, compactness and convergence properties of Möbius transformations

In this section we will discuss the convergence properties of sequences of Möbius transformations of hyperbolic n-space $\mathbb{H}^n$, the unit ball B^n and the n-sphere $\mathbb{S}^n$. Given the close relationships between these spaces, the compactness properties will also be similar. Further, it is to be expected that the results we achieve should be very similar to those well known in the theory of complex analysis and in particular the classical theory of normal families—a term coined by P. Montel in 1912 [**139**, p. 154]—and the key idea of equicontinuity. The theory of normal families—well covered in J. Schiff's book [**145**] on this topic—is important because it is closely connected with a standard proof of the Riemann mapping theorem, and with the circle of ideas surrounding basic theorems of Picard, Schottky, Landau and Bloch.

We will see a little later that the compactness results we achieve for families of Möbius transformations hold in much greater generality—in fact for families of quasiconformal mappings—and so the results obtained here foreshadow those.

The situation for families of Möbius transformations of $\mathbb{H}^n$ or B^n is a bit easier since we have already shown that these mappings act as isometries of these spaces when given their respective hyperbolic metrics. From this one can in fact easily deduce compactness results for sequences in $\mathrm{Con}(n)$, however it is possible to achieve interesting and useful distortion estimates which frame more general discussions later.

3.6.1. Distortion of chordal distances. With the aid of hyperbolic geometry one can give sharp bounds for the distortion of chordal distances under a Möbius transformation. These bounds stem from the following lemma relating spherical distortion and hyperbolic geometry.

LEMMA 3.6.1. *If f belongs to* $\mathrm{Möb}(B^n)$ *with $n \geq 2$, then*

$$\sup\left\{\frac{|f(y)-f(x)|}{|y-x|} : x,y \in \mathbb{S}^{n-1}\ ,\ x \neq y\right\} = e^{d_{\mathbb{H}}[0,f(0)]}\,. \tag{3.51}$$

PROOF. If f is in $\mathrm{O}(n)$, then (3.51) is a trivial statement as $|f(y)-f(x)| = |y-x|$ and $f(0) = 0$. Consider next the inversion R in a Euclidean sphere $\mathbb{S}^{n-1}(x_0, r)$ that is orthogonal to $\mathbb{S}^{n-1}$. Once we recall (3.29) and (3.21) we have $1+r^2 = |x_0|^2$ and

$$\frac{|R(x)-R(y)|}{|x-y|} = \frac{r^2}{|x-x_0|\cdot|y-x_0|}$$

for x and y on $\mathbb{S}^{n-1}$ with $x \neq y$. The right-hand side of this equation attains its maximum on $\mathbb{S}^{n-1}$ when $x = y = x_0/|x_0|$, in which case $|x-x_0| = |y-x_0| = |x_0|-1$. Remembering that $R(0) = x_0^*$, the image of x_0 under reflection in $\mathbb{S}^{n-1}$, and appealing to formula (3.47), we compute that

$$\begin{aligned}\sup\left\{\frac{|R(x)-R(y)|}{|x-y|} : x,y \in S^{n-1}\ ,\ x \neq y\right\} &= \frac{r^2}{(|x_0|-1)^2} = \frac{|x_0|^2-1}{(|x_0|-1)^2}\\ &= \frac{|x_0|+1}{|x_0|-1} = \frac{1+|x_0^*|}{1-|x_0^*|} = \frac{1+|R(0)|}{1-|R(0)|} = e^{d_{\mathbb{H}}[0,R(0)]}\,.\end{aligned}$$

Thus when $f = R$, relation (3.51) again holds true. Finally, from Theorem 3.3.15 we see that (3.51) is valid for an arbitrary f from $\mathrm{Möb}(B^n)$. □

As suggested, Lemma 3.6.1 is the stepping stone to a basic distortion theorem showing that Möbius transformations are bilipschitz in the spherical metric and giving the best possible estimate on the bilipschitz constant.

THEOREM 3.6.2. *Let f be a Möbius transformation of $\hat{\mathbb{R}}^n$, and let $\tilde{f}$ be its Poincaré extension. Let*

$$b = \exp\{d_{\mathbb{H}}[e_{n+1}, \tilde{f}(e_{n+1})]\}.$$

Then the bounds

$$b^{-1}q(x,y) \le q\big[f(x), f(y)\big] \le b\, q(x,y) \tag{3.52}$$

hold for all x and y in $\hat{\mathbb{R}}^n$. Moreover, the constant b is sharp.

PROOF. We establish the right-hand inequality in (3.52); this can subsequently be applied to f^{-1} to produce the lower bound once we observe that if $g = f^{-1}$, then $\tilde{g} = \tilde{f}^{-1}$ and $d_{\mathbb{H}}[e_{n+1}, \tilde{f}(e_{n+1})] = d_{\mathbb{H}}[e_{n+1}, \tilde{g}(e_{n+1})]$, which shows that f and f^{-1} give rise to the same constant b.

We once again call upon the transformation Φ in Möb$(n+1)$ that maps $\mathbb{H}^{n+1}$ to B^{n+1} and agrees with stereographic projection on $\hat{\mathbb{R}}^n$. Recall that Φ is an isometry between $(\mathbb{H}^{n+1}, d_{\mathbb{H}})$ and $(B^{n+1}, d_{\mathbb{H}})$, as well as an isometry between $(\hat{\mathbb{R}}^n, q)$ and $(\mathbb{S}^n, d)$, where d denotes the Euclidean metric.

Now $g = \Phi \circ \tilde{f} \circ \Phi^{-1}$ is a member of Möb(B^{n+1}) and, since $\Phi(e_{n+1}) = 0$,

$$d_{\mathbb{H}}\big[e_{n+1}, \tilde{f}(e_{n+1})\big] = d_{\mathbb{H}}\big[0, g(0)\big].$$

Lemma 3.6.1 thus yields

$$\begin{aligned}
&\sup\left\{\frac{q[f(x), f(y)]}{q(x,y)} : x, y \in \hat{\mathbb{R}}^n \ , \ x \neq y\right\} \\
&= \quad \sup\left\{\frac{|g(z) - g(w)|}{|z-w|} : z, w \in \mathbb{S}^n \ , \ z \neq w\right\} = b\,,
\end{aligned}$$

which confirms the upper estimate in (3.52) and at the same time demonstrates that the constant b cannot be replaced by any smaller constant. □

The next result furnishes a more localized method for gauging the chordal distortion of a Möbius transformation.

THEOREM 3.6.3. *If D is a domain in $\hat{\mathbb{R}}^n$ with at least two boundary points and if f belongs to* Möb(n), *then*

$$q\big[f(x), f(y)\big]\, q\big[f(D)^c\big] \le \frac{8\, q(x,y)}{\sqrt{q(x,\partial D)\, q(y,\partial D)}} \tag{3.53}$$

for all points x and y of D.

PROOF. Let x and y be arbitrary distinct points of D, and let u and v be points of the complement of D, that is, D^c, such that $q[f(u), f(v)] = q[f(D)^c]$. Put $c = 8/q[f(D)^c]$. Invoking Theorem 3.3.9, we calculate that

$$\begin{aligned}
&\frac{q(x,y)\, q(u,v)}{q(x,u)\, q(y,v)} \cdot \frac{q(x,y)\, q(u,v)}{q(x,v)\, q(y,u)} \\
&\quad = \quad [x,u,y,v][x,v,y,u] \\
&\quad = \quad \big[f(x), f(u), f(y), f(v)\big]\big[f(x), f(v), f(y), f(u)\big] \\
&\quad = \quad \frac{q[f(x), f(y)]\, q[f(u), f(v)]}{q[f(x), f(u)]\, q[f(y), f(v)]} \cdot \frac{q[f(x), f(y)]\, q[f(v), f(u)]}{q[f(x), f(v)]\, q[f(y), f(u)]}\,.
\end{aligned}$$

This therefore leads us to the estimates

$$\begin{aligned}
&\left\{\frac{q[f(x),f(y)]}{q(x,y)}\right\}^2 \\
&\quad = \left\{\frac{q(u,v)}{q[f(u),f(v)]}\right\}^2 \left\{\frac{q[f(x),f(u)]\,q[f(x),f(v)]\,q[f(y),f(u)]\,q[f(y),f(v)]}{q(x,u)\,q(x,v)\,q(y,u)\,q(y,v)}\right\} \\
&\quad \le \left\{\frac{q(u,v)}{q[f(u),f(v)]}\right\}^2 \left\{\frac{16}{q(x,u)\,q(x,v)\,q(y,u)\,q(y,v)}\right\} \\
&\quad = \left\{\frac{4}{q[f(u),f(v)]}\right\}^2 \left\{\frac{q(u,v)}{q(x,u)\,q(x,v)}\right\}\left\{\frac{q(u,v)}{q(y,u)\,q(y,v)}\right\} \\
&\quad \le \frac{c^2}{4}\left\{\frac{q(u,x)+q(x,v)}{q(x,u)\,q(x,v)}\right\}\left\{\frac{q(u,y)+q(y,v)}{q(y,u)\,q(y,v)}\right\} \\
&\quad \le \frac{c^2}{4}\left\{\frac{1}{q(x,v)}+\frac{1}{q(x,u)}\right\}\left\{\frac{1}{q(y,u)}+\frac{1}{q(y,v)}\right\} \\
&\quad \le \frac{c^2}{4}\left\{\frac{1}{q(x,D^c)}+\frac{1}{q(x,D^c)}\right\}\left\{\frac{1}{q(y,\partial D)}+\frac{1}{q(y,\partial D)}\right\} \\
&\quad = \frac{c^2}{q(x,\partial D)\,q(y,\partial D)}.
\end{aligned}$$

This then results in the estimate

$$\left\{\frac{q[f(x),f(y)]}{q(x,y)}\right\}^2 \le \frac{c^2}{q(x,\partial D)\,q(y,\partial D)}$$

that is the same as (3.53) which we were seeking. □

3.6.2. Normal families and convergence. Let $f_\nu : X \to Y$ $(\nu = 1, 2, \ldots)$ be a sequence of functions from a metric space (X, d) to a metric space (Y, d'). There are numerous ways in which to think of the sequence $\langle f_\nu \rangle_{\nu=1}^{\infty}$ (which for simplicity we denote by $\langle f_\nu \rangle$) converging to a function $f : X \to Y$. For starters, $\langle f_\nu \rangle$ could converge to f *pointwise in* X, meaning simply that $f_\nu(p) \to f(p)$ for each point p of X. A significantly stronger type of convergence is *uniform convergence on* X. This form of convergence demands that for each $\varepsilon > 0$ there should be an index N such that $d'[f_\nu(p), f(p)] < \varepsilon$ is true for every p in X as soon as $\nu \ge N$.

Between these extremes lies the mode of convergence appropriate to most of the convergence questions that come up in this book. It is called a *locally uniform convergence in* X and requires that each point p of X have a neighbourhood $U = U_p$ with the property that $f_\nu \to f$ uniformly on U. This is easily seen to be equivalent to the statement that for each compact subset F of X we have $f_\nu \to f$ uniformly on F. Another common name for this type of convergence is "uniform convergence on compact subsets".

Given a sequence $\langle f_\nu \rangle$ of mappings from X to Y, we would like to have a criterion telling us whether it is possible to extract from $\langle f_\nu \rangle$ a subsequence that converges in the locally uniform sense to some function $f : X \to Y$. More generally, we would like to characterize the families $\mathcal{F}$ of functions from X to Y that enjoy the following property: each sequence $\langle f_\nu \rangle$ from $\mathcal{F}$ has a locally uniformly convergent subsequence. Note that at this point we do not insist that the limit of such a subsequence be a member of $\mathcal{F}$. However this is clearly desirable and an intriguing separate problem.

In a classic 1907 paper [**118**], families fitting this description were dubbed *normal families* by the French analyst Paul Montel. The property of normality in a family $\mathcal{F}$ is closely allied with the notion of equicontinuity. We say that a family $\mathcal{F}$ of functions from X to Y is *equicontinuous at the point* p if for each $\varepsilon > 0$ there is a $\delta = \delta(p, \varepsilon) > 0$ such that $d'[f(q), f(p)] < \varepsilon$ holds for every f in $\mathcal{F}$ whenever $d(q, p) < \delta$.

A family $\mathcal{F}$ that is equicontinuous at each point of X is known as an *equicontinuous family*. The link between normality and equicontinuity is established in the following version of the Arzelà-Ascoli theorem that is general enough to meet all our needs in this book.

THEOREM 3.6.4. *Let $\mathcal{F}$ be a nonempty family of continuous functions from a separable and locally compact metric space X to a complete metric space Y. Then $\mathcal{F}$ is a normal family if and only if $\mathcal{F}$ is an equicontinuous family and $\mathcal{F}(p) = \{f(p) : f \in \mathcal{F}\}$ is a relatively compact subset of Y for each p in X.*

Without any explicit statement to the contrary, it will be a standing assumption that whenever the terms "normal" and "equicontinuous" are used in this book in reference to a family of mappings $f : A \to \hat{\mathbb{R}}^n$ with A a subset of $\hat{\mathbb{R}}^n$, the metric involved in both A and $\hat{\mathbb{R}}^n$ is the chordal metric.

Theorem 3.6.3 now leads to a very convenient method for detecting normality in a family of Möbius transformations.

THEOREM 3.6.5. *Suppose that D is a domain in $\hat{\mathbb{R}}^n$ and that $\mathcal{F}$ is a nonempty subfamily of* Möb(n). *If each point x of D has an open neighbourhood $U = U_x$ with the property that* $\inf\{q[f(U)^c] : f \in \mathcal{F}\} > 0$, *then $\mathcal{F}|D = \{f|D : f \in \mathcal{F}\}$ is a normal family. In particular, $\mathcal{F}|D$ is a normal family whenever*

$$\inf\{q[f(D)^c] : f \in \mathcal{F}\} > 0. \tag{3.54}$$

PROOF. The metric spaces $X = D$, $Y = \hat{\mathbb{R}}^n$, and the family $\mathcal{F}|D$ satisfy all the conditions laid down in Theorem 3.6.4. Moreover, since $\hat{\mathbb{R}}^n$ is compact, the set $\mathcal{F}(x)$ is relatively compact for every x in $\hat{\mathbb{R}}^n$. To establish the normality of $\mathcal{F}|D$ we have to check only that this family is equicontinuous. Given x_0 in D, we invoke the hypothesis of the theorem to choose $r > 0$ for which the chordal ball $B = \{x : q(x, x_0) < 2r\}$ is contained in D and has $d = \inf\{q[f(B)^c] : f \in \mathcal{F}\} > 0$. We elicit from Theorem 3.6.3 (applied in B rather than D) the information that

$$q\big[f(x), f(x_0)\big] \le \frac{8q(x, x_0)}{d\sqrt{q(x, B^c)\, q(x_0, B^c)}} \le \frac{8q(x, x_0)}{dr}$$

whenever f comes from $\mathcal{F}$ and $q(x, x_0) < r$. This inequality is more than adequate to guarantee the equicontinuity of $\mathcal{F}|D$ at x_0, an arbitrary point of D. □

It is worthwhile to cite a corollary of Theorem 3.6.5 that is frequently easier to use than the theorem itself.

COROLLARY 3.6.6. *Suppose that a nonempty subfamily $\mathcal{F}$ of* Möb(n) *enjoys the following property: there exist points a, b, and c of $\hat{\mathbb{R}}^n$ and a constant $m > 0$ such that*

$$q\big[f(a), f(b)\big] \ge m, \qquad q\big[f(a), f(c)\big] \ge m, \qquad q\big[f(b), f(c)\big] \ge m$$

for every f in $\mathcal{F}$. Then $\mathcal{F}$ is a normal family.

PROOF. If $D = \hat{\mathbb{R}}^n \setminus \{a,b\}$, then $q[f(D)^c] \geq q[f(a), f(b)] \geq m$ for every f in $\mathcal{F}$, so Theorem 3.6.5 ensures the normality of $\mathcal{F}|D$ and thus the equicontinuity of $\mathcal{F}$ at each point of D. The same argument applies to $\hat{\mathbb{R}}^n \setminus \{a,c\}$ and $\hat{\mathbb{R}}^n \setminus \{b,c\}$, revealing that $\mathcal{F}$ is equicontinuous at every point of $\hat{\mathbb{R}}^n$. An appeal to Theorem 3.6.4 finishes the proof. □

Next, having established a good general criterion to determine whether a limit function of a sequence $\langle f_\nu \rangle$ exists, we should decide what properties this limit has. The next result supplies the answer to this question.

THEOREM 3.6.7. *Let $\langle f_\nu \rangle$ be a sequence from* Möb(n) *that converges pointwise in $\hat{\mathbb{R}}^n$ to a function f. Then there are exactly three possibilities for the limit function:*

(1) *f is constant;*
(2) *f assumes exactly two values;*
(3) *f belongs to* Möb(n).

In the third case both $f_\nu \to f$ and $f_\nu^{-1} \to f^{-1}$ uniformly on $\hat{\mathbb{R}}^n$.

PROOF. Assuming that neither (1) nor (2) above describes the limit f, we prove that (3) offers the only alternative. We take as an assumption, therefore, the existence of points a, b, and c in $\hat{\mathbb{R}}^n$ such that the values $f(a)$, $f(b)$, and $f(c)$ are all different. We first observe that f is injective. If $f(x) = f(y)$ were to hold with $x \neq y$, we could choose distinct elements u and v from the set $\{a,b,c\}$ so that $f(u)$ and $f(v)$ both differ from $f(x)$ and compute

$$\begin{aligned} 0 &= \frac{q[f(x), f(y)]\, q[f(u), f(v)]}{q[f(x), f(u)]\, q[f(y), f(v)]} = \lim_{\nu\to\infty} \big[f_\nu(x), f_\nu(u), f_\nu(y), f_\nu(v)\big] \\ &= [x, u, y, v] > 0 \ , \end{aligned}$$

a contradiction. Thus f is injective. Furthermore,

$$\big[f(x), f(y), f(u), f(v)\big] = \lim_{\nu\to\infty} \big[f_\nu(x), f_\nu(y), f_\nu(u), f_\nu(v)\big] = [x, y, u, v]$$

whenever x, y, u, and v are distinct elements of $\hat{\mathbb{R}}^n$. Theorem 3.3.9 now certifies that f is a Möbius transformation of $\hat{\mathbb{R}}^n$.

Because $f_\nu(a) \to f(a)$, $f_\nu(b) \to f(b)$, and $f_\nu(c) \to f(c)$, it is apparent that there is a constant $m > 0$ such that

$$q\big[f_\nu(a), f_\nu(b)\big] \geq m, \qquad q\big[f_\nu(a), f_\nu(c)\big] \geq m, \qquad q\big[f_\nu(b), f_\nu(c)\big] \geq m$$

for all ν. Corollary 3.6.6 tells us that $\{f_\nu : \nu = 1, 2, \ldots\}$ is an equicontinuous family. Given $\varepsilon > 0$, we can choose for each x in $\hat{\mathbb{R}}^n$ a $\delta_x > 0$ with the feature that $q[f_\nu(y), f_\nu(x)] < \varepsilon/3$ for every $\nu \geq 1$ whenever y lies in $B_x = \{y : q(y,x) < \delta_x\}$. Next, we can select a finite number of these chordal balls—$B_{x_1}, B_{x_2}, \ldots, B_{x_r}$, say—to cover $\hat{\mathbb{R}}^n$ and then fix N large enough that $q[f_\nu(x_j), f_\mu(x_j)] < \varepsilon/3$ for $j = 1, 2, \ldots, r$ whenever $\mu \geq \nu \geq N$. Any y in $\hat{\mathbb{R}}^n$ belongs to B_{x_j} for some j, so

$$\begin{aligned} q\big[f_\nu(y), f_\mu(y)\big] &\leq q\big[f_\nu(y), f_\nu(x_j)\big] + q\big[f_\nu(x_j), f_\mu(x_j)\big] + q\big[f_\mu(x_j), f_\mu(y)\big] \\ &< \tfrac{\varepsilon}{3} + \tfrac{\varepsilon}{3} + \tfrac{\varepsilon}{3} = \varepsilon \end{aligned}$$

whenever $\mu \geq \nu \geq N$. Since y was an arbitrary point of $\hat{\mathbb{R}}^n$, we have just demonstrated that in case (3) the sequence $\langle f_\nu \rangle$ is a uniform Cauchy sequence with respect to the chordal metric on $\hat{\mathbb{R}}^n$. It therefore follows that f_ν converges uniformly to f on $\hat{\mathbb{R}}^n$.

To establish that $f_\nu^{-1} \to f^{-1}$ uniformly on $\hat{\mathbb{R}}^n$ in case (3), it suffices by what we have just shown to check that $f_\nu^{-1} \to f^{-1}$ pointwise in $\hat{\mathbb{R}}^n$. Given x in $\hat{\mathbb{R}}^n$, we consider the sequence $\langle y_\nu \rangle$ defined by $y_\nu = f_\nu^{-1}(x)$. Let y be an arbitrary accumulation point of $\langle y_\nu \rangle$. We can select a subsequence $\langle y_{\nu_k} \rangle$ of $\langle y_\nu \rangle$ such that $y_{\nu_k} \to y$. Because $f_\nu \to f$ uniformly on $\hat{\mathbb{R}}^n$, it is easy to see that $f_{\nu_k}(y_{\nu_k}) \to f(y)$. Consequently,

$$f(y) = \lim_{k\to\infty} f_{\nu_k}(y_{\nu_k}) = \lim_{k\to\infty} x = x\,.$$

We conclude that $y = f^{-1}(x)$ is the sole accumulation point of the sequence of points $\langle f_\nu^{-1}(x) \rangle$. In other words, $f_\nu^{-1}(x) \to f^{-1}(x)$ for each x in $\hat{\mathbb{R}}^n$, as desired. □

As examples, suppose that $f_\nu(x) = x + \nu e_1$. Then $f_\nu(x) \to \infty$ for each x in $\hat{\mathbb{R}}^n$; if $f_\nu(x) = \nu x$, then $f_\nu(x) \to \infty$ for $x \neq 0$ and $f_\nu(0) \to 0$. These examples illustrate that situations (1) and (2) in Theorem 3.6.7 really do arise.

In neither of these cases is convergence uniform on $\hat{\mathbb{R}}^n$: in case (1) uniform convergence is prohibited by the fact that $q[f_\nu(\hat{\mathbb{R}}^n)] = 2$ for every ν; in case (2) it is ruled out by the discontinuity of the limit mapping.

Theorem 3.6.7 implies the following lemma.

LEMMA 3.6.8. *If* $\langle f_\nu \rangle$ *and* $\langle g_\nu \rangle$ *are sequences in* Möb(n) *and if these sequences converge pointwise in* $\hat{\mathbb{R}}^n$ *(and hence, uniformly on* $\hat{\mathbb{R}}^n$*) to Möbius transformations* f *and* g*, respectively, then* $g_\nu \circ f_\nu \to g \circ f$ *uniformly on* $\hat{\mathbb{R}}^n$*.*

Next we establish the following useful result which we will use to identify the equivalence of various topologies on the Möbius group.

LEMMA 3.6.9. *Let* $\langle f_\nu \rangle$ *be a sequence from* Möb($\mathbb{H}^n$) *with* $n \geq 2$*, and let* f *belong to* Möb($\mathbb{H}^n$)*. The following statements are equivalent:*

(i) $f_\nu \to f$ *uniformly on* $\hat{\mathbb{R}}^n$*;*
(ii) $f_\nu \to f$ *uniformly on* $\hat{\mathbb{R}}^{n-1}$*;*
(iii) $f_\nu \to f$ *pointwise in* $\mathbb{H}^n$*.*

PROOF. The implications (i) ⇒ (ii) and (i) ⇒ (iii) are trivial. Assume that either (ii) or (iii) holds. It is clear from Theorem 3.6.5 that, in either situation, $\mathcal{F} = \{f_\nu : \nu = 1, 2, \ldots\}$ is a normal family. Therefore, every subsequence of $\langle f_\nu \rangle$ has a further subsequence that converges uniformly on $\hat{\mathbb{R}}^n$, necessarily to a member of Möb($\mathbb{H}^n$). If $\langle f_\nu \rangle$ failed to converge to f uniformly on $\hat{\mathbb{R}}^n$, there would be a subsequence $\langle f_{\nu_k} \rangle$ of $\langle f_\nu \rangle$ such that $f_{\nu_k} \to g$ uniformly on $\hat{\mathbb{R}}^n$, where g belonged to Möb($\mathbb{H}^n$) but was different from f. Given that (ii) is true, we would have $g = f$ on $\hat{\mathbb{R}}^{n-1}$, so $g^{-1} \circ f$ would be a Möbius transformation that fixed $\hat{\mathbb{R}}^{n-1}$ pointwise, yet was neither the identity transformation of $\hat{\mathbb{R}}^n$ nor the reflection in $\mathbb{R}^{n-1}$, contrary to Theorem 3.3.6. Assuming that (iii) holds, we would have $g = f$ in $\mathbb{H}^n$; hence, by continuity, $g = f$ on $\hat{\mathbb{R}}^{n-1}$, leading to precisely the same contradiction as above. In order for a contradiction not to follow, it must be true that $f_\nu \to f$ uniformly on $\hat{\mathbb{R}}^n$ whenever condition (ii) or (iii) holds. □

3.7. The Möbius group as a matrix group

We have already seen indications of the relationship between Möbius transformations, or linear fractional transformations, of the Riemann sphere $\hat{\mathbb{C}}$ and matrix groups. In this section we intend to clarify these relationships. As a first example of this, suppose that f and g are linear fractional transformations,

$$f(z) = \frac{az+b}{cz+d}, \quad g(z) = \frac{a'z+b'}{c'z+d'}$$

normalised so that both

$$ad - bc = 1, \quad a'd' - b'c' = 1.$$

We can identify f and g with the matrices

$$A = \begin{pmatrix} a & b \\ c & d \end{pmatrix}, \quad B = \begin{pmatrix} a' & b' \\ c' & d' \end{pmatrix}$$

respectively. The matrices A and B lie in the group $\mathrm{SL}(2,\mathbb{C})$ of 2×2 matrices with complex entries and determinant equal to 1. The group operation in the Möbius group is composition. We can compare the group operations with the following calculation:

$$(f\circ g)(z) = \frac{a\left(\frac{a'z+b'}{c'z+d'}\right)+b}{c\left(\frac{a'z+b'}{c'z+d'}\right)+d} = \frac{a(a'z+b')+b(c'z+d')}{c(a'z+b')+d(c'z+d')}$$

$$= \frac{(aa'+bc')z+(ab'+bd')}{(ca'+dc')z+(cb'+dd')}, \tag{3.55}$$

$$A.B = \begin{pmatrix} aa'+bc' & ab'+bd' \\ ca'+dc' & cb'+dd' \end{pmatrix}. \tag{3.56}$$

Then notice that (3.55) and (3.56) effect the same correspondence between matrices and Möbius transformations that we used between f and A and also between g and B. That is, the natural matrix representative for $f\circ g$ is simply the product of the matrix representation for f and that for g using the usual matrix multiplication. The correspondence $f \leftrightarrow A$ as above therefore describes an isomorphism of the groups $\mathrm{Möb}^+(2)$ and $\mathrm{PSL}(2,\mathbb{C}) = \mathrm{SL}(2,\mathbb{C})/\{\pm \mathit{identity}\}$—we have to projectivise as the matrices A and $-A$ give rise to the same linear fractional transformation.

We now consider to what extent this remarkable correspondence continues in higher dimensions. To do so we will have to develop a bit more general theory.

3.7.1. The Möbius group as a topological group. A *topological group* is a pair $\langle G,\mathcal{T}\rangle$, where G is a group and $\mathcal{T}$ is a Hausdorff topology on G with the property that the group operations of G are continuous relative to $\mathcal{T}$. Thus the function from $G\times G$ to G that sends (g_1,g_2) to g_1g_2 and the function from G to itself that takes g to g^{-1} are continuous. Here the topology on $G\times G$ is, as expected, the product topology associated with $\mathcal{T}$. One usually speaks of "the topological group G", with it being implicit that the underlying topology $\mathcal{T}$ has somehow been prescribed and is to remain fixed for the duration of the discussion.

The chordal metric q on $\hat{\mathbb{R}}^n$ induces in a natural way a metric q^* on the group $\mathrm{Möb}(n)$: for f and g in $\mathrm{Möb}(n)$,

$$q^*(f,g) = \max_{x\in\hat{\mathbb{R}}^n} q\big[f(x),g(x)\big].$$

The associated metric topology on Möb(n) is nothing but the topology of uniform convergence on $\hat{\mathbb{R}}^n$, which is to say that a sequence $\langle f_\nu \rangle$ from Möb(n) converges to a member f of Möb(n) in the metric q^* if and only if $f_\nu \to f$ uniformly on $\hat{\mathbb{R}}^n$. It was observed in the previous section that if sequences $\langle f_\nu \rangle$ and $\langle g_\nu \rangle$ from Möb(n) converge uniformly on $\hat{\mathbb{R}}^n$ to mappings f and g, then Theorem 3.6.7 and Lemma 3.6.8 imply that these limits are of necessity Möbius transformations and also that $g_\nu \circ f_\nu \to g \circ f$ and $f_\nu^{-1} \to f^{-1}$ uniformly on $\hat{\mathbb{R}}^n$. In other words, when endowed with the metric topology that derives from q^*, Möb(n) becomes a topological group. The groups Möb$(\mathbb{H}^n)$ and Möb(B^n) are closed topological subgroups of Möb(n).

Topological groups G and G' are *isomorphic* if there is a bijection $\varphi : G \to G'$ that is simultaneously a group homomorphism and a homeomorphism. Any such φ is called a *topological isomorphism* from G onto G'. For example, we know of a bijective homomorphism φ : Möb$(n) \to$ Möb$(\mathbb{H}^{n+1})$, the one given by $\varphi(f) = \tilde{f}$, where $\tilde{f}$ is the Poincaré extension of f. We assert that φ is a topological isomorphism. Indeed, that φ is also a homeomorphism follows instantly from Lemma 3.6.9.

Of course Lemma 3.6.9 has an obvious analogue for the group Möb(B^n), and this shows that the topological groups Möb$(\mathbb{H}^n)$ and Möb(B^n) are isomorphic: the function φ defined by $\varphi(f) = g^{-1} \circ f \circ g$, where g is any member of Möb(n) that maps B^n onto $\mathbb{H}^n$, delivers a topological isomorphism of Möb$(\mathbb{H}^n)$ onto Möb(B^n). We infer that Möb(n) and Möb(B^{n+1}) are isomorphic as topological groups.

3.7.2. Another model for hyperbolic space. To identify the matrix groups we need, we shall need another description of hyperbolic space and in particular its group of isometries.

We have already pointed out that a number of realizations exist for hyperbolic n-space $\mathbb{H}^n$. In the Poincaré half-space and ball models we have already met two of these. We now consider a third. The new model has the advantage that its group of isometries is a subgroup of GL$(n+1)$. Since this isometry group is also isomorphic as a topological group to Möb$(\mathbb{H}^n)$, we arrive at a means to represent the group Möb$(\mathbb{H}^n)$—and, ultimately, Möb(n)—as a group of matrices.

For reasons that will soon become apparent we shall, in this subsection and the next, write vectors x in $\mathbb{R}^{n+1}$ in the manner $x = (x_0, x_1, \ldots, x_n)$ and, with the obvious meaning, use $e_0, e_1, \ldots, e_n$ for the standard basis of $\mathbb{R}^{n+1}$.

For $n \geq 2$, let B_n denote the nondegenerate, symmetric, bilinear form defined on $\mathbb{R}^{n+1}$ by

$$B_n(x, y) = x_0 y_0 - x_1 y_1 - x_2 y_2 - \cdots - x_n y_n$$

or, to express it differently,

$$B_n(x, y) = \langle J(x), y \rangle ,$$

where $J : \mathbb{R}^{n+1} \to \mathbb{R}^{n+1}$ is the symmetric linear involution given by

$$J(x) = (x_0, -x_1, -x_2, \ldots, -x_n) .$$

With B_n is associated a quadratic form Q_n,

$$Q_n(x) = B_n(x, x) = x_0^2 - x_1^2 - x_2^2 - \cdots - x_n^2 .$$

We define a subset $\mathbb{H}^n$ of $\mathbb{R}^{n+1}$ as follows (we use the notation $\mathbb{H}^n$ since it will turn out to be a model for hyperbolic space when endowed with the correct metric of course):

$$\mathbb{H}^n = \big\{x \in \mathbb{R}^{n+1} : Q_n(x) = 1 \ , \ x_0 > 0\big\}.$$

Thus $\mathbb{H}^n$ is the "positive" sheet of the hyperboloid of two sheets in $\mathbb{R}^{n+1}$ determined by the equation

$$x_0^2 - x_1^2 - x_2^2 - \cdots - x_n^2 = 1 \,.$$

Observe that $x_0 \geq 1$ for every point x of $\mathbb{H}^n$. If $\gamma : [a, b] \to \mathbb{R}^{n+1}$ is a smooth path whose trajectory lies on $\mathbb{H}^n$, say $\gamma = (\gamma_0, \gamma_1, \ldots, \gamma_n)$, then

$$\big[\gamma_0(t)\big]^2 = 1 + \big[\gamma_1(t)\big]^2 + \big[\gamma_2(t)\big]^2 + \cdots + \big[\gamma_n(t)\big]^2$$

for every t in $[a, b]$. It follows from differentiation of this identity that

$$\gamma_0(t)\dot\gamma_0(t) = \gamma_1(t)\dot\gamma_1(t) + \gamma_2(t)\dot\gamma_2(t) + \cdots + \gamma_n(t)\dot\gamma_n(t)$$

for all such t. From these relations we deduce that

$$\begin{aligned} Q_n\big[\dot\gamma(t)\big] &= \big[\dot\gamma_0(t)\big]^2 - \sum_{i=1}^n \big[\dot\gamma_i(t)\big]^2 = \frac{\Big[\sum_{i=1}^n \gamma_i(t)\dot\gamma_i(t)\Big]^2}{\big[\gamma_0(t)\big]^2} - \sum_{i=1}^n \big[\dot\gamma_i(t)\big]^2 \\ &\leq \frac{\Big\{\sum_{i=1}^n \big[\gamma_i(t)\big]^2\Big\}\Big\{\sum_{i=1}^n \big[\dot\gamma_i(t)\big]^2\Big\}}{\big[\gamma_0(t)\big]^2} - \sum_{i=1}^n \big[\dot\gamma_i(t)\big]^2 \\ &= -\frac{\sum_{i=1}^n \big[\dot\gamma_i(t)\big]^2}{\big[\gamma_0(t)\big]^2} \leq 0 \end{aligned}$$

for each t in $[a, b]$, with strict inequality unless $\dot\gamma_1(t) = \dot\gamma_2(t) = \cdots = \dot\gamma_n(t) = 0$ (hence, also $\dot\gamma_0(t) = 0$). As a result, we see that the form Q_n is negative definite on tangent vectors to $\mathbb{H}^n$.

Mimicking what we previously did to construct the hyperbolic distances, we can use Q_n to define a distance $d_{\mathbb{H}}$ in $\mathbb{H}^n$: for any piecewise smooth path $\gamma : [a, b] \to \mathbb{H}^n$, we set

$$\ell_{\mathbb{H}}(\gamma) = \int_a^b \sqrt{-Q_n[\dot\gamma(t)]}\, dt \ ;$$

for x and y in $\mathbb{H}^n$, we define

$$d_{\mathbb{H}}(x, y) = \inf_\gamma \ell_{\mathbb{H}}(\gamma) \ ,$$

where the infimum extends over all piecewise smooth paths γ on $\mathbb{H}^n$ with initial point x and terminal point y. In the language of differential geometry, $\ell_{\mathbb{H}}$ and $d_{\mathbb{H}}$ are the length and distance functions associated with the Riemannian metric on $\mathbb{H}^n$ given by the fundamental form $ds^2 = dx_1^2 + \cdots + dx_n^2 - dx_0^2$.

We now want to show that, equipped with the above distance, the metric space $\mathbb{H}^n$ is isometric to the Poincaré ball B^n.

THEOREM 3.7.1. *The mapping* $F : \mathbb{H}^n \to B^n$ *defined by*

$$F(x) = \left(\frac{x_1}{1 + x_0}, \frac{x_2}{1 + x_0}, \ldots, \frac{x_n}{1 + x_0}\right)$$

is an isometry between $(\mathbb{H}^n, d_{\mathbb{H}})$ *and* $(B^n, d_{\mathbb{H}})$.

PROOF. Write $y = F(x)$ for x in $\mathbb{H}^n$. A computation gives

$$|y|^2 = \frac{x_0 - 1}{x_0 + 1} < 1 \ ,$$

so $F(\mathbb{H}^n)$ is contained in B^n. An inverse for F is easily exhibited:

$$F^{-1}(y) = \left(\frac{1+|y|^2}{1-|y|^2}, \frac{2y_1}{1-|y|^2}, \dots, \frac{2y_n}{1-|y|^2}\right)$$

for y in B^n. This now establishes that F is a bijection between $\mathbb{H}^n$ and B^n. To prove that F is an isometry it suffices to show that $\ell_{\mathbb{H}}(F \circ \gamma) = \ell_{\mathbb{H}}(\gamma)$ whenever γ is a smooth path on $\mathbb{H}^n$. Fix γ and write $\beta = F \circ \gamma$. Then

$$\beta_i(t) = \frac{\gamma_i(t)}{1 + \gamma_0(t)}$$

and

$$\dot{\beta}_i(t) = \frac{\dot{\gamma}_i(t)}{1 + \gamma_0(t)} - \frac{\gamma_i(t)\dot{\gamma}_0(t)}{[1 + \gamma_0(t)]^2}$$

for $i = 1, 2, \dots, n$.

Remembering that $\gamma_0\dot{\gamma}_0 = \gamma_1\dot{\gamma}_1 + \gamma_2\dot{\gamma}_2 + \cdots + \gamma_n\dot{\gamma}_n$, we calculate that

$$\begin{aligned}
&\frac{4|\dot{\beta}(t)|^2}{(1 - |\beta(t)|^2)^2} \\
&= 4\left(\frac{1+\gamma_0(t)}{2}\right)^2 \sum_{i=1}^{n} \left(\frac{\dot{\gamma}_i(t)}{1+\gamma_0(t)} - \frac{\gamma_i(t)\dot{\gamma}_0(t)}{[1+\gamma_0(t)]^2}\right)^2 \\
&= \sum_{i=1}^{n} \big[\dot{\gamma}_i(t)\big]^2 - \frac{2\dot{\gamma}_0(t)}{1+\gamma_0(t)} \sum_{i=1}^{n} \gamma_i(t)\dot{\gamma}_i(t) + \frac{[\dot{\gamma}_0(t)]^2}{[1+\gamma_0(t)]^2} \sum_{j=1}^{n} \big[\gamma_i(t)\big]^2 \\
&= \sum_{i=1}^{n} \big[\dot{\gamma}_i(t)\big]^2 - \frac{2\gamma_0(t)[\dot{\gamma}_0(t)]^2}{1+\gamma_0(t)} + \frac{[\dot{\gamma}_0(t)]^2[\gamma_0(t) - 1]}{1+\gamma_0(t)} \\
&= \sum_{i=1}^{n} \big[\dot{\gamma}_i(t)\big]^2 - \big[\dot{\gamma}_0(t)\big]^2 = -Q_n\big[\dot{\gamma}(t)\big] \, .
\end{aligned}$$

If $\gamma : [a, b] \to \mathbb{H}^n$, we thus obtain

$$\ell_{\mathbb{H}}(\beta) = \int_a^b \frac{2|\dot{\beta}(t)|\,dt}{1 - |\beta(t)|^2} = \int_a^b \sqrt{-Q_n[\dot{\gamma}(t)]}\,dt = \ell_{\mathbb{H}}(\gamma) \ ,$$

as desired. □

The isometry F from $\mathbb{H}^n$ to B^n can be described in geometric terms as a stereographic projection with respect to the point $-e_0$: for x in $\mathbb{H}^n$, $F(x)$ is the point where B^n, here identified with $\{x \in B^{n+1} : x_0 = 0\}$, is intersected by the ray in $\mathbb{R}^{n+1}$ that issues from $-e_0$ and passes through x. On the basis of Theorem 3.7.1 we realize $(\mathbb{H}^n, d_{\mathbb{H}})$ as a third model for hyperbolic n-space—the hyperboloid model. The correspondence $f \mapsto F^{-1} \circ f \circ F$ sets up an algebraic isomorphism between Möb(B^n) and Isom$(\mathbb{H}^n)$, the group of isometries of $(\mathbb{H}^n, d_{\mathbb{H}})$. We shall exploit this isomorphism to identify the members of Isom$(\mathbb{H}^n)$. It is through this process that we shall establish a topological isomorphism between Möb(B^n) and one of the classical matrix groups.

3.7.3. The isometry group $\mathrm{Isom}^+(\mathbb{H}^n)$**.** For $n \geq 2$ we denote by $\mathrm{O}(1,n)$ the subgroup of $\mathrm{GL}(n+1)$ consisting of all transformations T that preserve the bilinear form B_n: a member T of $\mathrm{GL}(n+1)$ belongs to $\mathrm{O}(1,n)$ if and only if

$$B_n(x,y) = B_n[T(x),T(y)]$$

for all x and y in $\mathbb{R}^{n+1}$. Recalling that $B_n(x,y) = \langle J(x), y\rangle$, where $J(x) = (x_0, -x_1, -x_2, \ldots, -x_n)$, the last condition can be rewritten as

$$\big\langle J(x), y\big\rangle = \big\langle [JT](x), T(y)\big\rangle = \big\langle [T^*JT](x), y\big\rangle$$

for all x and y in $\mathbb{R}^{n+1}$. It follows that a linear transformation $T : \mathbb{R}^{n+1} \to \mathbb{R}^{n+1}$ is in $\mathrm{O}(1,n)$ if and only if $T^*JT = J$. This is equivalent to the requirement $T^{-1} = JT^*J$.

Any transformation T from $\mathrm{O}(1,n)$ preserves the quadratic form Q_n, so T leaves invariant the hyperboloid $S : Q_n(x) = 1$, although it may interchange the two sheets of S. We use $\mathrm{O}^+(1,n)$ to denote the set

$$\mathrm{O}^+(1,n) = \{T \in \mathrm{O}(1,n) : T(\mathbb{H}^n) = \mathbb{H}^n\}\ .$$

Then $\mathrm{O}^+(1,n)$ is a subgroup of $\mathrm{O}(1,n)$. In fact, $\mathrm{O}^+(1,n)$ has index two in $\mathrm{O}(1,n)$, the left cosets of $\mathrm{O}^+(1,n)$ in $\mathrm{O}(1,n)$ being $\mathrm{O}^+(1,n)$ and $R\mathrm{O}^+(1,n)$, where R is the reflection in the hyperplane $P : x_0 = 0$. It is evident that a transformation T from $\mathrm{O}(1,n)$ belongs to $\mathrm{O}^+(1,n)$ if and only if $T(e_0)$ lies in $\mathbb{H}^n$, which happens if and only if $\langle T(e_0), e_0\rangle > 0$.

The relation $T^*JT = J$ for T in $\mathrm{O}(1,n)$ implies that $\det(T) = \pm 1$. Those T in $\mathrm{O}(1,n)$ with $\det(T) = 1$ form an index-two subgroup of $\mathrm{O}(1,n)$, a subgroup that we designate $\mathrm{SO}(1,n)$. The notation $\mathrm{SO}^+(1,n)$ stands for the group $\mathrm{O}^+(1,n) \cap \mathrm{SO}(1,n)$, which is of index two in $\mathrm{O}^+(1,n)$. All of the above groups are closed subgroups of the topological group $\mathrm{GL}(n+1)$, whose topology we take to be the metric topology associated with the operator norm. Of course, a sequence of linear transformations $T_\nu : \mathbb{R}^n \to \mathbb{R}^m$ converges in operator norm to $T : \mathbb{R}^n \to \mathbb{R}^m$ if and only if $T_\nu(x) \to T(x)$ for each x in $\mathbb{R}^n$, or, for that matter, for each x in some basis for $\mathbb{R}^n$.

The groups just introduced can be viewed more concretely as groups of $(n+1) \times (n+1)$ invertible matrices through the customary device of identifying a linear transformation $T : \mathbb{R}^{n+1} \to \mathbb{R}^{n+1}$ with its matrix A_T relative to the basis $e_0, e_1, \ldots, e_n$:

$$A_T = \begin{bmatrix} a_{00} & a_{01} & \cdots & a_{0n} \\ a_{10} & a_{11} & \cdots & a_{1n} \\ \vdots & \vdots & \ddots & \vdots \\ a_{n0} & a_{n1} & \cdots & a_{nn} \end{bmatrix},$$

where $a_{ij} = \langle T(e_j), e_i\rangle$ for $0 \leq i,j \leq n$. The correspondence $T \to A_T$ provides a topological isomorphism of $\mathrm{GL}(n+1)$ onto the group of nonsingular $(n+1)\times(n+1)$ real matrices.

In general the algebra of $m \times n$ real matrices is topologized by regarding it as a copy of $\mathbb{R}^{mn}$ and giving it the standard metric topology of this Euclidean space. Thus, a sequence $\langle A_\nu\rangle$ of $m \times n$ matrices converges to an $m \times n$ matrix A if and only if for each i and j the $(i,j)^{th}$-entry of A_ν converges to the $(i,j)^{th}$-entry of A as $\nu \to \infty$. Because of the way in which we have associated matrices with linear transformations, the inverse isomorphism is given by $A \mapsto T_A$, in which $T_A(x) = xA^*$. In this context A^* means the transpose of A; the multiplication is ordinary matrix multiplication in which a vector x in $\mathbb{R}^{n+1}$ is regarded as a $1 \times (n+1)$ matrix.

It is traditional to let notation such as $\mathrm{GL}(n)$, $\mathrm{O}(n)$, $\mathrm{O}(1,n)$, and so forth, serve for both a group of linear transformations and for its associated matrix group. For instance, $\mathrm{O}(1,n)$ can be thought of as the group of $(n+1) \times (n+1)$ real matrices A that satisfy the identity $A^*JA = J$ (or $A^{-1} = JA^*J$), where

$$J = \begin{bmatrix} 1 & 0 & 0 & \cdots & 0 \\ 0 & -1 & 0 & \cdots & 0 \\ 0 & 0 & -1 & \cdots & 0 \\ & \vdots & & & \vdots \\ 0 & 0 & 0 & \cdots & -1 \end{bmatrix}.$$

The group $\mathrm{O}^+(1,n)$ consists of the matrices A in $\mathrm{O}(1,n)$ with $a_{00} > 0$.

Let T be a linear transformation from the group $\mathrm{O}^+(1,n)$. Then $f = T|\mathbb{H}^n$ maps $\mathbb{H}^n$ bijectively to itself. Moreover, if $\gamma : [a,b] \to \mathbb{R}^{n+1}$ is an arbitrary smooth path whose trajectory lies on $\mathbb{H}^n$ and if $\beta = f \circ \gamma$, then $\dot\beta(t) = T[\dot\gamma(t)]$ for each t in $[a,b]$. Because T belongs to $\mathrm{O}^+(1,n)$, we discover that

$$\ell_{\mathbb{H}}(\beta) = \int_a^b \sqrt{-Q_n[\dot\beta(t)]}\, dt = \int_a^b \sqrt{-Q_n[\dot\gamma(t)]}\, dt = \ell_{\mathbb{H}}(\gamma)\,.$$

It follows that any mapping f of this kind is an isometry of $\mathbb{H}^n$ relative to its hyperbolic metric. There are no others, as we now learn.

THEOREM 3.7.2. *Each member of the group* $\mathrm{Isom}(\mathbb{H}^n)$ *is the restriction to* $\mathbb{H}^n$ *of a uniquely determined transformation from* $\mathrm{O}^+(1,n)$.

PROOF. Since $\mathbb{H}^n$ contains a basis for $\mathbb{R}^{n+1}$, for example, $\mathbb{H}^n$ includes the vectors $e_0, \sqrt{2}\, e_0 + e_1, \sqrt{2}\, e_0 + e_2, \ldots, \sqrt{2}\, e_0 + e_n$, two linear transformations of $\mathbb{R}^{n+1}$ that share the same restriction to $\mathbb{H}^n$ must coincide everywhere in $\mathbb{R}^{n+1}$. This fact shows the uniqueness assertion to be trivial. If $f = T|\mathbb{H}^n$ and $g = S|\mathbb{H}^n$, where T and S belong to $\mathrm{O}^+(1,n)$, then $g \circ f = ST|\mathbb{H}^n$. To complete the proof, therefore, it suffices to check that $\mathrm{Isom}(\mathbb{H}^n)$ has a set of generators of the stated type. It has already been observed that the correspondence $g \mapsto F^{-1} \circ g \circ F$, where $F : \mathbb{H}^n \to B^n$ is the mapping in Theorem 3.7.1, gives us an isomorphism of $\mathrm{M\ddot{o}b}(B^n)$ onto $\mathrm{Isom}(\mathbb{H}^n)$.

We shall identify two particular types of matrices A in the matrix group $\mathrm{O}^+(1,n)$ and verify that the associated isometries $f_A = T_A|\mathbb{H}^n$ generate $\mathrm{Isom}(\mathbb{H}^n)$. We do so by making certain that $\mathrm{M\ddot{o}b}(B^n)$ is generated by the mappings $g_A = F \circ f_A \circ F^{-1}$ that correspond to the aforementioned f_A under the above isomorphism. To this end, it will be useful to have a concrete representation of the function g_A for arbitrary A in $\mathrm{O}^+(1,n)$.

Given A in $\mathrm{O}^+(1,n)$ and z in B^n, write $x = F^{-1}(z)$, $y = f_A(x)$, and $w = F(y) = g_A(z)$. Then $y = xA^*$, so

$$y_i = a_{i0}x_0 + a_{i1}x_1 + \cdots + a_{in}x_n$$

for $i = 0, 1, \ldots, n$. From the formula for F^{-1} we obtain

$$(1 - |z|^2)y_i = (1 + |z|^2)a_{i0} + 2(a_{i1}z_1 + a_{i2}z_2 + \cdots + a_{in}z_n)\,.$$

It is also true that

$$w_i = \frac{y_i}{1+y_0} = \frac{(1-|z|^2)y_i}{(1-|z|^2)+(1-|z|^2)y_0}$$

for $i = 1, 2, \ldots, n$. As a result, we arrive at the expression

$$w_i = \frac{(1+|z|^2)a_{i0} + 2(a_{i1}z_1 + a_{i2}z_2 + \cdots + a_{in}z_n)}{|z|^2(a_{00}-1) + 2(a_{01}z_1 + a_{02}z_2 + \cdots + a_{0n}z_n) + a_{00} + 1}$$

for z belonging to B^n.

Consider the matrix

$$A = \begin{bmatrix} 1 & 0 & \cdots & 0 \\ 0 & & & \\ \vdots & & U & \\ 0 & & & \end{bmatrix},$$

where U comes from $\mathrm{O}(n)$. Then A is in $\mathrm{O}^+(1,n)$ and has $\det(A) = \det(U)$. The computation above shows us that

$$w = g_A(z) = zU^* = T_U(z)$$

for every z in B^n, which implies that $g_A = T_U$ is an orthogonal linear transformation. Here we recall Corollary 3.3.11 and note that g_A belongs to $\text{Möb}^+(B^n)$ if and only if U is in $\mathrm{SO}(n)$.

Next, let R_s be the reflection in the sphere $\Sigma = \mathbb{S}^{n-1}(se_1, r)$, where $s > 1$ and $r = \sqrt{s^2-1}$. Since Σ is orthogonal to $\mathbb{S}^{n-1}$, R_s is in $\text{Möb}(B^n)$. Indeed, a moment's thought reveals that $\text{Möb}(B^n)$ is generated by $\mathrm{O}(n)$ and $\{R_s : s > 1\}$, since any reflection in a Euclidean sphere orthogonal to $\mathbb{S}^{n-1}$ is conjugate via a rotation to R_s for some $s > 1$. We claim that $R_s = g_A$ for

$$A = \begin{bmatrix} \cosh(2t) & -\sinh(2t) & 0 & \cdots & 0 \\ \sinh(2t) & -\cosh(2t) & 0 & \cdots & 0 \\ 0 & 0 & & & \\ \vdots & \vdots & & I_{n-1} & \\ 0 & 0 & & & \end{bmatrix},$$

where t satisfies $\coth t = s$. (Then $\sinh t = r^{-1}$.) Certainly A is in $\mathrm{O}^+(1,n)$, with $\det(A) = -1$. Let $c = se_1$. To determine $w = g_A(z)$ for z in B^n, we first remark that

$$\begin{aligned} &|z|^2(a_{00}-1) + 2\left(\sum_{i=1}^{n} a_{0i}z_i\right) + a_{00} + 1 \\ &= |z|^2\big[\cosh(2t) - 1\big] - 2z_1\sinh(2t) + \cosh(2t) + 1 \\ &= 2|z|^2\sinh^2(t) - 2z_1\sinh(2t) + 2\cosh^2 t = 2|z-c|^2\sinh^2 t = \frac{2|z-c|^2}{r^2}, \end{aligned}$$

whence

$$w_i = \frac{r^2 z_i}{|z-c|^2}$$

for $i = 2, 3, \ldots, n$. When $i = 1$ we have

$$\begin{aligned} w_1 &= \frac{(1+|z|^2)\sinh(2t) - 2z_1\cosh(2t)}{2|z-c|^2\sinh^2(t)} \\ &= \frac{(|z-c|^2 + 1 - |c|^2 + 2\langle z, c\rangle)(2\sinh^2 t + 1) - 2z_1(2\cosh^2 t - 1)}{2|z-c|^2\sinh^2 t} \\ &= \coth t + \frac{r^2(z_1 - \coth t)}{|z-c|^2} = s + \frac{r^2(z_1 - s)}{|z-c|^2}. \end{aligned}$$

We conclude that

$$w = g_A(z) = c + \frac{r^2(z-c)}{|z-c|^2} = R_s(z)$$

for every z in B^n, which means that $g_A = R_s$ (Corollary 3.3.11). It follows that the group $\mathrm{Isom}(\mathbb{H}^n)$ is generated by the mappings f_A, where A ranges over all matrices of the two types we have just considered. The proof of the theorem is thus complete. □

We can define an algebraic isomorphism φ of the matrix group $\mathrm{O}^+(1,n)$ onto $\mathrm{Möb}(B^n)$ by setting $\varphi(A) = g_A$, where g_A is the unique member of $\mathrm{Möb}(B^n)$ satisfying $F^{-1} \circ g_A \circ F = f_A = T_A|\mathbb{H}^n$. The upshot of the preceding deliberations is encapsulated in the following theorem.

THEOREM 3.7.3. *The function φ is a topological isomorphism of the matrix group* $\mathrm{O}^+(1,n)$ *onto* $\mathrm{Möb}(B^n)$. *Under this isomorphism,* $\mathrm{SO}^+(1,n)$ *corresponds to* $\mathrm{Möb}^+(B^n)$.

PROOF. We must verify that φ is a homeomorphism. For an arbitrary sequence $\langle A_\nu \rangle$ from $\mathrm{O}^+(1,n)$, write $f_\nu = f_{A_\nu}$ and $g_\nu = g_{A_\nu}$. If $A_\nu \to A$, then $xA_\nu^* \to xA^*$ for every x in $\mathbb{R}^{n+1}$, so $f_\nu \to f = f_A$ pointwise in $\mathbb{H}^n$. It follows that $g_\nu \to g = g_A$ pointwise in B^n. The analogue of Lemma 3.6.9 for $\mathrm{Möb}(B^n)$ shows that $g_\nu \to g$ uniformly on $\hat{\mathbb{R}}^n$. We infer that φ is continuous. Conversely, if $g_\nu \to g$ uniformly on $\hat{\mathbb{R}}^n$, then $f_\nu \to f = F^{-1} \circ g \circ F$ pointwise on $\mathbb{H}^n$. The limit f must belong to $\mathrm{Isom}(\mathbb{H}^n)$, whence $f = f_A$ for a unique A in $\mathrm{O}^+(1,n)$. Naturally, $\varphi(A) = g$. Since $T_{A_\nu} \to T_A$ pointwise on $\mathbb{H}^n$ and since $\mathbb{H}^n$ contains a basis for $\mathbb{R}^{n+1}$, $T_{A_\nu} \to T_A$ in the operator norm. Therefore $A_\nu \to A$ in $\mathrm{O}^+(1,n)$. This shows that φ^{-1} is also continuous, making φ a homeomorphism.

The function $\psi(g) = \det[\varphi^{-1}(g)]$ is a homomorphism of $\mathrm{Möb}(B^n)$ onto $\{+1, -1\}$ with kernel $K = \varphi[\mathrm{SO}^+(1,n)]$, an index-two subgroup of $\mathrm{Möb}(B^n)$. An arbitrary member g of $\mathrm{Möb}^+(B^n)$ either belongs to $\mathrm{SO}(n)$ or has the structure $g = h \circ R_\Sigma$, where h is a sense-reversing member of $\mathrm{O}(n)$ and Σ is a Euclidean sphere that is orthogonal to $\mathbb{S}^{n-1}$. The proof of Theorem 3.7.2 demonstrates that $\varphi(g)$ is in $\mathrm{SO}^+(1,n)$ for such g; that is, $\mathrm{Möb}^+(B^n)$ lies in K. But $\mathrm{Möb}^+(B^n)$ is also of index two in $\mathrm{Möb}(B^n)$, so $\mathrm{Möb}^+(B^n) = K = \varphi[\mathrm{SO}^+(1,n)]$. □

Recalling that $\mathrm{Möb}(n)$ is topologically isomorphic to $\mathrm{Möb}(B^{n+1})$ via an isomorphism that maps $\mathrm{Möb}^+(n)$ to $\mathrm{Möb}^+(B^{n+1})$, we record a corollary of Theorem 3.7.3.

COROLLARY 3.7.4. *The Möbius groups* $\mathrm{Möb}(n)$ *and* $\mathrm{Möb}^+(n)$ *are isomorphic as topological groups to the matrix groups* $\mathrm{O}^+(1,n+1)$ *and* $\mathrm{SO}^+(1,n+1)$, *respectively.*

One implication of Corollary 3.7.4 is that Möb(n) carries the structure of a Lie group. For more about this side of the Möbius group we refer the reader to the recent book of D. Bump on this subject, [**26**].

3.8. Liouville's theorem

Just as conformal mappings in the plane are homeomorphic solutions to a system of partial differential equations (the Cauchy-Riemann equations in fact), so too are conformal mappings in space—the Cauchy-Riemann systems which we will come to in a moment. The Looman-Menchoff theorem describes the precise hypotheses on the derivatives of a solution to the Cauchy-Riemann equations under which one can assert the mapping is holomorphic. This is proved in R. Narasimhan's book [**132**]. In higher dimensions, $n \geq 3$, questions still remain as to whether all sufficiently regular solutions to the natural equations defining a conformal mapping are Möbius transformations. Of course this is demonstrably false in two dimensions—there are plenty of conformal mappings which are not Möbius, although entire (defined in $\mathbb{C}$ or $\hat{\mathbb{C}}$) holomorphic homeomorphisms are Möbius transformations. However in dimensions $n \geq 3$ this is basically true and reflects a remarkable higher–dimensional rigidity, with important applications which in many ways underpins the utility of the theory of quasiconformal mappings. We will explore these things not only in this section, but elsewhere in the book.

3.8.1. A little history. In 1850, the celebrated French mathematician Joseph Liouville added a short note [**99**] to a new edition of Gaspard Monge's classic work [**117**] *Application de l'Anallyse à la Géometrie*, whose publication Liouville was overseeing. The note was prompted by a series of three letters that Liouville had received in the years 1845 and 1846 from the renowned British physicist William Thomson. Thomson, better known today as Lord Kelvin, had studied in Paris under Liouville's tutelage for a period of time in the mid-1840s, so these two giants of nineteenth century science were well acquainted. In his letters, Thomson posed to Liouville a number of questions concerning inversions in spheres, questions that had arisen in conjunction with Thomson's research in electrostatics, in particular, with the so-called principle of electrical images. (It might be pointed out that the reflection in the unit sphere $\mathbb{R}^3$ is often referred to in physics circles as the "Kelvin transform".) To learn more about the interesting relationship between Thomson and Liouville, the reader is invited to consult Jesper Lützen's magnificient biography of Liouville [**101**].

Paraphrased in the language of the present chapter, the substance of Liouville's note is conveyed by the following remarkable assertion:

if D is a domain in $\mathbb{R}^n$ with $n \geq 3$, then any conformal mapping $f : D \to \mathbb{R}^n$ is the restriction to D of a Möbius transformation.

Befitting his motivation for writing the article, Liouville couched his discussion in the language of differential forms rather than mappings. As a consequence, his original formulation of the result bears scant resemblance to the preceding statement, although the relationship between the two formulations is quite transparent to anyone moderately well versed in differential geometry. Indeed, there is nothing in Liouville's work [**99**] that even has the label "theorem"! And the title of the note, "Extension au cas de trois dimensions de la question du tracé géographique"

[Extension to the case of three dimensions of the question of drawing geographic maps], gives no hint whatsoever as to its stunning contents. (The reference is to the "map projection problem" in two dimensions, whose solution Gauss had found but never published. Liouville had rediscovered the solution in 1847.) It was only later that Liouville published his theorem in a form approximating the statement of it that we have given [**100**].

Of greater importance here is the fact that the proof which Liouville outlined for his theorem makes use of certain implicit smoothness hypotheses. His argument thus translates to a proof of the stated conformal mapping interpretation of his result only under the added assumption that f be a mapping of class C^3 or better.

We initially draw attention to one significant corollary of Liouville's theorem:

the only domains in $\mathbb{R}^n$ *with* $n \geq 3$ *that are conformally equivalent to the unit ball* B^n *are round Euclidean balls and half-spaces.*

This stands in stark contrast to the marvelous discovery announced in 1851, a year after Liouville's note was published, by Riemann:

A simply connected proper subdomain D *of the complex plane* $\mathbb{C}$ *can be transformed via a (complex analytic) conformal mapping to the unit disk* $\mathbb{D}$.

Indeed it is the contrast between these two italicized statements immediately above which motivates the mapping problem for quasiconformal mappings. Although an arbitrary domain homeomorphic to B^n in $\mathbb{R}^n$, $n \geq 3$, cannot be mapped conformally to the unit ball, is there a "nearly conformal" mapping which effects this? The restriction to domains homeomorphic to the ball removes elementary topological obstructions, however it is fair to say that progress on this problem has been slow. We will discuss what is known in Chapter 7.

It is our intention to present here an adaptation of Rolf Nevanlinna's elementary proof, vintage 1960, for the C^4-version of Liouville's theorem [**133, 134**].

The theorem will resurface in a much more general form later in the book. In the subsection at hand we prepare the way for Nevanlinna's proof by introducing some convenient notation and by attending to a few preliminary technical details. The proof (as with even the more general approaches) is basically to find differential identities satisfied by the Jacobian determinant of any Möbius transformation. Similarities and Euclidean isometries have constant Jacobian, while we have already identified a formula for the Jacobian of an arbitrary reflection in Lemma 3.2.1. Thus we can reasonably expect to find these equations, but it is certainly not a triviality. Next one shows that a conjectured conformal mapping, say f, has a determinant which also satisfies these equations and consequently has the same form as the Jacobian of a Möbius transformation, say φ. The chain rule will show $J_{f\circ\varphi^{-1}}$ is constant while $f \circ \varphi^{-1}$ is certainly conformal. It is not too difficult to see that a conformal mapping with constant determinant is an isometry or similarity, and this implies f is Möbius. In fact we have made quite similar arguments in Lemma 3.4.2. What this discussion omits are the deep technical questions of regularity, as the equations found are of higher order than one might like, requiring higher degrees of differentiability for f. This considerable technical difficulty needs to be overcome to get the best result.

3.8.2. Technical considerations. Let U be an open set in $\mathbb{R}^n$, and let g belong to the class $C^2(U,\mathbb{R}^m)$. Then g is differentiable throughout U, and for each vector u in $\mathbb{R}^n$ we obtain a function $\partial_u g$ in $C^1(U,\mathbb{R}^m)$ by setting

$$\partial_u g(x) = g'(x)u = \sum_{i=1}^{n} u_i \partial_i g(x)\,. \tag{3.57}$$

In particular, we see that $\partial_{e_i} g = \partial_i g$, the partial derivative of g with respect to x_i. Now the function $\partial_u g$ is itself differentiable in U, so we can form $\partial_v(\partial_u g)$ for any vector v in $\mathbb{R}^n$. This produces a function that lies in $C(U,\mathbb{R}^m)$. We denote it by $\partial_{v,u} g$. From (3.57) we deduce that

$$\partial_{v,u} g = \sum_{i,j=1}^{n} u_i v_j \partial_{i,j} g\ , \tag{3.58}$$

which implies among other things that $\partial_{v,u} g = \partial_{u,v} g$. Should g be a C^k-function with $k > 2$, we could continue this process to obtain higher order "directional derivatives". (Since we do not insist that vectors involved in these differential operators be unit vectors, they are not directional derivatives in the strict sense of the term.)

Suppose now that $f : D \to \mathbb{R}^n$ is a conformal mapping, where D is again a domain in $\mathbb{R}^n$. For the remainder of this section we shall assume that f is a diffeomorphism of the differentiability class $C^4(D,\mathbb{R}^n)$. We write

$$\lambda = \|f'\| = |J_f|^{1/n}, \quad \text{and} \quad \rho = 1/\lambda.$$

Here J_f is the Jacobian determinant which is positive as f is assumed to be a diffeomorphism. Note that the function λ belongs to $C^3(D)$. Since $\lambda > 0$, ρ is a C^3-function in D as well. Moreover, Theorem 3.1.1 implies that

$$\langle f'(x)u, f'(x)v\rangle = \langle f'(x)^* f'(x)u, v\rangle = \langle \lambda^2(x)u, v\rangle = \lambda^2(x)\langle u, v\rangle \tag{3.59}$$

for each point x of D and for all vectors u and v in $\mathbb{R}^n$. Notice that if λ is constant, then (3.59) has $h(x) = \frac{1}{\lambda} f(x)$ and $\langle h'(x)u, h'(x)v\rangle = \langle u, v\rangle$, and as a special case of what follows we shall show that h is an isometry. The reader should compare this with what we have already established in the closely related Lemma 3.4.3.

The first step in Nevanlinna's proof is supplied by the following lemma.

LEMMA 3.8.1. *If D is a domain in $\mathbb{R}^n$ with $n \geq 3$ and $f : D \to \mathbb{R}^n$ is a conformal mapping of class C^4, then $\partial_{u,v}\rho$ vanishes identically in D whenever u and v are orthogonal vectors in $\mathbb{R}^n$.*

PROOF. Since plainly $\partial_{u,v}\rho = 0$ when $u = 0$ or $v = 0$ and since the correspondence $(u, v) \mapsto \partial_{u,v}\rho$ is bilinear, it suffices to treat the case in which both u and v are unit vectors. For any pair of orthogonal (unit) vectors u and v, we have

$$\big\langle \partial_u f(x), \partial_v f(x)\big\rangle = \big\langle f'(x)u, f'(x)v\big\rangle = \lambda^2(x)\langle u, v\rangle = 0$$

for every x in D. Thus $\langle \partial_u f, \partial_v f\rangle = 0$ in D.

If w is an arbitrary vector in $\mathbb{R}^n$, we can apply the operator ∂_w to this relation in order to obtain

$$0 = \partial_w\langle \partial_u f, \partial_v f\rangle = \langle \partial_{w,u} f, \partial_v f\rangle + \langle \partial_u f, \partial_{w,v} f\rangle$$

and conclude that

$$\langle \partial_{w,u} f, \partial_v f \rangle = -\langle \partial_{w,v} f, \partial_u f \rangle$$

in D. In particular, given that three unit vectors u, v, and w in $\mathbb{R}^n$ are mutually orthogonal (a possibility since $n \geq 3$), we find that

$$\langle \partial_{u,v} f, \partial_w f \rangle = -\langle \partial_{u,w} f, \partial_v f \rangle = \langle \partial_{v,w} f, \partial_u f \rangle = -\langle \partial_{u,v} f, \partial_w f \rangle ,$$

with the consequence that

$$\langle \partial_{u,v} f, \partial_w f \rangle = 0 \tag{3.60}$$

in D for every such choice of u, v, and w.

Consider orthogonal unit vectors u and v in $\mathbb{R}^n$, and fix a point x of D. If W denotes the subspace of $\mathbb{R}^n$ spanned by u and v and if $T = f'(x)$, then (3.60) and the conformality of T tell us that

$$\partial_{u,v} f(x) \in T(W^\perp)^\perp = T(W)^{\perp\perp} = T(W) .$$

Of course, $T(W)$ is spanned by the nonzero orthogonal vectors $\partial_u f(x) = T(u)$ and $\partial_v f(x) = T(v)$. Accordingly, the relation

$$\partial_{u,v} f = \frac{\langle \partial_{u,v} f, \partial_u f \rangle}{|\partial_u f|^2} \, \partial_u f + \frac{\langle \partial_{u,v} f, \partial_v f \rangle}{|\partial_v f|^2} \, \partial_v f$$

holds throughout D. However,

$$|\partial_u f|^2 = \langle \partial_u f, \partial_u f \rangle = \lambda^2 |u|^2 = \lambda^2$$

and

$$\langle \partial_{u,v} f, \partial_u f \rangle = \frac{\partial_v \langle \partial_u f, \partial_u f \rangle}{2} = \frac{\partial_v (\lambda^2)}{2} = \lambda \partial_v \lambda .$$

Similarly,

$$|\partial_v f|^2 = \lambda^2, \qquad \langle \partial_{u,v} f, \partial_v f \rangle = \lambda \partial_u \lambda .$$

It follows that

$$\partial_{u,v} f = \left(\frac{\partial_v \lambda}{\lambda} \right) \partial_u f + \left(\frac{\partial_u \lambda}{\lambda} \right) \partial_v f . \tag{3.61}$$

Remembering that $\rho = 1/\lambda$, we observe that

$$\frac{\partial_u \rho}{\rho} = -\frac{\partial_u \lambda}{\lambda} \quad , \quad \frac{\partial_v \rho}{\rho} = -\frac{\partial_v \lambda}{\lambda} .$$

Thus (3.61) can be rephrased as

$$\partial_v \rho \cdot \partial_u f + \partial_u \rho \cdot \partial_v f + \rho \cdot \partial_{u,v} f = 0 \tag{3.62}$$

in D, whenever u and v are orthogonal unit vectors in $\mathbb{R}^n$. (The multiplication used here is just ordinary multiplication of a vector by a scalar.)

As a next step, we look at three unit vectors u, v, and w in $\mathbb{R}^n$ that are mutually orthogonal. We apply the operator ∂_w to (3.62) and learn that

$$\partial_{w,v} \rho \cdot \partial_u f + \partial_{w,u} \rho \cdot \partial_v f + \partial_v \rho \cdot \partial_{w,u} f + \partial_u \rho \cdot \partial_{w,v} f + \partial_w \rho \cdot \partial_{u,v} f + \rho \cdot \partial_{u,v,w} f = 0$$

or, equivalently, that

$$\partial_{w,v}\rho\cdot\partial_u f+\partial_{w,u}\rho\cdot\partial_v f = -\partial_v\rho\cdot\partial_{w,u}f-\partial_u\rho\cdot\partial_{w,v}f-\partial_w\rho\cdot\partial_{u,v}f-\rho\cdot\partial_{u,v,w}f \tag{3.63}$$

in D. Because the dependence of the right-hand side of (3.63) on u, v, and w is symmetric in these quantities, the same must be true of the other side. Therefore, for instance,

$$\partial_{w,v}\rho\cdot\partial_u f+\partial_{w,u}\rho\cdot\partial_v f = \partial_{u,v}\rho\cdot\partial_w f+\partial_{w,u}\rho\cdot\partial_v f\ ,$$

demonstrating that

$$\partial_{u,v}\rho\cdot\partial_w f = \partial_{w,v}\rho\cdot\partial_u f$$

in D. Finally, for u, v, and w as specified, we can once more exploit the conformality of f to arrive at

$$\begin{aligned}\lambda^2\partial_{u,v}\rho = \partial_{u,v}\rho\,\langle\partial_w f,\partial_w f\rangle = \langle\partial_{u,v}\rho\cdot\partial_w f,\partial_w f\rangle &= \langle\partial_{w,v}\rho\cdot\partial_u f,\partial_w f\rangle \\ &= \partial_{w,v}\rho\cdot\langle\partial_u f,\partial_w f\rangle = 0\ ,\end{aligned}$$

which demands that $\partial_{u,v}\rho = 0$ throughout D. □

The function ρ satisfies another significant differential equation.

LEMMA 3.8.2. *Under the hypotheses of Lemma* 3.8.1,

$$n|\nabla\rho|^2 = 2\rho\Delta\rho \tag{3.64}$$

in D, where $\Delta\rho$ denotes the Laplacian of ρ; $\Delta\rho = \sum_{i=1}^n \partial_{i,i}\rho$.

It is possible to verify Lemma 3.8.2 by means of brute force calculation, an exercise that obscures the geometric significance of (3.64) and is bound to leave readers groping in the dark for motivation. A more illuminating approach to (3.64) takes advantage of a geometric fact that is of interest in its own right—namely, that the function $k = 2n^{-1}\rho\Delta\rho-|\nabla\rho|^2$ represents the scalar curvature of a certain metric to which the conformal mapping f naturally gives rise. Rather than plunge into a lengthy series of computations, we prefer to take the differential geometric route to (3.64), even though this means that the proof we ultimately give for Lemma 3.8.2 draws on information of an elementary character from outside sources.

3.8.2.1. *Elements of Riemannian geometry.* Let D be a domain in $\mathbb{R}^n$ with $n \geq 2$, and let k be a positive integer. A *Riemannian C^k-structure* in D is a symmetric and positive definite $(0,2)$-tensor field g of class C^k in D. To express this in more mundane language, g is a rule that assigns to each point x of D an inner product $g(x) = \langle\cdot,\cdot\rangle_x$ on $\mathbb{R}^n$ and does it in such a way that $x \mapsto \langle u(x), v(x)\rangle_x$ describes a C^k-function in D whenever u and v are of the class $C^k(D,\mathbb{R}^n)$. (When $k = \infty$, the accepted terminology is that g is a *Riemannian metric* in D.)

Given a structure g of this kind, one can perform yet another variation on (3.33) and (3.34) by defining the g-length $\ell_g(\gamma)$ of any piecewise smooth path $\gamma : [a,b] \to D$ through the formula

$$\ell_g(\gamma) = \int_a^b |\dot\gamma(t)|_{\gamma(t)}\,dt\ ,$$

in which $|\cdot|_x$ denotes the norm associated with $\langle\cdot,\cdot\rangle_x$, and using this to determine the g-distance $d_g(x,y)$ between points x and y of D:

$$d_g(x,y) = \inf_\gamma \ell_g(\gamma)\ ,$$

the infimum being taken over all piecewise smooth paths γ in D that start at x and terminate at y. The pair $\langle D, d_g\rangle$ is then a metric space. Moreover, the metric topology in D corresponding to the distance d_g is the same as the relative topology induced in D by the standard topology of $\mathbb{R}^n$. A Riemannian C^k-structure g carries with it a lot of other geometric structure. These include notions of g-volume, g-surface area, g-measurement of angles, and so forth, which we only allude to in passing.

Suppose that D and D' are domains in $\mathbb{R}^n$ and that g and g_0 are Riemannian C^k-structures in D and D', respectively (the possibility that $D = D'$ and $g = g_0$ is not excluded here). By an *isometry* of the pair $\langle D, g\rangle$ onto the pair $\langle D', g_0\rangle$ is meant a C^{k+1}-diffeomorphism f of D onto D' with the property that, giving the notation its obvious interpretation,

$$\langle u, v\rangle_x = \langle f'(x)u, f'(x)v\rangle_{f(x)} \tag{3.65}$$

for each point x of D and for all vectors u and v in $\mathbb{R}^n$. When a mapping of this kind exists, we say that $\langle D, g\rangle$ and $\langle D', g_0\rangle$ are *isometric*. Indeed, if f is such an isometry, then it is easy to see that

$$d_g(x, y) = d_{g_0}\big[f(x), f(y)\big]$$

for all x and y in D.

Of course the basic Riemannian metric of a subdomain D of $\mathbb{R}^n$ is the restricted Euclidean metric g_e: it has $g_e(x) = \langle\cdot,\cdot\rangle$, the standard Euclidean inner product on $\mathbb{R}^n$, for every x in D.

If we set

$$g = e^{2\varphi} g_e, \tag{3.66}$$

where φ comes from the class $C^k(D)$, we obtain a Riemannian C^k-structure in D. A structure g given by (3.66) is said to be *conformally Euclidean with metric density* e^{φ} (cf. Section 3.1, where $\rho = e^{\varphi}$). A second method of constructing Riemannian C^k-structure in D is via "pull-back".

Let f be a C^{k+1}-diffeomorphism of D onto a domain D' in which a Riemannian C^k-structure g_0 is assumed to be defined. For each x in D we can use (3.65) to determine an inner product $\langle\cdot,\cdot\rangle_x$ on $\mathbb{R}^n$. Then $g(x) = \langle\cdot,\cdot\rangle_x$ describes a Riemannian C^k-structure in D. We refer to g as the *pull-back of* g_0 *under* f, a relationship that is traditionally indicated by writing $g = f^*(g_0)$. Quite clearly the diffeomorphism f is an isometry of $\langle D, g\rangle$ onto $\langle D', g_0\rangle$ precisely when $g = f^*(g_0)$.

For technical reasons we shall henceforth assume that $k \geq 3$. With a Riemannian C^k-structure g in D are associated several different types of "curvature", each of which conveys information about the geometry introduced on D by g. Among these is the scalar curvature of g. This provides a local measure of the deviation of g-surface area from the Euclidean surface area. To be slightly more explicit, given a point x of D and $r > 0$ we let $S_r = \mathbb{S}^{n-1}(x, r)$ and $S'_r = \{y \in D : d_g(y, x) = r\}$. If σ_g and σ denote g-surface area and Euclidean surface area, respectively, it can be shown that

$$\frac{\sigma_g(S'_r)}{\sigma(S_r)} = 1 - cr^2 + o(r^2) \tag{3.67}$$

as $r \to 0$ for some constant c. The *scalar curvature* $k_g(x)$ of g at x is defined by $k_g(x) = c/c_n$, where $c_n > 0$ is a certain normalization factor that depends only on the dimension n.

C. Taubes' book [**151**] is a good place to start learning more about the differential geometry here and in particular for more precise details on the subject of scalar curvature.

Two facts are readily apparent from (3.67): first, the restricted Euclidean metric g_e in D has $k_{g_e} = 0$ throughout D; second, scalar curvature is invariant under isometry—that is, if f is an isometry of $\langle D, g\rangle$ onto $\langle D', g_0\rangle$, then $k_g(x) = k_{g_0}[f(x)]$ for every x in D.

In the event that g is a conformally Euclidean C^k-structure in D, say $g = e^{2\varphi}g_e$ with φ a function from $C^k(D)$, there is a standard formula giving its scalar curvature:

$$k_g = -n^{-1}e^{-2\varphi}\left[2\Delta\varphi + (n-2)|\nabla\varphi|^2\right]. \tag{3.68}$$

Accepting the foregoing discussion, we proceed to prove Lemma 3.8.2.

PROOF OF LEMMA 3.8.2. It has already been noted that, under the assumptions in force in Lemma 3.8.2, the statement $\langle f'(x)u, f'(x)v\rangle = \lambda^2(x)\langle u, v\rangle$ holds for each point x of D and for all vectors u and v in $\mathbb{R}^n$. We can rephrase this by saying that $g = f^*(g_e)$, the pull-back under f of the Riemannian metric g_e in $D' = f(D)$, is just the conformally Euclidean C^3-structure in D whose metric density is e^{φ}, with $\varphi = \log\lambda = -\log\rho$. Since

$$\nabla\varphi = -\rho^{-1}\nabla\rho \quad , \quad \Delta\varphi = -\rho^{-1}\Delta\rho + \rho^{-2}|\nabla\rho|^2 ,$$

and $n \geq 3$, we infer from (3.68) that

$$k_g = 2n^{-1}\rho\Delta\rho - |\nabla\rho|^2 .$$

On the other hand, f is an isometry of $\langle D, g\rangle$ onto $\langle D', g_e\rangle$, whence $k_g = k_{g_e}\circ f = 0$. Identity (3.64) follows. □

With Lemma 3.8.2 in hand, we return to Liouville's theorem, but first remind the reader of an elementary fact from algebra.

LEMMA 3.8.3. *If $B : \mathbb{R}^n \times \mathbb{R}^n \to \mathbb{R}$ is a symmetric bilinear mapping with the property that $B^n(u, v) = 0$ whenever $\langle u, v\rangle = 0$, then there is a real number σ such that $B(u, v) = \sigma\langle u, v\rangle$ for all u and v in $\mathbb{R}^n$.*

PROOF. Let u and v be vectors in $\mathbb{R}^n$. We have

$$B^n(u, v) = \sum_{i,j=1}^{n} u_i v_j B^n(e_i, e_j) = \sum_{i=1}^{n} u_i v_i B^n(e_i, e_i) ,$$

for $\langle e_i, e_j\rangle = 0$ if $i \neq j$. It is also true that $\langle e_i - e_j, e_i + e_j\rangle = 0$, whence

$$0 = B^n(e_i - e_j, e_i + e_j) = B^n(e_i, e_i) - 2B^n(e_i, e_j) - B^n(e_j, e_j) = B^n(e_i, e_i) - B^n(e_j, e_j) .$$

The inference is that $B^n(e_i, e_i) = B^n(e_j, e_j)$ when $1 \leq i, j \leq n$. Letting $\sigma = B^n(e_1, e_1)$, we see that $B^n(u, v) = \sigma\sum_{i=1}^{n} u_i v_i = \sigma\langle u, v\rangle$ for all u and v in $\mathbb{R}^n$. □

Lemmas 3.8.1, 3.8.2, and 3.8.3 now combine to supply an important ingredient in the proof of Liouville's theorem for C^4-mappings. It establishes that the Jacobian of our conformal mapping f has the same form as the Jacobian of a Möbius transformation.

LEMMA 3.8.4. *Under the hypotheses of Lemma* 3.8.1 *there exist a point b in $\mathbb{R}^n$ and real numbers α and β with $\alpha^2+\beta^2>0$ such that $\rho(x)=\beta|x-b|^2+\alpha$ for every x in D.*

PROOF. For fixed x in D the function $B:\mathbb{R}^n\times\mathbb{R}^n\to\mathbb{R}$ defined by $B^n(u,v)=\partial_{u,v}\rho(x)$ is a symmetric bilinear mapping with the property that $B^n(u,v)=0$ whenever $\langle u,v\rangle=0$ (Lemma 3.8.1). We apply Lemma 3.8.3 at each point of D to obtain a function $\sigma:D\to\mathbb{R}$ such that $\partial_{u,v}\rho=\sigma\langle u,v\rangle$ in D for all u and v in $\mathbb{R}^n$. In particular, taking $u=v=e_i$ discloses that $\partial_{i,i}\rho=\sigma$ for $1\le i\le n$, while the choice $u=e_i$, $v=e_j$ reveals that $\partial_{i,j}\rho=0$ for $i\ne j$. The identity $\partial_{i,i}\rho=\sigma$ shows that σ belongs to $C^1(D)$. Furthermore, given $1\le i\le n$, we can choose j different from i and compute that

$$\partial_i\sigma=\partial_i\partial_{j,j}\rho=\partial_{i,j,j}\rho=\partial_{j,i,j}\rho=\partial_j\partial_{i,j}\rho=0\,.$$

It is precisely at this point where the assumption that f is of class C^4 comes into play, for the calculation requires that ρ be a C^3-function.

It follows that σ is a constant function, say $\sigma(x)=2\beta$ for every x in D. Thus $\partial_{i,i}\rho=2\beta$ for $i=1,2,\ldots,n$. Assume first that $\beta\ne 0$. In conjunction with the knowledge that $\partial_{i,j}\rho=0$ when $i\ne j$, the relation $\partial_{i,i}\rho=2\beta$ forces $\nabla\rho$ to have the form

$$\nabla\rho(x)=(2\beta x_1+c_1,2\beta x_2+c_2,\ldots,2\beta x_n+c_n)=2\beta x+c$$

for some $c=(c_1,c_2,\ldots,c_n)$ in $\mathbb{R}^n$. This can be rewritten as $\nabla\rho(x)=2\beta(x-b)$ by setting $b=-(2\beta)^{-1}c$. In this case it becomes clear that ρ takes the form $\rho(x)=\beta|x-b|^2+\alpha$ for some constant α.

If, on the other hand, $\beta=0$, then we find that $\Delta\rho=0$ in D. Lemma 3.8.2 mandates that $\nabla\rho=0$ in D, which tells us that ρ is constant there. Suppose that $\rho(x)=\alpha$ for all x in D. Then $\alpha>0$ and, once again, $\rho(x)=\beta|x-b|^2+\alpha$ for any b in $\mathbb{R}^n$. □

The following simple, but interesting, geometric observation plays a key role in Nevanlinna's argument.

LEMMA 3.8.5. *Let c be a point of $\mathbb{R}^n$, and let $\gamma:[t_0,t_1]\to\mathbb{R}^n\setminus\{c\}$ be a C^1-path with the property that for each t in the interval (t_0,t_1) the vector $\dot\gamma(t)$ is a normal vector to the sphere $\mathbb{S}^{n-1}(c,|\gamma(t)-c|)$. Then the trajectory of γ lies on a ray issuing from c.*

PROOF. Since the hypotheses and conclusion of the lemma are translation invariant, there is no loss of generality in assuming that $c=0$. The radial projection $P:\mathbb{R}^n\setminus\{0\}\to\mathbb{S}^{n-1}$ given by $P(x)=|x|^{-1}x$ is a C^∞-function. Moreover, for any x in $\mathbb{R}^n\setminus\{0\}$ and any scalar multiple h of x it is true that $P(x+sh)=P(x)$ whenever $|s|$ is sufficiently small, so

$$P'(x)h=\lim_{s\to 0}\frac{P(x+sh)-P(x)}{s}=0\,.$$

Consider the path $\beta=P\circ\gamma$. By assumption $\dot\gamma(t)$ is a scalar multiple of $\gamma(t)$ for each t in (t_0,t_1). As a result, $\dot\beta(t)=P'[\gamma(t)]\dot\gamma(t)=0$ on (t_0,t_1). This makes β constant on $[t_0,t_1]$. This simply means that every point on the trajectory of γ projects to the same point of $\mathbb{S}^{n-1}$. The stated conclusion follows. □

The final lemma in this section represents a minor extension of Lemmas 3.4.2 and 3.4.3. It is simply a formulation of analytic continuation for Möbius transformations, while the infinitesimal version relies on Lemma 3.4.2.

LEMMA 3.8.6. *Let D be a domain in $\mathbb{R}^n$, and let $f : D \to \mathbb{R}^n$ be a function with the following property: for each point x of D there exist an open ball B_x centered at x and a transformation g_x in $\boldsymbol{E}(n)$ such that f and g_x coincide in B_x. Then f is the restriction to D of a unique member g of $\boldsymbol{E}(n)$.*

Consequently f is the restriction to D of a unique $g \in \boldsymbol{E}(n)$ if f is an injective mapping such that

$$\lim_{y \to x} |f(y) - f(x)|/|y - x| = 1$$

for each x in D.

PROOF. We initially prove the first part of the lemma. Fix x_0 in D and write g for g_{x_0}. We claim that $f = g$ in D. To see this consider $U = \{x \in D : g_x = g\}$. The set U is nonempty, for x_0 belongs to U. If x is in U and y in B_x, then $g_y = f = g_x = g$ in $B_y \cap B_x$, a nonempty open set. By Corollary 3.3.11, $g_y = g$. This implies that y is an element of U, i.e., B_x is contained in U. We have just demonstrated that U is an open set in which f and g coincide. If x is a point of $\bar{U} \cap D$, then $B_x \cap U$ is a nonempty open set in which $g_x = f = g$. Calling on Corollary 3.3.11 a second time, we conclude that $g_x = g$, so x lies in U. In other words, U is relatively closed in D. The connectedness of D dictates that $U = D$, whence $f = g|D$.

To see the second part of the lemma we simply appeal to Lemma 3.4.2. □

3.8.3. Liouville's theorem for C^4-mappings. All necessary pieces are now in place for the proof of the C^4-smooth version of Liouville's theorem. We recall that a conformal mapping $f : D \to \mathbb{R}^n$ is simply an injective function whose derivative is pointwise a scalar multiple of an orthogonal transformation, $f'(x) = \lambda(x)O(x)$, $\lambda(x) \geq 0$, and $O : D \to \mathbf{SO}(n)$.

THEOREM 3.8.7. *If D is a domain in $\mathbb{R}^n$ with $n \geq 3$ and $f : D \to \mathbb{R}^n$ is a conformal diffeomorphism that belongs to the class $C^4(D, \mathbb{R}^n)$, then f is the restriction to D of a Möbius transformation of $\hat{\mathbb{R}}^n$.*

PROOF. According to Lemma 3.8.4 there exist a point b in $\mathbb{R}^n$ and real constants α and β, not both zero, such that

$$\|f'(x)\| = \frac{1}{\beta|x - b|^2 + \alpha}$$

for every x in D. Lemma 3.8.4 applies equally well to $h = f^{-1}$, which is a conformal C^4-mapping of the domain $D' = f(D)$ onto D. Thus we can assert the existence of c in $\mathbb{R}^n$ and real numbers γ and δ with $\gamma^2 + \delta^2 > 0$ such that

$$\|h'(y)\| = \frac{1}{\gamma|y - c|^2 + \delta}$$

for every y in D'. Because $\|h'[f(x)]\| = \|f'(x)\|^{-1}$, we deduce that

$$\big(\beta|x - b|^2 + \alpha\big)\big(\gamma|f(x) - c|^2 + \delta\big) = 1 \tag{3.69}$$

whenever x belongs to D. We now distinguish two cases.

Case 1: $\beta = 0$. Here $\|f'(x)\| = \alpha^{-1}$ is constant in D. From (2.22) and the conformality of f we deduce that $L_f(x) = \ell_f(x) = \alpha^{-1}$ for each x in D, which implies that

$$\lim_{y \to x} \frac{|f(y) - f(x)|}{|y - x|} = \frac{1}{\alpha}$$

for such x. Lemma 3.8.6 tells us that αf must be the restriction to D of a Euclidean isometry from $\boldsymbol{E}(n)$. As a result, we discover that f is the restriction to D of a similarity transformation of $\mathbb{R}^n$, and hence the restriction to D of a Möbius transformation.

Case 2: $\beta \neq 0$. Identity (3.69) requires that $\gamma \neq 0$. This identity has another crucial implication: if $\mathbb{S}^{n-1}(b, t) \cap D \neq \emptyset$, then the image of this intersection under f lies on the sphere $\mathbb{S}^{n-1}(c, r)$, where r is an algebraic function of t (to be precise, $r = r(t) = \{[(\beta t^2 + \alpha)^{-1} - \delta]/\gamma\}^{1/2}$). We now fix a point x_1 of D such that $x_1 \neq b$ and $f(x_1) \neq c$. We then fix a point x_0 on the open line segment between x_1 and b, doing it in such a way that the closed line segment I with endpoints x_0 and x_1 is contained in D and that $f(I)$ does not contain c. Setting $t_0 = |x_0 - b|$, $t_1 = |x_1 - b|$, and $u = (x_0 - b)/|x_0 - b|$, we parametrize I as follows: $x = x(t) = b + tu$ for $t_0 \le t \le t_1$. From this we retrieve a smooth, injective parametrization of $f(I)$: $y = y(t) = f(b + tu)$ for $t_0 \le t \le t_1$. Now $\dot{x}(t) = u$ is a normal vector to $\mathbb{S}^{n-1}(b, t)$ for each t in (t_0, t_1), so the conformality of f ensures that the vector $\dot{y}(t) = f'(b + tu)u$ is normal to the sphere $\mathbb{S}^{n-1}(c, r(t)) = \mathbb{S}^{n-1}(c, |y(t) - c|)$ for such t. By Lemma 3.8.5, $f(I)$ lies on some ray emanating from c. We infer that $f(I)$ is the line segment whose endpoints are $y_0 = f(x_0)$ and $y_1 = f(x_1)$. Write $r_0 = |y_0 - c|$ and $r_1 = |y_1 - c|$. We proceed under the assumption that $r_0 < r_1$ (the case $r_1 < r_0$ is handled similarly). For $t_0 \le t \le t_1$ we obtain

$$r(t) - r_0 = |y(t) - y_0| = \int_{t_0}^{t} |\dot{y}(\tau)|\, d\tau = \int_{t_0}^{t} \|f'(b + \tau u)\|\, d\tau = \int_{t_0}^{t} \frac{d\tau}{\beta\tau^2 + \alpha}\,.$$

If $\alpha \neq 0$, the last integral does not define an algebraic function of t on $[t_1, t_2]$, whereas the dependence of $r(t)$ (hence, of $r(t) - r_0$) on t is known to be algebraic. We are thus led to conclude: in Case 2 we must have $\alpha = 0$, which in conjunction with (3.69) implies that b cannot be a point of D. By the same argument applied to $h = f^{-1}$, we see that $\delta = 0$ and that c is not a point of D'. Finally, we consider $H = R \circ f$, where R is an inversion in $\mathbb{S}^{n-1}(c, 1)$. Then $H : D \to \mathbb{R}^n$ is a conformal mapping, H belongs to $C^4(D, \mathbb{R}^n)$, and

$$\|H'(x)\| = \|R'[f(x)]f'(x)\| = \|R'[f(x)]\| \cdot \|f'(x)\| = \frac{1}{|f(x) - c|^2}\, \frac{1}{\beta|x - b|^2} = \gamma$$

for all x in D. To see this, use (3.1), (3.10), and (3.69), not forgetting that $\alpha = \delta = 0$. In other words, H is a mapping to which Case 1 applies. Accordingly, H is the restriction to D of a Möbius transformation, so $f = R^{-1} \circ H = R \circ H$ has the same property. □

For the sake of completeness, we should at least sketch an argument showing why the function defined on $[t_0, t_1]$ by $t \mapsto \int_{t_0}^{t} (\beta\tau^2 + \alpha)^{-1}\, d\tau$ is not an algebraic function when both α and β are nonzero and $\beta\tau^2 + \alpha$ has no roots in $[t_0, t_1]$. (Of course, if one were prepared to accept the fact that logarithm and inverse trigonometric functions are transcendental functions, such an argument would be unnecessary.)

Perhaps the easiest way to demonstrate this is to consider the analytic function ζ defined in an open disk B centered at t_0 and containing neither root of $\alpha z^2 + \beta$ by $\zeta(z) = \int_{t_0}^{z} (\beta\zeta^2 + \alpha)^{-1}\, d\zeta$, where the integration is carried out along any smooth path in B that has initial point t_0 and terminal point z. The function element (ζ, B) can be continued analytically along all paths in the complex plane that avoid the singularities of $(\beta z^2 + \gamma)^{-1}$. These singularities are exactly two simple poles at points z_1 and z_2. The continuation of (ζ, B) along a smooth path γ that starts and ends at t_0 and has winding numbers $n(\gamma, z_1) = k$ and $n(\gamma, z_2) = 0$, with k an arbitrary integer, results in a function element (ζ_k, B_k) for which $\zeta_k(t_0) = ka$, where a is the residue of $(\alpha z^2 + \beta)^{-1}$ at z_1. This implies that the Riemann surface S of the complete analytic function $\underline{\zeta}$ determined by (ζ, B) is infinitely sheeted over the complex plane.

If there were a polynomial $P(z, w)$ of positive degree in w such that $P(t, \zeta(t)) = 0$ for all t in some interval $[t_0, t_0+\epsilon]$ with $\epsilon > 0$, then $\underline{\zeta}$ would be a complete algebraic function, and S would be finitely sheeted over the complex plane. Therefore, no such polynomial P can exist under the conditions stated, so $t \mapsto \int_0^t \frac{d\tau}{\alpha\tau^2+\beta}$ is not an algebraic function.

Over the years Liouville's theorem has yielded only grudgingly to refinement in the form of relaxed smoothness hypotheses. In 1958, more than a century after Liouville's paper, Hartman established the result for conformal mappings of class C^1 [**71**]. The theorem for conformal mappings as we have defined them was not proven until 1962, at which time F.W. Gehring showed that, when formulated in terms of conformal invariants such as moduli, the result is true without any a priori differentiability assumptions whatsoever [**46**].

Gehring's work was further generalized by Yu.G. Reshetnyak in 1967 [**141**], who removed the injectivity assumption in the result we have presented.

The best possible version of Liouville's theorem, in a well-defined sense that we shall not go into here, was proved in 1993 by T. Iwaniec and G.J. Martin, but only for even dimensions [**85**]. Whether or not their result remains valid in odd dimensions is an intriguing open question to which there are only partial answers; see [**82, 84**].

To round out the present discussion we shall state one variant of Gehring's theorem and an extremely surprising recent refinement of it due to J. Heinonen and P. Koskela [**75**], although we shall not have the means to prove it for some chapters to come.

Let D be a domain in $\mathbb{R}^n$, and let $f : D \to \mathbb{R}^n$ be a continuous injection. The *linear dilatation* of f at the point x of D is the quantity $H_f(x)$ defined by

$$H_f(x) = \limsup_{r\to 0} \frac{L_f(x, r)}{\ell_f(x, r)}\,, \tag{3.70}$$

where for $0 < r < \operatorname{dist}(x, \partial D)$ we set

$$L_f(x, r) = \max_{|h|=r} |f(x+h) - f(x)|, \qquad \ell_f(x, r) = \min_{|h|=r} |f(x+h) - f(x)|\,.$$

Plainly $1 \le H_f(x) \le \infty$. If f happens to be differentiable at x and if $f'(x)$ is nonsingular, then from (2.16) we derive the inequalities

$$L\big[f'(x)\big] - \max_{|h|=r} |\varepsilon(x+h)| \le \frac{L_f(x,r)}{r} \le L\big[f'(x)\big] + \max_{|h|=r} |\varepsilon(x+h)|$$

and

$$\ell\big[f'(x)\big] - \max_{|h|=r} |\varepsilon(x+h)| \le \frac{\ell_f(x,r)}{r} \le \ell\big[f'(x)\big] + \max_{|h|=r} |\varepsilon(x+h)|\,,$$

which are valid whenever $0 < r < \operatorname{dist}(x, \partial D)$. From this it is relatively straightforward to establish that

$$H_f(x) = \frac{L[f'(x)]}{\ell[f'(x)]} = H\big[f'(x)\big]. \tag{3.71}$$

In particular, $H_f(x) = 1$ would hold for every x in D, should f be a conformal diffeomorphism of D. The simplest rendering of Gehring's theorem reads:

THEOREM 3.8.8. *If D is a domain in $\mathbb{R}^n$ with $n \ge 3$ and $f : D \to \mathbb{R}^n$ is a continuous injection with the property that $H_f(x) = 1$ for every x in D, then f is the restriction to D of a Möbius transformation of $\hat{\mathbb{R}}^n$.*

The addition that the aforementioned work of Heinonen and Koskela would make would be to replace the linear distortion function $H_f(x)$ in the above theorem by the possibly smaller function

$$\underline{H}_f(x) = \liminf_{r \to 0} \frac{L_f(x,r)}{\ell_f(x,r)}\,. \tag{3.72}$$

The paucity of conformal mappings in dimensions three and above dictates that a reasonable surrogate for conformality be found, lest the incredible richness, flexibility, and utility of plane conformal mapping theory be totally lost in the passage to higher dimensions. The trick is to produce a theory with an abundance of mappings, yet not to retreat so far from conformality that many of the desirable geometric and analytic features of the plane theory are sacrificed in the process. The theory of quasiconformal mappings, of which we shall give an account in the remaining chapters of this book, would seem to strike a very happy medium in this regard.

CHAPTER 4

The Moduli of Curve Families

One of the consequences of conformality for a mapping f is that the directional derivatives in directions v, $\|v\| = 1$, all have the same magnitude since $\|f'(x)v\| = \|\lambda(x)O(x)v\| = \lambda(x)\|v\| = \lambda(x)$. Since $\lambda^n(x) = J_f(x)$, the Jacobian determinant, there is a straightforward functional relationship between the infinitesimal length distortion (in any direction) and the local distortion of volume. Integrating this local relationship leads to the "length–area" method—so named because it was first used in two dimensions—to identify geometrically natural conformal invariants related to the configurations of sets in the domain D and their images in the range D' of f. We will see that these ideas give rise to very powerful tools to aid in the study of the geometry of mappings.

Let D and D' be domains in $\mathbb{R}^n$ with $n \geq 2$, and let f be a homeomorphism of D onto D'. There is an obvious standard by which to gauge how far f deviates from being a conformal mapping. Taking a cue from the final pages of the previous chapter, we can consider the quantity

$$H(f) = \sup\{H_f(x) : x \in D\}, \tag{4.1}$$

where $H_f(x) \geq 1$ is the linear dilatation of f at x as defined at (3.70). Theorem 3.8.8, as yet unproven, suggests this as an intuitive measure of "closeness to conformality" on the part of f since $H(f) = 1$ if only if f is a conformal mapping.

A mapping for which $H(f)$ is finite is called *quasiconformal.* When $n = 2$ and f is a diffeomorphism, this definition goes back in spirit to one formulated by Grötzsch in the 1920s, although he did not use the expression "quasiconformal", which owes its coinage to Ahlfors.

This definition of quasiconformality has some nice features. It is simple, direct, very general, and quite appealing from a geometric-aesthetic vantage point. Further, it requires no regularity of differentiability properties of the mapping to formulate. Indeed one can see clear generalisations to abstract metric spaces—a study undertaken to great effect in recent years after the foundational papers of Heinonen and Koskela.

Unfortunately the above definition of quasiconformality suffers from a serious drawback that renders it a less-than-ideal starting point from which to build up a theory: many basic facts concerning these mappings are extremely awkward to establish, proceeding solely from the knowledge that their linear dilatations are bounded.

Fortunately, there are more efficient ways to develop the theory. There is, for instance, the approach initiated by Yu G. Reshetnyak and outlined in his book [**140**] basically deriving information about quasiconformal mappings from the theory of partial differential equations. This is because such mappings naturally arise as homeomorphic solutions to certain types of nonlinear partial differential equations, these days called Beltrami systems.

This approach employs some fairly heavy duty analytical machinery—which necessitates the imposition of a priori regularity conditions on the mappings considered—while placing only token emphasis on the geometric features of quasiconformal mappings, seeking to derive them as a consequence of the analysis. The modern approach to the higher-dimensional theory of quasiconformal mappings from this point of view is fully covered in the monograph of T. Iwaniec and G. Martin [**84**] which also contains a fairly extensive bibliography should the reader care to explore further.

The tack we choose to follow in the present text, adhering to the precedent set by the well-known books of O. Lehto and K. Virtanen [**97**] and J. Väisälä [**160**], does just the opposite. From the outset, our development of the theory of quasiconformal mappings highlights the geometric aspects of the theory, delaying any examination of subtle analytical issues until we have developed the requisite theory of the geometry of mappings.

Inspired by the work of A. Beurling and L. Ahlfors [**10**], this treatment rests on the study of a conformal invariant known as the "conformal modulus of a curve family" or "extremal length". There are few books about extremal length as a topic in its own right, and so for the interested reader we mention L.V. Ahlfors' book on conformal invariants [**8**] and for a more recent and far deeper treatment M. Ohtsuka's book [**135**].

Over the next few pages we present a heuristic introduction to the concept in an effort to motivate it.

4.0.4. A conformal invariant. Suppose that $f : D \to D'$ is a diffeomorphism between domains D and D' in $\mathbb{R}^n$, where $n \geq 2$, and that Γ is a family of smooth paths in D. We are really thinking here of a "large" family of paths. A good example to keep in mind would be the family of all smooth paths in D that connect a given pair of nonempty, disjoint compact subsets E and F of D.

We write $f(\Gamma)$ for $\{f \circ \gamma : \gamma \in \Gamma\}$. Then $f(\Gamma)$ is also a family of smooth paths in D'. Earlier, in a different context, we indicated how, given a continuous positive function $\rho : D \to [0, \infty)$, to assign a ρ-length $\ell_\rho(\gamma)$ to any smooth path γ in D:

$$\ell_\rho(\gamma) = \int_\gamma \rho(x)\,|dx|\,.$$

There is an analogous notion of ρ-volume: the *ρ-volume* $v_\rho(A)$ of a Lebesgue measurable subset A of D is given by

$$v_\rho(A) = \int_A \rho^n(x)\,dx\ ,$$

dx denoting integration with respect to Lebesgue measure in $\mathbb{R}^n$.

We must now introduce a normalization into the picture by restricting our attention to functions ρ that satisfy

$$\ell_\rho(\gamma) \geq 1, \qquad \text{for every } \gamma \text{ in } \Gamma. \tag{4.2}$$

Any continuous function $\rho : D \to [0, \infty)$ endowed with this extra property (4.2) is called an *admissible density* for Γ, and the collection of all such admissible densities is symbolized by $\operatorname{Adm}(\Gamma)$.

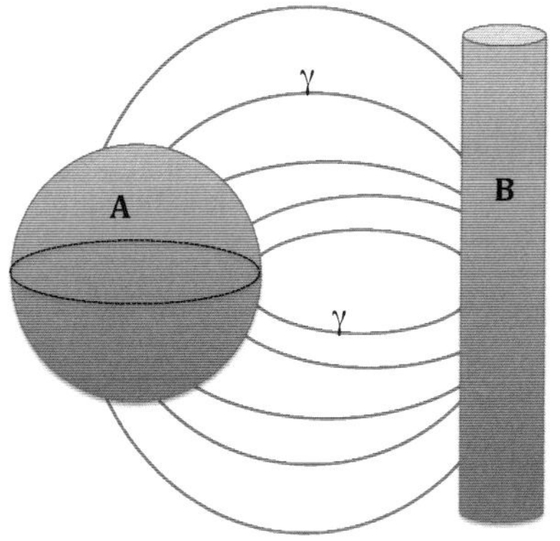

The family of curves $\gamma \in \Gamma$ joining continua A (a sphere) to B (a cylinder) in $\mathbb{R}^n$. In this case $\rho(x) = 1/\mathrm{dist}(A, B)$ is an admissible function.

Intuitively, the constraint $\ell_\rho(\gamma) \geq 1$ prevents ρ from being, on average, excessively small along the trajectory of any $\gamma \in \Gamma$. Thus, if ρ is an admissible density for Γ and if Γ contains sufficiently many paths that their collective trajectories fill up a substantial portion of D (in the measure theoretic sense), one might expect that the associated volume $v_\rho(D)$ could not itself be small.

These thoughts prompt one to ask how small a ρ-volume might actually be achieved through the proper choice of ρ. We have bounded from below the length and now wish to control the volume, and so we are led to consider the quantity

$$M(\Gamma) = \inf\{v_\rho(D) : \rho \in \mathrm{Adm}(\Gamma)\},$$

called the modulus of the curve family Γ. It may happen that $\mathrm{Adm}(\Gamma) = \emptyset$. If so, we agree to set $M(\Gamma) = \infty$. It is also possible that $\mathrm{Adm}(\Gamma) \neq \emptyset$, but that $v_\rho(D) = \infty$ for all ρ in $\mathrm{Adm}(\Gamma)$. In this event we again have $M(\Gamma) = \infty$.

Of course, we can go through the same process for $\tilde{\Gamma} = f(\Gamma)$ and then pose the question: How do $M(\Gamma)$ and $M(\tilde{\Gamma})$ compare?

Fortunately there is an easy way of passing from an admissible density $\tilde{\rho}$ for $\tilde{\Gamma}$ to a density ρ in $\mathrm{Adm}(\Gamma)$; namely, we can define $\rho : D \to [0, \infty)$ by the "push forward",

$$\rho(x) = \tilde{\rho}[f(x)]\,\|f'(x)\|\,.$$

To check that ρ does belong to $\mathrm{Adm}(\Gamma)$ we make the following calculation. Let $\gamma : [a, b] \to D$ be a path in Γ. Then $\beta = f \circ \gamma$ is a member of $\tilde{\Gamma}$, whence

$$\begin{aligned} 1 \;&\leq\; \ell_{\tilde{\rho}}(\beta) = \int_\beta \tilde{\rho}(x)\,|dx| = \int_a^b \tilde{\rho}[\beta(t)]\,|\dot{\beta}(t)|\,dt = \int_a^b \tilde{\rho}[\beta(t)]\,\big|f'[\gamma(t)]\dot{\gamma}(t)\big|\,dt \\ &\leq\; \int_a^b \tilde{\rho}[\beta(t)]\|f'[\gamma(t)]\| \cdot |\dot{\gamma}(t)|dt = \int_a^b \rho[\gamma(t)]\,|\dot{\gamma}(t)|dt = \int_\gamma \rho(x)|dx| = \ell_\rho(\gamma). \end{aligned}$$

Moreover, we recall the definition of $H_O(T)$ for a linear transformation T given at (2.9) and we further define the *outer distortion*

$$K_O(f) = \sup_{x \in D} H_O[f'(x)] = \sup_{x \in D} \frac{\|f'(x)\|^n}{|J_f(x)|} \tag{4.3}$$

and assume that $K_O(f) < \infty$. Then

$$v_\rho(D) \le K_O(f)\, v_{\tilde\rho}(D') \,, \tag{4.4}$$

by the standard change of variable formula

$$\begin{aligned} v_\rho(D) &= \int_D \rho^n(x)\,dx = \int_D \tilde\rho^n\big[f(x)\big]\|f'(x)\|^n\,dx \\ &\le K_O(f)\int_D \tilde\rho^n\big[f(x)\big]|J_f(x)|\,dx = K_O(f)\int_{D'} \tilde\rho^n(y)\,dy = K_O(f)v_{\tilde\rho}(D')\,. \end{aligned}$$

Given that $\mathrm{Adm}(\tilde\Gamma) \neq \emptyset$, it follows from (4.4) that

$$M(\Gamma) \le v_\rho(D) \le K_O(f) v_{\tilde\rho}(D')$$

for every $\tilde\rho$ in $\mathrm{Adm}(\tilde\Gamma)$, and hence that

$$M(\Gamma) \le K_O(f) M(\tilde\Gamma)\,. \tag{4.5}$$

Inequality (4.5) holds trivially when $\mathrm{Adm}(\tilde\Gamma) = \emptyset$, a case in which $M(\tilde\Gamma) = \infty$ by definition.

Naturally, the same game can be played for the mapping f^{-1} to produce a companion inequality to (4.5):

$$M(\tilde\Gamma) \le K_I(f) M(\Gamma)\,, \tag{4.6}$$

provided $K_I(f) = K_O(f^{-1}) < \infty$. Incidentally, it is a consequence of (2.10) that

$$K_I(f) = \sup_{y\in D'} H_O\big[(f^{-1})'(y)\big] = \sup_{x\in D} H_I\big[f'(x)\big] = \sup_{x\in D} \frac{|J_f(x)|}{\ell[f'(x)]^n}\,.$$

If the diffeomorphism f is a conformal mapping of D onto D', we have

$$H_O\big[f'(x)\big] = H_I\big[f'(x)\big] = 1$$

for every x in D, in which event we obtain $K_O(f) = K_I(f) = 1$.

On the strength of (4.5) and (4.6), we can therefore assert: *if $f : D \to D'$ is a conformal diffeomorphism between domains D and D' in $\mathbb{R}^n$, then $M(\Gamma) = M[f(\Gamma)]$ whenever Γ is a family of smooth paths in D.* In short, $M(\Gamma)$ is a conformally invariant quantity. It is called the *conformal modulus* of Γ.

The number $\lambda(\Gamma) = M(\Gamma)^{1/(1-n)}$ is known as the *conformal extremal length* of Γ.

As we mentioned earlier, this concept is an outgrowth of the so-called length-area method popularized in the first half of the last century by Radó, Grötzsch, and other stalwarts of classical complex function theory. In a landmark 1950 paper [**10**], Beurling and Ahlfors began by distilling the principles that underlie length-area arguments into the notion of extremal length, and then proceeded to mold this invariant into a powerful geometric and analytic tool.

Our chief objective in the present chapter is to establish the fundamental properties of the conformal modulus in n-dimensional Euclidean space. In order to maximize our gain from this endeavour, it will be necessary to abandon the category of smooth paths and to make sense of $M(\Gamma)$ for an arbitrary family of paths. This is since even if γ is a smooth path, $f(\gamma)$ may not be without a priori assumptions on f which we will not want to make.

Passage to the more general context therefore requires that we address certain technical issues that do not come up in the smooth setting.

Let us return for a moment to the diffeomorphism $f : D \to D'$. Under the assumption that the numbers $K_O(f)$ and $K_I(f)$ are finite, we have verified that

$$\frac{1}{K_O(f)}\, M(\Gamma) \le M\big[f(\Gamma)\big] \le K_I(f)\, M(\Gamma) \tag{4.7}$$

for every family Γ of smooth paths in D. When both $K_O(f)$ and $K_I(f)$ are finite, we declare f to be a *quasiconformal diffeomorphism* with *outer dilatation* $K_O(f)$, *inner dilatation* $K_I(f)$, and *maximal dilatation*

$$K(f) = \max\{K_O(f), K_I(f)\}. \tag{4.8}$$

It follows from (2.13) and (3.71) that

$$H(f)^{n/2} \le K(f) \le H(f)^{n-1},$$

so f is quasiconformal if and only if its linear dilatation H_f remains bounded in D.

From (2.12) we learn that $K(f) \ge 1$, with equality holding precisely when f is a conformal mapping. The point of all this is that $K(f)$ represents another global measure of the "deviation from conformality" of the mapping f.

When $K(f)$ is small, f is in some sense "not far from being conformal", whence the expression "quasiconformal". Unfortunately, the class of quasiconformal diffeomorphisms does not inherit enough of the properties of the class of conformal mappings to provide by itself an adequate framework in which to generalize the theory of plane conformal mappings. For instance one shortcoming of this class is that a sequence of quasiconformal diffeomorphisms $f_\nu : D \to D'_\nu$ may converge locally uniformly in D to a homeomorphism $f : D \to D'$ that is nowhere differentiable! Even if the maximal distortion were to remain uniformly bounded along this sequence, this limit may not be differentiable everywhere—but basically the limit will satisfy (4.7) if nonconstant.

As it turns out—this is a topic we shall take up in earnest later—inequality (4.7) provides the key to a quite satisfactory substitute for the classical conformal mapping theory in Euclidean spaces of arbitrary dimension, a substitute whose desirability is made necessary by Liouville's theorem.

The theory to which we allude embraces a wealth of mappings yet retains many of the driving features of the two-dimensional theory. It springs from the study of homeomorphisms $f : D \to D'$ that are not presumed to be even minimally differentiable, but that satisfy the geometric condition

$$\frac{M(\Gamma)}{K} \le M\big[f(\Gamma)\big] \le K M(\Gamma)$$

for every family Γ of (not necessarily smooth) paths in D, where $1 \le K < \infty$ is a constant. A homeomorphism of this sort will later be called a *K-quasiconformal mapping* of D onto D'.

If G and G' are domains in the complex plane and if f is a sense-preserving diffeomorphism of G onto G', then according to (2.13), (3.71), and (2.23),

$$H_O\big[Df(z)\big] = H_I\big[Df(z)\big] = H\big[Df(z)\big] = H_f(z) = \frac{|\partial f(z)| + |\overline{\partial} f(z)|}{|\partial f(z)| - |\overline{\partial} f(z)|}$$

for every z in G. It is not difficult to see that

$$\frac{|\partial f(z)| + |\overline{\partial} f(z)|}{|\partial f(z)| - |\overline{\partial} f(z)|} = \frac{\max_\theta |\partial_\theta f(z)|}{\min_\theta |\partial_\theta f(z)|},$$

where for θ in $[0, 2\pi]$ we set

$$\partial_\theta f(z) = \lim_{t\to 0} \frac{f(z+te^{i\theta}) - f(z)}{t}.$$

If we write

$$Q_f(z) = \frac{\max_\theta |\partial_\theta f(z)|}{\min_\theta |\partial_\theta f(z)|}, \tag{4.9}$$

then

$$K(f) = K_O(f) = K_I(f) = \sup_D Q_f.$$

In other words, f is a quasiconformal diffeomorphism if and only if the function Q_f is bounded in G. In fact arguably the first appearance of quasiconformal mappings in the literature occurred in a 1928 paper of Grötzsch [**66**], wherein he studied sense-preserving diffeomorphisms f for which the quantity Q_f is bounded.

Assuming that we have a mapping f with Q_f bounded, we can define another complex-valued function μ_f in D by

$$\mu_f(z) = \frac{\overline{\partial} f(z)}{\partial f(z)}.$$

The function μ_f is known as the *complex dilatation* of f. We can easily verify that

$$\|\mu_f\|_\infty = \sup_D \mu_f = \frac{K(f)-1}{K(f)+1} < 1.$$

We can then ask what properties f acquires as a solution in G of the tautological partial differential equation

$$\overline{\partial} f = \mu\, \partial f, \tag{4.10}$$

where $\mu = \mu_f$. An equation of the general type (2.5), but in which $\mu : G \to \mathbb{C}$ is an arbitrary Lebesgue measurable function whose L^∞-norm $\|\mu\|_\infty$ satisfies $\|\mu\|_\infty < 1$, is called a *Beltrami equation*.

Virtually the entire subject of plane quasiconformal mappings can be developed through the theory of Beltrami equations. However we shall not pursue this line of inquiry here. Instead, we refer the reader to [**15**] for a thorough modern discussion of Beltrami equations.

4.0.5. Some real analysis. We use m_n to denote Lebesgue measure in $\mathbb{R}^n$ and m_n^* to denote the associated outer measure. In situations where it is important to keep track of the variable of integration, x say, we may choose to write $dm_n(x)$—or just dx—in place of dm_n.

Let A be a Lebesgue measurable subset of $\mathbb{R}^n$. In the literature it is now commonplace to find the symbol $|A|$ employed as a substitute for $m_n(A)$. We, too, shall use this notation from time to time. If $f : A \to [0, \infty]$ is a Lebesgue measurable function, then

$$\int_A f\, dm_n, \quad \int_A f(x)\, dm_n(x), \quad \text{and} \quad \int_A f(x)\, dx$$

are various notation for the Lebesgue integral of f over A. This integral certainly exists in the circumstances described, though it may be infinite. The integral also exists if $f : A \to [-\infty, \infty]$ belongs to the class $L^1(A)$;

$$L^1(A) = \{f : A \to [-\infty, \infty] \text{ is Lebesgue measurable and } \int_A |f|\, dm_n < \infty\}.$$

Under the assumption that $0 < |A| < \infty$ and that f is a member of $L^1(A)$, the average value of f over A is symbolized by f_A:

$$f_A = \frac{1}{|A|} \int_A f\, dm_n\,. \tag{4.11}$$

Let X be a topological space. By a *Borel subset of* X (or a *Borel set in* X) is meant a member of the σ-algebra of subsets of X generated by the family of open sets in X. A function $\varphi : A \to Y$, where A is contained in X and Y is a topological space, is called a *Borel function* provided that $\varphi^{-1}(F)$ is a Borel set in X whenever F is a Borel set in Y. (Implicit here is that the domain of φ, $A = \varphi^{-1}(Y)$, is a Borel set in X.) This is the case if and only if $\varphi^{-1}(V)$ is a Borel set in X for every open subset V of Y. For example, any continuous function $\varphi : A \to Y$ whose domain A is a Borel set in X certainly qualifies as a Borel function. If $\{A_\nu\}$ is a countable family of pairwise disjoint Borel sets in X and $\varphi_\nu : A_\nu \to Y$ is a Borel function, then the function $\varphi : \bigcup_\nu A_\nu \to Y$ defined by $\varphi(x) = \varphi_\nu(x)$ for x in A_ν is a Borel function. The composition $\psi \circ \varphi$ of Borel functions $\varphi : A \to Y$ and $\psi : B \to Z$, where A is a subset of X, B is a subset of Y, and $\varphi(A)$ is contained in B, is a Borel function. If $\varphi : X \to Y$ is a homeomorphism between X and Y, then $\varphi(E)$ is a Borel set in Y whenever E is a Borel set in X. We could not draw the same conclusion if we knew only that φ belonged to $C(X, Y)$. If, for instance, U is an open set in $\mathbb{R}^n$, φ is in the class $C(U, \mathbb{R}^m)$, and E is a Borel set in U, then the most that can normally be said about the set $\varphi(E)$ is that it is Lebesgue measurable. We recall that every Borel set in $\mathbb{R}^n$ is Lebesgue measurable and every Lebesgue measurable set A in $\mathbb{R}^n$ has the structure $A = E \cup N$, where E is a Borel set and $|N| = 0$.

Consider once again a Borel set A in a topological space X. A necessary and sufficient condition for $f : A \to [-\infty, \infty]$ to be a Borel function is that $\{x \in A : f(x) < \alpha\}$ be a Borel set in X for every real number α. The sum, difference, product, and quotient of Borel functions $f, g : A \to [-\infty, \infty]$ are Borel functions (though the domains of these functions need not coincide with A). If $f_\nu : A \to [-\infty, \infty]$ $(\nu = 1, 2, \ldots)$ is a sequence of Borel functions, then $\sup_\nu f_\nu$, $\inf_\nu f_\nu$, $\limsup_{\nu\to\infty} f_\nu$, and $\liminf_{\nu\to\infty} f_\nu$ (all defined pointwise) are Borel functions, as is $\lim_{\nu\to\infty} f_\nu$, defined on the set where this pointwise limit exists. If A is a Borel set in $\mathbb{R}^n$ and $f : A \to [-\infty, \infty]$ is a Borel function, then f is Lebesgue measurable. For a function $f : A \to \mathbb{R}^n$ to be a Borel function it is necessary and sufficient that each of its coordinate functions be a real-valued Borel function.

LEMMA 4.0.9. *Suppose that U is an open set in $\mathbb{R}^n$ and that f is a continuous function defined on U and valued in $\mathbb{R}^m$. Then the functions $L_f : U \to [0, \infty]$ and $\ell_f : U \to [0, \infty]$ defined by*

$$L_f(x) = \limsup_{h\to 0} \frac{|f(x+h) - f(x)|}{|h|} = \lim_{\delta\to 0} \sup_{0<|h|<\delta} \frac{|f(x+h) - f(x)|}{|h|}$$

and

$$\ell_f(x) = \liminf_{h\to 0} \frac{|f(x+h)-f(x)|}{|h|} = \lim_{\delta\to 0} \inf_{0<|h|<\delta} \frac{|f(x+h)-f(x)|}{|h|}$$

are Borel functions.

We check this assertion for L_f by exhibiting it as the pointwise limit of a sequence $\langle f_\nu \rangle$ of Borel functions (an analogous argument handles ℓ_f). For $\nu = 1, 2, \dots$ let $E_\nu = \{x \in U : \operatorname{dist}(x, \mathbb{R}^n \backslash U) > \nu^{-1}\}$, and for $h \in \mathbb{Q}^n$ with $0 < |h| < \nu^{-1}$ define $f_{\nu,h} : U \to [0, \infty)$ by

$$f_{\nu,h}(x) = \begin{cases} \frac{|f(x+h)-f(x)|}{|h|} & \text{if } x \in E_\nu\ , \\ 0 & \text{if } x \in U \setminus E_\nu\ . \end{cases}$$

The continuity of f ensures that $f_{\nu,h}$ is a Borel function, a fact which enables us to conclude that the function $f_\nu = \sup\{f_{\nu,h} : h \in \mathbb{Q}^n,\ 0 < |h| < \nu^{-1}\}$ is also Borel. Plainly $f_\nu \to L_f$ pointwise in U.

Reasoning similar to the above shows that the various partial derivatives $\partial_1 f, \partial_2 f, \dots, \partial_n f$ of f are likewise Borel functions; to be more precise, assuming that $F_i = \{x \in U : \partial_i f(x) \text{ exists}\}$ is nonempty, F_i is a Borel set and $\partial_i f : F_i \to \mathbb{R}^m$ is a Borel function.

An important result from real analysis that we shall occasionally invoke is the following version of the Rademacher-Stepanov theorem. A thorough account of this can be found in either of the references [**73**] or [**98**]. The classical Rademacher theorem from 1919 asserts that each real-valued Lipschitz function on $\mathbb{R}^n$ is almost everywhere Fréchet differentiable. The more general Stepanov theorem from 1923 says that an arbitrary real function on $\mathbb{R}^n$ is Fréchet differentiable at almost all points at which f is Lipschitz. In fact, Stepanov gave the proof for measurable f only. This result is traditionally proved by the application of Rademacher's theorem to Lipschitz extensions of Lipschitz functions; a straightforward proof can be found in [**103**].

THEOREM 4.0.10. *If U is an open set in $\mathbb{R}^n$ and $f : U \to \mathbb{R}^m$ is a continuous function with the property that $L_f(x) < \infty$ for almost every x in U, then f is differentiable at almost every point of U.*

We shall frequently have cause to use the change of variable formula for locally Lipschitz homeomorphisms. A function $f : U \to \mathbb{R}^m$, where U is an open set in $\mathbb{R}^n$, is *locally Lipschitz* if for each point x of U there exist a constant $\lambda = \lambda_x$ and an open neighbourhood $V = V_x$ of x such that

$$|f(z) - f(y)| \leq \lambda |z - y|$$

holds for all y and z belonging to V. If, for example, f is a function from the class $C^1(U, \mathbb{R}^m)$, then f is locally Lipschitz.

If the inequality $|f(z) - f(y)| \leq \lambda|z - y|$ is actually true for all y and z in U, then we say that f is a *λ-Lipschitz mapping* or that f satisfies a *Lipschitz condition with constant* λ. It is a consequence of Theorem 4.0.10 that a locally Lipschitz function $f : U \to \mathbb{R}^n$ is differentiable almost everywhere in U, so its Jacobian J_f is a Borel function whose domain includes almost every point of U. A proof of the next theorem can be found in H. Federer's exhaustive monograph [**37**] as the "change of variables formula".

THEOREM 4.0.11. *If $f : D \to D'$ is a locally Lipschitz homeomorphism between domains D and D' in $\mathbb{R}^n$, then*

$$\int_{D'} \rho \, dm_n = \int_D (\rho \circ f)|J_f| \, dm_n$$

for every Borel function $\rho : D' \to [0, \infty]$.

4.0.6. Hausdorff measure. Let V be an affine subspace of $\mathbb{R}^n$ of dimension p, where $1 \leq p \leq n-1$. Then Lebesgue measure m_p on $\mathbb{R}^p$ can be transported by means of a Euclidean isometry $\varphi : \mathbb{R}^p \to V$ to a measure on V, a measure that we still call Lebesgue measure and continue to symbolize by m_p. More generally, any smooth p-dimensional submanifold of $\mathbb{R}^n$ supports a natural analogue of p-dimensional Lebesgue measure. The most efficient way to describe this measure is to do so in the context of Hausdorff measure.

Let $0 \leq p < \infty$. For each subset A of $\mathbb{R}^n$ and each $t > 0$ we set

$$\mathcal{H}_t^p(A) = \alpha(p) 2^{-p} \inf \sum_{\nu} \operatorname{diam}(E_\nu)^p \,,$$

where $\alpha(p) = \Gamma(1/2)^p / \Gamma[(p+2)/2]$ and where the infimum is extended over all countable covers $\{E_\nu\}$ of A whose members E_ν are nonempty subsets of $\mathbb{R}^n$ with $\operatorname{diam}(E_\nu) < t$. Note that $\mathcal{H}_t^p(A)$ is a nonincreasing function of t, so

$$\mathcal{H}^p(A) = \lim_{t \to 0} \mathcal{H}_t^p(A) = \sup_{t > 0} \mathcal{H}_t^p(A) \tag{4.12}$$

exists in $[0, \infty]$. The extended real number $\mathcal{H}^p(A)$ is called the *normalized p-dimensional Hausdorff outer measure of A*.

In defining $\mathcal{H}^p(A)$ we could have restricted our attention to coverings of A by compact (or open or convex) sets without affecting the value obtained for $\mathcal{H}^p(A)$. (If, on the other hand, we were to consider only coverings of A by Euclidean balls, whether they were open or closed, we would in general obtain a slightly different quantity for the limit in (4.12); namely, we would arrive at the *p-dimensional spherical outer measure of A*, which is denoted by $\sigma_{\mathbb{S}}^p(A)$. It can be demonstrated that

$$\mathcal{H}^p(A) \leq \sigma_{\mathbb{S}}^p(A) \leq \left(\frac{2n}{n+1} \right)^{p/2} \mathcal{H}^p(A)$$

for every subset A of $\mathbb{R}^n$ and for every $p \geq 0$.

It is easy to see that $\mathcal{H}_t^p$ and $\mathcal{H}^p$ are outer measures on $\mathbb{R}^n$. True, but less straightforward to verify, is the fact that every Borel set in $\mathbb{R}^n$ is $\mathcal{H}^p$-measurable. Thus the restriction of $\mathcal{H}^p$ to the σ-algebra of Borel subsets of $\mathbb{R}^n$ is a measure. It follows directly from the above definitions that if $\mathcal{H}^p(A) < \infty$ and if $q > p$, then $\mathcal{H}^q(A) = 0$.

In fact,

$$\mathcal{H}_t^q(A) \leq c \, t^{q-p} \mathcal{H}_t^p(A)$$

for every $t > 0$, where the constant $c = 2^{q-p}\alpha(p)/\alpha(q)$.

One can confirm without difficulty that $\mathcal{H}^p(A) = 0$ for $p > n$. The *Hausdorff dimension of A*, denoted by $\dim_{\mathcal{H}}(A)$, is defined as the infimum of $\{p > 0 : \mathcal{H}^p(A) = 0\}$. Accordingly, $0 \leq \dim_{\mathcal{H}}(A) \leq n$. Of course, this number need not be an integer, hence it may differ from the topological dimension of A. It is readily checked that $\mathcal{H}^p(A) = \infty$ whenever p satisfies $0 \leq p < \dim_{\mathcal{H}}(A)$.

Observe that $\mathcal{H}^0$ is just counting measure: $\mathcal{H}^0(A) = \#(A)$, the cardinality of A, if $\#(A) < \infty$, and $\mathcal{H}^0(A) = \infty$ otherwise. It can be shown that $\mathcal{H}^n = \sigma_{\mathbb{S}}^n = m_n^*$, the Lebesgue outer measure on $\mathbb{R}^n$. In fact, the restriction of $\mathcal{H}^p$—or $\sigma_{\mathbb{S}}^p$—to any p-dimensional plane S in $\mathbb{R}^n$ coincides with a p-dimensional Lebesgue outer measure on S.

If a function $f : \mathbb{R}^n \to \mathbb{R}^m$ satisfies a Lipschitz condition, say $|f(y) - f(x)| \leq \lambda|y - x|$ for all x and y in $\mathbb{R}^n$, then $\operatorname{diam}(f(E)) \leq \lambda \operatorname{diam}(E)$ for every subset E of $\mathbb{R}^n$, with the result that

$$\mathcal{H}^p(f(A)) \leq \lambda^p \mathcal{H}^p(A)$$

for any set A in $\mathbb{R}^n$ and any $p \geq 0$. This implies, in particular, that $\mathcal{H}^p(f(A)) = \lambda^p \mathcal{H}^p(A)$ whenever f is a similarity transformation of $\mathbb{R}^n$ with dilation factor λ. Note that $\mathcal{H}^p(A) = 0$ if A is a countable subset of $\mathbb{R}^n$ and $p > 0$.

For $n \geq 2$ we follow the reasonably widespread practice of using σ_{n-1} to denote the completion of the measure obtained when $\mathcal{H}^{n-1}$ is restricted to the collection of Borel sets in $\mathbb{R}^n$. We refer to σ_{n-1} as *surface area measure* in $\mathbb{R}^n$. There are two standard situations in which integrals of the type $\int_A f\, d\sigma_{n-1}$ can be reduced with a minimum of fuss to integrals involving Lebesgue measure m_{n-1}. The first occurs when A is the graph of a real-valued, locally Lipschitz function of $n - 1$ of the variables $x_1, x_2, \ldots, x_n$.

Suppose for simplicity's sake that A is the set of points in $\mathbb{R}^n$ described by the equation $x_n = \varphi(x')$ for $x' = (x_1, x_2, \ldots, x_{n-1})$ in A', where A' is a Lebesgue measurable subset of an open set U in $\mathbb{R}^{n-1}$ and $\varphi : U \to \mathbb{R}$ is a locally Lipschitz function. Then

$$\int_A f\, d\sigma_{n-1} = \int_{A'} f\big(x', \varphi(x')\big)\sqrt{1 + |\nabla\varphi(x')|^2}\, dm_{n-1}(x')$$

whenever $f : A \to [0, \infty]$ is σ_{n-1}-measurable or $f : A \to [-\infty, \infty]$ is σ_{n-1}-integrable.

Secondly, an integral $\int_{\mathbb{S}^{n-1}} f\, d\sigma_{n-1}$ can be transformed to an integral over $\mathbb{R}^{n-1}$ by means of stereographic projection $\pi : \hat{\mathbb{R}}^{n-1} \to \mathbb{S}^{n-1}$; to wit,

$$\int_{\mathbb{S}^{n-1}} f\, d\sigma_{n-1} = 2^{n-1} \int_{\mathbb{R}^{n-1}} \frac{f[\pi(x)]\, dm_{n-1}(x)}{(1 + |x|^2)^{n-1}}, \tag{4.13}$$

a formula valid for any σ_{n-1}-measurable function $f : \mathbb{S}^{n-1} \to [0, \infty]$ or any extended real-valued function f that is integrable over $\mathbb{S}^{n-1}$ with respect to σ_{n-1}.

A result that will be called upon repeatedly in this book is the "spherical coordinate" version of Fubini's theorem, which is a simple case of the Federer-Young coarea formula [**37**].

THEOREM 4.0.12. *If $f : \mathbb{R}^n \to [0, \infty]$ is Lebesgue measurable or if $f : \mathbb{R}^n \to [-\infty, \infty]$ belongs to $L^1(\mathbb{R}^n)$, then*

$$\int_{\mathbb{R}^n} f\, dm_n = \int_0^\infty \left[\int_{S^{n-1}(r)} f\, d\sigma_{n-1}\right] dm_1 \tag{4.14}$$

$$= \int_0^\infty \left[\int_{S^{n-1}} f(ru)\, d\sigma_{n-1}(u)\right] r^{n-1}\, dr\,. \tag{4.15}$$

We now introduce some convenient notation. For $n = 1, 2, \ldots$ we write

$$\Omega_n = m_n(B^n), \qquad \omega_n = \sigma_n(\mathbb{S}^n)\,.$$

For later purposes it will prove convenient to adopt the convention that $\omega_0 = 2$. We infer from (4.14) that

$$\omega_{n-1} = n\Omega_n \tag{4.16}$$

for $n \geq 2$. It is known that

$$\omega_{2n-1} = \frac{2\pi^n}{(n-1)!}, \qquad \omega_{2n} = \frac{2^{n+1}\pi^n}{1 \cdot 3 \cdot 5 \cdots (2n-1)} \tag{4.17}$$

for $n \geq 1$.

4.1. Path integrals

Here we collect some of the ideas and tools we will need in order to develop the theory of the modulus of path families.

4.1.1. Paths, curves, and arcs.

By a *path in* $\hat{\mathbb{R}}^n$ we understand a continuous function γ of the type $\gamma : [a, b] \to \hat{\mathbb{R}}^n$, with $-\infty < a < b < \infty$. Under the more general heading of a *curve in* $\hat{\mathbb{R}}^n$ we include any continuous function $\gamma : I \to \hat{\mathbb{R}}^n$ in which I is a nondegenerate subinterval of $[-\infty, \infty]$.

Should I be an open (respectively, a half-open) interval in $(-\infty, \infty)$, we may refer to γ as an *open* (respectively, a *half-open*) *path*. The range $\gamma(I)$ of a curve $\gamma : I \to \hat{\mathbb{R}}^n$ is known as its *trajectory*, a set we denote by $|\gamma|$.

We speak of γ as a *curve in* A, where A is a subset of $\hat{\mathbb{R}}^n$, under the condition that $|\gamma|$ is contained in A.

The symbol $\mathcal{C}(A)$ stands for the collection of all such curves,

$$\mathcal{C}(A) = \{\gamma : I \to A \ \text{ is continuous}\}.$$

To say that a curve $\beta : J \to \hat{\mathbb{R}}^n$ is a *subcurve* of a curve $\gamma : I \to \hat{\mathbb{R}}^n$ means that J is a subinterval of I and β is the restriction of γ to J. The notation $\beta \prec \gamma$ will be used to describe the fact that such a relationship exists between curves β and γ.

Let $\gamma : [a, b] \to \hat{\mathbb{R}}^n$ and $\beta : [c, d] \to \hat{\mathbb{R}}^n$ be paths. We use $-\gamma$ to denote the *reverse path* of γ, the path defined on $[a, b]$ by $[-\gamma](t) = \gamma(b + a - t)$. Under the assumption that $\gamma(b) = \beta(c)$, γ and β can be strung together to form their *path sum* $\gamma + \beta$:

$$[\gamma + \beta](t) = \begin{cases} \gamma(t) & \text{if } a \leq t \leq b\,, \\ \beta(t + c - b) & \text{if } b \leq t \leq b + d - c\,. \end{cases}$$

By induction we define the path sum $\gamma_1 + \gamma_2 + \cdots + \Gamma_{rect}$ whenever $\gamma_j : [a_j, b_j] \to \hat{\mathbb{R}}^n$ are paths satisfying $\gamma_j(b_j) = \gamma_{j+1}(a_{j+1})$ for $j = 1, 2, \ldots, r-1$: $\gamma_1 + \gamma_2 + \cdots + \Gamma_{rect} = (\gamma_1 + \gamma_2 + \cdots + \gamma_{r-1}) + \Gamma_{rect}$.

A subset J of $\hat{\mathbb{R}}^n$ is called an *arc* if J is homeomorphic to the interval $[0, 1]$, an *open arc* if it is homeomorphic to $(0, 1)$, and a *half-open arc* if it is homeomorphic to $[0, 1)$. Sets of these three types are by definition the trajectories of curves. In the other direction, it is a basic fact of topology that the trajectory $|\gamma|$ of a nonconstant curve γ in $\hat{\mathbb{R}}^n$ is arcwise connected. That is, for each pair of distinct points x and y of $|\gamma|$ there exists an arc J in $|\gamma|$ whose endpoints are x and y.

4.1.2. Rectifiable curves. Let $\gamma : [a,b] \to \hat{\mathbb{R}}^n$ be a path. A *length* $\ell(\gamma)$ is assigned to γ as follows: if ∞ is not a point of $|\gamma|$, we define

$$\ell(\gamma) = \sup_P \ell(\gamma, P),$$

where for each partition $P : a = t_0 < t_1 < \cdots < t_r = b$ of $[a,b]$ the approximating polygonal length $\ell(\gamma, P)$ is given by

$$\ell(\gamma, P) = \sum_{j=1}^{r} |\gamma(t_j) - \gamma(t_{j-1})|$$

and the supremum extends over all such partitions; if $|\gamma| = \{\infty\}$, we set $\ell(\gamma) = 0$; if γ is a nonconstant path with ∞ in its trajectory, we declare that $\ell(\gamma) = \infty$. In all cases it is clear that $0 \leq \ell(\gamma) \leq \infty$, with $\ell(\gamma) = 0$ if and only if the trajectory of γ reduces to a point. We say that γ is *rectifiable* provided $\ell(\gamma) < \infty$. This being the situation, it is a simple matter to show that each subpath of γ is rectifiable and that

$$\ell(\gamma) = \sum_{j=1}^{r} \ell\big(\gamma|[t_{j-1}, t_j]\big) \tag{4.18}$$

for every partition $P : a = t_0 < t_1 < \cdots < t_r = b$ of $[a,b]$. More generally, if we can express γ as a path sum $\gamma = \gamma_1 + \gamma_2 + \cdots + \Gamma_{rect}$, then

$$\ell(\gamma) = \ell(\gamma_1) + \ell(\gamma_2) + \cdots + \ell(\Gamma_{rect}),$$

which shows that γ is rectifiable precisely when every one of its "summands" enjoys this property. For the reverse path $-\gamma$ of γ it is true that $\ell(-\gamma) = \ell(\gamma)$. Thus $-\gamma$ is rectifiable whenever γ is rectifiable.

Consider a rectifiable path $\gamma : [a,b] \to \mathbb{R}^n$. The *length function* of γ is the function $\sigma_\gamma : [a,b] \to [0, \ell(\gamma)]$ defined by $\sigma_\gamma(a) = 0$ and

$$\sigma_\gamma(t) = \ell(\gamma|[a,t])$$

if $a < t \leq b$. In the next lemma we list some key properties of this function. Before doing so, we recall for the reader some relevant items from real analysis. The rectifiability of γ dictates that its coordinate functions $\gamma_i : [a,b] \to \mathbb{R}$ $(1 \leq i \leq n)$ be functions of bounded variation on $[a,b]$. It then follows from Lebesgue's differentiation theorem that $\dot{\gamma}(t) = (\gamma_1'(t), \gamma_2'(t), \ldots, \gamma_n'(t))$ exists for almost every t in $[a,b]$ and that the function $|\dot{\gamma}|$ belongs to $L^1([a,b])$.

We also remind the reader of what it means for a function $\varphi : [a,b] \to \mathbb{R}$ to be *absolutely continuous*: corresponding to each $\varepsilon > 0$ there is a $\delta > 0$ such that

$$\sum_{j=1}^{r} |\varphi(b_j) - \varphi(a_j)| < \varepsilon$$

holds for each finite collection $[a_1, b_1], [a_2, b_2], \ldots, [a_r, b_r]$ of nonoverlapping, nondegenerate subintervals of $[a,b]$ with $\sum_{j=1}^{r}(b_j - a_j) < \delta$. Every such function φ is of bounded variation on $[a,b]$ (but not conversely) and satisfies

$$\varphi(t) = \varphi(a) + \int_a^t \varphi'(u)\, du$$

for each t in $[a,b]$. The definition of absolute continuity for a real-valued function on $[a,b]$ carries over in essentially the same manner to a definition of absolute continuity for a path $\gamma : [a,b] \to \mathbb{R}^n$. Alternatively, γ is absolutely continuous if and only if each of its coordinate functions has this property.

LEMMA 4.1.1. *Let $\gamma : [a,b] \to \mathbb{R}^n$ be a rectifiable path of length ℓ, and let $\sigma = \sigma_\gamma$ be its length function. Then:*

(i) $|\gamma(t) - \gamma(u)| \le \sigma(t) - \sigma(u)$ *when* $a \le u < t \le b$;
(ii) σ *is nondecreasing and continuous;*
(iii) σ *is absolutely continuous if and only if γ is as well;*
(iv) $\sigma'(t)$ *and* $\dot\gamma(t)$ *exist and satisfy* $\sigma'(t) = |\dot\gamma(t)|$ *for almost every t in* $[a,b]$;
(v) *both σ' and $|\dot\gamma|$ belong to $L^1([a,b])$ and*

$$\int_a^b |\dot\gamma(t)|\,dt = \int_a^b \sigma'(t)\,dt \le \ell\,, \tag{4.19}$$

with equality holding precisely when γ is absolutely continuous.

PROOF. If $a \le u < t \le b$, then (4.18) implies that

$$\sigma(t) - \sigma(u) = \ell\big(\gamma|[u,t]\big) \ge |\gamma(t) - \gamma(u)|\,,$$

proving (i). This tells us, among other things, that $\sigma(t) \ge \sigma(u)$ for $t \ge u$; that is σ is nondecreasing. To finish the proof of (ii) it is enough to verify that $\sigma(t^+) = \lim_{u\to t^+} \sigma(u) = \sigma(t)$ holds for each t in $[a,b)$ and $\sigma(t^-) = \lim_{u\to t^-}\sigma(u) = \sigma(t)$ holds for each t in $(a,b]$. We supply the details of the argument in the first case only. Fix t in $[a,b)$. Because σ is nondecreasing, $\sigma(t^+) \ge \sigma(t)$. Assuming that $3d = \sigma(t^+) - \sigma(t) > 0$, we shall construct a strictly decreasing sequence $\langle t_\nu\rangle$ in (t,b) such that $\sigma(t_\nu) - \sigma(t_{\nu+1}) \ge d$ for every ν. This results in a contradiction, since it forces

$$\ell \ge \ell\big(\gamma|[t_{\nu+1},t_1]\big) = \sum_{j=1}^{\nu}\big[\sigma(t_j) - \sigma(t_{j+1})\big] \ge \nu d$$

to hold for every $\nu \ge 1$. The construction of the desired sequence is as follows. We first fix t_1 in the interval (t,b) with the property that $|\gamma(u) - \gamma(t)| \le d$ whenever $t \le u \le t_1$. As

$$\ell(\gamma|[t,t_1]) = \sigma(t_1) - \sigma(t) \ge \sigma(t^+) - \sigma(t) \ge 3d,$$

we can find a partition P: $t = u_0 < u_1 < \cdots < u_r = t_1$ of the interval $[t,t_1]$ such that $r > 1$ and

$$\sum_{k=1}^{r} |\gamma(u_k) - \gamma(u_{k-1})| \ge 2d\,.$$

We then set $t_2 = u_1$. Because $|\gamma(u_1) - \gamma(t)| \le d$, we see that

$$\begin{aligned}\sigma(t_1) - \sigma(t_2) &\ge \sum_{k=2}^{r} |\gamma(u_k) - \gamma(u_{k-1})| \\ &= \sum_{k=1}^{r} |\gamma(u_k) - \gamma(u_{k-1})| - |\gamma(u_1) - \gamma(t)| \ge 2d - d = d\,.\end{aligned}$$

We now repeat the same argument, this time starting with t_2 instead of t_1, to produce an element t_3 of (t,t_2) for which $\sigma(t_2) - \sigma(t_3) \ge d$. Inductively, we generate a sequence $\langle t_\nu\rangle$ with the stated property. Assertion (ii) is thereby established.

The inequality in (i) ensures that γ is absolutely continuous whenever σ has this property. Under the converse assumption that γ is absolutely continuous, let $\varepsilon > 0$.

Choose $\delta > 0$ so that $\sum_{j=1}^{r} |\gamma(b_j) - \gamma(a_j)| < \varepsilon$ whenever $[a_1, b_1], [a_2, b_2], \ldots, [a_r, b_r]$ are nonoverlapping, nondegenerate subintervals of $[a, b]$ with $\sum_{j=1}^{r}(b_j - a_j) < \delta$. Given such a collection of intervals, we can take a partition P_j of $[a_j, b_j]$ for which

$$\sigma(b_j) - \sigma(a_j) = \ell(\gamma_j) \leq \ell(\gamma_j, P_j) + \frac{\varepsilon}{r} ,$$

where γ_j is the restriction of γ to $[a_j, b_j]$. The partitioning of the intervals $[a_j, b_j]$ creates a new collection of nonoverlapping closed intervals in $[a, b]$ ($[c_1, d_1]$, $[c_2, d_2], \ldots, [c_p, d_p]$, say) whose length-sum is smaller than δ. Accordingly,

$$\begin{aligned}\sum_{j=1}^{r}[\sigma(b_j) - \sigma(a_j)] &\leq \sum_{j=1}^{r}\Big[\ell(\gamma_j, P_j) + \frac{\varepsilon}{r}\Big] = \varepsilon + \sum_{j=1}^{r}\ell(\gamma_j, P_j) \\ &= \varepsilon + \sum_{k=1}^{p}|\gamma(d_k) - \gamma(c_k)| < 2\varepsilon .\end{aligned}$$

Because this is true for each collection $[a_1, b_1], [a_2, b_2], \ldots, [a_r, b_r]$ of the type in question, σ is seen to be absolutely continuous. We have now confirmed (iii).

The Lebesgue differentiation theorem ensures that the derivatives $\sigma'(t)$ and $\dot{\gamma}(t)$ both exist at almost every point t of $[a, b]$, that σ' and $|\dot{\gamma}|$ belong to $L^1([a, b])$, and that

$$\int_a^b \sigma'(t) \leq \sigma(b) - \sigma(a) = \ell ,$$

with equality in force if and only if the function σ (hence, by (iii), the path γ) is absolutely continuous. In order to complete the proofs of (iv) and (v) we need demonstrate only that $|\dot{\gamma}(t)| = \sigma'(t)$ is true for almost every t in (a, b). If t is a point of (a, b) for which both $\dot{\gamma}(t)$ and $\sigma'(t)$ exist, then (i) implies that

$$|\dot{\gamma}(t)| = \lim_{h\to 0^+}\left|\frac{\gamma(t+h) - \gamma(t)}{h}\right| \leq \lim_{h\to 0^+}\frac{\sigma(t+h) - \sigma(t)}{h} = \sigma'(t) ,$$

so at least $|\dot{\gamma}(t)| \leq \sigma'(t)$ holds almost everywhere in (a, b). Let

$$E = \{t \in (a, b) : |\dot{\gamma}(t)| < \sigma'(t)\} .$$

We maintain that $m_1(E) = 0$. For $\nu = 1, 2, \ldots$ let E_ν be the set of t in E such that

$$\frac{\sigma(v) - \sigma(u)}{v - u} \geq \left|\frac{\gamma(v) - \gamma(u)}{v - u}\right| + \frac{1}{\nu}$$

holds whenever u and v in (a, b) satisfy the conditions $u \leq t \leq v$ and $0 < v - u < \nu^{-1}$. Because

$$\lim_{u\to t^- v\to t^+}\frac{\sigma(v) - \sigma(u)}{v - u} = \sigma'(t), \qquad \lim_{u\to t^- v\to t^+}\frac{\gamma(v) - \gamma(u)}{v - u} = \dot{\gamma}(t)$$

for any t at which both σ and γ are differentiable, it becomes clear that $E = \bigcup_{\nu=1}^{\infty} E_\nu$. We prove that $m_1(E_\nu) = 0$ for each ν. To this end, fix ν and let $\eta > 0$. Choose a partition $P : a = t_0 < t_1 < \cdots < t_r = b$ of $[a, b]$ such that $t_j - t_{j-1} < \nu^{-1}$ for $1 \leq j \leq r$ and $\ell \leq \ell(\gamma, P) + \nu^{-1}\eta$. Writing $I_j = [t_{j-1}, t_j]$, we deduce from the definition of E_ν that

$$\sigma(t_j) - \sigma(t_{j-1}) > |\gamma(t_j) - \gamma(t_{j-1})| + \frac{t_j - t_{j-1}}{\nu}$$

for each j with $E_\nu \cap I_j \neq \emptyset$. Assuming $E_\nu \neq \emptyset$, we get

$$\begin{aligned} m_1(E_\nu) &\leq \sum_{E_\nu \cap I_j \neq \emptyset} (t_j - t_{j-1}) \\ &\leq \nu \sum_{E_\nu \cap I_j \neq \emptyset} \Big[\sigma(t_j) - \sigma(t_{j-1}) - |\gamma(t_j) - \gamma(t_{j-1})|\Big] \\ &\leq \nu \sum_{j=1}^{r} \big[\sigma(t_j) - \sigma(t_{j-1}) - |\gamma(t_j) - \gamma(t_{j-1})|\big] \\ &= \nu\big[\ell - \ell(\gamma, P)\big] \leq \eta\,. \end{aligned}$$

Since $\eta > 0$ was arbitrary, $m_1(E_\nu) = 0$, and this is the case for $\nu = 1, 2, \ldots$. As a consequence, $m_1(E) = 0$, whence $|\dot{\gamma}(t)| = \sigma'(t)$ almost everywhere in $[a, b]$. □

We remark that a Lipschitz path $\gamma : [a, b] \to \mathbb{R}^n$, meaning a path for which there is a constant $\lambda > 0$ such that $|\gamma(u) - \gamma(t)| \leq \lambda|u - t|$ for all t and u in $[a, b]$, is absolutely continuous. Thus the length of such a path can be computed by means of the formula $\ell(\gamma) = \int_a^b |\dot{\gamma}(t)|\,dt$. In particular, this comment applies to any piecewise C^1-path. We also wish to point out that if $\gamma : [a, b] \to \hat{\mathbb{R}}^n$ is a rectifiable path and if $\beta = \gamma|[c, d]$ with $a \leq c < d \leq b$, then $\sigma_\beta(t) = \sigma_\gamma(t) - \sigma_\gamma(c)$ for $c \leq t \leq d$.

The length $\ell(\gamma)$ of an arbitrary curve $\gamma : I \to \hat{\mathbb{R}}^n$ is defined by

$$\ell(\gamma) = \sup_\beta \ell(\beta)\ ,$$

in which the supremum is taken over all paths β with $\beta \prec \gamma$. When $\ell(\beta) < \infty$ for every such β we call the curve γ *locally rectifiable*. Once more γ is said to be rectifiable if $\ell(\gamma) < \infty$. As previously, $\ell(\gamma) = \infty$ if ∞ is a point of $|\gamma|$, save when $|\gamma| = \{\infty\}$, in which event $\ell(\gamma) = 0$. Rectifiable curves are "nicely behaved" at the endpoints of their parameter intervals, a fact to which the next result attests.

LEMMA 4.1.2. *If $\gamma : I \to \hat{\mathbb{R}}^n$ is a rectifiable curve, then γ has a unique extension to a rectifiable curve $\gamma^* : \bar{I} \to \hat{\mathbb{R}}^n$, where $\bar{I}$ indicates the closure of I in $[-\infty, \infty]$. Furthermore, $\ell(\gamma) = \ell(\gamma^*)$.*

PROOF. The case $|\gamma| = \{\infty\}$ is trivial. We can therefore assume that $|\gamma|$ lies in $\mathbb{R}^n$. If we fix a point t_0 of I, then for any t in I we have $|\gamma(t) - \gamma(t_0)| \leq \ell(\gamma)$, so the trajectory $|\gamma|$ must be bounded. We further assume that $\bar{I} \neq I$, there being nothing to prove otherwise.

Let c be an endpoint of $\bar{I}$ in $[-\infty, \infty]$ that is not already included in I. We shall show that $\lim_{t \to c} \gamma(t)$ exists in $\mathbb{R}^n$. Once this is established, we can safely define $\gamma^*(c) = \lim_{t \to c} \gamma(t)$ and be certain that γ^* is continuous at c. We deal with the case of a right endpoint (a left endpoint can be treated similarly). If the limit in question failed to exist, there would be sequences $\langle t_\nu \rangle$ and $\langle s_\nu \rangle$ in I such that $t_\nu \to c$, $s_\nu \to c$, $\gamma(t_\nu) \to x$, and $\gamma(s_\nu) \to y$, where x and y were different points of $\mathbb{R}^n$. Passing to subsequences if necessary, we could presume in addition that $t_1 < s_1 < t_2 < s_2 < \cdots$ and that $|\gamma(s_\nu) - \gamma(t_\nu)| \geq d > 0$ for all ν. But then we would have

$$\ell(\gamma) \geq \ell\big(\gamma|[t_1, s_\nu]\big) \geq \sum_{j=1}^{\nu} |\gamma(s_j) - \gamma(t_j)| \geq \nu d$$

for all ν, which would contradict the rectifiability of γ. The existence of an extension curve $\gamma^* : \bar{I} \to \mathbb{R}^n$ is thereby established. Its uniqueness is evident.

What remains for us to show is that $\ell(\gamma^*) = \ell(\gamma)$. The inequality $\ell(\gamma) \le \ell(\gamma^*)$ is a trivial one. To obtain the reverse inequality, let $\beta : [a,b] \to \mathbb{R}^n$ be an arbitrary subpath of γ^*, let $\varepsilon > 0$, and let $P : a = u_0 < u_1 < \cdots < u_r = b$ be an arbitrary partition of $[a,b]$. We demonstrate that $\ell(\beta, P) \le \ell(\gamma) + 2\varepsilon$. Subject to a possible refinement of P, we are free to suppose that $r \ge 2$, that $|\beta(u_1) - \beta(a)| < \varepsilon$, and that $|\beta(b) - \beta(u_{r-1})| < \varepsilon$. Now $\beta|[u_1, u_{r-1}]$ is a subpath of γ, whence

$$\ell(\beta, P) \le |\beta(u_1) - \beta(a)| + \ell(\gamma) + |\beta(b) - \beta(u_{r-1})| \le \ell(\gamma) + 2\varepsilon\,.$$

Taking the supremum over P yields $\ell(\beta) \le \ell(\gamma) + 2\varepsilon$; letting $\varepsilon \to 0$ then gives $\ell(\beta) \le \ell(\gamma)$. Since β was an arbitrary subpath of γ^*, we conclude that $\ell(\gamma^*) \le \ell(\gamma)$. This shows that γ^* is rectifiable and has the same length as γ. □

Let $\gamma : I \to \hat{\mathbb{R}}^n$ be a curve, and let I have endpoints a and b ($a < b$) in $[-\infty, \infty]$. An equivalent alternative to the definition we have given for $\ell(\gamma)$ is

$$\ell(\gamma) = \lim_{d \to b^-\ c \to a^+} \ell\big(\gamma|[c,d]\big)\,. \tag{4.20}$$

If γ happens to be rectifiable, then in the notation of Lemma 4.1.2

$$\lim_{c \to a^+} \ell\big(\gamma^*|[a,c]\big) = \lim_{d \to b^-} \ell\big(\gamma^*|[d,b]\big) = 0\,. \tag{4.21}$$

This follows from (4.20), Lemma 4.1.2, and the fact, not difficult to verify, that

$$\ell(\gamma) \ge \sum_\nu \ell(\gamma_\nu)$$

for any curve $\gamma : I \to \hat{\mathbb{R}}^n$ and any collection of subcurves $\gamma_\nu : I_\nu \to \hat{\mathbb{R}}^n$ of γ with pairwise nonoverlapping domains I_ν.

We say a curve $\gamma : I \to \mathbb{R}^n$ is *locally absolutely continuous* under the condition that every subpath of γ is absolutely continuous. Lemma 4.1.1(v) and (4.20) justify the formula

$$\ell(\gamma) = \int_a^b |\dot{\gamma}(t)|\, dt$$

for any such curve. Here a and b ($a < b$) are the endpoints of I in $[-\infty, \infty]$.

4.1.3. Normal representation of paths. To state that a path $\gamma : [a,b] \to \hat{\mathbb{R}}^n$ can be obtained from a path $\beta : [c,d] \to \hat{\mathbb{R}}^n$ through the *change of parameter* h means that h is a nondecreasing continuous function on $[a,b]$ with $h([a,b]) = [c,d]$ and with the property that $\gamma = \beta \circ h$. It is straightforward to check that $\ell(\gamma) = \ell(\beta)$ when γ and β are related in this way, so γ is rectifiable if and only if β is rectifiable. Any nonconstant rectifiable path can be obtained by making a change of parameter in a very special kind of path. This fact is the content of the following theorem establishing an arc length parameterisation.

THEOREM 4.1.3. *Let $\gamma : [a,b] \to \mathbb{R}^n$ be a rectifiable path of positive length ℓ. There exists a unique rectifiable path $\gamma_0 : [0,\ell] \to \mathbb{R}^n$ endowed with the following two features:*

(i) *γ is obtainable from γ_0 through a change of parameter, and*
(ii) *the length function σ_0 of γ_0 is given by $\sigma_0(s) = s$ for $0 \le s \le \ell$.*

PROOF. Suppose first that a path γ_0 with properties (i) and (ii) exists. Let γ arise from γ_0 via the change of parameter $h : [a, b] \to [0, \ell]$. We claim that $h = \sigma_\gamma$, the length function of the path γ; i.e., $\gamma = \gamma_0 \circ \sigma_\gamma$. If this is true, the uniqueness of γ_0 is an immediate consequence. Write $\sigma = \sigma_\gamma$. Consider t in $[a, b]$. If $h(t) = 0$, then $h([a, t]) = \{0\}$ by reason of the monotonicity of h. It follows that $\gamma = \gamma_0 \circ h$ is constant on $[a, t]$, which in turn implies that $\sigma(t) = 0$. Assuming next that $h(t) > 0$, we extract from (ii) the information that

$$\sigma(t) = \ell(\gamma|[a, t]) = \ell(\gamma_0 \circ h|[a, t]) = \ell\Big(\gamma_0|\big[0, h(t)\big]\Big) = \sigma_0\big[h(t)\big] = h(t)\,.$$

Therefore $h = \sigma$, so there can be at most one path γ_0 enjoying properties (i) and (ii). It remains to establish the existence of such a path. Notice that the condition $\sigma(t) = \sigma(u)$ for $a \le u < t \le b$ forces γ to be constant on the interval $[u, t]$, whence $\gamma(t) = \gamma(u)$. As a result, we obtain a well-defined function $\gamma_0 : [0, \ell] \to \mathbb{R}^n$ by insisting that $\gamma_0[\sigma(t)] = \gamma(t)$ for t in $[a, b]$. Let $\langle s_\nu \rangle$ be a strictly monotone sequence in $[0, \ell]$ such that $s_\nu \to s$. If t_ν in $[a, b]$ satisfies $\sigma(t_\nu) = s_\nu$, then $\langle t_\nu \rangle$ is a strictly monotone sequence in $[a, b]$ and $t = \lim_{\nu\to\infty} t_\nu$ has $\sigma(t) = s$. Therefore, $\gamma_0(s_\nu) = \gamma(t_\nu) \to \gamma(t) = \gamma_0(s)$. This makes certain that γ_0 is continuous. Since the paths γ and γ_0 are related by the change of parameter σ, γ_0 is rectifiable with $\ell(\gamma_0) = \ell(\gamma) = \ell$. Lastly, for $s = \sigma(t) > 0$ we have

$$\sigma_0(s) = \ell(\gamma_0|[0, s]) = \ell(\gamma|[0, t]) = \sigma(t) = s\ ,$$

which furnishes confirmation of (iii). □

The path γ_0 whose existence and uniqueness is assured by Theorem 4.1.3 is known as the *normal* (or *arclength*) *representation* of the given rectifiable path $\gamma : [a, b] \to \mathbb{R}^n$. A path $\gamma : [a, b] \to \mathbb{R}^n$ whose length function $\sigma = \sigma_\gamma$ satisfies $\sigma(t) = t - a$ for t in $[a, b]$ is said to be *parametrized by arclength.*

If another path $\beta : [c, d] \to \mathbb{R}^n$ is related to γ through a change of parameter $h : [a, b] \to [c, d]$, then $\sigma_\gamma = \sigma_\beta \circ h$ and

$$\gamma = \beta \circ h = \beta_0 \circ \sigma_\beta \circ h = \beta_0 \circ \sigma_\gamma\,.$$

The uniqueness part of Theorem 4.1.3 thus informs us that $\beta_0 = \gamma_0$.

On the strength of Lemma 4.1.1 and Theorem 4.1.3, we can assert that

$$|\gamma_0(s) - \gamma_0(u)| \le \sigma_0(s) - \sigma_0(u) \le s - u$$

whenever $0 \le u \le s \le \ell$, which makes the path γ_0 a Lipschitz (hence, an absolutely continuous) path. Furthermore, $|\dot\gamma_0(s)| = \sigma_0'(s) = 1$ for almost every s in $[0, \ell]$.

Let $\gamma : [a, b] \to \mathbb{R}^n$ be a nonconstant rectifiable path, and let $\sigma = \sigma_\gamma$. It is not difficult to show that $(-\gamma)_0 = -\gamma_0$ and that $\gamma_0 = \beta_0 + \alpha_0$ when $\gamma = \beta + \alpha$ (assuming, of course, that $\ell(\alpha) > 0$ and $\ell(\beta) > 0$). Also, if $\beta = \gamma|[c, d]$ and if $\ell(\beta) = \sigma(d) - \sigma(c) > 0$, then $\beta_0(s) = \gamma_0[s + \sigma(c)]$ for $0 \le s \le \ell(\beta)$.

Consider an arc J in $\mathbb{R}^n$. If γ and β are homeomorphisms of $[0, 1]$ onto J with $\gamma(0) = \beta(0)$ and $\gamma(1) = \beta(1)$, then γ is obtainable from β by the change of parameter $h = \beta^{-1} \circ \gamma : [0, 1] \to [0, 1]$. As a result, $\ell(\gamma) = \ell(\beta)$.

There is an important point to be made here. Rectifiability in this situation depends on the point set J, not in a particular homeomorphism of $[0, 1]$ onto J. As a matter of fact, $\ell(\gamma) = \mathcal{H}^1(J)$ for any homeomorphism γ of $[0, 1]$ onto J, where $\mathcal{H}^1$ is the normalized one-dimensional Hausdorff measure in $\mathbb{R}^n$. We are therefore entitled to speak of a *rectifiable arc* J and its length $\ell(J)$ without reference to any parametrization. A rectifiable arc can be parametrized by arclength starting from either of its endpoints.

4.1.4. Integration along curves. Suppose that A is a Borel set in $\hat{\mathbb{R}}^n$, that $\rho : A \to [0, \infty]$ is a Borel function, and that $\gamma : I \to A$ is a locally rectifiable curve.

We wish to define

$$\int_\gamma \rho\, ds = \int_\gamma \rho(x)|dx|\,,$$

the *integral of* ρ *along* γ *with respect to arclength*.

We consider first the case where $\gamma : [a, b] \to A$ is a rectifiable path of length ℓ. If $\ell = 0$, that is, if γ is a constant path, we decree that $\int_\gamma \rho\, ds = 0$; if $\ell > 0$, we define

$$\int_\gamma \rho\, ds = \int_0^\ell \rho\big[\gamma_0(s)\big]\, ds\ , \tag{4.22}$$

where $\gamma_0 : [0, \ell] \to A$ is the normal representation of γ.

Notice that the integrand $\rho \circ \gamma_0$ appearing on the right-hand side of (4.22) is a nonnegative Borel function, so this Lebesgue integral is meaningful and takes its value in $[0, \infty]$. If γ is related to a second path β through a change of parameter, then $\int_\gamma \rho\, ds = \int_\beta \rho\, ds$. In the nontrivial case, by which we mean that β and γ are nonconstant, this follows directly from the fact that $\gamma_0 = \beta_0$.

Let $\gamma : [a, b] \to A$ be a rectifiable path. It is straightforward to verify that

$$\int_{-\gamma} \rho\, ds = \int_\gamma \rho\, ds \tag{4.23}$$

and that

$$\int_\gamma \rho\, ds = \int_\beta \rho\, ds + \int_\alpha \rho\, ds \tag{4.24}$$

if $\gamma = \beta + \alpha$. If $\beta = \gamma|[c, d]$ with $a \le c < d \le b$, then since $\rho \ge 0$,

$$\int_\gamma \rho\, ds \ge \int_\beta \rho\, ds\,. \tag{4.25}$$

This is a trivial assertion if β is constant. In the nonconstant case, we make a linear change of variable in computing

$$\begin{aligned}\int_\gamma \rho\, ds \ge \int_{\sigma(c)}^{\sigma(d)} \rho\big[\gamma_0(s)\big]\, ds &= \int_0^{\sigma(d)-\sigma(c)} \rho\Big[\gamma_0[s + \sigma(c)]\Big]\, ds \\ &= \int_0^{\ell(\beta)} \rho[\beta_0(s)]\, ds = \int_\beta \rho\, ds\ ,\end{aligned}$$

in which $\sigma = \sigma_\gamma$.

For an arbitrary locally rectifiable curve $\gamma : I \to A$ we set

$$\int_\gamma \rho\, ds = \sup_\beta \int_\beta \rho\, ds\ ,$$

the supremum extending over all paths β such that $\beta \prec \gamma$. When J is a rectifiable arc in A, we define $\int_J \rho\, ds$ by

$$\int_J \rho\, ds = \int_\gamma \rho\, ds,$$

where γ is any homeomorphism of $[0, 1]$ onto J. The resulting quantity is independent of the choice of γ. In fact it can be shown that

$$\int_J \rho\, ds = \int_J \rho\, d\mathcal{H}^1.$$

More generally, if E is a locally rectifiable open (or half-open) arc in A, we define

$$\int_E \rho\, ds = \sup_J \int_J \rho\, ds,$$

the supremum now being taken over subarcs J of E.

If $\gamma : I \to A$ is a locally absolutely continuous curve, then it is possible to evaluate $\int_\gamma \rho\, ds$ without first passing to the normal representation for γ. We preface the result that tells us how to do so with the following remark: if f and g are nonnegative functions and if the product fg appears in an integrand, say in an integral $\int_A fg\, dm_n$ or $\int_\gamma fg\, ds$, we shall observe the usual convention in dealing with the typical indeterminate forms that crop up in this context, agreeing to set $0 \cdot \infty = \infty \cdot 0 = 0$.

LEMMA 4.1.4. *If A is a Borel set in $\hat{\mathbb{R}}^n$, $\rho : A \to [0, \infty]$ is a Borel function, and $\gamma : I \to A$ is a locally absolutely continuous curve, then*

$$\int_\gamma \rho\, ds = \int_a^b \rho[\gamma(t)]\,|\dot{\gamma}(t)|\, dt\ ,$$

where a and b are the endpoints of I in $[-\infty, \infty]$.

PROOF. By the definition of the arclength integral for an arbitrary locally rectifiable curve γ and by elementary properties of the Lebesgue integral, it suffices to prove the lemma in the case of an absolutely continuous path $\gamma : [a, b] \to \mathbb{R}^n$ of length $\ell > 0$. Then $\dot{\gamma}$ is defined almost everywhere in $[a, b]$ and belongs to $L^1[a, b]$, $|\dot{\gamma}(t)| = \sigma_\gamma'(t)$ for almost every t in $[a, b]$, and $\sigma_\gamma : [a, b] \to [0, \ell]$ is absolutely continuous. We may assume that the function ρ is bounded. (If not, we apply the result for the bounded case to the functions $\rho_\nu = \min\{\rho, \nu\}$ for $\nu = 1, 2, \ldots$, and appeal to the monotone convergence theorem.) Noting that $\rho \circ \gamma_0$ is then integrable over $[0, \ell]$, we invoke the change of variable formula for Lebesgue integrals over closed intervals in $\mathbb{R}$ and calculate

$$\int_a^b \rho[\gamma(t)]\,|\dot{\gamma}(t)|\, dt = \int_a^b \rho\{\gamma_0[\sigma_\gamma(t)]\}\sigma_\gamma'(t)\, dt = \int_0^\ell \rho[\gamma_0(s)]\, ds = \int_\gamma \rho\, ds\ ,$$

as asserted. □

Let $\gamma : I \to A$ be a locally rectifiable curve in a Borel subset A of $\hat{\mathbb{R}}^n$, and let $\rho : A \to [0, \infty]$ be a Borel function. If β is a subcurve of γ, then it follows immediately from definitions that (4.25) remains valid. Thus

$$\int_\gamma \rho\, ds \geq \int_\beta \rho\, ds\,. \tag{4.26}$$

More generally, if $\gamma_\nu : I_\nu \to \hat{\mathbb{R}}^n$ are subcurves of γ whose domains I_ν are pairwise nonoverlapping, then a variation of the argument that established (4.25) can be used to justify the statement

$$\int_\gamma \rho\, ds \geq \sum_\nu \int_{\gamma_\nu} \rho\, ds\,. \tag{4.27}$$

Arclength integrals exhibit the following expected absolute continuity property, which the reader can easily verify:

LEMMA 4.1.5. *If $\int_\gamma \rho\, ds < \infty$, then corresponding to each $\varepsilon > 0$ there is a $\delta > 0$ such that $\int_\beta \rho\, ds < \varepsilon$ whenever $\beta \prec \gamma$ and $\ell(\beta) < \delta$.*

Together with (4.20) and (4.21), this fact enters into the proof (also left to the reader) that

$$\int_\gamma \rho\, ds = \int_{\gamma^*} \rho\, ds \tag{4.28}$$

in case γ is rectifiable. To see this, recall Lemma 4.1.2. In the integral on the right in (4.28) one can define $\rho(x)$ arbitrarily at any endpoint x of γ^* that does not already belong to A.

4.1.5. Transformation of arclength integrals. We being by considering the following situation: γ is a locally rectifiable curve in $\hat{\mathbb{R}}^n$, $f : |\gamma| \to \hat{\mathbb{R}}^m$ is a continuous function, A is a Borel subset of $\hat{\mathbb{R}}^m$ containing $|f \circ \gamma|$, and $\rho : A \to [0, \infty]$ is a Borel function. It is natural to seek an expression for the integral

$$\int_{f\circ\gamma} \rho\, ds$$

—or, at the very least, an estimate for this integral—in terms of the integral along γ of some integrand that involves the Borel function $\rho \circ f$.

Unfortunately, this process might not even get off the ground as the integral $\int_{f\circ\gamma} \rho\, ds$ may fail to be meaningful. For instance, under the conditions specified, the curve $f \circ \gamma$ need not be locally rectifiable.

One way to circumvent this problem is to insist at the outset that f be reasonably well behaved on γ, an idea we make precise with the following definitions.

In the special case where γ is a rectifiable path of length ℓ, we say that f is *absolutely continuous on* γ if either γ (and, with it, $f \circ \gamma$) is a constant path, or γ is nonconstant and the path $f \circ \gamma_0 : [0, \ell] \to A$ is absolutely continuous.

In the case of an arbitrary locally rectifiable curve γ, we declare f to be *locally absolutely continuous on* γ provided it is absolutely continuous on every subpath of γ. For example, if γ is a rectifiable path and if the function f satisfies a Lipschitz condition on $|\gamma|$—say $|f(y)-f(x)| \leq \lambda|y-x|$ for all points x and y of $|\gamma|$, where $\lambda > 0$ is a constant—then f is absolutely continuous on γ. Assuming that $\ell = \ell(\gamma) > 0$, we see this by remarking that

$$|f \circ \gamma_0(s) - f \circ \gamma_0(u)| \leq \lambda|\gamma_0(s) - \gamma_0(u)| \leq \lambda\big[\sigma_0(s) - \sigma_0(u)\big] = \lambda(s - u)$$

whenever $0 \leq u \leq s \leq \ell$, which makes $f \circ \gamma_0$ a Lipschitz (hence, absolutely continuous) path. In particular, if U is an open set in $\mathbb{R}^n$ and f is a member of the class $C^1(U, \mathbb{R}^m)$, then f is locally absolutely continuous on every locally rectifiable curve in U, for such a function satisfies a Lipschitz condition on each compact set in U. The local absolute continuity of a function f on a locally rectifiable curve

γ guarantees that the curve $f \circ \gamma$ is also locally rectifiable. In fact, if β is any nonconstant subpath of γ, then $f \circ \beta = f \circ \beta_0 \circ \sigma_\beta$. As a consequence, $f \circ \beta$ is obtained from the absolutely continuous (therefore, rectifiable) path $f \circ \beta_0$ by the change of parameter $h = \sigma_\beta$, rendering $f \circ \beta$ rectifiable as well. We can now state a change of variable theorem for arclength integrals.

THEOREM 4.1.6. *Let U be an open set in $\mathbb{R}^n$, and let f belong to the class $C(U, \mathbb{R}^m)$. If γ is a locally rectifiable curve in U on which f is locally absolutely continuous, then*

$$\int_\gamma (\rho \circ f)\, \ell_f \, ds \le \int_{f\circ\gamma} \rho\, ds \le \int_\gamma (\rho \circ f)\, L_f \, ds \tag{4.29}$$

for every Borel function $\rho : A \to [0, \infty]$ whose domain A includes the trajectory of $f \circ \gamma$.

PROOF. As in the proof of Lemma 4.1.4, it is enough to verify (4.29) in the case where $\gamma : [a, b] \to U$ is a rectifiable path and ρ is a bounded function. The discussion preceding the statement of the theorem shows that the path $\beta = f \circ \gamma$ is rectifiable. Should γ happen to be a constant path, (4.29) would be a completely trivial statement. Moreover, if γ were nonconstant but β were constant, then the inequalities in (4.29) would still be trivial, for $\ell_f(x) = 0$ would be true for all x in $|\gamma|$. We proceed assuming that $\ell = \ell(\gamma)$ and $\ell^* = \ell(\beta)$ are positive. Let $\alpha = f \circ \gamma_0$. We know that $\beta_0 = \alpha_0$, because β is related to α through the change of parameter $h = \sigma_\gamma$. By assumption the path α is absolutely continuous, so its length function $\sigma_\alpha : [0, \ell] \to [0, \ell^*]$ is also absolutely continuous and has $\sigma_\alpha'(s) = |\dot\alpha(s)|$ for almost every s is $[0, \ell]$ (Lemma 4.4). Furthermore, $\beta_0 \circ \sigma_\alpha = \alpha_0 \circ \sigma_\alpha = \alpha = f \circ \gamma_0$.

As a bounded function, $\rho \circ \beta_0$ belongs to the class $L^1[0, \ell^*]$. The formula for an absolutely continuous change of variable in a one-dimensional Lebesgue integral now provides the information that

$$\begin{aligned} \int_\beta \rho\, ds &= \int_0^{\ell^*} \rho[\beta_0(u)]\, du = \int_0^\ell \rho[\beta_0[\sigma_\alpha(s)]]\sigma_\alpha'(s)\, ds \\ &= \int_0^\ell \rho[f[\gamma_0(s)]]\, |\dot\alpha(s)|\, ds\,. \end{aligned} \tag{4.30}$$

Now consider a point s in $(0, \ell)$ for which both $\dot\alpha(s)$ and $\dot\gamma_0(s)$ exist and for which $|\dot\gamma_0(s)| = 1$—this holds at almost every s in $(0, \ell)$. The path γ_0 cannot be constant on any interval $[s, s+h]$ with $h > 0$ and $s + h < \ell$, for $\ell(\gamma_0|[s, s+h]) = h > 0$. This fact implies the existence of a sequence $\langle h_\nu \rangle$ of positive numbers such that $s + h_\nu < \ell$, $h_\nu \to 0$, and $x_\nu = \gamma_0(s + h_\nu) \ne \gamma_0(s)$. Consequently,

$$\begin{aligned} |\dot\alpha(s)| &= \lim_{\nu\to\infty} \frac{|\alpha(s+h_\nu) - \alpha(s)|}{h_\nu} \\ &= \lim_{\nu\to\infty} \left(\frac{|\gamma_0(s+h_\nu) - \gamma_0(s)|}{h_\nu} \cdot \frac{|f(x_\nu) - f[\gamma_0(s)]|}{|x_\nu - \gamma_0(s)|} \right) \\ &\le \lim_{\nu\to\infty} \frac{|\gamma_0(s+h_\nu) - \gamma_0(s)|}{h_\nu} \limsup_{\nu\to\infty} \frac{|f(x_\nu) - f[\gamma_0(s)]|}{|x_\nu - \gamma_0(s)|} \\ &\le |\dot\gamma_0(s)|\, L_f[\gamma_0(s)] = L_f[\gamma_0(s)]\,. \end{aligned}$$

Thus $|\dot{\alpha}(s)| \leq L_f[\gamma_0(s)]$ holds almost everywhere in $(0,\ell)$. Similarly, $|\dot{\alpha}(s)| \geq \ell_f[\gamma_0(s)]$ for almost every s in $(0,\ell)$. Together with (4.30) and the definitions of the integrals at the extremes of (4.29), these estimates complete the proof. □

As a simple illustration of the use of Theorem 4.1.6 we show that the chordal length $\ell_q(\gamma)$ of a rectifiable path $\gamma : [a,b] \to \mathbb{R}^n$—by definition

$$\ell_q(\gamma) = \sup_P \ell_q(\gamma, P) = \sup_P \sum_{j=1}^{r} q\big[\gamma(t_j), \gamma(t_{j-1})\big] ,$$

the supremum ranging over all partitions $P : a = t_0 < t_1 < \cdots < t_r = b$ of $[a,b]$—is given by

$$\ell_q(\gamma) = 2\int_\gamma \frac{|dx|}{1+|x|^2} . \tag{4.31}$$

Clearly $\ell_q(\gamma) = \ell(\pi \circ \gamma)$, where $\pi : \hat{\mathbb{R}}^n \to S^n$ is a stereographic projection. We know that π is the restriction to $\hat{\mathbb{R}}^n$ of R, the inversion in the sphere $\mathbb{S}^n(e_{n+1}, \sqrt{2}\,)$. As R is conformal,

$$L_\pi(x) = \ell_\pi(x) = \|R'(x)\| = \frac{2}{1+|x|^2}$$

for every x in $\mathbb{R}^n$. Applying Theorem 4.1.6 with $f = \pi$ and $\rho = 1$ yields (4.31).

Theorem 4.1.6 delivers a useful estimate for arclength integrals of radial functions.

LEMMA 4.1.7. *If $\gamma : I \to \mathbb{R}^n$ is a locally rectifiable curve, if $J = \{|\gamma(t)| : t \in I\}$, and if $\rho : J \to [0,\infty]$ is a Borel function, then*

$$\int_\gamma \rho(|x|)|dx| \geq \int_J \rho(r)\, dr . \tag{4.32}$$

PROOF. It suffices to treat the case of a rectifiable path $\gamma : [a,b] \to \mathbb{R}^n$ such that $J = [c,d]$ is a nondegenerate interval. Again, we need only consider bounded functions ρ. Select a_1 and b_1 in $[a,b]$ for which $|\gamma(a_1)| = c$ and $|\gamma(b_1)| = d$. Replacing γ by its reverse path $-\gamma$ if need be, we may assume that $a_1 < b_1$. Moreover, by considering $\gamma|[a_1,b_1]$ in place of γ—this step does not increase the left-hand side of (4.32) and leaves the right-hand side unchanged—we reduce matters to the situation in which $|\gamma(a)| = c$ and $|\gamma(b)| = d$. The function $f : \mathbb{R}^n \to [0,\infty)$ defined by $f(x) = |x|$ is a Lipschitz function (in fact, $|f(y) - f(x)| \leq |y - x|$ holds for all x and y) with $L_f(x) = 1$ for every x in $\mathbb{R}^n$. Theorem 4.1.6 tells us that

$$\int_\gamma \rho(|x|)|dx| = \int_\gamma \rho(|x|)L_f(x)|dx| \geq \int_{f\circ\gamma} \rho(y)|dy| .$$

Write $\beta = f \circ \gamma$, $\ell = \ell(\beta)$, and $h = \beta_0$. Then $h : [0,\ell] \to [c,d]$ is an absolutely continuous function with $h(0) = c$, $h(\ell) = d$, and $|h'(s)| = 1$ almost everywhere in $[0,\ell]$. Since ρ belongs to $L^1(J)$, the change of variable $r = h(s)$ is justified in the computation

$$\begin{aligned}
\int_J \rho(r)\, dr &= \int_c^d \rho(r)\, dr = \int_0^\ell \rho\big[h(s)\big]h'(s)\, ds \leq \int_0^\ell \rho\big[h(s)\big]|h'(s)|\, ds \\
&= \int_0^\ell \rho\big[\beta_0(s)\big]\, ds = \int_{f\circ\gamma} \rho(y)|dy| \leq \int_\gamma \rho(|x|)|dx| ,
\end{aligned}$$

which completes the proof. □

4.2. Moduli of curve families

We now come to the central point of this chapter, namely the definition of modulus in a general setting.

4.2.1. The p-modulus M_p. Let Γ be a family of curves in $\hat{\mathbb{R}}^n$, where $n \geq 2$. We shall henceforth use the expression *admissible density for* Γ to denote a Borel function $\rho : \hat{\mathbb{R}}^n \to [0, \infty]$ with the property that

$$\int_\gamma \rho \, ds \geq 1$$

for every locally rectifiable member γ of Γ.

The notation $\operatorname{Adm}(\Gamma)$ indicates the collection of all such densities. For p in $[1, \infty]$ we set

$$M_p(\Gamma) = \inf\left\{ \int_{\mathbb{R}^n} \rho^p(x) \, dm_n(x) : \rho \in \operatorname{Adm}(\Gamma) \right\}$$

if $\operatorname{Adm}(\Gamma) \neq \emptyset$. Otherwise $M_p(\Gamma) = \infty$ if $\operatorname{Adm}(\Gamma) = \emptyset$. Of course $\operatorname{Adm}(\Gamma)$ is empty if and only if Γ contains a constant curve, for the function $\rho \equiv \infty$ belongs to $\operatorname{Adm}(\Gamma)$ in every other case.

Obviously $0 \leq M_p(\Gamma) \leq \infty$. The quantity $M_p(\Gamma)$ is known as the *p-modulus* of the family Γ, while $\lambda_p(\Gamma) = M_p(\Gamma)^{1/(1-p)}$ is termed its *p-extremal length.*

In situations where no confusion about the dimension is likely to arise, it is customary to abbreviate $M_n(\Gamma)$ and $\lambda_n(\Gamma)$ simply to $M(\Gamma)$ and $\lambda(\Gamma)$, respectively, and to speak of them as the *conformal modulus* and *conformal extremal length of* Γ.

In point of fact, it is only the conformal modulus that is needed for the development of the theory of quasiconformal mappings. Nevertheless, the p-modulus turns up in a number of contexts closely related to this theory, so there is some benefit to discussing moduli of the more general type.

The informational content of $M_p(\Gamma)$ is not easy to fathom. It suffices to say that this number aggregates in some subtle way data about the abundance of curves in Γ, the relative lengths of these curves, and the amount of space their collective trajectories occupy. In this way it is not unlike an isoperimetric quantity. Notice, for example, that any family Γ with no locally rectifiable curves in it has $M_p(\Gamma) = 0$, because the function $\rho \equiv 0$ is trivially admissible for such a family.

By definition, the p-modulus is a set function whose domain consists of the subsets of $\mathcal{C}(\hat{\mathbb{R}}^n)$, the family of all curves in $\hat{\mathbb{R}}^n$, and whose values lie in $[0, \infty]$. We can say more.

LEMMA 4.2.1. *The p-modulus is an outer measure on $\mathcal{C}(\hat{\mathbb{R}}^n)$. That is:*

(i) $M_p(\emptyset) = 0$;

(ii) *if Γ_1 and Γ_2 are subsets of $\mathcal{C}(\hat{\mathbb{R}}^n)$ with Γ_1 contained in Γ_2, then*

$$M_p(\Gamma_1) \leq M_p(\Gamma_2);$$

(iii) *for any countable collection $\{\Gamma_\nu\}$ of subsets of $\mathcal{C}(\hat{\mathbb{R}}^n)$,*

$$M_p\Big(\bigcup_\nu \Gamma_\nu\Big) \leq \sum_\nu M_p(\Gamma_\nu).$$

PROOF. For (i) simply note that the empty family includes no locally rectifiable curves. Assertion (ii) follows from the fact that $\mathrm{Adm}(\Gamma_2)$ is contained in $\mathrm{Adm}(\Gamma_1)$ when Γ_1 is a subset of Γ_2. As for (iii), we may assume that $\sum_\nu M_p(\Gamma_\nu) < \infty$ (then $M_p(\Gamma_\nu) < \infty$ for each ν). Let $\varepsilon > 0$ be given. Choose ρ_ν in $\mathrm{Adm}(\Gamma_\nu)$ for which

$$\int_{\mathbb{R}^n} \rho_\nu^p \, dm_n \leq M_p(\Gamma_\nu) + 2^{-\nu}\varepsilon \, .$$

Clearly $\rho = (\sum_\nu \rho_\nu^p)^{1/p}$ is a Borel function with the property that $\rho \geq \rho_\nu$ for every ν. If γ is a locally rectifiable member of $\Gamma = \bigcup_\nu \Gamma_\nu$, then γ belongs to Γ_ν for some ν, so

$$\int_\gamma \rho \, ds \geq \int_\gamma \rho_\nu \, ds \geq 1 \, .$$

This places ρ in $\mathrm{Adm}(\Gamma)$. We infer that

$$M_p(\Gamma) \leq \int_{\mathbb{R}^n} \rho^p \, dm_n = \int_{\mathbb{R}^n} \sum_\nu \rho_\nu^p \, dm_n$$

$$= \sum_\nu \int_{\mathbb{R}^n} \rho_\nu^p \, dm_n \leq \sum_\nu \big(M_p(\Gamma_\nu) + 2^{-\nu}\varepsilon\big) \leq \sum_\nu M_p(\Gamma_\nu) + \varepsilon \, .$$

Letting $\varepsilon \to 0$ establishes (iii). □

In view of Lemma 4.2.1, an obvious question raises itself: Which curve families Γ in $\hat{\mathbb{R}}^n$ are M_p-measurable in the sense of Carathéodory? We are in fact asking which Γ meet the requirement that

$$M_p(\Gamma_0) = M_p(\Gamma_0 \cap \Gamma) + M_p(\Gamma_0 \setminus \Gamma)$$

for every subset Γ_0 of $\mathcal{C}(\hat{\mathbb{R}}^n)$.

The answer to this question reveals that the p-modulus is not terribly interesting from a purely measure theoretic standpoint. To be specific, the following is known:

(1) if $M_p(\Gamma) = 0$, then Γ is M_p-measurable;
(2) if $0 < M_p(\Gamma) < \infty$, then Γ is not M_p-measurable;
(3) if $M_p(\Gamma) = \infty$, then Γ may or not be M_p-measurable.

In particular, there is no significant σ-algebra of subsets of $\mathcal{C}(\hat{\mathbb{R}}^n)$ on which M_p is a nontrivial measure. Nevertheless, some of the language of measure theory retains its usefulness in conjunction with the p-modulus. For instance, one speaks of a property of curves holding M_p-*almost everywhere* when the family Γ of curves that lack the property in question has $M_p(\Gamma) = 0$.

Let Γ_1 and Γ_2 be families of curves in $\hat{\mathbb{R}}^n$ that are related in the following way: each curve in Γ_1 has a subcurve belonging to Γ_2. We then say that Γ_2 *minorizes* Γ_1 and express this relationship between the two curve families symbolically by writing $\Gamma_2 \prec \Gamma_1$. We record a simple yet consequential observation.

LEMMA 4.2.2. *If* $\Gamma_2 \prec \Gamma_1$, *then* $M_p(\Gamma_1) \leq M_p(\Gamma_2)$.

PROOF. The assertion is trivial if $M_p(\Gamma_2) = \infty$. Under the assumption that $M_p(\Gamma_2) < \infty$, consider a function ρ from $\mathrm{Adm}(\Gamma_2)$. If a curve γ from Γ_1 is locally rectifiable and if β in Γ_2 is chosen so that $\beta \prec \gamma$, then by (4.26),

$$\int_\gamma \rho \, ds \geq \int_\beta \rho \, ds \geq 1 \, .$$

This means that ρ belongs to $\mathrm{Adm}(\Gamma_1)$. Therefore

$$M_p(\Gamma_1) \le \int_{\mathbb{R}^n} \rho^p \, dm_n \, .$$

Since ρ was an arbitrary member of $\mathrm{Adm}(\Gamma_2)$, the inequality $M_p(\Gamma_1) \le M_p(\Gamma_2)$ follows when the infimum ρ is taken over all such ρ. □

Lemma 4.2.2 lends quantitative substance to a rule of thumb well worth bearing in mind when dealing with moduli of curve families: *the shorter the curves in a family, the larger its p-modulus is likely to be.* In particular, we have seen that a family Γ with no "short" curves (meaning no locally rectifiable curves) has $M_p(\Gamma) = 0$.

A couple of examples will serve to elucidate the concept of the p-modulus. Before turning to these, we fix convenient notation for several types of curve families that will surface again and again throughout this book.

Let E, F, and G be nonempty subsets of $\hat{\mathbb{R}}^n$ (in most applications G will be a domain, and E and F disjoint subsets of $\bar{G}$). The notation $\Delta(E, F : G)$ stands for the family of "open paths with endpoints" that connect E and F through G; thus for a curve $\gamma : (a, b) \to \hat{\mathbb{R}}^n$ to be a member of $\Delta(E, F : G)$, it is required that $|\gamma|$ be a subset of G, that both $\gamma(a^+) = \lim_{t\to a^+} \gamma(t)$ and $\gamma(b^-) = \lim_{t\to b^-} \gamma(t)$ exist, and that one of these limits belong to E and the other to F. We use $\Delta_0(E, F; G)$ to indicate the family of all open paths γ in G whose trajectories cluster at both E and F, in the sense that $\bar{\gamma} \cap E \ne \emptyset$ and $\bar{\gamma} \cap F \ne \emptyset$. Lastly, the notation $\bar{\Delta}(E, F : G)$ is reserved for the family of all paths that join E and F through G. To be more precise, a path $\gamma : [a, b] \to \hat{\mathbb{R}}^n$ belongs to $\bar{\Delta}(E, F : G)$ provided that $\gamma(t)$ lies in G when $a < t < b$, while $\gamma(a)$ is an element of E and $\gamma(b)$ an element of F, or vice versa. Notice that $\Delta(E, F : G)$ minorizes $\bar{\Delta}(E, F : G)$ and is contained in $\Delta_0(E, F : G)$. Now we come to the promised examples.

THEOREM 4.2.3. *Let E be a nonempty Borel set in $\mathbb{R}^{n-1}$ viewed as a subspace of $\mathbb{R}^n$, with $n \ge 2$. For $h > 0$, let $G = \{x + te_n : x \in E,\ 0 < t < h\}$ and $F = E + he_n = \{x + he_n : x \in E\}$. For $1 < p < \infty$ the family $\Gamma = \Delta(E, F : G)$ has*

$$M_p(\Gamma) = \frac{m_{n-1}(E)}{h^{p-1}} \, . \tag{4.33}$$

PROOF. Since G is a Borel subset of $\hat{\mathbb{R}}^n$, $\rho = h^{-1}\chi_G$ is a Borel function, and plainly

$$\int_\gamma \rho \, ds = \frac{\ell(\gamma)}{h} \ge 1$$

for every locally rectifiable γ in Γ. Thus ρ finds itself in $\mathrm{Adm}(\Gamma)$. According to Fubini's theorem,

$$M_p(\Gamma) \le \int_{\mathbb{R}^n} \rho^p \, dm_n = \frac{m_n(G)}{h^p} = \frac{h m_{n-1}(E)}{h^p} = \frac{m_{n-1}(E)}{h^{p-1}} \, . \tag{4.34}$$

On the other hand, for each x in E the path $\gamma_x : (0, h) \to \mathbb{R}$ given by $\gamma_x(t) = x + te_n$ is a locally absolutely continuous member of Γ. Lemma 4.1.4 and Hölder's inequality ensure that for any ρ in $\mathrm{Adm}(\Gamma)$ we have

$$1 \le \int_{\gamma_x} \rho \, ds = \int_0^h \rho(x, t) \, dt \le \left(\int_0^h \rho^p(x, t) \, dt \right)^{1/p} h^{(p-1)/p} \, .$$

We conclude that

$$\frac{1}{h^{p-1}} \le \int_0^h \rho^p(x,t)\,dt$$

for each x in E. Integrating both sides of this inequality over E and invoking Fubini's theorem one more time, we discover that

$$\frac{m_{n-1}(E)}{h^{p-1}} \le \int_E \left[\int_0^h \rho^p(x,t)\,dt\right] dm_{n-1}(x) = \int_G \rho^p\,dm_n \le \int_{\mathbb{R}^n} \rho^p\,dm_n\,.$$

Since ρ in $\mathrm{Adm}(\Gamma)$ was arbitrary, we deduce that

$$\frac{m_{n-1}(E)}{h^{p-1}} \le M_p(\Gamma)\,.$$

In tandem with (4.34) this inequality establishes (4.33). □

In the next example we observe the conventions $\overline{B}^n(x_0,0) = \mathbb{S}^{n-1}(x_0,0) = \{x_0\}$, $B^n(x_0,\infty) = \mathbb{R}^n$ and $\mathbb{S}^{n-1}(x_0,\infty) = \{\infty\}$.

THEOREM 4.2.4. *Let x_0 be a point of $\mathbb{R}^n$, let $0 \le a < b \le \infty$, and let $\Gamma = \Delta(E,F : G)$, where $E = \mathbb{S}^{n-1}(x_0,a)$, $F = \mathbb{S}^{n-1}(x_0,b)$, and $G = B^n(x_0,b) \setminus \overline{B}^n(x_0,a)$. Then for $1 < p < \infty$ it is true that*

$$M_p(\Gamma) = \omega_{n-1}\left(\int_a^b r^{m-1}\,dr\right)^{1-p}, \tag{4.35}$$

in which $m = (p-n)/(p-1)$. In particular, the conformal modulus of Γ is given by

$$M(\Gamma) = \omega_{n-1}\left(\log\frac{b}{a}\right)^{1-n}. \tag{4.36}$$

PROOF. We assume initially that

$$c = \int_a^b r^{m-1}\,dr < \infty, \qquad m = (p-n)/(p-1).$$

For the sake of simplicity we assume, in addition, that $x_0 = 0$. Since $M_p(\Gamma)$ is readily seen to be translation invariant, this entails no loss of generality. Fix an admissible density ρ for Γ. For each u in $\mathbb{S}^{n-1}$ the curve γ_u defined on (a,b) by $\gamma_u(r) = ru$ is a locally absolutely continuous member of Γ. Lemma 4.1.4 and Hölder's inequality yield

$$\begin{aligned} 1 &\le \left(\int_{\gamma_u} \rho\,ds\right)^p = \left(\int_a^b \rho(ru)\,dr\right)^p = \left(\int_a^b \rho(ru)\,r^{n-1/p}\,r^{1-n/p}\,dr\right)^p \\ &\le \left(\int_a^b \rho^p(ru)\,r^{n-1}\,dr\right)\left(\int_a^b r^{m-1}\,dr\right)^{p-1} = c^{p-1}\int_a^b \rho^p(ru)\,r^{n-1}\,dr\,, \end{aligned}$$

with the result that

$$0 < c^{1-p} \le \int_a^b \rho^p(ru)\,r^{n-1}\,dr\,.$$

Because this inequality holds for every u in $\mathbb{S}^{n-1}$, we can integrate it over $\mathbb{S}^{n-1}$ with respect to surface area. In fact, recalling (4.14), we may compute that

$$\begin{aligned}\int_{\mathbb{R}^n} \rho^p \, dm_n &\geq \int_G \rho^p \, dm_n = \int_a^b \left[\int_{S^{n-1}} \rho^p(ru) \, d\sigma_{n-1}(u) \right] r^{n-1} \, dr \\ &= \int_{S^{n-1}} \left[\int_a^b \rho^p(ru) \, r^{n-1} \, dr \right] d\sigma_{n-1}(u) \geq \omega_{n-1} c^{1-p} .\end{aligned}$$

We thus secure a lower bound for $M_p(\Gamma)$ in the form of

$$M_p(\Gamma) \geq c^{1-p} \omega_{n-1} . \tag{4.37}$$

Next, let $\rho : \hat{\mathbb{R}}^n \to [0, \infty]$ be defined by

$$\rho(x) = \begin{cases} c^{-1}|x|^{m-1} & \text{if } x \in G , \\ 0 & \text{if } x \notin G . \end{cases}$$

Then ρ is a radial Borel function. Lemma 4.1.7 gives

$$\int_\gamma \rho \, ds \geq \int_a^b \rho(r) \, dr = c^{-1} \int_a^b r^{m-1} \, dr = 1$$

for every locally rectifiable γ in Γ. This means ρ is a member of $\mathrm{Adm}(\Gamma)$. Consequently,

$$\begin{aligned} M_p(\Gamma) &\leq \int_{\mathbb{R}^n} \rho^p \, d\mu_n = \int_a^b \left[\int_{S^{n-1}} \rho^p(ru) \, d\sigma_{n-1}(u) \right] r^{n-1} \, dr \\ &= c^{-p} \int_a^b \left[\int_{S^{n-1}} r^{p(m-1)} \, d\sigma_{n-1}(u) \right] r^{n-1} \, dr \\ &= c^{-p} \omega_{n-1} \int_a^b r^{pm-1+n-1} \, dr = c^{-p} \omega_{n-1} \int_a^b r^{m-1} \, dr = c^{1-p} \omega_{n-1} . \end{aligned}$$

In view of (4.37), we obtain

$$M_p(\Gamma) = \omega_{n-1} \left(\int_a^b r^{m-1} \, dr \right)^{1-p} ,$$

which establishes (4.35) whenever the integral involved is finite (which is certainly the case when $0 < a < b < \infty$).

It remains to check (4.35) when the integral on the right is infinite, that is, when the right-hand side of (4.35) is 0, a situation that does not arise unless $a = 0$ or $b = \infty$ or both. For this, consider a_1 and b_1 satisfying $a < a_1 < b_1 < b$. The family Γ is minorized by $\Gamma_1 = \Delta(E_1, F_1 : G_1)$, with $E_1 = \mathbb{S}^{n-1}(x_0, a_1)$, $F_1 = \mathbb{S}^{n-1}(x_0, b_1)$, and $G_1 = B^n(x_0, b_1) \setminus \overline{B}^n(x_0, a_1)$. Also, $\int_{a_1}^{b_1} r^{m-1} \, dr < \infty$. By Lemma 4.2.2,

$$M_p(\Gamma) \leq M_p(\Gamma_1) = \omega_{n-1} \left(\int_{a_1}^{b_1} r^{m-1} \, dr \right)^{1-p} .$$

Letting $a_1 \to a$ and $b_1 \to b$, we learn that $M_p(\Gamma) = 0$, as desired. □

We can generalise this result using conformal invariance to obtain a useful corollary which gives us a formula for the modulus of the curve family in $\mathbb{R}^n$ which connects two disjoint balls.

COROLLARY 4.2.5. *Let $x, y \in \mathbb{R}^n$ and $r, s > 0$ with $r + s < |x-y|$. Then*

$$M\big[\bar{B}^n(x,r), \bar{B}^n(y,s); \mathbb{R}^n \setminus \{\bar{B}^n(x,r) \cup \bar{B}^n(y,s)\}\big]$$

$$(4.38)\qquad = \omega_{n-1}\left[\log \frac{\sqrt{|x-y|^2-(s-r)^2}+\sqrt{|x-y|^2-(s+r)^2}}{\sqrt{|x-y|^2-(s-r)^2}-\sqrt{|x-y|^2-(s+r)^2}}\right]^{1-n}$$

$$= \omega_{n-1}\left[2\log \frac{2\sqrt{rs}}{\sqrt{|x-y|^2-(s-r)^2}-\sqrt{|x-y|^2-(s+r)^2}}\right]^{1-n}.$$

PROOF. We are computing a conformal invariant so we may adjust the situation by a similarity. In particular we may replace x by 0 and r by 1 and rotate so $y-x$ is a positive scalar multiple of e_1. Then the two balls (in terms of the original data) are $B = B^n(0,1)$ and $B^n(\frac{|x-y|}{r}\, e_1, \frac{s}{r})$. Let $R_{\mathbb{S}}$ be inversion in the unit sphere. We put $B^* = R_{\mathbb{S}}(B^n(\frac{|x-y|}{r}\, e_1, \frac{s}{r}))$, and conformal invariance again shows us that

$$M\big[\bar{B}^n(x,r), \bar{B}^n(y,s); \mathbb{R}^n \setminus (\bar{B}^n(x,r) \cup \bar{B}^n(y,s))\big] = M\big[\bar{B}^*, \hat{\mathbb{R}}^n \setminus B, B \setminus \bar{B}^*\big].$$

As a round Euclidean ball in B^n, B^* is also a hyperbolic ball. There is Möbius transformation ϕ of B to itself so that $\phi(B^*) = B^n(0,s)$. Then

$$(4.39)\qquad M\big[\bar{B}^*, \hat{\mathbb{R}}^n \setminus B, B \setminus \bar{B}^*\big] = \omega_{n-1}\Big(\log \frac{1}{s}\Big)^{1-n}$$

by (4.36). We now need to compute s. The hyperbolic radius of $B^n(0,s)$ is $\log \frac{1+s}{1-s}$. This must be the same as that of B^*, as ϕ is an isometry. We consider the action on the reflection $R_{\mathbb{S}}$ on the e_1-axis to see that the hyperbolic diameter of B^* is the hyperbolic distance between the two (diametrically opposite on $B^* \cap \mathbb{R}\, e_1$) points

$$R_{\mathbb{S}}\Big[\big(\frac{|x-y|}{r}+\frac{s}{r}\big)e_1\Big] = \frac{r\, e_1}{|x-y|+s}, \quad \text{and} \quad R_{\mathbb{S}}\Big[\big(\frac{|x-y|}{r}-\frac{s}{r}\big)e_1\Big] = \frac{r\, e_1}{|x-y|-s}.$$

Then $\operatorname{diam}_{\mathbb{H}}(B^n(0,s)) = \operatorname{diam}_{\mathbb{H}}(B^*)$ reads as

$$\begin{aligned}
2\log\frac{1+s}{1-s} &= \int_{\frac{r}{|x-y|+s}}^{\frac{r}{|x-y|-s}} \frac{2dt}{1-t^2} = -\log\frac{1+\frac{r}{|x-y|+s}}{1-\frac{r}{|x-y|+s}} + \log\frac{1+\frac{r}{|x-y|-s}}{1-\frac{r}{|x-y|-s}} \\
&= \log\frac{(|x-y|+s-r)(|x-y|-s+r)}{(|x-y|+s+r)(|x-y|-s-r)} \\
&= \log\frac{|x-y|^2-(s-r)^2}{|x-y|^2-(s+r)^2}.
\end{aligned}$$

Then

$$(4.40)\qquad s = \frac{\sqrt{|x-y|^2-(s-r)^2}-\sqrt{|x-y|^2-(s+r)^2}}{\sqrt{|x-y|^2-(s-r)^2}+\sqrt{|x-y|^2-(s+r)^2}},$$

which together with (4.39) finally gives the equation (4.38). □

We can check Corollary 4.2.5 by considering the special case of two balls with the same radius. First note that if $r = s$, then (4.38) reads as

$$(4.41)\quad M\big[\bar{B}^n(x,r), \bar{B}^n(y,r); \hat{\mathbb{R}}^n\big] = \omega_{n-1}\left[2\log\frac{2r}{|x-y|-\sqrt{|x-y|^2-4r^2}}\right]^{1-n}.$$

Notice that as $r \to 0$, (4.41) has the asymptotics

$$M\big[\bar{B}^n(x,r), \bar{B}^n(y,r); \hat{\mathbb{R}}^n\big] \approx \omega_{n-1}\, 2^{1-n} \left[\log \frac{|x-y|}{r}\right]^{1-n}.$$

Another way to see (4.41) is to note that when the two balls have the same radius there is a hyperplane in which reflection across this hyperplane moves one to the other. After rotation, this hyperplane can be assumed to be the $\{x_n = 0\}$-plane. We may translate so $y-x$ lies on the x_n-axis. Following the earlier argument, except using the upper half-space model for hyperbolic geometry, we see that for $r < t$, $M\big[\{x_n\!=\!0\}, \bar{B}^n(te_n, r), \mathbb{H}^n\big] \!=\! \omega_{n-1}\big[\log 1/s\big]^{1-n}$, and $\log \frac{1+s}{1-s} = \frac{1}{2}\mathrm{diam}_{\mathbb{H}}\big[B^n(te_n, r)\big]$ since we can isometrically map $\mathbb{H}^n$ to $B^n(0,1)$ in their hyperbolic metrics. This latter number, with $t = |x-y|/2$, is

$$\mathrm{diam}_{\mathbb{H}}\Big[B^n\Big(\frac{|x-y|}{2}\, e_n, r\Big)\Big] = \int_{\frac{1}{2}|x-y|-r}^{\frac{1}{2}|x-y|+r} \frac{dt}{t} = \log \frac{|x-y|+2r}{|x-y|-2r} = \log \frac{1+\frac{2r}{|x-y|}}{1-\frac{2r}{|x-y|}}.$$

Hence

$$s = \frac{\sqrt{1+\frac{2r}{|x-y|}} - \sqrt{1-\frac{2r}{|x-y|}}}{\sqrt{1+\frac{2r}{|x-y|}} + \sqrt{1-\frac{2r}{|x-y|}}} = \frac{2r}{|x-y| + \sqrt{|x-y|^2 - 4r^2}}.$$

Now $M\big[\{x_n = 0\}, \bar{B}^n(te_n, r), \mathbb{H}^n\big]$ is only half of the story—accounting for the missing factor of 2—as

$$2M\big[\{x_n = 0\}, \bar{B}^n(te_n, r), \mathbb{H}^n\big] = M\big[\bar{B}^n(-te_n, r), \bar{B}^n(te_n, r), \mathbb{H}^n\big]$$

is a property of disjoint curve families that we will soon prove. Now the two formulas coincide, giving and alternate verification of the corollary in this special case.

Regrettably, the preceding examples are abberations, for it is rarely possible to compute the p-modulus of a curve family exactly. The utility of the concept lodges ultimately in one's ability to derive good bounds for $M_p(\Gamma)$. Upcoming sections will reveal some general procedures for accomplishing this. In the meantime, we must content ourselves with several elementary illustrations.

LEMMA 4.2.6. *If E is a Borel set in $\mathbb{R}^n$ with $|E| < \infty$ and Γ is a family of curves in E for which $d = \inf\{\ell(\gamma) : \gamma \in \Gamma\} > 0$, then*

$$M_p(\Gamma) \le \frac{|E|}{d^p} < \infty\,.$$

This is a direct consequence of the fact that $\rho = d^{-1}\chi_E$ is an admissible density for Γ.

LEMMA 4.2.7. *If Γ is a family of curves in $\hat{\mathbb{R}}^n$ with the property that*

$$d = \inf\{q(|\gamma|) : \gamma \in \Gamma\} > 0$$

and if $p > n/2$, then

$$M_p(\Gamma) \le \frac{2^p \omega_{n-1}}{d^p} \int_0^\infty \frac{r^{n-1}dr}{(1+r^2)^p} < \infty\,. \tag{4.42}$$

PROOF. It follows from (4.31) that

$$d \le q(|\gamma|) \le \ell_q(\gamma) = 2\int_\gamma \frac{|dx|}{1+|x|^2}$$

for every locally rectifiable member γ of Γ. Therefore, $\rho(x) = (2/d)(1+|x|^2)^{-1}$ for x in $\mathbb{R}^n$ and $\rho(\infty) = 0$ defines a function ρ belonging to $\mathrm{Adm}(\Gamma)$, which means that

$$M_p(\Gamma) \le \frac{2^p}{d^p}\int_{\mathbb{R}^n} \frac{dx}{(1+|x|^2)^p} = \frac{2^p\omega_{n-1}}{d^p}\int_0^\infty \frac{r^{n-1}\,dr}{(1+r^2)^p}\,.$$

For the integral in (4.42) to be finite it is necessary that $2p - n + 1 > 1$; this is the condition $p > n/2$. □

Lemma 4.2.7 implies, for instance, that $M_p[\Delta(E, F : \hat{\mathbb{R}}^n)] < \infty$ whenever E and F are nonempty sets in $\hat{\mathbb{R}}^n$ with $q(E, F) > 0$ and $p > n/2$. As a matter of fact, it is true that $M_p[\Delta(E, F : \hat{\mathbb{R}}^n)] < \infty$ for every $p > 1$ when $q(E, F) > 0$.

LEMMA 4.2.8. *If E and F are nonempty subsets of $\hat{\mathbb{R}}^n$ for which $q(E, F) > 0$, then $M_p[\Delta(E, F : \hat{\mathbb{R}}^n)] < \infty$ for every $p > 1$.*

PROOF. The point ∞ does not belong to both $\bar{E}$ and $\bar{F}$. Without loss of generality we may presume that $\bar{E}$ lies in $\mathbb{R}^n$. Choose r obeying the condition $0 < 2r < q(E, F)$, and let

$$G = \{x \in \mathbb{R}^n : \mathrm{dist}(x, E) < r\}\,.$$

Then G is a bounded open set that contains E. Moreover, since $q(x, y) \le 2|x - y|$ for all x and y in $\mathbb{R}^n$, the sets G and F are disjoint. It follows that $\Delta(E, F : \hat{\mathbb{R}}^n)$ is minorized by the family $\Delta(E, \partial G : G)$. Because every curve in the latter family has length no smaller than r, Lemma 4.2.2 and Lemma 4.2.7 combine to inform us that

$$M_p[\Delta(E, F : \hat{\mathbb{R}}^n)] \le M_p[\Delta(E, \partial G : G)] \le \frac{|G|}{r^p} < \infty$$

for every $p > 1$. □

4.2.2. Separated curve famlies. Although we pointed out that the outer measure M_p generally has little to offer as far as interesting measure theory goes, there are isolated instances in which some measure-like behavior on the part of M_p does appear. We describe here a property of M_p that is reminiscent of a Carathéodory outer measure.

Let $\{\Gamma_\nu\}$ be a countable collection of curve families in $\hat{\mathbb{R}}^n$ with $n \ge 2$. We say that the members of this collection are *separated* if there is an associated family of pairwise disjoint Borel sets E_ν in $\hat{\mathbb{R}}^n$ such that

$$\int_\gamma \chi_{E_\nu^c}\,ds = 0 \tag{4.43}$$

whenever γ is a locally rectifiable curve from Γ_ν. In the interest of brevity we call $\{E_\nu\}$ a *separating collection of Borel sets for* $\{\Gamma_\nu\}$. Condition (4.43) would be fulfilled, for example, should $|\gamma|$ be a subset of E_ν for every γ in Γ_ν.

We now have the following result.

THEOREM 4.2.9. *If the members of a countable collection $\{\Gamma_\nu\}$ of curve families in $\hat{\mathbb{R}}^n$ are separated, then $M_p(\bigcup_\nu \Gamma_\nu) = \sum_\nu M_p(\Gamma_\nu)$.*

PROOF. Due to the subadditivity of M_p, we only need to demonstrate that $\Gamma = \bigcup_\nu \Gamma_\nu$ has

$$M_p(\Gamma) \geq \sum_\nu M_p(\Gamma_\nu)\,. \tag{4.44}$$

We may assume that $M_p(\Gamma) < \infty$, (4.44) holding trivially otherwise.

Let ρ in $\mathrm{Adm}(\Gamma)$ be given. Set $\rho_\nu = \rho\chi_{E_\nu}$, where $\{E_\nu\}$ is a separating collection of Borel sets for $\{\Gamma_\nu\}$. Because

$$\int_\gamma \chi_{E_\nu^c}\, ds = 0$$

for every locally rectifiable γ in Γ_ν, the monotone convergence theorem allows us to conclude that

$$0 \leq \int_\gamma \rho\,\chi_{E_\nu^c}\, ds = \lim_{k\to\infty} \int_\gamma \min\{\rho, k\}\,\chi_{E_\nu^c}\, ds \leq \lim_{k\to\infty} k \int_\gamma \chi_{E_\nu^c}\, ds = 0\,.$$

Thus

$$\int_\gamma \rho\,\chi_{E_\nu^c}\, ds = 0$$

for such γ. It follows that

$$1 \leq \int_\gamma \rho\, ds = \int_\gamma \rho(\chi_{E_\nu} + \chi_{E_\nu^c})\, ds = \int_\gamma \rho_\nu\, ds + \int_\gamma \rho\,\chi_{E_\nu^c}\, ds = \int_\gamma \rho_\nu\, ds\,.$$

Thus ρ_ν is an admissible density for Γ_ν. As a result,

$$\sum_\nu M_p(\Gamma_\nu) \leq \sum_\nu \int_{\mathbb{R}^n} \rho_\nu^p\, dm_n = \sum_\nu \int_{E_\nu} \rho^p\, dm_n = \int_{\bigcup_\nu E_\nu} \rho^p\, dm_n \leq \int_{\mathbb{R}^n} \rho^p\, dm_n\,.$$

Taking the infimum over $\mathrm{Adm}(\Gamma)$ leads to (4.44). □

Lower bounds for moduli can sometimes be obtained by employing Lemma 4.2.2 and Theorem 4.2.9 in combination.

THEOREM 4.2.10. *If the members of a countable collection $\{\Gamma_\nu\}$ of curve families in $\hat{\mathbb{R}}^n$ are separated and if $\Gamma \prec \Gamma_\nu$ for every ν, then $M_p(\Gamma) \geq \sum_\nu M_p(\Gamma_\nu)$.*

PROOF. Since $\Gamma \prec \bigcup_\nu \Gamma_\nu$, the assertion is an immediate consequence of the propositions cited prior to the statement of the theorem. □

For example, let $\Gamma = \Delta(E, F : G)$. If $\Gamma_\nu = \Delta(E_\nu, F_\nu : G_\nu)$, where $G_1, G_2, \ldots$ is a (finite or infinite) sequence of pairwise disjoint Borel sets that lie in G and where E_ν is contained in E and F_ν in F for every ν, then Theorem 4.2.10 applies.

A useful source of upper estimates for moduli is the following companion to the preceding theorem.

THEOREM 4.2.11. *If the members of a countable collection $\{\Gamma_\nu\}$ of curve families in $\hat{\mathbb{R}}^n$ are separated and if $\Gamma_\nu \prec \Gamma$ for every ν, then*

$$M_p(\Gamma) \leq \left[\sum_\nu \lambda_p(\Gamma_\nu)\right]^{1-p}. \tag{4.45}$$

PROOF. We may suppose that $M_p(\Gamma) > 0$ and that $M_p(\Gamma_\nu) < \infty$ for each ν. Indeed, the content of the theorem is trivial when $M_p(\Gamma) = 0$, while any family Γ_ν for which $M_p(\Gamma_\nu) = \infty$ (that is, $\lambda_p(\Gamma_\nu) = 0$) can simply be deleted from the collection without effect on the sum in (4.45). We may further assume that $\{\Gamma_\nu\}$ is a finite collection—$\Gamma_1, \Gamma_2, \ldots, \Gamma_N$, say—for letting $N \to \infty$ in the finite version of (4.45) would establish the same estimate for an infinite collection. For $\nu = 1, 2, \ldots, N$ let ρ_ν belong to $\mathrm{Adm}(\Gamma_\nu)$. If $E_1, E_2, \ldots, E_N$ is a separating family of Borel sets for $\{\Gamma_\nu\}$, then the function $\tilde{\rho}_\nu = \rho_\nu \chi_{E_\nu}$ is another admissible density for Γ_ν (recall the proof of Theorem 4.2.9). Consider $\rho = \sum_{\nu=1}^N a_\nu \tilde{\rho}_\nu$, where the a_ν are constants satisfying $0 < a_\nu < 1$ and $\sum_{\nu=1}^N a_\nu = 1$. If γ is a locally rectifiable curve in Γ, we can select for each ν a curve γ_ν from Γ_ν that is a subcurve of γ and compute

$$\int_\gamma \rho\, ds = \sum_{\nu=1}^N a_\nu \int_\gamma \tilde{\rho}_\nu\, ds \geq \sum_{\nu=1}^N a_\nu \int_{\gamma_\nu} \tilde{\rho}_\nu\, ds \geq \sum_{\nu=1}^N a_\nu = 1\,.$$

Accordingly, the function ρ belongs to $\mathrm{Adm}(\Gamma)$.

Because the sets E_ν are pairwise disjoint, we obtain

$$\begin{aligned} M_p(\Gamma) &\leq \int_{\mathbb{R}^n} \rho^p\, dm_n = \int_{\mathbb{R}^n} \Big(\sum_{\nu=1}^N a_\nu \tilde{\rho}_\nu\Big)^p dm_n = \int_{\mathbb{R}^n} \sum_{\nu=1}^N a_\nu^p\, \tilde{\rho}_\nu^p\, dm_n \\ &= \sum_{\nu=1}^N a_\nu^p \int_{\mathbb{R}^n} \tilde{\rho}_\nu^p\, dm_n \leq \sum_{\nu=1}^N a_\nu^p \int_{\mathbb{R}^n} \rho_\nu^p\, dm_n\,. \end{aligned}$$

Taking successive infima over $\rho_1, \rho_2, \ldots$, and ρ_N, we learn that

$$M_p(\Gamma) \leq \sum_{\nu=1}^N a_\nu^p\, M_p(\Gamma_\nu)\,.$$

Finally, we make a specific choice for the numbers a_ν:

$$a_\nu = \frac{\lambda_p(\Gamma_\nu)}{\sum\limits_{\nu=1}^N \lambda_p(\Gamma_\nu)}$$

for $\nu = 1, 2, \ldots, N$. Then by the definition of p-extremal length,

$$M_p(\Gamma) \leq \frac{\sum\limits_{\nu=1}^N \lambda_p(\Gamma_\nu)^p M_p(\Gamma_\nu)}{\Big[\sum\limits_{\nu=1}^N \lambda_p(\Gamma_\nu)\Big]^p} = \frac{\sum\limits_{\nu=1}^N \lambda_p(\Gamma_\nu)}{\Big[\sum\limits_{\nu=1}^N \lambda_p(\Gamma_\nu)\Big]^p} = \Big[\sum_{\nu=1}^N \lambda_p(\Gamma_\nu)\Big]^{1-p}\,,$$

as claimed. □

To give a concrete illustration of a situation in which Theorem 4.2.11 can prove quite handy, suppose that G is a set which includes the spherical rings $R_\nu = B^n(x_0, b_\nu) \setminus \overline{B}^n(x_0, a_\nu)$, where $0 < a \leq a_1 < b_1 < a_2 < b_2 < \cdots \leq b < \infty$. By virtue of Theorems 4.2.4 and 4.2.11, it is true of the family $\Gamma = \Delta(E, F : G)$ that

$$M_p(\Gamma) \leq \omega_{n-1}\Big(\sum_\nu \int_{a_\nu}^{b_\nu} r^{m-1}\, dr\Big)^{1-p}$$

whenever E is contained in $B^n(x_0, a)$ and F is a subset of $\hat{\mathbb{R}}^n \setminus \overline{B}^n(x_0, b)$. Here, as earlier, $m = (p-n)/(p-1)$. Indeed, the conditions stated ensure that $\Gamma_\nu \prec \Gamma$ for every ν, where Γ_ν is the family $\Delta[\mathbb{S}^{n-1}(x_0, a_\nu), \mathbb{S}^{n-1}(x_0, b_\nu) : R_\nu]$.

4.2.3. The influence of nonrectifiable curves. Since the members of a curve family Γ that are not locally rectifiable form a subfamily Γ' of Γ with $M_p(\Gamma') = 0$ for every $p > 1$, it is natural to ask whether the same holds true for the subfamily of nonrectifiable curves. Or, to put the question differently:

Is $M_p(\Gamma) = M_p(\Gamma_{rect})$ for every curve family Γ in $\hat{\mathbb{R}}^n$, where Γ_{rect} denotes the family of all rectifiable curves in Γ?

When $1 < p < n$ the answer is in the negative. Theorem 4.2.4 shows, for instance, that the family $\Gamma = \Delta(\mathbb{S}^{n-1}, \{\infty\}, \mathbb{R}^n \setminus \overline{B}^n)$ has $M_p(\Gamma) = \omega_{n-1}[(n-p)/(p-1)]^{p-1} > 0$ when $1 < p < n$, whereas $M_p(\Gamma_{rect}) = 0$ as every curve in Γ is nonrectifiable.

The situation changes when $p \geq n$, as we shall now see. We note especially that the nonrectifiable curves in any curve family exert no influence on its conformal modulus, a fact of paramount importance for our approach to the subject of quasiconformal mappings.

THEOREM 4.2.12. *If $p \geq n$, then $M_p(\Gamma) = M_p(\Gamma_{rect})$ for every family Γ of curves in $\hat{\mathbb{R}}^n$.*

PROOF. Obviously $M_p(\Gamma_{rect}) \leq M_p(\Gamma)$ for a given family Γ. We must establish the truth of the reverse inequality. To this end, we introduce an auxiliary function $\rho_0 : \hat{\mathbb{R}}^n \to [0, \infty]$ as follows. We define

$$\rho_0(x) = \begin{cases} e^{-1} & \text{if } |x| \leq e\,, \\ \frac{1}{|x| \log |x|} & \text{if } e < |x| < \infty\,, \\ 0 & \text{if } x = \infty\,. \end{cases}$$

Then ρ_0 is a radial Borel function. Because $p \geq n \geq 2$, an easy computation reveals that

$$a = \int_{\mathbb{R}^n} \rho_0^p \, dm_n = e^{n-p}\Omega_n + \omega_{n-1} \int_e^\infty r^{n-p-1}(\log r)^{-p} \, dr < \infty\,.$$

Furthermore, for any locally rectifiable but nonrectifiable curve γ in $\hat{\mathbb{R}}^n$ it is the case that

$$\int_\gamma \rho_0 \, ds = \infty\,. \tag{4.46}$$

This is clear if $|\gamma|$ is bounded, since then $b = \inf\{\rho_0(x) : x \in |\gamma|\} > 0$ and

$$\int_\gamma \rho_0 \, ds \geq \int_\gamma b \, ds = b\ell(\gamma) = \infty\,.$$

Assuming that $|\gamma|$ is unbounded, we fix x_1 on $|\gamma|$ with $|x_1| = c \geq e$ and choose for $\nu = 2, 3, \ldots$ a point x_ν on $|\gamma|$ such that $|x_\nu| = \nu c$. The curve γ has a subpath γ_ν whose endpoints are x_1 and x_ν, so we see with the aid of Lemma 4.1.7 that

$$\int_\gamma \rho_0 \, ds \geq \int_{\gamma_\nu} \rho_0 \, ds \geq \int_c^{\nu c} \frac{dr}{r \log r}\,.$$

The last term tends to ∞ with ν. This shows the correctness of (4.46) for any locally rectifiable curve γ in $\hat{\mathbb{R}}^n$ whose trajectory is unbounded.

Now consider a function ρ in $\mathrm{Adm}(\Gamma_{rect})$. First note that if $\mathrm{Adm}(\Gamma_{rect})$ is empty, then $M_p(\Gamma) = M_p(\Gamma_{rect}) = \infty$. Fixing $\varepsilon > 0$, we set $\rho_\varepsilon = (\rho^p + \varepsilon^p \rho_0^p)^{1/p}$. We claim that ρ_ε is a member of $\mathrm{Adm}(\Gamma)$. For any locally rectifiable γ from Γ it is true that either γ belongs to Γ_{rect}, in which event

$$\int_\gamma \rho_\varepsilon \, ds \geq \int_\gamma \rho \, ds \geq 1 \ ,$$

or that $\ell(\gamma) = \infty$, a case in which (4.46) gives

$$\int_\gamma \rho_\varepsilon \, ds \geq \varepsilon \int_\gamma \rho_0 \, ds = \infty \, .$$

Thus ρ_ε is a function from $\mathrm{Adm}(\Gamma)$.

We infer from this that

$$M_p(\Gamma) \leq \int_{\mathbb{R}^n} \rho_\varepsilon^p \, dm_n = \int_{\mathbb{R}^n} \rho^p \, dm_n + \varepsilon^p a \, .$$

As ρ was an arbitrary member of $\mathrm{Adm}(\Gamma_{rect})$, it follows that

$$M_p(\Gamma) \leq M_p(\Gamma_{rect}) + \varepsilon^p a \, .$$

Since this holds for every $\varepsilon > 0$, $M_p(\Gamma_{rect}) \leq M_p(\Gamma)$. □

Let Γ be a family of curves in $\hat{\mathbb{R}}^n$. By modifying the proof of Theorem 4.2.12 one can check that $M_p(\Gamma) = M_p(\Gamma_{rect})$ for each $p > 1$, provided that every locally rectifiable curve in Γ has a bounded trajectory: the role of auxiliary function can then be taken over by any Borel function $\rho_0 : \hat{\mathbb{R}}^n \to [0, \infty)$ that belongs to $L^p(\mathbb{R}^n)$ and has $\inf_K \rho_0 > 0$ for each compact set K in $\mathbb{R}^n$. Thus it is not the breakdown of rectifiability per se, but rather a special kind of nonrectifiability (the presence in Γ of curves with unbounded trajectories) that appears to cause problems when $1 < p < n$. However, even unbounded trajectories can sometimes be tolerated in this range of p. For example, if Γ is a family of curves in a Borel set E in $\mathbb{R}^n$ with $|E| < \infty$, then $M_p(\Gamma) = M_p(\Gamma_{rect})$ for all $p > 1$. In this case $\rho_0 = \chi_E$ can serve as the auxiliary function in the proof of Theorem 4.2.12.

One consequence of Theorem 4.2.12, Lemma 4.2.2, and Theorem 4.2.4 that will later prove to be of considerable value is the following.

THEOREM 4.2.13. *Let Γ be a family of nonconstant curves in $\hat{\mathbb{R}}^n$. Suppose there is a point x_0 of $\hat{\mathbb{R}}^n$ that belongs to $\bar{\gamma}$ for every γ in Γ. If $x_0 = \infty$, then $M_p(\Gamma) = 0$ whenever $p \geq n$; if $x_0 \neq \infty$, then $M_p(\Gamma) = 0$ whenever $1 < p \leq n$.*

PROOF. If $x_0 = \infty$, then Theorem 4.2.12 tells us immediately that $M_p(\Gamma) = M_p(\Gamma_{rect}) = M_p(\emptyset) = 0$ for $p \geq n$. Assuming next that x_0 is a finite point and that $1 < p \leq n$, let $\Gamma_\nu = \{\gamma \in \Gamma : |\gamma| \cap S^{n-1}(x_0, 1/\nu) \neq \emptyset\}$ for $\nu = 1, 2, \ldots$. Because each curve γ in Γ is nonconstant and has x_0 in $\bar{\gamma}$, $\Gamma = \bigcup_{\nu=1}^\infty \Gamma_\nu$. Therefore $M_p(\Gamma) \leq \sum_{\nu=1}^\infty M_p(\Gamma_\nu)$. We shall be finished with the proof if we can verify that $M_p(\Gamma_\nu) = 0$ for each ν. Fix ν. The fact that x_0 belongs to $\bar{\gamma}$ for every γ in Γ implies that Γ_ν is minorized by the family

$$\Gamma_\nu^\varepsilon = \Delta\big[\mathbb{S}^{n-1}(x_0, \varepsilon), \ \mathbb{S}^{n-1}(x_0, 1/\nu) : B^n(x_0, 1/\nu) \setminus \overline{B}^n(x_0, \varepsilon)\big]$$

whenever $0 < \varepsilon < \nu^{-1}$. By Theorem 4.2.4 and Lemma 4.2.2,

$$M_p(\Gamma_\nu) \le M_p(\Gamma_\nu^\varepsilon) = \omega_{n-1}\left(\int_\varepsilon^{1/\nu} r^{m-1}\,dr\right)^{1-p},$$

where $m = (p-n)/(p-1) < 0$. Now $m - 1 < -1$, so $\int_0^{1/\nu} r^{m-1}\,dr = \infty$. Consequently,

$$M_p(\Gamma_\nu) \le \lim_{\varepsilon\to 0} M_p(\Gamma_\nu^\varepsilon) = 0\,.$$

The conclusion announced by the theorem when $x_0 \ne \infty$ now follows. □

If $\Gamma = \Delta(\mathbb{S}^{n-1}, \{\infty\} : \mathbb{R}^n \setminus \overline{B}^n)$, then

$$M_p(\Gamma) = \omega_{n-1}[(n-p)/(p-1)]^{p-1} > 0$$

whenever $1 < p < n$, even though ∞ belongs to $\bar{\gamma}$ for every γ in Γ.

If $\Gamma = \Delta(\mathbb{S}^{n-1}, \{0\} : B^n \setminus \{0\})$, then

$$M_p(\Gamma) = \omega_{n-1}[(p-n)/(p-1)]^{p-1} > 0$$

for $p > n$ despite the fact that 0 is in $\bar{\gamma}$ for every γ in Γ. We infer that Theorem 4.2.13 is not, in general, subject to improvement.

COROLLARY 4.2.14. *Let Γ be a family of nonconstant curves in $\hat{\mathbb{R}}^n$. If $\bigcap_{\gamma\in\Gamma}\bar{\gamma}$ is not empty, then $M(\Gamma) = 0$. More generally, if there is a countable subset E of $\hat{\mathbb{R}}^n$ such that $\bar{\gamma}\cap E$ is nonempty for every γ in Γ, then $M(\Gamma) = 0$.*

Corollary 4.2.14 ensures that any countable family Γ of nonconstant curves in $\hat{\mathbb{R}}^n$ must have $M(\Gamma) = 0$. A second, somewhat technical, application of Theorem 4.2.12 will enable us to pass back and forth between curve families that are characterized by different "endpoint requirements".

THEOREM 4.2.15. *Let E, F, and G be nonempty subsets of $\hat{\mathbb{R}}^n$, with E and F disjoint. If $p \ge n$, then*

$$M_p[\Delta(E, F : G)] = M_p[\Delta_0(E, F : G)] = M_p[\bar{\Delta}(E, F : G)].$$

PROOF. Write $\Gamma = \Delta(E, F : G)$, $\Gamma_0 = \Delta_0(E, F : G)$, and $\bar{\gamma} = \bar{\Delta}(E, F : G)$. There is no harm in assuming that Γ_0 is nonempty: if $\Gamma_0 = \emptyset$, then $\Gamma = \bar{\gamma} = \emptyset$ and all the moduli in question are 0. Since Γ is a subfamily of Γ_0, $M_p(\Gamma) \le M_p(\Gamma_0)$. To establish the opposite inequality, we shall check that Γ minorizes $(\Gamma_0)_{rect}$. If so, Lemma 4.2.2 and Theorem 4.2.12 combine to inform us that $M_p(\Gamma_0) = M_p[(\Gamma_0)_r] \le M_p(\Gamma)$. Let $\gamma : (a, b) \to G$ be a rectifiable member of Γ_0. According to Lemma 4.1.2, γ admits an extension to a rectifiable curve $\gamma^* : [a, b] \to \hat{\mathbb{R}}^n$. Because $|\gamma^*| = \bar{\gamma}$ meets both of the disjoint sets E and F, there exist c and d with $a \le c < d \le b$ such that one of the points $\gamma^*(c)$ or $\gamma^*(d)$ lies in E and the other in F. Thus $\beta = \gamma|(c, d)$ belongs to Γ and hence $\Gamma \prec (\Gamma_0)_{rect}$.

Due to the fact that $\Gamma \prec \bar{\gamma}$, $M_p(\bar{\gamma}) \le M_p(\Gamma)$. To finish the proof, we must demonstrate that $M_p(\Gamma) \le M_p(\bar{\gamma})$. We assume that $M_p(\bar{\gamma}) < \infty$ and consider ρ from $\mathrm{Adm}(\bar{\gamma})$. Let $\gamma : (a, b) \to \hat{\mathbb{R}}^n$ be a rectifiable curve in Γ. Its extension $\gamma^* : [a, b] \to \hat{\mathbb{R}}^n$ is rectifiable with $\ell(\gamma^*) = \ell(\gamma)$. Of course, γ^* may not qualify as member of $\bar{\gamma}$, for it might happen that $a = -\infty$ or $b = \infty$, in which event γ^* would not technically fit our definition of a path. We are, however, free to choose a homeomorphism $h : [0, 1] \to [a, b]$ with $h(0) = a$ and $h(1) = b$ and to define a path $\tilde{\gamma}$ by $\tilde{\gamma} = \gamma^* \circ h$. The path $\tilde{\gamma}$ is a rectifiable member of $\bar{\gamma}$. Furthermore, h sets up a one-to-one correspondence between the subpaths $\tilde{\beta}$ of the curve $\alpha = \tilde{\gamma}|(0,1)$ and the subpaths β of γ, $\tilde{\beta} = \alpha|[c,d]$ corresponding to $\beta = \gamma|[h(c), h(d)]$. Clearly $\tilde{\beta}$ arises from β via a change of parameter, so

$$\int_\gamma \rho\, ds = \sup_\beta \int_\beta \rho\, ds = \sup_{\tilde{\beta}} \int_{\tilde{\beta}} \rho\, ds = \int_\alpha \rho\, ds = \int_{\alpha^*} \rho\, ds = \int_{\tilde{\gamma}} \rho\, ds \geq 1,$$

where we recall (4.28).

We have just verified that ρ is in $\mathrm{Adm}(\Gamma_{rect})$. Therefore,

$$M_p(\Gamma) = M_p(\Gamma_{rect}) \leq \int_{\mathbb{R}^n} \rho^p\, dm_n\,.$$

But ρ was an arbitrary admissible density for $\bar{\gamma}$, and hence $M_p(\Gamma) \leq M_p(\bar{\gamma})$. □

In conjunction with the comments made subsequent to the proof of Theorem 4.2.12, the proof given for Theorem 4.2.15 shows that its conclusions sometimes remain valid for every $p > 1$. This is the case, for example, if G is a Borel set in $\mathbb{R}^n$ with $|G| < \infty$. For the same reasons it is true that $M_p[\Delta(E, F : G)] = M_p[\bar{\Delta}(E, F : G)]$ for all $p > 1$ whenever the sets E and F lie in $\mathbb{R}^n$: in this instance all the locally rectifiable curves involved have bounded trajectories.

Having gained access to the material in the present section, we are in a position to refine and expand upon the discussion with which we opened this chapter.

THEOREM 4.2.16. *Let $f : D \to D'$ be a homeomorphism between domains D and D' in $\hat{\mathbb{R}}^n$ with $n \geq 2$. Suppose that the restriction of f to*

$$D_0 = \{x \in D : x \neq \infty,\ f(x) \neq \infty\}$$

is a quasiconformal diffeomorphism. Then

$$\frac{1}{K_O}\, M(\Gamma) \leq M\big[f(\Gamma)\big] \leq K_I\, M(\Gamma) \tag{4.47}$$

for every family Γ of curves in D, where $K_O = K_O(f|D_0)$ and $K_I = K_I(f|D_0)$.

PROOF. Let Γ be a family of curves in D. We verify the first inequality in (4.47); as usual the second follows by applying the first to f^{-1}.

Write $D_0' = f(D_0)$, $\tilde{\Gamma} = f(\Gamma)$, $\Gamma_0 = \{\gamma \in \Gamma : |\gamma| \subset D_0\}$, and $\tilde{\Gamma}_0 = \{\beta \in \tilde{\Gamma} : |\beta| \subset D_0'\} = f(\Gamma_0)$. In proving that $M(\Gamma) \leq K_O M(\tilde{\Gamma})$ we may presume that $M(\tilde{\Gamma}) < \infty$. This means, in particular, that neither Γ nor $\tilde{\Gamma}$ contains a constant curve. By Corollary 4.2.14, $M(\Gamma \setminus \Gamma_0) = 0$ as every curve γ in $\Gamma \setminus \Gamma_0$ passes through either ∞ or $f^{-1}(\infty)$, which implies that $M(\Gamma) = M(\Gamma_0)$. Similarly, $M(\tilde{\Gamma}) = M(\tilde{\Gamma}_0)$. Therefore, to complete the proof it is enough to verify that

$$M(\Gamma_0) \leq K_O\, M(\tilde{\Gamma}_0)\,. \tag{4.48}$$

Given $\tilde{\rho}$ in $\mathrm{Adm}(\tilde{\Gamma}_0)$, we define $\rho : \hat{\mathbb{R}}^n \to [0, \infty]$ by

$$\rho(x) = \begin{cases} \tilde{\rho}[f(x)] L_f(x) & \text{if } x \in D_0, \\ 0 & \text{if } x \notin D_0. \end{cases}$$

Then ρ is a Borel function. If γ is a locally rectifiable curve from Γ_0, then f, being of class C^1 in D_0, is locally absolutely continuous on γ. This ensures that the curve $f \circ \gamma$ is locally rectifiable, whence by Theorem 4.1.6

$$1 \leq \int_{f\circ\gamma} \tilde{\rho}\, ds \leq \int_\gamma (\tilde{\rho} \circ f)\, L_f\, ds = \int_\gamma \rho\, ds\,.$$

This tells us that ρ is an admissible density for Γ_0.

Now

$$L_f^n(x) = \|f'(x)\|^n \le K_O |J_f(x)|$$

for every x in D_0, so by Theorem 4.0.11

$$\begin{aligned} M(\Gamma_0) &\le \int_{\mathbb{R}^n} \rho^n \, dm_n = \int_{D_0} \rho^n \, dm_n = \int_{D_0} (\tilde{\rho} \circ f)^n \, L_f^n \, dm_n \\ &\le K_O \int_{D_0} (\tilde{\rho} \circ f)^n |J_f| \, dm_n = K_O \int_{D_0'} \tilde{\rho}^n \, dm_n \le K_O \int_{R^n} \tilde{\rho}^n \, dm_n \, . \end{aligned}$$

Taking the infimum over $\tilde{\rho}$ produces inequality (4.48). □

One special case of Theorem 4.2.16 deserves a separate statement.

COROLLARY 4.2.17. *Let $f : D \to D'$ be a homeomorphism between domains D and D' in $\hat{\mathbb{R}}^n$ with $n \ge 2$. Suppose that the restriction of f to D_0 (as above) is a conformal diffeomorphism. Then*

$$M(\Gamma) = M\big(f(\Gamma)\big)$$

for every family Γ of curves in D.

We shall learn later that the conclusion of Corollary 4.2.17 remains valid if $f|D_0$ is conformal, but not a priori of class C^1. Thus $M(\Gamma)$ is truly a conformal invariant. The same cannot be said about $M_p(\Gamma)$ for $p \ne n$. To see this, consider the inversion R in $\mathbb{S}^{n-1}$ and the curve family $\Gamma = \Delta(\mathbb{S}^{n-1}, \{\infty\} : R^n \setminus \overline{B}^n)$. Then $R(\Gamma) = \Delta(\mathbb{S}^{n-1}, \{0\} : B^n \setminus \{0\})$. If $p > n$, $M_p(\Gamma) = 0$, whereas $M_p[R(\Gamma)] > 0$; if $1 < p < n$, $M_p(\Gamma) > 0$, whereas $M_p[R(\Gamma)] = 0$.

The p-modulus does behave nicely, however, under similarity transformations.

THEOREM 4.2.18. *Let $n \ge 2$ and let $1 < p < \infty$. If f is a similarity transformation of $\mathbb{R}^n$ with dilation factor λ, then $M_p\big(f(\Gamma)\big) = \lambda^{n-p} M_p(\Gamma)$ for every family Γ of curves in $\hat{\mathbb{R}}^n$.*

PROOF. The key point here is that $\ell_f(x) = L_f(x) = \|f'(x)\| = \lambda$ and $|J_f(x)| = \lambda^n$ for every x in $\mathbb{R}^n$. Let Γ be a family of curves in $\hat{\mathbb{R}}^n$, and let $\tilde{\Gamma} = f(\Gamma)$. It suffices to show that $M_p(\Gamma) \le \lambda^{p-n} M_p(\tilde{\Gamma})$—the inequality in the other direction follows by considering f^{-1}—under the assumption that $M_p(\tilde{\Gamma}) < \infty$. If $\tilde{\rho}$ is a member of $\mathrm{Adm}(\tilde{\Gamma})$, we define a function ρ by $\rho(x) = \lambda \tilde{\rho}(f(x))$. Since f is a Lipschitz mapping of $\mathbb{R}^n$ that fixes ∞ and since the condition $M_p(\tilde{\Gamma}) < \infty$ prevents any curve in Γ from having trajectory $\{\infty\}$, f is locally absolutely continuous on every locally rectifiable curve γ in Γ.

By Theorem 4.1.6,

$$1 \le \int_{f \circ \gamma} \tilde{\rho} \, ds \le \int_\gamma (\tilde{\rho} \circ f) \, L_f \, ds = \int_\gamma (\tilde{\rho} \circ f) \lambda \, ds = \int_\gamma \rho \, ds \, .$$

This shows that ρ is admissible for Γ, from which we infer that

$$\begin{aligned} M_p(\Gamma) &\le \int_{\mathbb{R}^n} \rho^p \, dm_n = \lambda^{p-n} \int_{\mathbb{R}^n} (\tilde{\rho} \circ f)^p \lambda^n \, dm_n \\ &= \lambda^{p-n} \int_{\mathbb{R}^n} (\tilde{\rho} \circ f)^p |J_f| \, dm_n = \lambda^{p-n} \int_{\mathbb{R}^n} \tilde{\rho}^p \, dm_n \, , \end{aligned}$$

and hence that $M_p(\Gamma) \le \lambda^{p-n} M_p(\tilde{\Gamma})$. □

We note, in particular, that $M_p(\Gamma) = M_p\big(f(\Gamma)\big)$ for every p in $(1, \infty)$ and every curve family Γ in $\hat{\mathbb{R}}^n$, whenever f belongs to the Euclidean group $\boldsymbol{E}(n)$.

4.2.4. The cap inequality. In this section we derive an inequality for the modulus of an important specific type of curve family which will be useful later on. The point of this section is that using this modulus inequality we will achieve *lower bounds* on the modulus of all curve families given fairly general geometric information. It is often quite straightforward to obtain upper bounds and we have already seen examples above. One could for instance identify an admissible function or one could identify a special family of shorter curves. To obtain lower bounds, one must be able to say something generic about all admissible functions. This is not so easy as we shall see.

By an *open* (respectively, a *closed*) *cap* of a Euclidean sphere $S = \mathbb{S}^{n-1}(x_0, r)$ with $n \geq 2$ we mean a nonempty set C of the form $C = S \cap H$, where H is an open (respectively, a closed) half-space in $\mathbb{R}^n$. This definition allows for the eventuality that $C = S$.

Our next objective is the derivation of a lower bound for a class of integrals of the general type

$$\int_C \rho^p \, d\sigma_{n-1},$$

an estimate that will pay dividends in the form of sharp lower bounds for the moduli of certain curve families.

THEOREM 4.2.19. *Let C be an open cap of a Euclidean sphere S of radius r in $\mathbb{R}^n$, and let x and y be different points of $\bar{C}$. If a Borel function $\rho : C \to [0, \infty]$ enjoys the property that*

$$\int_\alpha \rho \, ds \geq 1$$

whenever α is an open circular arc in C whose endpoints are x and y, then

$$\int_C \rho^p \, d\sigma_{n-1} \geq b(n,p)\, r^{n-p-1} \tag{4.49}$$

for each $p > n - 1$, where

$$b(n,p) = 2^{n-2p-1}\omega_{n-2}\left[\int_0^\infty t^{(2-n)/(p-1)} \; (1+t^2)^{(n-p-1)/(p-1)} \, dt\right]^{1-p}. \tag{4.50}$$

PROOF. No loss of generality is incurred by supposing that $S = \mathbb{S}^{n-1}$, $x = e_n$, and $y = \pi(ce_1)$ for some $c \geq 0$, where $\pi : \hat{\mathbb{R}}^{n-1} \to S^{n-1}$ is a stereographic projection. If not, we could select a similarity transformation g of $\mathbb{R}^n$ that transforms $\mathbb{S}^{n-1}$ to S, e_n to x, and y_0 to y, where y_0 is a point of the form $y_0 = \pi(ce_1)$ with $c \geq 0$. Then $\ell_g(z) = L_g(z) = \|g'(z)\| = r$ would be true for every z in $\mathbb{R}^n$. The set $C_0 = g^{-1}(C)$ would be an open cap of $\mathbb{S}^{n-1}$ with e_n and y_0 in its closure, and the Borel function $\rho_0 : C_0 \to [0, \infty]$ defined by $\rho_0(z) = r\rho[g(z)]$ would have by Theorem 4.1.6,

$$\int_A \rho_0 \, ds = r \int_A \rho \circ g \, ds = \int_A (\rho \circ g)\, L_g \, ds = \int_{g(A)} \rho \, ds \geq 1$$

for every open circular arc A in C_0 with endpoints e_n and y_0.

If (4.49) were known to hold in the normalized situation, then it would follow that

$$\begin{aligned} b(n,p) &\leq \int_{C_0} \rho_0^p \, d\sigma_{n-1} = \int_{C_0} (\rho^p \circ g)\, r^p \, d\sigma_{n-1} \\ &= r^{p-n+1} \int_{C_0} (\rho^p \circ g)\, r^{n-1} \, d\sigma_{n-1} = r^{p-n+1} \int_C \rho^p \, d\sigma_{n-1} \,. \end{aligned}$$

In other words, (4.49) would hold in the general case.

We therefore proceed assuming that $S = \mathbb{S}^{n-1}$, that $x = e_n$ and $y = \pi(ce_1)$ with $c \geq 0$, and temporarily, that $n \geq 3$. Let R be the reflection in the sphere $\mathbb{S}^{n-1}(e_n, \sqrt{2})$. We remind the reader that $R = R^{-1}$ and that $R|\hat{\mathbb{R}}^{n-1}$ gives the stereographic projection π. Since $\infty = R(e_n)$ belongs to $R(\bar{C})$, the set $E = \hat{\mathbb{R}}^{n-1} \setminus R(C)$ is (depending upon the exact nature and location of C) the empty set, a point, a closed Euclidean ball in $\mathbb{R}^{n-1}$, or a closed half-space in $\hat{\mathbb{R}}^{n-1}$.

By assumption, $z_0 = R(y) = ce_1$ is not an interior point of E, so there must be an open hemisphere of $\mathbb{S}^{n-2}$—choose one and call it K—with the feature that $z_0 + tu$ lies in $R(C)$ whenever u comes from K and $t > 0$. Then for each u in K the injective C^∞-path $\gamma_u : (0, \infty) \to \mathbb{R}^n$ defined by $\gamma_u(t) = R(z_0 + tu)$ parameterizes an open circular arc α_u in C with endpoints $\lim_{t\to\infty} \gamma_u(t) = R(\infty) = e_n$ and $\lim_{t \to 0} \gamma_u(t) = R(z_0) = y$. As R is a conformal diffeomorphism of $\mathbb{R}^n \setminus \{e_n\}$ onto itself and has $\|R'(z)\| = 2(1 + |z|^2)^{-1}$ for z in $\mathbb{R}^{n-1}$, we find that

$$|\dot{\gamma}_u(t)| = \|R'(z_0 + tu)\| \cdot |u| = \frac{2}{1 + |z_0 + tu|^2} \,.$$

From the information about ρ supplied in the hypotheses we extract the inequality

$$1 \leq \int_{A_u} \rho \, ds = 2 \int_0^\infty \frac{\rho[R(z + tu)]\, dt}{1 + |z_0 + tu|^2}$$

for each u in K. By integrating both sides of this inequality over K with respect to surface measure σ_{n-2}, we are led to the estimate

$$\begin{aligned} \frac{\omega_{n-2}}{2} &\leq 2 \int_K \left\{ \int_0^\infty \frac{\rho[R(z_0 + tu)]\, dt}{1 + |z_0 + tu|^2} \right\} d\sigma_{n-2}(u) \\ &= 2 \int_0^\infty \left\{ \int_K \frac{\rho[R(z_0 + tu)] t^{n-2} \, d\sigma_{n-2}(u)}{t^{n-2}(1 + |z_0 + tu|)^2} \right\} dt \\ &= 2 \int_0^\infty \left\{ \int_{H \cap S^{n-2}(z_0, t)} \frac{\rho[R(z)]\, d\sigma_{n-2}(z)}{|z - z_0|^{n-2}(1 + |z|^2)} \right\} dt \\ &= 2 \int_H \frac{\rho[R(z)]\, dm_{n-1}(z)}{|z - z_0|^{n-2}(1 + |z|^2)} \,, \end{aligned}$$

where $H = \{z_0 + tu : u \in K, \, t > 0\}$. Next, we want to use Hölder's inequality to isolate a term of the form

$$\int_H \frac{\rho^p[R(z)]\, dm_{n-1}(z)}{(1 + |z|^2)^{n-1}}$$

so that we can apply (4.13)—the form of spherical integration — to the function $f = \rho^p \chi_C$. To do this we write the integrand as

$$\frac{\rho[R(z)]}{|z - z_0|^{n-2}(1 + |z|^2)} = \frac{\rho[R(z)]}{(1 + |z|^2)^{(n-1)/p}} \frac{(1 + |z|^2)^{(n-1)/(p-1)}}{|z - z_0|^{n-2}}$$

and the Hölder conjugate to p is $q = p/(p-1)$. Then

$$\begin{aligned}
\left(\frac{\omega_{n-2}}{4}\right)^p &\le \left[\int_H \frac{\rho[R(z)]}{(1+|z|^2)^{(n-1)/p}} \times \frac{(1+|z|^2)^{(n-1)/p-1}}{|z-z_0|^{n-2}}\, dm_{n-1}(z)\right]^p \\
&\le \left[\int_H \frac{\rho^p[R(z)]\, dm_{n-1}(z)}{(1+|z|^2)^{n-1}}\right] \\
&\quad \times \left[\int_H |z-z_0|^{p(2-n)/(p-1)}\, (1+|z|^2)^{(n-1-p)/(p-1)}\, dm_{n-1}(z)\right]^{p-1} \\
&\le \left[\int_{R(C)} \frac{\rho^p[R(z)]\, dm_{n-1}(z)}{(1+|z|^2)^{n-1}}\right] \\
&\quad \times \left[\int_{\mathbb{R}^{n-1}} |z-z_0|^{p(2-n)/(p-1)} (1+|z|^2)^{(n-1-p)/(p-1)}\, dm_{n-1}(z)\right]^{p-1} \\
&= \left[2^{1-n}\int_C \rho^p\, d\sigma_{n-1}\right] \times \left[J(z_0)\right]^{p-1},
\end{aligned}$$

in which we have introduced the notation

$$J(z_0) = \int_{\mathbb{R}^{n-1}} |z-z_0|^{p(2-n)/(p-1)}\, \left(1+|z|^2\right)^{(n-1-p)/(p-1)}\, dm_{n-1}(z)$$

for $z_0 = ce_1$ with $c \ge 0$. Consequently,

$$\int_C \rho^p\, d\sigma_{n-1} \ge 2^{n-2p-1}\omega_{n-2}^p J(z_0)^{1-p}. \tag{4.51}$$

All that remains for us to do is to find the appropriate upper bound for $J(z_0)$. We can express any z in $\mathbb{R}^{n-1}$ uniquely in the manner $z = te_1 + \zeta$ with t a real number and ζ in P, the hyperplane in $\mathbb{R}^{n-1}$ that passes through the origin and has normal e_1. For the purpose of integration we shall think of P as $\mathbb{R}^{n-2}$. Because $z_0 = ce_1$, we can use Fubini's theorem to write

$$\begin{aligned}
J(z_0) &= \int_{\mathbb{R}^{n-1}} |z-z_0|^{p(2-n)/(p-1)}\, \left(1+|z|^2\right)^{(n-1-p)/(p-1)}\, dm_{n-1}(z) \\
&= \int_{\mathbb{R}^{n-2}} \left[\int_{-\infty}^{\infty} |\zeta+(t-c)e_1|^{\frac{p(2-n)}{p-1}}\, (1+|\zeta+te_1|^2)^{\frac{n-1-p}{p-1}}\, dt\right] dm_{n-2}(\zeta) \\
&= \int_{\mathbb{R}^{n-2}} \left[\int_{-\infty}^{\infty} [|\zeta|^2+(t-c)^2]^{\frac{p(2-n)}{2(p-1)}} (1+|\zeta|^2+t^2)^{\frac{n-1-p}{p-1}}\, dt\right] dm_{n-2}(\zeta) \\
&= \int_{\mathbb{R}^{n-2}} I(c,|\zeta|)\, dm_{n-2}(\zeta).
\end{aligned}$$

Here $I(v,w)$ is the quantity defined for $v \ge 0$ and $w \ge 0$ by

$$I(v,w) = \int_{-\infty}^{\infty} \left[w^2+(t-v)^2\right]^{\frac{p(2-n)}{2(p-1)}}\, \left(1+w^2+t^2\right)^{\frac{n-1-p}{p-1}}\, dt.$$

By a result of Hardy, Littlewood, and Pólya [**70**], $I(v,w) \le I(0,w)$. This implies that

$$\begin{aligned} J(z_0) &= \int_{\mathbb{R}^{n-2}} I\big(c,|\zeta|\big)\, dm_{n-2}(\zeta) \le \int_{\mathbb{R}^{n-2}} I\big(0,|\zeta|\big)\, dm_{n-2}(\zeta) \\ &= \int_{\mathbb{R}^{n-1}} |\zeta|^{p(2-n)/(p-1)} \,\big(1+|\zeta|^2\big)^{(n-1-p)/(p-1)}\, dm_{n-1}(\zeta) \\ &= \omega_{n-2}\int_0^\infty r^{p(2-n)/(p-1)}\,\big(1+r^2\big)^{(n-1-p)/(p-1)}\, r^{n-2}\, dr \\ &= \omega_{n-2}\int_0^\infty r^{(2-n)/(p-1)}\,\big(1+r^2\big)^{(n-1-p)/(p-1)}\, dr\,. \end{aligned}$$

This integral is finite as long as the exponent of r is smaller than -1 for large r and bigger than -1 for small r. This follows from the observation that

$$\frac{2-n}{p-1} > -1, \qquad \frac{2-n}{p-1} + 2\left(\frac{n-1-p}{p-1}\right) = \frac{n-2}{p-1} - 2 < -1$$

when $p > n-1$. On the basis of inequality (4.51), we arrive at (4.49) with

$$b(n,p) = 2^{n-2p+1}\omega_{n-2}\left(\int_0^\infty r^{(2-n)/(p-1)}\,\big(1+r^2\big)^{(n-1-p)/(p-1)}\, dr\right)^{p-1}$$

when $n \ge 3$.

Finally, let $n=2$. In this case, either C is an open arc on $\mathbb{S}^1$ or $C = \mathbb{S}^1$. If we let α denote an open subarc of C whose endpoints are e_2 and y, then

$$1 \le \int_\alpha \rho\, ds \le \int_C \rho\, ds = \int_C \rho\, d\sigma_1\,,$$

so Hölder's inequality leads to

$$1 \le \int_C \rho^p\, d\sigma_1 \cdot \left(\int_C d\sigma_1\right)^{p-1} \le (2\pi)^{p-1}\int_C \rho^p\, d\sigma_1$$

for $p > 1$. We have thus confirmed (4.49) with $b(2,p) = (2\pi)^{1-p}$. Moreover, since we have agreed to set $\omega_0 = 2$, formula (4.50) describes the constant $b(n,p)$ for all $n \ge 2$ and $p > n-1$. □

The numbers $b(n,p)$. These numbers will come up a few times in what follows and so we say a few words about them. We first consider the integral

$$I_{n,p} = \int_0^\infty r^{(2-n)/(p-1)}\big(1+r^2\big)^{(n-1-p)/(p-1)}\, dr.$$

This integral has a closed form expression in terms of the usual Gamma function as

$$I_{n,p} = \frac{1}{2}\,\frac{\Gamma\left[\frac{1-n+p}{2(p-1)}\right]^2}{\Gamma\left[\frac{1-n+p}{p-1}\right]}, \qquad p > n-1.$$

The most common use of these numbers is when $p = n$, associated with the conformal modulus. Then

$$I_{n,n} = \frac{1}{2}\,\frac{\Gamma\left[\frac{1}{2(n-1)}\right]^2}{\Gamma\left[\frac{1}{n-1}\right]}, \qquad I_{2,2} = \frac{\pi}{2}, \qquad I_{3,3} = \frac{\Gamma\left[\frac{1}{4}\right]^2}{2\sqrt{\pi}} \approx 3.62561.$$

Also

$$\lim_{n\to\infty} \frac{1}{n}\, I_{n,n} = 2.$$

Then, for large n

$$b(n,n) = \frac{\omega_{n-2}}{2^{n+1}}\, I_{n,n} \approx \frac{n\,\omega_{n-2}}{2^n}.$$

Under closer scrutiny, the proof of Theorem 4.2.19 is seen in the special case $C = S$ to yield a lower bound for $\int_C \rho^p\, d\sigma_{n-1}$ that is substantially better than the one in (4.49).

THEOREM 4.2.20. *Let S be a Euclidean sphere of radius r in $\mathbb{R}^n$, and let x and y be different points of S. If a Borel function $\rho : S \to [0,\infty]$ has the property that $\int_\alpha \rho\, ds \geq 1$ whenever A is a circular arc on S whose endpoints are x and y, then*

$$\int_S \rho^p\, d\sigma_{n-1} \geq 2^p b(n,p)\, r^{n-p-1} \tag{4.52}$$

for $p > n-1$, with $b(n,p)$ given by (4.50)*. Furthermore, the lower bound in* (4.52) *is sharp.*

PROOF. At one point in the proof of Theorem 4.2.19 for the case $n \geq 3$ we were forced to pick a hemisphere K of $\mathbb{S}^{n-2}$ such that $z_0 + tu$ belonged to $R(C)$ for every u in K and $t > 0$. When $C = S$ no such choice is necessary: we automatically have $z_0 + tu$ in $R(\mathbb{S}^{n-1})$ for each u in $\mathbb{S}^{n-2}$ and $t > 0$. Whereas in the proof of Theorem 4.2.19 we integrated a certain inequality over K, in the present situation we can do the same integration over all of $\mathbb{S}^{n-2}$. This change accounts for the improved bound when $n \geq 3$.

If $n = 2$, the points e_2 and y partition S^1 into a pair of circular arcs α_1 and α_2. Therefore

$$\int_{S^1} \rho\, d\sigma_1 = \int_{\alpha_1} \rho\, ds + \int_{\alpha_2} \rho\, ds \geq 2,$$

and again the original proof adjusts the lower bound in (4.49) accordingly.

The lower bound in (4.52) is actually attained for some function ρ_0 in the case where x and y are antipodal points of S. For instance, when $S = \mathbb{S}^{n-1}$, $x = e_n$, and $y = -e_n$, we shall check that equality holds in (4.52) for ρ_0 defined as follows. Assuming that $n \geq 3$, we let R be the reflection that entered into the proof of Theorem 4.2.19 and set $\rho_0 = \varphi \circ R$, in which $\varphi : \hat{\mathbb{R}}^{n-1} \to [0,\infty]$ is the function defined by

$$\varphi(z) = a|z|^{2-n/p-2} \left(1 + |z|^2\right)^{n-2/p-1}$$

for z in $\mathbb{R}^n$ and $\varphi(\infty) = 0$, with

$$a = \left[2\int_0^\infty r^{(2-n)/(p-1)} \left(1+r^2\right)^{(n-p-1)/(p-1)}\, dr\right]^{-1}.$$

If γ is any locally rectifiable member of the family $\Delta(\{e_n\},\{-e_n\}:\mathbb{S}^{n-1})$, then by Lemma 4.1.7

$$\begin{aligned}\int_\gamma \rho_0\,ds &= \int_{R\circ\gamma}(\rho_0\circ R)\|R'\|\,ds = \int_{R\circ\gamma}\varphi\|R'\|\,ds\\ &\geq 2a\int_0^\infty r^{(2-n)/(p-1)}\big(1+r^2\big)^{(n-2)/(p-1)}\big(1+r^2\big)^{-1}\,dr\\ &= 2a\int_0^\infty r^{(2-n)/(p-1)}\big(1+r^2\big)^{(n-p-1)/(p-1)}\,dr = 1\,.\end{aligned}$$

In particular, $\int_\alpha \rho_0\,ds \geq 1$ for any circular arc α on $\mathbb{S}^{n-1}$ with endpoints e_n and $-e_n$—equality actually holds in this case. On the other hand,

$$\begin{aligned}&\int_{\mathbb{S}^{n-1}}\rho_0\,d\sigma_{n-1}\\ &= 2^{n-1}\int_{\mathbb{R}^{n-1}}\frac{\rho_0[\pi(z)]\,dm_{n-1}(z)}{(1+|z|^2)^{n-1}}\\ &= a^p2^{n-1}\int_{\mathbb{R}^{n-1}}|z|^{p(2-n)/(p-1)}\,\big(1+|z|^2\big)^{p(n-2)/(p-1)}\,\big(1+|z|^2\big)^{-(n-1)}\,dm_{n-1}(z)\\ &= \omega_{n-2}a^p2^{n-1}\int_0^\infty r^{(2-n)/(p-1)}\,\big(1+r^2\big)^{(n-p-1)/(p-1)}\,dr\\ &= \omega_{n-2}2^{n-p-1}\bigg(\int_0^\infty r^{(2-n)/(p-1)}\,\big(1+r^2\big)^{(n-p-1)/(p-1)}\,dr\bigg)^{1-p} = 2^pb(n,p)\ ,\end{aligned}$$

giving equality in (4.52). When $n=2$, the function $\rho_0\equiv 1/\pi$ produces equality in (4.52) for $S=\mathbb{S}^1$, $x=e_2$, and $y=-e_2$. □

Notice that

$$c_n = 2^nb(n,n) = \frac{1}{4}\,\omega_{n-2}\,\frac{\Gamma\left[\frac{1}{2(n-1)}\right]^2}{\Gamma\left[\frac{1}{n-1}\right]}. \tag{4.53}$$

With a little effort one can prove

$$n\,\frac{\pi}{16}\leq\frac{1}{4}\,\frac{\Gamma\left[\frac{1}{2(n-1)}\right]^2}{\Gamma\left[\frac{1}{n-1}\right]}\leq\frac{n}{2},$$

$$c_n \geq n\,\omega_{n-2}\,\frac{\pi}{16}. \tag{4.54}$$

The formula for ω_{n-2} can be found at (4.17), and so one can determine the rough order of c_n should one need it.

One direct consequence of Theorem 4.2.19 that we shall later have reason to exploit is the following bound for the diameter of the image of a spherical cap under a smooth mapping.

THEOREM 4.2.21. *Let C be an open cap of a Euclidean sphere S of radius r in $\mathbb{R}^n$, and let f be a function from $C^1(U,\mathbb{R}^m)$, where U is an open set in $\mathbb{R}^n$ that* *contains C. Then*

$$\operatorname{diam}\big[f(C)\big] \le \left[a_n r \int_C \|f'\|^n \, d\sigma_{n-1}\right]^{1/n}$$

with $a_n = b(n,n)^{-1}$. If $C = S$, this estimate holds with a_n replaced by $A_n = 2^{-n}a_n$.

PROOF. Fix x and y in C. We wish to check that

$$|f(x) - f(y)|^n \le a_n r \int_C \|f'\|^n \, d\sigma_{n-1}\,. \tag{4.55}$$

We may clearly suppose that $d = |f(x) - f(y)| > 0$. Consider $\rho = d^{-1}L_f = d^{-1}\|f'\|$. Let γ be any locally rectifiable curve from the family $\Delta(\{x\},\{y\} : C)$. Since f is locally absolutely continuous on γ and since $\ell(f \circ \gamma) \ge d$, we have

$$1 \le \frac{\ell(f\circ\gamma)}{d} = \int_{f\circ\gamma} \frac{ds}{d} \le \frac{1}{d}\int_\gamma L_f \, ds = \int_\gamma \rho \, ds\,.$$

The restriction of ρ to C thus satisfies the hypotheses of Theorem 4.2.19. Therefore

$$\frac{1}{d^n}\int_C \|f'\|^n \, d\sigma_{n-1} = \int_C \rho^n \, d\sigma_{n-1} \ge \frac{b(n,n)}{r} = \frac{1}{a_n r}\,,$$

which establishes (4.55). Should $C = S$, Theorem 4.2.20 would allow us to replace the constant a_n with the smaller one A_n. □

Our real motive for introducing Theorems 4.2.19 and 4.2.20 was to obtain the following important conformal modulus estimate.

THEOREM 4.2.22. *Let E and F be nonempty, disjoint sets in $\hat{\mathbb{R}}^n$ with $n \ge 2$, and let G be a subset of $\hat{\mathbb{R}}^n$ with the following property: there exist a point x_0 of $\mathbb{R}^n$ and positive real numbers a and b ($a < b$) such that, for each r in the interval (a,b), G contains an open cap C_r of $\mathbb{S}^{n-1}(x_0,r)$ whose closure intersects both E and F. Then*

$$M\big[\Delta(E,F:G)\big] \ge b_n \log\frac{b}{a} \tag{4.56}$$

with $b_n = b(n,n)$. Furthermore, if we can take $C_r = \mathbb{S}^{n-1}(x_0,r)$ for each r in (a,b), then

$$M\big[\Delta(E,F:G)\big] \ge c_n \log\frac{b}{a} \tag{4.57}$$

with $c_n = 2^n b_n$. The lower bound in (4.57) is sharp.

PROOF. Write $\Gamma = \Delta(E,F:G)$. If ρ belongs to $\operatorname{Adm}(\Gamma)$, then ρ is also an admissible density for $\Delta(E\cap\bar{C}_r, F\cap\bar{C}_r : C_r)$ whenever $a < r < b$. This implies that $\rho|C_r$ fulfills the conditions of Theorem 3.13 for any x in $E\cap\bar{C}_r$ and y in $F\cap\bar{C}_r$, whence

$$\begin{aligned}\int_{\mathbb{R}^n} \rho^n \, dm_n &= \int_0^\infty \left(\int_{S^{n-1}(x_0,r)} \rho^n \, d\sigma_{n-1}\right) dr \ge \int_a^b \left(\int_{C_r} \rho^n \, d\sigma_{n-1}\right) dr \\ &\ge b(n,n)\int_a^b \frac{dr}{r} = b_n \log\frac{b}{a}\,.\end{aligned}$$

Estimate (4.56) follows by taking the infimum over ρ. If, in fact, we can take $C_r = \mathbb{S}^{n-1}(x_0,r)$ for every r in (a,b), then Theorem 4.2.20 enables us to replace b_n with $c_n = 2^n b_n$ in the above argument, which results in (4.57).

To see that (4.57) is not, in general, subject to improvement, we assess the situation for the configuration

$$G = B^n(b) \setminus \overline{B}^n(a), \qquad E = \{te_n : a < t < b\}, \qquad F = \{-te_n : a < t < b\}.$$

In the course of the proof of Theorem 4.2.20 we established the existence of a Borel function $\rho_0 : S^{n-1} \to [0,\infty]$ that has $\int_\gamma \rho_0\, ds \geq 1$ for each locally rectifiable member γ of the family $\Delta(\{e_n\},\{-e_n\} : S^{n-1})$ and satisfies

$$\int_{\mathbb{S}^{n-1}} \rho_0^n \, d\sigma_{n-1} = c_n \, .$$

Define $\rho : \hat{\mathbb{R}}^n \to [0,\infty]$ by

$$\rho(x) = \begin{cases} |x|^{-1}\rho_0(|x|^{-1}x) & \text{if } x \text{ lies in } G, \\ 0 & \text{otherwise.} \end{cases}$$

Let $f(x) = |x|^{-1}x$ for x in G. Then f belongs to $C^1(G,\mathbb{R}^n)$, $f(G) = \mathbb{S}^{n-1}$, $f(E) = \{e_n\}$, $f(F) = \{-e_n\}$, and $L_f(x) = |x|^{-1}$ throughout G.

If a curve γ in $\Gamma = \Delta(E,F : G)$ is locally rectifiable, then $f \circ \gamma$ is a locally rectifiable member of $\Delta(\{e_n\},\{-e_n\} : S^{n-1})$. Hence,

$$\int_\gamma \rho \, ds = \int_\gamma (\rho_0 \circ f)\, L_f \, ds \geq \int_{f\circ\gamma} \rho_0 \, ds \geq 1 \, .$$

This shows us that ρ is among the admissible densities for Γ. We conclude that

$$\begin{aligned} M(\Gamma) &\leq \int_{\mathbb{R}^n} \rho^n \, dm_n = \int_a^b \left(\int_{S^{n-1}(r)} \rho^n \, d\sigma_{n-1} \right) dr \\ &= \int_a^b \left[\int_{S^{n-1}} \rho^n(ru) r^{n-1} \, d\sigma_{n-1}(u) \right]^r = \int_a^b \left(\int_{S^{n-1}} \rho_0^n \, d\sigma_{n-1} \right) \frac{dr}{r} \\ &= c_n \log \frac{b}{a} \, . \end{aligned}$$

Therefore we see that in this instance equality holds in (4.57). □

We note that the estimates (4.56) and (4.57) have p-modulus analogues, provided $p > n-1$. For example, with the other hypotheses in the theorem remaining as they are, we obtain in place of (4.56) the lower bound

$$M_p\big[\Delta(E,F : G)\big] \geq b(n,p) \int_a^b r^{n-p-1} \, dr \tag{4.58}$$

for $p > n-1$. The changes needed in the proofs to achieve these results are straightforward.

Estimate (4.56) is often referred to in the literature as the "cap inequality" for the conformal modulus. It has a multitude of implications. We close this section by pointing out a few of them.

THEOREM 4.2.23. *If D is an open half-space or an open Euclidean ball in $\mathbb{R}^n$ with $n \geq 2$ and if E and F are disjoint, nondegenerate, connected subsets of $\bar{D}$ for which $\bar{E} \cap \bar{F} \neq \emptyset$, then $M[\Delta(E,F : D)] = \infty$.*

PROOF. Because the conformal modulus is invariant under Möbius transformations, it suffices to consider the situation when $\bar{E} \cap \bar{F}$ contains some finite point, say x_0. Since E and F are nondegenerate, we can fix $b > 0$ so that neither E nor F is contained in $B^n(x_0, b)$. The connectedness of E and F, together with the fact that x_0 belongs to $\bar{E} \cap \bar{F}$, then ensures that $\mathbb{S}^{n-1}(x_0, r)$ meets both E and F for every r in $(0, b)$. This implies that the closure of the cap $C_r = \mathbb{S}^{n-1}(x_0, r) \cap D$ intersects each of these sets for every r in that interval. Theorem 3.13 tells us that

$$M\big[\Delta(E, F : D)\big] \geq b_n \log \frac{b}{a}$$

whenever $0 < a < b$. Letting $a \to 0$, we draw the stated conclusion. □

There are many domains D other than half-spaces and balls for which Theorem 3.17 remains valid. These include $D = \hat{\mathbb{R}}^n$ and $D = \mathbb{R}^n$, to which the proof of Theorem 4.2.23 carries over without essential change. As we shall see later, the theorem is true for any bounded domain D in $\mathbb{R}^n$ fitting the following description: $D = \{x : \varphi(x) < 0\}$, where φ is a function from the class $C^1(\mathbb{R}^n)$ that has $\nabla\varphi(x) \neq 0$ for every x in ∂D. Theorem 4.2.23 does not hold in arbitrary domains. For example, if $D = B^{n-1} \times (-\infty, \infty)$, $E = \{te_n : 1 \leq t < \infty\}$, and $F = \{te_n : -\infty < t \leq 0\}$, then ∞ is in $\bar{E} \cap \bar{F}$, yet $M[\Delta(E, F : D)] \leq \Omega_{n-1}$. Indeed, $\Delta(E, F : D)$ is minorized by $\Delta(B^{n-1}, e_n + B^{n-1} : B^{n-1} \times (0, 1))$.

The p-modulus counterpart of Theorem 4.2.23 holds when $p > n$, provided $\bar{E} \cap \bar{F}$ contains a finite point. To see this, in the proof of Theorem 4.2.23 just use (4.58) in place of (4.56).

The corresponding assertion when $p > n$ and $\bar{E} \cap \bar{F} = \{\infty\}$ need not be true. To see this, let $E = \{te_n : 1 \leq t < \infty\}$, let $F = \{te_n : -\infty < t \leq -1\}$, and consider $\Gamma = \bar{\Delta}(E, F : \hat{\mathbb{R}}^n)$. It is not difficult to verify that the function $\rho : \hat{\mathbb{R}}^n \to [0, \infty)$ defined by $\rho(x) = (2|x|)^{-1}$ when $e^{-1} \leq |x| < \infty$ and $\rho(x) = 0$ otherwise is an admissible density for Γ. As a result,

$$M_p\big[\Delta(E, F : \hat{\mathbb{R}}^n)\big] = M_p(\Gamma) \leq \int_{\mathbb{R}^n} \rho^p \, dm_n = \frac{\omega_{n-1} e^{p-n}}{2^p(p-n)} < \infty$$

if $p > n$, even though $\bar{E} \cap \bar{F} = \{\infty\}$.

Similar reasoning shows that $M_p[\Delta(E, F : \hat{\mathbb{R}}^n)]$ is finite for $E = \{te_n : 0 < t \leq 1\}$ and $F = \{te_n : -1 \leq t < 0\}$ when $1 < p < n$, so the p-modulus analogue of Theorem 4.2.23 is generally false for this range of p.

Here is a second application of the cap inequality and Theorem 4.2.19.

THEOREM 4.2.24. *If D is an open half-space or an open Euclidean ball in $\mathbb{R}^n$ with $n \geq 2$ and if E and F are disjoint, nondegenerate, connected sets in $\bar{D}$, then*

$$M[\Delta(E, F : D)] = M[\Delta(\bar{E}, \bar{F} : D)]\,.$$

PROOF. Because of Corollary 4.2.17, it is sufficient to treat the half-space case. By virtue of Theorem 4.2.15, we need to check only that $M(\Gamma) = M(\Gamma')$ for $\Gamma = \bar{\Delta}(E, F : D)$ and $\Gamma' = \bar{\Delta}(\bar{E}, \bar{F} : D)$. Since Γ is a subfamily of Γ', $M(\Gamma) \leq M(\Gamma')$. In establishing the reverse inequality, we may assume that $M(\Gamma) < \infty$. In light of Theorems 4.2.15 and 4.2.23, this forces $\bar{E}$ and $\bar{F}$ to be disjoint. Let ρ in $\mathrm{Adm}(\Gamma)$

have $\int_{\mathbb{R}^n} \rho^n \, dm_n < \infty$. We maintain that ρ must also be admissible for Γ'. Fix γ, a rectifiable path in Γ', and let $0 < \varepsilon < 1/2$. We shall demonstrate that

$$\int_\gamma \rho \, ds \geq 1 - 2\varepsilon \,. \tag{4.59}$$

Letting $\varepsilon \to 0$, we conclude that ρ is in $\mathrm{Adm}(\Gamma')$. It then follows that $M(\Gamma') \leq \int_{\mathbb{R}^n} \rho^n \, dm_n$, and hence, by taking the infimum over ρ, that $M(\Gamma') \leq M(\Gamma)$.

In verifying (4.59) we may suppose that γ actually belongs to

$$\bar{\Delta}(\,\bar{E}, \bar{F} : D \setminus (\,\bar{E} \cup \bar{F}\,))$$

(if not, we simply replace γ by a subpath belonging to the stated family), that γ has for its domain the unit interval $[0,1]$, and that $x_0 = \gamma(0)$ belongs to $\bar{E}$, $y_0 = \gamma(1)$ to $\bar{F}$. Since γ is rectifiable, both x_0 and y_0 are finite points. Define $\delta > 0$ by $\delta = 2^{-1} \min\{|x_0 - y_0|, \mathrm{diam}(E), \mathrm{diam}(F)\}$. For r in $(0, \delta)$ the closure of the cap $C_r = D \cap \mathbb{S}^{n-1}(x_0, r)$ intersects both $|\gamma|$ and E (neither of these two connected subsets of $\bar{D}$ lies in $B^n(x_0, r)$, and each has x_0 in its closure). For some r in $(0, \delta)$ there must exist an injective path $\alpha : [0,1] \to \bar{C}_r$ such that $|\alpha|$ is a circular arc with $\alpha(0)$ in E, $\alpha(1)$ in $|\gamma|$, $\alpha(t)$ in C_r for t in $(0,1]$, and $\int_\alpha \rho \, ds < \varepsilon$. Otherwise, since $E \cap \bar{C}_r$ and $|\gamma| \cap \bar{C}_r$ are nonempty, disjoint sets when $0 < r < \delta$, Theorem 4.2.19 (applied to the function $\varepsilon^{-1}\rho$ and to points x of $E \cap \bar{C}_r$ and y of $|\gamma| \cap \bar{C}_r$) would bring the inequality

$$\int_{C_r} \rho^n \, d\sigma_{n-1} \geq \frac{b_n \varepsilon^n}{r}$$

into play for every r in $(0, \delta)$. This would have the consequence that

$$\int_{\mathbb{R}^n} \rho^n \, dm_n \geq \int_0^\delta \left(\int_{C_r} \rho^n \, d\sigma_{n-1} \right) dr \geq b_n \varepsilon^n \int_0^\delta \frac{dr}{r} = \infty \,,$$

contrary to our assumption concerning ρ. Fix r and α as indicated, say with $\alpha(1) = \gamma(c)$. As $\alpha(1)$ is neither x_0 nor y_0, we have $0 < c < 1$. A similar argument applied to the path $\alpha + \gamma|[c,1]$ at its endpoint y_0 will produce a rectifiable path $\beta : [0,1] \to \bar{D}$ with $\beta(1)$ in F, $\beta(0) = \gamma(d)$ for some d in $(c,1)$, $\beta(t)$ in D when $0 \leq t < 1$, and $\int_\beta \rho \, ds < \varepsilon$. Therefore $\gamma_1 = \alpha + \gamma|[c,d] + \beta$ is a rectifiable path in Γ, which means that

$$1 \leq \int_{\gamma_1} \rho \, ds \leq \int_\alpha \rho \, ds + \int_\gamma \rho \, ds + \int_\beta \rho \, ds \leq 2\varepsilon + \int_\gamma \rho \, ds \,.$$

This shows that (4.59) holds. □

The connectedness of E and F is crucial to the validity of Theorem 4.2.24. For instance, if $D = H^n$, $E = \partial D$, and $F = \{te_n : t \in \mathbb{Q},\, t > 0\}$, then

$$M[\Delta(E, F : D)] = 0$$

(Corollary 4.2.14 applies, for F is a countable set), whereas $M[\Delta(\,\bar{E}, \bar{F} : D)] = \infty$ by Theorem 4.2.23. Theorem 4.2.24 holds for certain domains D different from half-spaces and balls (for instance, $D = \hat{\mathbb{R}}^n$, $D = \mathbb{R}^n$) but is not true for every domain. A moment's reflection on the above proof reveals, however, that the analogue of Theorem 4.2.24 for an arbitrary domain D does hold under the additional hypothesis that both $\bar{E}$ and $\bar{F}$ are subsets of D.

Finally, it is worthwhile exploring the limitations of the cap inequality. To do this we consider a fairly natural example. Consider the parallel plates

$$\begin{array}{rcl} E & = & \{x \in \mathbb{R}^n : x_n = +1,\ |x_i| < M, i = 1, 2, \dots, n-1\}, \\ F & & \{x \in \mathbb{R}^n : x_n = -1,\ |x_i| < M, i = 1, 2, \dots, n-1\}. \end{array}$$

Then

$$\mathbb{S}^{n-1}(0,r) \cap E \neq \emptyset, \quad \mathbb{S}^{n-1}(0,r) \cap F \neq \emptyset$$

as long as $1 \leq r \leq \sqrt{1+(n-1)M^2}$.

The cap inequality (4.56) gives

$$M(E, F; \mathbb{R}^n) \geq \frac{c_n}{2} \log[1 + (n-1)M^2]. \tag{4.60}$$

However, if $\rho > 0$ is an admissible function, then for $x \in \mathbb{R}^n$ write $x = (x', x_n)$ and $Q = \{|x_i| \leq M, i = 1, 2, \dots, n-1\} \subset \mathbb{R}^{n-1}$ to see that $\{(t, x') : -1 \leq t \leq 1\}$ is a path from E to F, and so

$$\begin{array}{rcl} \displaystyle\int_{\mathbb{R}^n} \rho^n(x)\, dx & = & \displaystyle\int_{\mathbb{R}^n} \rho^n(x_n, x')\, dx_n\, dx' \geq \int_Q \Big[\int_{-1}^{+1} \rho^n(x_n, x')\, dx_n \Big]\, dx' \\ & \geq & \displaystyle 2^{1-n} \int_Q \Big(\int_{-1}^{+1} \rho(x_n, x')\, dx_n \Big)^n\, dx' \quad \text{(Hölder's inequality)} \\ & \geq & \displaystyle 2^{1-n} \int_Q dx' = 2^{1-n} M^{n-1} \quad (\rho \text{ is admissible}), \end{array}$$

which gives a substantially better lower bound—at least for M large. Actually a trivial upper bound is obtained as

$$\rho(x) = \frac{1}{2} \chi_{\tilde{Q}}(x)$$

is an admissible function, where the box $\tilde{Q} = \{x : |x_n| < 2, |x_i| < M+1, i \neq n\}$. Then

$$M(E, F; \mathbb{R}^n) \leq 2^{-n} |\tilde{Q}| = 2^{2-n}(M+1)^{n-1}.$$

The two estimates we have found together show that

$$2^{1-n} M^{n-1} \leq M(E, F; \mathbb{R}^n) \leq 2^{2-n}(M+1)^{n-1}. \tag{4.61}$$

Now the cap inequality gives a better bound when M is small as

$$\frac{c_n}{2} \log[1 + (n-1)M^2] \approx \frac{1}{2}(n-1) c_n M^2, \quad M \to 0.$$

If M is small, then we can surround E and F by the balls $B^n(\pm 1, \sqrt{n-1}M)$. Then

$$M(E, F; \mathbb{R}^n) \leq M(B^n(+1, \sqrt{n-1}M), B^n(-1, \sqrt{n-1}M); \mathbb{R}^n)$$

as the latter has shorter curves. We have a formula for the modulus above, derived at (4.41), and this gives

$$\begin{aligned} & M[B^n(+1, \sqrt{n-1}\,M), B^n(-1, \sqrt{n-1}\,M); \mathbb{R}^n] \\ = \;& \omega_{n-1} \left[2 \log \frac{\sqrt{n-1}\,M}{1 - \sqrt{1-(n-1)\,M^2}} \right]^{1-n} \\ = \;& \omega_{n-1} \left[\log 4 - \log[(n-1)\,M^2] - \frac{(n-1)\,M^2}{2} + O[M^4] \right]^{1-n} \\ \approx \;& \omega_{n-1} \log^{1-n} \frac{1}{(n-1)\,M^2}, \end{aligned}$$

which has much slower decay than that given by the cap inequality. When $M \ll 1$, is it not hard to construct a quasiconformal map $\mathbb{R}^n \setminus B^n(\pm 1, \sqrt{n-1}\,M) \mapsto \mathbb{R}^n \setminus (E \cup F)$; this last estimate will be correct up to a uniformly bounded (but depending on n) multiplicative factor.

While the capacity of two plates seems a natural problem, it will be a lot more work to significantly improve these estimates—at least in dimension $n \geq 3$ where conformal mapping techniques fail.

4.3. Technical properties of moduli

In this section we collect an assortment of technical facts about moduli which we can use to understand the geometry of mappings which distort moduli by a bounded multiplicative factor—the quasiconformal mappings.

We begin with a discussion of several situations in which symmetry enhances our ability to compute or estimate moduli.

4.3.1. Symmetry principles for moduli. The first step on the way to these symmetry principles is the following lemma.

LEMMA 4.3.1. *Suppose that D is an open half-space in $\mathbb{R}^n$ with $n \geq 2$ and that E and F are nonempty, disjoint, compact subsets of $\bar{D}$. Then*

$$M_p[\Delta(E, F : D)] = M_p[\Delta(E, F : \bar{D}\,)]$$

for every $p > 1$.

PROOF. Since $\Gamma_1 = \Delta(E, F : D)$ is contained in $\Gamma_2 = \Delta(E, F : \bar{D}\,)$, the only thing we need to prove is that $M_p(\Gamma_2) \leq M_p(\Gamma_1)$. We may assume that $M_p(\Gamma_1) < \infty$. The hypotheses ensure that the compact set $E \cup F$ has at least one finite point. Given λ in $(0, 1)$, we define a function $f : \hat{\mathbb{R}}^n \to \hat{\mathbb{R}}^n$ as follows: letting ν denote the unit normal to ∂D that points into D and writing $A = (E \cup F) \cap \mathbb{R}^n$, we set

$$f(x) = x + \lambda\, \mathrm{dist}(x, A)\, \nu$$

for x in $\mathbb{R}^n$ and $f(\infty) = \infty$. Notice that $f(x) = x$ if x belongs to $E \cup F$. As

$$|\mathrm{dist}(x, A) - \mathrm{dist}(y, A)| \leq |x - y|$$

for x and y in $\mathbb{R}^n$, we see that for such points

$$(1 - \lambda)|x - y| \leq |f(x) - f(y)| \leq (1 + \lambda)|x - y|\ ,$$

which implies that f is continuous on $\hat{\mathbb{R}}^n$. Moreover,

$$1 - \lambda \le \ell_f(x) \le L_f(x) \le 1 + \lambda$$

holds throughout $\mathbb{R}^n$. The Rademacher-Stepanov theorem informs us that f is differentiable almost everywhere in $\mathbb{R}^n$. At any point x where f is differentiable, we infer from (2.8) and (2.22) that

$$|J_f(x)| \ge \left[\ell_f(x)\right]^n \ge (1-\lambda)^n .$$

Now let ρ_1 belong to $\mathrm{Adm}(\Gamma_1)$ and define a Borel function $\rho_2 : \hat{\mathbb{R}}^n \to [0, \infty]$ by

$$\rho_2(x) = (1+\lambda)\rho_1\left[f(x)\right] .$$

Let $\gamma_2 : (a,b) \to \bar{D}$ be a locally rectifiable curve from Γ_2, say with $\gamma_2(a^+)$ in E and $\gamma_2(b^-)$ in F, and let γ_2^* denote the continuous extension of γ_2 to $[a,b]$. There exist c and d satisfying $a \le c < d \le b$ with $\gamma_2^*(c)$ in E, $\gamma_2^*(d)$ in F, and $\gamma_2(t)$ in $[\bar{D} \setminus (E \cup F)] \cap \mathbb{R}^n$ whenever $c < t < d$. Set $\gamma = \gamma_2|(c,d)$ and $\gamma_1 = f \circ \gamma$. Then γ_1 is locally rectifiable, $\gamma_1^*(c) = \gamma^*(c)$ lies in E, $\gamma_1^*(d) = \gamma^*(d)$ belongs to F, and $\gamma_1(t) = \gamma(t) + \lambda d[\gamma(t), A]\nu$ is in D for $c < t < d$; in other words, γ_1 is a curve from Γ_1. It follows that

$$\int_{\gamma_2} \rho_2 \, ds = (1+\lambda) \int_{\gamma_2} \rho_1 \circ f \, ds \ge \int_{\gamma} (\rho_1 \circ f) \, L_f \, ds \ge \int_{\gamma_1} \rho_1 \, ds \ge 1 ,$$

which marks ρ_2 as a member of $\mathrm{Adm}(\Gamma_2)$. We conclude that

$$\begin{aligned} M_p(\Gamma_2) &\le \int_{\mathbb{R}^n} \rho_2^p \, dm_n = (1+\lambda)^p \int_{\mathbb{R}^n} (\rho_1 \circ f)^p \, dm_n \\ &\le \frac{(1+\lambda)^p}{(1-\lambda)^n} \int_{\mathbb{R}^n} (\rho_1 \circ f)^p |J_f| \, dm_n = \frac{(1+\lambda)^p}{(1-\lambda)^n} \int_{\mathbb{R}^n} \rho_1^p \, dm_n . \end{aligned}$$

Here we have used the change of variable formula for a Lipschitz mapping (Theorem 4.0.11). Taking the infimum over ρ_1 and then letting $\lambda \to 0$, we arrive at $M_p(\Gamma_2) \le M_p(\Gamma_1)$. □

Because the conformal modulus is a Möbius invariant, we can assert on the strength of Lemma 4.3.1 that $M[\Delta(E, F : D)] = M[\Delta(E, F : \bar{D})]$ whenever D is an open Euclidean ball in $\mathbb{R}^n$ with $n \ge 2$ and the sets E and F are nonempty, disjoint, compact subsets of $\bar{D}$. In tandem with Theorem 4.2.24, Lemma 4.3.1 thus leads to:

THEOREM 4.3.2. *If D is an open half-space or an open Euclidean ball in $\mathbb{R}^n$ with $n \ge 2$ and if E and F are disjoint, nondegenerate, connected subsets of $\bar{D}$, then*

$$\Delta(E, F : D)] = M[\Delta(\bar{E}, \bar{F} : \bar{D})] .$$

PROOF. We may suppose that D is a half-space. If $\bar{E} \cap \bar{F}$ is not empty, then the two moduli in question are infinite; if $\bar{E}$ and $\bar{F}$ are disjoint, then Theorem 4.2.24 and Lemma 4.3.1 together give

$$M[\Delta(E, F : D)] = M[\Delta(\bar{E}, \bar{F} : D)] = M[\Delta(\bar{E}, \bar{F} : \bar{D})].$$

□

Our first symmetry principle for moduli reads as follows.

THEOREM 4.3.3. *Suppose that D is an open half-space in $\mathbb{R}^n$ with $n \geq 2$ and that E and F are nonempty, disjoint, compact subsets of $\bar{D}$. If E^* and F^* denote the images of E and F under reflection in the sphere or hyperplane ∂D, then*

$$M_p\big[\Delta(E \cup E^*, F \cup F^* : \hat{\mathbb{R}}^n)\big] = 2M_p\big[\Delta(E, F : D)\big]$$

for every $p > 1$.

PROOF. Write $\Gamma = \Delta(E, F : D)$ and $\Gamma_0 = \Delta(E \cup E^*, F \cup F^* : \hat{\mathbb{R}}^n)$. Let $f : \hat{\mathbb{R}}^n \to \hat{\mathbb{R}}^n$ denote the reflection in ∂D, and let $\Gamma^* = f(\Gamma)$. Then

$$\Gamma^* = \Delta(E^*, F^* : D^*),$$

where $D^* = f(D)$. Since f is a Euclidean isometry, it follows that $M_p(\Gamma) = M_p(\Gamma^*)$ for $p > 1$ (Theorem 4.2.18). Because the families Γ and Γ^* are separated and $\Gamma \cup \Gamma^*$ is a subfamily of Γ_0, we extract from Theorem 4.2.9 the information that

$$2M_p(\Gamma) = M_p(\Gamma) + M_p(\Gamma^*) \leq M_p(\Gamma_0)\,.$$

To derive the reverse inequality, we consider the continuous mapping $g : \hat{\mathbb{R}}^n \to \hat{\mathbb{R}}^n$ given by

$$g(x) = \begin{cases} x & \text{if } x \in \bar{D}\,, \\ f(x) & \text{if } x \in D^*\,. \end{cases}$$

Then $|g(y) - g(x)| \leq |y - x|$ throughout $\mathbb{R}^n$, and $\ell_g(x) = L_g(x) = 1$ for every x in $\mathbb{R}^n$. Let $\Gamma_1 = \Delta(E, F : \bar{D})$, and let ρ_1 belong to $\mathrm{Adm}(\Gamma_1)$. If γ_0 in Γ_0 is locally rectifiable, then $\gamma_1 = g \circ \gamma_0$ is a locally rectifiable member of Γ_1. Therefore $\rho_0 = \rho_1 \circ g$ satisfies

$$\int_{\gamma_0} \rho_0 \; ds = \int_{\gamma_0} \rho_1 \circ g \; ds = \int_{\gamma_0} (\rho_1 \circ g)\; L_g \; ds \geq \int_{\gamma_1} \rho_1 \; ds \geq 1\,,$$

putting ρ_0 in $\mathrm{Adm}(\Gamma_0)$. Hence, since $\rho_0 = \rho_1$ in D, $\rho_0 = \rho_1 \circ f$ in D^*, and $|J_f(x)| = 1$ for every x in D^*,

$$\begin{aligned} M_p(\Gamma_0) &\leq \int_{\mathbb{R}^n} \rho_0^p \; dm_n = \int_D \rho_1^p \; dm_n + \int_{D^*} (\rho_1 \circ f)^p |J_f| \; dm_n \\ &= \int_D \rho_1^p \; dm_n + \int_D \rho_1^p \; dm_n = 2\int_D \rho_1^p \; dm_n \leq 2\int_{\mathbb{R}^n} \rho_1^p \; dm_n\,. \end{aligned}$$

Taking the infimum over ρ_1 and then invoking Lemma 4.3.1 yields

$$M_p(\Gamma_0) \leq 2M_p(\Gamma_1) = 2M_p(\Gamma)\,,$$

which completes the proof. □

When dealing with the conformal modulus we can improve upon Theorem 4.3.3 slightly.

COROLLARY 4.3.4. *Suppose that D is an open half-space or an open Euclidean ball in $\mathbb{R}^n$ with $n \geq 2$ and that E and F are either disjoint, compact sets in $\bar{D}$ or disjoint, connected sets in $\bar{D}$. If E^* and F^* are the images of E and F under reflection in ∂D, then*

$$M\big[\Delta(E \cup E^*, F \cup F^* : \hat{\mathbb{R}}^n)\big] = 2M\big[\Delta(E, F : D)\big]\,.$$

PROOF. We may apply a preliminary Möbius transformation and then recall Theorem 3.3.7, to reduce the discussion to the case where D is a half-space.

If E and F are compact sets, then Corollary 4.3.4 is just a special instance of Theorem 4.3.3. Assume now that E and F are disjoint, connected sets in $\bar{D}$. If either is degenerate, the assertion holds trivially, with both moduli in question being zero. Taking E and F to be nondegenerate, we may assume that $\bar{E} \cap \bar{F} = \emptyset$, for both of the indicated moduli are infinite otherwise. Then $\bar{E}$ and $\bar{F}$ are disjoint compacts sets in $\bar{D}$. By Theorems 4.3.3 and 4.2.24,

$$\begin{aligned} M\big[\Delta(E \cup E^*, F \cup F^* : \hat{\mathbb{R}}^n)\big] &\leq M\big[\Delta(\bar{E} \cup \bar{E}^*, \bar{F} \cup \bar{F}^* : \hat{\mathbb{R}}^n)\big] \\ &= 2M\big[\Delta(\bar{E}, \bar{F} : D)\big] = 2M\big[\Delta(E, F : D)\big] . \end{aligned}$$

On the other hand, $\Delta(E, F : D)$ and $\Delta(E^*, F^* : D^*)$ are separated and Möbius equivalent curve families whose union is contained in $\Delta(E \cup E^*, F \cup F^* : \hat{\mathbb{R}}^n)$. Therefore

$$\begin{aligned} 2M\big[\Delta(E, F : D)\big] &= M\big[\Delta(E, F : D)\big] + M\big[\Delta(E^*, F^* : D^*)\big] \\ &\leq M\big[\Delta(E \cup E^*, F \cup F^* : \hat{\mathbb{R}}^n)\big] . \end{aligned}$$

The theorem thus follows for connected sets E and F. □

Here is a second symmetry result for moduli.

THEOREM 4.3.5. *Suppose that D is an open half-space in $\mathbb{R}^n$ with $n \geq 2$ and that E in $\mathbb{R}^n \cap \bar{D}$ and F in $\mathbb{R}^n \cap \partial D$ are nonempty, disjoint sets. If E^* and D^* are the images of E and D under reflection in ∂D, then*

$$M_p\big[\Delta(E, F : D)\big] = 2^{p-1} M_p\big[\Delta(E, E^* : D \cup F \cup D^*)\big]$$

for every $p > 1$.

PROOF. Write $\Gamma = \Gamma(E, F : D)$, $\Gamma^* = \Delta(E^*, F : D^*)$, and $\Gamma_0 = \Delta(E, E^* : D \cup F \cup D^*)$. Plainly $\Gamma^* = f(\Gamma)$, where f is the reflection in ∂D. Since f is a Euclidean isometry, $M_p(\Gamma) = M_p(\Gamma^*)$ for $p > 1$. Now Γ and Γ^* are separated families for which $\Gamma \prec \Gamma_0$ and $\Gamma^* \prec \Gamma_0$, so by Theorem 4.2.11

$$M_p(\Gamma_0) \leq \big[\lambda_p(\Gamma) + \lambda_p(\Gamma^*)\big]^{1-p} = 2^{1-p} \lambda_p(\Gamma)^{1-p} = 2^{1-p} M_p(\Gamma)$$

whenever $p > 1$; that is, $M_p(\Gamma) \geq 2^{p-1} M_p(\Gamma_0)$ for every $p > 1$.

In attempting to establish the reverse inequality, $M_p(\Gamma) \leq 2^{p-1} M_p(\Gamma_0)$, we may presume that $M_p(\Gamma_0)$ is finite. We also draw attention to the fact that $M_p(\Gamma) = M_p(\Gamma_{rect})$ for all $p > 1$, true because every γ in Γ has a bounded trajectory. (Recall Theorem 4.2.12 and the remarks following its proof.) Given ρ_0 in $\mathrm{Adm}(\Gamma_0)$, we define a Borel function $\rho : \hat{\mathbb{R}}^n \to [0, \infty]$ by

$$\rho(x) = \begin{cases} \rho_0(x) + \rho_0[f(x)] & \text{if } x \in D , \\ 0 & \text{if } x \notin D . \end{cases}$$

We claim that ρ belongs to $\mathrm{Adm}(\Gamma_{rect})$. Let γ be a rectifable curve from Γ. We may assume that $\gamma : (0,1) \to D$ and that its extension $\gamma^* : [0,1] \to \bar{D}$ has $\gamma^*(0)$ in E and $\gamma^*(1)$ in F. Clearly $\beta = f \circ \gamma$ is a rectifiable curve belonging to Γ^*.

The path $\alpha = \gamma^* + (-\beta^*)$ is a well-defined, rectifiable path. To be explicit, α is the path defined on $[0,2]$ by

$$\alpha(t) = \begin{cases} \gamma^*(t) & \text{if } 0 \le t \le 1, \\ f\big[\gamma^*(2-t)\big] & \text{if } 1 \le t \le 2 . \end{cases}$$

Thus $\gamma_0 = \alpha|(0,2)$ is a rectifiable member of Γ_0. Using the fact that f is a diffeomorphism with $\ell_f = L_f = 1$, we compute

$$\begin{aligned} \int_\gamma \rho \, ds &= \int_\gamma (\rho_0 + \rho_0 \circ f) \, ds = \int_\gamma \rho_0 \, ds + \int_\gamma (\rho_0 \circ f) \, L_f \, ds \\ &= \int_\gamma \rho_0 \, ds + \int_\beta \rho_0 \, ds = \int_{\gamma^*} \rho_0 \, ds + \int_{\beta^*} \rho_0 \, ds = \int_\alpha \rho_0 \, ds \\ &= \int_{\gamma_0} \rho_0 \, ds \ge 1 . \end{aligned}$$

Therefore ρ is an admissible density for the family Γ_{rect}. Finally, exploiting the fact that $|J_f| = 1$, we obtain

$$\begin{aligned} M_p(\Gamma) &= M_p(\Gamma_{rect}) \le \int_{\mathbb{R}^n} \rho^p \, dm_n = \int_D (\rho_0 + \rho_0 \circ f)^p \, dm_n \\ &\le 2^{p-1} \int_D \big[\rho_0^p + (\rho_0 \circ f)^p\big] \, dm_n \\ &= 2^{p-1} \int_D \rho_0^p \, dm_n + 2^{p-1} \int_D (\rho_0 \circ f)^p |J_f| \, dm_n \\ &= 2^{p-1} \int_D \rho_0^p \, dm_n + 2^{p-1} \int_{D^*} \rho_0^p \, dm_n = 2^{p-1} \int_{\mathbb{R}^n} \rho_0^p \, dm_n . \end{aligned}$$

We now take the infimum over all admissible ρ_0 to arrive at the inequality which finishes the proof. □

When $p \ge n$, Theorem 4.3.5 remains true even if one of the sets E or F contains the point ∞. This is so because the family Γ of all nonconstant curves in γ in $\hat{\mathbb{R}}^n$ with ∞ in $\bar{\gamma}$ has $M_p(\Gamma) = 0$ for $p \ge n$. The hypothesis in the theorem that E and F be sets in $\mathbb{R}^n$ cannot be weakened, however, when $1 < p < n$: if $D = \mathbb{H}^n$, $E = \mathbb{S}^{n-1} \cap \bar{D}$, and $F = \{\infty\}$, then $M_p[\Delta(E, F : D)] > 0$ when $1 < p < n$, whereas $M_p[\Delta(E, E^* : D \cup F \cup D^*)] = 0$ for all $p > 1$ as no curve in $\Delta(E, E^* : D \cup F \cup D^*)$ is locally rectifiable.

In the Möbius invariant case $p = n$, we can state a somewhat nicer result.

COROLLARY 4.3.6. *Suppose that D is an open half-space or an open Euclidean ball in $\mathbb{R}^n$ with $n \ge 2$ and that E in $\bar{D}$ and F in ∂D are nonempty, disjoint sets. If E^* and D^* are the images of E and D under reflection in ∂D, then*

$$M\big[\Delta(E, F : D)\big] = 2^{n-1} M\big[\Delta(E, E^* : D \cup F \cup D^*)\big] .$$

In particular, it is true that

$$M[\Delta(E, \partial D : D)] = 2^{n-1} M[\Delta(E, E^* : \hat{\mathbb{R}}^n)]$$

for any nonempty subset E of D.

4.3.2. Continuity properties of moduli. This section addresses, among others, the following question: Is the conformal modulus of $\Delta(E, F : \hat{\mathbb{R}}^n)$, where E and F are disjoint continua in $\hat{\mathbb{R}}^n$ (so nondegenerate, connected, compact subsets of $\hat{\mathbb{R}}^n$), sensitive to small perturbations of E and F? If E' and F' are continua that are "close" to E and F, respectively, could there be a drastic difference between $M[\Delta(E, F : \hat{\mathbb{R}}^n)]$ and $M[\Delta(E', F' : \hat{\mathbb{R}}^n)]$?

Of course, to turn this into a well-posed question we must be specific about the way in which the proximity of sets is measured. We shall ultimately take "close" in the present context to mean "close in the Hausdorff metric on the space of nonempty compact subsets of $\hat{\mathbb{R}}^n$".

In order to handle problems of this kind, it will be necessary to lay a bit of technical groundwork. We do so in a series of lemmas. The first of these merely quotes the Clarkson inequalities, a pair of classical inequalities from real analysis used to establish the uniform convexity of L^p. These inequalities were first identified in 1936 by J.A. Clarkson, [**29**], but a more elementary treatment is given by K.O. Friedrichs [**40**].

In what follows, the symbol $\|g\|_p$ indicates the L^p-norm of the function g.

LEMMA 4.3.7. *Let f and g be functions in $L^p(\mathbb{R}^n)$, where $1 < p < \infty$. If $p \geq 2$, then*

$$\|f + g\|_p^p + \|f - g\|_p^p \leq 2^{p-1}(\|f\|_p^p + \|g\|_p^p) ,$$

and if $1 < p < 2$, then

$$\|f + g\|_p^q + \|f - g\|_p^q \leq 2(\|f\|_p^p + \|g\|_p^p)^{q/p}$$

with $q = p/(p-1)$.

The second lemma establishes a connection between L^p-convergence in $\mathbb{R}^n$ and what might be termed "L^1-convergence along curves in $\mathbb{R}^n$".

LEMMA 4.3.8. *Assume that $n \geq 2$ and $1 < p < \infty$. If $\langle f_\nu \rangle$ is a sequence of Borel functions from $L^p(\mathbb{R}^n)$ that converges in L^p-norm to a Borel function $f : \mathbb{R}^n \to \mathbb{R}$, then there exist a subsequence $\langle f_{\nu_k} \rangle$ of $\langle f_\nu \rangle$ and a subset Γ of $\mathcal{C}(\mathbb{R}^n)$ with $M_p(\Gamma) = 0$ such that*

$$\lim_{k \to \infty} \int_\gamma |f_{\nu_k} - f| \, ds = 0 \tag{4.62}$$

for every curve γ in $\mathcal{C}(\mathbb{R}^n) \setminus \Gamma$.

PROOF. We may assume that each of the functions f_ν is defined everywhere in $\mathbb{R}^n$. We choose $\nu_1 < \nu_2 < \cdots$ in such a way that

$$\int_{\mathbb{R}^n} |f_{\nu_k} - f|^p \, dm_n < 2^{-(p+1)k} .$$

For $k = 1, 2, \ldots$ we define a Borel function $g_k : \hat{\mathbb{R}}^n \to [0, \infty]$ by insisting that g_k agree in $\mathbb{R}^n$ with $|f_{\nu_k} - f|$ and have $g_k(\infty) = 0$. Let Γ denote the subset of $\mathcal{C}(\mathbb{R}^n)$ made up of all curves in $\mathbb{R}^n$ that are not locally rectifiable, together with every locally rectifiable member γ of $\mathcal{C}(\mathbb{R}^n)$ for which

$$\limsup_{k \to \infty} \int_\gamma g_k \, ds > 0 .$$

Then (4.62) is certainly true for each curve γ in the family $\mathcal{C}(\mathbb{R}^n) \setminus \Gamma$. The only question is whether or not $M_p(\Gamma) = 0$.

For $k = 1, 2, \ldots$ let Γ_k designate the family of curves γ in $\mathbb{R}^n$ that either fail to be locally rectifiable or are locally rectifiable with $\int_\gamma g_k \, ds \geq 2^{-k}$. It is apparent that the function $\rho_k = 2^k g_k$ is an admissible density for Γ_k, so by the construction of g_k

$$M_p(\Gamma_k) \leq 2^{pk} \int_{\mathbb{R}^n} g_k^p \, dm_n < 2^{pk} 2^{-(p+1)k} = 2^{-k} .$$

If γ is a locally rectifiable curve from Γ, there must be a $\delta > 0$ such that $\int_\gamma g_k \, ds \geq \delta > 2^{-k}$ holds for infinitely many k. As a result, we see that Γ is a subfamily of $\bigcup_{k=\ell}^\infty \Gamma_k$ for each $\ell \geq 1$. Therefore,

$$M_p(\Gamma) \leq \sum_{k=\ell}^\infty M_p(\Gamma_k) \leq \sum_{k=\ell}^\infty 2^{-k} = 2^{-\ell+1}$$

whenever $\ell \geq 1$, which implies that $M_p(\Gamma) = 0$. □

The next result exposes a nontrivial convergence property of the p-modulus, one with obvious measure theoretic overtones.

A sequence of curve families $\langle \Gamma_\nu \rangle$ is a nondecreasing sequence if $\Gamma_\nu \subset \Gamma_{\nu+1}$, so additional curves are being included.

LEMMA 4.3.9. *If $\langle \Gamma_\nu \rangle$ is a nondecreasing sequence of curve families in $\hat{\mathbb{R}}^n$ with $n \geq 2$ and if $\Gamma = \bigcup_{\nu=1}^\infty \Gamma_\nu$, then $M_p(\Gamma) = \lim_{\nu\to\infty} M_p(\Gamma_\nu)$ for every $p > 1$.*

PROOF. Since $\langle M_p(\Gamma_\nu) \rangle$ is a monotone sequence in $[0, \infty]$ and since Γ_ν is contained in Γ for every ν, $M = \lim_{\nu\to\infty} M_p(\Gamma_\nu)$ exists in $[0, \infty]$ and satisfies $M \leq M_p(\Gamma)$. We need to show that $M_p(\Gamma) \leq M$. For this it may be assumed that $M < \infty$. Because $M_p(\Gamma_\nu) \leq M$ for every ν, we can rule out the presence of constant curves in Γ_ν and therefore be certain that every locally rectifiable curve from Γ has its trajectory in $\mathbb{R}^n$. Choose for each ν a function ρ_ν in $\mathrm{Adm}(\Gamma_\nu)$ such that

$$\|\rho_\nu\|_p^p = \int_{\mathbb{R}^n} \rho_\nu^p \, dm_n \leq M_p(\Gamma_\nu) + \frac{1}{\nu} \leq M + \frac{1}{\nu} .$$

We claim that this selection process generates a Cauchy sequence $\langle \rho_\nu \rangle$ in $L^p(\mathbb{R}^n)$. If $\mu \geq \nu$ and if γ is a locally rectifiable member of Γ_ν, then γ also belongs to Γ_μ. Accordingly, we are able to infer that

$$\frac{1}{2} \int_\gamma (\rho_\mu + \rho_\nu) \, ds = \frac{1}{2} \int_\gamma \rho_\mu \, ds + \frac{1}{2} \int_\gamma \rho_\nu \, ds \geq \frac{1}{2} + \frac{1}{2} = 1 ,$$

whence $(\rho_\mu + \rho_\nu)/2$ is an admissible density for Γ_ν. It follows that

$$M_p(\Gamma_\nu) \leq \|(\rho_\mu + \rho_\nu)/2\|_p^p = 2^{-p} \|\rho_\mu + \rho_\nu\|_p^p$$

whenever $\mu \geq \nu$. If $p \geq 2$, Clarkson's inequalities yield

$$\begin{aligned} \|\rho_\mu - \rho_\nu\|_p^p &\leq 2^{p-1}(\|\rho_\mu\|_p^p + \|\rho_\nu\|_p^p) - \|\rho_\mu + \rho_\nu\|_p^p \\ &\leq 2^{p-1}\left(2M + \frac{1}{\mu} + \frac{1}{\nu}\right) - 2^p \min\{M_p(\Gamma_\mu), M_p(\Gamma_\nu)\} \end{aligned}$$

for all μ and ν. Similarly, when $1 < p < 2$ and $q = p/(p-1)$, Clarkson's inequalities tell us that

$$\begin{aligned}\|\rho_\mu - \rho_\nu\|_p^p &\le 2(\|\rho_\mu\|_p^p + \|\rho_\nu\|_p^p)^{q/p} - \|\rho_\mu + \rho_\nu\|^q \\ &\le 2\left(2M + \frac{1}{\mu} + \frac{1}{\nu}\right)^{q/p} - 2^q \min\bigl\{M_p(\Gamma_\mu)^{q/p}, M_p(\Gamma_\nu)^{q/p}\bigr\}\end{aligned}$$

for all μ and ν. As $\lim_{\nu\to\infty} M_p(\Gamma_\nu) = M$, it becomes clear that, whatever the value of p might be, $\|\rho_\mu - \rho_\nu\|_p \to 0$ as μ and ν tend to ∞, which shows $\langle\rho_\nu\rangle$ is a Cauchy sequence in $L^p(\mathbb{R}^n)$.

We may thus assume that $\rho_\nu \to \rho$ in an L^p-norm, where $\rho : \mathbb{R}^n \to [0,\infty)$ is a Borel function. (Recall that every function in $L^p(\mathbb{R}^n)$ coincides almost everywhere in $\mathbb{R}^n$ with a Borel function taking values in $(-\infty,\infty)$.) We set $\rho(\infty) = 0$.

Finally, we apply Lemma 4.3.8 to produce a subfamily Γ_0 of Γ with $M_p(\Gamma_0) = 0$ and a subsequence $\langle\rho_{\nu_k}\rangle$ of $\langle\rho_\nu\rangle$ with the property that

$$\lim_{k\to\infty} \int_\gamma |\rho_{\nu_k} - \rho|\, ds = 0$$

for every curve γ in the family $\Gamma \setminus \Gamma_0$. (Implicit here is that all curves in $\Gamma\setminus\Gamma_0$ are locally rectifiable.) Fix such a γ. We assert that $\int_\gamma \rho\, ds \ge 1$. This is a trivial statement when $\int_\gamma \rho\, ds = \infty$, so we suppose that $\int_\gamma \rho\, ds < \infty$. Then $\int_\gamma \rho_{\nu_k}\, ds$ must be finite for all large k and, since γ belongs to Γ_{ν_k} once k becomes suitably large, we get

$$\int_\gamma \rho\, ds = \lim_{k\to\infty} \int_\gamma \rho_{\nu_k}\, ds \ge 1\,.$$

We reach the conclusion that ρ is a member of $\mathrm{Adm}(\Gamma\setminus\Gamma_0)$, which means that

$$M_p(\Gamma\setminus\Gamma_0) \le \|\rho\|_p^p = \lim_{k\to\infty} \|\rho_{\nu_k}\|_p^p \le M\,.$$

Because M_p is subadditive, we thus discover that

$$M_p(\Gamma) \le M_p(\Gamma\setminus\Gamma_0) + M_p(\Gamma_0) \le M\ ,$$

completing the proof. □

Surprisingly the direct analogue of Lemma 4.3.9 for nonincreasing sequences turns out to be false. To see this, consider for $k = 1, 2, \ldots$ the family

$$\Delta_k = \Delta\bigl(\mathbb{S}^{n-1}, \mathbb{S}^{n-1}(r_k) : B^n(r_k)\setminus\overline{B}^n\bigr)\ ,$$

where $r_k = 2 - (1/2)^k \to 2$. Set $\Gamma_\nu = \bigcup_{k=\nu}^\infty \Delta_k$. Then $\Gamma_1 \supset \Gamma_2 \supset \cdots$ and $\Gamma = \bigcap_{\nu=1}^\infty \Gamma_\nu = \emptyset$. Thus $M_p(\Gamma) = 0$. However, $\Delta_\nu \subset \Gamma_\nu$ and $\Delta_\nu \prec \Gamma_\nu$, so

$$M_p(\Gamma_\nu) = M_p(\Delta_\nu) = \omega_{n-1}\biggl(\int_1^{r_\nu} t^{q-1}\, dt\biggr)^{1-p} \to \omega_{n-1}\biggl(\int_1^2 t^{q-1}\, dt\biggr)^{1-p} > 0\ ,$$

where $q = (p-n)/(p-1)$. In other words, $M_p(\Gamma_\nu) \not\to M_p(\Gamma)$.

The lemma that follows next has consequences far beyond the limited application we intend to make of it. It can, for instance, be used to establish the existence of geodesics for certain conformally Euclidean metrics in subdomains of $\mathbb{R}^n$.

LEMMA 4.3.10. *Let G be a subset of $\hat{\mathbb{R}}^n$, and let $\rho : \bar{G} \to [0, \infty]$ be a lower semicontinuous function. Suppose that $\langle \gamma_\nu \rangle$ is a sequence of rectifiable curves in G and that x_ν and y_ν are the initial and terminal points, respectively, of γ_ν^*. Given that the sequence $\langle \ell(\gamma_\nu) \rangle$ is bounded and that the sequences $\langle x_\nu \rangle$ and $\langle y_\nu \rangle$ are both convergent—say, $x_\nu \to x$ and $y_\nu \to y$—there exists a rectifiable path γ in $\bar{G}$ with initial point x, with terminal point y, and with the property that*

$$\int_\gamma \rho\, ds \leq \liminf_{\nu \to \infty} \int_{\gamma_\nu} \rho\, ds\,. \tag{4.63}$$

PROOF. If $x = y$, the assertion is a triviality: any constant path γ whose trajectory is $\{x\}$ meets the requirements. We may assume, therefore, that $x \neq y$ and, by discarding finitely many terms, that $x_\nu \neq y_\nu$ for $\nu = 1, 2, \ldots$. The boundedness of $\langle \ell(\gamma_\nu) \rangle$ guarantees that both x and y are finite points. We may further suppose that $\beta_\nu = \gamma_\nu^*$ is parametrized by arclength on the interval $[0, \ell_\nu]$, where $\ell_\nu = \ell(\gamma_\nu)$. We write $\ell_0 = \sup_\nu \ell_\nu$ and extend β_ν to $[0, \ell_0]$ by setting $\beta_\nu(s) = y_\nu$ when $\ell_\nu < s \leq \ell_0$. Then $|\beta_\nu(t) - \beta_\nu(s)| \leq |t - s|$ holds for every ν whenever t and s belong to $[0, \ell_0]$. Notice especially that $\{\beta_\nu\}$ is an equicontinuous family on $[0, \ell_0]$. Since $x_\nu \to x \neq \infty$, the sequence $\langle x_\nu \rangle$ is bounded; since $|\beta_\nu(s)| \leq |x_\nu| + \ell_0$ for every s in $[0, \ell_0]$, the family $\{\beta_\nu\}$ is uniformly bounded on this interval.

Choose a subsequence $\langle \gamma_{\nu_k} \rangle$ of $\langle \gamma_\nu \rangle$ for which $\ell = \lim_{k \to \infty} \ell_{\nu_k}$ exists and

$$\lim_{k \to \infty} \int_{\gamma_{\sigma_{\mathbb{S}} \nu_k}} \rho\, ds = \liminf_{\nu \to \infty} \int_{\gamma_{\sigma_{\mathbb{S}} \nu}} \rho\, ds\,.$$

Necessarily, $0 < \ell \leq \ell_0$. Through an appeal to the Arzelà-Ascoli theorem we may suppose, by passing to a further subsequence and afterwards relabeling, that $\beta_{\nu_k} \to \beta$ uniformly on $[0, \ell_0]$, where $\beta : [0, \ell_0] \to \bar{G}$ is a path. Then $\gamma = \beta | [0, \ell]$ is also a path in $\bar{G}$. Moreover,

$$\gamma(0) = \lim_{k \to \infty} \beta_{\nu_k}(0) = \lim_{k \to \infty} x_{\nu_k} = x\,,$$

and, because $|y_{\nu_k} - \beta_{\nu_k}(\ell)| = |\beta_{\nu_k}(\ell_{\nu_k}) - \beta_{\nu_k}(\ell)| \leq |\ell_{\nu_k} - \ell| \to 0$,

$$\gamma(\ell) = \lim_{k \to \infty} \beta_{\nu_k}(\ell) = \lim_{k \to \infty} y_{\nu_k} = y\,.$$

This shows that γ has initial point x and terminal point y. In addition,

$$|\gamma(t) - \gamma(s)| = \lim_{k \to \infty} |\beta_{\nu_k}(t) - \beta_{\nu_k}(s)| \leq |t - s|$$

for all t and s in $[0, \ell]$, which reveals that the path γ is absolutely continuous (in particular, it is rectifiable) and has $|\dot{\gamma}(s)| \leq 1$ for almost every s in $[0, \ell]$.

There remains the verification of (4.63). Let b be a point of $(0, \ell)$. Then b belongs to $(0, \ell_{\nu_k})$ for all sufficiently large k. The combination of Fatou's lemma and the lower semicontinuity of ρ gives for each such b:

$$\begin{aligned}
\int_0^b \rho\big[\gamma(s)\big]\, ds &\leq \int_0^b \liminf_{k \to \infty} \rho\big[\beta_{\nu_k}(s)\big]\, ds \leq \liminf_{k \to \infty} \int_0^b \rho\big[\beta_{\nu_k}(s)\big]\, ds \\
&\leq \liminf_{k \to \infty} \int_0^{\ell_{\sigma_{\mathbb{S}} \nu_k}} \rho\big[\beta_{\nu_k}(s)\big]\, ds = \lim_{k \to \infty} \int_{\beta_{\sigma_{\mathbb{S}} \nu_k}} \rho\, ds \\
&= \lim_{k \to \infty} \int_{\gamma_{\sigma_{\mathbb{S}} \nu_k}} \rho\, ds = \liminf_{\nu \to \infty} \int_{\gamma_{\sigma_{\mathbb{S}} \nu}} \rho\, ds\,.
\end{aligned}$$

Therefore, calling to mind Lemma 4.1.4, we conclude that

$$\begin{aligned}\int_\gamma \rho\, ds &= \int_0^\ell \rho\big[\gamma(s)\big]\,|\dot\gamma(s)|\, ds \le \int_0^\ell \rho\big[\gamma(s)\big]\, ds \\ &= \lim_{b\to\ell^-}\int_0^b \rho\big[\gamma(s)\big]\, ds \le \liminf_{\nu\to\infty}\int_{\gamma_\nu}\rho\, ds\,,\end{aligned}$$

which confirms (4.63) and completes the proof. □

In order to retrieve information about the moduli of curve families from Lemma 4.3.10, we shall need to know that $M_p(\Gamma)$ is completely determined by the lower semicontinuous members of $\mathrm{Adm}(\Gamma)$. We make this fact explicit in another lemma.

LEMMA 4.3.11. *If Γ is a family of nonconstant curves in $\hat{\mathbb{R}}^n$ with $n \ge 2$, then*

$$M_p(\Gamma) = \inf\left\{\int_{\mathbb{R}^n}\rho^p\, dm_n\ : \rho \in \mathrm{Adm}(\Gamma)\cap \mathcal{L}(\hat{\mathbb{R}}^n)\right\}, \tag{4.64}$$

where $\mathcal{L}(\hat{\mathbb{R}}^n)$ is the class of extended real-valued functions on $\hat{\mathbb{R}}^n$ that are lower semicontinuous.

PROOF. Let M denote the infimum that appears on the right in (4.64). It is obvious that $M_p(\Gamma) \le M$. Assuming that $M_p(\Gamma) < \infty$, we shall verify that $M \le M_p(\Gamma)$. Let $\varepsilon > 0$ be given. Choose ρ in $\mathrm{Adm}(\Gamma)$ for which

$$\int_{\mathbb{R}^n}\rho^p\, dm_n < M_p(\Gamma) + \varepsilon\,.$$

The Vitali-Carathéodory theorem from real analysis [**142**, pp. 56–57] implies the existence of a lower semicontinuous function $\rho_0 : \mathbb{R}^n \to [0,\infty]$ satisfying $\rho_0 \ge \rho$ throughout $\mathbb{R}^n$ and enjoying the property that

$$\int_{\mathbb{R}^n}\rho_0^p\, dm_n \le \int_{\mathbb{R}^n}\rho^p\, dm_n + \varepsilon\,.$$

Each locally rectifiable curve γ from Γ has its trajectory in $\mathbb{R}^n$, so

$$\int_\gamma \rho_0\, ds \ge \int_\gamma \rho\, ds \ge 1$$

for any such curve. If we define $\rho_0(\infty) = \liminf_{x\to\infty}\rho_0(x)$, then ρ_0 becomes a member of the class $\mathrm{Adm}(\Gamma)\cap\mathcal{L}(\hat{\mathbb{R}}^n)$. Accordingly,

$$M \le \int_{\mathbb{R}^n}\rho_0^p\, dm_n \le \int_{\mathbb{R}^n}\rho^p\, dm_n + \varepsilon \le M_p(\Gamma) + 2\varepsilon\,.$$

Letting $\varepsilon \to 0$, we obtain $M \le M_p(\Gamma)$. □

In concrete geometric situations one can often do even better than Lemma 4.3.11. One example is the family $\Delta_0(E,F:D)$, where D is a domain in $\hat{\mathbb{R}}^n$ and where E and F are nonempty, disjoint, compact subsets of D. It can be shown that for such a configuration

$$M_p\big[\Delta_0(E,F:D)\big] = \inf\int_{\mathbb{R}^n}\rho^p\, dm_n\ ,$$

in which the infimum is taken over the collection of lower semicontinuous admissible densities ρ for $\Delta_0(E,F:D)$ that are actually real-valued and continuous in $D\backslash\{\infty\}$.

Having completed the necessary preliminaries, we can finally begin to discuss the continuity properties of moduli. Our first step in this direction is to consider the effect on $M_p[\Delta(E,F:G)]$ of "thickening" the sets E and F incrementally. If A is a nonempty subset of $\hat{\mathbb{R}}^n$ and $0 \le r \le 2$, we employ the notation $A(r)$ to denote

$$A(r) = \{x \in \hat{\mathbb{R}}^n : q(x,A) \le r\}.$$

We next establish:

LEMMA 4.3.12. *Let G be a nonempty set in $\hat{\mathbb{R}}^n$ with $n \ge 2$, and let E and F be nonempty, disjoint, compact subsets of $\hat{\mathbb{R}}^n$. Then*

$$M_p\big[\Delta(E,F:G)\big] \le \lim_{r\to 0} M_p\big[\Delta\big(E(r),F(r):G\big)\big] \le M_p\big[\Delta(E,F:\bar{G})\big] \tag{4.65}$$

for $p \ge n$. If G is an open set in $\mathbb{R}^n$ of finite volume, $|G| < \infty$, then (4.65) *remains true when $1 < p < n$.*

PROOF. Write $\Gamma(r) = \Delta(E(r),F(r):G)$—thus, $\Gamma(0) = \Delta(E,F:G)$—and let $\Gamma = \Delta(E,F:\bar{G})$. The monotonicity of M_p ensures that $M = \lim_{r\to 0} M_p[\Gamma(r)]$ exists and that $M_p[\Delta(E,F:G)] \le M$ for every $p > 1$. We now suppose that $p \ge n$ and demonstrate that $M \le M_p(\Gamma)$. It is only necessary to treat the case in which $M > 0$ and $M_p(\Gamma) < \infty$. This means, in particular, that $\Gamma(r)$ is nonempty for all $r > 0$. In the ensuing argument we again take advantage of the auxiliary function $\rho_0 : \hat{\mathbb{R}}^n \to [0,\infty)$ defined by

$$\rho_0(x) = \begin{cases} e^{-1} & \text{if } |x| \le e\ , \\ \frac{1}{|x|\log|x|} & \text{if } e < |x| < \infty\ , \\ 0 & \text{if } x = \infty\ . \end{cases}$$

(Recall the proof of Theorem 4.2.12.) This function is continuous and belongs to $L^p(\mathbb{R}^n)$ whenever $p \ge n$. Fixing $p \ge n$, we set $a = \int_{\mathbb{R}^n} \rho_0^p \, dm_n$. Given a function ρ from the class $\mathrm{Adm}(\Gamma) \cap \mathcal{L}(\hat{\mathbb{R}}^n)$ and given $\varepsilon > 0$, we let $\rho_\varepsilon = (\rho^p + \varepsilon^p \rho_0^p)^{1/p}$. Then ρ_ε is another member of the class $\mathrm{Adm}(\Gamma) \cap \mathcal{L}(\hat{\mathbb{R}}^n)$. We assert that there exists a $\delta > 0$ such that $(1+\varepsilon)\rho_\varepsilon$ is an admissible density for $\Gamma(r)$ whenever $0 < r < \delta$.

If the assertion is false, then there has to be a sequence $\langle r_\nu \rangle$ in $(0,1)$ with $r_\nu \to 0$ such that the family $\Gamma(r_\nu)$ contains a locally rectifiable curve γ_ν for which

$$\int_{\gamma_\nu} \rho_\varepsilon \, ds < \frac{1}{1+\varepsilon}\,. \tag{4.66}$$

We may assume that $E(r_\nu)$ and $F(r_\nu)$ are disjoint for every ν, that the initial point x_ν of the curve γ_ν^* lies in $E(r_\nu)$, and that its terminal point y_ν is a point of $F(r_\nu)$. Passing to subsequences if need be, we may suppose additionally that $x_\nu \to x$ and $y_\nu \to y$, where x belongs to E and y to F. Since E and F are disjoint, either $x \ne \infty$ or $y \ne \infty$. For definiteness, say that $x \ne \infty$. Fix a number $b > \max\{e,|x|\}$. No generality is lost by assuming that $|x_\nu| < b$ for all ν. We shall show that the curves γ_ν $(\nu = 1,2,\dots)$ are all rectifiable and that the sequence $\langle \ell(\gamma_\nu) \rangle$ is bounded. To this end, we introduce the constant $c_\nu = \sup\{|z| : z \in |\gamma_\nu|\}$. We first observe that

$$c_\nu \le c = b^{\exp(1/\varepsilon)} \tag{4.67}$$

for every ν. This is trivially the case if $c_\nu \le b$. If $c_\nu > b$, then γ_ν has a subcurve β_ν in the family $\Delta[\mathbb{S}^{n-1}(b), \mathbb{S}^{n-1}(c_\nu) : \mathbb{R}^n \setminus \overline{B}^n(b)]$. By (4.66) and Lemma 4.1.7,

$$1 \ge \int_{\gamma_\nu} \rho_\varepsilon \, ds \ge \int_{\gamma_\nu} \varepsilon\rho_0 \, ds \ge \varepsilon \int_{\beta_\nu} \rho_0 \, ds \ge \varepsilon \int_b^{c_\nu} \frac{dr}{r\log r} = \varepsilon \log\left(\frac{\log c_\nu}{\log b}\right)\,,$$

from which (4.67) quickly follows. We note especially that

$$\rho_\varepsilon \geq \varepsilon\rho_0 \geq \frac{\varepsilon}{c\log c} > 0$$

on $|\gamma_\nu|$, implying that

$$\ell(\gamma_\nu) = \int_{\gamma_\nu} ds \leq \frac{c\log c}{\varepsilon}\int_{\gamma_\nu} \rho_\varepsilon\, ds < \frac{c\log c}{\varepsilon} < \infty$$

for all ν. Appealing to Lemma 4.3.10, we can assert the existence of a rectifiable curve γ in Γ with x and y as endpoints and with the property that

$$\int_\gamma \rho_\varepsilon\, ds \leq \liminf_{\nu\to\infty}\int_{\gamma_\nu} \rho_\varepsilon\, ds \leq \frac{1}{1+\varepsilon} < 1\,.$$

This contradicts the fact that ρ_ε originated in $\mathrm{Adm}(\Gamma)\cap\mathcal{L}(\hat{\mathbb{R}}^n)$. The only alternative is that a $\delta > 0$ fitting the description in the claim exists. Thus, for r in $(0,\delta)$ we have

$$M_p\bigl[\Gamma(r)\bigr] \leq (1+\varepsilon)^p\int_{\mathbb{R}^n} \rho_\varepsilon^p\, dm_n = (1+\varepsilon)^p\biggl(\int_{\mathbb{R}^n}\rho^p\, dm_n + \varepsilon^p a\biggr)\,.$$

Since ρ in $\mathrm{Adm}(\Gamma)\cap\mathcal{L}(\hat{\mathbb{R}}^n)$ was arbitrary, Lemma 4.3.11 informs us that

$$M_p\bigl[\Gamma(r)\bigr] \leq (1+\varepsilon)^p\bigl[M_p(\Gamma) + \varepsilon^p a\bigr]$$

for all r such that $0 < r < \delta$. By letting $r \to 0$ we conclude that

$$M \leq (1+\varepsilon)^p\bigl[M_p(\Gamma) + \varepsilon^p a\bigr]\,.$$

If we now let $\varepsilon \to 0$, we get $M \leq M_p(\Gamma)$—provided, of course, that $p \geq n$.

Finally, if G is an open set in $\mathbb{R}^n$ with $|G| < \infty$, the above argument can be carried out with the role of auxiliary function played by $\rho_0 = \chi_G$. In this case we find that

$$\ell(\gamma_\nu) = \int_{\gamma_\nu} \rho_0\, ds \leq \frac{1}{\varepsilon}\int_{\gamma_\nu}\rho_\varepsilon\, ds < \frac{1}{\varepsilon}$$

for every ν, so Lemma 4.3.10 can again be invoked. □

As a special case, Lemma 4.3.12 yields our first continuity result for moduli.

THEOREM 4.3.13. *If E and F are nonempty compact subsets of $\hat{\mathbb{R}}^n$ with $n \geq 2$, then*

$$M_p\bigl[\Delta(E,F:\hat{\mathbb{R}}^n)\bigr] = \lim_{r\to 0} M_p\bigl[\Delta\bigl(E(r),F(r):\hat{\mathbb{R}}^n\bigr)\bigr]$$

for every $p \geq n$.

The conclusion of Lemma 4.3.12 is actually true for all $p > 1$, provided that Δ is changed to Δ_0. Indeed, an even more general fact was established by J. Hesse [**77, 78**].

THEOREM 4.3.14. *If D is a domain in $\hat{\mathbb{R}}^n$ with $n \geq 2$ and if $\langle E_\nu\rangle$ and $\langle F_\nu\rangle$ are non-increasing sequences of nonempty compact sets in D, then*

$$\lim_{\nu\to\infty} M_p\bigl[\Delta_0(E_\nu,F_\nu:D)\bigr] = M_p\bigl[\Delta_0\bigl(\bigcap_{\nu=1}^{\infty}E_\nu,\bigcap_{\nu=1}^{\infty}F_\nu:D\bigr)\bigr]$$

for every $p > 1$.

We shall not give the proof of Hesse's theorem, for it involves ideas and techniques that we cannot develop here. Theorem 4.3.13 for the case $p = n$ provides us with all the information we shall really need in the sequel.

The *Hausdorff distance* $q_\sigma(E, F)$ between nonempty compact sets E and F in $\hat{\mathbb{R}}^n$ is defined by

$$q_\sigma(E, F) = \max\{\ \max_{x \in E} q(x, F)\ ,\ \max_{x \in F} q(x, E)\ \}$$

or, equivalently, by

$$q_\sigma(E, F) = \inf\{r > 0 : E \subset F(r) \ \text{ and } \ F \subset E(r)\}\,.$$

It is not too difficult to show that q_σ defines a complete metric on the space of all nonempty compact subsets of $\hat{\mathbb{R}}^n$. In particular, it is meaningful to speak of a sequence $\langle E_\nu \rangle$ of compact sets in $\hat{\mathbb{R}}^n$ converging in the Hausdorff metric to a compact set E; this simply demands that

$$q_\sigma(E_\nu, E) \to 0, \qquad \text{as } \nu \to \infty.$$

As an example, if E is a nonempty compact set in $\hat{\mathbb{R}}^n$, it is easy to see that $E(r) \to E$, where $E(r)$ denotes the spherical r neighbourhood of E, in the Hausdorff metric as $r \to 0$. Another illustration is that if E is a nonempty compact set in $\hat{\mathbb{R}}^n$ and if $\langle f_\nu \rangle$ is a sequence from $C(E, \hat{\mathbb{R}}^n)$ that converges uniformly on E to a function f, then $f_\nu(E) \to f(E)$ in the Hausdorff metric.

Many real-valued set functions defined in $\hat{\mathbb{R}}^n$ are continuous relative to the Hausdorff metric. For instance, if sequences $\langle E_\nu \rangle$ and $\langle F_\nu \rangle$ of compact sets in $\hat{\mathbb{R}}^n$ converge in the Hausdorff metric to compact sets E and F, then $q(E_\nu) \to q(E)$ and $q(E_\nu, F_\nu) \to q(E, F)$. Observe, also, that $q_\sigma(E, F) = q_\sigma[g(E), g(F)]$ for all nonempty compact sets E and F in $\hat{\mathbb{R}}^n$ whenever g is a chordal isometry of $\hat{\mathbb{R}}^n$.

The proof of a key result pertaining to the continuity of conformal moduli is our next task.

THEOREM 4.3.15. *Let $\langle E_\nu \rangle$ and $\langle F_\nu \rangle$ be sequences of continua in $\hat{\mathbb{R}}^n$ with $n \geq 2$. Assume that $E_\nu \to E$ and $F_\nu \to F$ in the Hausdorff metric, where the compact sets E and F are disjoint. Then*

$$M\big[\Delta(E, F : \hat{\mathbb{R}}^n)\big] = \lim_{\nu \to \infty} M\big[\Delta(E_\nu, F_\nu : \hat{\mathbb{R}}^n)\big]\,. \tag{4.68}$$

PROOF. We first observe that the sets E and F are connected. To see this, if E, say, could be covered by a pair of disjoint open sets U and V, each of which meets E, then for large ν we would have $E_\nu \subset U \cup V$, $E_\nu \cap U \neq \emptyset$, and $E_\nu \cap V \neq \emptyset$, which would contradict the hypothesised connectedness of E_ν.

We have $q(E, F) > 0$, so $M = M[\Delta(E, F : \hat{\mathbb{R}}^n)] < \infty$ by Lemma 4.2.8. By performing a chordal isometry that moves some point of $(E \cup F)^c$ to ∞, we reduce matters to the case where E and F are compact sets in $\mathbb{R}^n$.

Then E_ν and F_ν are disjoint subsets of $K = \{x \in \mathbb{R}^n : \operatorname{dist}(x, E \cup F) \leq 1\}$ once ν is sufficiently large—we shall suppose this to be true for every ν—and if we set

$$\delta_\nu = \max\{\max_{x \in E} \operatorname{dist}(x, E_\nu)\ ,\ \max_{x \in E_\nu} \operatorname{dist}(x, E)\ ,\ \max_{x \in F} \operatorname{dist}(x, F_\nu)\ ,\ \max_{x \in F_\nu} \operatorname{dist}(x, F)\},$$

then

(4.69) $$\delta_\nu \to 0, \quad \text{as } \nu \to \infty$$

since $|x-y| \le 2(1+c^2)q(x,y)$ for x and y in K, where $c = \max\{|x| : x \in K\}$.

Let $r > 0$. For large ν we find that E_ν is contained in $E(r)$ and F_ν in $F(r)$, a fact which carries with it the implication that

$$\limsup_{\nu\to\infty} M\big[\Delta(E_\nu, F_\nu : \hat{\mathbb{R}}^n)\big] \le M\big[\Delta\big(E(r), F(r) : \hat{\mathbb{R}}^n\big)\big] .$$

Letting $r \to 0$ and referring to Lemma 4.3.12, we conclude that

(4.70) $$\limsup_{\nu\to\infty} M\big[\Delta(E_\nu, F_\nu : \hat{\mathbb{R}}^n)\big] \le M .$$

If $M = 0$, (4.68) follows, and so we proceed under the assumption that $M > 0$, which implies that neither E nor F can be degenerate. What remains is the problem of showing that

$$\liminf_{\nu\to\infty} M\big[\Delta(E_\nu, F_\nu : \hat{\mathbb{R}}^n)\big] \ge M .$$

By Theorem 4.2.15, we are at liberty to work with $\Gamma_\nu = \bar{\Delta}(E_\nu, F_\nu : \hat{\mathbb{R}}^n)$ and $\Gamma = \bar{\Delta}(E, F : \hat{\mathbb{R}}^n)$, and to verify that $\liminf_{\nu\to\infty} M(\Gamma_\nu) \ge M(\Gamma) = M$. Given the truth of inequality (4.70), there is no harm in assuming that $M(\Gamma_\nu) < 2M$ for every ν. Choose ρ_ν in $\mathrm{Adm}(\Gamma_\nu)$ such that

(4.71) $$\int_{\mathbb{R}^n} \rho_\nu^n \, dm_n \le M(\Gamma_\nu) + 2^{-\nu} M < 3M .$$

Let $0 < \varepsilon < 1/2$. We make the claim that for all suitably large ν, the function $(1-2\varepsilon)^{-1}\rho_\nu$ is an admissible density for Γ. Assuming this to be the case, we deduce that

$$M = M(\Gamma) \le (1-2\varepsilon)^{-n} \int_{\mathbb{R}^n} \rho_\nu^n \, dm_n \le (1-2\varepsilon)^{-n}\big[M(\Gamma_\nu) + 2^{-\nu}M\big]$$

for large ν. Accordingly,

$$M \le (1-2\varepsilon)^{-n} \liminf_{\nu\to\infty} M(\Gamma_\nu) .$$

Letting $\varepsilon \to 0$, we infer that

$$M \le \liminf_{\nu\to\infty} M(\Gamma_\nu) = \liminf_{\nu\to\infty} M\big[\Delta(E_\nu, F_\nu : \hat{\mathbb{R}}^n)\big] ,$$

which is enough to finish the proof of the theorem.

To establish our claim, we fix a real number δ satisfying

$$0 < \delta < 4^{-1} \min\{\mathrm{diam}(E), \mathrm{diam}(F), \mathrm{dist}(E, F)\} ,$$

after which we fix an index ν_0 such that the following four conditions are met whenever $\nu \ge \nu_0$:

$$\delta_\nu < \delta, \qquad \mathrm{diam}(E_\nu) > 2\delta, \qquad \mathrm{diam}(F_\nu) > 2\delta, \qquad c_n \varepsilon^n \log\frac{\delta}{\delta_\nu} > 3M .$$

Here δ_ν is defined by (4.69) and c_n is the constant that appeared in Theorem 4.2.22. The choice of ν_0 is possible because $\delta_\nu \to 0$ and because $\mathrm{diam}(E_\nu) \to \mathrm{diam}(E)$ and $\mathrm{diam}(F_\nu) \to \mathrm{diam}(F)$, as one easily checks.

We maintain that $(1-2\varepsilon)\rho_\nu$ is in $\mathrm{Adm}(\Gamma)$ whenever $\nu \ge \nu_0$. Fix $\nu \ge \nu_0$ and consider a rectifiable path γ from Γ. Assume that $\gamma : [a, b] \to \mathbb{R}^n$, with $x_0 = \gamma(a)$

in E and $y_0 = \gamma(b)$ in F. We are going to use γ to construct a rectifiable path β belonging to the family $\bar{\Delta}(E_\nu, F : \mathbb{R}^n)$ and with the property that

$$\int_\gamma \rho_\nu \, ds \geq \int_\beta \rho_\nu \, ds - \varepsilon \, .$$

If $|\gamma|$ and E_ν intersect, this is not difficult as our conditions show that E_ν and F are disjoint, and we can take β to be $\gamma|[c, b]$ for some c in $[a, b)$.

We shall therefore make the assumption that E_ν and $|\gamma|$ are disjoint. For each r satisfying $\delta_\nu < r < \delta$ the sphere $\mathbb{S}^{n-1}(x_0, r)$ meets both E_ν and $|\gamma|$. By Theorems 4.2.15 and 4.2.22,

$$M\big[\bar{\Delta}(E_\nu, |\gamma| : B)\big] \geq c_n \log \frac{\delta}{\delta_\nu} \ ,$$

where $B = B^n(x_0, \delta)$. There must be a rectifiable path γ_1 in $\bar{\Delta}(E_\nu, |\gamma|, B)$ for which $\int_{\gamma_1} \rho_\nu \, ds < \varepsilon$, otherwise $\varepsilon^{-1}\rho_\nu$ would be an admissible density for $\bar{\Delta}(E_\nu, |\gamma| : B)$, which in combination with (4.71) would lead to a contradiction; namely the inequality

$$3M > \int_{\mathbb{R}^n} \rho_\nu^n \, dm_n \geq \varepsilon^n M\big[\bar{\Delta}(E_\nu, |\gamma| : \mathbb{R}^n)\big] \geq c_n \varepsilon^n \log \frac{\delta}{\delta_\nu} > 3M$$

would hold.

We may assume that $\gamma_1 : [a_1, b_1] \to \mathbb{R}^n$ has $\gamma_1(a_1)$ in E_ν and $\gamma_1(b_1) = \gamma(c)$, where $a \leq c < b$. Then $\beta = \gamma_1 + \gamma|[c, b]$ delivers a rectifiable path in $\bar{\Delta}(E_\nu, F : \hat{\mathbb{R}}^n)$ with

$$\int_\gamma \rho_\nu \, ds \geq \int_\beta \rho_\nu \, ds - \int_{\gamma_1} \rho_\nu \, ds \geq \int_\beta \rho_\nu \, ds - \varepsilon \, .$$

We can now perform an analogous procedure on the path β near its terminal point y_0 so as to produce a rectifiable path α in $\Gamma_\nu = \bar{\Delta}(E_\nu, F_\nu : \mathbb{R}^n)$ for which

$$\int_\beta \rho_\nu \, ds \geq \int_\alpha \rho_\nu \, ds - \varepsilon \geq 1 - \varepsilon \, ,$$

where ρ_ν is the admissible density for Γ_ν. As a consequence,

$$\int_\gamma \rho_\nu \, ds \geq 1 - 2\varepsilon \ ,$$

which is to say that $(1 - 2\varepsilon)^{-1}\rho_\nu$ belongs to $\mathrm{Adm}(\Gamma)$. This is true whenever $\nu \geq \nu_0$, so the proof of the theorem is complete. □

In conjunction with Theorem 4.3.14, the proof of Theorem 4.3.15 (modulo a few minor alterations) establishes a more general result:

THEOREM 4.3.16. *If D is a domain in $\hat{\mathbb{R}}^n$ with $n \geq 2$ and if $\langle E_\nu \rangle$ and $\langle F_\nu \rangle$ are sequences of continua in D such that $E_\nu \to E$ and $F_\nu \to F$ in the Hausdorff metric, where E and F are disjoint compact subsets of D, then*

$$M\big[\Delta(E, F : D)\big] = \lim_{\nu \to \infty} M\big[\Delta(E_\nu, F_\nu : D)\big] \, .$$

Indeed, subject to more substantial modifications, the argument shows that

$$M_p\big[\Delta(E, F : D)\big] = \lim_{\nu \to \infty} M_p\big[\Delta(E_\nu, F_\nu : D)\big]$$

for every $p \geq n$, provided the limit sets E and F are nondegenerate continua in $D \cap \mathbb{R}^n$.

The assumption that E and F be disjoint is essential in Theorem 4.3.15. If, for example, $E_\nu = \{te_n : 2^{-\nu} \le t \le 2^{-\nu+1}\}$ and $F_\nu = \{-te_n : 2^{-\nu} \le t \le 2^{-\nu+1}\}$ for $\nu = 1, 2, \ldots$, then $E_\nu \to \{0\}$ and $F_\nu \to \{0\}$ in the Hausdorff metric. On the other hand, $M[\Delta(E_\nu, F_\nu : \hat{\mathbb{R}}^n)] = m$, a finite constant, for every ν, whereas $M[\Delta(\{0\}, \{0\} : \hat{\mathbb{R}}^n)] = \infty$. (We can, however, dispense with the disjointedness of E and F when both these limits are assumed to be nondegenerate: if E and F are intersecting continua, it is a simple exercise to show that $M[\Delta(E_\nu, F_\nu : \hat{\mathbb{R}}^n)] \to \infty = M[\Delta(E, F : \hat{\mathbb{R}}^n)]$.) The connectedness of E_ν and F_ν is likewise critical to the truth of Theorem 4.3.15. To see this, consider the sets

$$E = \{te_n : 1 \le t \le 2\}, \qquad F = \{-te_n : 1 \le t \le 2\}.$$

By virtue of the cap inequality, $M[\Delta(E, F : \hat{\mathbb{R}}^n)] \ge c_n \log 2 > 0$. If

$$E_\nu = \{(k/\nu)e_n : k = \nu, \nu+1, \ldots, 2\nu\}$$

and $F_\nu = F$ for $\nu = 1, 2, \ldots$, then $E_\nu \to E$ and $F_\nu \to F$ in the Hausdorff metric, while $M[\Delta(E_\nu, F_\nu : \hat{\mathbb{R}}^n)] = 0$ for every ν.

4.4. Extremal metrics

Given that a curve family Γ in $\hat{\mathbb{R}}^n$ has $M_p(\Gamma) < \infty$, it is natural to pose the following question: Is there an admissible density ρ for Γ such that

$$M_p(\Gamma) = \int_{\mathbb{R}^n} \rho^p \, dm_n \ ?$$

To see that in general the answer is no, consider the situation when $M_p(\Gamma) = 0$. What we are looking for in this case is a Borel function $\rho : \hat{\mathbb{R}}^n \to [0, \infty]$ satisfying $\rho = 0$ almost everywhere in $\mathbb{R}^n$, yet having the feature that $\int_\gamma \rho \, ds \ge 1$ for every locally rectifiable member γ of Γ. When Γ is "large enough", no such function need exist of course. For example, there is obviously no density of this type when $1 < p \le n$ and $\Gamma = \Delta(\mathbb{S}^{n-1}, \{0\} : B^n \setminus \{0\})$, nor when $p \ge n$ and $\Gamma = \Delta(\mathbb{S}^{n-1}, \{\infty\} : \mathbb{R}^n \setminus \overline{B}^n)$.

In order to get a more satisfactory response to our question we must cast our nets a bit wider and take into consideration densities ρ_0 that are *M_p-almost admissible for* Γ, in the following sense:

there is a subfamily Γ_0 of Γ with $M_p(\Gamma_0) = 0$ such that ρ_0 is an admissible density for the family $\Gamma \setminus \Gamma_0$.

We shall now prove that, when $M_p(\Gamma) < \infty$, there exists an M_p-almost admissible density ρ_0 for Γ with the property that

$$M_p(\Gamma) = \int_{\mathbb{R}^n} \rho_0^p \, dm_n.$$

We refer to any function meeting these qualifications as a *p-extremal metric* Γ. Furthermore, assuming that both ρ_0 and ρ_1 are p-extremal metrics for Γ, we shall demonstrate that $\rho_0 = \rho_1$ almost everywhere in $\mathbb{R}^n$. The existence and essential uniqueness of p-extremal metrics are established in the ensuing set of theorems.

THEOREM 4.4.1. *If Γ is a curve family in $\hat{\mathbb{R}}^n$ with $n \ge 2$ and if $M_p(\Gamma) < \infty$, then Γ has a p-extremal metric.*

PROOF. Choose a sequence $\langle\rho_\nu\rangle$ from $\mathrm{Adm}(\Gamma)\cap L^p(\mathbb{R}^n)$ such that

$$\int_{\mathbb{R}^n}\rho_\nu^p\,dm_n\to M_p(\Gamma).$$

We show that $\langle\rho_\nu\rangle$ is a Cauchy sequence with respect to the L^p-norm. For all ν and μ the function $(\rho_\mu+\rho_\nu)/2$ is an admissible density for Γ, the implication being that

$$M_p(\Gamma)\le 2^{-p}\|\rho_\mu+\rho_\nu\|_p^p\,.$$

Clarkson's inequalities inform us that

$$\begin{aligned}\|\rho_\mu-\rho_\nu\|_p^p &\le 2^{p-1}(\|\rho_\mu\|_p^p+\|\rho_\nu\|_p^p)-\|\rho_\mu+\rho_\nu\|_p^p\\ &\le 2^{p-1}(\|\rho_\mu\|_p^p+\|\rho_\nu\|_p^p)-2^pM_p(\Gamma)\end{aligned}$$

when $p\ge 2$ and that, with $q=p/(p-1)$,

$$\begin{aligned}\|\rho_\mu-\rho_\nu\|_p^q &\le 2(\|\rho_\mu\|_p^p+\|\rho_\nu\|_p^p)^q-\|\rho_\mu+\rho_\nu\|_p^q\\ &\le 2(\|\rho_\mu\|_p^p+\|\rho_\nu\|_p^p)^{q/p}-2^qM_p(\Gamma)^{q/p}\end{aligned}$$

when $1<p\le 2$.

In either case, it is clear from the selection of $\langle\rho_\nu\rangle$ that $\|\rho_\mu-\rho_\nu\|_p\to 0$ as $\mu,\nu\to\infty$. Let $\rho_0:\hat{\mathbb{R}}^n\to[0,\infty)$ be a Borel function with the property that $\rho_\nu\to\rho_0$ in $L^p(\mathbb{R}^n)$. According to Lemma 4.3.8 we may assume, at the possible expense of replacing $\langle\rho_\nu\rangle$ with some subsequence, that there is a subfamily Γ_0 of Γ having $M_p(\Gamma_0)=0$ such that $\int_\gamma|\rho_\nu-\rho_0|\,ds\to 0$ for every curve γ in $\Gamma\setminus\Gamma_0$. For each γ in $\Gamma\setminus\Gamma_0$ it is either true that $\int_\gamma\rho_0\,ds=\infty$ or that ρ_0 is integrable along γ, with

$$\int_\gamma\rho_0\,ds=\lim_{\nu\to\infty}\int_\gamma\rho_\nu\,ds\ge 1\,.$$

Thus ρ_0 is seen to be an M_p-almost admissible density for Γ, and

$$\int_{\mathbb{R}^n}\rho_0^p\,dm_n=\lim_{\nu\to\infty}\int_{\mathbb{R}^n}\rho_\nu^p\,dm_n=M_p(\Gamma)\,.$$

By definition, ρ_0 is a p-extremal metric for Γ. □

The uniqueness of p-extremal metrics can be derived from the following variational principle.

THEOREM 4.4.2. *Suppose that Γ is a family of curves in $\hat{\mathbb{R}}^n$ with $n\ge 2$, that $M_p(\Gamma)<\infty$, and that ρ_0 is a p-extremal metric for Γ. If Γ_1 is a subfamily of Γ for which $M_p(\Gamma_1)=M_p(\Gamma)$, then*

$$M_p(\Gamma)\le\int_{\mathbb{R}^n}\rho_0^{p-1}(x)\,\rho_1(x)\,dm_n(x)\tag{4.72}$$

whenever ρ_1 is a member of $\mathrm{Adm}(\Gamma_1)$ *that belongs to* $L^p(\mathbb{R}^n)$.

PROOF. Let Γ_0 be a subfamily of Γ with $M_p(\Gamma_0)=0$ such that ρ_0 is an admissible density for the family $\Gamma\setminus\Gamma_0$. If $\Gamma_2=\Gamma_1\setminus\Gamma_0$, then $M_p(\Gamma_2)=M_p(\Gamma_1)=M_p(\Gamma)$. Fix ρ_1, an L^p-integrable member of $\mathrm{Adm}(\Gamma_1)$. For t in $(0,1)$, set $\rho=\rho_0+t\rho_1$. Then ρ is a Borel function on $\hat{\mathbb{R}}^n$. Moreover,

$$\int_\gamma\rho\,ds=\int_\gamma\rho_0\,ds+t\int_\gamma\rho_1\,ds\ge 1+t$$

for each locally rectifiable curve γ in Γ_2. It follows that $(1+t)^{-1}\rho$ is an admissible density for Γ_2. We infer that

$$M_p(\Gamma) = M_p(\Gamma_2) \le (1+t)^{-p} \int_{\mathbb{R}^n} \rho^p \, dm_n .$$

Next, because $M_p(\Gamma) = \int_{\mathbb{R}^n} \rho_0^p \, dm_n$, we see that

$$\big[(1+t)^p - 1\big] M_p(\Gamma) \le \int_{\mathbb{R}^n} \big[(\rho_0 + t\rho_1)^p - \rho_0^p\big] \, dm_n \tag{4.73}$$

whenever $0 < t < 1$. Write

$$g_t = t^{-1}[(\rho_0 + t\rho_1)^p - \rho_0^p], \quad \text{for } t \in (0,1).$$

Then

$$\lim_{t \to 0} g_t(x) = p \, \rho_0^{p-1}(x) \, \rho_1(x) \tag{4.74}$$

for every x in $\mathbb{R}^n$ at which both $\rho_0(x) < \infty$ and $\rho_1(x) < \infty$. Since ρ_0 and ρ_1 are L^p-functions, this is the case for almost every x in $\mathbb{R}^n$. Furthermore,

$$0 \le g_t(x) = \frac{p}{t} \int_{\rho_0(x)}^{\rho_0(x)+t\rho_1(x)} u^{p-1} \, du \le p\big[\rho_0(x) + \rho_1(x)\big]^{p-1} \rho_1(x)$$

for each such x. The function $(\rho_0 + \rho_1)^{p-1}$ belongs to $L^{p/(p-1)}(\mathbb{R}^n)$. Accordingly, the function $p(\rho_0 + \rho_1)^{p-1}\rho_1$ finds itself in $L^1(\mathbb{R}^n)$. Once we recall (4.73) and (4.74), and then appeal to the dominated convergence theorem, we are able to claim that

$$\begin{aligned} pM_p(\Gamma) &= \lim_{t \to 0} \frac{(1+t)^p - 1}{t} M_p(\Gamma) \le \lim_{t \to 0} \int_{\mathbb{R}^n} g_t \, dm_n \\ &= \int_{\mathbb{R}^n} \lim_{t \to 0} g_t \, dm_n = p \int_{\mathbb{R}^n} \rho_0^{p-1} \rho_1 \, dm_n \ , \end{aligned}$$

which clearly implies (4.72) and completes the proof. □

From Theorem 4.4.2 it is an easy step to the uniqueness result we seek.

THEOREM 4.4.3. *Let Γ be a family of curves in $\hat{\mathbb{R}}^n$ with $n \ge 2$. If $M_p(\Gamma) < \infty$ and if ρ_0 and ρ_1 are p-extremal metrics for Γ, then $\rho_0 = \rho_1$ almost everywhere in $\mathbb{R}^n$.*

PROOF. Choose a subfamily Γ_0 of Γ with $M_p(\Gamma_0) = 0$ such that ρ_1 is an admissible density for the family $\Gamma_1 = \Gamma \backslash \Gamma_0$. Since $M_p(\Gamma_1) = M_p(\Gamma)$, Theorem 4.4.2 and Hölder's inequality yield

$$M_p(\Gamma) \le \int_{\mathbb{R}^n} \rho_0^{p-1} \rho_1 \, dm_n \le \left(\int_{\mathbb{R}^n} \rho_0^p \, dm_n \right)^{(p-1)/p} \left(\int_{\mathbb{R}^n} \rho_1^p \, dm_n \right)^{1/p} = M_p(\Gamma) \, .$$

Equality at the extremes forces equality to hold throughout this string of inequalities, which in turn dictates that $\rho_0 = \rho_1$ almost everywhere in $\mathbb{R}^n$. □

We close this section by recording a property of p-extremal metrics that will later turn out to be of some value. The following terminology is used: when E and F are nonempty subsets of a set G in $\hat{\mathbb{R}}^n$, we say that a subset S of $\hat{\mathbb{R}}^n$ *separates* E *and* F *relative to* G provided S is disjoint from $E \cup F$ and no connected component of $G \setminus S$ contains points from both E and F.

THEOREM 4.4.4. *Let G be a set in $\hat{\mathbb{R}}^n$ with $n \geq 2$, let E and F be nonempty subsets of G, and let $\Gamma = \Delta(E, F : G)$. Assume the existence of a bounded set S in $\mathbb{R}^n$ that separates E and F relative to G and has $d = \operatorname{dist}(S, A) > 0$, where $A = (E \cup F) \cap \mathbb{R}^n$. Then $M_p(\Gamma) < \infty$ for every $p > 1$. Moreover, for any r in $(0, d)$ and for any p-extremal metric ρ_0 for Γ, it is the case that*

$$\int_{U_r} \rho_0^{p-1}(x)\, dm_n(x) \geq 2r M_p(\Gamma)\ , \tag{4.75}$$

in which $U_r = \{x \in \mathbb{R}^n : \operatorname{dist}(x, S) < r\}$.

PROOF. Fix r in $(0, d)$ and write $U = U_r$. Then U is a bounded open set in $\mathbb{R}^n$, so $\rho_1 = (2r)^{-1}\chi_U$ is a Borel function and

$$\int_{\mathbb{R}^n} \rho_1^p\, dm_n = (2r)^{-p}|U| < \infty$$

for each $p > 1$. We assert that ρ_1 is an admissible density for Γ. If this is true, then $M_p(\Gamma) < \infty$ and Theorem 4.4.2 (with $\Gamma_1 = \Gamma$) lets us know that

$$\frac{1}{2r}\int_U \rho_0^{p-1}\, dm_n = \int_{\mathbb{R}^n} \rho_0^{p-1}\rho_1\, dm_n \geq M_p(\Gamma)$$

whenever ρ_0 is a p-extremal metric for Γ. Thus (4.75) holds. To confirm that ρ_1 is a member of $\operatorname{Adm}(\Gamma)$, consider a locally rectifiable curve γ from Γ, say with endpoints x in E and y in F. Because S separates E and F relative to G, $|\gamma|$ must intersect S. Choose a point z of $S \cap |\gamma|$. Since $|x - z| > r$ and $|y - z| > r$, we see that γ must have a subcurve β such that

$$z \in |\beta| \subset B^n(z, r) \subset U$$

and such that both endpoints of β lie on the sphere $\mathbb{S}^{n-1}(z, r)$. We conclude that

$$\int_\gamma \rho_1\, ds \geq \int_\beta \rho_1\, ds = \frac{1}{2r}\int_\beta ds = \frac{\ell(\beta)}{2r} \geq \frac{2r}{2r} = 1\ ,$$

which identifies ρ_1 as a function from $\operatorname{Adm}(\Gamma)$. □

4.5. ACL-functions and Fuglede's theorem

Let $n \geq 2$. In the discussion that follows we use Π_i to denote the orthogonal projection of $\mathbb{R}^n$ onto the hyperplane $\mathbb{R}_i^{n-1} = \{x \in \mathbb{R}^n : x_i = 0\}$. Thus if $x = (x_1, x_2, \ldots, x_n)$, then $\Pi_i(x) = x - x_i e_i$.

Let Q be a closed n-interval in $\mathbb{R}^n$, meaning a set of the form $[a_1, b_1] \times [a_2, b_2] \times \cdots \times [a_n, b_n]$ with $a_i < b_i$ for each i.

A function $u : Q \to \mathbb{R}$ is said to have the ACL-*property* or to be an ACL-*function*—ACL stands for "absolutely continuous on lines"—if u is continuous and if for $i = 1, 2, \ldots, n$ the following condition is met: there is a subset E_i of $\Pi_i(Q)$ such that $m_{n-1}(E_i) = 0$ and such that for every point y of $\Pi_i(Q) \setminus E_i$ the function of x_i defined on the interval $[a_i, b_i]$ by $x_i \mapsto u(y + x_i e_i)$ is absolutely continuous. Note the implication that, whenever y belongs to the set $\Pi_i(Q) \setminus E_i$, the partial derivative $\partial_i u(x)$ exists at almost every point x of the line segment $Q \cap \Pi_i^{-1}(\{y\})$. Thus, letting A_i designate the (Borel) set of x in Q at which $\partial_i u(x)$ fails to exist, Fubini's theorem ensures that

$$m_n(A_i) = \int_{\Pi_i(Q)} m_1\big[A_i \cap \Pi_i^{-1}(\{y\})\big]\, dm_{n-1}(y) = 0\,.$$

It follows that the partial derivatives $\partial_1 u, \partial_2 u, \ldots, \partial_n u$ of an ACL-function $u : Q \to \mathbb{R}$ are Borel functions that are defined almost everywhere in Q.

It is a common practice in this context—and one that we shall follow—to assign $\partial_i u(x)$ the value 0 at any x in A_i. Then under this convention the symbol $\partial_i u$ stands for a Borel function that is defined everywhere in Q. In particular, the formal gradient

$$\nabla u(x) = (\partial_1 u(x), \partial_2 u(x), \ldots, \partial_n u(x))$$

exists for almost every x in Q. We stress, however, that ∇u is only the formal gradient of u: we are not claiming that u is actually differentiable anywhere in Q.

In the added event that the partials $\partial_1 u, \partial_2 u, \ldots, \partial_n u$ all belong to $L^p(Q)$, we call u an ACL^p*-function*. Here $1 \le p \le \infty$.

For example, a function $u : Q \to \mathbb{R}$ that satisfies a Lipschitz condition on Q is certainly an ACL-function—indeed, an ACL$^\infty$-function. It is easy to see that the sum $u + v$ and product uv of ACL-functions u and v defined on Q are themselves ACL-functions. If $u : Q \to \mathbb{R}$ is an ACL-function (respectively, an ACL^p-function) and $\varphi : u(Q) \to \mathbb{R}$ is a Lipschitz function, then $v = \varphi \circ u$ is an ACL-function (respectively, an ACL^p-function). This follows from the fact that $|v(y) - v(x)| \le \lambda |u(y) - u(x)|$ for all x and y in Q—hence, that $|\partial_i v(x)| \le \lambda |\partial_i u(x)|$ for every x in Q at which both these partials exist—where λ is a Lipschitz constant for φ.

It is a simple exercise to check that an ACL-function $u : Q \to \mathbb{R}$ whose formal gradient ∇u vanishes almost everywhere in Q is constant in this n-interval.

We draw our attention to an important feature of ACL^1-functions.

LEMMA 4.5.1. *If Q is a closed n-interval in $\mathbb{R}^n$ and $u : Q \to \mathbb{R}$ is an* ACL^1*-function, then for $i = 1, 2, \ldots, n$*

$$\int_Q \varphi \cdot \partial_i u \, dm_n = -\int_Q u \cdot \partial_i \varphi \, dm_n \tag{4.76}$$

whenever $\varphi : Q \to \mathbb{R}$ is an ACL^1*-function that vanishes on ∂Q.*

PROOF. Fix i and let φ be as indicated. For almost every y in $\Pi_i(Q)$ we can exploit the absolute continuity of the functions $x_i \mapsto u(y+x_i e_i)$ and $x_i \mapsto \varphi(y+x_i e_i)$ on $[a_i, b_i]$, the integration by parts formula, and the information that φ vanishes on ∂Q to conclude that

$$\int_{a_i}^{b_i} \varphi(y + x_i e_i) \partial_i u(y + x_i e_i) \, dx_i = -\int_{a_i}^{b_i} u(y + x_i e_i) \partial_i \varphi(y + x_i e_i) \, dx_i \, .$$

Since $\partial_i u$ and $\partial_i \varphi$ belong to $L^1(Q)$, while u and φ are plainly members of $L^\infty(Q)$, we can invoke Fubini's theorem to justify the computation

$$\begin{aligned} \int_Q \varphi \cdot \partial_i u \, dm_n &= \int_{\Pi_i(Q)} \left[\int_{a_i}^{b_i} \varphi(y + x_i e_i) \partial_i u(y + x_i e_i) \, dx_i \right] dm_{n-1}(y) \\ &= -\int_{\Pi_i(Q)} \left[\int_{a_i}^{b_i} u(y + x_i e_i) \partial_i \varphi(y + x_i e_i) \, dx_i \right] dm_{n-1}(y) \\ &= -\int_Q u \cdot \partial_i \varphi \, dm_n \, , \end{aligned}$$

which confirms (4.76). □

Let U be an open set in $\mathbb{R}^n$ with $n \geq 2$. To state that a function $u : U \to \mathbb{R}$ belongs to the class $\mathrm{ACL}(U)$—one speaks of u being *absolutely continuous on lines in* U—means that $u|Q$ is an ACL-function for each closed n-interval Q in U. Note that this definition requires u to be at least a continuous function.

The property of a function being in $\mathrm{ACL}(U)$ is clearly a local one: a function $u : U \to \mathbb{R}$ is a member of $\mathrm{ACL}(U)$ if and only if each point x of U has an open neighbourhood $V = V_x$ in U such that $u|V$ belongs to $\mathrm{ACL}(V)$.

The class $\mathrm{ACL}^p(U)$, where $1 \leq p \leq \infty$, is defined in an analogous way, and membership in $\mathrm{ACL}^p(U)$ is likewise a local property of a function.

Since U can be expressed as a countable union of closed n-intervals, we infer from our earlier discussion that the partial derivatives $\partial_1 u, \partial_2 u, \ldots, \partial_n u$ of a function u from the class $\mathrm{ACL}(U)$ are Borel functions each of whose domains includes almost every point of U. As earlier, we stick to the convention of setting $\partial_i u(x) = 0$ for any point x in U at which this partial derivative fails to exist, thereby ensuring that $\partial_i u$ is a Borel function defined everywhere in u. Thus, u has a formal gradient $\nabla u(x)$ for almost every x in U. Should U be connected and $\nabla u = 0$ almost everywhere in U, then u would be constant there. If u is in $\mathrm{ACL}^p(U)$, then the functions $\partial_1 u, \partial_2 u, \ldots, \partial_n u$ and $|\nabla u|$ are locally L^p-functions in U, which is to say that their restrictions to any compact subset A of U belong to $L^p(A)$. Finally, a function $f : U \to \mathbb{R}^m$ is a member of the class $\mathrm{ACL}(U, \mathbb{R}^m)$ (respectively, $\mathrm{ACL}^p(U, \mathbb{R}^m)$) provided that each of its coordinate functions belongs to $\mathrm{ACL}(U)$ (respectively, $\mathrm{ACL}^p(U)$). Such a function f has a formal derivative matrix $f'(x)$ at almost every point x of U.

Let U again be an open set in $\mathbb{R}^n$—here $n = 1$ is allowed—and let $u : U \to \mathbb{R}$ be a function that is locally integrable in U. We say that a locally integrable function $v : U \to \mathbb{R}$ is a *weak* (or *distributional*) *partial derivative of* u *in* U *with respect to the variable* x_i if it is true that

$$\int_U u \cdot \partial_i \varphi \, dm_n = -\int_U \varphi v \, dm_n$$

for every φ in $C_0^\infty(U)$.

If $\tilde{v}$ is a second function satisfying these conditions, then it is not difficult to show that $\tilde{v} = v$ almost everywhere in U, so v is essentially unique. Given that a weak i^{th}-partial of u exists, we use $D_i u$ to denote a generic representative of this derivative (modulo almost everywhere equality). Of course, if u comes from $C^1(U)$, we can take $D_i u = \partial_i u$.

Note that there are continuous functions $u : U \to \mathbb{R}$—the Cantor function $u : (0, 1) \to [0, 1]$ gives a one-dimensional example—such that $\partial_i u$ exists almost everywhere in U, is locally integrable in U, but is not a weak i^{th}-partial of u in U.

For p in $[1, \infty]$ the *Sobolev space* $W^{1,p}(U)$ consists of all functions $u : U \to \mathbb{R}$ belonging to $L^p(U)$ and possessing weak partials $D_1 u, D_2 u, \ldots, D_n u$ that are themselves in $L^p(U)$. For u in $W^{1,p}(U)$ we set

$$\|u\|_{W^{1,p}(U)} = \|u\|_{L^p(U)} + \sum_{i=1}^{n} \|D_i u\|_{L^p(U)} \, .$$

This defines a seminorm on $W^{1,p}(U)$, one for which $\|u\|_{W^{1,p}(U)} = 0$ if and only if $u = 0$ almost everywhere in U. The space $W^{1,p}(U)$ is complete with respect to this seminorm. To learn more about Sobolev spaces, see [**114**].

Focusing our attention now on a function u from the class $\mathrm{ACL}^1(U)$, we let φ belong to $C_0^\infty(U)$. We select a finite collection $Q_1, Q_2, \ldots, Q_p$ of closed n-intervals in U whose interiors cover the support of φ. Via a standard partition of unity construction we can express φ as a sum $\varphi = \varphi_1 + \varphi_2 + \cdots + \varphi_p$, where φ_k is a function in $C_0^\infty(U)$ whose support lies in the interior of Q_k. With the aid of Lemma 4.1.1, we are then able to compute

$$\int_U u \cdot \partial_i \varphi \, dm_n = \sum_{k=1}^{p} \int_{Q_k} u \cdot \partial_i \varphi_k \, dm_n = -\sum_{k=1}^{p} \int_{Q_k} \varphi_k \cdot \partial_i u \, dm_n = -\int_U \varphi \cdot \partial_i u \, dm_n$$

for $i = 1, 2, \ldots, n$. The inference:

LEMMA 4.5.2. *If u is a member of $\mathrm{ACL}^1(U)$, then the actual partial derivative $\partial_1 u, \partial_2 u, \ldots, \partial_n u$ of u also defines weak partial derivatives for u in U.*

Accordingly, any function u from $\mathrm{ACL}^p(U)$ is locally a $W^{1,p}$-function in U, in the sense that $u|V$ belongs to $W^{1,p}(V)$ for every bounded open set V whose closure is contained in U.

Any function u in $\mathrm{ACL}^p(U)$ is nicely approximable by functions from $C_0^\infty(U)$. In order to give this statement precise meaning, we recall for the reader some facts about the smoothing of functions. Fix a nonnegative function ψ in $C_0^\infty(\mathbb{R}^n)$ that has support in B^n and satisfies $\int_{\mathbb{R}^n} \psi \, dm_n = 1$. For $\nu = 1, 2, \ldots$ define ψ_ν by $\psi_\nu(x) = \nu^n \psi(\nu x)$. Then ψ_ν is a function in $C_0^\infty(\mathbb{R}^n)$, its support lies in $B^n(\nu^{-1})$, and $\int_{\mathbb{R}^n} \psi_\nu \, dm_n = 1$. If $f : \mathbb{R}^n \to \mathbb{R}$ is a locally integrable function, we can form the convolution $f_\nu = f * \psi_\nu$:

$$f_\nu(x) = \int_{\mathbb{R}^n} f(x-y)\psi_\nu(y) \, dm_n(y) = \int_{\mathbb{R}^n} f(y)\psi_\nu(x-y) \, dm_n(y) .$$

The function f_ν is a member of the class $C^\infty(\mathbb{R}^n)$. Its partial derivative $\partial_i f_\nu$ is given by $\partial_i f_\nu = f * \partial_i \psi_\nu$. (Under the extra assumption that f has a weak partial derivative $D_i f$ in $\mathbb{R}^n$, $\partial_i f_\nu$ can also be represented in the manner $\partial_i f_\nu = D_i f * \psi_\nu$.) Were f to vanish almost everywhere off a compact set K in $\mathbb{R}^n$, f_ν would belong to $C_0^\infty(\mathbb{R}^n)$; in fact, the support of f_ν would be contained in $\{x : \mathrm{dist}(x, K) \le \nu^{-1}\}$. We list for reference three well-known convergence properties of the sequence $\langle f_\nu \rangle$:

(i) If x is a *Lebesgue point* for f, by which we understand a point with the property that

$$\lim_{r \to 0} \frac{1}{|B^n(x,r)|} \int_{B^n(x,r)} |f(y) - f(x)| \, dm_n(y) = 0 ,$$

then $f_\nu(x) \to f(x)$. Since almost every x in $\mathbb{R}^n$ is a Lebesgue point for f, it follows that $f_\nu \to f$ pointwise almost everywhere in $\mathbb{R}^n$.

(ii) If f is continuous in a neighbourhood of a compact subset K of $\mathbb{R}^n$, then $f_\nu \to f$ uniformly on K.

(iii) If f belongs to $L^p(\mathbb{R}^n)$ with $1 \le p < \infty$, then $f_\nu \to f$ in L^p-norm.

As we suggested earlier, the foregoing comments enter into the proof of a useful approximation result for ACL^p-functions.

THEOREM 4.5.3. *Let U be an open set in $\mathbb{R}^n$ with $n \geq 2$, and let $1 \leq p < \infty$. If a function u belongs to the class $\mathrm{ACL}^p(U)$, then there exists a sequence $\langle u_\nu \rangle$ in $C_0^\infty(U)$ with the following property: for each compact set K in U, $u_\nu \to u$ uniformly on K and $\partial_i u_\nu \to \partial_i u$ in L^p-norm on K for $i = 1, 2, \ldots, n$.*

PROOF. It suffices to show that, given a compact set K in U and $\varepsilon > 0$, we can find a function v in $C_0^\infty(U)$ with $|v(x) - u(x)| < \varepsilon$ for every x in K and

$$\int_K |\partial_i v - \partial_i u|^p \, dm_n < \varepsilon, \qquad \text{for } i = 1, 2, \ldots, n.$$

Once this is demonstrated, we can write $U = \bigcup_{\nu=1}^\infty K_\nu$, where $\langle K_\nu \rangle$ is a sequence of compact sets in U such that K_ν lies in the interior of $K_{\nu+1}$, and choose u_ν in $C_0^\infty(U)$ such that $|u_\nu - u| < \nu^{-1}$ on K_ν and $\int_{K_\nu} |D_i u_\nu - D_i u|^p \, dm_n < \nu^{-1}$ for $i = 1, 2, \ldots, n$. Then $\langle u_\nu \rangle$ will be a sequence with the desired characteristics.

We now fix K and ε as indicated. Define a function $f : \mathbb{R}^n \to \mathbb{R}$ by $f(x) = \varphi(x)u(x)$ for x in U, where φ is a function from $C_0^\infty(U)$ such that $\varphi \equiv 1$ in an open neighbourhood V of K, and $f(x) = 0$ for x in $\mathbb{R}^n \setminus U$. Then f is a continuous function whose support is contained in that of φ. The function f also belongs to $\mathrm{ACL}(\mathbb{R}^n)$, a fact which is made evident by looking at the restriction of f to each of the sets U and $\mathbb{R}^n \setminus \mathrm{supp}(\varphi)$. Plainly $\partial_i f = u\partial_i \varphi + \varphi \partial_i u$ almost everywhere in U and $\partial_i f = 0$ in $\mathbb{R}^n \setminus U$. In particular, $\partial_i f$ belongs to $L^p(\mathbb{R}^n)$ and $\partial_i f = \partial_i u$ almost everywhere in V. We make the following remarks concerning the mollified sequence $f_\nu = f * \psi_\nu$ that was discussed above: the function f_ν is in $C_0^\infty(\mathbb{R}^n)$ and, for ν suitably large, its support lies in U; $f_\nu \to f$ uniformly on K, a set on which f and u agree; since $\partial_i f$ is a weak partial derivative for f in $\mathbb{R}^n$, we have $\partial_i f_\nu = \partial_i f * \psi_\nu$; since $\partial_i f$ belongs to $L^p(\mathbb{R}^n)$, $\partial_i f_\nu \to \partial_i f$ on $\mathbb{R}^n$ in L^p-norm—hence, $\partial_i f_\nu \to \partial_i u$ in $L^p(K)$. By taking $v = f_\nu | U$ for large ν, we obtain a function in $C_0^\infty(U)$ with $|v - u| < \varepsilon$ on K and $\int_K |\partial_i v - \partial_i u|^p \, dm_n < \varepsilon$ for $i = 1, 2, \ldots, n$. □

Theorem 4.5.3 has a converse:

THEOREM 4.5.4. *If $u : U \to \mathbb{R}$ is a function for which there exists a sequence $\langle u_\nu \rangle$ in $C_0^\infty(U)$ such that $u_\nu \to u$ locally uniformly in U and such that $\partial_i u_\nu \to v_i$ locally in L^p-norm in U for $i = 1, 2, \ldots, n$, then u is a member of the class $\mathrm{ACL}^p(U)$ and $\partial_i u = v_i$ almost everywhere in U.*

We shall not need this fact, so we supply no proof for it.

The main result in this section is a theorem due to [**42**]. It states, albeit in more exact terms, that an ACL^p-function is reasonably well behaved on M_p-almost every curve in its domain.

THEOREM 4.5.5. *Let U be an open set in $\mathbb{R}^n$ with $n \geq 2$, and let u be a function from the class $\mathrm{ACL}^p(U)$ with $1 < p < \infty$. If Γ consists of the family of curves in U on which u fails to be locally absolutely continuous, then $M_p(\Gamma) = 0$.*

PROOF. Write $U = \bigcup_{\mu=1}^\infty U_\mu$, where $U_1 \subset U_2 \subset \cdots$ are bounded open sets with $\bar{U}_\mu \subset U$. Let Γ_μ denote the set of rectifiable paths in U_μ on which u fails to be absolutely continuous, and let Γ_0 be the family of curves in U that are not locally rectifiable. Then $\bigcup_{\mu=0}^\infty \Gamma_\mu \prec \Gamma$, whence $M_p(\Gamma) \leq \sum_{\mu=0}^\infty M_p(\Gamma_\mu)$. Since $M_p(\Gamma_0) = 0$, it suffices to verify that $M_p(\Gamma_\mu) = 0$ for $\mu = 1, 2, \ldots$. Fix such an index μ. By Theorem 4.5.3, we can pick a sequence $\langle u_\nu \rangle$ in $C_0^\infty(U)$ such that

$u_\nu \to u$ uniformly on $K = \bar{U}_\mu$ and $\partial_i u_\nu \to \partial_i u$ in L^p-norm on K for $i = 1, 2, \ldots, n$. We apply Lemma 4.3.8 successively to the sequences $\langle \partial_1 u_\nu \rangle, \langle \partial_2 u_\nu \rangle, \ldots, \langle \partial_n \varphi_\nu \rangle$—to be certain that all the technical requirements of the lemma are fulfilled, we agree for the moment to set $\partial_i u_\nu \equiv \partial_i u \equiv 0$ in $\mathbb{R}^n \setminus K$ and, per convention, to write $\partial_i u(x) = 0$ for any x in U at which this partial fails to exist—in order to arrive at a subfamily Γ'_μ of Γ_μ with $M_p(\Gamma'_\mu) = 0$ and a subsequence $\langle u_{\nu_k} \rangle$ of $\langle u_\nu \rangle$ such that

$$\lim_{k \to \infty} \int_\gamma |\partial_i u_{\nu_k} - \partial_i u| \, ds = 0$$

for each path γ in $\Gamma_\mu \setminus \Gamma'_\mu$ and for $i = 1, 2, \ldots, n$. By using $\langle u_{\nu_k} \rangle$ in place of the original sequence $\langle u_\nu \rangle$, we reduce matters to the situation where

$$\lim_{\nu \to \infty} \int_\gamma |\partial_i u_\nu - \partial_i u| \, ds = 0 \tag{4.77}$$

for all such γ and i.

We now assert that $\Gamma_\mu = \Gamma'_\mu$. If not, consider a path β from $\Gamma_\mu \setminus \Gamma'_\mu$. By the definition of Γ_μ, u fails to be absolutely continuous on β, which mandates that $\ell = \ell(\beta) > 0$. Thus, if $\gamma : [0, \ell] \to U_\mu$ is the normal representation of β, we are told that $u \circ \gamma$ is not an absolutely continuous function on $[0, \ell]$. We shall see that (4.77) leads to contradiction in this situation.

Write $\gamma = (\gamma_1, \gamma_2, \ldots, \gamma_n)$. Because u_ν belongs to $C^\infty(U_\mu)$, the function $u_\nu \circ \gamma$ is absolutely continuous on $[0, \ell]$ and

$$u_\nu[\gamma(s)] - u_\nu[\gamma(0)] = \int_0^s [u_\nu \circ \gamma]'(t) \, dt = \int_0^s \left\{ \sum_{i=1}^n \partial_i u_\nu[\gamma(t)] \gamma'_i(t) \right\} dt$$

for each s in this interval. Letting $\nu \to \infty$, we obtain

$$u[\gamma(s)] - u[\gamma(0)] = \int_0^s \left\{ \sum_{i=1}^n \partial_i u[\gamma(t)] \gamma'_i(t) \right\} dt \tag{4.78}$$

whenever $0 \le s \le \ell$. Indeed, $u_\nu \to u$ uniformly on $|\gamma|$ and, since $|\gamma'_i| \le |\dot{\gamma}| = 1$ almost everywhere in $[0, \ell]$, (4.77) implies that

$$\begin{aligned}
&\left| \int_0^s \left\{ \sum_{i=1}^n \partial_i u_\nu[\gamma(t)] \gamma'_i(t) - \sum_{i=1}^n \partial_i u[\gamma(t)] \gamma'_i(t) \right\} dt \right| \\
&\qquad \le \sum_{i=1}^n \int_0^s |\partial_i u_\nu[\gamma(t)] - \partial_i u[\gamma(t)]| \cdot |\gamma'_i(t)| \, dt \\
&\qquad \le \sum_{i=1}^n \int_0^\ell |\partial_i u_\nu[\gamma(t)] - \partial_i u[\gamma(t)]| \, dt \\
&\qquad = \sum_{i=1}^n \int_\gamma |\partial_i u_\nu - \partial_i u| \, ds \to 0
\end{aligned}$$

as $\nu \to \infty$.

Notice that the continuous function $\partial_i u_\nu \circ \gamma$ is in $L^1([0, \ell])$. In conjunction with (4.77), this fact implies that $\partial_i u \circ \gamma$ also belongs to $L^1([0, \ell])$.

Therefore $\sum_{i=1}^n (\partial_i u \circ \gamma) \gamma'_i$ is an L^1-function on $[0, \ell]$, and so the right-hand side (4.78) really does make sense. This formula reveals that $u \circ \gamma$ is absolutely continuous, contradicting the assumption that β comes from Γ_μ. Consequently, $\Gamma_\mu = \Gamma'_\mu$ and $M_p(\Gamma_\mu) = M_p(\Gamma'_\mu) = 0$. The proof is complete. $\square$

We close this section by establishing a result of Gehring [**46**] on the oscillation of ACL-functions over spheres. Recall that the *oscillation* of a function $u : A \to \mathbb{R}^n$ over a subset S of A is the quantity $\operatorname{osc}_S u$ defined by

$$\operatorname{osc}_S u = \sup_{x,y\in S} |u(y) - u(x)|$$

or, stated differently, $\operatorname{osc}_S u = \operatorname{diam}[u(S)]$. If $n = 1$, $\operatorname{osc}_S u = \sup u(S) - \inf u(S)$.

THEOREM 4.5.6. *Let x_0 be a point of $\mathbb{R}^n$ with $n \geq 2$, let $0 \leq a < b \leq \infty$, and let $U = B^n(x_0, b) \setminus \overline{B}^n(x_0, a)$. If a function u is a member of the class* $\mathrm{ACL}(U)$, *then*

$$\int_a^b \left(\operatorname{osc}_{S_r} u\right)^n \frac{dr}{r} \leq A_n \int_U |\nabla u|^n \, dm_n \ ,$$

where $S_r = \mathbb{S}^{n-1}(x_0, r)$ and $A_n = 2^{-n} b(n,n)^{-1}$.

PROOF. We may assume that $|\nabla u|$ belongs to $L^n(U)$, hence that u is a function from the class $\mathrm{ACL}^n(U)$. Choose a sequence $\langle u_\nu \rangle$ from $C_0^\infty(U)$ such that $u_\nu \to u$ uniformly on compact subsets of U and $|\nabla u_\nu| \to |\nabla u|$ in L^n-norm on such sets.

Fix numbers a_1 and b_1 satisfying $a < a_1 < b_1 < b$ and write $K = \overline{B}^n(b_1) \setminus B^n(a_1)$. It is evident that $\operatorname{osc}_{S_r} u_\nu$ converges uniformly to $\operatorname{osc}_{S_r} u$ on the interval $[a_1, b_1]$. Because $|\nabla u_\nu| \to |\nabla u|$ in $L^n(K)$, we can invoke Theorem 4.2.21 and conclude that

$$\begin{aligned}
\int_{a_1}^{b_1} \left(\operatorname{osc}_{S_r} u\right)^n \frac{dr}{r} &= \lim_{\nu\to\infty} \int_{a_1}^{b_1} \left(\operatorname{osc}_{S_r} u_\nu\right)^n \frac{dr}{r} \\
&\leq A_n \lim_{\nu\to\infty} \int_{a_1}^{b_1} \left(\int_{S_r} |\nabla u_\nu|^n \, d\sigma_{n-1}\right) dr \\
&= A_n \lim_{\nu\to\infty} \int_K |\nabla u_\nu|^n \, dm_n \\
&= A_n \int_K |\nabla u|^n \, dm_n \leq A_n \int_U |\nabla u|^n \, dm_n \, .
\end{aligned}$$

To finish the proof, let $a_1 \to a$ and $b_1 \to b$. $\square$

We also record in passing a variant of Theorem 4.5.6 that deals with the oscillation of ACL-functions over hemispheres. Its proof, which differs only slightly from that presented for Theorem 4.5.6, is left to the reader as an exercise.

THEOREM 4.5.7. *Let $n \geq 2$, let $D = B^n \cap \mathbb{H}^n$, and let u be a function from the class* $\mathrm{ACL}(D)$. *Then*

$$\int_a^b \left(\operatorname{osc}_{C_r} u\right)^n \frac{dr}{r} \leq a_n \int_{D(a,b)} |\nabla u|^n \, dm_n$$

whenever $0 < a < b < 1$, where $C_r = \mathbb{S}^{n-1}(r) \cap H^n$, $D(a,b) = \{x \in D : a < |x| < b\}$, and $a_n = b(n,n)^{-1}$.

CHAPTER 5

Rings and Condensers

In the previous chapter the reader will have noted that much of the discussion of the modulus of curve families centered around various families of curves in $\mathbb{R}^n$ or $\hat{\mathbb{R}}^n$ joining two continua, say E and F, with perhaps the only relevant exception being that curve family which exhibited the sharpness of the cap inequality. In any case, this situation is the most geometrically appealing and, it turns out, the most useful. This chapter will expose this utility and the interesting applications one obtains once we have reasonable estimates for moduli in terms of the underlying geometry, along with an understanding of various extremal configurations.

5.1. Rings

A ring R will be a domain in $\hat{\mathbb{R}}^n$ with the property that $\hat{\mathbb{R}}^n \setminus R$ has exactly two components, say E and F. Thus, E and F are compact in $\hat{\mathbb{R}}^n$, although one of them might be unbounded as a subset of $\mathbb{R}^n$. In order to proceed, we need to establish the topological invariance of rings. Thus we will preface the forthcoming discussion of rings and condensers with a brief synopsis of pertinent background information from topology. The reader is referred to any basic graduate text on topology, for instance [**127**], for details.

5.1.1. A touch of topology. Suppose first that D is a domain in $\hat{\mathbb{R}}^n$ with $n \geq 2$ and that C is a component of $D^c = \hat{\mathbb{R}}^n \setminus D$. Then $C \cap \bar{D}$ is a component of ∂D. Furthermore, the correspondence $C \mapsto C \cap \bar{D}$ is a bijection between the collection of components of D^c and the collection of components of ∂D.

Next, let $f : D \to D'$ be a homeomorphism between domains D and D' in $\hat{\mathbb{R}}^n$, where again $n \geq 2$. Assume that $D \neq \hat{\mathbb{R}}^n$. If x is a point of ∂D, then the *cluster set of f at x*, denoted by $C_f(x)$, is defined as follows:

$$C_f(x) = \bigcap_{r>0} \overline{f\big[D \cap B^n(x,r)\big]}\,.$$

For a point y of $\hat{\mathbb{R}}^n$ to belong to $C_f(x)$ it is thus necessary and sufficient that there be a sequence $\langle x_\nu \rangle$ in D such that $x_\nu \to x$ and $f(x_\nu) \to y$. The set $C_f(x)$ is easily seen to be a nonempty compact subset of $\partial D'$. If E is a nonempty subset of ∂D, we use $C_f(E)$ to indicate the cluster set of E,

$$C_f(E) = \bigcup_{x \in E} C_f(x).$$

Suppose that B is a component of ∂D. It can be shown that $B' = C_f(B)$ is of necessity a component of $\partial D'$ and that $B \mapsto C_f(B)$ establishes a one-to-one correspondence between the collection of components of ∂D and the collection of components of $\partial D'$.

For $n \geq 2$ we employ the term *ring* in $\hat{\mathbb{R}}^n$ to denote a domain R in $\hat{\mathbb{R}}^n$ with the property that R^c has exactly two components. If C_0 and C_1 are these two components, then we may sometimes identify R by making use of the notation $R = \mathcal{R}(C_0, C_1)$ to denote the ring

$$\mathcal{R}(C_0, C_1) = \hat{\mathbb{R}}^n \setminus \{C_0 \cup C_1\}.$$

In the case that R lies in $\mathbb{R}^n$, that is, $\infty \notin R$, we shall always choose the labeling so that ∞ belongs to C_0.

We say that R is *nondegenerate* when both C_0 and C_1 are continua, so both have infinite cardinality. According to a comment made above, $R = \mathcal{R}(C_0, C_1)$ has precisely two boundary components—namely, $B_0 = C_0 \cap \bar{R}$ and $B_1 = C_1 \cap \bar{R}$. If F_0 and F_1 are nonempty, disjoint, compact, connected sets in $\hat{\mathbb{R}}^n$, then elementary topology tells us that the open set $(F_0 \cup F_1)^c$ has a unique component R which is a ring, say $R = \mathcal{R}(C_0, C_1)$, satisfying the conditions $B_0 \subset F_0 \subset C_0$ and $B_1 \subset F_1 \subset C_1$. We refer to R as the *ring in* $\hat{\mathbb{R}}^n$ *associated with the pair* (F_0, F_1).

Let R be a ring in $\hat{\mathbb{R}}^n$, and let $f : R \to \hat{\mathbb{R}}^n$ be a continuous injection. On the strength of the invariance of domain theorem, we can assert that f is a homeomorphism of R onto the domain $R' = f(R)$. Since R has exactly two boundary components, the same must be true of R'. In other words, R' is itself a ring.

5.1.2. Capacities of rings. Given a ring $R = \mathcal{R}(C_0, C_1)$ in $\hat{\mathbb{R}}^n$ we can naturally associate the curve family $\Gamma_R = \Delta(C_0, C_1 : \hat{\mathbb{R}}^n)$.

For $1 < p < \infty$ the extended real number $M_p(\Gamma_R)$ is known as the *p-capacity of* R. The symbol $\operatorname{Cap}_p(R)$ serves to denote this quantity:

$$\operatorname{Cap}_p(R) = M_p(\Delta(C_0, C_1 : \hat{\mathbb{R}}^n)), \quad R = \hat{\mathbb{R}}^n \setminus \{C_0 \cup C_1\}.$$

When $p = n$, the dimension of the ambient space, this gives the *conformal capacity* of R, for which it is customary to drop the subscript from the notation. Thus we write $\operatorname{Cap}(R)$ in place of $\operatorname{Cap}_n(R)$, allowing context to specify the dimension.

The *modulus of* R, designated by $\operatorname{Mod}(R)$, is defined as follows:

$$\operatorname{Mod}(R) = \left[\frac{\omega_{n-1}}{\operatorname{Cap}(R)}\right]^{1/(n-1)}.$$

Notice that $\operatorname{Cap}(R) = 0$, or, equivalently, $\operatorname{Mod}(R) = \infty$, if the ring R is degenerate. Further, as we shall find out later, the converse is also true, but not quite so obvious.

It is a consequence of the elementary properties of the p-modulus M_p discussed earlier that various other curve families naturally related to R could have been employed in place of Γ_R to define the capacity $\operatorname{Cap}_p(R)$, for they share the same p-modulus with Γ_R. For example,

$$\operatorname{Cap}_p(R) = M_p\big[\Delta(B_0, B_1 : \hat{\mathbb{R}}^n)\big] = M_p\big[\Delta(B_0, B_1 : R)\big]\,, \tag{5.1}$$

where B_0 and B_1 are the boundary components of R. If either $p \geq n$ or R is a bounded ring in $\mathbb{R}^n$, then

$$\operatorname{Cap}_p(R) = M_p\big[\Delta_0(B_0, B_1 : R)\big] = M_p\big[\bar{\Delta}(B_0, B_1 : R)\big]\,. \tag{5.2}$$

Moreover, should R be the ring in $\hat{\mathbb{R}}^n$ associated with a pair (F_0, F_1) of compact and connected sets, it would also be true that

$$\text{Cap}_p(R) = M_p\big[\Delta(F_0, F_1 : \hat{\mathbb{R}}^n)\big]\,. \tag{5.3}$$

The prototypical rings are, of course, the Euclidean spherical rings, meaning rings of the type $R = B^n(x_0, b) \setminus \overline{B}^n(x_0, a)$ with x_0 in $\mathbb{R}^n$ and $0 \le a < b \le \infty$. It follows from (5.1) and Theorem 4.2.4 that any such ring has

$$\text{Cap}_p(R) = \omega_{n-1}\left(\int_a^b r^{m-1}\,dr\right)^{1-p}$$

for every $p > 1$, where $m = (p-n)/(p-1)$. In particular,

$$\text{Cap}(R) = \omega_{n-1}\left(\log\frac{b}{a}\right)^{1-n}, \qquad \text{Mod}(R) = \log\frac{b}{a}\,. \tag{5.4}$$

Let $R = \mathcal{R}(C_0, C_1)$ be a ring in $\hat{\mathbb{R}}^n$, and let f be a homeomorphism of R onto the ring $R' = \mathcal{R}(C_0', C_1')$. Assume that the restriction of f to $R_0 = \{x \in R : x \ne \infty,\ f(x) \ne \infty\}$ is a conformal diffeomorphism. Since f is easily seen to transform the curve family $\Delta_0(B_0, B_1 : R)$ to $\Delta_0(B_0', B_1' : R')$, we infer from (5.2) and Corollary 4.2.17 that $\text{Cap}(R) = \text{Cap}(R')$ and $\text{Mod}(R) = \text{Mod}(R')$.

In the classical setting of complex analysis, $\hat{\mathbb{C}}$, generalisations of the Riemann mapping theorem show that it is always possible to produce a mapping f of the above description so that R' is a Euclidean annulus.

We therefore are able to note that the modulus of a ring is invariant under Möbius transformations. If, for instance, R is the chordal spherical ring defined by

$$R = \{x \in \hat{\mathbb{R}}^n : a < q(x, x_0) < b\}, \tag{5.5}$$

where x_0 belongs to $\hat{\mathbb{R}}^n$ and $0 \le a < b \le 2$, then any chordal isometry that maps x_0 to 0 transforms R to $R' = \{x \in \hat{\mathbb{R}}^n : a < q(x, 0) < b\}$, which can be described in Euclidean terms as

$$R' = \{x : a(4-a^2)^{-1/2} < |x| < b(4-b^2)^{-1/2}\}.$$

As chordal isometries are Möbius transformations, we conclude that with R as defined at (5.5),

$$\text{Cap}(R) = \omega_{n-1}\left(\log\frac{b\sqrt{4-a^2}}{a\sqrt{4-b^2}}\right)^{1-n}, \qquad \text{Mod}(R) = \log\frac{b\sqrt{4-a^2}}{a\sqrt{4-b^2}}\,. \tag{5.6}$$

The following result is often helpful when it comes to finding bounds for the modulus of a ring.

LEMMA 5.1.1. *If $\{R_\nu\}$ is a collection of pairwise disjoint rings in $\hat{\mathbb{R}}^n$ and if R is a ring in $\hat{\mathbb{R}}^n$ whose complementary components are separated in $\hat{\mathbb{R}}^n$ by each R_ν, then* $\text{Mod}(R) \ge \sum_\nu \text{Mod}(R_\nu)$.

PROOF. Let $B_{\nu 0}$ and $B_{\nu 1}$ (resp., B_0 and B_1) be the boundary components of R_ν (resp., R). Set

$$\Gamma_\nu = \Delta(B_{\nu 0}, B_{\nu 1} : R_\nu) \quad\text{and}\quad \Gamma = \Delta(B_0, B_1 : R).$$

The members of $\{\Gamma_\nu\}$ are separated and $\Gamma_\nu \prec \Gamma$ for each ν. In view of (5.1) and Theorem 4.2.11,

$$\begin{aligned}\operatorname{Mod}(R) &= \omega_{n-1}^{1/(n-1)} M(\Gamma)^{1/(1-n)} \geq \omega_{n-1}^{1/(n-1)} \sum_\nu \lambda(\Gamma_\nu) \\ &= \sum_\nu \left[\frac{\omega_{n-1}}{M(\Gamma_\nu)}\right]^{1/(n-1)} = \sum_\nu \operatorname{Mod}(R_\nu)\,,\end{aligned}$$

as asserted. □

Here is a simple symmetry principle for rings. We say the sets C_0 and C_1 are symmetric in a chordal sphere Σ if the reflection $R_\Sigma(C_0) = C_1$ (and therefore $R(C_1) = C_0$).

LEMMA 5.1.2. *Suppose that a ring* $R = \mathcal{R}(C_0, C_1)$ *in* $\hat{\mathbb{R}}^n$ *contains a chordal sphere* Σ *with respect to which the sets* C_0 *and* C_1 *are symmetric. Then* $R\setminus\Sigma$ *is the disjoint union of rings* R_0 *and* R_1 *such that* $\operatorname{Mod}(R_0) = \operatorname{Mod}(R_1) = \operatorname{Mod}(R)/2$.

PROOF. It is an easy fact of topology that Σ splits R into a pair of rings R_0 and R_1, the labels being chosen so that C_0 is a component of R_0^c and C_1 is a component of R_1^c. Thus the boundary components of R_0 are Σ and $B_0 = C_0 \cap \bar{R}$; those of R_1 are Σ and $B_1 = C_1 \cap \bar{R}$. By assumption $C_0 = C_1^*$, the reflection of C_1 in Σ. Choose a component D of Σ^c that is a half-space or Euclidean ball in $\mathbb{R}^n$. We may suppose that C_1 is the component of R^c which lies in D. Then R_1 is contained in D. Appealing to elementary properties of curve family moduli, to (5.1) and to Corollary 4.3.6, we obtain

$$\begin{aligned}\operatorname{Cap}(R_1) &= M\big[\Delta(\Sigma, B_1 : R_1)\big] = M\big[\Delta(\Sigma, C_1 : D)\big] \\ &= 2^{n-1} M\big[\Delta(C_1^*, C_1 : \hat{\mathbb{R}}^n)\big] = 2^{n-1} M\big[\Delta(C_0, C_1 : \hat{\mathbb{R}}^n)\big] \\ &= 2^{n-1}\operatorname{Cap}(R)\,,\end{aligned}$$

whence $\operatorname{Mod}(R_1) = \operatorname{Mod}(R)/2$. Finally, since the modulus of a ring is a Möbius invariant and since R_0 is obtained from R_1 via reflection in Σ, $\operatorname{Mod}(R_0) = \operatorname{Mod}(R_1)$. □

5.1.3. The Grötzsch and Teichmüller rings. Here we introduce two extremely important rings. They will have very strong extremal properties from which many key geometric estimates in the theory of quasiconformal mappings will ultimately depend.

For $n \geq 2$ and $1 < r \leq \infty$, the ring $R_G(n, r)$ in $\hat{\mathbb{R}}^n$ whose complementary components are $C_0 = \{te_1 : r \leq t \leq \infty\}$ and $C_1 = \overline{B}^n$ is called the *n-dimensional Grötzsch ring* corresponding to r,

$$R_G(n, r) = \mathbb{R}^n \setminus \{[re_1, \infty) \cup \bar{B}^n\}, \quad r > 1.$$

We define a function $\Phi_n : (1, \infty] \to (1, \infty]$ by requiring that

$$\log \Phi_n(r) = \operatorname{Mod}\big[R_G(n, r)\big]\,.$$

It follows from Theorem 4.3.15 and general properties of the moduli of curve families that Φ_n is a nondecreasing, continuous function with $\Phi_n(\infty) = \infty$.

The cap inequality delivers the lower bound

$$\operatorname{Cap}\big[R_G(n, r)\big] = M\big[\Delta(C_0, C_1 : \hat{\mathbb{R}}^n\big] \geq c_n \log \frac{2}{r-1}$$

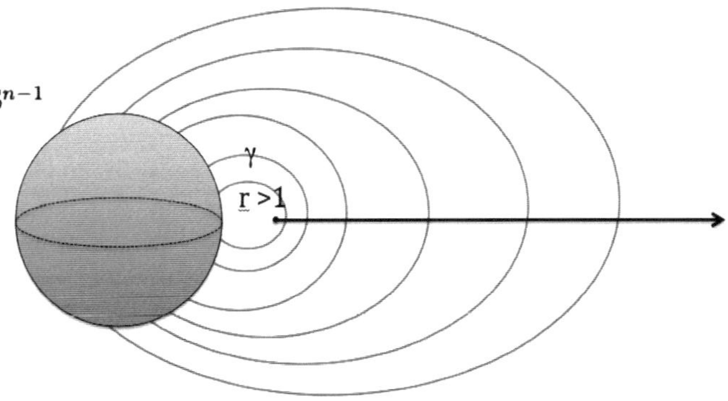

The Grötzsch ring. This ring consists of those curves connecting $\mathbb{S}^{n-1}$ and the infinite line segment $[re_1, +\infty]$, $r > 1$.

when $1 < r \leq 2$, which implies that $\Phi_n(r) \to 1$ as $r \to 1$.

For $n \geq 2$ and $0 < r \leq \infty$, we use $R_T(n, s)$ to denote the ring in $\hat{\mathbb{R}}^n$ with complementary components $C_0 = \{te_1 : s \leq t \leq \infty\}$ and $C_1 = \{te_1 : -1 \leq t \leq 0\}$,

$$R_T(n, s) = \mathbb{R}^n \setminus \{[-e_1, 0] \cup [se_1, \infty)\}, \quad s > 0.$$

This ring is known as the *n-dimensional Teichmüller ring* corresponding to s.

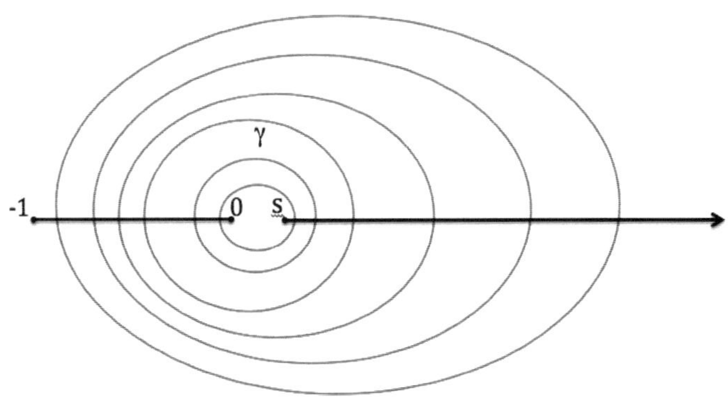

The Teichmüller ring. This ring consists of those curves connecting the finite segment $[-e_1, 0]$ and the infinite line segment $[se_1, +\infty]$, $s > 0$.

Paralleling the situation with Φ_n, a function $\Psi_n : (0, \infty] \to (1, \infty]$ is defined by insisting that

$$\log \Psi_n(s) = \mathrm{Mod}\big[R_T(n, s)\big] .$$

As was the case for Φ_n, the function Ψ_n is nondecreasing and continuous. It satisfies $\Psi_n(\infty) = \infty$ and $\lim_{s \to 0} \Psi_n(r) = 1$.

The next lemma describes the relationship between the two functions Φ_n and Ψ_n.

LEMMA 5.1.3. *$\Psi_n(r) = [\Phi_n(\sqrt{1+r}\,)]^2$ for $r > 0$.*

PROOF. Fix r in $(0, \infty)$, and let R_0 and R_1 be the rings into which $R = R_T(n, r)$ is divided by the sphere $\Sigma = \mathbb{S}^{n-1}(-e_1, \sqrt{1+r}\,)$, labeled so that ∞ lies on the boundary of R_0. Lemma 5.1.2 implies that $\mathrm{Mod}(R) = 2\mathrm{Mod}(R_0)$, for the components of R^c are symmetric with respect to Σ. On the other hand, the similarity transformation $f(x) = (1+r)^{-1/2}(x + e_1)$ maps R_0 to $R_G(n, \sqrt{1+r}\,)$. Therefore, by the conformal invariance of the modulus of a ring,

$$\log \Psi_n(r) = \mathrm{Mod}(R) = 2\mathrm{Mod}(R_0) = 2\log\left[\Phi_n(\sqrt{1+r}\,)\right] ,$$

which produces the desired identity for finite r. Since $\Psi_n(\infty) = \Phi_n(\infty) = \infty$, it holds when $r = \infty$ as well. □

The function Φ_n has another noteworthy feature.

LEMMA 5.1.4. *The function $r \mapsto \Phi_n(r)/r$ is nondecreasing on the interval* $(1, \infty)$.

PROOF. Suppose that $1 < r < s < \infty$. Let R_0 and R_1 be the rings into which $R = R_G(n, s)$ is partitioned by the sphere $\mathbb{S}^{n-1}(s/r)$, labeled so that ∞ belongs to ∂R_0. Then R_0 and R_1 are disjoint, and each of these rings separates the components of R^c. Now R_1 is just a spherical ring, while the similarity transformation $f(x) = (r/s)x$ maps R_0 to $R_0' = R_G(n, r)$. Recalling Lemma 5.1.1, we obtain

$$\begin{aligned} \log \Phi_n(s) &= \mathrm{Mod}(R) \geq \mathrm{Mod}(R_0) + \mathrm{Mod}(R_1) = \mathrm{Mod}(R_0') + \mathrm{Mod}(R_1) \\ &= \log \Phi_n(r) + \log \frac{s}{r} . \end{aligned}$$

This plainly leads to the conclusion that $\Phi_n(r)/r \leq \Phi_n(s)/s$, which is what we set out to show. □

It is an immediate consequence of Lemma 5.1.4 that Φ_n is strictly increasing on $(1, \infty)$. This lemma also guarantees that

$$\lambda_n = \lim_{r\to\infty} \frac{\Phi_n(r)}{r}$$

exists in $(0, \infty]$. A detailed study of the function Φ_n would disclose the fact that

$$4 \leq \lambda_n \leq 2^{n/(n-1)} e^{n(n-2)/(n-1)} . \tag{5.7}$$

We shall see that $\lambda_2 = 4$, but the exact value of λ_n for any $n \geq 3$ remains a longstanding open question.

We shall not attempt to derive (5.7), being content to achieve a more modest goal, which is merely to demonstrate that λ_n is finite. We shall be aided in this endeavor by the next lemma.

LEMMA 5.1.5. *Let f be a diffeomorphism of the spherical ring $R_0 = B^n(b)\setminus\overline{B}^n(a)$ onto a ring R in $\mathbb{R}^n$, and let $N_f : R_0 \to (0, \infty)$ be the magnitude of its radial derivative*

$$N_f(x) = |f'(x)x|/|x|.$$

Then

$$\mathrm{Mod}(R) \leq \int_a^b \frac{Q(r)\,dr}{r} ,$$

where

$$Q(r) = \max_{|x|=r} \left\{ \frac{[N_f(x)]^n}{|J_f(x)|} \right\}^{1/(n-1)}$$

for $a < r < b$.

PROOF. Write $\Gamma = \Delta_0(B_0, B_1 : R)$, where B_0 and B_1 are the components of ∂R. For each u in $\mathbb{S}^{n-1}$ define $\gamma_u : (a,b) \to \mathbb{R}^n$ by $\gamma_u(r) = ru$. Then $f \circ \gamma_u$ is a locally absolutely continuous curve in Γ. If ρ belongs to $\mathrm{Adm}(\Gamma)$, we can use Hölder's inequality to obtain

$$\begin{aligned}
1 &\le \int_{f\circ\gamma_u} \rho\, ds = \int_a^b \rho\big[f(ru)\big]|f'(ru)u|\, dr = \int_a^b \rho\big[f(ru)\big] N_f(ru)\, dr \\
&= \int_a^b \rho\big[f(ru)\big]|J_f(ru)|^{1/n} r^{(n-1)/n} N_f(ru)|J_f(ru)|^{-1/n} r^{(1-n)/n}\, dr \\
&\le \int_a^b \rho\big[f(ru)\big]|J_f(ru)|^{1/n} r^{(n-1)/n}\big[Q(r)\big]^{(n-1)/n} r^{(1-n)/n}\, dr \\
&\le \left\{\int_a^b \rho^n\big[f(ru)\big]|J_f(ru)| r^{n-1}\, dr\right\}^{1/n} \left\{\int_a^b \frac{Q(r)\, dr}{r}\right\}^{(n-1)/n}.
\end{aligned}$$

This leads to the estimate

$$\left[\int_a^b \frac{Q(r)\, dr}{r}\right]^{1-n} \le \int_a^b \rho^n\big[f(ru)\big]|J_f(ru)| r^{n-1}\, dr\,.$$

Integration of this inequality over the sphere $\mathbb{S}^{n-1}$ yields

$$\omega_{n-1}\left[\int_a^b \frac{Q(r)\, dr}{r}\right]^{1-n} \le \int_{R_0} (\rho^n \circ f)|J_f|\, dm_n = \int_R \rho^n\, dm_n \le \int_{\mathbb{R}^n} \rho^n\, dm_n\,.$$

Taking the infimum over ρ and recalling (5.2), we arrive at

$$\omega_{n-1}\left[\int_a^b \frac{Q(r)\, dr}{r}\right]^{1-n} \le M(\Gamma) = \mathrm{Cap}(R)\,.$$

The conclusion of the lemma now follows from the definition of $\mathrm{Mod}(R)$. □

We can now apply Lemma 5.1.5 to derive information about a specific ring.

LEMMA 5.1.6. *Let* $a > 1$. *If* $R = R_E(n,a)$ *is the ring in* $\hat{\mathbb{R}}^n$ *whose boundary components are the ellipsoid* B_0 *with equation*

$$B_0 \;:\; (x_1/\sqrt{a^2+1}\,)^2 + (x_2/a)^2 + \cdots + (x_n/a)^2 = 1$$

and the line segment $B_1 = \{te_1 : -1 \le t \le 1\}$, *then*

$$\mathrm{Mod}(R) \le \int_1^b \left(\frac{r^2+1}{r^2-1}\right)^{n-2/n-1} \frac{dr}{r}\,, \tag{5.8}$$

in which $b = a + \sqrt{a^2+1}$. *Equality holds in* (5.8) *when* $n = 2$; *that is,*

$$\mathrm{Mod}[R_E(2,a)] = \log b, \qquad \textit{for } a > 1.$$

PROOF. We shall first construct a C^∞-diffeomorphism f of $R_0 = B^n(b) \setminus \overline{B}^n$ onto R, where b is the number indicated above. To do so, we shall take advantage of a standard conformal mapping φ in the complex plane, namely, the function φ defined for $|z| > 1$ by $\varphi(z) = (z + z^{-1})/2$.

We remark that φ maps the domain $D = \mathbb{C} \setminus \bar{B}^2$ conformally onto $D' = \mathbb{C} \setminus [-1, 1]$, that it transforms the circle $C_r = S^1(r)$ with $r > 1$ to the ellipse E_r whose equation is

$$\frac{x^2}{\left(\frac{r+r^{-1}}{2}\right)^2} + \frac{y^2}{\left(\frac{r-r^{-1}}{2}\right)^2} = 1 ,$$

and that $\Im\varphi(z) > 0$ if and only if $\Im z > 0$. In particular, since it is true that $(b + b^{-1})/2 = \sqrt{a^2+1}$ and $(b - b^{-1})/2 = a$, we note that φ maps the annulus $G_0 = B^2(b) \setminus \bar{B}^2$ to the plane ring G whose boundary components are the real interval $[-1, 1]$ and the ellipse with equation $(x/\sqrt{a^2+1}\,)^2 + (y/a)^2 = 1$. The conformal invariance of the modulus and (5.4) therefore ensure equality in (5.8) for $n = 2$. Further, φ carries $G_0^+ = \{z \in G_0 : \Im z \geq 0\}$ to $G^+ = \{z \in G : \Im z \geq 0\}$.

The construction of the diffeomorphism f will be accomplished through the process of "rotating $\varphi|G_0^+$ about the real axis". Having dealt with the case $n = 2$ above, we assume that $n > 2$ for the remainder of the proof.

If $x = (x_1, x_2, \ldots, x_n)$ is a point of $\mathbb{R}^n$ for which $x_2^2 + x_3^2 + \cdots + x_n^2 > 0$, then we can give a unique description of x in terms of its cylindrical coordinates relative to the x_1-axis: $x = (x_1, \rho, \omega)$, where $\rho = (x_2^2 + x_3^2 + \cdots + x_n^2)^{1/2}$ and where $\omega = \rho^{-1}(x_2, x_3, \ldots, x_n)$ belongs to $\mathbb{S}^{n-2}$ ($\omega = \pm 1$ if $n = 2$). To define f, we consider $x = (x_1, x_2, \ldots, x_n)$ in R_0. If $x_2 = x_3 = \cdots = x_n = 0$, we set $f(x) = (\varphi(x_1), 0, \ldots, 0)$; otherwise, let x have cylindrical coordinates (x_1, ρ, ω) and define $y = f(x)$ to be the point with cylindrical coordinates (y_1, σ, ω), where $y_1 + i\sigma = \varphi(x_1 + i\rho)$, that is, where

$$y_1 = \frac{x_1(|x|^2+1)}{2|x|^2}, \qquad \sigma = \frac{\rho(|x|^2-1)}{2|x|^2} .$$

In other words, each x in R_0 has its image $y = f(x)$ given by

$$y_1 = \frac{x_1(|x|^2+1)}{2|x|^2} , \; y_2 = \frac{\rho x_2(|x|^2-1)}{2|x|^2} , \; \ldots , \; y_n = \frac{\rho x_n(|x|^2-1)}{2|x|^2} ,$$

with $\rho = (x_2^2 + \cdots + x_n^2)^{1/2}$. So defined, f is a C^∞-diffeomorphism of R_0 onto R. Owing to the rotational symmetry of f about the x_1-axis, a moment's thought should reveal that

$$N_f(x) = |\varphi'(x_1 + i\rho)| = \frac{|(x_1+i\rho)^2 - 1|}{2|x|^2}$$

for every x in R_0. An elementary but somewhat tedious calculation shows that

$$J_f(x) = |\varphi'(x_1 + i\rho)|^2 \left(\frac{|x|^2-1}{2|x|^2}\right)^{n-2} = \frac{|(x_1+i\rho)^2 - 1|^2(|x|^2-1)^{n-2}}{(2|x|^2)^n}$$

for such x. As a result,

$$\frac{[N_f(x)]^n}{|J_f(x)|} = \left[\frac{|(x_1+i\rho)^2 - 1|}{|x|^2 - 1}\right]^{n-2} .$$

The last expression attains its maximum on $\mathbb{S}^{n-1}(r)$, $1 < r < b$, when $x_1 = 0$ and $\rho = r$. It follows that

$$Q(r) = \max_{|x|=r} \left\{ \frac{[N_f(x)]^n}{|J_f(x)|} \right\}^{1/(n-1)} = \left(\frac{r^2+1}{r^2-1} \right)^{(n-2)/(n-1)}$$

for r in $(1, b)$. Lemma 5.1.5 now contributes the estimate (5.8) for $\mathrm{Mod}(R)$. □

We are now in a position to say something about the constant λ_n.

LEMMA 5.1.7. *If $R_E(n,a)$ is the ring in Lemma 5.1.6, then*

$$\log \lambda_n = \lim_{a\to\infty} \left\{ \mathrm{Mod}\big[R_E(n,a)\big] - \log \frac{a}{2} \right\}. \tag{5.9}$$

PROOF. Fix $a > 4$. Define a pair of rings R' and R'' by $R' = \mathcal{R}(C_0', C_1')$ and $R'' = \mathcal{R}(C_0'', C_1'')$, where $C_0' = \hat{\mathbb{R}}^n \setminus B^n(-e_1, a-2)$, $C_0'' = \hat{\mathbb{R}}^n \setminus B^n(-e_1, a+2)$, and $C_1' = C_1'' = \{te_1 : -1 \le t \le 1\}$. Also, let C_0 and $C_1 = C_1'$ designate the complementary components of $R = R_E(n,a)$.

If x lies in C_0'', then

$$|x - e_1| + |x + e_1| \ge 2|x + e_1| - 2 \ge 2(a+2) - 2 = 2(a+1) \ge 2(a^2+1)^{1/2},$$

which puts x in C_0. Thus C_0'' is a subset of C_0. Similarly, for x in C_0 we have

$$2|x + e_1| \ge |x + e_1| + |x - e_1| - 2 \ge 2(a^2+1)^{1/2} - 2 > 2(a-1) - 2 = 2(a-2),$$

showing that x belongs to C_0'. Therefore C_0 is a subset of C_0'. Lemma 5.1.1 implies that

$$\mathrm{Mod}(R') \le \mathrm{Mod}(R) \le \mathrm{Mod}(R'').$$

Observe that the Möbius transformations

$$f(x) = \frac{(a-2)(x+e_1)}{|x+e_1|^2}, \qquad g(x) = \frac{(a+2)(x+e_1)}{|x+e_1|^2}$$

map R' and R'' onto $R_G[n, (a-2)/2]$ and $R_G[n, (a+2)/2]$, respectively.

Because of the conformal invariance of the modulus, we see that

$$\mathrm{Mod}(R') = \log \Phi_n\left(\frac{a-2}{2}\right), \qquad \mathrm{Mod}(R'') = \log \Phi_n\left(\frac{a+2}{2}\right).$$

Assuming that $\lambda_n < \infty$, we conclude that

$$\begin{aligned}
\log \lambda_n &= \lim_{a\to\infty} \left\{ \log \Phi_n\left(\frac{a-2}{2}\right) - \log \frac{a-2}{2} \right\} \\
&= \lim_{a\to\infty} \left\{ \log \Phi_n\left(\frac{a-2}{2}\right) - \log \frac{a}{2} \right\} \\
&\le \liminf_{a\to\infty} \left\{ \mathrm{Mod}\big[R_E(n,a)\big] - \log \frac{a}{2} \right\} \\
&\le \limsup_{a\to\infty} \left\{ \mathrm{Mod}\big[R_E(n,a)\big] - \log \frac{a}{2} \right\} \\
&\le \lim_{a\to\infty} \left\{ \log \Phi_n\left(\frac{a+2}{2}\right) - \log \frac{a}{2} \right\} \\
&= \lim_{a\to\infty} \left\{ \log \Phi_n\left(\frac{a+2}{2}\right) - \log \frac{a+2}{2} \right\} = \log \lambda_n.
\end{aligned}$$

This yields (5.9). Should $\lambda_n = \infty$, the first part of the above computation would certify that the limit in (5.9) is ∞ as well. □

Putting the above pieces together, we demonstrate that λ_n is finite by producing an upper bound for it.

THEOREM 5.1.8. *The constant λ_n is finite. In fact, $\lambda_2 = 4$ and*

$$\lambda_n \le 4\exp\left\{\int_1^\infty \left[\left(\frac{r^2+1}{r^2-1}\right)^{(n-2)/(n-1)} - 1\right]\frac{dr}{r}\right\}$$

when $n > 2$.

The upper bounds here grow relatively quickly with $\lambda_3 \le 12.407\ldots$, $\lambda_4 \le 35.538\ldots$, $\lambda_5 \le 99.363\ldots$, and so forth. Estimates can be obtained by analysing the singularities at $r = 1$ since

$$4\exp\left\{\int_2^\infty \left[\left(\frac{r^2+1}{r^2-1}\right)^{(n-2)/(n-1)} - 1\right]\frac{dr}{r}\right\} \le 4\exp\left\{\int_2^\infty \frac{2}{r^2-1}\,\frac{dr}{r}\right\} = \frac{16}{3},$$

but these will not be useful in what follows.

PROOF. Lemma 5.1.7 and the fact that (5.8) holds with equality when $n = 2$ imply that $\lambda_2 = 4$. We proceed under the assumption that $n > 2$.

Setting $b = b(a) = a + \sqrt{a^2+1}$ for $a > 1$, we compute

$$\begin{aligned}
&\lim_{a\to\infty}\left\{\operatorname{Mod}\big[R_E(n,a)\big] - \log\frac{a}{2}\right\}\\
&\le \lim_{a\to\infty}\left\{\int_1^b \left(\frac{r^2+1}{r^2-1}\right)^{(n-2)/(n-1)}\frac{dr}{r} - \log\frac{a}{2}\right\}\\
&= \lim_{a\to\infty}\left\{\int_1^b \left[\left(\frac{r^2+1}{r^2-1}\right)^{(n-2)/(n-1)} - 1\right]\frac{dr}{r} + \log\frac{2b}{a}\right\}\\
&= \log 4 + \int_1^\infty \left[\left(\frac{r^2+1}{r^2-1}\right)^{(n-2)/(n-1)} - 1\right]\frac{dr}{r} < \infty\,.
\end{aligned}$$

The integrand in this last integral grows like $(r-1)^{-(n-2)/(n-1)}$ as $r \to 1$ and like r^{-3} as $r \to \infty$. An appeal to Lemma 5.1.7 completes the proof. □

We summarize the preceding considerations as follows:

THEOREM 5.1.9. *The function Φ_n is strictly increasing and continuous on the interval $(1,\infty)$, and we have the estimates*

$$r \le \Phi_n(r) \le \lambda_n r\,.$$

The function Ψ_n is strictly increasing and continuous on the interval $(0,\infty)$, and we have the estimates

$$r + 1 \le \Psi_n(r) \le \lambda_n^2(r+1)\,.$$

PROOF. We have verified the assertions concerning Φ_n. The statements regarding Ψ_n follow by application of Lemma 5.1.3. □

5.2. Condensers

The concept of a "condenser" is one that comes up quite naturally in the theory of elliptic partial differential equations. This arises through the introduction of various quantities that measure the "capacity" of a condenser, in analogy with the physical notion of capacitance. In fact the study of such PDEs planted the seeds from which modern potential theory has grown.

5.2.1. Capacities of condensers. By definition, a *condenser* in $\mathbb{R}^n$, where $n \geq 2$, is a pair $C = (U, F)$, where U is an open set in $\mathbb{R}^n$ and F is a nonempty compact subset of U. (If the set $U \setminus F$ is a ring in $\hat{\mathbb{R}}^n$, then we call C a *ringlike condenser*.)

With the condenser C we associate the class $\mathcal{A}(C)$ of continuous functions $u : \hat{\mathbb{R}}^n \to [0,1]$ that belong to $\mathrm{ACL}(\mathbb{R}^n)$ and have $u(x) = 1$ for every x in F and $u(x) = 0$ for every x in $\hat{\mathbb{R}}^n \setminus U$. For instance any function $u \in C_0^\infty(\mathbb{R}^n)$ that has support in U and is identically 1 on F belongs to $\mathcal{A}(C)$, provided we agree to set $u(\infty) = 0$.

A function u that belongs to $\mathcal{A}(C)$ thus has a formal gradient $\nabla u(x)$ at almost every point x of $\mathbb{R}^n$. For $1 < p < \infty$, we define the *p-capacity of* C, denoted $\mathrm{Cap}_p(C)$, by

$$\mathrm{Cap}_p(C) = \inf\left\{ \int_{\mathbb{R}^n} |\nabla u|^p \, dm_n : u \in \mathcal{A}(C) \right\}. \tag{5.10}$$

Then $0 \leq \mathrm{Cap}_p(C) < \infty$.

Below we illustrate a ringlike condenser. It will turn out that for most p and condensers C there is an optimal choice of u which achieves the infimum in (5.10) above; this is discussed in Section 5.6 below. This u will be a C^∞-function. We illustrate the level lines of this "extremal function" u when $p = 2$.

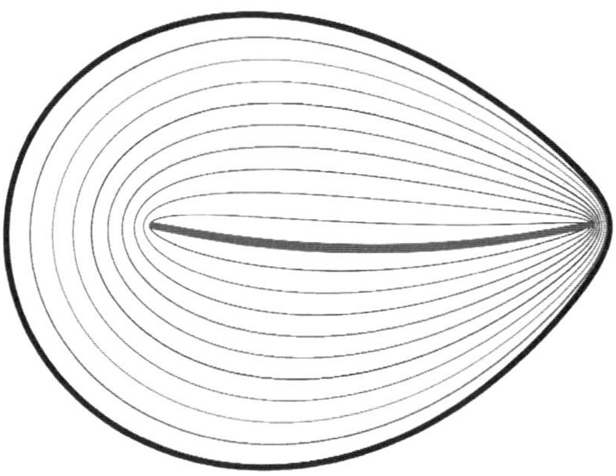

The level lines of the extremal function u for a condenser when $p = 2$.

When no confusion about the dimension n is possible, we mimic an earlier practice by abbreviating $\mathrm{Cap}_n(C)$ with $\mathrm{Cap}(C)$ and referring to this quantity as the *conformal capacity of* C. There are a number of other function classes that would serve just as well as $\mathcal{A}(C)$ in the definition of $\mathrm{Cap}_p(C)$. We indicate several of these in a lemma.

LEMMA 5.2.1. *If $C = (U, F)$ is a condenser in $\mathbb{R}^n$ and $1 < p < \infty$, then*

$$\mathrm{Cap}_p(C) = \inf\left\{ \int_{\mathbb{R}^n} |\nabla u|^p \, dm_n : u \in \mathcal{A}_k(C) \right\} \tag{5.11}$$

for any of the following classes of functions:

(i) $\mathcal{A}_0(C)$: *the class of functions in* $\mathcal{A}(C)$ *that are compactly supported in* U*;*

(ii) $\mathcal{A}_1(C) = \{u \in \mathcal{A}_0(C) : u \equiv 1$ *in a neighbourhood of* $F\}$*;*

(iii) $\mathcal{A}_2(C) = \mathcal{A}_1(C) \cap C^\infty(\mathbb{R}^n)$.

PROOF. Consider a function u from $\mathcal{A}(C)$. For $0 < \varepsilon < 1/2$ let $u_\varepsilon = \varphi_\varepsilon \circ u$, where $\varphi_\varepsilon : \mathbb{R} \to [0,1]$ is given by

$$\varphi_\varepsilon(t) = \begin{cases} 0 & \text{if } t < \varepsilon\,, \\ (1-2\varepsilon)^{-1}(t-\varepsilon) & \text{if } \varepsilon \le t \le 1-\varepsilon\,, \\ 1 & \text{if } t > 1-\varepsilon\,. \end{cases}$$

Since φ_ε satisfies a Lipschitz condition on $\mathbb{R}$ with constant $(1-2\varepsilon)^{-1}$, u_ε is in the class ACL$(\mathbb{R}^n)$ and has $|\nabla u_\varepsilon| \le (1-2\varepsilon)^{-1}|\nabla u|$ almost everywhere in $\mathbb{R}^n$. Also, u_ε is supported in $\{x : u(x) \ge \varepsilon\}$, which is a compact subset of U, while $u_\varepsilon(x) = 1$ for every x in $\{x : u(x) > 1-\varepsilon\}$, which is a neighbourhood of F; i.e., u_ε is a member of the family $\mathcal{A}_1(C)$. Because $\mathcal{A}_1(C)$ is contained in $\mathcal{A}(C)$, we have

$$\begin{aligned}\operatorname{Cap}_p(C) &\le \inf\left\{\textstyle\int_{\mathbb{R}^n} |\nabla v|^p\,dm_n : v \in \mathcal{A}_1(C)\right\} \le \liminf_{\varepsilon\to 0} \textstyle\int_{\mathbb{R}^n} |\nabla u_\varepsilon|^p\,dm_n \\ &\le \lim_{\varepsilon\to 0}(1-2\varepsilon)^{-p} \textstyle\int_{\mathbb{R}^n} |\nabla u|^p\,dm_n = \int_{\mathbb{R}^n} |\nabla u|^p\,dm_n\,.\end{aligned}$$

As u was an arbitrary member of $\mathcal{A}(C)$, (5.11) follows in the case $k = 1$, and hence, since $\mathcal{A}_0(C)$ lies between $\mathcal{A}_1(C)$ and $\mathcal{A}(C)$, in the case $k = 0$ as well.

Next, let u be a member of $\mathcal{A}_1(C)$ with $\int_{\mathbb{R}^n} |\nabla u|^p\,dm_n < \infty$. Define u_ν for $\nu = 1, 2, \ldots$ by $u_\nu = u * \psi_\nu$, where $\langle\psi_\nu\rangle$ is the standard sequence of mollifiers associated with a nonnegative C^∞-function ψ that is supported in B^n and has $\int_{\mathbb{R}^n} \psi\,dm_n = 1$. Then u_ν is a member of $C_0^\infty(\mathbb{R}^n)$ and

$$0 \le u_\nu(x) = \int_{\mathbb{R}^n} u(x-y)\psi_\nu(y)\,dm_n(y) \le \int_{\mathbb{R}^n} \psi_\nu(y)\,dm_n(y) = 1\,.$$

For large ν, the support of u_ν lies in U and $u_\nu \equiv 1$ in a neighbourhood of F. Therefore, u_ν belongs to the family $\mathcal{A}_2(C)$ once ν is sufficiently large. Because $|\nabla u_\nu| \to |\nabla u|$ in L^p-norm, we have

$$\begin{aligned}\operatorname{Cap}_p(C) &\le \inf\left\{\int_{\mathbb{R}^n} |\nabla v|^p\,dm_n : v \in \mathcal{A}_2(C)\right\} \le \lim_{\nu\to\infty} \int_{\mathbb{R}^n} |\nabla u_\nu|^p\,dm_n \\ &= \int_{\mathbb{R}^n} |\nabla u|^p\,dm_n\,.\end{aligned}$$

The above inequality holds trivially for any u in $\mathcal{A}_1(C)$ with $\int_{\mathbb{R}^n} |\nabla u|^p\,dm_n = \infty$, so we can take the infimum over u in $\mathcal{A}_1(C)$, a class for which (5.11) has already been established, to confirm (5.11) for $k = 2$. □

Further variations on Lemma 5.2.1 are possible. To mention one,

$$\operatorname{Cap}_p(C) = \inf\left\{\int_V |\nabla u|^p\,dm_n : u \in \mathcal{A}_3(C)\right\}\,, \tag{5.12}$$

where $V = U \setminus F$ and $\mathcal{A}_3(C)$ denotes the class of ACL-functions $u : V \to [0,1]$ such that $u(x) \to 0$ as $x \to \partial U \cap \partial V$ and $u(x) \to 1$ as $x \to F \cap \partial V$.

The proof of (5.12) is similar to the first part of the proof of Lemma 5.2.1. Another variant of (5.11), which will come into play when we consider spherical symmetrization of condensers, demands some preparatory comments.

By a k-*simplex* in $\mathbb{R}^n$, where $0 \le k \le n$, is meant a subset of $\mathbb{R}^n$ of the type $T = f(\Delta_k)$, in which Δ_k is the *standard* k-*simplex in* $\mathbb{R}^n$—Δ_k is the convex hull of the points $0, e_1, e_2, \ldots, e_k$ —and f is a transformation from the affine group of $\mathbb{R}^n$. The points $p_0 = f(0)$, $p_1 = f(e_1), \ldots, p_k = f(e_k)$ are called the *vertices of* T. Thus T comprises precisely those x in $\mathbb{R}^n$ that can be expressed in the form

$$x = t_0 p_0 + t_1 p_1 + \cdots + t_k p_k$$

with $0 \le t_i \le 1$ and $\sum_{i=0}^k t_i = 1$. Let $\mathcal{T} = \{T\}$ be a simplicial tesselation of $\mathbb{R}^n$ whose tiles T are congruent n-simplexes: the union of all the tiles in $\mathcal{T}$ is $\mathbb{R}^n$, and distinct members T and T' of $\mathcal{T}$ meet, if at all, in a lower-dimensional simplex whose vertices are common vertices of T and T'. Such tesselations are easy to construct in all dimensions.

We say that a function $u : \mathbb{R}^n \to \mathbb{R}$ is *piecewise linear relative to* $\mathcal{T}$ provided that u is continuous and that for each simplex T from $\mathcal{T}$ the function u_T obtained by restricting u to T is linear:

$$u_T(x) = a_1 x_1 + a_2 x_2 + \cdots + a_n x_n + b$$

for x in T, where $a_1, a_2, \ldots, a_n$, and b real numbers (which, of course, vary with T).

Notice that u_T is completely determined by the values of u at the vertices of T.

A function u of this sort is plainly differentiable, with constant gradient, in the interior of each n-simplex from $\mathcal{T}$. Moreover, u_T satisfies a Lipschitz condition with constant $\lambda_T = (a_1^2 + a_2^2 + \cdots + a_n^2)^{1/2}$ in T, so u will satisfy a Lipschitz condition with constant $\lambda = \sup_T \lambda_T$ in the event that λ is finite. In particular, this will always be the case when u has compact support.

We shall speak of a function $u : \mathbb{R}^n \to \mathbb{R}$ as *piecewise linear* if it is piecewise linear relative to some tesselation $\mathcal{T}$ of the type under consideration. (We warn the reader that this definition is somewhat more restrictive than the standard definition found in topology texts for a piecewise linear function.)

For any given $\mathcal{T}$ and continuous function $v : \mathbb{R}^n \to \mathbb{R}$, there is a unique piecewise linear function $u : \mathbb{R}^n \to \mathbb{R}$ with the property that $u(p) = v(p)$ whenever p is the vertex of an n-simplex belonging to $\mathcal{T}$. If v belongs to the class $C_0^\infty(\mathbb{R}^n)$, then a straightforward calculation shows that an inequality

$$|\nabla u(x) - \nabla v(x)| \le 2c\lambda\delta\sqrt{n}$$

holds in the interior of any tile T from $\mathcal{T}$. Here c is a constant that depends only on the similarity class of T, λ is a Lipschitz constant for the mapping $\nabla v : \mathbb{R}^n \to \mathbb{R}^n$, and δ is the diameter of T. (The case $T = \Delta_n$ is easy to handle; the general case reduces to the special one after an affine change of variable.) This leads directly to the L^∞-estimate

$$\big\| \, |\nabla u| - |\nabla v| \, \big\|_\infty \le 2c\lambda\delta\sqrt{n}\,. \tag{5.13}$$

The support of u is a compact subset of $\{x : \operatorname{dist}[x, \operatorname{supp}(v)] \le \delta\}$.

With these remarks behind us, we add a postscript to Lemma 5.2.1.

LEMMA 5.2.2. *If* $C = (U, F)$ *is a condenser in* $\mathbb{R}^n$ *and* $1 < p < \infty$, *then*

$$\operatorname{Cap}_p(C) = \inf\left\{ \int_{\mathbb{R}^n} |\nabla u|^p \, dm_m : u \in \mathcal{A}_4(C) \right\},$$

where $\mathcal{A}_4(C)$ *is the class of piecewise linear functions in* $\mathcal{A}_0(C)$.

PROOF. Fix a simplicial tesselation $\mathcal{T} = \{T\}$ of $\mathbb{R}^n$ with tiles that are congruent n-simplexes of unit diameter. For δ in $(0, 1]$ the dilation $x \mapsto \delta x$ transforms $\mathcal{T}$ into a tesselation $\mathcal{T}_\delta$ of $\mathbb{R}^n$ that differs from $\mathcal{T}$ only in scale. Given a function v from the class $\mathcal{A}_2(C)$, we let u_δ be the piecewise linear function on $\mathbb{R}^n$ that agrees with v on the set of vertices of $\mathcal{T}_\delta$. Then u_δ belongs to the class $\mathcal{A}_4(C)$ for all sufficiently small δ, so

$$\begin{aligned}\operatorname{Cap}_p(C) &\leq \inf\left\{\int_{\mathbb{R}^n} |\nabla u|^p\, dm_n : u \in \mathcal{A}_4(C)\right\} \leq \lim_{\delta\to 0} \int_{\mathbb{R}^n} |\nabla u_\delta|^p\, dm_n \\ &= \int_{\mathbb{R}^n} |\nabla v|^p\, dm_n\,.\end{aligned}$$

The last equality is justified by (5.13), which implies that $|\nabla u_\delta| \to |\nabla v|$ in L^∞-norm as $\delta \to 0$ —hence, since all functions concerned are supported in some fixed compact set, in L^p-norm as well. Taking the infimum over v leads to the stated conclusion. □

The link between the present discussion and our earlier discussion concerning the moduli of curve familes is provided by the following result.

THEOREM 5.2.3. *Let $C = (U, F)$ be a condenser in $\mathbb{R}^n$. If $\Gamma_C = \Delta(U^c, F : \hat{\mathbb{R}}^n)$, then $\operatorname{Cap}_p(C) = M_p(\Gamma_C)$ for every $p \geq n$.*

PROOF. Let $\mathcal{A}_2(C)$ be the family of functions identified in Lemma 5.2.1. Given u in $\mathcal{A}_2(C)$, we define a Borel function $\rho : \hat{\mathbb{R}}^n \to [0, \infty)$ by setting $\rho(x) = |\nabla u(x)| = L_u(x)$ if x is in $\mathbb{R}^n$ and $\rho(\infty) = 0$. Consider a locally rectifiable curve γ from the family Γ_C. The trajectory of γ lies in $\mathbb{R}^n$, so the C^∞-function u is locally absolutely continuous on γ, making $\beta = u \circ \gamma$ a locally rectifiable curve with trajectory in the real interval $[0, 1]$. Because $\bar{\gamma}$ meets both U^c, in a neighbourhood of which $u \equiv 0$, and F, in a neighbourhood of which $u \equiv 1$, the trajectory of β is the entire interval $[0, 1]$. As a result, $\ell(\beta) \geq 1$. By Theorem 4.1.6,

$$\int_\gamma \rho\, ds = \int_\gamma L_u\, ds \geq \int_\beta ds = \ell(\beta) \geq 1\,.$$

The consequence is that ρ belongs to $\operatorname{Adm}(\Gamma_C)$, whence

$$M_p(\Gamma_C) \leq \int_{\mathbb{R}^n} \rho^p\, dm_n = \int_{\mathbb{R}^n} |\nabla u|^p\, dm_n\,.$$

Taking the infimum over $\mathcal{A}_2(C)$, we learn from Lemma 5.2.1 that $M_p(\Gamma_C) \leq \operatorname{Cap}_p(C)$. Indeed, the above argument establishes this inequality for every $p > 1$.

The inequality $\operatorname{Cap}_p(C) \leq M_p(\Gamma_C)$ for $p \geq n$ is more delicate. Fix such a p. We know from Lemma 4.2.7 that $M_p(\Gamma_C) < \infty$. Writing $E = U^c$, we can assert on the strength of Theorem 4.3.13 that $M_p(\Gamma_C) = \lim_{r\to 0} M_p[\Gamma(r)]$, where $\Gamma(r) = \Delta[E(r), F(r) : \hat{\mathbb{R}}^n]$ for $0 < r \leq 2$. Given $\varepsilon > 0$, we choose r so that $M_p[\Gamma(r)] < M_p(\Gamma_C) + \varepsilon$ and then select an admissible density ρ for $\Gamma(r)$ having

$$\int_{\mathbb{R}^n} \rho^p\, dm_n < M_p(\Gamma_C) + \varepsilon.$$

We thereupon fix s in $(0, r/2)$ and define a function $\tilde{\rho} : \mathbb{R}^n \to [0, \infty)$ as follows:

$$\tilde{\rho}(x) = \frac{1}{|B^n(x, s)|} \int_{B^n(x,s)} \rho\, dm_n\,.$$

Because it belongs to the class $L^p(\mathbb{R}^n)$ with $p > 1$, ρ is locally integrable in $\mathbb{R}^n$, so the function $\tilde{\rho}$ is well defined. Hölder's inequality tells us that

$$\begin{aligned} 0 \; &\le \; \tilde{\rho}(x) = \frac{1}{|B^n(x,s)|}\int_{B^n(x,s)} \rho\, dm_n \le \left[\frac{1}{|B^n(x,s)|}\int_{B^n(x,s)} \rho^p\, dm_n\right]^{1/p} \\ &\le \; (s^n\Omega_n)^{-1/p}\|\rho\|_p = \lambda < \infty\,. \end{aligned}$$

The absolute continuity property of Lebesgue integrals ensures that $\tilde{\rho}$ is a continuous function. We will shortly find it advantageous to express $\tilde{\rho}(x)$ differently, namely, in the manner

$$\tilde{\rho}(x) = c\int_{B^n(s)} \rho(x+y)\, dm_n(y) \tag{5.14}$$

with $c = (s^n\Omega_n)^{-1}$.

We are going to use the function $\tilde{\rho}$ to construct a member u of $\mathcal{A}(C)$. We start by defining a function $w : \mathbb{R}^n \to [0,\infty)$ through the formula

$$w(x) = \inf_\gamma \int_\gamma \tilde{\rho}\, ds,$$

in which the infimum is taken over all rectifiable paths γ with initial point in F and terminal point x. It is easy to see that for x and y in $\mathbb{R}^n$ we have

$$|w(y) - w(x)| \le \int_x^y \tilde{\rho}\, ds \le \lambda|y-x|\;,$$

where the integration is performed along the line segment between x and y. Thus w is a Lipschitz function in $\mathbb{R}^n$. As such, w is differentiable almost everywhere in $\mathbb{R}^n$. At any point x where w is differentiable we can exploit the continuity of $\tilde{\rho}$ to conclude that

$$|\nabla w(x)| = L_w(x) \le \lim_{y\to x}\frac{1}{|y-x|}\int_x^y \tilde{\rho}\, ds = \tilde{\rho}(x)\,. \tag{5.15}$$

Notice that $w(x) = 0$ whenever x belongs to F. Consider a finite point x of the set $E(r-2s)$ and an arbitrary rectifiable path $\gamma : [a,b] \to \mathbb{R}^n$ with $\gamma(a)$ in F and $\gamma(b) = x$. For any y in $B^n(s)$ we see that

$$q\big(y+\gamma(a),F\big) \le 2d\big(y+\gamma(a),F\big) \le 2|y| < 2s < r$$

and that

$$q\big(y+\gamma(b),E\big) = q(y+x,E) \le q(y+x,x) + q(x,E) \le 2|y| + r - 2s < 2s + r - 2s = r\,.$$

As a result, we obtain for each y in $B^n(s)$ a rectifiable path γ_y from the family $\Gamma(r)$ by setting $\gamma_y(t) = y + \gamma(t)$ for $a \le t \le b$. Because ρ originated in $\mathrm{Adm}[\Gamma(r)]$, we have

$$\int_\gamma \rho(y+z)\,|dz| = \int_{\gamma_y} \rho\, ds \ge 1$$

for every y in $B^n(s)$. It follows from (5.14) and Fubini's theorem that

$$\begin{aligned} \int_\gamma \tilde{\rho}\, ds \; &= \; \int_\gamma \tilde{\rho}(z)\,|dz| = c\int_\gamma \left[\int_{B^n(s)} \rho(y+z)\, dm_n(y)\right]|dz| \\ &= \; c\int_{B^n(s)}\left[\int_\gamma \rho(y+z)\,|dz|\right] dm_n(y) \ge c\int_{B^n(s)} dm_n(y) = 1\,. \end{aligned}$$

Since γ was an arbitrary rectifiable path joining F to x, we infer that $w(x) \geq 1$. This is true of every x in $E(r-2s) \cap \mathbb{R}^n$.

We next define $v : \hat{\mathbb{R}}^n \to [0,1]$ by $v(x) = \min\{1, w(x)\}$ if x is in $\mathbb{R}^n$ and $v(\infty) = 1$. Now v is continuous in $\mathbb{R}^n$ and $v \equiv 1$ in $E(r-2s) \cap \mathbb{R}^n$, a set that includes a punctured neighbourhood of ∞, so v is a continuous function. Note that $v(x) = 0$ whenever x is an element of F and that $v(x) = 1$ whenever x belongs to $E = U^c$. Clearly

$$|v(y) - v(x)| \leq |w(y) - w(x)|$$

for all x and y in $\mathbb{R}^n$. Thus v is a Lipschitz function in $\mathbb{R}^n$, and $|\nabla v(x)| \leq |\nabla w(x)|$ is true for any x at which both v and w are differentiable. By virtue of (5.15), we can assert that $|\nabla v(x)| \leq \tilde{\rho}(x)$ holds for almost every x in $\mathbb{R}^n$. We conclude that $u = 1 - v$ is a function from the class $\mathcal{A}(C)$ and satisfies $|\nabla u| \leq \tilde{\rho}$ almost everywhere in $\mathbb{R}^n$.

Therefore,

$$\begin{aligned}
\operatorname{Cap}_p(C) &\leq \int_{\mathbb{R}^n} |\nabla u|^p \, dm_n \\
&\leq \int_{\mathbb{R}^n} \tilde{\rho}^p \, dm_n = \int_{\mathbb{R}^n} \left[\frac{1}{|B^n(x,s)|} \int_{B^n(x,s)} \rho(y) \, dm_n(y) \right]^p dm_n(x) \\
&\leq \int_{\mathbb{R}^n} \left[\frac{1}{|B^n(x,s)|} \int_{B^n(x,s)} \rho^p(y) \, dm_n(y) \right] dm_n(x) \\
&= c \int_{\mathbb{R}^n} \left[\int_{B^n(s)} \rho^p(x+y) \, dm_n(y) \right] dm_n(x) \\
&= c \int_{B^n(s)} \left[\int_{\mathbb{R}^n} \rho^p(x+y) \, dm_n(x) \right] dm_n(y) \\
&= \int_{\mathbb{R}^n} \rho^p \, dm_n < M_p(\Gamma_C) + \varepsilon \, .
\end{aligned}$$

Finally, we let $\varepsilon \to 0$ to get $\operatorname{Cap}_p(C) \leq M_p(\Gamma_C)$.

This completes the proof of the theorem. □

We stress here that, because we have invoked Theorem 4.3.13, we have established this inequality only when $p \geq n$.

Consider a domain D in $\hat{\mathbb{R}}^n$ with $n \geq 2$ and a pair F_0 and F_1 of nonempty, disjoint, compact subsets of $\bar{D}$. Theorem 5.2.3 is a special case of a principle that spells out the relationship between the modulus $M_p[\Delta_0(F_0, F_1 : D)]$ and the *variational p-capacity* $\operatorname{Cap}_p(F_0, F_1 : D)$ of the configuration $(F_0, F_1 : D)$. The latter quantity is defined by

$$\operatorname{Cap}_p(F_0, F_1 : D) = \inf \left\{ \int_{D \cap \mathbb{R}^n} |\nabla u|^p \, dm_n : u \in \mathcal{A}(F_0, F_1 : D) \right\} ,$$

where $\mathcal{A}(F_0, F_1 : D)$ denotes the class of continuous functions $u : D \cup F_0 \cup F_1 \to [0,1]$ that have the ACL-property in $D \cap \mathbb{R}^n$ and satisfy $u \equiv 0$ on F_0 and $u \equiv 1$ on F_1.

Lemma 5.2.1 has a natural analogue here. The role of $\mathcal{A}_3(C)$, for instance, is played by the class $\mathcal{A}_3(F_0, F_1 : D)$, which consists of those u in $\mathcal{A}(F_0, F_1 : D)$ that are identically 0 and 1, respectively, on neighbourhoods U_0 and U_1 of F_0 and F_1 in the relative topology of $D \cup F_0 \cup F_1$.

The principle to which we just alluded asserts that

$$M_p\big[\Delta_0(F_0, F_1 : D)\big] = \mathrm{Cap}_p(F_0, F_1 : D)$$

for every $p > 1$. This was shown for certain configurations by W. Ziemer [**170**] and was established by J. Hesse [**78**] for the case where F_0 and F_1 are contained in D. The result in the generality stated is due to V.A. Shlyk [**146, 147**].

Theorem 5.2.3 for the case $p = n$ is perfectly adequate to meet our needs in the present book.

The identification of $M_p[\Delta_0(F_0, F_1 : D)]$ with $\mathrm{Cap}_p(F_0, F_1 : D)$ has many interesting consequences. To cite one example, it is the mechanism that enabled Hesse to prove Theorem 4.3.14, a result we quoted without proof. Indeed, let $\langle F_{0\nu}\rangle$ and $\langle F_{1\nu}\rangle$ be nonincreasing sequences of nonempty, compact subsets of $\bar{D}$, and let $F_0 = \bigcap_{\nu=1}^{\infty} F_{0\nu}$ and $F_1 = \bigcap_{\nu=1}^{\infty} F_{1\nu}$. We assume that the sets F_0 and F_1 are disjoint, which forces $F_{0\nu}$ and $F_{1\nu}$ to be disjoint once ν is sufficiently large. Writing $\Gamma_\nu = \Delta_0(F_{0\nu}, F_{1\nu} : D)$ and $\Gamma = \Delta_0(F_0, F_1 : D)$, we claim that $M_p(\Gamma) = \lim_{\nu\to\infty} M_p(\Gamma_\nu)$. The monotonicity of M_p ensures that the limit in question exists and is no smaller than $M_p(\Gamma)$. If u is a function from $\mathcal{A}_3(F_0, F_1 : D)$, then u belongs to $\mathcal{A}(F_{0\nu}, F_{1\nu} : D)$ for all suitably large ν, from which we infer that

$$\lim_{\nu\to\infty} M_p(\Gamma_\nu) = \lim_{\nu\to\infty} \mathrm{Cap}_p(F_{0\nu}, F_{1\nu} : D) \le \int_{D\cap\mathbb{R}^n} |\nabla u|^p \, dm_n\,.$$

Taking the infimum over u leads to the inequality

$$\lim_{\nu\to\infty} M_p(\Gamma_\nu) \le \mathrm{Cap}_p(F_0, F_1 : D) = M_p(\Gamma)\,.$$

Accordingly, Hesse's theorem—as a matter of fact, a more general version of it—follows from Shlyk's.

Let $R = \mathcal{R}(C_0, C_1)$ be a ring in $\hat{\mathbb{R}}^n$ with ∞ in C_0. Theorem 5.2.3 implies that $\mathrm{Cap}_p(R) = \mathrm{Cap}_p(C)$ for $p \ge n$, where C is the ringlike condenser $(R \cup C_1, C_1)$. Indeed, $\Gamma_C = \Gamma_{rect}$ in this case.

5.3. Spherical symmetrization of condensers

For the duration of the present section we fix a point x_0 of $\mathbb{R}^n$ with $n \ge 2$ and a ray L^* that emanates from x_0. We also write $S_r = \mathbb{S}^{n-1}(x_0, r)$ for $0 \le r < \infty$.

Let U be an open set in $\mathbb{R}^n$. The *spherical symmetrization of U with respect to L^** is the subset U^* of $\mathbb{R}^n$ determined by the following two requirements:

(1) the set S_r has nonempty intersection with (resp., is contained in) U^* precisely when it enjoys the same property relative to U;
(2) if $U \cap S_r$ is a nonempty proper subset of S_r, then $U^* \cap S_r$ is the unique proper open cap of S_r that is centered on L^* and has the same σ_{n-1}-measure as $U \cap S_r$.

In particular, when S_r is not a subset of U but $U \cap S_r$ has surface area $\omega_{n-1} r^{n-1}$, $U^* \cap S_r$ is just the set obtained from S_r by removing its point of intersection with $L^\star$, which for the remainder of the section will serve to indicate the ray; dual to L^*; i.e., the ray issuing from x_0 in the direction opposite to that of L^*.

This construction is designed to ensure the axial symmetry of U^* with respect to L^*, meaning that $g(U^*) = U^*$ for every Euclidean isometry g of $\mathbb{R}^n$ which fixes L^* pointwise.

It is a simple matter to check that U^* is itself an open set, that U^* is connected if U is, and that $(U^*)^c$ is connected when U^c has this feature. On the basis of Theorem 4.0.12, applied separately to $f = \mathcal{X}_U$ and $f = \mathcal{X}_{U^*}$, we can assert that the sets U and U^* have the identical Lebesgue measure. If V is an open subset of U, then V^* plainly lies in U^*.

Suppose next that F is a closed subset of $\mathbb{R}^n$. The construction of the spherical symmetrization of F with respect to L^*, again denoted by F^*, involves only a slight deviation from the procedure for an open set. To be precise, in the case of a closed set F condition (2) above changes to: if $F \cap S_r$ is a nonempty proper subset of S_r, then $F^* \cap S_r$ is the unique closed cap of S_r whose center is on L^* and whose σ_{n-1}-measure equals that of $F \cap S_r$.

Thus, in the event that $F \cap S_r$ is a nonempty set with $\sigma_{n-1}(F \cap S_r) = 0$, $F^* \cap S_r$ consists of the single point at which S_r meets L^*. Again the set F^* is axially symmetric relative to L^*. Observing that

$$F^* = \mathbb{R}^n \setminus (\mathbb{R}^n \setminus F)^\star , \tag{5.16}$$

where $\star$ denotes spherical symmetrization with respect to $L^\star$, we see that F^* is closed in $\mathbb{R}^n$. When F is connected, so is F^*; when F is bounded, F^* is also bounded. As was true for an open set, a closed set F and its symmetrized relative F^* have equal Lebesgue measure. It is evident that E^* lies in F^* whenever E is a closed set that is contained in F.

Finally, let $C = (U, F)$ be a condenser in $\mathbb{R}^n$. Then F^* is a nonempty, compact subset of U^*. The condenser $C^* = (U^*, F^*)$ is called the *spherical symmetrization of C with respect to L^**. We further note here that if the original condenser C is ringlike, then C^* will be ringlike as well.

A symmetrization result that is central to the understanding of condensers and their capacities was established in the plane by W.K. Hayman [**68, 69**] and in 3-space by F.W. Gehring [**46**], whose approach was subsequently extended to higher dimensions by G.D. Mostow [**125**].

The book [**41**] of V.N. Dubinin contains the most recent treatment of the topic of symmetrization and its various applications.

The symmetrization result we want asserts that

$$\operatorname{Cap}_p(C^*) \le \operatorname{Cap}_p(C) \tag{5.17}$$

for every $p > 1$.

For a systematic overview of inequalities of the type (5.17) arising from various kinds of symmetrization, the reader should consult a paper of J. Sarvas [**143**].

In order to confirm (5.17) we must devise a scheme for symmetrizing continuous functions that is adapted to the foregoing method of symmetrizing open and closed sets. It will greatly streamline our treatment of the symmetrization of functions if we are at liberty to quote certain standard theorems from the subject of geometric measure theory. For ease of reference we state the results in question, begging the indulgence of readers who would prefer a totally self-contained presentation.

(The approach in [**143**] is more elementary in this respect, but pursuing it would entail a major digression in order to establish the necessary notation, teminology, and so forth.)

For a subset A of $\mathbb{R}^n$ with $n \geq 2$ and for $\rho > 0$, we call $T_\rho(A) = \{x \in \mathbb{R}^n : \operatorname{dist}(x, A) \leq \rho\}$ the *tubular neighbourhood of* A *with radius* ρ. Similarly, if A lies on a Euclidean sphere S in $\mathbb{R}^n$, we refer to $\widetilde{T}_\rho(A) = \{x \in S : d_S(x, A) \leq \rho\}$, where d_S indicates the geodesic (= great circle) distance on S, as the *spherical tubular neighbourhood of* A *with radius* ρ. We have of course earlier encountered tubular neighbourhoods as in the sets $A(\rho) = \{x \in \hat{\mathbb{R}}^n : q(x, A) \leq \rho\}$, which are nothing but chordal tubular neighbourhoods of A.

The first result we wish to cite is a version of the Brunn-Minkowski theorem—specifically, the Brunn-Minkowski theorem for spherical geometry [**67**].

THEOREM 5.3.1. *If* A *is a* σ_{n-1}*-measurable subset of a Euclidean sphere* S *in* $\mathbb{R}^n$ *and if* K *is a cap of* S *with* $\sigma_{n-1}(K) = \sigma_{n-1}(A)$*, then*

$$\sigma_{n-1}\big[\widetilde{T}_\rho(K)\big] \leq \sigma_{n-1}\big[\widetilde{T}_\rho(A)\big] \tag{5.18}$$

for every $\rho > 0$.

A subset A of $\mathbb{R}^n$ is said to be m*-rectifiable* if A can be represented in the manner $A = \Phi(B)$, where B is a bounded set in $\mathbb{R}^m$ and Φ is a Lipschitz mapping of B into $\mathbb{R}^n$. For example, any bounded subset of a hyperplane in $\mathbb{R}^n$ is $(n-1)$-rectifiable, as is any subset of a Euclidean sphere in $\mathbb{R}^n$. The union of a finite collection of m-rectifiable sets in $\mathbb{R}^n$ is itself m-rectifiable. Obviously $\sigma_{n-1}(A) < \infty$ whenever A is an $(n-1)$-rectifiable Borel set in $\mathbb{R}^n$. A second proposition from geometric measure theory to which we shall appeal relates surface area to volume. As such, it belongs to the realm of ideas surrounding the isoperimetric inequality.

THEOREM 5.3.2. *If* F *is a compact subset of* $\mathbb{R}^n$ *whose interior is dense in* F *and whose boundary is* $(n-1)$*-rectifiable, then*

$$\sigma_{n-1}(\partial F) = \lim_{\rho \to 0} \frac{m_n[T_\rho(F)] - m_n(F)}{\rho}\,. \tag{5.19}$$

Another well-known result that we shall have occasion to invoke is the Federer-Young co-area formula [**37**].

THEOREM 5.3.3. *If* D *is an open set in* $\mathbb{R}^n$ *and if* $u : D \to \mathbb{R}$ *satisfies a Lipschitz condition, then for any Lebesgue measurable function* $f : D \to [0, \infty]$ *it is true that*

$$\int_D f|\nabla u|\, dm_n = \int_I \left(\int_{\Sigma_t} f\, d\sigma_{n-1}\right) dt\,,$$

where $I = u(D)$ *and where* $\Sigma_t = \{x : u(x) = t\}$ *for* t *in* I.

We have already made rather extensive use of a simple case of the co-area formula, under the name "Fubini's theorem in spherical coordinates" in Theorem 4.0.12.

Suppose that A is a Lebesgue measurable subset of $\mathbb{R}^n$ and that ν is a unit vector in $\mathbb{R}^n$. A point x of $\mathbb{R}^n$ is said to be a *point of density for* A *in the direction* ν if

$$\lim_{r \to 0} \frac{m_1[\{t \in [-r, r] : x + t\nu \in A\}]}{2r} = 1\,.$$

We shall exploit the following piece of information which follows from elementary Lebesgue measure theory.

THEOREM 5.3.4. *If A is a Lebesgue measurable subset of $\mathbb{R}^n$ and if ν is a unit vector in $\mathbb{R}^n$, then almost every point of A is a point of density for A in the direction ν.*

We begin to apply the foregoing results to matters at hand by proving a few preparatory lemmas that will have significant consequences for the symmetrization of functions. The first of these reads:

LEMMA 5.3.5. *If F is a closed subset of $\mathbb{R}^n$, then $T_\rho(F^*)$ is contained in $T_\rho(F)^*$ for every $\rho > 0$.*

PROOF. Fix $\rho > 0$. We need only demonstrate that $T_\rho(F^*) \cap S_r$ is a subset of $T_\rho(F)^* \cap S_r$ for each $r \geq 0$. As this is readily checked when $r = 0$, we consider the situation for a fixed $r > 0$. We assume that $T_\rho(F^*) \cap S_r$ is nonempty, for there is nothing to prove otherwise.

Note that the set $T_\rho(F^*)$ shares with F^* the property of axial symmetry about L^*, a fact which makes it clear that $T_\rho(F^*) \cap S_r$ is a closed cap of S_r centered on L^*.

We claim that $T_\rho(F)^* \cap S_r$ is also nonempty. If this is true, then $T_\rho(F^*) \cap S_r$ is a second closed cap of S_r with center on L^*, so the proof of the lemma reduces to the problem of verifying the inequality

$$\sigma_{n-1}\big[T_\rho(F^*) \cap S_r\big] \leq \sigma_{n-1}\big[T_\rho(F)^* \cap S_r\big] . \tag{5.20}$$

Let x and y be points of S_r and F^*, respectively, for which $|x-y| \leq \rho$. If $t = |y-x_0|$, then

$$|r - t| = \big|\, |x - x_0| - |y - x_0| \,\big| \leq |x - y| \leq \rho .$$

Now S_t intersects F^*, hence it must intersect F. We infer that

$$\operatorname{dist}(S_r, F) \leq \operatorname{dist}(S_r, F \cap S_t) = |r - t| \leq \rho .$$

Taking into account our knowledge that F is closed, we conclude that S_r has a nonempty intersection with $T_\rho(F)$. Therefore, $T_\rho(F)^* \cap S_r$ is likewise nonempty, as asserted. We now turn to the verification of (5.20).

Let t be any positive number for which $|r - t| \leq \rho$, and let $P : S_t \to S_r$ be dilation by the factor f/t. A bit of elementary trigonometry reveals that, for any point z of S_t, the set $\overline{B}^n(z,\rho) \cap S_r$ is the closed cap of S_r whose center is $P(z)$ and whose radius in the spherical geometry of S_r is αr, where α in $[0,\pi]$ satisfies $\rho^2 = r^2 + t^2 - 2rt\cos\alpha$. Since F is closed,

$$T_\rho(F \cap S_t) \cap S_r = \bigcup_{z \in F \cap S_t} \big[\overline{B}^n(z,\rho) \cap S_r\big] .$$

We deduce that, under the constraints on t as outlined above,

$$T_\rho(F \cap S_t) \cap S_r = \widetilde{T}_{\alpha r}\big[P(F \cap S_t)\big] \ ,$$

where spherical tubular neighbourhoods are taken with respect to $S = S_r$. Similarly, we see that for t in the stated range

$$T_\rho(F^* \cap S_t) \cap S_r = \widetilde{T}_{\alpha r}\big[P(F^* \cap S_t)\big] .$$

Now $P(F^* \cap S_t)$ is a closed cap of S_r that has the same σ_{n-1}-measure as $P(F \cap S_t)$, for under the mapping P the surface area just scales by the factor $(r/t)^{n-1}$. Inequality (5.18) thus informs us that

$$\sigma_{n-1}\big[T_\rho(F^* \cap S_t) \cap S_r\big] \le \sigma_{n-1}\big[T_\rho(F \cap S_t) \cap S_r\big] \tag{5.21}$$

whenever $t \ge 0$ and $|r - t| \le \rho$. Let I denote the set of all these t; the special case $t = 0$ is trivial.

The fact that F^* is a closed set allows us to write

$$T_\rho(F^*) \cap S_r = T_\rho\big[\bigcup_{t\in I}(F^* \cap S_t)\big] \cap S_r = \bigcup_{t\in I}\big[T_\rho(F^* \cap S_t) \cap S_r\big]\,.$$

As the sets in the last union are all closed caps of S_r centered on L^*, we find with the help of (5.21) that

$$\begin{aligned}\sigma_{n-1}\big[T_\rho(F^*) \cap S_r\big] &= \sup_{t\in I}\sigma_{n-1}\big[T_\rho(F^* \cap S_t) \cap S_r\big] \\ &\le \sup_{t\in I}\sigma_{n-1}\big[T_\rho(F \cap S_t) \cap S_r\big] \le \sigma_{n-1}\big[T_\rho(F) \cap S_r\big] \\ &= \sigma_{n-1}\big[T_\rho(F)^* \cap S_r\big]\,,\end{aligned}$$

which establishes (5.20) and thereby completes the proof. □

Another important aspect of spherical symmetrization that will be vital to our applications can be stated in heuristic terms as follows: although volume is preserved when a set undergoes spherical symmetrization, boundary surface area tends to be compressed.

In fact, we shall need to make this statement precise only for a particular class of sets, namely, for polyhedral sets. We interpret the term "polyhedral set" very loosely, using it merely to indicate a subset P of $\mathbb{R}^n$ with $n \ge 2$, connected or otherwise, that can be expressed as the union of finitely many nonoverlapping (not necessarily congruent) n-simplexes. No special combinatorial requirements are imposed on P. Naturally, the class of polyhedral sets includes any closed, convex polyhedron in $\mathbb{R}^n$. If A is a Borel subset of $\mathbb{R}^n$, we shall employ the notation σ_A for the function defined on $[0, \infty)$ by $\sigma_A(r) = \sigma_{n-1}(A \cap B_r)$, where $B_r = \overline{B}^n(x_0, r)$. Then $\sigma_A(0) = 0$, σ_A is nondecreasing, and σ_A is bounded if and only if $\sigma_{n-1}(A)$ is finite. In case $\sigma_{n-1}(A) < \infty$, σ_A is continuous if and only if $\sigma_{n-1}(A \cap S_r) = 0$ for every $r \ge 0$.

We state the next lemma for every dimension $n \ge 2$, but only present a proof that is valid when $n \ge 3$, the range of n of greatest importance for this book. The case $n = 2$, where $\sigma_{n-2} = \mathcal{H}^1$ measures arc length, can be treated directly using more elementary arguments which we encourage the reader to develop.

LEMMA 5.3.6. *Suppose that P is a polyhedral set in $\mathbb{R}^n$. Write $\sigma = \sigma_{\partial P}$ and $\sigma^* = \sigma_{\partial P^*}$. If $0 \le r_1 < r_2 \le \infty$, then*

$$\sigma^*(r_2) - \sigma^*(r_1) \le \sigma(r_2) - \sigma(r_1)\,. \tag{5.22}$$

In particular, $\sigma_{n-1}(\partial P^*) \le \sigma_{n-1}(\partial P)$.

PROOF FOR $n \ge 3$. Because ∂P is covered by the union of finitely many hyperplanes, it is clear that $\sigma_{n-1}(\partial P \cap S_r) = 0$ for each $r \ge 0$. Since this forces $P^* \cap S_r$ and $\mathrm{int}(P^*) \cap S_r$ to have equal surface area, it implies that $\sigma_{n-1}(\partial P^* \cap S_r) = 0$ for $r \ge 0$. Define $A(r)$ for $r \ge 0$ by $A(r) = \sigma_{n-1}(P \cap S_r) = \sigma_{n-1}(P^* \cap S_r)$.

When $0 < A(r) < \omega_{n-1}r^{n-1}$, the set $\partial P^* \cap S_r$ is just the $(n-2)$-sphere on S_r that borders the closed cap $P^* \cap S_r$. Now $A(r) > 0$ if and only if the point $L^* \cap S_r$ lies in $\mathrm{int}(P)^*$, while $A(r) < \omega_{n-1}r^{n-1}$ precisely when $L^\star \cap S_r$ is a point of the open set $(\mathbb{R}^n \setminus P)^\star$. As each of the sets $\mathrm{int}(P)$ and $\mathbb{R}^n \setminus P$ has but a finite number of components and as spherical symmetrization preserves connectedness, each of the sets $\{r : A(r) > 0\}$ and $\{r : A(r) < \omega_{n-1}r^{n-1}\}$ is a union of a finite number of disjoint open intervals. We conclude that $\{r : 0 < A(r) < \omega_{n-1}r^{n-1}\}$ can be written as the disjoint union of finitely many open intervals I_j, say $I_j = (a_j, b_j)$ for $j = 1, 2, \ldots, N$. Moreover, if J is a component of $[0,\infty) \setminus \bigcup_{j=1}^N I_j$, then $\{x \in \partial P^* : |x - x_0| \in J\}$ is contained in the line $L^* \cup L^\star$. Therefore, such a set has no impact on $\sigma_{n-1}(\partial P^*)$. In other words,

$$\sigma_{n-1}(\partial P^*) = \sum_{j=1}^{N} \sigma_{n-1}(\Sigma_j') ,$$

where $\Sigma_j' = \{x \in \partial P^* : |x - x_0| \in I_j\}$. The boundary of P is obviously $(n-1)$-rectifiable. As a result, $\sigma_{n-1}(\partial P) < \infty$ and the function σ is seen to be bounded and continuous. Assume next that $a_j < r_1 < r_2 < b_j$ for some j, and write $\Sigma' = \{x \in \partial P^* : r_1 < |x - x_0| \le r_2\}$. Then $\sigma_{n-1}(\Sigma') = \sigma^*(r_2) - \sigma^*(r_1)$. We shall prove that $\sigma_{n-1}(\Sigma') \le \sigma(r_2) - \sigma(r_1)$ in this situation. By letting $r_1 \to a_j$ and $r_2 \to b_j$, we are able to conclude that $\sigma_{n-1}(\Sigma_j') \le \sigma(b_j) - \sigma(a_j) < \infty$. Such being the case for $j = 1, 2, \ldots, N$, we can infer that $\sigma_{n-1}(\partial P^*) \le \sigma_{n-1}(\partial P) < \infty$. Consequently, σ^* is also a bounded and continuous function, one for which (5.22) holds whenever r_1 and r_2 lie in the same interval I_j. However, because σ^* remains constant in each component of $[0,\infty) \setminus \bigcup_{j=1}^N I_j$ and because both σ and σ^* are continuous functions, this is actually sufficient to establish (5.22) in general.

Consider $F = \{x \in P : r_1 \le |x - x_0| \le r_2\}$ and $\Sigma = \{x \in \partial P : r_1 < |x - x_0| \le r_2\}$. We remark that

$$\partial F = (P \cap S_{r_1}) \cup \Sigma \cup (P \cap S_{r_2})$$

and that any two of the three sets in the right-hand union intersect in a set of σ_{n-1}-measure zero. Thus

$$\begin{aligned} \sigma_{n-1}(\partial F) &= \sigma_{n-1}(P \cap S_{r_1}) + \sigma_{n-1}(\Sigma) + \sigma_{n-1}(P \cap S_2) \\ &= \sigma_{n-1}(P \cap S_{r_1}) + \sigma(r_2) - \sigma(r_1) + \sigma_{n-1}(P \cap S_2) . \end{aligned}$$

Since it can be covered by a finite union of hyperplanes and spheres, the set ∂F is $(n-1)$-rectifiable. It is quite possible that F has a finite number of isolated points. If any such points are disregarded— their presence would not invalidate (5.19)— F satisfies the hypotheses of Theorem 5.3.2.

We observe that $F^* = \{x \in P^* : r_1 \le |x - x_0| \le r_2\}$ and that

$$\partial F^* = (P^* \cap S_{r_1}) \cup \Sigma' \cup (P^* \cap S_{r_2}) ,$$

which leads to

$$\begin{aligned} \sigma_{n-1}(\partial F^*) &= \sigma_{n-1}(P^* \cap S_{r_1}) + \sigma_{n-1}(\Sigma') + \sigma_{n-1}(P^* \cap S_{r_2}) \\ &= \sigma_{n-1}(P \cap S_{r_1}) + \sigma_{n-1}(\Sigma') + \sigma_{n-1}(P \cap S_{r_2}) . \end{aligned}$$

Comparing the last line to the earlier expression for $\sigma_{n-1}(\partial F)$, we see that the desired estimate $\sigma_{n-1}(\Sigma') \le \sigma(r_2) - \sigma(r_1)$ will indeed hold if only $\sigma_{n-1}(\partial F^*) \le \sigma_{n-1}(\partial F)$.

For $r_1 < r \leq r_2$, $\Sigma' \cap S_r$ is the $(n-2)$-sphere on S_r that borders the cap $P^* \cap S_r$. For such r, the quantity $A(r) = \sigma_{n-1}(P^* \cap S_r)$ can be expressed as

$$A(r) = \omega_{n-2} r^{n-1} \int_0^{\varphi} \sin^{n-2}\theta \, d\theta \ ,$$

in which $\varphi = \varphi(r)$ is the angle between the unit direction vector u of L and a radial vector from x_0 to a point of $\Sigma' \cap S_r$. The integral in the above formula has a value between 0 and $\omega_{n-1}/\omega_{n-2}$, and it describes an increasing function of φ that is bilipschitz on any closed subinterval of $(0, \pi)$. Of course, we also have $A(r) = \sigma_{n-1}(P \cap S_r)$. Because P is a polyhedral set and $n \geq 3$, it is not difficult to show —it suffices to consider the case where P is an n-simplex— that A is Lipschitz. We infer that the function $\psi(r) = r^{1-n} A(r)/\omega_{n-2}$ is bounded away from 0 and $\omega_{n-1}/\omega_{n-2}$ on $(r_1, r_2]$ and satisfies a Lipschitz condition there. The upshot of this observation is that for r in $(r_1, r_2]$, we can represent φ in the form $\varphi = f(r)$, with f a Lipschitz function. If $u_1, u_2, \ldots, u_{n-1}$ is an orthonormal basis for the subspace of $\mathbb{R}^n$ orthogonal to u, then the function Φ defined on $(r_1, r_2] \times S^{n-2}$ by

$$\Phi(r, y) = x_0 + r\cos\big[f(r)\big]u + \sum_{i=1}^{n-1} r\sin\big[f(r)\big] y_i u_i$$

for $y = (y_1, y_2, \ldots, y_{n-1})$ in $\mathbb{S}^{n-2}$ is a Lipschitz mapping whose image is Σ'. Therefore Σ'—and, along with it, the set ∂F^*— is $(n-1)$-rectifiable. This means that F^* also fulfills the hypotheses of Theorem 5.3.2. Using Lemma 5.3.5 and recalling that Lebesgue measure is preserved under spherical symmetrization, we compute

$$\begin{aligned}
\sigma_{n-1}(\partial F^*) &= \lim_{\rho \to 0} \frac{m_n[T_\rho(F^*)] - m_n(F^*)}{\rho} \leq \lim_{\rho \to 0} \frac{m_n[T_\rho(F)^*] - m_n(F^*)}{\rho} \\
&= \lim_{\rho \to 0} \frac{m_n[T_\rho(F)] - m_n(F)}{\rho} = \sigma_{n-1}(\partial F)\,.
\end{aligned}$$

The proof of (5.22) is thus complete. □

We now focus our attention on a continuous function $u : \mathbb{R}^n \to \mathbb{R}$. For each real number t we define sets F_t and U_t by

$$F_t = \{x : u(x) \leq t\}, \qquad U_t = \{x : u(x) < t\}\,.$$

Then F_t is closed in $\mathbb{R}^n$, while U_t is an open subset of $\mathbb{R}^n$. We note that

$$F_t = \bigcap_{s>t} F_s, \qquad U_t = \bigcup_{s<t} F_s\,. \tag{5.23}$$

Let F_t^* and U_t^* be the spherical symmetrizations of F_t and U_t with respect to our given ray L^*.

The *spherical symmetrization of u with respect to L^** is the function $u^* : \mathbb{R}^n \to \mathbb{R}$ defined by

$$u^*(x) = \inf\{t : x \in F_t^*\}\,.$$

We note that if x is a point of $\mathbb{R}^n$, then x certainly belongs to F_t^* for $t = \max\{u(y) : |y - x_0| \leq |x - x_0|\}$ but is not an element of F_t^* when $t < \min\{u(y) : |y - x_0| \leq |x - x_0|\}$, so u^* is well defined.

In view of (5.23) and the nature of the spherical symmetrization procedure, it becomes apparent that

$$F_t^* = \bigcap_{s>t} F_s^*, \qquad U_t^* = \bigcup_{s<t} F_s^* \,,$$

from which it immediately follows that

$$\{x : u^*(x) \le t\} = F_t^*, \qquad \{x : u^*(x) < t\} = U_t^* \tag{5.24}$$

for every real t. The statements in (5.24) have several implications deserving of comment. First, they imply that u^* is a continuous function, for $\{x : a < u^*(x) < b\} = U_b^* \setminus F_a^*$ is an open set whenever $a < b$. A second consequence of (5.24) is that $u^*(S_r) = u(S_r)$ for each $r \ge 0$ and hence that $u^*(x_0) = u(x_0)$ and $u^*(\mathbb{R}^n) = u(\mathbb{R}^n)$. To see this, note that

$$\{x \in S_r : u(x) = t\} = (F_t \cap S_r) \setminus (U_t \cap S_r)\,.$$

Because U_t is a subset of F_t, $\{x \in S_r : u(x) = t\}$ is empty precisely when the sets $F_t \cap S_r$ and $U_t \cap S_r$ coincide. In light of the fact that the former is a relatively closed subset of S_r and the latter is relatively open in S_r, this happens if and only if either

$$F_t \cap S_r = U_t \cap S_r = S_r \quad \text{or} \quad F_t \cap S_r = U_t \cap S_r = \emptyset.$$

By the definition of the spherical symmetrization of u, this is equivalent to saying that either $F_t^* \cap S_r = U_t^* \cap S_r = S_r$ or $F_t^* \cap S_r = U_t^* \cap S_r = \emptyset$. In other words, $\{x \in S_r : u^*(x) = t\}$ is empty precisely when $\{x \in S_r : u(x) = t\}$ is, whence $u^*(S_r) = u(S_r)$.

A third noteworthy consequence of (5.24) is the following observation:

$$\sigma_{n-1}\big[\{x \in S_r : u^*(x) \in I\}\big] = \sigma_{n-1}\big[\{x \in S_r : u(x) \in I\}\big] \tag{5.25}$$

for any $r \ge 0$ and any real interval I. For instance, in the case $I = [a, b]$ we find that

$$\begin{aligned}
\sigma_{n-1}\big[\{x \in S_r : a \le u^*(x) \le b\}\big] &= \sigma_{n-1}(F_b^* \cap S_r) - \sigma_{n-1}(U_a^* \cap S_r) \\
&= \sigma_{n-1}(F_b \cap S_r) - \sigma_{n-1}(U_a \cap S_r) \\
&= \sigma_{n-1}\big[\{x \in S_r : a \le u(x) \le b\}\big]\,.
\end{aligned}$$

Since each of the sets F_t^* involved in the definition of u^* exhibits axial symmetry about L^*, this property is passed along to u^*, in the sense that $u^*[g(x)] = u^*(x)$ whenever x is in $\mathbb{R}^n$ and g is a Euclidean isometry of $\mathbb{R}^n$ that leaves L^* pointwise fixed. Stated differently, for $x \ne x_0$ the value $u^*(x)$ depends only on $r = |x - x_0|$ and the angle φ at which the vector $x - x_0$ meets a direction vector of L^*.

The next result shows that one smoothness property of u is always inherited by u^*.

LEMMA 5.3.7. *If $u : \mathbb{R}^n \to \mathbb{R}$ satisfies a Lipschitz condition with constant λ, then u^* satisfies a Lipschitz condition with the same constant.*

PROOF. The Lipschitz condition on u implies that the set $T_\rho(F_t)$ is contained in $F_{t+\lambda\rho}$ for every t in $\mathbb{R}$ and $\rho > 0$. According to Lemma 5.3.5, we thus have

$$T_\rho(F_t^*) \subset T_\rho(F_t)^* \subset F_{t+\lambda\rho}^*$$

for such t and ρ. Let x and y be different points of $\mathbb{R}^n$ and set $\rho = |y - x|$. If x belongs to F_t^*, then y is in $T_\rho(F_t^*)$ and, therefore, in $F_{t+\lambda\rho}^*$. It follows that

$$u^*(y) \le t + \lambda\rho = t + \lambda|y - x|\,.$$

Taking the infimum over t yields

$$u^*(y) \le u^*(x) + \lambda|y - x|\,.$$

Since the roles of x and y here are interchangeable, we discover that

$$|u^*(y) - u^*(x)| \le \lambda|y - x|\,,$$

which is what we sought to prove. □

We now further explore the relationship between a continuous function $u : \mathbb{R}^n \to \mathbb{R}$ and its spherical symmetrization u^* with respect to L^*. Let $I = (a, b)$ be an open interval in $\mathbb{R}$ such that the open set $D = \{x : u(x) \in I\}$ is not empty, and let $D' = \{x : u^*(x) \in I\}$. Then D' is nonempty—in fact, as noted in (5.25), $D \cap S_r$ and $D' \cap S_r$ have the same surface area for each $r \ge 0$.

For t in I we write

$$\Sigma_t = \{x : u(x) = t\}, \qquad \Sigma_t' = \{x : u^*(x) = t\}\,.$$

It was remarked earlier that $\Sigma_t \cap S_r$ is empty if and only if $\Sigma_t' \cap S_r$ is empty. Now let $f : D \to [0, \infty)$ and $g : D' \to [0, \infty)$ be functions related to each other as follows: for each $r \ge 0$ and each t in I such that $\Sigma_t \cap S_r$ is nonempty, f and g are constant and assume the same value on the sets $\Sigma_t \cap S_r$ and $\Sigma_t' \cap S_r$, respectively. We record in the following two lemmas information about f and g that will contribute to the proof of (5.17).

LEMMA 5.3.8. *Let D, D', f, and g be as described above, with g continuous. Then f is also continuous and*

$$\int_D f^q\, dm_n = \int_{D'} g^q\, dm_n \tag{5.26}$$

for every $q > 0$.

PROOF. If f fails to be continuous, we can produce a point x of D and a sequence $\langle x_\nu \rangle$ in D such that $x_\nu \to x$, yet $f(x_\nu) \to c$ with $c \ne f(x)$. Let $r_\nu = |x_\nu - x_0|$. Because u^* assumes the same values on $D' \cap S_{r_\nu}$ that u does on $D \cap S_{r_\nu}$, we can select a sequence $\langle y_\nu \rangle$ in D' with y_ν on S_{r_ν} and $u^*(y_\nu) = u(x_\nu)$. Thus $g(y_\nu) = f(x_\nu)$. Through passage to a subsequence if need be, we may assume that $y_\nu \to y$, where y belongs to S_r for $r = |x - x_0|$. Then $t = u^*(y) = u(x)$ is in I, so y is a point of D'. Since x belongs to $\Sigma_t \cap S_r$ and y to $\Sigma_t' \cap S_r$, we obtain

$$f(x) = g(y) = \lim_{\nu\to\infty} g(y_\nu) = \lim_{\nu\to\infty} f(x_\nu) = c\,,$$

contrary to our assumption. The contradiction shows that f must, in fact, be continuous.

It remains to prove (5.26). To this end, fix a real number β, let

$$U = \{x \in D : f(x) < \beta\}, \qquad U' = \{x \in D' : g(x) < \beta\}\,,$$

and write $E_r = u(U \cap S_r)$ for $r \ge 0$. The set $U \cap S_r$ is a relatively open subset of S_r on which u is continuous, except when $U \cap S_r = U' \cap S_r = \emptyset$. Therefore, we can express E_r as the disjoint union of countably many (possibly degenerate) intervals, say $E_r = \bigcup I_k$. Consider an element t of I_k. Since $\Sigma_t \cap S_r$ meets U and since f is constant on $\Sigma_t \cap S_r$, $\Sigma_t \cap S_r$ is contained in U. Because the value of g on $\Sigma'_t \cap S_r$ coincides with that of f on $\Sigma_t \cap S_r$, $\Sigma'_t \cap S_r$ lies in U'. As a result, we learn that

$$U \cap S_r = \bigcup \{x \in S_r : u(x) \in I_k\}, \qquad U' \cap S_r = \bigcup \{x \in S_r : u^*(x) \in I_k\}.$$

By virtue of (5.25) we can assert that

$$\sigma_{n-1}(U \cap S_r) = \sigma_{n-1}(U' \cap S_r)$$

for each $r \geq 0$. From this we quickly deduce that the sets

$$A = \{x \in D : \alpha \leq f(x) < \beta\}, \qquad A' = \{x \in D' : \alpha \leq g(x) < \beta\}$$

satisfy

$$\sigma_{n-1}(A \cap S_r) = \sigma_{n-1}(A' \cap S_r)$$

whenever $\alpha < \beta$ and $r \geq 0$. Finally, (4.14) ensures that $m_n(A) = m_n(A')$, whatever the values of α and β. We infer that f and g induce the same measure on $\mathbb{R}$, call it μ. The validity of (5.26) is then guaranteed by the following result in elementary real analysis: for $q > 0$,

$$\int_D f^q \, dm_n = \int_{D'} g^q \, dm_n = \int_0^\infty t^q \, d\mu.$$

This completes the proof. □

There is a companion result to Lemma 5.3.8 that compares the area integrals of the functions f and g over the level sets Σ_t and Σ'_t, respectively, albeit in the presence of certain extra restrictions on the function u and the value t.

LEMMA 5.3.9. *Let D, D', f, and g be as in Lemma 5.3.8, but relative to an underlying function u that is nonconstant, nonnegative, piecewise linear, and of compact support. Then*

$$\int_{\Sigma'_t} g \, d\sigma_{n-1} \leq \int_{\Sigma_t} f \, d\sigma_{n-1} \tag{5.27}$$

for every t in (a, b) such that Σ_t is nonempty and does not pass through any vertex of an n-simplex from the tesselation $\mathcal{T}$ of $\mathbb{R}^n$ with which u is associated.

PROOF. Fix t as described and write $\Sigma = \Sigma_t$, $\Sigma' = \Sigma'_t$. Since $u \geq 0$ and has compact support, the condition on Σ implies that $t > 0$. Because u is nonconstant and piecewise linear, it further implies that Σ is contained in a finite union of hyperplanes in $\mathbb{R}^n$, with the result that $\sigma_{n-1}(\Sigma \cap S_r) = 0$ for each $r \geq 0$. Referring to (5.25), we conclude that $\sigma_{n-1}(\Sigma' \cap S_r) = 0$ for all such r. If $\Sigma' \cap S_r$ is not empty, then (5.24) shows that

$$\Sigma' \cap S_r = (F_t^* \cap S_r) \setminus (U_t^* \cap S_r),$$

whence $\Sigma' \cap S_r$ either reduces to a point or is a closed annulus in the spherical geometry of S_r. The added knowledge that $\sigma_{n-1}(\Sigma' \cap S_r) = 0$ thus ensures that $\Sigma' \cap S_r$ be either a point of S_r or a degenerate annulus, which is to say an $(n-2)$-sphere, on S_r. In the latter case, $\Sigma' \cap S_r$ is the common boundary relative to S_r of two nonempty open caps, namely, $\{x \in S_r : u^*(x) < t\}$ and $\{x \in S_r : u^*(x) > t\}$.

The set Σ partitions each of the finitely many n-simplexes T in $\mathcal{T}$ with which it has contact into two nondegenerate convex polyhedra. On one of these $u \geq t$, and on the other $u \leq t$. It follows that $P = \{x : u(x) \geq t\}$ is a polyhedral set whose boundary is Σ. We claim that Σ' is the boundary of $\{x : u^*(x) \geq t\}$ or, equivalently, the boundary of $P^\star$, for by the counterpart of (5.16) dealing with open sets we have

$$P^\star = (\mathbb{R}^n \setminus U_t)^\star = \mathbb{R}^n \setminus U_t^* = \{x : u^*(x) \geq t\}.$$

Consider a point x of Σ', say in $\Sigma' \cap S_{r_0}$. The fact that $u(S_{r_0}) = u^*(S_{r_0})$ means that $\Sigma \cap S_{r_0}$ is not empty. It is then easy to see that there must be an open interval J with r_0 as an endpoint and with the property that S_r meets both $\{x : u(x) < t\}$ and $\{x : u(x) > t\}$ for every r in J. The remarks above imply that $\Sigma' \cap S_r$ is an $(n-2)$-sphere for such r, the common boundary relative to S_r of $\{x \in S_r : u^*(x) < t\}$ and $\{x \in S_r : u^*(x) > t\}$. In particular, $\Sigma' \cap S_r$ lies on the boundary of $P^\star$ for every r in J. Since $\Sigma' \cap S_{r_0}$ is either a point of S_{r_0} or an $(n-2)$-sphere on S_{r_0}, it is clear on geometric grounds that $\Sigma' \cap S_r$ approaches $\Sigma' \cap S_{r_0}$ as $r \to r_0$. Accordingly, $\partial P^\star$ includes $\Sigma' \cap S_{r_0}$, hence, includes the point x. We have now demonstrated that $\partial P^\star$ contains Σ'. The opposite containment is trivial, so $\partial P^\star = \Sigma'$.

Lemma 5.3.6 is applicable to the pair P and $P^\star$. Thus (5.22) holds, where now $\sigma = \sigma_\Sigma$ and $\sigma^* = \sigma_{\Sigma'}$.

Let $R > 0$ be the largest radius for which $\Sigma \cap S_R$ is not empty (and thus also for which $\Sigma' \cap S_R$ is not empty). For $\nu = 1, 2, \ldots$ and for $j = -1, 0, 1, \ldots, \nu$, set $r_{\nu,j} = jR/\nu$ and define f_ν by

$$f_\nu = \sum_{j=0}^{\nu} m_{\nu,j}(f)\chi_{A_{\nu,j}},$$

where

$$A_{\nu,j} = \{x \in \Sigma : r_{\nu,j-1} < |x - x_0| \leq r_{\nu,j}\}$$

and

$$m_{\nu,j}(f) = \min\{f(x) : x \in \bar{A}_{\nu,j}\}$$

for $j = 0, 1, \ldots, \nu$ (take $m_{\nu,j}(f) = 0$ if the set $A_{\nu,j}$ is empty). Then $\langle f_\nu \rangle$ is uniformly bounded on Σ and, owing to the special character of f, $f_\nu \to f$ pointwise on Σ. Because $\sigma_{n-1}(\Sigma)$ is finite, the dominated convergence theorem now implies that

$$\int_\Sigma f\, d\sigma_{n-1} = \lim_{\nu\to\infty} \int_\Sigma f_\nu\, d\sigma_{n-1} = \lim_{\nu\to\infty} \sum_{j=1}^{\nu} m_{\nu,j}(f)\big[\sigma(r_{\nu,j}) - \sigma(r_{\nu,j-1})\big].$$

Similarly, with $m_{\nu,j}(g)$ defined in an analogous way relative to Σ',

$$\int_{\Sigma'} g\, d\sigma_{n-1} = \lim_{\nu\to\infty} \sum_{j=1}^{\nu} m_{\nu,j}(g)\big[\sigma^*(r_{\nu,j}) - \sigma^*(r_{\nu,j-1})\big].$$

The structures of f and g demand that $m_{\nu,j}(f) = m_{\nu,j}(g)$, so (5.22) then implies the inequality (5.27). □

We come at last to the crux of the proof of (5.17).

THEOREM 5.3.10. *If $u : \mathbb{R}^n \to [0, \infty)$ is a piecewise linear function with compact support and u^* is its spherical symmetrization with respect to the ray L^*, then*

$$\int_{\mathbb{R}^n} |\nabla u^*|^p\, dm_n \leq \int_{\mathbb{R}^n} |\nabla u|^p\, dm_n \tag{5.28}$$

for every $p > 1$.

PROOF. The function u^* is continuous and, because $u^*(S_r) = u(S_r)$ for every $r \geq 0$, is seen to be compactly supported. As a piecewise linear function of compact support, u satisfies a Lipschitz condition on $\mathbb{R}^n$. By Lemma 5.3.7, the same is true of u^*. It follows from the Rademacher-Stepanov theorem that u^* is differentiable almost everywhere in $\mathbb{R}^n$. If we agree to set $\nabla u^*(x) = 0$ for any x at which u^* fails to be differentiable, we obtain a Borel function $|\nabla u^*| : \mathbb{R}^n \to [0, \lambda]$, where λ is a Lipschitz constant for u. (Naturally $|\nabla u| : \mathbb{R}^n \to [0, \lambda]$ is itself a Borel function, provided we observe the same convention for points of nondifferentiability.) Since both $|\nabla u|$ and $|\nabla u^*|$ vanish off some fixed compact set, these functions belong to $L^p(\mathbb{R}^n)$ for every $p > 1$. Moreover, the axial symmetry of u^* about L^* transfers to $|\nabla u^*|$: if g is a Euclidean isometry of $\mathbb{R}^n$ that fixes L^* pointwise, then the chain rule lets us know that $u^* = u^* \circ g^{-1}$ is differentiable at the point $g(x)$ if and only if u^* is differentiable at x, in which event

$$|\nabla u^*[g(x)]| = |\nabla u^*[g(x)]| \, \|g'(x)\| = |\nabla[u^* \circ g](x)| = |\nabla u^*(x)| \, .$$

We may assume that u is nonconstant, for (5.28) holds trivially when $u = u^* = 0$. Let $0 = t_0 < t_1 < \cdots < t_N = \|u\|_\infty$ list the finite set of values taken by u on the set composed of all vertices of n-simplexes from the tesselation $\mathcal{T}$ of $\mathbb{R}^n$ with which u is associated. For $i = 1, 2, \ldots, N$ we define

$$D_i = \{x : t_{i-1} < u(x) < t_i\}, \qquad D_i' = \{x : t_{i-1} < u^*(x) < t_i\}$$

and set

$$\Sigma_t = \{x : u(x) = t\}, \qquad \Sigma_t' = \{x : u^*(x) = t\}$$

for any $t \geq 0$. Because of Theorem 5.3.4 almost every point x of $\Sigma_i' = \Sigma_{t_i}'$ is a point of density for Σ_i' in the direction of each of the basic vectors e_i and because $\nabla u^*(x) = 0$ for any such x where u^* is also differentiable, we infer that $\nabla u^* = 0$ almost everywhere in Σ_i'.

This tells us that

$$\int_{\Sigma_i'} |\nabla u^*|^p \, dm_n = 0$$

for $i = 0, 1, \ldots, N$. To establish (5.28) it thus suffices to verify the following inequality:

$$\int_{D_i'} |\nabla u^*|^p \, dm_n \leq \int_{D_i} |\nabla u|^p \, dm_n \tag{5.29}$$

for $i = 1, 2, \ldots, N$. Indeed, knowing (5.29) and bearing in mind that $\mathbb{R}^n$ is the union of the sets D_i' and Σ_i'—recall that $u^*(\mathbb{R}^n) = u(\mathbb{R}^n)$— we obtain

$$\int_{\mathbb{R}^n} |\nabla u^*|^p \, dm_n = \sum_{i=1}^N \int_{D_i'} |\nabla u^*|^p \, dm_n \leq \sum_{i=1}^N \int_{D_i} |\nabla u|^p \, dm_n \leq \int_{\mathbb{R}^n} |\nabla u|^p \, dm_n \, .$$

Fixing i, we proceed with the verification of (5.29).

We can choose a sequence $\langle g_\nu \rangle$ of continuous functions in D_i' such that $0 \leq g_\nu \leq \lambda^{p-1}$, g_ν is axially symmetric with respect to L^*, and $g_\nu \to |\nabla u^*|^{p-1}$ almost everywhere in D_i'. As we discovered in the proof of Lemma 5.3.9, for $r \geq 0$ and t in (t_{i-1}, t_i) the set $\Sigma_t' \cap S_r$ is empty, is a singleton, or is the boundary relative to S_r of the closed cap $F_t^* \cap S_r$. By virtue of its axial symmetry, therefore, the function g_ν is constant on $\Sigma_t' \cap S_r$. If we define f_ν in D_i by insisting that f_ν be constant on $\Sigma_t \cap S_r$ and take the same value there that g_ν takes on $\Sigma_t' \cap S_r$—this for every

$r \geq 0$ and t in (t_{i-1}, t_i) with $\Sigma'_t \cap S_r$ nonempty—we obtain a continuous function by Lemma 5.3.8.

Lemma 5.3.9 shows that

$$\int_{\Sigma'_t} g_\nu \, d\sigma_{n-1} \leq \int_{\Sigma_t} f_\nu \, d\sigma_{n-1}$$

whenever $t_{i-1} < t < t_i$. On the strength of Theorem 5.3.3 we can state that

$$\begin{aligned} \int_{D'_t} g_\nu |\nabla u^*| \, dm_n &= \int_{t_{i-1}}^{t_i} \left(\int_{\Sigma'_t} g_\nu \, d\sigma_{n-1} \right) dt \\ &\leq \int_{t_{i-1}}^{t_i} \left(\int_{\Sigma_t} f_\nu \, d\sigma_{n-1} \right) dt = \int_{D_i} f_\nu |\nabla u| \, dm_n \,. \end{aligned} \tag{5.30}$$

In combination with Lemma 5.3.8, Hölder's inequality yields

$$\begin{aligned} \int_{D_i} f_\nu |\nabla u| \, dm_n &\leq \left(\int_{D_i} |\nabla u|^p \, dm_n \right)^{1/p} \left(\int_{D_i} f_\nu^q \, dm_n \right)^{1/q} \\ &= \left(\int_{D_i} |\nabla u|^p \, dm_n \right)^{1/p} \left(\int_{D'_i} g_\nu^q \, dm_n \right)^{1/q} , \end{aligned} \tag{5.31}$$

where q is the exponent conjugate to p. Now $0 \leq g_\nu |\nabla u^*| \leq \lambda^p$ almost everywhere and D'_i is bounded, so the dominated convergence theorem is applicable. Taking into account (5.30) and (5.31), this leads to

$$\begin{aligned} \int_{D'_i} |\nabla u^*|^p \, dm_n &= \lim_{\nu \to \infty} \int_{D'_i} g_\nu |\nabla u^*| \, dm_n \leq \liminf_{\nu \to \infty} \int_{D_i} f_\nu |\nabla u| \, dm_n \\ &\leq \left(\int_{D_i} |\nabla u|^p \, dm_n \right)^{1/p} \lim_{\nu \to \infty} \left(\int_{D'_i} g_\nu^q \, dm_n \right)^{1/q} \\ &= \left(\int_{D_i} |\nabla u|^p \, dm_n \right)^{1/p} \left(\int_{D_i} |\nabla u^*|^p \, dm_n \right)^{1/q} . \end{aligned}$$

In other words,

$$\left(\int_{D'_i} |\nabla u^*|^p \, dm_n \right)^p \leq \left(\int_{D_i} |\nabla u|^p \, dm_n \right) \left(\int_{D'_i} |\nabla u^*|^p \, dm_n \right)^{p/q} .$$

Assuming, as we may, that $I = \int_{D'_i} |\nabla u^*|^p \, dm_n \neq 0$, we divide by $I^{p/q}$ to obtain (5.29). This finishes the proof. □

There are, to be sure, other classes of functions u for which (5.28) holds. For example, this estimate is valid for any C_0^∞-function $u : \mathbb{R}^n \to \mathbb{R}$. The proofs hinge on technical points that do not come up in the relatively simple situation we have treated. Alternatively, one could consider establishing the necessary density results from which these more general results would follow from Theorem 5.3.10. We now have a result that we have been seeking.

THEOREM 5.3.11. *If C is a condenser in $\mathbb{R}^n$ and C^* is its spherical symmetrization with respect to the ray L^*, then $\operatorname{Cap}_p(C^*) \leq \operatorname{Cap}_p(C)$ for every $p > 1$.*

PROOF. Suppose that $C = (U, F)$. Let $u : \mathbb{R}^n \to [0, 1]$ be a piecewise linear function with compact support in U that is identically 1 on F. Consider $v = u^\star$, the spherical symmetrization of u with respect to the ray $L^\star$. Then $v : \mathbb{R}^n \to [0, 1]$ is a continuous function of compact support. Moreover, v satisfies a Lipschitz condition on $\mathbb{R}^n$ (Lemma 5.3.7). Next, we apply (5.24) and (5.16) to conclude that

$$\begin{aligned} F^* \subset \{x : u(x) = 1\}^* &= \mathbb{R}^n \setminus \{x : u(x) < 1\}^\star = \mathbb{R}^n \setminus \{x : v(x) < 1\} \\ &= \{x : v(x) = 1\}. \end{aligned}$$

Similarly, using the analogue of (5.16) for open sets, we obtain

$$\{x : v(x) > 0\} = \mathbb{R}^n \setminus \{x : v(x) \le 0\} = \mathbb{R}^n \setminus \{x : u(x) \le 0\}^\star = \{x : u(x) > 0\}^\star.$$

Since $\{x : u(x) > 0\}$ is a subset of U, $\{x : v(x) > 0\}$ is seen to be a subset of U^*. If we set $v(\infty) = 0$, then v becomes a function from the class $\mathcal{A}(C^*)$. Theorem 5.3.10 permits us to assert that

$$\mathrm{Cap}_p(C^*) \le \int_{\mathbb{R}^n} |\nabla v|^p \, dm_n \le \int_{\mathbb{R}^n} |\nabla u|^p \, dm_n .$$

To complete the proof, we invoke Lemma 5.2.2. □

Let $R = \mathcal{R}(C_0, C_1)$ be a ring in $\mathbb{R}^n$ (by our notational convention this requires ∞ to be in C_0). The spherical symmetrization of R with respect to the ray L is the ring $R^* = (R \cup C_1)^* \setminus C_1^*$.

In tandem with Theorem 5.2.3, Theorem 5.3.11 yields the following important result.

THEOREM 5.3.12. *If R is a ring in $\mathbb{R}^n$ and R^* is its spherical symmetrization with respect to the ray L^*, then*

$$\mathrm{Mod}(R^*) \ge \mathrm{Mod}(R).$$

5.4. Estimating the moduli of rings

When knowledge of the Grötzsch and Teichmüller rings is combined with the process of spherical symmetrization, we are able to develop an effective tool for estimating the moduli of various rings from geometric data. This section seeks to outline these sorts of estimates.

5.4.1. The Grötzsch and Teichmüller bounds. We first consider the extremal properties of the Grötzsch ring in the following theorem.

THEOREM 5.4.1. *If a ring R in $\hat{\mathbb{R}}^n$ separates a closed ball $\overline{B}^n(x, r)$ from both ∞ and a finite point y, then*

$$\mathrm{Mod}(R) \le \log \Phi_n \left(\frac{|y - x|}{r} \right).$$

PROOF. When we spherically symmetrize R with respect to the ray L^* originating at x and passing through y, we obtain a ring R^* that separates the complementary components of a ring R' which is similar to $R_G(n, |y - x|/r)$. According to Theorem 5.3.12 and Lemma 5.1.1,

$$\mathrm{Mod}(R) \le \mathrm{Mod}(R^*) \le \mathrm{Mod}(R') = \mathrm{Mod}\left[R_G\Big(n, \frac{|y-x|}{r}\Big)\right] = \log \Phi_n \left(\frac{|y - x|}{r} \right),$$

as claimed. □

After seeing this result, the reader should be well aware that there must be a similar result for the Teichmüller ring. Thus, in the next estimate it is the function Ψ_n that surfaces in the key role.

THEOREM 5.4.2. *If a ring R in $\hat{\mathbb{R}}^n$ separates a pair of distinct points x and y from both ∞ and a finite point z, then*

$$\operatorname{Mod}(R) \leq \log \Psi_n \left(\frac{|z-x|}{|y-x|} \right).$$

PROOF. Let R^* be the spherical symmetrization of R with respect to the ray L^* that emanates from x and goes through y. Then R^* separates the complementary components of a ring R' that is similar to $R_T(n, |z-x|/|y-x|)$. Therefore

$$\operatorname{Mod}(R) \leq \operatorname{Mod}(R^*) \leq \operatorname{Mod}(R') = \operatorname{Mod}\left[R_T\left(n, \frac{|z-x|}{|y-x|}\right)\right] = \log \Psi_n \left(\frac{|z-x|}{|y-x|} \right),$$

the promised upper bound. □

Theorem 5.4.2 admits a Möbius invariant formulation.

THEOREM 5.4.3. *If a ring R in $\hat{\mathbb{R}}^n$ separates one pair of distinct points x and y from another pair of distinct points u and v, then*

$$\operatorname{Mod}(R) \leq \log \Psi_n \left[\frac{q(x,u)q(y,v)}{q(x,y)q(u,v)} \right].$$

PROOF. Let f be a Möbius transformation that sends v to ∞. The ring $f(R)$ separates $\{f(x), f(y)\}$ from $\{f(u), \infty\}$, and

$$\frac{|f(u)-f(x)|}{|f(y)-f(x)|} = [f(x), f(y), f(u), \infty] = [x, y, u, v] = \frac{q(x,u)q(y,v)}{q(x,y)q(u,v)}.$$

Since $\operatorname{Mod}(R) = \operatorname{Mod}[f(R)]$, the stated estimate is a consequence of Theorem 5.4.2. □

We note the following corollary of Theorem 5.4.3.

COROLLARY 5.4.4. *Let R be the ring associated with a pair (F_0, F_1) of disjoint continua in $\hat{\mathbb{R}}^n$. Then*

$$\operatorname{Mod}(R) \leq \log \Psi_n \left[\frac{4}{q(F_0)q(F_1)} \right] \tag{5.32}$$

and

$$\operatorname{Mod}(R) \leq \log \Psi_n \left[\frac{8q(F_0, F_1)}{q(F_0)q(F_1)} \right]. \tag{5.33}$$

In particular, $\operatorname{Mod}(R) < \infty$, so $M[\Delta(F_0, F_1 : \hat{\mathbb{R}}^n)] = \operatorname{Cap}(R) > 0$.

PROOF. Choose points x and y of F_0 such that $q(x,y) = q(F_0)$, and points u and v of F_1 with $q(u,v) = q(F_1)$. Now R separates F_0 and F_1 in $\hat{\mathbb{R}}^n$, and

$$\frac{q(x,u)q(y,v)}{q(x,y)q(u,v)} \leq \frac{4}{q(F_0)q(F_1)}.$$

Since Ψ_n is an increasing function, (5.32) follows from Theorem 5.4.3. As for (5.33), we first choose x in F_0 and u in F_1 for which $q(x,u) = q(F_0, F_1)$ and afterwards select y in F_0 and v in F_1 such that $q(x,y) = q(F_0)/2$ and $q(u,v) = q(F_1)/2$.

In this instance,

$$\frac{q(x,u)q(y,v)}{q(x,y)q(u,v)} \le \frac{8q(F_0,F_1)}{q(F_0)q(F_1)}\,.$$

Theorem 5.4.3 again yields the desired bound. □

5.5. Sets of capacity zero

Let $n \ge 2$ and let E be a compact subset of $\hat{\mathbb{R}}^n$, with $E \ne \hat{\mathbb{R}}^n$. We declare E to have *(conformal) capacity zero* provided that $M[\Delta(E,F : \hat{\mathbb{R}}^n)] = 0$ for every continuum F in E^c.

We express this state of affairs by writing $\mathrm{Cap}(E) = 0$, although in this context we assign no numerical meaning to the symbol $\mathrm{Cap}(E)$ per se.

Corollary 5.4.4 implies that the components of a set E of capacity zero must be points. That is, E must be totally disconnected. Corollary 4.2.14 quickly shows that every countable compact subset of $\hat{\mathbb{R}}^n$ has capacity zero.

We shall shortly construct a Cantor set—hence, an uncountable set—of capacity zero.

By most conventional standards of size (apart from cardinality), sets of capacity zero should be considered exceedingly small; for instance, if $\mathrm{Cap}(E) = 0$, then the Hausdorff dimension of E is certainly zero [**108**].

To test whether a compact set E is of capacity zero it actually suffices to check that $M[\Delta(E,F : \hat{\mathbb{R}}^n)] = 0$ for some continuum F in E^c. The truth of this statement becomes clear once a basic comparison principle for conformal moduli is established. In formulating this principle we shall make use of the quantity $\delta_D(\Gamma_1,\Gamma_2)$, defined for nonempty curve families Γ_1 and Γ_2 in a subdomain D of $\hat{\mathbb{R}}^n$ by

$$\delta_D(\Gamma_1,\Gamma_2) = \inf\Big\{M\big[\bar{\Delta}(|\gamma_1|,|\gamma_2| : D)\big] : \gamma_1 \in \Gamma_1\,,\ \gamma_2 \in \Gamma_2\Big\}\,.$$

When $D = \hat{\mathbb{R}}^n$ we drop the subscript and simply write $\delta(\Gamma_1,\Gamma_2)$. If there is a constant $c > 0$ such that $q(|\gamma_1|) \ge c$ and $q(|\gamma_2|) \ge c$ for every γ_1 in Γ_1 and γ_2 in Γ_2, then

$$\delta(\Gamma_1,\Gamma_2) \ge \omega_{n-1}\big[\log \Psi_n(4c^{-2})\big]^{1-n} > 0\,. \tag{5.34}$$

This is a consequence of Corollary 5.4.4.

THEOREM 5.5.1. *Let F_1, F_2, and F_3 be nonempty subsets of a domain D in $\hat{\mathbb{R}}^n$ with $n \ge 2$, and let $\Gamma_{ij} = \bar{\Delta}(F_i,F_j : D)$ for $1 \le i < j \le 3$. Then*

$$M(\Gamma_{13}) \ge 3^{-n}\min\{M(\Gamma_{12}), M(\Gamma_{23}), \delta_D(\Gamma_{12},\Gamma_{23})\}\,.$$

PROOF. We may assume that $M(\Gamma_{13}) < \infty$. Let ρ be an admissible density for Γ_{13}. Suppose that $\int_\gamma \rho\,ds \ge 1/3$ for every rectifiable path γ in Γ_{12}. Then 3ρ belongs to $\mathrm{Adm}(\Gamma_{12})$, in which event

$$\int_{\mathbb{R}^n} \rho^n\,dm_n \ge 3^{-n}M(\Gamma_{12})\,.$$

Similarly, if $\int_\gamma \rho\,ds \ge 1/3$ for every rectifiable path γ in Γ_{23}, then

$$\int_{\mathbb{R}^n} \rho^n\,dm_n \ge 3^{-n}M(\Gamma_{23})\,.$$

In all remaining cases there must exist rectifiable paths γ_1 in Γ_{13} and γ_2 in Γ_{23} for which $\int_{\gamma_1} \rho\, ds < 1/3$ and $\int_{\gamma_2} \rho\, ds < 1/3$. Fix such paths, say $\gamma_1 : [a_1, b_1] \to D$ with $\gamma_1(a_1)$ in F_1 and $\gamma_2 : [a_2, b_2] \to D$ with $\gamma_2(b_2)$ in F_3. Consider an arbitrary rectifiable path $\gamma : [a, b] \to D$ that belongs to the family $\bar{\Delta}(|\gamma_1|, |\gamma_2| : D)$. We may presume that $\gamma(a)$ belongs to $|\gamma_1|$ and $\gamma(b)$ belongs to $|\gamma_2|$. Choose c_1 in $[a_1, b_1]$ and c_2 in $[a_2, b_2]$ for which $\gamma_1(c_1) = \gamma(a)$ and $\gamma_2(c_2) = \gamma(b)$. The path $\beta = \gamma_1|[a_1, c_1] + \gamma + \gamma_2|[c_2, b_2]$ is a rectifiable member of Γ_{13}, so

$$1 \le \int_\beta \rho\, ds \le \int_{\gamma_1} \rho\, ds + \int_\gamma \rho\, ds + \int_{\gamma_2} \rho\, ds < \frac{2}{3} + \int_\gamma \rho\, ds\,,$$

which implies that $\int_\gamma \rho\, ds \ge 1/3$. We thus discover that 3ρ is an admissible density for $\bar{\Delta}(|\gamma_1|, |\gamma_2| : D)$, the inference being that

$$\int_{\mathbb{R}^n} \rho^n\, dm_n \ge 3^{-n} M\big[\bar{\Delta}(|\gamma_1|, |\gamma_2| : D)\big] \ge 3^{-n} \delta_D(\Gamma_{12}, \Gamma_{23})\,.$$

In all possible cases, therefore,

$$\int_{\mathbb{R}^n} \rho^n\, dm_n \ge 3^{-n} \min\big\{M(\Gamma_{12}), M(\Gamma_{23}), \delta_D(\Gamma_{12}, \Gamma_{23})\big\}\,.$$

Since ρ in $\mathrm{Adm}(\Gamma_{13})$ was arbitrary, the conclusion of the theorem follows. □

We are actually interested in the following corollary of Theorem 5.5.1.

COROLLARY 5.5.2. *If E is a nonempty compact subset of $\hat{\mathbb{R}}^n$ with $n \ge 2$ and if there exists a continuum F in E^c such that $M[\Delta(E, F : \hat{\mathbb{R}}^n)] = 0$, then* $\mathrm{Cap}(E) = 0$.

PROOF. Write $F_1 = F$. Suppose that E^c were to contain a continuum F_2 for which $M[\Delta(E, F_2 : \hat{\mathbb{R}}^n)] > 0$. We could choose a third continuum F_3 in $(E \cup F_1 \cup F_2)^c$ and set $\Gamma_j = \bar{\Delta}(E, F_j : \hat{\mathbb{R}}^n)$ for $1 \le j \le 3$, and $\Gamma_{jk} = \bar{\Delta}(F_j, F_k : \hat{\mathbb{R}}^n)$ for $1 \le j < k \le 3$. Since $q(E, F_2) > 0$ and $q(F_2, F_3) > 0$, we could infer from (5.34) that $\delta(\Gamma_2, \Gamma_{23}) > 0$. Invoking Theorem 4.2.15, Corollary 5.4.4, and Theorem 5.5.1, we could then assert that

$$M(\Gamma_3) \ge 3^{-n} \min\big\{M(\Gamma_2), M(\Gamma_{23}), \delta(\Gamma_2, \Gamma_{23})\big\} > 0\,.$$

By a second appeal to the same set of facts we could then conclude that

$$M(\Gamma_1) \ge 3^{-n} \min\big\{M(\Gamma_3), M(\Gamma_{13}), \delta(\Gamma_3, \Gamma_{13})\big\} > 0\,,$$

which is a contradiction. Therefore no such continuum F_2 can exist and hence $\mathrm{Cap}(E) = 0$. □

We close this section by constructing a Cantor set of capacity zero in $\mathbb{R}^n$. Let $E_0 = \{te_1 : 0 \le t \le 1\}$. We mimic Cantor's classical "middle third" construction to obtain a sequence $E_0 \supset E_1 \supset E_2 \supset \cdots$ of compact sets, but with the following twist: at each stage of the construction we remove a "midsection" of such a length as to render E_ν the union of 2^ν closed intervals $I_{\nu j}$ of length 2^{-3^ν}. Let $E = \bigcap_{\nu=1}^\infty E_\nu$ and $F = \mathbb{S}^{n-1}(2)$. Since $I_{\nu j}$ is separated from F by a spherical ring with radii 2^{-3^ν} and 1, it follows easily that

$$M\big[\Delta(E_\nu, F : \hat{\mathbb{R}}^n)\big] \le \frac{2^\nu \omega_{n-1}}{(3^\nu \log 2)^{n-1}} \le \left(\frac{2}{3}\right)^\nu \frac{\omega_{n-1}}{(\log 2)^{n-1}}\,.$$

Thus E is a Cantor set for which

$$M\big[\Delta(E, F : \hat{\mathbb{R}}^n)\big] \le \lim_{\nu\to\infty} M\big[\Delta(E_\nu, F : \hat{\mathbb{R}}^n)\big] = 0\,.$$

According to Corollary 5.5.2, $\mathrm{Cap}(E) = 0$.

5.6. Extremal functions for condensers

Given a condenser $C = (U, F)$ and a number $p \ge 1$, and following our discussion of extremal densities for the moduli of curve families, we again consider if there is a function $w : \mathbb{R}^n \to \mathbb{R}$ with $w|F = 1$ and $w|U^c = 0$ with the property that equality holds in the definition of the p-capacity of the condenser. We would also like to ensure w is from the class $\mathcal{A}(C)$ and that it is the unique function for which equality might hold. Such a function will be called an extremal function. The reader familiar with these sorts of variation problems may well expect that $w|U \setminus F$ is a much nicer function—at least continuously differentiable.

This section takes a somewhat selective look at the questions of existence, uniqueness, and regularity of extremal functions for condensers in $\mathbb{R}^n$. The emphasis will be on extremal functions for the conformal capacity of ringlike condensers, since this is the only case that has a direct bearing on later developments in the text. For a full-blown treatment of the topics in the present section, along with an abundance of related material, we urge the reader to consult the book of Heinonen, Kilpeläinen, and Martio [**74**].

By a *p-extremal function* for the condenser $C = (U, F)$ we mean a function w from the class $\mathcal{A}(C)$ which enjoys the property that

$$\mathrm{Cap}_p(C) = \int_{\mathbb{R}^n} |\nabla w|^p \, dm_n\,. \tag{5.35}$$

Let $V = U \setminus F$. Theorem 5.3.4 tells us that almost every point of $\mathbb{R}^n \setminus V$ is a point of density for $\mathbb{R}^n \setminus V$ in the direction of each of the basic vectors e_i, and because a function u from $\mathcal{A}(C)$ has $\nabla u(x) = 0$ for any point x at which the formal gradient $\nabla u(x)$ exists and is such a point of density, we see that each u in $\mathcal{A}(C)$ has $\nabla u = 0$ almost everywhere in $\mathbb{R}^n \setminus V$. In particular, (5.35) can be restated as

$$\mathrm{Cap}_p(C) = \int_V |\nabla w|^p \, dm_n\,. \tag{5.36}$$

It is clearly in the set V that any interesting behavior on the part of w is going to be observed. In fact, as the following lemma makes plain, in the study of p-extremal functions it would suffice to concentrate our attention on the class $\mathcal{A}_3(C)$. Incidentally, (5.12) is an immediate corollary of this lemma.

LEMMA 5.6.1. *Suppose that $C = (U, F)$ is a condenser in $\mathbb{R}^n$ and that v is a member of the class $\mathcal{A}_3(C)$ for which*

$$\int_V |\nabla v|^p \, dm_n < \infty.$$

Define $u : \hat{\mathbb{R}}^n \to [0, 1]$ by $u(x) = v(x)$ for x in V, $u(x) = 1$ for x in F and $u(x) = 0$ for $x \in U^c$. Then u belongs to the class $\mathcal{A}(C)$ and

$$\int_{\mathbb{R}^n} |\nabla u|^p \, dm_n = \int_V |\nabla v|^p \, dm_n.$$

PROOF. It follows straight from the definition of the class $\mathcal{A}_3(C)$ that the function u is continuous. We must check that u has the ACL-property in $\mathbb{R}^n$. Fix a closed n-interval $Q = [a_1, b_1] \times [a_2, b_2] \times \cdots \times [a_n, b_n]$ in $\mathbb{R}^n$ and a basic unit vector e_i. We must show that for the m_{n-1}-almost every point y of the projection $p_i(Q)$ the correspondence $t \mapsto u(y + te_i)$ describes a function which is absolutely continuous on $[a_i, b_i]$. Fubini's theorem allows us to write

$$\int_V |\nabla v|^p \, dm_n = \int_{p_i(V)} \left(\int_{A_y} |\nabla v|^p \, dm_1 \right) dm_{n-1} ,$$

in which $A_y = \{x \in V : p_i(x) = y\}$. Since $\int_V |\nabla v|^p \, dm_n < \infty$, this ensures the existence of a set E_0 in $p_i(Q)$ such that $m_{n-1}(E_0) = 0$ and such that $\int_{A_y} |\nabla v|^p \, dm_1 < \infty$ for every y in $p_i(Q) \setminus E_0$. Next, we choose a sequence $\langle Q_\nu \rangle$ of closed n-intervals in V, say $Q_\nu = [a_{1\nu}, b_{1\nu}] \times [a_{2\nu}, b_{2\nu}] \times \cdots \times [a_{n\nu}, b_{n\nu}]$, for which

$$Q \cap V \subset \bigcup_{\nu=1}^{\infty} \operatorname{int}(Q_\nu) .$$

Because v is an ACL-function in V we can select for each ν a set E_ν in $p_i(Q_\nu)$ with $m_{n-1}(E_\nu) = 0$ and with the feature that $t \mapsto v(y + te_i)$ is absolutely continuous on $[a_{\nu i}, b_{\nu i}]$ for each y from $p_i(Q_\nu) \setminus E_\nu$. Let $E = \bigcup_{\nu=0}^{\infty} E_\nu$. Then $m_{n-1}(E) = 0$. Fix a point y of $p_i(Q) \setminus E$ and define w on $[a_i, b_i]$ by $w(t) = u(y + te_i)$. We shall verify that

$$\left[\sum_{j=1}^{r} |w(d_j) - w(c_j)| \right]^p \leq \left[\sum_{j=1}^{r} (d_i - c_i) \right]^{p-1} \int_{A_y} |\nabla v|^p \, dm_1 \tag{5.37}$$

whenever $[c_1, d_1], [c_2, d_2], \ldots, [c_r, d_r]$ are nonoverlapping intervals in $[a_i, b_i]$. Since y does not lie in E_0, $\int_{A_y} |\nabla v|^p \, dm_1 < \infty$. Thus (5.37) implies that w is an absolutely continuous function. But y in $p_i(Q) \setminus E$ is arbitrary, so the confirmation of (5.37) is enough to complete the proof of the lemma.

Consider a nondegenerate subinterval $[a, b]$ of $[a_i, b_i]$. Set

$$I = \{y + te_i : a \leq t \leq b\}, \qquad \alpha = y + ae_i \text{ and } \beta = y + be_i.$$

The key element in the verification of (5.37) is the recognition that

$$|w(b) - w(a)| \leq \int_{I \cap V} |\nabla v| \, dm_1 . \tag{5.38}$$

Inequality (5.38) holds trivially if both α and β lie in F or if both points lie in $\mathbb{R}^n \setminus U$. We therefore disregard these cases in what follows and assume $I \cap V$ is not empty. Assume initially that I is contained in V. We can find a partition $P : a = t_0 < t_1 < \cdots < t_r = b$ of $[a, b]$ such that each segment $\{y + te_i : t_{k-1} \leq t \leq t_k\}$ lies in $\operatorname{int}[Q_{\nu(k)}]$ for some $\nu(k)$. As y is not an element of the set $E_{\nu(k)}$, w is absolutely continuous on $[t_{k-1}, t_k]$ for each k, and hence, is absolutely continuous on $[a, b]$. Accordingly,

$$|w(b) - w(a)| \leq \int_a^b |w'(t)| \, dt = \int_I |\partial_i v| \, dm_1 \leq \int_I |\nabla v| \, dm_1 .$$

By continuity, (5.38) holds when the open segment (α, β) is in V, but I itself meets ∂V.

All remaining cases fall into one of three categories, depending on the location of α and β: both α and β in V^c; both points in V; one of the points in V^c and the other in V. For example, if α lies in F and β in $\mathbb{R}^n \setminus U$, then we can find a subinterval $[c,d]$ of $[a,b]$ such that $\gamma = y + ce_i$ belongs to $F \cap \partial V$, $\delta = y + de_i$ belongs to $\partial U \cap \partial V$, and $J = (\gamma, \delta)$ is a subset of V. In this instance,

$$1 = |w(b) - w(a)| = |w(d) - w(c)| \le \int_J |\nabla v| \, dm_1 \le \int_{I \cap V} |\nabla v| \, dm_1 .$$

The other situations are handled in essentially the same way. Thus (5.38) has been established.

Finally, let $[c_1, d_1], [c_2, d_2], \ldots, [c_r, d_r]$ be an arbitrary finite collection of non-overlapping intervals in $[a_i, b_i]$. Writing $I_k = \{y + te_i : c_k \le t \le d_k\}$, we deduce from (5.38) and Hölder's inequality—both the integral and summation versions are needed—that

$$\begin{aligned} \left[\sum_{k=1}^{r} |w(d_k) - w(c_k)|\right]^p &\le \left[\sum_{k=1}^{r} \int_{I_k \cap V} |\nabla v| \, dm_1\right]^p \\ &\le \left[\sum_{k=1}^{r} \ell(I_k \cap V)^{1/q} \left(\int_{I_k \cap V} |\nabla v|^p \, dm_1\right)^{1/p}\right]^p \\ &\le \left[\sum_{k=1}^{r} \ell(I_k)\right]^{p-1} \left(\sum_{k=1}^{r} \int_{I_k \cap V} |\nabla v|^p \, dm_1\right) \\ &\le \left[\sum_{k=1}^{r} (d_k - c_k)\right]^{p-1} \int_{A_y} |\nabla v|^p \, dm_1 , \end{aligned}$$

in which q is the exponent conjugate to p. The validity of (5.38) is now established and, with that, the proof is finished. □

5.6.1. Monotone functions. Let V be an open set in $\mathbb{R}^n$, and let v be a real-valued continuous function whose domain is a subset of $\hat{\mathbb{R}}^n$ that includes V. We define v to be *monotone* in V provided that

$$\inf_{x \in D} v(x) = \min_{x \in \partial D} v(x), \qquad \sup_{x \in D} v(x) = \max_{x \in \partial D} v(x) \tag{5.39}$$

for every domain D such that $\bar{D}$ is contained in V.

This particular notion of monotonicity owes its definition to Lebesgue. It is intended to capture the notions of the maximum and minimum principle familiar to us from complex analysis.

An equivalent formulation is the following: v is monotone in V if and only if v is constant in any domain D such that $\bar{D}$ lies in V and v is constant on ∂D. We shall need yet a third characterisation of monotonicity. In order to formulate it in an efficient way, we must introduce some new terminology.

If x_0 is a point of V, then by an *asymptote* to ∂V from x_0 we understand a half-open arc A in V whose endpoint is x_0 and whose closure has a nonempty intersection with ∂V.

It is not hard to see that for each point x of ∂D, where D is the component of V that includes x_0, there is an asymptote A to ∂V from x_0 such that x belongs to $\bar{A}$.

A continuous function $v : V \to \mathbb{R}$ is said to be *asymptotically regulated above* (resp., *below*) when for each point x_0 of V and each $\varepsilon > 0$ there exists an asymptote

A to ∂V from x_0 such that $v(x) \leq v(x_0) + \varepsilon$ (respectively, $v(x) \geq v(x_0) - \varepsilon$) holds for each x on A.

We declare v to be *asymptotically regulated* if it is asymptotically regulated above and below.

LEMMA 5.6.2. *Let V be an open set in $\mathbb{R}^n$. A continuous function $v : V \to \mathbb{R}$ is monotone in V if and only if v is asymptotically regulated.*

PROOF. Assume first that v is monotone in V. Given x_0 in V and $\varepsilon > 0$, we let D denote the component of $\{x \in V : v(x) < v(x_0) + \varepsilon\}$ containing x_0. If $\bar{D}$ were included in V, then v would have the constant value $v(x_0) + \varepsilon$ on ∂D, yet would not be constant in D, a state of affairs that would conflict with the requirements of monotonicity. Therefore ∂D must intersect ∂V, so D contains an asymptote A to ∂V from x_0. Obviously the inequality $v(x) \leq v(x_0) + \varepsilon$ is satisfied along A. This proves that v is asymptotically regulated above. In a similar fashion v is shown to be asymptotically regulated below.

For the converse, assume that the function v is asymptotically regulated. Consider a domain D whose closure $\bar{D}$ is a subset of V. Let x_0 be an arbitrary point of D and let $\varepsilon > 0$. Since v is asymptotically regulated above, we can select an asymptote A to ∂V from x_0 such that $v(x) \leq v(x_0) + \varepsilon$ is true along A. Plainly A must meet ∂D. If x_1 is a point of $A \cap \partial D$, we have

$$\min_{x \in \partial D} v(x) \leq v(x_1) \leq v(x_0) + \varepsilon \,.$$

Letting $\varepsilon \to 0$, we find that $\min_{x \in \partial D} v(x) \leq v(x_0)$, a statement that holds for every x_0 in D. Thus, because v is continuous on $\bar{D}$,

$$\min_{x \in \partial D} v(x) \leq \inf_{x \in D} v(x) = \min_{v \in \bar{D}} v(x) \leq \min_{x \in \partial D} v(x) \,.$$

In other words, the first statement in (5.39) holds for D. The second half of (5.39) is obtained by an analogous argument, starting from the information that v is asymptotically regulated below. □

Suppose that V is an open set in $\mathbb{R}^n$ and that $v : \bar{V} \to \mathbb{R}$ is a continuous function. Let t be a value assumed by v, and let G be any subset of V that can be expressed as a union of components D of $\{x \in V : v(x) \neq t\}$ for which $\bar{D}$ lies in V. Noting that $v(x) = t$ for every x in ∂G, we are able to define a continuous function $u : \bar{V} \to \mathbb{R}$ by insisting that $u(x) = v(x)$ for x in $\bar{V} \setminus G$ and $u(x) = t$ for x in G. For want of a better expression, we shall say in these circumstances that u is obtained from v by means of a *Lebesgue modification* (*with respect to* G).

For technical reasons we also define this process for the case $G = \emptyset$, doing so in the obvious way—namely, setting $u = v$. Observe that $u = v$ on ∂V. The foregoing procedure is "oscillation reducing" in the following sense made precise by our next lemma.

LEMMA 5.6.3. *Let V be an open set in $\mathbb{R}^n$, and let $v : \bar{V} \to \mathbb{R}$ be a continuous function. If u is a function that arises from v via a Lebesgue modification, then*

$$\operatorname{osc}_I u \leq \operatorname{osc}_I v$$

for every closed line segment I in V.

PROOF. Assume that the modification on v is performed with respect to a set G which is a union of components D of $\{x \in V : v(x) \neq t\}$ with $\bar{D}$ contained in V.

We may suppose that G is nonempty. Let I be a closed line segment in V, and let x and y be points of I for which

$$\operatorname{osc}_I u = |u(y) - u(x)|.$$

If both x and y lie in G, then $u(x) = u(y) = t$ and $\operatorname{osc}_I u = 0 \leq \operatorname{osc}_I v$. If both x and y are points of $V \setminus G$, a set in which $u = v$, then

$$\operatorname{osc}_I u = |u(y) - u(x)| = |v(y) - v(x)| \leq \operatorname{osc}_I v\,.$$

Lastly, if one of the points x or y belongs to G and the other to $V \setminus G$ (x is in G and y in $V \setminus G$, say), then we can pick a point z of $I \cap \partial G$. In this case $u(x) = u(z) = v(z) = t$ and $u(y) = v(y)$, so again

$$\operatorname{osc}_I u = |u(y) - u(x)| = |v(y) - v(z)| \leq \operatorname{osc}_I v\,.$$

Now all possible cases have been accounted for and the assertion of the lemma follows. □

In the proofs of two upcoming results we shall make use of the following observation, whose straightforward verification is left to the reader.

LEMMA 5.6.4. *If v is an* ACL*-function in an open subset V of $\mathbb{R}^n$ and if $u : V \to \mathbb{R}$ is a function with the feature that $\operatorname{osc}_I u \leq \operatorname{osc}_I v$ for every closed line segment I in V, then u has the* ACL*-property in V and $|\nabla u| \leq |\nabla v|$ holds almost everywhere in that set.*

The next theorem gives us our first glimpse into the structure of a p-extremal function for a condenser.

THEOREM 5.6.5. *If w is a p-extremal function for the condenser $C = (U, F)$ in $\mathbb{R}^n$, then w is monotone in the set $V = U \setminus F$.*

PROOF. Assume that w fails to be monotone in V. Then there is a domain D with $\bar{D}$ in V such that w is constant on ∂D but not constant in D. If t is the value assumed by w on ∂D, then D is a component of $\{x \in V : w(x) \neq t\}$. Let $u : \bar{V} \to [0, 1]$ be the function obtained from the restriction of w to $\bar{V}$ by carrying out the Lebesgue modification procedure with respect to D. The function u is continuous and $u = w$ on ∂V. Thus $u \equiv 1$ on $F \cap \partial V$ and $u \equiv 0$ on $\partial U \cap \partial V$. According to Lemmas 5.6.3 and 5.6.4, u is an ACL-function in V, one for which $|\nabla u| \leq |\nabla w|$ holds almost everywhere in V. We note especially that u belongs to the class $\mathcal{A}_3(C)$. Now the fact that the ACL-function w is not constant in the domain D means that ∇w cannot vanish almost everywhere in D, whence $\int_D |\nabla w|^p \, dm_n > 0$. On the other hand, $\int_D |\nabla u|^p \, dm_n = 0$, because u is constant in D by construction. Recalling (5.12) and (5.36), we compute that

$$\begin{aligned}
\operatorname{Cap}_p(C) &\leq \int_V |\nabla u|^p \, dm_n = \int_{V\sigma_{\mathbb{S}} \setminus D} |\nabla u|^p \, dm_n \\
&\leq \int_{V\sigma_{\mathbb{S}} \setminus D} |\nabla w|^p \, dm_n < \int_V |\nabla w|^p \, dm_n = \operatorname{Cap}_p(C)\,.
\end{aligned}$$

In order to avoid this contradiction, we must abandon the assumption that a domain D exists as described. The alternative is for w to be monotone in V. □

To close this subsection, we append one final supplement to Lemma 5.2.1.

LEMMA 5.6.6. *If $C = (U, F)$ is a condenser in $\mathbb{R}^n$ and $1 < p < \infty$, then*

$$\text{Cap}_p(C) = \inf\left\{\int_V |\nabla u|^p \, dm_n : u \in \mathcal{A}_5(C)\right\}, \tag{3.38}$$

where $V = U \setminus F$ and $\mathcal{A}_5(C)$ is the class of functions u belonging to $\mathcal{A}(C)$ that are monotone in V.

PROOF. Let u_0 be a member of the class $\mathcal{A}_2(C)$, and let w_0 designate the restriction of u_0 to $\bar{V}$. We choose an enumeration $r_1, r_2, \ldots$ of the rational numbers in the interval $[0, 1]$ and inductively define a sequence of functions $w_1, w_2, \ldots$ on $\bar{V}$ as follows: for $\nu \geq 1$, let G_ν be the union of all components D of $\{x \in V : w_{\nu-1}(x) > r_\nu\}$ for which $\bar{D}$ is contained in V, and let w_ν be the function obtained from $w_{\nu-1}$ by means of the Lebesgue modification with respect to G_ν. Then $w_\nu : \bar{V} \to [0, 1]$ is continuous, $w_\nu \equiv w_0 \equiv 1$ on $F \cap \partial V$, and $w_\nu \equiv w_0 \equiv 0$ on $\partial U \cap \partial V$. Furthermore, the sequence $\langle w_\nu \rangle$ is nondecreasing on $\bar{V}$. Since the set $\bar{V}$ is compact, Dini's theorem shows that $w_\nu \to w$ uniformly on $\bar{V}$, where $w : \bar{V} \to [0, 1]$ is a continuous function that is identically 1 on $F \cap \partial V$ and identically 0 on $\partial U \cap \partial V$. Also, Lemma 5.6.3 implies that the sequence $\langle \text{osc}\,_I w_\nu \rangle$ is nonincreasing whenever I is a closed line segment in V. In view of the uniform convergence of $\langle w_\nu \rangle$, therefore, we see that

$$\text{osc}\,_I w = \lim_{\nu \to \infty} \text{osc}\,_I w_\nu \leq \text{osc}\,_I w_0$$

for each such I.

We now repeat the preceding construction, this time starting with the function $v_0 = 1 - w$. The outcome is a nonincreasing sequence $\langle v_\mu \rangle$ of continuous functions on $\bar{V}$ that converges uniformly to a continuous function $v : \bar{V} \to [0, 1]$ which satisfies $v \equiv 0$ on $F \cap \partial V$, $v \equiv 1$ on $\partial U \cap \partial V$, and

$$\text{osc}\,_I \, v \leq \text{osc}\,_I \, v_0 = \text{osc}\,_I \, w \leq \text{osc}\,_I \, w_0,$$

for all closed line segments I in V.

Finally, the function $u = 1 - v$ is continuous, it has $u \equiv 1$ on $F \cap \partial V$ and $u \equiv 0$ on $\partial U \cap \partial V$, and

$$\text{osc}\,_I u = \text{osc}\,_I v \leq \text{osc}\,_I v_0 = \text{osc}\,_I w \leq \text{osc}\,_I w_0$$

for all segments I in V. Since w_0 is an ACL-function in V, we see that u is an ACL-function in V for which $|\nabla u| \leq |\nabla w_0| = |\nabla u_0|$ is true almost everywhere. In particular,

$$\int_V |\nabla u|^p \, dm_n \leq \int_V |\nabla u_0|^p \, dm_n < \infty.$$

If we extend u to $\hat{\mathbb{R}}^n$ by setting $u(x) = 1$ for x in F and $u(x) = 0$ for x in U^c, we produce a function belonging to $\mathcal{A}(C)$. We claim that u is monotone in V, which places u in $\mathcal{A}_5(C)$. Accepting the monotonicity of u in V at face value for a moment and remembering that $\nabla u = 0$ almost everywhere in $\mathbb{R}^n \setminus V$, we obtain

$$\begin{aligned}
\text{Cap}_p(C) &\leq \inf\left\{\int_{\mathbb{R}^n} |\nabla \tilde{u}|^p \, dm_n : \tilde{u} \in \mathcal{A}_5(C)\right\} \leq \int_{\mathbb{R}^n} |\nabla u|^p \, dm_n \\
&= \int_V |\nabla u|^p \, dm_n \leq \int_{\mathbb{R}^n} |\nabla u_0|^p \, dm_n \, .
\end{aligned}$$

Taking the infimum over u_0, we confirm (3.38).

It remains to establish our claim that u is monotone in V. The argument that we give, which relies on Lemma 5.6.2, is borrowed from [**74**].

We first demonstrate that the method employed to construct the sequences $\langle w_\nu \rangle$ and $\langle v_\mu \rangle$ invariably produces a limit function which is asymptotically regulated below. In particular, the function v has this property, from which it follows immediately that $u = 1 - v$ is asymptotically regulated above. Since the construction of $\langle w_\nu \rangle$ typifies what is going on here, we work with it.

Let x_0 in V and $\varepsilon > 0$ be prescribed. Choose an index ν such that $r_{\nu+1}$ satisfies $w(x_0) - \varepsilon < r_{\nu+1} < w(x_0)$, and consider the function w_ν. If D denotes the component of $\{x \in V : w_\nu(x) > r_{\nu+1}\}$ containing x_0, then $\bar{D}$ does not lie in V, for otherwise D would be a subset of $G_{\nu+1}$, which would lead to a contradiction: $w(x_0) \le w_{\nu+1}(x_0) = r_{\nu+1} < w(x_0)$.

Let A be an asymptote to ∂V from x_0. Then A lies in D. More significantly here, A is a subset of $V \setminus G_{\nu+1}$. For x on A we thus have

$$w_{\nu+1}(x) = w_\nu(x) \ge r_{\nu+1}\,.$$

If $r_{\nu+2} < r_{\nu+1}$, then A must also be in $V \setminus G_{\nu+2}$, which implies that

$$w_{\nu+2}(x) = w_{\nu+1}(x) \ge r_{\nu+1}$$

for all x on A; if $r_{\nu+2} > r_{\nu+1}$, then

$$w_{\nu+2}(x) \ge \min\{r_{\nu+2}, w_{\nu+1}(x)\} \ge \min\{r_{\nu+2}, r_{\nu+1}\} = r_{\nu+1}$$

for such x. In other words, $w_{\nu+2} \ge r_{\nu+1}$ holds along A. Through repetition of the foregoing reasoning, we discover that $w_{\nu+\lambda} \ge r_{\nu+1} > w(x_0) - \varepsilon$ holds on the arc A for every $\lambda > 0$. Letting $\lambda \to \infty$, we deduce that the inequality $w(x) \ge w(x_0) - \varepsilon$ is satisfied for every point x of A. Accordingly, w is seen to be asymptotically regulated below. As noted, the same is true of v.

The function $v_0 = 1 - w$ is asymptotically regulated above. We now check by induction that each term from the sequence $\langle v_\mu \rangle$ inherits this property as well. Assuming this to be true for v_μ, we must verify it for $v_{\mu+1}$. Given x_0 in V and $\varepsilon > 0$, choose an asymptote A_0 to ∂V from x_0 along which v_μ obeys the inequality

$$v_\mu(x) \le v_\mu(x_0) + \varepsilon\,.$$

Let $G'_{\mu+1}$ be the union of all components D of $\{x \in V : v_\mu(x) > r_{\mu+1}\}$ with $\bar{D}$ contained in V. If x_0 is not an element of $G'_{\mu+1}$, then

$$v_{\mu+1}(x) \le v_\mu(x) \le v_\mu(x_0) + \varepsilon = v_{\mu+1}(x_0) + \varepsilon$$

holds for all x on the asymptote $A = A_0$. In the case where x_0 does belong to $G'_{\mu+1}$, A_0 must intersect $\partial G'_{\mu+1}$. It follows that A_0 has a subarc A'_0 with endpoints x_0 and a point x_1 of $\partial G'_{\mu+1}$ such that $A'_0 \setminus \{x_1\}$ is a subset of $G'_{\mu+1}$. Now $v_{\mu+1}(x_1) = v_\mu(x_1) = r_{\mu+1}$. By the induction hypothesis applied at the point x_1, V contains an asymptote A_1 to ∂V from x_1 with the feature that

$$v_\mu(x) \le v_\mu(x_1) + \varepsilon$$

whenever x is a point of A_1. In this situation $A'_0 \cup A_1$ contains an asymptote A to ∂V from x_0. Since $v_{\mu+1}(x) = v_{\mu+1}(x_0)$ whenever x lies on A'_0 and

$$v_{\mu+1}(x) \le v_\mu(x) \le v_\mu(x_1) + \varepsilon = v_{\mu+1}(x_1) + \varepsilon = v_{\mu+1}(x_0) + \varepsilon$$

for points x of A_1, $v_{\mu+1}(x) \leq v_{\mu+1}(x_0) + \varepsilon$ is valid along A in this case as well. Therefore $v_{\mu+1}$ is asymptotically regulated above. By induction, each of the functions v_μ also has this property. Finally, presented with x_0 in V and $\varepsilon > 0$, we can fix an index μ such that $v_\mu(x_0) \leq v(x_0) + (\varepsilon/2)$ and then pick an asymptote A to ∂V from x_0 along which $v_\mu(x) \leq v_\mu(x_0) + (\varepsilon/2)$ holds. The inequality

$$v(x) \leq v_\mu(x) \leq v_\mu(x_0) + \varepsilon \leq v(x_0) + \varepsilon$$

is true for all x on A. As a result, v is asymptotically regulated above, which makes u asymptotically regulated below. Since we saw earlier that u is asymptotically regulated above, Lemma 5.6.2 tells us that u is indeed monotone in V, as maintained. □

5.6.2. Existence and uniqueness of extremal functions. It is apparent that there can be no p-extremal function for a condenser $C = (U, F)$ in $\mathbb{R}^n$ whose p-capacity is zero. This is because a supposed extremal function w would be an ACL-function whose formal gradient vanished almost everywhere in $\mathbb{R}^n$. This would make w constant in $\mathbb{R}^n$ and it therefore could not in fact lie in $\mathcal{A}(C)$.

In other words, $\mathrm{Cap}_p(C) > 0$ is a necessary condition for the existence of a p-extremal function. In the case of the conformal capacity of a ringlike condenser, the only case of any real consequence for applications in this book, the condition $\mathrm{Cap}(C) > 0$ is also sufficient to ensure the existence (and uniqueness) of an extremal function.

Moreover, the argument for existence given by Gehring in [**46**] rests entirely on ideas and estimates already found in the text. Accordingly, we shall consider the existence question only in this limited context. For a discussion of the existence of p-extremal functions for more general condensers and for p different from n, situations in which potential theoretic complications inevitably surface, we again refer the reader to the book of Heinonen, Kilpeläinen, and Martio cited earlier.

In the following lemma we will make tacit use of the information, a direct consequence of Theorem 5.2.3, that a condenser $C = (U, F)$ in $\mathbb{R}^n$ with $\mathrm{Cap}(C) > 0$ has $\mathrm{diam}(F) > 0$ and $\mathrm{dist}(F, U^c) < \infty$.

LEMMA 5.6.7. *Let $C = (U, F)$ be a ringlike condenser in $\mathbb{R}^n$ with $\mathrm{Cap}(C) > 0$, let $b = \mathrm{diam}(F)$ and $c = \mathrm{dist}(F, U^c)$, and let $A = A_n$ be the constant from Theorem 4.5.6.*

If M is a real number greater than $\mathrm{Cap}(C)$ and $\mathcal{F}$ is the family of functions u from the class $\mathcal{A}(C)$ that are monotone in $V = U \setminus F$ and satisfy $\int_{\mathbb{R}^n} |\nabla u|^n \, dm_n \leq M$, then

$$|u(y) - u(x)|^n \leq AM \left(\log \frac{b}{a}\right)^{-1} \tag{5.40}$$

whenever u belongs to $\mathcal{F}$ and $|y - x| \leq a < b$, while

$$|u(y) - u(x)|^n \leq AM \left(\log \frac{a}{c}\right)^{-1} \tag{5.41}$$

whenever u belongs to $\mathcal{F}$ and $\min\{\mathrm{dist}(x, F), \mathrm{dist}(y, F)\} \geq a > c$.

In particular, the family $\mathcal{F}$ is equicontinuous at each point of $\hat{\mathbb{R}}^n$.

PROOF. We begin with the verification of (5.40). Fix points x and y such that $|y - x| \leq a$, with $0 < a < b$. To make statement (5.40) nontrivial, we assume that $AM[\log(b/a)]^{-1} < 1$, for $|u(y) - u(x)| \le 1$ is always true of the functions u in question. Let $z = (x + y)/2$. Theorem 4.5.6 implies the existence of a radius r in the interval $(a/2, b/2)$ for which

$$(\text{osc}\ _S u)^n \le AM \left(\log \frac{b}{a}\right)^{-1} < 1\ , \tag{5.42}$$

where $S = \mathbb{S}^{n-1}(z, r)$. Suppose first that S lies in the ring V. Since $\text{diam}(S) < b = \text{diam}(F)$ and since $\mathbb{R}^n \setminus U$ is unbounded, V must include the ball $B = B^n(z, r)$. The monotonicity of u in V thus dictates that

$$|u(y) - u(x)|^n \le (\text{osc}\ _B u)^n = (\text{osc}\ _S u)^n \le AM \left(\log \frac{b}{a}\right)^{-1} .$$

Next, if S meets both V and F, then by (5.42)

$$\min_S u = \max_S u - \text{osc}\ _S u = 1 - \text{osc}\ _S u > 0\ ,$$

whence S does not intersect U^c.

It now follows that $\bar{B}$ lies in U. Then any component D of $V \cap B$ has ∂D contained in $F \cup S$, so the monotonicity condition leads in this instance to the inequality

$$\begin{aligned} |u(y) - u(x)|^n &\le (\text{osc}\ _B u)^n = (1 - \inf_{V \cap B} u)^n \le (1 - \min_S u)^n \\ &= (\text{osc}\ _S u)^n \le AM \left(\log \frac{b}{a}\right)^{-1} . \end{aligned}$$

Should S meet both V and U^c, a similar argument would confirm (5.40).

Finally, if S is disjoint from V, then $\bar{B}$ must be a subset of either F or U^c. In each of these two cases, u is constant in $\bar{B}$ and (5.40) holds trivially. As a result, (5.40) is valid in all cases that arise.

We next turn to (5.41). Consider x and y such that

$$\min\{\text{dist}(x, F), \text{dist}(y, F)\} \ge a,$$

where now $c < a < \infty$. We may assume that x and y are finite points: once (5.41) is verified for finite x and y, the continuity of u at ∞ ensures that it remains valid when $x = \infty$ or $y = \infty$. We assume, in addition, that $AM[\log(a/c)]^{-1} < 1$, (5.41) being a trivial assertion otherwise. We choose a point z_0 of F for which $d(z_0, U^c) = c$ and, again invoking Theorem 4.5.6, we fix a radius r_1 in (c, a) such that

$$(\text{osc}\ _{S_1} u)^n \le AM \left(\log \frac{a}{c}\right)^{-1} < 1$$

holds for $S_1 = \mathbb{S}^{n-1}(z_0, r_1)$.

The sphere S_1 necessarily meets the connected set U^c. As $\text{osc}\ _{S_1} u < 1$, however, S_1 must be disjoint from F. The points x and y plainly lie outside S. Let $d = \max\{\text{dist}(x, F), \text{dist}(y, F)\}$. We appeal to Theorem 4.5.6 one last time to obtain a radius r_2 in $(d, ad/c)$ such that

$$(\text{osc}\ _{S_2} u)^n \le AM \log \left(\frac{a}{c}\right)^{-1} < 1\ ,$$

where $S_2 = \mathbb{S}^{n-1}(z_0, r_2)$. The sphere S_2 intersects U^c, is disjoint from F, and encloses both x and y. Consequently, the spherical ring bounded by S_1 and S_2, call it R, does not meet F but includes x and y. If D is any component of $V \cap R$, then

∂D lies in $S_1 \cup S_2 \cup U^c$. Because $u \equiv 0$ in U^c and u is monotone in V, we observe that

$$\text{osc}_{\,R}u = \sup_{V \cap R} u \leq \max\{\max_{S_1} u, \max_{S_2} u\} = \max\{\text{osc}_{\,S_1}u, \text{osc}_{\,S_2}u\}$$

and conclude that

$$|u(y) - u(x)|^n \leq (\text{osc}_{\,R}u)^n \leq AM\left(\log\frac{a}{c}\right)^{-1},$$

as demanded by (5.41). □

We now establish the existence of an extremal function for the conformal capacity of a ringlike condenser. The matter of its uniqueness will be dealt with separately.

THEOREM 5.6.8. *If $C = (U, F)$ is a ringlike condenser in $\mathbb{R}^n$ with $\text{Cap}(C) > 0$, then there exists a function w in $\mathcal{A}(C)$ for which*

$$\text{Cap}(C) = \int_{\mathbb{R}^n} |\nabla w|^n \, dm_n.$$

PROOF. Let $\langle w_\nu \rangle$ be an arbitrary sequence of functions from the class $\mathcal{A}(C)$ such that w_ν is monotone in the ring $V = U \setminus F$, such that $\int_{\mathbb{R}^n} |\nabla w_\nu|^n \, dm_n \leq M = 2\text{Cap}(C)$, and such that

$$\text{Cap}(C) = \lim_{\nu \to \infty} \int_{\mathbb{R}^n} |\nabla w_\nu|^n \, dm_n\,. \tag{5.43}$$

Lemma 5.6.4 attests to the existence of such sequences. Because of Lemma 5.6.7 and the Arzelà-Ascoli theorem, Theorem 3.6.4, we may assume, after possibly passing to a subsequence and relabelling, that $w_\nu \to w$ uniformly on $\hat{\mathbb{R}}^n$. Then $w : \mathbb{R}^n \to [0, 1]$ is a continuous function, $w(x) = 1$ for every x in F, and $w(x) = 0$ for every x in U^c. We shall prove that w is an extremal function for C.

Clarkson's inequality, found in Lemma 4.3.7, for the case $p = n$ (or, to be more accurate, the analogue of this result for vector-valued functions), in the presence of the information that $(w_\nu + w_\mu)/2$ is a member of the class $\mathcal{A}(C)$ for all ν and μ, gives

$$\begin{aligned}
&\int_{\mathbb{R}^n} |\nabla w_\nu - \nabla w_\mu|^n \, dm_n \\
&\leq\; 2^{n-1}\int_{\mathbb{R}^n} |\nabla w_\nu|^n \, dm_n + 2^{n-1}\int_{\mathbb{R}^n} |\nabla w_\mu|^n \, dm_n - \int_{\mathbb{R}^n} |\nabla w_\nu + \nabla w_\mu|^n \, dm_n \\
&\leq\; 2^{n-1}\int_{\mathbb{R}^n} |\nabla w_\nu|^n \, dm_n + 2^{n-1}\int_{\mathbb{R}^n} |\nabla w_\mu|^n \, dm_n - 2^n\text{Cap}(C)\,.
\end{aligned}$$

As a consequence, we discover that any sequence $\langle w_\nu \rangle$ of the type under discussion,

$$\limsup_{\nu,\mu \to \infty} \int_{\mathbb{R}^n} |\nabla w_\nu - \nabla w_\mu|^n \, dm_n \leq 2^{n-1}\text{Cap}(C) + 2^{n-1}\text{Cap}(C) - 2^n\text{Cap}(C) = 0,$$

that is, $\langle \nabla w_\nu \rangle$, forms a Cauchy sequence in $\mathbb{R}^n$ with respect to the L^n-norm. It thus follows that $\nabla w_\nu \to f = (f_1, f_2, \ldots, f_n)$ in the L^n-norm, where we may take $f : \mathbb{R}^n \to \mathbb{R}^n$ to be a Borel function. Of course, (5.43) implies that $\text{Cap}(C) = \int_{\mathbb{R}^n} |f|^n \, dm_n$. All that is needed now to finish the proof is a demonstration that w belongs to the class $\text{ACL}(\mathbb{R}^n)$ and has $\nabla w = f$ almost everywhere in $\mathbb{R}^n$.

Fix i in $\{1,2,\ldots,n\}$, let p_i be the projection of $\mathbb{R}^n$ onto $\mathbb{R}_i^{n-1}$, and for y in $\mathbb{R}^{n-1}$ let $L_y = \{x \in \mathbb{R}^n : p_i(x) = y\}$. Together, Fatou's lemma and Fubini's theorem imply that

$$\int_{\mathbb{R}_i^{n-1}} \left[\liminf_{\nu\to\infty} \int_{L_y} |\partial_i w_\nu - f_i|^n \, dm_1 \right] dm_{n-1} \le \liminf_{\nu\to\infty} \int_{\mathbb{R}^n} |\partial_i w_\nu - f_i|^n \, dm_n = 0 \,,$$

from which we infer that

$$\begin{aligned} &\liminf_{\nu\to\infty} \int_{L_y} |\partial_i w_\nu - f_i|^n \, dm_1 \\ (5.44) \qquad &= \liminf_{\nu\to\infty} \int_{-\infty}^{\infty} |\partial_i w_\nu(y+te_i) - f_i(y+te_i)|^n \, dt = 0 \end{aligned}$$

for m_{n-1}-almost everywhere y in $\mathbb{R}_i^n$. Since each term of the sequence $\langle w_\nu \rangle$ is an ACL-function in $\mathbb{R}^n$, m_{n-1}-almost every point y of $\mathbb{R}_i^{n-1}$ has the property that each one of the functions $t \mapsto w_\nu(y+te_i)$ is absolutely continuous on every closed subinterval $[a,b]$ of $\mathbb{R}$. Consider a point y of $\mathbb{R}_i^{n-1}$ which possesses this property and for which (5.44) is also true. Passing to a subsequence (depending on y, of course), we may assume that $\partial_i w_\nu \to f_i$ on $L = L_y$ with respect to the L^n-norm for the measure m_1. Then $\partial_i w_\nu \to f_i$ on each closed line segment I in L in the L^1-norm for m_1. Hence, f_i is locally m_1-integrable on L. For any closed subinterval $[a,b]$ of $\mathbb{R}$, we obtain

$$\begin{aligned} w(y+be_i) - w(y+ae_i) &= \lim_{\nu\to\infty} \big[w_\nu(y+be_i) - w_\nu(y+ae_i) \big] \\ &= \lim_{\nu\to\infty} \int_a^b \partial_i w_\nu(y+te_i)\, dt = \int_a^b f_i(y+te_i)\, dt \,. \end{aligned}$$

It thus becomes apparent that $t \mapsto w(y+te_i)$ describes a function which is locally absolutely continuous on the real line and whose derivative at t is $f_i(y+te_i)$ for almost every t. Because this is the case for almost every y in $\mathbb{R}_i^{n-1}$ and for $i = 1,2,\ldots,n$, it is evident that w is an ACL-function in $\mathbb{R}^n$, one whose formal gradient coincides with f almost everywhere. □

We next address the question of the uniqueness of a p-extremal function for a condenser $C = (U,F)$ in $\mathbb{R}^n$, assuming that such a function w is known to exist. We shall arrive at the uniqueness of w by exploiting a variational condition to which each p-extremal function for C is subject, which is quite similar to our approach to establish uniqueness for extremal admissible functions for the modulus.

LEMMA 5.6.9. *If w is a p-extremal function for a condenser $C = (U,F)$ in $\mathbb{R}^n$, then*

$$(5.45) \qquad \int_{\mathbb{R}^n} |\nabla w|^{p-2} \langle \nabla w, \nabla \varphi \rangle \, dm_n = 0$$

for every continuous function $\varphi : \hat{\mathbb{R}}^n \to \mathbb{R}$ that vanishes identically in $F \cup U^c$, enjoys the ACL*-property in $\mathbb{R}^n$, and has $\int_{\mathbb{R}^n} |\nabla\varphi|^p \, dm_n < \infty$.*

In particular, if $V = U \setminus F$, then (5.45) holds with $\mathbb{R}^n$ replaced by V whenever φ belongs to $C_0^\infty(V)$.

PROOF. We first note that because ∇w may well vanish on a set of positive measure in $\mathbb{R}^n$, we take for granted that the integrand in (5.45) is governed by the usual convention that $\infty \cdot 0 = 0$. This implicit assumption comes into play when $1 < p < 2$.

Fix φ as specified and consider for real t the function $w_t = w + t\varphi$. If $\psi : \mathbb{R} \to \mathbb{R}$ is the function defined by $\psi(s) = 0$ when $s \leq 0$, $\psi(s) = s$ when $0 < s \leq 1$, and $\psi(s) = 1$ when $s > 1$, then ψ satisfies a Lipschitz condition with constant 1 on $\mathbb{R}$. It follows that $u_t = \psi \circ w_t$ is a member of the class $\mathcal{A}(C)$. Since $|\nabla u_t| \leq |\nabla w_t|$ almost everywhere in $\mathbb{R}^n$, we find that the function E given on $\mathbb{R}^n$ by

$$E(t) = \int_{\mathbb{R}^n} |\nabla w_t|^p \, dm_n = \int_{\mathbb{R}^n} |\nabla w + t\nabla \varphi|^p \, dm_n$$

has

$$E(0) = \int_{\mathbb{R}^n} |\nabla w|^p \, dm_n = \mathrm{Cap}_p(C) \leq \int_{\mathbb{R}^n} |\nabla u_t|^p \, dm_n \leq \int_{\mathbb{R}^n} |\nabla w_t|^p \, dm_n = E(t)$$

for all t. Thus $E(t)$ attains its minimum value when $t = 0$. We shall show that E is differentiable at 0, with

$$E'(0) = p \int_{\mathbb{R}^n} |\nabla w|^{p-2} \langle \nabla w, \nabla \varphi \rangle \, dm_n \,. \tag{5.46}$$

As 0 is an extreme point for E, it must be the case that $E'(0) = 0$. This yields (5.45).

Let b and c be arbitrary points of $\mathbb{R}^n$. A straightforward calculation reveals that

$$\lim_{t \to 0} \frac{|b + tc|^p - |b|^p}{t} = \begin{cases} 0 & \text{if } b = 0 \,, \\ p|b|^{p-2}\langle b, c \rangle & \text{if } b \neq 0 \,. \end{cases} \tag{5.47}$$

Furthermore, when $0 < |t| \leq 1$ we have

$$\left| \frac{|b + tc|^p - |b|^p}{t} \right| \leq p2^{p-1}(|c| \cdot |b|^{p-1} + |c|^p) \tag{5.48}$$

for

$$\begin{aligned} \left| \frac{|b + tc|^p - |b|^p}{t} \right| &= \frac{p}{|t|} \left| \int_{|b|}^{|b+tc|} s^{p-1} \, ds \right| \leq \frac{p}{|t|} \int_{|b|}^{|b|+|t||c|} s^{p-1} \, ds \\ &\leq p|c|(|b| + |c|)^{p-1} \leq p2^{p-1}(|c| \cdot |b|^{p-1} + |c|^p) \,. \end{aligned}$$

Both $|\nabla w|$ and $|\nabla \varphi|$ are L^p-functions on $\mathbb{R}^n$. Hölder's inequality thus shows us that the function $|\nabla \varphi| \cdot |\nabla w|^{p-1}$ is integrable over $\mathbb{R}^n$. Therefore the function $p2^{p-1}(|\nabla \varphi| \cdot |\nabla w|^{p-2} + |\nabla \varphi|^p)$ is also integrable over $\mathbb{R}^n$.

For each point x of $\mathbb{R}^n$ at which both $\nabla w(x)$ and $\nabla \varphi(x)$ exist—this accounts for almost every point of $\mathbb{R}^n$ —we can assert on the basis of (5.47) that

$$\begin{aligned} &\lim_{t \to 0} \frac{|\nabla w(x) + t\nabla \varphi(x)|^p - |\nabla w(x)|^p}{t} \\ &= \begin{cases} 0 & \text{if } \nabla w(x) = 0 \,, \\ p|\nabla w(x)|^{p-2}\langle \nabla w(x), \nabla \varphi(x) \rangle & \text{if } \nabla w(x) \neq 0 \,. \end{cases} \end{aligned} \tag{5.49}$$

In view of (5.48), the estimate

$$\left| \frac{|\nabla w + t\nabla \varphi|^p - |\nabla w|^p}{t} \right| \leq p2^{p-1}(|\nabla \varphi| \cdot |\nabla w|^{p-1} + |\nabla \varphi|^p) \tag{5.50}$$

is in force almost everywhere in $\mathbb{R}^n$ whenever $0 < |t| \leq 1$. This, together with (5.49) and (5.50), and the dominated convergence theorem, tells us that

$$E'(0) = \lim_{t \to 0} \int_{\mathbb{R}^n} \frac{|\nabla w + t\nabla\varphi|^p - |\nabla w|^p}{t}\, dm_n = p \int_{\mathbb{R}^n} |\nabla w|^{p-2} \langle \nabla w, \nabla \varphi \rangle \, dm_n$$

as stated in (5.46), provided we interpret the integrand in (5.46) in accordance with the stated convention when $1 < p < 2$. □

Let V be an open set in $\mathbb{R}^n$, and let $f : V \to \mathbb{R}^n$ be a mapping whose coordinate functions $f_1, f_2, \ldots, f_n$ are locally integrable in V. In the context of the following discussion such a function is aptly called a *locally integrable vector field* in V. We say that a locally integrable function $g : V \to \mathbb{R}$ is a *weak divergence of* f under the condition that

$$\int_V g\varphi \, dm_n = -\int_V \langle f, \nabla\varphi \rangle \, dm_n \tag{5.51}$$

for every function φ from the class $C_0^\infty(V)$. If a function g with this property exists (it need not), then g is unique up to modification on a set of measure zero. We use the notation $\operatorname{Div} f$ to denote a generic representative of the class of functions that g determines modulo such modification. For example, if the coordinate functions of f are all locally $W^{1,p}$-functions in V for some $p \geq 1$, then we can take

$$\operatorname{Div} f = \sum_{i=1}^n D_i f_i,$$

where $D_i f_i$ denotes the weak derivative of f_i with respect to x_i. Moreover, in case each f_i is an ACL^p-function in V for some $p \geq 1$, this translates to

$$\operatorname{Div} f = \sum_{i=1}^n \partial_i f_i$$

(to have $\partial_i f_i$ globally defined, we set $\partial_i f_i(x) = 0$ for any x in V at which this partial derivative fails to exist). Now $\operatorname{Div} f$ is almost everywhere in V the divergence of f in the classical sense of the term.

When a locally integrable vector field $f : V \to \mathbb{R}^n$ has $\operatorname{Div} f = 0$ almost everywhere in V, we declare f to be *weakly divergence-free* there. In light of (5.51), this happens if and only if

$$\int_V \langle f, \nabla\varphi \rangle \, dm_n = 0 \tag{5.52}$$

whenever φ belongs to $C_0^\infty(V)$.

Now let w be a p-extremal function for a condenser $C = (U, F)$ in $\mathbb{R}^n$, and let $V = U \setminus F$. One implication of Lemma 5.6.9 is that the vector field $f : V \to \mathbb{R}^n$ defined by $f = |\nabla w|^{p-2}\nabla w$—for definiteness, define $f(x) = 0$ if $\nabla w(x)$ fails to exist or if $1 < p < 2$ and $\nabla w(x) = 0$—is weakly divergence-free in V. This is more commonly expressed by saying that w furnishes a weak solution in V to the partial differential equation

$$\operatorname{Div}(|\nabla w|^{p-2}\nabla w) = 0\,. \tag{5.53}$$

Equation (5.53), which is the Euler-Lagrange equation for the variational problem under consideration here, is known as the *p-Laplace equation* (or the *p-harmonic equation*), and any continuous weak solution of (5.53) in an open subset V of $\mathbb{R}^n$ is said to be a *(weakly) p-harmonic function in* V.

Of course when $p = 2$, (5.53) plainly reduces to Laplace's equation.

Many basic questions surrounding the p-Laplace operator $u \mapsto |\nabla u|^{p-2}\nabla u$ (and others of its genre) have yet to be answered. They provide fertile ground for current and future research. Indeed, recent years have already witnessed a surge of activity in the general area of degenerate, nonlinear, elliptic partial differential equations, under which heading (5.53) falls. For more information about the p-Laplace equation itself, see [**74**].

We now establish the uniqueness of extremal functions using Lemma 5.6.9.

THEOREM 5.6.10. *Let $C = (U, F)$ be a condenser in $\mathbb{R}^n$. If a p-extremal function for C exists, then it is unique.*

PROOF. Consider arbitrary p-extremal functions w and v for C. Then $\varphi = w - v$ fits the description of functions to which (5.45) is applicable.

Since $\nabla\varphi = \nabla w - \nabla v$ almost everywhere in $\mathbb{R}^n$, this variational conditional tells us that

$$\operatorname{Cap}_p(C) = \int_{\mathbb{R}^n} |\nabla w|^p \, dm_n = \int_{\mathbb{R}^n} |\nabla w|^{p-2}\langle \nabla w, \nabla v\rangle \, dm_n \, .$$

The Cauchy-Schwartz inequality implies that

$$\operatorname{Cap}_p(C) = \int_{\mathbb{R}^n} |\nabla w|^{p-2}\langle \nabla w, \nabla v\rangle \, dm_n \le \int_{\mathbb{R}^n} |\nabla w|^{p-1}|\nabla v| \, dm_n \, . \tag{5.54}$$

An application of Hölder's inequality then gives

$$\begin{aligned} \int_{\mathbb{R}^n} |\nabla w|^{p-1}|\nabla v| \, dm_n &\le \left(\int_{\mathbb{R}^n} |\nabla w|^p \, dm_n\right)^{(p-1)/p} \left(\int_{\mathbb{R}^n} |\nabla v|^p \, dm_n\right)^{1/p} \\ &= \operatorname{Cap}_p(C) \, . \end{aligned} \tag{5.55}$$

As a result, we see that equality must hold throughout (5.54) and (5.55).

For equality to hold in (5.55), $|\nabla w|$ and $|\nabla v|$ must be proportional almost everywhere in $\mathbb{R}^n$. But the L^p-norms of these functions are the same, so we actually have $|\nabla w| = |\nabla v|$ almost everywhere. From equality in (5.54) we infer that $\langle \nabla w, \nabla v\rangle = |\nabla w|\cdot|\nabla v|$ almost everywhere in $\mathbb{R}^n$. In combination with $|\nabla w| = |\nabla v|$, this ensures that $\nabla w = \nabla v$—or, as is the same, that $\nabla\varphi = 0$ —almost everywhere.

Because φ is an ACL-function, we deduce that φ is constant in $\mathbb{R}^n$. However, $\varphi(x) \to 0$ as $x \to \infty$, whence φ vanishes identically in $\hat{\mathbb{R}}^n$. In other words, w and v coincide in $\hat{\mathbb{R}}^n$, and this establishes the uniqueness of the p-extremal function. □

As might be anticipated from our experience with curve families, it is rarely possible to give an explicit formula for the p-extremal function of a condenser. One simple case in which this can be done is that of a spherical ringlike condenser, which for the sake of simplicity we assume to be centered at the origin:

$$C = (B^n(b), \overline{B}^n(a)), \qquad \text{where } 0 < a < b < \infty.$$

We recall from the proof of Theorem 5.2.3 that

$$\operatorname{Cap}_p(C) \geq M_p(\Gamma_C) = \omega_{n-1}\left(\int_a^b r^{m-1}\,dr\right)^{1-p}$$

for $1 < p < \infty$, where $m = (p-n)/(p-1)$.

The C^∞-function w is defined in the ring $R = B^n(b) \setminus \bar{B}^n(a)$ by

$$w(x) = \frac{\int\limits_{|x|}^{b} t^{m-1}\,dt}{\int\limits_{a}^{b} t^{m-1}\,dt},$$

so for $p \neq n$

$$w(x) = \frac{b^{(p-n)/(p-1)} - |x|^{(p-n)/(p-1)}}{b^{(p-n)/(p-1)} - a^{(p-n)/(p-1)}}$$

and for $p = n$

$$w(x) = \frac{\log b - \log|x|}{\log b - \log a}$$

belongs to the class $\mathcal{A}_3(C)$, and an easy calculation reveals that $\int_R |\nabla w|^p\,dm_n = M_p(\Gamma_C)$. Thus

$$\operatorname{Cap}_p(C) = \int_R |\nabla w|^p\,dm_n\,.$$

By extending w to be 1 in $\overline{B}^n(a)$ and 0 in $\hat{\mathbb{R}}^n \setminus B^n(b)$, we obtain the p-extremal function for C. In particular, the extremal function for $\operatorname{Cap}(C)$ is given in R by

$$w(x) = \frac{\log b - \log|x|}{\log b - \log a}\,.$$

5.6.3. The regularity of extremal functions. The precise degree of Sobolev regularity of the p-extremal function for a condenser C in $\mathbb{R}^n$ is, if anything, even more of a mystery than the exact conditions under which the extremal is certain to exist. For this reason, in addressing the question of regularity of extremal functions we return the discussion to the context of conformal capacity for ringlike condensers, where we have already established existence and uniqueness and where a regularity theorem can be presented without introducing an excess of material not germane to the text.

There is a common strategy employed to attack questions of regularity for weak solutions of nonlinear, elliptic partial differential equations—such as the n-harmonic equation we have already run in to. This proceeds as follows. One first exposes a modicum of smoothness in the solution—Hölder continuity, to be specific—by appealing to a well-known theorem of de Giorgi [**32**] or, which amounts to the same thing in the case of interest here, to Moser's "iteration" method [**123, 124**] (see also [**56, 55**]).

Next, by taking advantage of some special feature of the equation under investigation such as an a priori estimate that allows one to invoke the Sobolev embedding theorem, one determines that the solution actually belongs to a slightly better regularity class—finds it to be a C^1-function, say.

If one is extremely fortunate, more subtle equation-specific information can come into play and disclose an even greater degree of smoothness by virtue of standard regularity theorems found in the literature of partial differential equations such as those of C.B. Morrey [**122**] and E. Hopf [**79**].

We shall roughly follow this approach in deriving a regularity result that will be needed for the proof of Liouville's theorem in the setting of quasiconformal mappings. To get started, we state de Giorgi's theorem in a form suited to our purposes.

THEOREM 5.6.11. *Let D be a domain in $\mathbb{R}^n$, and let A be a symmetric $n \times n$ matrix whose entries are Lebesgue measurable functions $a_{ij} : D \to \mathbb{R}$. Assume the existence of positive constants M_1 and M_2 such that the ellipticity condition*

$$M_1|k|^2 \le \langle Ak, k\rangle \le M_2|k|^2 \tag{5.56}$$

is satisfied almost everywhere in D for each fixed k in $\mathbb{R}^n$. If $v : D \to \mathbb{R}$ is an ACL*-function for which*

$$\int_D v^2\, dm_n \le M_3 < \infty \quad , \quad \int_D |\nabla v|^2\, dm_n < \infty \tag{5.57}$$

and if

$$\int_D \langle A\nabla v, \nabla\varphi\rangle\, dm_n = 0 \tag{5.58}$$

for every function φ from the class $C_0^\infty(D)$, then for each compact subset K of D there exist constants $a > 0$ and $\alpha > 0$, which depend only on the numbers M_j and the set K, such that

$$|v(y) - v(x)| \le a|y - x|^\alpha$$

for all points x and y of K.

Suppose now that w is the extremal function for the conformal capacity of a ringlike condenser $C = (U, F)$ in $\mathbb{R}^n$. Assume additionally that we have the following information about w: for each compact set E in the ring $R = U \setminus F$, there is a positive constant $M = M(E)$ such that

$$M^{-1} \le |\nabla w(x)| \le M \tag{5.59}$$

holds for almost every x in E. We describe this state of affairs by saying that w is *essentially nonsingular*.

Consider a domain D whose closure lies in R. Set $d = d(\bar{D}, \partial R)$, fix c in the interval $(0, d)$, and let M be a constant for which (5.59) holds almost everywhere in $E = \{x : \operatorname{dist}(x, D) \le c\}$. For each vector h in $\mathbb{R}^n$ satisfying $0 < |h| \le c$, define a function $v = v_h : D \to \mathbb{R}$ by

$$v(x) = \frac{w(x+h) - w(x)}{|h|}.$$

Because w is an ACL-function in $\mathbb{R}^n$ and $|\nabla w|$ has a finite L^n-norm, v plainly belongs to the class $\mathrm{ACL}^n(D)$. Since (5.59) is satisfied almost everywhere in E, it is not difficult to show that $|v| \le M$ throughout D (Theorem 4.5.5 is relevant here), whence

$$\int_D |v|^2\, dm_n \le M^2|D| = M_3 < \infty\,. \tag{5.60}$$

Keeping in mind that $n \geq 2$, we apply Hölder's inequality to see that

$$
\begin{aligned}
\int_D |\nabla v|^2 \, dm_n \quad &\leq \quad |D|^{(n\ \ 2)/2} \int_D |\nabla v|^n \, dm_n \\
&\leq \quad (2/|h|)^n |D|^{(n-2)/n} \int_{\mathbb{R}^n} |\nabla w|^n \, dm_n < \infty \,.
\end{aligned} \tag{5.61}
$$

If $\varphi : \mathbb{R}^n \to \mathbb{R}$ is a C_0^∞-function with support in D and if $0 < |h| \leq c$, then the function ψ defined by $\psi(x) = \varphi(x-h)$ is a C_0^∞-function whose support lies in R. From Lemma 5.6.9 we are assured that

$$
\int_R |\nabla w|^{n-2} \langle \nabla w, \nabla \varphi \rangle \, dm_n = \int_R |\nabla w|^{n-2} \langle \nabla w, \nabla \psi \rangle \, dm_n = 0 \,.
$$

By making a change of variable in the second integral and then subtracting the first, we find that

$$
\int_D \big\langle |\nabla w(x+h)|^{n-2} \nabla w(x+h) - |\nabla w(x)|^{n-2} \nabla w(x), \nabla \varphi(x) \big\rangle \, dm_n(x) = 0 \tag{5.62}
$$

whenever φ comes from $C_0^\infty(D)$ and $0 < |h| \leq c$.

Next, let g be the function given in $\mathbb{R}^n$ by $g(y) = n^{-1}|y|^n$. Then g is a C^1-function. Let $g_i = \partial_i g$ for $i = 1, 2, \ldots, n$. To be explicit,

$$
g_i(y) = |y|^{n-2} y_i \,.
$$

The functions g_i are differentiable at every point of $\mathbb{R}^n$, and $g_{ij} = \partial_j g_i = \partial_{i,j} g$ has the rule of correspondence

$$
g_{ij}(y) = |y|^{n-2} \delta_{ij} + (n-2)|y|^{n-4} y_i y_j
$$

if $y \neq 0$, while $g_{ij}(0) = \delta_{ij}$ when $n = 2$ and $g_{ij}(0) = 0$ when $n > 2$. Here δ_{ij} is the Kronecker delta. Fix a vector h in $\mathbb{R}^n$ for which $0 < |h| \leq c$ and define $\gamma : D \times [0,1] \to \mathbb{R}^n$ by

$$
\gamma(x,t) = t \nabla w(x+h) + (1-t) \nabla w(x) \,,
$$

where as usual we agree to set $\nabla w(z) = 0$ for any point z of R at which this formal gradient fails to exist. Then γ is a Borel function. Both of the formal gradients $\nabla w(x)$ and $\nabla w(x+h)$ exist for almost every point x of D. Furthermore, in view of (5.59), for almost every x at which both do exist, we have

$$
M^{-1} \leq |\nabla w(x)| \,, \ |\nabla w(x+h)| \leq M \,. \tag{5.63}
$$

We can rephrase (5.62) by stating that

$$
\int_D \sum_{i=1}^n \Big\{ g_i\big[\gamma(x,1)\big] - g_i\big[\gamma(x,0)\big] \Big\} \, \partial_i \varphi(x) \, dm_n(x) = 0 \tag{5.64}
$$

for every φ in $C_0^\infty(D)$. On the other hand, an elementary computation shows us that $v = v_h$ satisfies

$$
g_i\big[\gamma(x,1)\big] - g_i\big[\gamma(x,0)\big] = |h| \sum_{j=1}^n \Big\{ \int_0^1 g_{ij}\big[\gamma(x,t)\big] \, dt \Big\} \partial_j v(x) \tag{5.65}
$$

for any x in D such that the two formal gradients $\nabla w(x)$ and $\nabla w(x+h)$ exist, hence, for almost every x in D. Combining (5.65) with (5.64), we can thus assert that v satisfies (5.58), in which A is now the matrix whose $(i,j)^{th}$ entry is the function $a_{ij} : D \to \mathbb{R}^n$ defined by

$$a_{ij}(x) = \int_0^1 g_{ij}\big[\gamma(x,t)\big]\, dt\,.$$

As the function h_{ij} given in $D \times [0,1]$ by $h_{ij}(x,t) = g_{ij}[\gamma(x,t)]$ is a Borel function which is integrable over $D \times [0,1]$ with respect to the measure m_{n+1}, it is a consequence of Fubini's theorem that the function a_{ij} is Lebesgue measurable. Plainly $a_{ij} = a_{ji}$. Finally, for any vector k from $\mathbb{R}^n$ and any x from D we have

$$\langle A(x)k, k\rangle = \int_0^1 |\gamma(x,t)|^{n-2}|k|^2 + (n-2)|\gamma(x,t)|^{n-4}\langle\gamma(x,t), k\rangle^2\, dt\,,$$

so

$$|k|^2 \int_0^1 |\gamma(x,t)|^{n-2}\, dt \le \langle A(x)k,k\rangle \le (n-1)|k|^2 \int_0^1 |\gamma(x,t)|^{n-2}\, dt\,.$$

It is readily verified that

$$\frac{1}{4M} \le \int_0^1 |\gamma(x,t)|^{n-2}\, dt \le M^{n-2}$$

whenever x is a point of D for which (5.63) is true. It follows that (5.56) holds almost everywhere in D, with $M_1 = (4M)^{-1}$ and $M_2 = (n-1)M^{n-2}$.

The simple conclusion we draw from Theorem 5.6.11 on the basis of (5.60), (5.61), and the somewhat long-winded deliberations we have just gone through is best summarized in the form of a lemma.

LEMMA 5.6.12. *Suppose that the extremal function w for the conformal capacity of a ringlike condenser $C = (U,F)$ in $\mathbb{R}^n$ is essentially nonsingular. Let D be a domain whose closure lies in the ring $R = U \setminus F$, and let $0 < c < d(\bar{D}, \partial R)$. Then for each compact set K in D there exist constants $a > 0$ and $\alpha > 0$, which depend only on w, n, c, D, and K, with the following property: for each vector h in $\mathbb{R}^n$ satisfying $0 < |h| \le c$, the function $v = v_h$ defined in D by*

$$v(x) = \frac{w(x+h) - w(x)}{|h|}$$

satisfies the Hölder estimate

$$|v(y) - v(x)| \le a|y - x|^\alpha \tag{5.66}$$

for all x and y in K.

Lemma 5.6.12 represents the first step in upgrading our knowledge of the smoothness of the function w. The next step is taken in the following theorem.

THEOREM 5.6.13. *If the extremal function w for the conformal capacity of a ringlike condenser $C = (U,F)$ in $\mathbb{R}^n$ is essentially nonsingular, then w is a C^1-function in the ring $R = U \setminus F$ and is free of critical points there.*

PROOF. It suffices to show that w is of class C^1 and has no critical points in each domain D for which $\bar{D}$ lies in R. Fix D and let $c = d(\bar{D}, \partial R)/2$. Fixing a basic unit vector e_i, we denote by $\mathcal{V}$ the family of all functions v_{te_i} (as per Lemma 5.6.12) with $0 < t \le c$. To avoid cumbersome notation, we write v_t in place of v_{te_i}.

Estimate (5.66) implies that $\mathcal{V}$ is equicontinuous at each point of D. Since $v_t(x) \to \partial_i w(x)$ as $t \to 0$ for any x in D at which $\partial_i w(x)$ exists, (5.66) also implies that the set $\mathcal{V}(y) = \{v_t(y) : 0 < t \leq c\}$ is bounded for each y in D. The Arzelà-Ascoli theorem, Theorem 3.6.4, tells us that $\mathcal{V}$ is a normal family. Therefore, any sequence $\langle t_\nu \rangle$ from $(0, c]$ such that $t_\nu \to 0$ has a subsequence $\langle t_{\nu_k} \rangle$ for which $v_{t_{\nu_k}} \to v_0$ uniformly on compact sets in D, where $v_0 : D \to \mathbb{R}$ is a continuous function.

Now $v_0(x) = \partial_i w(x)$ for every x in D where $\partial_i w(x)$ exists, which means almost everywhere in D. We conclude that, regardless of the original sequence $\langle t_\nu \rangle$, the same limit function v_0 is obtained for all subsequences $\langle t_{\nu_k} \rangle$ with the property described.

It therefore follows that

$$v_0(x) = \lim_{t \to 0^+} \frac{w(x + te_i) - w(x)}{t}$$

for every x in D.

A similar argument starting with $-e_i$ in place of e_i shows that

$$v_0(x) = \lim_{t \to 0^-} \frac{w(x + te_i) - w(x)}{t}$$

for all such x. Thus $\partial_i w(x)$ exists at each point x of D, and $\partial_i w = v_0$ is a continuous function in D. As this is the case for $i = 1, 2, \ldots, n$, w is a C^1-function in D. The fact that w is essentially nonsingular makes certain that $|\nabla w| \geq m > 0$ almost everywhere in D. Because w is of class C^1 in D, this inequality must hold throughout D. In particular, ∇w vanishes at no point of D, confirming the absence of critical points for w there. □

A real-valued function u whose domain includes an open set U in $\mathbb{R}^n$ is said to be *real-analytic in* U if for each point c of U there is an open ball $B = B^n(c, r)$ in which u can be represented as the sum of a convergent power series centered at c: for x in B,

$$u(x) = \sum_{\alpha} a_\alpha (x - c)^\alpha \,, \tag{5.67}$$

where the sum extends over all multi indices $\alpha = (\alpha_1, \alpha_2, \ldots, \alpha_n)$, the coefficients a_α are real numbers, and $(x - c)^\alpha = (x_1 - c_1)^{\alpha_1}(x_2 - c_2)^{\alpha_2} \cdots (x_n - c_n)^{\alpha_n}$. Any u of this type is of class C^∞ in U. Moreover, the coefficients a_α are related to u by

$$a_\alpha = \frac{\partial^\alpha u(c)}{\alpha!} = \frac{\partial_1^{\alpha_1} \partial_2^{\alpha_2} \cdots \partial_n^{\alpha_n} u(c)}{\alpha_1! \alpha_2! \ldots ! \alpha_n!} \,.$$

Real-analyticity on the part of u in U is often indicated by $u \in C^\omega(U)$.

To give an example that came up in the discussion preparatory to Lemma 5.6.12 and will resurface shortly, the function $g(x) = n^{-1}|x|^n$ is real-analytic in $\mathbb{R}^n \setminus \{0\}$. We are now ready to state the regularity theorem that has been our objective from the beginning of this section.

THEOREM 5.6.14. *If the extremal function w for the conformal capacity of a ringlike condenser $C = (U, F)$ in $\mathbb{R}^n$ is essentially nonsingular, then w is real-analytic in the ring $R = U \setminus F$, where it satisfies the n-Harmonic equation in R,*

$$\operatorname{Div}(|\nabla w|^{n-2} \nabla w) = 0.$$

PROOF. It is enough to show that w is real-analytic and solves the n-Harmonic equation in each domain D whose closure is contained in R.

Fix D as indicated. According to Lemma 5.6.9,

$$\int_D |\nabla w|^{n-2}\langle \nabla w, \nabla \varphi\rangle \, dm_n = 0$$

for every function φ from $C_0^\infty(D)$. This can be restated as

$$\int_D \big\langle \nabla g(\nabla w), \nabla \varphi \big\rangle \, dm_n = 0$$

in which $g(x) = n^{-1}|x|^n$.

If G denotes the Hessian matrix of g, the $n \times n$ matrix whose $(i,j)^{th}$ entry is $g_{ij} = \partial_{i,j} g$, then

$$\big\langle G(\nabla w)k, k \big\rangle = |\nabla w|^{n-1}|k|^2 + (n-2)|\nabla w|^{n-4}\langle \nabla w, k\rangle^2$$

in D. From this it is clear that the ellipticity condition

$$M_1|k|^2 \le \big\langle G(\nabla w)k, k\big\rangle \le M_2|k|^2$$

is satisfied pointwise in D for each k in $\mathbb{R}^n$, with

$$M_1 = \min_{x\in\bar{D}} |\nabla w(x)|^{n-2}, \qquad M_2 = (n-1)\max_{x\in\bar{D}} |\nabla w(x)|^{n-2},$$

and Theorem 5.6.13 guarantees that $0 < M_1 \le M_2 < \infty$. Since g is real-analytic in $\mathbb{R}^n \setminus \{0\}$, the set where $\nabla w(x)$ lies for x in D, we find ourselves in precisely the situation covered by a well-known regularity theorem due to Morrey [**122**, Theorem 9.2]. It informs us that w is a C^∞-function in D.

Given an open ball B such that $\bar{B}$ is a subset of D, we invoke the classical divergence theorem to justify the computation

$$\int_B \operatorname{Div}(\varphi|\nabla w|^{n-2}\nabla w)\, dm_n = \int_{\partial B} \varphi|\nabla w|^{n-2}\frac{\partial w}{\partial \nu}\, d\sigma = 0 \tag{5.68}$$

for any C^∞-function φ in D with support in B, where ν denotes the outward directed unit normal to ∂B. Since

$$\operatorname{Div}(\varphi|\nabla w|^{n-2}\nabla w) = |\nabla w|^{n-2}\langle \nabla w, \nabla\varphi\rangle + \operatorname{Div}(|\nabla w|^{n-2}\nabla w)\varphi$$

we infer from (5.45) and (5.68) that

$$\int_B \operatorname{Div}(|\nabla w|^{n-2}\nabla w)\varphi \, dm_n = 0 \tag{5.69}$$

whenever φ belongs to $C_0^\infty(B)$. Now the function $\operatorname{Div}(|\nabla w|^{n-2}\nabla w)$ is continuous in B, so the validity of (5.69) for all φ in $C_0^\infty(B)$ forces the conclusion that $\operatorname{Div}(|\nabla w|^{n-2}\nabla w) \equiv 0$ in B. Because this is the case for each open ball B with $\bar{B}$ contained in D, w satisfies $\operatorname{Div}(|\nabla w|^{n-2}\nabla w) = 0$ everywhere in D. When $n = 2$ this is nothing other than Laplace's equation, in which event the function w is harmonic in D and is thus real-analytic there. When $n > 2$, we have

$$\operatorname{Div}(|\nabla w|^{n-2}\nabla w) = |\nabla w|^{n-4}\bigg(|\nabla w|^2\Delta w + \sum_{i,j=1}^{n} \partial_i w\cdot\partial_j w\cdot\partial_{i,j} w\bigg).$$

We recall that our hypothesis implies $|\nabla w|$ is never zero in D, so w gives us a C^∞-solution in D to the equation

$$|\nabla w|^2 \Delta w + \sum_{i,j=1}^{n} \partial_i w \cdot \partial_j w \cdot \partial_{i,j} w = 0\,.$$

The last equation is again of a type to which a standard regularity theorem addresses itself, in this instance a result of E. Hopf [**79, 80**] which stipulates that, under the conditions present here, w is a real-analytic function in D. □

CHAPTER 6

Quasiconformal Mappings

In this chapter we begin our investigation into the theory and application of quasiconformal mappings in higher dimensions. We have developed numerous tools to help us and it is interesting to see the interplay between them. One of the striking features of the theory—alongside the rigidity apparent in Liouville's theorem—is the local to global properties of these mappings. This is seen most easily through the various equivalent definitions for a quasiconformal mapping—some are infinitesimal, such as that through the linear dilatation, and some are global, such as the definition via the moduli of curve families. Thus the initial parts of this chapter are in large part concerned with to giving the various definitions of a quasiconformal mapping and establishing their equivalence.

We then turn to issues around the boundary behavior of quasiconformal mappings—seeking to establish a higher-dimensional analogue of the Carathéodory theorem for conformal mappings between Jordan domains, and the reflection principle. Subsequently we discuss the compactness and normal families properties for quasiconformal mappings from which we will discover that they are about as good as one could possibly hope. Finally, we return to the Liouville theorem since we have considerably more information about the problem which can now be phrased in terms of 1-quasiconformal mappings.

6.1. The definition of quasiconformality via conformal moduli

We of course begin with the definition of quasiconformality that we have already investigated in the setting of diffeomorphisms—through the modulus of curve families.

6.1.1. Dilatations of homeomorphisms. Suppose that $n \geq 2$ and that f is a homeomorphism of a domain D in $\hat{\mathbb{R}}^n$ onto a second such domain D'. Let $\mathcal{C}(D)$ denote the family of all curves in D. The *inner dilatation* $K_I(f)$, *outer dilatation* $K_O(f)$, and *maximal dilatation* $K(f)$ of f are the quantities defined as follows:

$$\begin{aligned} K_I(f) &= \inf_{K\in(0,\infty)} \big\{K : M\big[f(\Gamma)\big] \leq KM(\Gamma) \text{ for every subfamily } \Gamma \text{ of } \mathcal{C}(D)\big\}\,, \\ K_O(f) &= \inf_{K\in(0,\infty)} \big\{K : M(\Gamma) \leq KM\big[f(\Gamma)\big] \text{ for every subfamily } \Gamma \text{ of } \mathcal{C}(D)\big\}\,, \\ K(f) &= \max\big\{K_I(f), K_O(f)\big\}\,. \end{aligned}$$

We observe the accepted convention in defining $K_I(f) = \infty$ or $K_O(f) = \infty$ whenever the corresponding set of numbers K is empty.

Naturally, the above definitions must be reconciled with the ones presented earlier for the dilatations of a diffeomorphism between domains in $\mathbb{R}^n$. This will be accomplished in due course. Since curve families Γ in D are readily exhibited for which $0 < M(\Gamma) < \infty$ and $0 < M[f(\Gamma)] < \infty$, it is easy to see that $K_I(f) > 0$ and $K_O(f) > 0$, and as with the case of diffeomorphisms we obtain

$$K(f) \geq 1.$$

Corollary 4.2.14 implies that if D_0 is a domain obtained from D by removing a countable set of points, then f and $f|D_0$ have the same dilatations. In particular, such is the case for $D_0 = \{x \in D : x \neq \infty,\, f(x) \neq \infty\}$, which f maps homeomorphically to $D_0' = \{x \in D' : x \neq \infty,\, f^{-1}(x) \neq \infty\}$. As far as dilatations are concerned, therefore, it matters little whether quasiconformal mappings are considered in the context of $\hat{\mathbb{R}}^n$ or in the context of $\mathbb{R}^n$. Therefore throughout this chapter we shall freely shift the emphasis back and forth between $\hat{\mathbb{R}}^n$ and $\mathbb{R}^n$, depending on the topic under study: when we discuss the boundary behavior of a mapping, for instance, the compactness of $\hat{\mathbb{R}}^n$ will be of importance; on the other hand, we investigate the potential differentiability of a function f only at finite points x for which $f(x)$ is likewise finite, and hence are more likely to state a result in the finite setting.

The homeomorphism f is said to be a *quasiconformal mapping* of D onto D' provided $K(f) < \infty$. If, in fact, $K(f) \leq K < \infty$, then we call f a *K-quasiconformal mapping*. For this to be true it is necessary and sufficient that

$$\frac{1}{K}\, M(\Gamma) \leq M\big[f(\Gamma)\big] \leq K M(\Gamma) \tag{6.1}$$

hold for every family Γ of curves in D.

For example, every Möbius transformation of $\hat{\mathbb{R}}^n$ is a 1-quasiconformal self-mapping of $\hat{\mathbb{R}}^n$ by Corollary 4.2.17.

For technical reasons it will later prove convenient to associate with the homeomorphism $f : D \to D'$ a second set of dilatations. Denoted $K_I^*(f)$, $K_O^*(f)$, and $K^*(f)$, these are called the *inner, outer*, and *maximal ring dilatations* of f. They are defined by

$$\begin{aligned}
K_I^*(f) &= \inf_{K \in (0,\infty)} \big\{K : \mathrm{Cap}\big[f(R)\big] \leq K\mathrm{Cap}(R) \ \text{ for every ring } R,\, \bar{R} \subset D\big\}\,, \\
K_O^*(f) &= \inf_{K \in (0,\infty)} \big\{K : \mathrm{Cap}(R) \leq K\mathrm{Cap}\big[f(R)\big] \ \text{ for every ring } R,\, \bar{R} \subset D\big\}\,, \\
K^*(f) &= \max\big\{K_I^*(f),\, K_O^*(f)\big\}\,.
\end{aligned}$$

Recall that a ring R whose boundary components are B_0 and B_1 has $\mathrm{Cap}(R) = M(\Gamma)$ for $\Gamma = \Delta(B_0, B_1 : R)$. Assuming that $\bar{R}$ lies in D, we see that the boundary components of $f(R)$ are just $f(B_0)$ and $f(B_1)$, so $\mathrm{Cap}[f(R)] = M[f(\Gamma)]$. If, in addition, R is nondegenerate, then $f(R)$ is also nondegenerate, whence $\mathrm{Cap}(R) > 0$ and $\mathrm{Cap}[f(R)] > 0$ by Corollary 5.4.4. We deduce from these observations that

$$0 < K_I^*(f) \leq K_I(f), \quad 0 < K_O^*(f) \leq K_O(f), \quad 1 \leq K^*(f) \leq K(f)\,. \tag{6.2}$$

We shall eventually prove that $1 \leq K_I^*(f) = K_I(f)$, $1 \leq K_O^*(f) = K_O(f)$, and $K^*(f) = K(f)$.

We list for future reference several elementary properties of dilatations.

LEMMA 6.1.1. *Let $f : D \to D'$ and $g : D' \to D''$ be homeomorphisms, where D, D' and D'' are domains in $\hat{\mathbb{R}}^n$ with $n \geq 2$. Then*

$$K_I(f^{-1}) = K_O(f), \quad K_O(f^{-1}) = K_I(f), \quad K(f^{-1}) = K(f) \tag{6.3}$$

and

$$K_I(g\circ f)\le K_I(g)K_I(f),\qquad K_O(g\circ f)\le K_O(g)K_O(f)$$

so that

$$K(g\circ f)\le K(g)K(f)\ , \tag{6.4}$$

with equality holding throughout when $K(f)=1$ *or* $K(g)=1$. *In particular, if both* f *and* g *are quasiconformal mappings, so are* f^{-1} *and* $g\circ f$.

PROOF. Statement (6.3) follows straight from the definitions of the dilatations concerned. The first inequality in (6.4) is trivial if either $K_I(f)=\infty$ or $K_I(g)=\infty$. Assuming that both of these dilatations are finite, we clearly have

$$M\big[g\circ f(\Gamma)\big]\le K_I(g)M\big[f(\Gamma)\big]\le K_I(g)K_I(f)M(\Gamma)$$

for every family Γ of curves in D. This ensures that $K_I(g\circ f)\le K_I(g)K_I(f)$.

The other parts of (6.4) are just as easy. If $K(f)=1$, then $K(f^{-1})=1$ by (6.3). Therefore $K_I(f)\le 1$, $K_I(f^{-1})\le 1$, and, by (6.4),

$$\begin{aligned} K_I(g) &= K_I(g\circ f\circ f^{-1})\le K_I(g\circ f)K_I(f^{-1})\le K_I(g\circ f)\\ &\le K_I(g)K_I(f)\le K_I(g)\ ; \end{aligned} \tag{6.5}$$

thus, $K_I(g\circ f)=K_I(g)$ when $K(f)=1$. For similar reasons, equality would obtain elsewhere in (6.4) should $K(f)=1$. The same is true when $K(g)=1$. □

In general, the inequality (6.4) cannot be improved. To see this, consider the two mappings $f,g:\mathbb{R}^n\to\mathbb{R}^n$ defined by

$$(x_1,x_2,\dots,x_n)\overset{f}{\mapsto}(Ax_1,x_2,\dots,x_n),\quad (x_1,x_2,\dots,x_n)\overset{g}{\mapsto}(Bx_1,x_2,\dots,x_n).$$

Using our formulas for the dilatations of diffeomorphisms at (4.3) (or see (2.9)), we compute that if $A,B\ge 1$, then

$$K_I(f)=A,\quad K_I(g)=B,\quad K_O(f)=A^{n-1},\quad K_O(g)=B^{n-1},$$

$$K_I(g\circ f)=AB,\quad K_O(g\circ f)=A^{n-1}B^{n-1}$$

so that equality can hold in (6.4).

Needless to say, there is a direct analogue of Lemma 6.1.1 for ring dilatations as well.

Consider now a diffeomorphism $f:D\to D'$ between domains D and D' in $\mathbb{R}^n$, where again $n\ge 2$. It follows from Theorem 4.2.16 that the (newly defined) dilatations $K_I(f)$ and $K_O(f)$ satisfy

$$K_I(f)\le\sup_{x\in D}H_I\big[f'(x)\big],\qquad K_O(f)\le\sup_{x\in D}H_O\big[f'(x)\big]\,.$$

We assert that equality holds in both cases. In other words, as regards a diffeomorphism, its "generalized" dilatations coincide with the dilatations introduced for it earlier in the text. We shall extract this information from the following theorem.

THEOREM 6.1.2. *Suppose that* $f:D\to D'$ *is a homeomorphism between domains* D *and* D' *in* $\mathbb{R}^n$ *with* $n\ge 2$ *and that* $K_O(f)<\infty$. *If* f *is differentiable at a point* x_0 *of* D, *then*

$$\|f'(x_0)\|^n\le K_O(f)|J_f(x_0)|\,.$$

PROOF. We may assume that $x_0 = f(x_0) = 0$. If not, simply consider in place of f the mapping g defined in the domain $\{x \in \mathbb{R}^n : x + x_0 \in D\}$ by $g(x) = f(x + x_0) - f(x_0)$, noting that $g'(0) = f'(x_0)$ and that, by Lemma 6.1.1, $K_O(g) = K_O(f)$. Moreover, there is no loss of generality in assuming that $f'(0)$ takes the form

$$f'(0)x = (\lambda_1 x_1, \lambda_2 x_2, \ldots, \lambda_n x_n)$$

in which $\lambda_1 \geq \lambda_2 \geq \cdots \geq \lambda_n \geq 0$, since Theorem 2.3.3 asserts the existence of orthogonal linear transformations U and V so that $U \circ f'(0) \circ V$ has this form. If $h = U \circ f \circ V$ in the domain $V^{-1}(D)$, then $h'(0) = U \circ f'(0) \circ V$, $\|h'(0)\| = \|f'(0)\|$, $|J_h(0)| = |J_f(0)|$, and, once again by virtue of Lemma 6.1.1, $K_O(h) = K_O(f)$. Passing from f to h we achieve the desired simplification.

Thus $\|f'(0)\| = \lambda_1$ and $|J_f(0)| = \lambda_1 \lambda_2 \cdots \lambda_n$. We wish to verify that

$$\lambda_1^n \leq K_O(f)\, \lambda_1 \lambda_2 \cdots \lambda_n \,. \tag{6.6}$$

Since this would trivially be the case were $\lambda_1 = 0$, we may suppose that $\lambda_1 > 0$.

For any x in D we can express $f(x)$ in the manner

$$f(x) = f'(0)x + |x| \varepsilon(x) \,,$$

where $\lim_{x \to 0} \varepsilon(x) = 0$. Let η in $(0, \lambda_1/2)$ be given. Choose $\delta > 0$ so that the n-interval $Q = [0, \delta] \times [0, \delta] \times \cdots \times [0, \delta]$ lies in D and so that $|\varepsilon(x)| < n^{-1/2}\eta$ holds for every x in Q. Then

$$|f(x) - f'(0)x| = |x| \cdot |\varepsilon(x)| < \eta\delta \tag{6.7}$$

for all such x. Using E and F to denote the faces of Q on which $x_1 = 0$ and $x_1 = \delta$, respectively, we consider the curve family $\Gamma = \Delta(E, F; \mathrm{int}[Q])$.

By Theorem 4.2.3, $M(\Gamma) = 1$. Using (6.7) we can assert that $f(Q)$ is contained in

$$G = \{x : -\eta\delta \leq x_i \leq (\lambda_i + \eta)\delta, \ \ 1 \leq i \leq n\}.$$

Furthermore, the set $f(E)$ lies between the hyperplanes with equations $x_1 = -\eta\delta$ and $x_1 = \eta\delta$, while $f(F)$ lies between the hyperplanes described by $x_1 = (\lambda_1 - \eta)\delta$ and $x_1 = (\lambda_1 + \eta)\delta$. It follows that any curve from $f(\Gamma)$ must have length at least $(\lambda_1 - 2\eta)\delta$. Remembering Lemma 4.2.6, we are led to the estimate

$$M\big[f(\Gamma)\big] \leq \frac{|G|}{(\lambda_1 - 2\eta)^n \delta^n} = \frac{(\lambda_1 + 2\eta)(\lambda_2 + 2\eta) \cdots (\lambda_n + 2\eta)}{(\lambda_1 - 2\eta)^n} \,.$$

Because $1 = M(\Gamma) \leq K_O(f) M[f(\Gamma)]$, we infer that

$$(\lambda_1 - 2\eta)^n \leq K_O(f)(\lambda_1 + 2\eta)(\lambda_2 + 2\eta) \cdots (\lambda_n + 2\eta) \,.$$

We arrive at (6.6) by letting $\eta \to 0$. □

We take special note of one implication of Theorem 6.1.2: under the conditions stated in the theorem, $f'(x_0) = 0$ whenever $J_f(x_0) = 0$. We are now prepared to state the following.

COROLLARY 6.1.3. *Let $f : D \to D'$ be a homeomorphism between domains D and D' in $\hat{\mathbb{R}}^n$ with $n \geq 2$. Suppose that the restriction of f to $D_0 = \{x \in D : x \neq \infty$, $f(x) \neq \infty\}$ is a diffeomorphism. Then*

$$K_I(f) = \sup_{x \in D_0} H_I\big[f'(x)\big], \qquad K_O(f) = \sup_{x \in D_0} H_O\big[f'(x)\big] \,.$$

In particular,

$$1 \le K_O(f) \le \left[K_I(f)\right]^{n-1}, \qquad 1 \le K_I(f) \le \left[K_O(f)\right]^{n-1}.$$

PROOF. Earlier observations made it clear that $K_O(f) \le \sup_{x \in D_0} H_O[f'(x)]$. The reverse inequality is trivial when $K_O(f) = \infty$ and follows from Theorem 6.1.2 otherwise. The statement regarding $K_I(f)$ is obtained by applying the foregoing to f^{-1}, using the facts that $K_I(f) = K_O(f^{-1})$ and that $H_I[f'(x)] = H_O[(f^{-1})'(y)]$ for $y = f(x)$. The final inequalities then come as consequences of the estimates

$$1 \le H_O(T) \le \left[H_I(T)\right]^{n-1}, \qquad 1 \le H_I(T) \le \left[H_O(T)\right]^{n-1},$$

valid for any non-singular linear transformation $T : \mathbb{R}^n \to \mathbb{R}^n$ once we recall (2.12). □

Observe that in the setting of Corollary 6.1.3 the mapping f can be declared quasiconformal as soon as either $K_O(f) < \infty$ or $K_I(f) < \infty$.

The following refinement of Theorem 6.1.2 will play a significant role in future considerations.

THEOREM 6.1.4. *Under the hypotheses of Theorem* 6.1.2,

$$\|f'(x_0)\|^n \le K_O^*(f)|J_f(x_0)|.$$

PROOF. If $J_f(x_0) = 0$, then Theorem 6.1.2 ensures that $f'(x_0) = 0$, so the assertion becomes trivial. We proceed under the assumption that $J_f(x_0) \ne 0$. As in the proof of Theorem 6.1.2, we are at liberty to suppose that $x_0 = f(x_0) = 0$ and that $f'(0)$ takes the form $f'(0)x = (\lambda_1 x_1, \lambda_2 x_2, \dots, \lambda_n x_n)$, where now $\lambda_1 \ge \lambda_2 \ge \cdots \ge \lambda_n > 0$. The desired inequality thus translates to

$$\lambda_1^n \le K_O^*(f)\,\lambda_1\lambda_2\cdots\lambda_n. \tag{6.8}$$

Again we can write $f(x)$ for x in D as

$$f(x) = f'(0)x + |x|\varepsilon(x),$$

where $\varepsilon(x) \to 0$ as $x \to 0$. Let $h > 0$ and $\delta > 0$ be such that the closed n-interval $Q = \{x : |x_1| \le \lambda_1^{-1}\delta h,\ |x_i| \le h + \lambda_i^{-1}\delta h \text{ for } i = 2, 3, \dots, n\}$ is a subset of D. The ring R whose boundary components are $B_0 = \partial Q$ and $B_1 = \{x : x_1 = 0, |x_i| \le h \text{ for } i = 2, 3, \dots, n\}$ has its closure in D. Let $\Gamma = \Delta(B_0, B_1 : R)$. The family Γ contains the curves γ_y and β_y for each y in B_1, where $\gamma_y(t) = y + te_1$ and $\beta_y(t) = y - te_1$ for $0 < t < \lambda_1^{-1}\delta h$. Moreover, the curve families $\Gamma_1 = \{\gamma_y : y \in B_1\}$ and $\Gamma_2 = \{\beta_y : y \in B_1\}$ are separated. Arguing as we did in Theorem 4.2.3, we find that

$$M(\Gamma_1) = M(\Gamma_2) = \frac{m_{n-1}(B_1)}{(\lambda_1^{-1}\delta h)^{n-1}} = \left(\frac{2\lambda_1}{\delta}\right)^{n-1},$$

which leads by way of Theorem 4.2.9 to the estimate

$$\mathrm{Cap}R = M(\Gamma) \ge M(\Gamma_1) + M(\Gamma_2) = 2\left(\frac{2\lambda_1}{\delta}\right)^{n-1}. \tag{6.9}$$

We now impose further restrictions on h and δ. First, we fix δ in $(0, 1)$. Then, given η in $(0, \delta/2)$, we choose $h > 0$ so as to be certain that

$$|f(x) - f'(0)x| < \eta h \tag{6.10}$$

holds for every x in Q. Remarking that $f'(0)$ transforms R to the ring with boundary components $B_0' = \partial Q'$, where $Q' = \{x : |x_1| \leq \delta h,\ |x_i| \leq (\lambda_i + \delta)h$ for $i = 2, 3, \ldots, n\}$ and $B_1' = \{x : x_1 = 0,\ |x_i| \leq \lambda_i h$ for $i = 2, 3, \ldots, n\}$, we see that $\operatorname{dist}(B_0', B_1') \geq \delta h$. In light of (6.10) we conclude that

$$\operatorname{dist}[f(B_0), f(B_1)] \geq (\delta - 2\eta)h,$$

which has the consequence that $\ell(\gamma) \geq (\delta - 2\eta)h$ for every γ in

$$f(\Gamma) = \Delta[f(B_0), f(B_1), f(R)].$$

Referring once more to (6.10), we note that $f(R)$ is contained in the n-interval $Q'' = \{x : |x_1| \leq (\delta + \eta)h,\ |x_i| \leq (\lambda_i + \delta + \eta)h$ for $i = 2, 3, \ldots, n\}$, so

$$|f(R)| \leq 2^n h^n (\delta + \eta)(\lambda_2 + \delta + \eta) \cdots (\lambda_n + \delta + \eta)\,.$$

Calling again upon Lemma 4.2.6, we conclude that

$$\operatorname{Cap}\big[f(R)\big] = M\big[f(\Gamma)\big] \leq \frac{2^n (\delta + \eta)(\lambda_2 + \delta + \eta) \cdots (\lambda_n + \delta + \eta)}{(\delta - 2\eta)^n}\,. \tag{6.11}$$

Since $\operatorname{Cap} R \leq K_O^*(f)\operatorname{Cap}[f(R)]$, (6.9) and (6.11) together give us the inequality

$$\lambda_1^{n-1}(\delta - 2\eta)^n \leq K_O^*(f)\delta^{n-1}(\delta + \eta)(\lambda_2 + \delta + \eta) \cdots (\lambda_n + \delta + \eta)\,.$$

Then we let $\eta \to 0$ to obtain

$$\lambda_1^{n-1}\delta^n \leq K_O^*(f)\delta^n(\lambda_2 + \delta)(\lambda_3 + \delta) \cdots (\lambda_n + \delta)\ ,$$

which simplifies to

$$\lambda_1^{n-1} \leq K_O^*(f)(\lambda_2 + \delta)(\lambda_3 + \delta) \cdots (\lambda_n + \delta)\,.$$

Finally, we let $\delta \to 0$ to yield

$$\lambda_1^{n-1} \leq K_O^*(f)\lambda_2\lambda_3 \cdots \lambda_n\,.$$

This obviously implies (6.8). □

We remark that the proof presented for Theorem 6.1.4 shows that its conclusion remains valid when $J_f(x_0) \neq 0$, even without the assumption that $K_O(f) < \infty$. We also record a consequence of Corollary 6.1.3 and Theorem 6.1.4.

THEOREM 6.1.5. *Under the hypotheses of Corollary 6.1.3,*

$$K_I^*(f) = K_I(f), \qquad K_O^*(f) = K_O(f), \qquad K^*(f) = K(f)\,.$$

6.2. Examples and the computation of dilatation

Here we give some elementary, but nevertheless important, examples along with some techniques useful for calculating the dilatation of mappings.

We first remind the reader of the fact that a nonsingular linear transformation $T : \mathbb{R}^n \to \mathbb{R}^n$, where $n \geq 2$, maps the unit sphere $\mathbb{S}^{n-1}$ to an ellipsoid $E(T)$, which we now call the *dilatation ellipsoid* of T. If $\lambda_1 \geq \lambda_2 \geq \cdots \geq \lambda_n > 0$ denote the semi-axis lengths of $E(T)$, we recall that

$$\|T\| = \lambda_1, \quad \ell(T) = \lambda_n, \quad |\det(T)| = \lambda_1\lambda_2 \cdots \lambda_n.$$

It is not always an easy task to determine the λ_i. If, however, T transforms some orthogonal basis $v_1, v_2, \ldots, v_n$ of $\mathbb{R}^n$ to another such basis $v_1' = T(v_1), v_2' = T(v_2), \ldots, v_n' = T(v_n)$, matters simplify greatly. Namely, writing $u_i = v_i/|v_i|$, $u_i' = v_i'/|v_i'|$ and letting U denote the orthogonal transformation that takes u_i'

to u_i for $i = 1, 2, \ldots, n$, we note that $U \circ T$ transforms u_i to $(|v_i'|/|v_i|)u_i$, from which it follows that $E(U \circ T)$—hence, also $E(T)$—has semi-axis lengths $|v_i'|/|v_i|$ for $i = 1, 2, \ldots, n$. In particular,

$$(6.12) \qquad \|T\| = \max_i \frac{|v_i'|}{|v_i|}, \qquad \ell(T) = \min_i \frac{|v_i'|}{|v_i|}, \qquad |\det(T)| = \frac{|v_1'| \cdot |v_2'| \cdots |v_n'|}{|v_1| \cdot |v_2| \cdots |v_n|}.$$

It turns out that this observation is useful in the computation of the dilatation of mappings effecting change of coordinates as we shall see.

Suppose now that $f : D \to D'$ is a homeomorphism between domains D and D' in $\mathbb{R}^n$ and that f is differentiable at a point x_0 of D, with $J_f(x_0) \neq 0$. Let $\gamma : (a, b) \to D$ be a smooth curve through x_0—say, $\gamma(t_0) = x_0$ and $v = \dot{\gamma}(t_0) \neq 0$. If $\beta = f \circ \gamma$, then $f'(x_0)$ transforms v to $v' = \dot{\beta}(t_0)$. If we are able to identify n such smooth curves that have pairwise orthogonal trajectories at x_0 and that f maps to curves whose trajectories are mutually orthogonal at $f(x_0)$, we can determine the semi-axis lengths of the dilatation ellipsoid for $f'(x_0)$ and, more to the point, use (6.12) to calculate $\|f'(x_0)\|$, $\ell[f'(x_0)]$, and $|J_f(x_0)|$. Our first example illustrates this technique.

6.2.1. Radial stretchings. Let $n \geq 2$ and let α be a non-zero real number. If $f : \hat{\mathbb{R}}^n \to \hat{\mathbb{R}}^n$ is the homeomorphism defined by

$$(6.13) \qquad f(x) = |x|^{\alpha - 1} x$$

for $x = 0$ and $x = \infty$, we interpret this formula in the obvious way: $f(0) = 0$ and $f(\infty) = \infty$ if $\alpha > 0$, $f(0) = \infty$ and $f(\infty) = 0$ if $\alpha < 0$. Then

$$(6.14) \qquad \begin{cases} K_I(f) = |\alpha| \quad , & K_O(f) = |\alpha|^{n-1}, \quad \text{if } |\alpha| \geq 1, \\ K_I(f) = |\alpha|^{1-n}, & K_O(f) = |\alpha|^{-1} \quad , \quad \text{if } 0 < |\alpha| < 1. \end{cases}$$

Thus f is a quasiconformal mapping of $\hat{\mathbb{R}}^n$ onto itself. (This mapping is known as the *radial stretching of* $\hat{\mathbb{R}}^n$ *corresponding* to α.)

On the basis of our earlier comments, we can assert that f has the same dilatations as $f|D_0$, where $D_0 = \mathbb{R}^n \setminus \{0\}$. Since $f|D_0$ is a diffeomorphism of D_0 onto itself, we can determine its dilatations with the aid of Corollary 6.1.3.

Our first job is to compute the quantities $\|f'(x)\|$, $\ell[f'(x)]$, and $|J_f(x)|$ for x in D_0. The symmetry of f cuts down considerably on the work involved in doing so. It is straightforward to check that $U \circ f = f \circ U$ for every orthogonal linear transformation U, which fact implies that

$$U \circ f'(x) = f'\big[U(x)\big] \circ U$$

whenever x is in D_0 and U belongs to $\mathrm{O}(n)$. This dictates that $\|f'(x)\|$, $\ell[f'(x)]$, and $|J_f(x)|$ be radial functions of x in D_0; that is, these numbers depend only on $|x|$. Therefore we need only ascertain their values for $x = re_1$, where $0 < r < \infty$. Furthermore, given i satisfying $2 \leq i \leq n$, we can choose U so that $U(e_1) = e_1$ and $U(e_2) = e_i$ in order to conclude that

$$(6.15) \qquad |f'(re_1)e_i| = |f'(re_1)e_2|$$

for $i = 3, 4, \ldots, n$. Because f leaves the two-dimensional subspace spanned by e_1 and e_i ($i \geq 2$) invariant, maps the positive x_1-axis to itself, and carries $\mathbb{S}^{n-1}(r)$ to $\mathbb{S}^{n-1}(r^\alpha)$, $f'(re_1)$ must transform e_1 to a multiple of itself and e_i to a vector that is simultaneously a tangent vector to $\mathbb{S}^{n-1}(r^\alpha)$ at $r^\alpha e_1$ and a vector in the span of e_1 and e_i—in other words, to a multiple of itself. In view of (6.15) and the discussion leading up to (6.12), the semi-axis lengths of $E[f'(re_1)]$ will be completely determined as soon as $|f'(re_1)e_1|$ and $|f'(re_1)e_2|$ are known.

The curve $\gamma : (0,\infty) \to \mathbb{R}^n$ defined by $\gamma(r) = re_1$ has $\dot{\gamma}(r) = e_1$ for every $r > 0$. Its image $\beta = f \circ \gamma$ under f is given by $\beta(r) = r^\alpha e_1$, so $\dot{\beta}(r) = \alpha r^{\alpha-1} e_1$ for $r > 0$. It follows that

$$|f'(re_1)e_1| = |\dot{\beta}(r)| = |\alpha| r^{\alpha-1}.$$

Next, fix $r > 0$ and define $\gamma_1 : (-\pi,\pi) \to \mathbb{R}^n$ by $\gamma_1(\theta) = (r\cos\theta)e_1 + (r\sin\theta)e_2$. Then $\gamma_1(0) = re_1$ and $\dot{\gamma}_1(0) = re_2$. In this case $\beta_1 = f \circ \gamma_1$ is presented by the formula $\beta_1(\theta) = (r^\alpha \cos\theta)e_1 + (r^\alpha \sin\theta)e_2$, whence $\dot{\beta}_1(0) = r^\alpha$ and

$$|f'(re_1)e_2| = r^{-1}|f'(re_1)re_2| = r^{-1}|\dot{\beta}_1(0)| = r^{\alpha-1}.$$

We conclude from the above discussion that the dilatation ellipsoid of $f'(x)$ for arbitrary x in D_0 has semi-axis lengths $\lambda_1 = |\alpha| \cdot |x|^{\alpha-1}$, $\lambda_2 = |x|^{\alpha-1}, \dots, \lambda_n = |x|^{\alpha-1}$ if $|\alpha| \geq 1$, and $\lambda_1 = |x|^{\alpha-1}, \dots, \lambda_{n-1} = |x|^{\alpha-1}$, $\lambda_n = |\alpha| \cdot |x|^{\alpha-1}$ if $0 < |\alpha| < 1$. When $|\alpha| \geq 1$, Corollary 6.1.3 gives

$$K_I(f) = \sup_{x\in D_0} H_I\big[f'(x)\big] = \sup_{x \in D_0} \frac{|\alpha|(|x|^{\alpha-1})^n}{(|x|^{\alpha-1})^n} = |\alpha|$$

and

$$K_O(f) = \sup_{x\in D_0} H_O\big[f'(x)\big] = \sup_{x \in D_0} \frac{|\alpha|^n(|x|^{\alpha-1})^n}{|\alpha|(|x|^{\alpha-1})^n} = |\alpha|^{n-1},$$

as desired. The situation when $0 < |\alpha| < 1$ is handled similarly.

Notice that the radial stretching f is not differentiable at the origin when $0 < \alpha < 1$ and has $f'(0) = 0$ when $\alpha > 1$.

In order to simplify the discussion of subsequent examples, we review several of the common coordinate systems used to describe points in $\mathbb{R}^n$ with $n \geq 2$. If $1 \leq j < n$, we can think of $\mathbb{R}^n$ as $\mathbb{R}^j \times \mathbb{R}^{n-j}$ and identify a point $x = (x_1, \dots, x_j, x_{j+1}, \dots, x_n)$ by using its polar coordinates in $\mathbb{R}^j$ and its Euclidean coordinates in $\mathbb{R}^{n-j}$; i.e., we can write $x = (r, u, z)$, in which $r \geq 0$, u in $\mathbb{S}^{j-1}$, and z in $\mathbb{R}^{n-j}$ are related to the Euclidean coordinates of x through

$$r = (x_1^2 + x_2^2 + \cdots + x_j^2)^{1/2}, \quad x_i = ru_i \text{ for } 1 \leq i \leq j, \quad z = (x_{j+1}, x_{j+2}, \dots, x_n).$$

This description of x is unique if $r > 0$. When $j = 2$, it is more common to write $x = (r, \theta, z)$ with $0 \leq \theta < 2\pi$, it being implicit that $u = (\cos\theta, \sin\theta)$.

We refer to (r, u, z) as the *cylindrical coordinates of x with respect to* $\mathbb{R}^j$. The analogue of this setup for $j = n$ is the standard *polar coordinate system* for $\mathbb{R}^n$, in which x is written $x = (r, u)$ with $r \geq 0$ and u in $\mathbb{S}^{n-1}$.

When $n \geq 3$ we shall sometimes employ the *spherical coordinate system* (*with respect to the positive x_n-axis*). In this system a point x of $\mathbb{R}^n$ is represented as $x = (\rho, \varphi, u)$, in which $\rho \geq 0$, φ in $[0, \pi]$, and u in $\mathbb{S}^{n-2}$ bear the following relations to the Euclidean coordinates of x:

$$\rho = |x|, \qquad x_i = u_i \rho \sin\varphi \text{ for } 1 \leq i \leq n-1, \qquad x_n = \rho\cos\varphi.$$

Thus, for $x \neq 0$, φ is the angle between x and e_n. If x is not a point of the x_n-axis, it has a unique description in spherical coordinates.

Let $n \geq 3$ and $0 < \alpha \leq 2\pi$. The domain $W(n, \alpha)$ in $\mathbb{R}^n$ that is defined using cylindrical coordinates with respect to $\mathbb{R}^2$ by

$$W(n, \alpha) = \{x = (r, \theta, z) : r > 0 \ , \ 0 < \theta < \alpha\}$$

will be called the *standard open dihedral wedge of angle* α *in* $\mathbb{R}^n$.

In our next example we exhibit a quasiconformal diffeomorphism between two standard wedges.

6.2.2. Mapping between dihedral wedges. Let $0 < \alpha < \beta \leq 2\pi$. The diffeomorphism $f : W(n, \alpha) \to W(n, \beta)$ described in terms of cylindrical coordinates with respect to $\mathbb{R}^2$ by

$$(r, \theta, z) \mapsto (r, \beta\theta/\alpha, z) \tag{6.16}$$

has dilatations

$$K_I(f) = \frac{\beta}{\alpha}, \qquad K_O(f) = \left(\frac{\beta}{\alpha}\right)^{n-1}.$$

This mapping is called the *natural folding of* $W(n, \alpha)$ *to* $W(n, \beta)$.

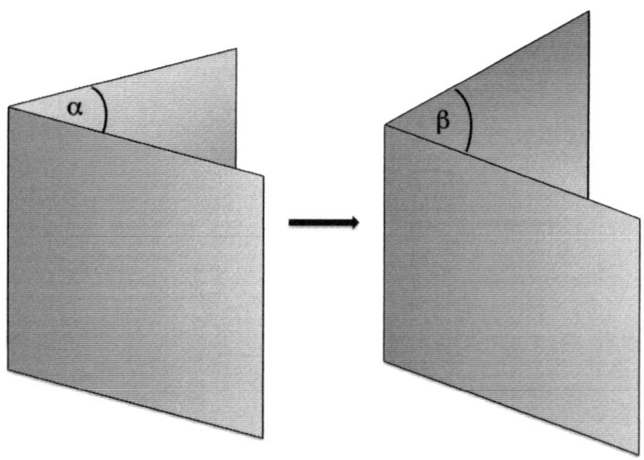

Mapping between dihedral wedges.

We determine the semi-axis lengths for the dilatation ellipsoid of $f'(x)$ at the point x of $W(n, \alpha)$. Since f commutes with translation by any vector belonging to the span of $\{e_3, e_4, \ldots, e_n\}$, it is enough to deal with x of the form $x = x_1 e_1 + x_2 e_2$, i.e., with x whose cylindrical coordinate representation is $x = (r, \theta, 0)$. Fix such a point x. It is obvious that $f'(x)e_i = e_i$ for $j = 3, 4, \ldots, n$. In order to pin down the behavior of $f'(x)$ in two additional (orthogonal) directions, we consider the curves $\gamma_1 : (0, \infty) \to W(n, \alpha)$ and $\gamma_2 : (0, \alpha) \to W(n, \alpha)$ that are given by

$$\gamma_1(r) = tx, \qquad \gamma_2(t) = (r \cos t)e_1 + (r \sin t)e_2 \,.$$

Clearly

$$\gamma_1(1) = x, \qquad v_1 = \dot{\gamma}_1(1) = x \ ,$$

while

$$\gamma_2(\theta) = x, \qquad v_2 = \dot{\gamma}_2(\theta) = (-r \sin \theta)e_1 + (r \cos \theta)e_2 \,.$$

The definition of f yields for $\zeta_1 = f \circ \gamma_1$ and $\zeta_2 = f \circ \gamma_2$ the formulas

$$\zeta_1(t) = t f(x)$$

and

$$\zeta_2(t) = \big[r\cos(\beta t/\alpha)\big]e_1 + \big[r\sin(\beta t/\alpha)\big]e_2 .$$

Therefore

$$v_1' = f'(x)v_1 = \dot\zeta_1(1) = f(x)$$

and

$$v_2' = f'(x)v_2 = \dot\zeta_2(\theta) = (\beta r/\alpha)\big[-\sin(\beta\theta/\alpha)\big]e_1 + (\beta r/\alpha)\big[\cos(\beta\theta/\alpha)\big]e_2 .$$

As a consequence, writing $v_i = v_i' = e_i$ when $3 \le i \le n$, we have

$$\frac{|v_1'|}{|v_1|} = 1\ ,\quad \frac{|v_2'|}{|v_2|} = \frac{\beta}{\alpha}\ ,\quad \frac{|v_3'|}{|v_3|} = 1\ ,\ \ldots\ ,\quad \frac{|v_n'|}{|v_n|} = 1 .$$

Thus, for any x in $W(n,\alpha)$ the dilatation ellipsoid of $f'(x)$ has semi-axis lengths $\lambda_1 = \beta/\alpha$, $\lambda_2 = \lambda_3 = \cdots = \lambda_n = 1$, so

$$\|f'(x)\| = \frac{\beta}{\alpha}, \qquad \ell\big[f'(x)\big] = 1, \qquad |J_f(x)| = \frac{\beta}{\alpha} .$$

We invoke Corollary 6.1.3 to get $K_I(f) = (\beta/\alpha)$ and $K_O(f) = (\beta/\alpha)^{n-1}$.

When $\beta < 2\pi$ the rule of correspondence for the natural folding f of $W(n,\alpha)$ to $W(n,\beta)$ defines a homeomorphism of $\overline{W(n,\alpha)}$ onto $\overline{W(n,\beta)}$, provided we set $f(\infty) = \infty$. Moreover, f fixes every point of the "face" of $W(n,\alpha)$ on which $\theta = 0$.

This example of course makes sense for $\mathbb{R}^2$, where $W(2,\alpha)$ is the set described in polar coordinates by $\{(r,\theta) : r > 0,\ 0 < \theta < \alpha\}$ and f is the mapping $(r,\theta) \mapsto (r,\ \beta\theta/\alpha)$. In this case the dilatations simplify to $K_I(f) = K_O(f) = K(f) = \beta/\alpha$.

Of course, in the plane one can also map $W(2,\alpha)$ conformally to $W(2,\beta)$ by means of a suitable branch of the complex (β/α)-power function.

For $n \ge 3$ and $0 < \alpha \le \pi$ we refer to $C(n,\alpha) = \{x \in \mathbb{R}^n : x_n > |x|\cos\alpha\}$ as the *standard open cone in* $\mathbb{R}^n$ *with vertex at the origin and vertex half-angle* α. This domain enjoys the spherical coordinate description $\{x = (r,\varphi,u) : r > 0,\ \varphi < \alpha\}$.

6.2.3. Mapping between cones. Let $0 < \alpha < \beta \le \pi$. The diffeomorphism $f : C(n,\alpha) \to C(n,\beta)$ given in spherical coordinates by

$$(\rho,\varphi,u) \mapsto (\rho,\ \beta\varphi/\alpha,\ u) \tag{6.17}$$

has dilatations

$$K_I(f) = \max\left\{\left(\frac{\beta}{\alpha}\right)^{n-1},\ \frac{\beta\sin^2\alpha}{\alpha\sin^2\beta}\right\}, \qquad K_O(f) = \frac{\beta^{n-1}\sin^{n-2}\alpha}{\alpha^{n-1}\sin^{n-2}\beta} .$$

This mapping is called the *natural folding of* $C(n,\alpha)$ *to* $C(n,\beta)$.

We determine the semi-axis lengths for the dilatation ellipsoid of $f'(x)$ at x in $C(n,\alpha)$. Owing to the rotational symmetry of f about the x_n-axis, we can restrict our attention to points x that lie in the (x_1,x_n)-plane and have $x_1 \ge 0$, i.e., that admit spherical coordinate representations of the sort $x = (\rho,\varphi,e_1)$. Fix such an x. Set $v_1 = \rho^{-1}x = \rho^{-1}(x_1e_1 + x_ne_n)$, $v_n = \rho^{-1}(x_ne_1 - x_1e_n)$, and $v_i = e_i$ for $i = 2,3,\ldots,n-1$. Let $v_1', v_2', \ldots, v_n'$ be the unit vectors constructed in an analogous way starting with $f(x)$ instead of x. It is geometrically clear that $f'(x)v_i$ must be a multiple of v_i'. In fact, by considering appropriate curves through x one sees that

$$f'(x)v_1 = v_1', \quad f'(x)v_i = \frac{\sin(\beta\varphi/\alpha)}{\sin\varphi} v_i' \text{ for } 2 \le i \le n-1, \quad f'(x)v_n = \frac{\beta}{\alpha} v_n' ,$$

where the expression $\sin(\beta\varphi/\alpha)/\sin\varphi$ is given the obvious interpretation for $\varphi = 0$, namely, β/α.

Adopting a similar convention for $t = 0$, we note that the inequality

$$\frac{\sin\beta}{\sin\alpha} \le \frac{\sin\beta t}{\sin\alpha t} \le \frac{\beta}{\alpha}$$

is valid when $0 < \alpha < \beta \le \pi$ and $0 \le t \le 1$, from which we infer that

$$\frac{\sin\beta}{\sin\alpha} \le \frac{\sin(\beta\varphi/\alpha)}{\sin\varphi} \le \frac{\beta}{\alpha}$$

for $0 \le \varphi < \alpha$. The upshot of these considerations is that for $x = (\rho, \varphi, u)$ in $C(n, \alpha)$ we have

$$\|f'(x)\| = \frac{\beta}{\alpha}, \quad \ell[f'(x)] = \min\left\{1, \frac{\sin(\beta\varphi/\alpha)}{\sin\alpha}\right\}, \quad |J_f(x)| = \frac{\beta}{\alpha}\left[\frac{\sin(\beta\varphi/\alpha)}{\sin\varphi}\right]^{n-2} .$$

This yields

$$H_I[f'(x)] = \frac{\beta}{\alpha} \max\left\{\left[\frac{\sin(\beta\varphi/\alpha)}{\sin\varphi}\right]^{n-2}, \left[\frac{\sin\varphi}{\sin(\beta\varphi/\alpha)}\right]^2\right\}$$

and

$$H_O[f'(x)] = \left(\frac{\beta}{\alpha}\right)^{n-1} \left[\frac{\sin\varphi}{\sin(\beta\varphi/\alpha)}\right]^{n-2} .$$

Taking suprema over φ leads by way of Corollary 6.1.3 to

$$K_I(f) = \max\left\{\left(\frac{\beta}{\alpha}\right)^{n-1}, \frac{\beta\sin^2\alpha}{\alpha\sin^2\beta}\right\}, \qquad K_O(f) = \frac{\beta^{n-1}\sin^{n-2}\alpha}{\alpha^{n-1}\sin^{n-2}\beta} ,$$

as maintained.

Observe that the natural folding of $C(n, \alpha)$ to $C(n, \beta)$ is a quasiconformal mapping, save when $\beta = \pi$. For β smaller than π this mapping clearly extends to a homeomorphism of $\overline{C(n, \alpha)}$ onto $\overline{C(n, \beta)}$ that fixes ∞. We next show how to map a half-space quasiconformally to a cylinder.

6.2.4. Mapping a half-space to a cylinder. Let $n \ge 3$, let $D = \mathbb{H}^n$, and let

$$D' = \{x \in \mathbb{R}^n : x_1^2 + \cdots + x_{n-1}^2 < 1\}. \tag{6.18}$$

The diffeomorphism $f : D \to D'$ that sends a point x which is given in spherical coordinates by $x = (\rho, \varphi, u)$ to the point described in cylindrical coordinates with respect to $\mathbb{R}^{n-1}$ by

$$f(x) = (2\varphi/\pi, u, (2/\pi)\log\rho) \tag{6.19}$$

has dilatations

$$K_I(f) = \left(\frac{\pi}{2}\right)^{n-2}, \qquad K_O(f) = \left(\frac{\pi}{2}\right)^2 .$$

By symmetry it suffices to determine the semi-axis lengths of $E[f'(x)]$ for x having spherical coordinates of the form (ρ, φ, e_1). Fix x of this type. We consider

$f'(x)v_i$, where $v_1, v_2, \ldots, v_n$ are the unit vectors assigned to x in the previous example. From the geometry we discern that $v_i' = f'(x)v_i$ is a multiple of e_i for $i = 1, 2, \ldots, n$. Calculations then reveal that $E[f'(x)]$ has two semi-axes of length $2/(\pi\rho)$ and $n-2$ of length $2\varphi/(\pi\rho\sin\varphi)$, the latter taken to mean $2/(\pi\rho)$ when $\varphi = 0$. Because $1 \le \varphi/\sin\varphi \le \pi/2$ when $0 \le \varphi < \pi/2$, we learn that

$$\|f'(x)\| = \frac{2\varphi}{\pi\rho\sin\varphi}, \qquad \ell\big[f'(x)\big] = \frac{2}{\pi\rho} \quad , \quad |J_f(x)| = \frac{2^n\varphi^{n-2}}{\pi^n\sin^{n-2}\varphi}.$$

As a result,

$$H_I\big[f'(x)\big] = \frac{\varphi^{n-2}}{\sin^{n-2}\varphi}, \qquad H_O\big[f'(x)\big] = \frac{\varphi^2}{\sin^2\varphi}$$

for any point $x = (\rho, \varphi, u)$ of D. We take suprema over φ in $[0, \pi/2)$ and then invoke Corollary 6.1.3 to conclude that

$$K_I(f) = \left(\frac{\pi}{2}\right)^{n-2}, \qquad K_O(f) = \left(\frac{\pi}{2}\right)^2.$$

This mapping is quasiconformal with $K(f) = \pi^2/4$ when $n = 3$ and with $K(f) = (\pi/2)^{n-2}$ when $n \ge 4$.

The definition of $f(x)$ actually makes sense for x in $\bar{D} \setminus \{0, \infty\}$ and shows that f can be extended to a homeomorphism of $\bar{D} \setminus \{0, \infty\}$ onto $\bar{D}' \setminus \{\infty\}$. If we set $f(0) = f(\infty) = \infty$, f becomes a continuous mapping of $\bar{H}^n$ onto $\bar{D}'$.

The previous mappings can be composed in a variety of ways to generate further examples of quasiconformal mappings. We obviously get a map from B^n to a cylinder.

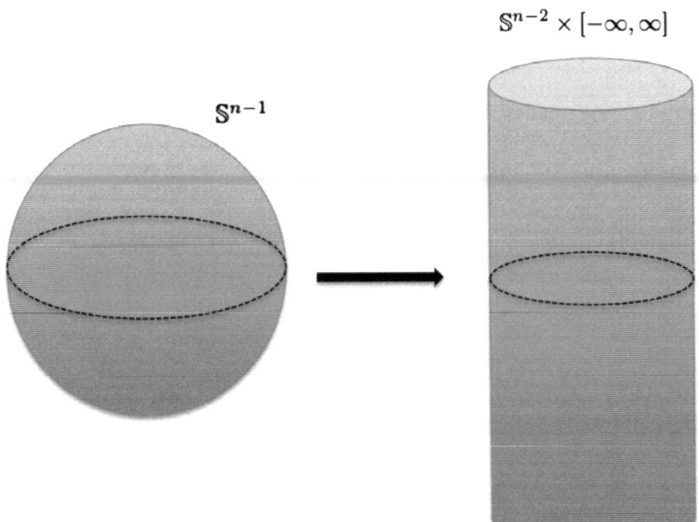

The mapping between $\mathbb{H}^n$ and the infinite cylinder $\mathbb{S}^{n-2} \times [-\infty, \infty]$ gives us a map from B^n to the infinite cylinder.

6.2.5. Mapping a half-cylinder to a cylinder.

LEMMA 6.2.1. *Let $n \geq 3$. If $D = \{x \in \mathbb{R}^n : x_1^2 + x_2^2 + \cdots + x_{n-1}^2 < 1,\ x_n < 0\}$ is a half-cylinder and $D' = \{x \in \mathbb{R}^n : x_1^2 + x_2^2 + \cdots + x_{n-1}^2 < 1\}$ is a cylinder, then there is a mapping $g : \bar{D} \to \bar{D}'$, continuous in the spherical metric, that maps D quasiconformally onto D' and fixes every point of*

$$\partial D \cap \partial D' = \{x \in \mathbb{R}^n : x_1^2 + x_2^2 + \cdots + x_{n-1}^2 - 1,\ x_n < 0\}.$$

PROOF. Let $f : \bar{\mathbb{H}}^n \to \bar{D}'$ be the mapping constructed previously. Write $D_1 = \mathbb{H}^n \cap B^n$.

We obtain a homeomorphism $g_1 : \bar{D} \to \bar{D}_1$ which is quasiconformal in D by noting that $f|\bar{D}_1$ is a homeomorphism of $\bar{D}_1$ onto $\bar{D}$ and taking $g_1 = (f|\bar{D}_1)^{-1}$. Observe that $g_1(\partial D \cap \partial D') = \bar{B}^{n-1}$. Next, let R be the reflection in the sphere $\mathbb{S}^{n-1}(-e_1, 2)$. The image of D_1 under R is the quarter-space $D_2 = \{x \in \mathbb{H}^n : x_1 > 1\}$, while $H' = R(B^{n-1}) = \{x \in \partial\mathbb{H}^n : x_1 > 1\}$. Thus $g_2 = R|\bar{D}_1$ is a homeomorphism of $\bar{D}_1$ onto $\bar{D}_2$ that transforms D_1 conformally to D_2.

Although D_2 is not in standard position, it is clear how to get a folding $g_3 : \bar{D}_2 \to \bar{H}^n$ that maps D_2 quasiconformally onto $\mathbb{H}^n$ and fixes $\bar{H}'$ pointwise. The homeomorphism $g_4 = R^{-1}|\bar{\mathbb{H}}^n = R|\bar{\mathbb{H}}^n$ maps $\mathbb{H}^n$ conformally onto itself and sends $\bar{H}'$ back to $\overline{B}^n$. Finally, let $g_5 = f$. The composition $g = g_5 \circ g_4 \circ g_3 \circ g_2 \circ g_1$ is a continuous mapping of $\bar{D}$ onto $\bar{D}'$, it fixes every point of $\partial D \cap \partial D'$, and its restriction to D is a quasiconformal diffeomorphism of D onto D'. □

Notice that if g is the mapping in Lemma 6.2.1 and if $D'' = \{x \in D' : x_n < 1\}$, then by setting $h(x) = e_n + g^{-1}[g(x) - e_n]$ for $x \in \bar{D} \setminus \{\infty\}$ and $h(\infty) = \infty$ we obtain a homeomorphism of $\bar{D}$ onto $\bar{D}''$ that maps D quasiconformally onto D'' and fixes $\partial D \cap \partial D''$ pointwise.

6.2.6. Mapping a half-space to a hairy half-space. The next example shows that a quasiconformal mapping of a half-space can be quite wild. Let $\mathcal{F}$ be a disjoint collection of balls $B^n(x_j, r_j) \subset \mathbb{R}^{n-1} = \partial\mathbb{H}^n$. Then define a "hairy" half-space as

$$H = \mathbb{H}^n \bigcup_{B \in \mathcal{F}} B \times [0, \infty) \subset \mathbb{R}^n.$$

THEOREM 6.2.2. *There is a quasiconformal mapping B^n onto the domain H.*

PROOF. Since B^n is conformally equivalent to $\mathbb{H}^n$ we may as well define the mapping on $\mathbb{H}^n$. Let $g : B^{n-1} \times (0, -\infty) \to B^{n-1} \times (-\infty, +\infty)$ be defined as in the previous section. Note that $g|\mathbb{S}^{n-2} \times (0, -\infty)$ is the identity. Define the mapping as follows:

$$f(x) = \begin{cases} g_j(x) = x_j + r_j g(\frac{x - x_j}{r_j}) & x \in B \times (0, -\infty),\ B = B^{n-1}(x_j, r_j) \in \mathcal{F}, \\ x & \text{otherwise.} \end{cases}$$

Then it is easy to see that f is a homeomorphism as $f|\mathbb{S}^{n-2}(x_j, r_j) \times (0, -\infty)$ is the identity mapping. The mapping $x \overset{g_j}{\mapsto} x_j + r_j g(\frac{x - x_j}{r_j})$ is quasiconformal with the same dilatation as g, as it can be written as a composition of g with similarities. Thus f has bounded dilatation except possibly on the sets $x \in \mathbb{S}^{n-2}(x_j, r_j) \times (0, -\infty)$, $B^{n-1}(x_j, r_j) \in \mathcal{F}$, where the identity and g_j are glued together. This raises a general question about when piecewise quasiconformal mappings are quasiconformal. This is an entirely local problem at the interface between regions where a mapping is quasiconformal. We shall address this later in the much more general situation of Theorem 6.4.22. It suffices to say that the smooth codimension 1 interfaces $\mathbb{S}^{n-2}(x_j, r_j) \times (0, \infty)$ (of sigma-finite $(n-1)$-measure 0) will be allowable. This completes the proof. □

6.2.7. Locally bilipschitz mappings are quasiconformal. The next lemma identifies an entire class of mappings that are quasiconformal. We say that a homeomorphism $f : D \to D'$ between domains D and D' in $\mathbb{R}^n$ is *locally λ-bilipschitz*, where $1 \leq \lambda < \infty$, provided each point x of D has an open neighbourhood $U = U_x$ with the property that

$$\lambda^{-1}|z - y| \leq |f(z) - f(y)| \leq \lambda|z - y|$$

whenever y and z belong to U. When the above inequality is true for all y and z in D, we call f a *λ-bilipschitz homeomorphism.*

LEMMA 6.2.3. *If $f : D \to D'$ is a locally λ-bilipschitz homeomorphism between domains D and D' in $\mathbb{R}^n$ with $n \geq 2$, then f is a λ^{2n}-quasiconformal mapping of D onto D'.*

PROOF. The assumptions ensure that f is locally absolutely continuous on every locally rectifiable curve in D. Certainly we have $\lambda^{-1} \leq \ell_f(x) \leq L_f(x) \leq \lambda$ for every x in D. According to the Rademacher-Stepanov theorem (Theorem 4.0.10), f is differentiable at almost every point of D. Furthermore, at any point of differentiability x of f we have

$$|J_f(x)| \geq \ell\big[f'(x)\big]^n = \ell_f(x)^n \geq \lambda^{-n}\,.$$

Let Γ be a family of curves in D. We shall verify that $M(\Gamma) \leq \lambda^{2n} M[f(\Gamma)]$. We assume that $M[f(\Gamma)] < \infty$—the inequality holds trivially otherwise—and consider an admissible density ρ for $f(\Gamma)$. Define $\tilde{\rho} : \hat{\mathbb{R}}^n \to [0, \infty]$ as follows: $\tilde{\rho}(x) = \lambda\rho[f(x)]$ if x is in D and $\tilde{\rho}(x) = 0$ if x belongs to D^c. For any locally rectifiable curve γ in Γ we have by Theorem 4.1.6

$$\int_\gamma \tilde{\rho}\, ds = \lambda \int_\gamma \rho \circ f\, ds \geq \int_\gamma (\rho \circ f) L_f\, ds \geq \int_{f\circ\gamma} \rho\, ds \geq 1\,.$$

Therefore $\tilde{\rho}$ belongs to $\mathrm{Adm}(\Gamma)$, and so

$$\begin{aligned} M(\Gamma) &\leq \int_{\mathbb{R}^n} \tilde{\rho}^n\, dm_n = \lambda^n \int_{\mathbb{R}^n} \rho^n \circ f\, dm_n \leq \lambda^{2n} \int_D (\rho^n \circ f)|J_f|\, dm_n \\ &= \lambda^{2n} \int_{D'} \rho^n\, dm_n \leq \lambda^{2n} \int_{\mathbb{R}^n} \rho^n\, dm_n\,. \end{aligned}$$

As the admissible density ρ for $f(\Gamma)$ was arbitrary, we see that

$$M(\Gamma) \leq \lambda^{2n} M[f(\Gamma)],$$

for every subfamily Γ of $\mathcal{C}(D)$. We infer that $K_O(f) \leq \lambda^{2n}$. The same argument applied to f^{-1} gives $K_I(f) = K_O(f^{-1}) \leq \lambda^{2n}$, so $K(f) \leq \lambda^{2n}$ as stated. □

We now describe a concrete situation to which we may apply Lemma 6.2.3.

Let $n \geq 2$ and let D be a bounded domain in $\mathbb{R}^n$ which is *strictly starlike with respect to the origin*, meaning that 0 lies in D and that each ray emanating from 0 intersects ∂D in exactly one point. A natural mapping $f : \mathbb{R}^n \to \mathbb{R}^n$ is defined by insisting that on each ray L from the origin f act as dilation by the factor $1/\ell(L \cap D)$. This mapping f, which is readily seen to be a homeomorphism, is known as the *linear radial stretching of* $\mathbb{R}^n$ *associated with* D. We wish to point out a geometric condition on D under which f is a quasiconformal homeomorphism. The notation $C(n, \alpha, x)$, where x belongs to $\mathbb{R}^n \setminus \{0\}$ and $0 < \alpha < \pi$, is used in what follows to indicate the standard open cone in $\mathbb{R}^n$ whose vertex is x, whose vertex half-angle is α, and whose axis is the ray from x that goes through the origin; in other words, $C(n, \alpha, x) = \{y \in \mathbb{R}^n : \langle x - y, x\rangle > |x - y| \cdot |x| \cos \alpha\}$.

LEMMA 6.2.4. *Let f be the linear radial stretching of $\mathbb{R}^n$ associated with a bounded domain D that is strictly starlike with respect to the origin. Assume the existence of an angle α in $(0, \pi/4]$ with the property that the truncated cone $C(n, \alpha, x) \cap B^n(x, |x|)$ is contained in D whenever x is a point of ∂D. Then f is a bilipschitz homeomorphism of $\mathbb{R}^n$ onto itself—hence, a quasiconformal mapping of $\mathbb{R}^n$ onto itself—that transforms D to B^n.*

PROOF. For nonzero z in $\mathbb{R}^n$, we use L_z to designate the ray from the origin that passes through z. Write $m = \min\{|x| : x \in \partial D\}$ and $M = \max\{|x| : x \in \partial D\}$. We first demonstrate that

$$\frac{|y - x| \sin \alpha}{M} \le |f(y) - f(x)| \le \frac{|y - x|}{m} \tag{6.20}$$

whenever x and y are boundary points of D. Fix such a pair of points, say with $|y| \le |x|$. Remarking that x lies in $L_x \setminus B^n(m)$ and y in $L_y \setminus B^n(m)$, we discover that

$$|f(y) - f(x)| = m^{-1}|mf(y) - mf(x)| \le \frac{|y - x|}{m},$$

where $mf(x)$ and $mf(y)$ are the points of $L_x \setminus B^n(m)$ and $L_y \setminus B^n(m)$ at minimal distance from each other. Let $\theta = \theta(x, y)$. If $\theta \ge \pi/2$, then plainly

$$|f(y) - f(x)| \ge \sqrt{2} \ge \frac{|y - x|\sqrt{2}}{2M} \ge \frac{|y - x| \sin \alpha}{M}.$$

Also, if $0 < \theta < \pi/2$, then the fact that $|y| \le |x|$ and y is not a point of $C(n, \alpha, x) \cap B^n(x, |x|)$, in combination with a little elementary trigonometry, reveals that

$$|f(y) - f(x)| \ge \sin \theta \ge \frac{|y| \sin \theta}{|x|} \ge \frac{|y - x| \sin \alpha}{|x|} \ge \frac{|y - x| \sin \alpha}{M}.$$

Finally $y = x$ (hence, $f(y) = f(x)$) when $\theta = 0$, so we have confirmed (6.20) in all cases.

Now let x and y be arbitrary points of $\mathbb{R}^n$. We wish to exhibit a constant $\lambda \ge 1$, independent of x and y, such that

$$\lambda^{-1}|y - x| \le |f(y) - f(x)| \le \lambda |y - x|. \tag{6.21}$$

The estimate $|f(x)| \le m^{-1}|x|$, valid for all x in $\mathbb{R}^n$, shows that f is continuous at 0. In establishing (6.21), therefore, we need only concern ourselves with the case where both x and y are different from zero. For nonzero z in $\mathbb{R}^n$ we let z^* denote the point where L_z meets ∂D. We are then entitled to write

$$f(z) = \frac{|z| f(z^*)}{|z^*|}.$$

Assume initially that x is a point for which $|x^*| = 1$—note that $f(x^*) = x^*$ and $f(x) = x$ for any such point—and let $\theta = \theta(x, y)$. Then $\theta[f(x^*), f(y^*)] = \theta$ as well, and by (6.20),

$$|f(y) - y| = \frac{|y| \cdot |f(y^*) - y^*|}{|y^*|} \leq \frac{|y| \cdot |f(y^*) - x^*| + |y| \cdot |y^* - x^*|}{|y^*|}$$

$$\leq \frac{|y| \cdot |f(y^*) - f(x^*)| + (M \csc \alpha)|y| \cdot |f(y^*) - f(x^*)|}{m}. \tag{6.22}$$

As $|f(y^*) - f(x^*)| \leq \theta$, we conclude that

$$|f(y) - y| \leq \theta m^{-1}(1 + M \csc \alpha)|y|. \tag{6.23}$$

If $\theta \geq \pi/2$, then $|y - x| \geq |y|$ and (6.23) gives (recall that $f(x) = x$ in the situation under consideration)

$$\begin{aligned} |f(y) - f(x)| &\leq |f(y) - y| + |y - x| \\ &\leq (1 + \theta m^{-1} + \theta m^{-1} M \csc \alpha)|y - x| \\ &\leq (1 + \pi m^{-1} + \pi m^{-1} M \csc \alpha)|y - x|\,; \end{aligned}$$

if $0 < \theta \leq \pi/2$, then $|y - x| \geq |y| \sin \theta$ and in this instance (6.23) yields

$$|f(y) - f(x)| \leq \left(1 + \frac{\theta}{m \sin \theta} + \frac{\theta M \csc \alpha}{m \sin \theta}\right)|y - x| \leq \left(1 + \frac{\pi}{2m} + \frac{\pi M \csc \alpha}{2m}\right)|y - x|\,;$$

if $\theta = 0$, we have both $f(y) = y$ and $f(x) = x$, so

$$|f(y) - f(x)| = |y - x|.$$

In every case which arises we have demonstrated that

$$|f(y) - f(x)| \leq (1 + \pi m^{-1} + \pi m^{-1} M \csc \alpha)|y - x|,$$

always presuming, of course, that $|x^*| = 1$. However, if $r = |x^*| \neq 1$, then the dilation $\varphi(z) = r^{-1}z$ maps D to a domain D_0 which satisfies all of the conditions that D satisfies. Its associated radial linear stretching is $f_0 = f \circ \varphi^{-1}$. Relative to D_0, $|\varphi(x)^*| = 1$. Since $m_0 = r^{-1}m$, $M_0 = r^{-1}M$, and $m \leq r \leq M$, we can use the normalized case to infer

$$\begin{aligned} |f(y) - f(x)| &= |f_0[\varphi(y)] - f_0[\varphi(x)]| \\ &\leq (1 + \pi r m^{-1} + \pi m^{-1} M \csc \alpha)|\varphi(y) - \varphi(x)| \\ &= (r^{-1} + \pi m^{-1} + \pi r^{-1} m^{-1} M \csc \alpha)|r\varphi(y) - r\varphi(x)| \\ &\leq (m^{-1} + \pi m^{-1} + \pi m^{-2} M \csc \alpha)|y - x|. \end{aligned}$$

We thus see that the inequality

$$|f(y) - f(x)| \leq \beta |y - x|$$

holds with $\beta = m^{-1} + \pi m^{-1} + \pi m^{-2} M \csc \alpha$ for all x and y in $\mathbb{R}^n$. Similar arguments confirm that there is a constant $\beta^* > 0$ so that

$$|f(y) - f(x)| \geq \beta^* |y - x|$$

for all x and y in $\mathbb{R}^n$. Therefore (6.21) is satisfied for $\lambda = \max\{\beta, 1/\beta^*\}$. □

Lemma 6.2.4 implies, for instance, that an arbitrary convex open polyhedron in $\mathbb{R}^n$ can be mapped to B^n via a quasiconformal self-mapping of $\mathbb{R}^n$. (By an *open polyhedron* is meant a bounded domain D in $\mathbb{R}^n$ with $n \geq 2$ such that $\partial D = \partial \bar{D}$ and such that ∂D can be covered by finitely many hyperplanes.)

6.2.8. Straightening hypersurfaces. Our final example in this section shows how a smooth hypersurface in $\mathbb{R}^n$ can locally be "straightened out" by means of a quasiconformal mapping.

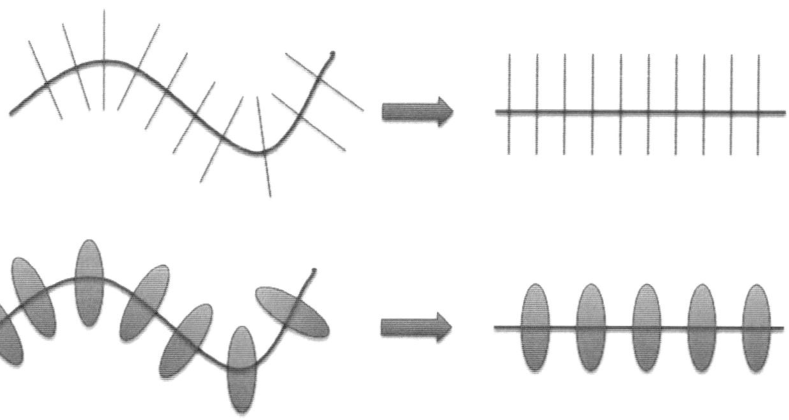

Straightening a smooth curve in two and three dimensions.

LEMMA 6.2.5. *Let G be a domain in $\mathbb{R}^{n-1}$ with $n \geq 2$, let $u : G \to \mathbb{R}$ be a C^1-function such that $|\nabla u(\xi)| \leq m < \infty$ for every ξ in G, and let $D = G \times (-\infty, \infty)$. The diffeomorphism $f : D \to D$ defined by $f(x) = x - u(x_1, x_2, \ldots, x_{n-1})e_n$ is a quasiconformal mapping that maps the graph of u, that is, the set*

$$S = \{x = \xi + x_n e_n : \xi \in G,\ \ x_n = u(\xi)\},$$

to G.

PROOF. The derivative matrix of f at $x = \xi + x_n e_n$ has the form

$$\begin{bmatrix} 1 & 0 & \cdots & 0 & 0 \\ 0 & 1 & \cdots & 0 & 0 \\ \vdots & \vdots & \cdots & \vdots & \vdots \\ 0 & 0 & & 1 & 0 \\ -D_1 u(\xi) & -D_2 u(\xi) & \cdots & -D_{n-1}u(\xi) & 1 \end{bmatrix},$$

which enables us to conclude that $J_f(x) = 1$ and that

$$\begin{aligned} &\|f'(x)\| \\ &= \max_{|h|=1} |f'(x)h| \\ &= \max_{|h|=1} \sqrt{h_1^2 + \cdots + h_{n-1}^2 + (-h_1 D_1 u(\xi) - \cdots - h_{n-1}D_{n-1}u(\xi) + h_n)^2} \\ &\leq \max_{|h|=1} \sqrt{|h|^2 + (1 + |\nabla u(\xi)|^2)|h|^2} = \sqrt{2 + |\nabla u(\xi)|^2} \leq \sqrt{2 + m^2} \end{aligned}$$

for every x in D. It follows from Corollary 6.1.3 that

$$K_O(f) \leq (2 + m^2)^{n/2}, \qquad K_I(f) \leq (2 + m^2)^{n(n-1)/2}.$$

Consequently, f is a quasiconformal mapping of D onto itself and transforms the graph of u to the domain G in $\mathbb{R}^{n-1}$. □

6.3. Some measure theory

In preparation for our discussion of the linear dilatation of a homeomorphism we will need to assemble some background information from measure theory. First, we discuss the differentiation of a measure μ with respect to Lebesgue measure.

6.3.1. The symmetric derivative of a measure.

The type of measure μ that we focus upon here will be locally finite Borel measures defined in an open subset U of $\mathbb{R}^n$; thus we shall assume that the domain of μ includes the σ-algebra of Borel subsets of U and that $\mu(K) < \infty$ for every compact set K in U.

To say that such a measure is *locally absolutely continuous with respect to Lebesgue measure* means the following: corresponding to each compact set K in U and each $\varepsilon > 0$ there exists a $\delta = \delta(\varepsilon, K) > 0$ such that $\mu(A) < \varepsilon$ for every Borel subset A of K with $|A| < \delta$. An obvious example of a measure which enjoys this property is one defined on the σ-algebra of Lebesgue measurable sets in U by

$$\mu(A) = \int_A \rho \, dm_n \ ,$$

where $\rho : U \to [0, \infty]$ is a function that is locally integrable in U.

The Radon-Nikodym theorem implies that any locally finite Borel measure in U which is locally absolutely continuous with respect to Lebesgue measure has the above structure for some function ρ. Moreover, ρ is uniquely determined up to a set of Lebesgue measure zero. The function ρ—or, to be more precise, the equivalence class of functions modulo almost everywhere equality that ρ represents—is called the *Radon-Nikodym derivative of μ with respect to m_n*.

We quote an alternative characterisation of local absolute continuity with respect to Lebesgue measure that will be of some use.

THEOREM 6.3.1. *A locally finite Borel measure μ in an open subset U of $\mathbb{R}^n$ is locally absolutely continuous with respect to Lebesgue measure if and only if $\mu(A) = 0$ whenever A is a Borel subset of U with $|A| = 0$.*

Given a locally finite Borel measure μ in an open set U in $\mathbb{R}^n$, we define a function $\mu' : U \to [0, \infty]$ by

$$\mu'(x) = \limsup_{r \to 0} \frac{\mu[B^n(x,r)]}{|B^n(x,r)|} = \limsup_{r \to 0} \frac{\mu[B^n(x,r)]}{\Omega_n r^n} \ .$$

Since $B^n(x,r) \subset \overline{B}^n(x,r) \subset B^n(x, \lambda r)$ for every $\lambda > 1$, it is an easy exercise to show that

$$\mu'(x) = \limsup_{r \to 0} \frac{\mu[\overline{B}^n(x,r)]}{|\overline{B}^n(x,r)|} \ .$$

Whenever it turns out that, in fact,

$$\mu'(x) = \lim_{r \to 0} \frac{\mu[B^n(x,r)]}{|B^n(x,r)|} < \infty \ ,$$

we shall refer to $\mu'(x)$ as the *symmetric derivative of* μ at x (*with respect to Lebesgue measure*). One version of the Lebesgue differentiation theorem which we shall have cause to use reads as follows.

THEOREM 6.3.2. *If μ is a locally finite Borel measure in an open subset U of $\mathbb{R}^n$, then μ possesses a symmetric derivative at almost every point of U. The function $\mu' : U \to [0, \infty]$ is Lebesgue measurable and obeys the inequality*

$$\mu(A) \geq \int_A \mu' \, dm_n \tag{6.24}$$

for each Borel set A in U. In particular, μ' is locally integrable in U. If, in addition, μ is locally absolutely continuous with respect to Lebesgue measure, then equality holds in (6.24), in which event μ' is the Radon-Nikodym derivative of μ with respect to m_n and

$$\int_U \varphi \, d\mu = \int_U \varphi \mu' \, dm_n$$

is true for every Borel function $\varphi : U \to [0, \infty]$.

As usual, the $0 \cdot \infty = \infty \cdot 0 = 0$ convention is in force for the integrand $\varphi\mu'$ in the last integral. Notice that if μ is given in U by

$$\mu(A) = \int_A \rho \, dm_n,$$

where $\rho : U \to [0, \infty]$ is a locally integrable function, then Theorem 6.3.2 ensures that

$$\lim_{r \to 0} \frac{1}{|B^n(x,r)|} \int_{B^n(x,r)} \rho \, dm_n = \rho(x)$$

for almost every point x of U. In particular, taking $\rho = \chi_E$ in U, where E is a Lebesgue measurable subset of U, we can conclude that almost every point x of E has the feature that

$$\lim_{r \to 0} \frac{|E \cap B^n(x,r)|}{|B^n(x,r)|} = 1 \,.$$

Any point x of $\mathbb{R}^n$ with this last property is called a *point of density for E.*

Suppose that $f : D \to D'$ is a homeomorphism between domains D and D' in $\mathbb{R}^n$. Then $f(A)$ is a Borel set in D' whenever A is a Borel set in D. This observation enables us to define a locally finite Borel measure μ_f in D by setting

$$\mu_f(A) = |f(A)|$$

for each Borel subset A of D. The measure μ_f is called the *measure induced by f*, while the function μ_f' is known as the *generalized Jacobian* of f (or the *volume derivative of f*).

It is a simple matter to see that if f happens to be differentiable at a point x, then μ_f has a symmetric derivative at x and $\mu_f'(x) = |J_f(x)|$. This fact is especially easy to verify when $f'(x) = 0$ or when $f'(x)$ is nonsingular, the cases of greatest interest in the sequel.

We can assert on the grounds of Theorem 6.3.2 that μ_f has a symmetric derivative almost everywhere in D and that

$$|f(A)| \geq \int_A \mu_f' \, dm_n \tag{6.25}$$

whenever A is a Borel subset of D. More generally, the monotone convergence theorem and (6.25) together imply that

$$\int_{D'} \varphi \, dm_n \geq \int_D (\varphi \circ f) \mu_f' \, dm_n \tag{6.26}$$

for every Borel function $\varphi : D' \to [0, \infty]$.

Consider further the homeomorphism $f : D \to D'$. We speak of f having the *Lusin property* under the condition that $|f(K)| = 0$ for each compact set K in D with $|K| = 0$.

If f has this property, then $|f(A)| = 0$ holds for an arbitrary subset A of D with $|A| = 0$. Indeed, given such a set A we can choose a Borel set E such that $A \subset E \subset D$ and $|E| = 0$. Should $|f(E)| > 0$, the Borel set $f(E)$ would contain a compact set K' with $|K'| > 0$. Then $K = f^{-1}(K')$ would be a compact subset of D with $|K| \leq |E| = 0$, yet with $|f(K)| = |K'| > 0$, contrary to the assumption about f. Accordingly, $|f(A)| \leq |f(E)| = 0$. In particular, when a homeomorphism f is endowed with the Lusin property, its induced measure is locally absolutely continuous with respect to Lebesgue measure. The important properties of homeomorphisms with the Lusin property are summarized in the next theorem.

THEOREM 6.3.3. *Suppose that a homeomorphism $f : D \to D'$ between domains D and D' in $\mathbb{R}^n$ has the Lusin property.*

If A is a Lebesgue measurable set in D, then $f(A)$ is a Lebesgue measurable set and

$$|f(A)| = \int_A \mu_f' \, dm_n \,. \tag{6.27}$$

If f^{-1} also has the Lusin property, then $\mu_f' > 0$ almost everywhere in D and the formula

$$\int_{D'} \varphi \, dm_n = \int_D (\varphi \circ f) \mu_f' \, dm_n \tag{6.28}$$

is valid for every Lebesgue measurable function $\varphi : D' \to [0, \infty]$.

PROOF. Because μ_f is locally absolutely continuous with respect to Lebesgue measure, Theorem 6.3.2 gives

$$|f(E)| = \int_E \mu_f' \, dm_n$$

for every Borel set E in D. If A is a Lebesgue measurable set in D, we can express A as a disjoint union $A = E \cup N$, where E is a Borel set and $|N| = 0$. Now $f(E)$ is a Borel set and $|f(N)| = 0$, so $f(A) = f(E) \cup f(N)$ is Lebesgue measurable and

$$|f(A)| = |f(E)| = \int_E \mu_f' \, dm_n = \int_A \mu_f' \, dm_n \,.$$

Let $S = \{x \in D : \mu_f'(x) = 0\}$. Then S is a Lebesgue measurable subset of D and (6.27) yields

$$|f(S)| = \int_S \mu_f' \, dm_n = 0 \,.$$

If f^{-1} also has the Lusin property, then we see that $|S| = |f^{-1}[f(S)]| = 0$, whence $\mu_f' > 0$ almost everywhere in D. In this event (6.28) is true when $\varphi = \chi_A$ in D', where A is a Lebesgue measurable subset of D', for the set $f^{-1}(A)$ is Lebesgue measurable and by (6.27)

$$\begin{aligned}\int_{D'} \varphi\, dm_n &= |A| = |f[f^{-1}(A)]| = \int_{f^{-1}(A)} \mu'_f\, dm_n = \int_D \chi_{f^{-1}(A)}\mu'_f\, dm_n \\ &= \int_D (\varphi \circ f)\mu'_f\, dm_n\,.\end{aligned}$$

Thus (6.28) holds for any Lebesgue measurable simple function $\varphi : D' \to [0, \infty]$; hence, after an appeal to the monotone convergence theorem, it holds for an arbitrary Lebesgue measurable function $\varphi : D' \to [0, \infty]$. □

We draw our attention to the following corollary of Theorem 6.3.3.

COROLLARY 6.3.4. *Let $f : D \to D'$ be a homeomorphism between domains D and D' in $\mathbb{R}^n$. If both f and f^{-1} have the Lusin property and if f is differentiable almost everywhere in D, then $f'(x)$ is nonsingular for almost every x in D and the standard change of variable formula,*

$$\int_{D'} \varphi\, dm_n = \int_D (\varphi \circ f)|J_f|\, dm_n\ ,$$

is valid for every Lebesgue measurable function $\varphi : D' \to [0, \infty]$.

Taking for granted the Rademacher-Stepanov theorem, it is not hard to check that a locally bilipschitz homeomorphism $f : D \to D'$ satisfies all the hypotheses of Corollary 6.3.4. Therefore the change of variable formula in the corollary holds true for such a mapping, a fact which we have already used on several occasions earlier in this book.

6.3.2. Linear measure and absolute continuity of paths. The Hausdorff $\mathcal{H}^1$-measure of a subset A of $\mathbb{R}^n$ is often referred to as its *linear measure*. Several elementary facts about linear measure will be needed in upcoming sections. We therefore incorporate these into the following series of lemmas.

LEMMA 6.3.5. *If A is a connected set in $\mathbb{R}^n$, then* $\operatorname{diam}(A) \le \mathcal{H}^1(A)$.

PROOF. We may assume that $\operatorname{diam}(A) > 0$. Let x and y be arbitrary distinct points of A, and let $f : \mathbb{R}^n \to \mathbb{R}^n$ be the orthogonal projection of $\mathbb{R}^n$ onto the line L through x and y. Then $f(A)$ contains the line segment having x and y for endpoints. Thus

$$|y - x| \le m_1^*\big[f(A)\big] = \mathcal{H}^1\big[f(A)\big] \le \mathcal{H}^1(A)\ ,$$

where $m_1^* = \mathcal{H}^1$ on the line L and f is nonexpansive (that is, $|f(w) - f(z)| \le |w - z|$ for all z and w). The stated conclusion follows directly. □

The next lemma establishes a relationship between the length of a curve and the linear measure of its trajectory.

LEMMA 6.3.6. *If γ is a curve in $\mathbb{R}^n$, then $\mathcal{H}^1(|\gamma|) \le \ell(\gamma)$. Equality holds when γ is injective.*

PROOF. It suffices to consider the case where γ is a path. In verifying the inequality we may further assume that $0 < \ell(\gamma) < \infty$ and that γ is presented to us in normal form, $\gamma = \gamma_0 : [0, \ell(\gamma)] \to \mathbb{R}^n$. Given $t > 0$, choose a partition

$0 = s_0 < s_1 < \cdots < s_p = \ell(\gamma)$ of $[0, \ell(\gamma)]$ with $s_j - s_{j-1} < t$ for $j = 1, 2, \ldots, p$. Writing $I_j = [s_{j-1}, s_j]$, we have

$$d\big[\gamma(I_j)\big] \le \ell(\gamma|I_j) = s_j - s_{j-1} < t\,.$$

Hence, since the sets $\gamma(I_j)$ cover $|\gamma|$ and since $\alpha(1)/2 = 1$,

$$\mathcal{H}^1_t(|\gamma|) \le \sum_{j=1}^{p} d\big[\gamma(I_j)\big] \le \sum_{j=1}^{p} \ell(\gamma|I_j) = \ell(\gamma)\,.$$

Letting $t \to 0$ yields $\mathcal{H}^1(|\gamma|) \le \ell(\gamma)$.

Let $\gamma : [a, b] \to \mathbb{R}^n$ be an injective path, and let $P : a = t_0 < t_1 < \cdots < t_p = b$ be an arbitrary partition of $[a, b]$. Setting $F_j = (t_{j-1}, t_j)$, we obtain with the help of Lemma 6.3.5 that

$$\begin{aligned} \ell(\gamma, P) &= \sum_{j=1}^{p} |\gamma(t_j) - \gamma(t_{j-1})| \le \sum_{j=1}^{p} d\big[\gamma(\bar{F}_j)\big] = \sum_{j=1}^{p} d\big[\gamma(F_j)\big] \\ &\le \sum_{j=1}^{p} \mathcal{H}^1\big[\gamma(F_j)\big] = \mathcal{H}^1\Bigg[\bigcup_{j=1}^{p} \gamma(F_j)\Bigg] \le \mathcal{H}^1(|\gamma|)\,. \end{aligned}$$

We note that the set $\gamma(\bar{F}_j)$ is compact, so $\gamma(F_j) = \gamma(\bar{F}_j) \setminus \{\gamma(t_{j-1}), \gamma(t_j)\}$ is a Borel set, and is thus $\mathcal{H}^1$-measurable. It follows that $\ell(\gamma) \le \mathcal{H}^1(|\gamma|)$. In view of the first part of the lemma, $\mathcal{H}^1(|\gamma|) = \ell(\gamma)$ for an injective path γ. □

Linear measure also has a bearing on the absolute continuity of paths.

LEMMA 6.3.7. *Suppose that $\gamma : [a, b] \to \mathbb{R}^n$ is an injective path with the following property: corresponding to each $\varepsilon > 0$ there is a $\delta > 0$ such that $\mathcal{H}^1[\gamma(F)] < \varepsilon$ whenever F is a compact set in (a, b) with $m_1(F) < \delta$. Then γ is absolutely continuous.*

PROOF. Let $\varepsilon > 0$. We produce a $\delta > 0$ so that $\sum_{j=1}^{p} |\gamma(b_j) - \gamma(a_j)| \le \varepsilon$ whenever $I_1 = [a_1, b_1], I_2 = [a_2, b_2], \ldots, I_p = [a_p, b_p]$ is a collection of nondegenerate, nonoverlapping intervals in $[a, b]$ with $\sum_{j=1}^{p} (b_j - a_j) < \delta$. Choose δ as specified in the hypothesis, and let $I_1, I_2, \ldots, I_p$ be as indicated.

Given $0 < \eta < (1/2) \min_j (b_j - a_j)$, we apply the hypothesis to the set $F = \bigcup_{j=1}^{p} F_j$, where $F_j = [a_j + \eta, b_j - \eta]$. Relying on Lemma 6.3.5, we compute

$$\begin{aligned} \sum_{j=1}^{p} |\gamma(b_j - \eta) - \gamma(a_j + \eta)| &\le \sum_{j=1}^{p} d\big[\gamma(F_j)\big] \le \sum_{j=1}^{p} \mathcal{H}^1\big[\gamma(F_j)\big] \\ &= \mathcal{H}^1\Bigg[\bigcup_{j=1}^{p} \gamma(F_j)\Bigg] = \mathcal{H}^1\big[\gamma(F)\big] < \varepsilon\,, \end{aligned}$$

since $m_1(F) < \delta$.

We let $\eta \to 0$ to get $\sum_{j=1}^{p} |\gamma(b_j) - \gamma(a_j)| \le \varepsilon$, the inequality desired. □

We shall need a second result that relates absolute continuity to linear measure.

LEMMA 6.3.8. *Let $\gamma : I \to \mathbb{R}^n$ be a path. Suppose there is a compact set F in I for which the following conditions are met:*

(i) *γ is absolutely continuous on each closed subinterval of $I \setminus F$;*
(ii) $\mathcal{H}^1[\gamma(F)] = 0$;

(iii) $\int_{I \setminus F} |\dot{\gamma}(t)| \, dt < \infty$.

Then γ is absolutely continuous.

PROOF. Define a function $\rho : I \to [0, \infty)$ by

$$\rho(t) = \begin{cases} |\dot{\gamma}(t)| & \text{if } t \in I \setminus F \text{ and } \gamma \text{ is differentiable at } t \, , \\ 0 & \text{if } t \in F \text{ or } \dot{\gamma}(t) \text{ fails to exist.} \end{cases}$$

Conditions (i) and (iii) above guarantee that ρ is a Borel function which is integrable over I.

We demonstrate that

$$d\big[\gamma(J)\big] \le \int_J \rho(t) \, dt$$

for each subinterval J of I, whose inequality clearly implies the absolute continuity of γ. Fix J and write $J = (J \cap F) \cup (\bigcup_\nu J_\nu)$, where the J_ν are nonoverlapping intervals in $I \setminus F$. (Should J be contained in F, we would have $\operatorname{diam}[\gamma(J)] \le \mathcal{H}^1[\gamma(F)] = 0$ by Lemma 6.3.5, so the stated inequality would hold trivially. We may thus assume that J meets $I \setminus F$.) Since $\gamma | J_\nu$ is locally absolutely continuous, Lemmas 6.3.5 and 6.3.6 give

$$\begin{aligned} d\big[\gamma(J)\big] \le \mathcal{H}^1\big[\gamma(J)\big] &\le \mathcal{H}^1\big[\gamma(J \cap F)\big] + \sum_\nu \mathcal{H}^1\big[\gamma(J_\nu)\big] = \sum_\nu \mathcal{H}^1\big[\gamma(J_\nu)\big] \\ &\le \sum_\nu \ell(\gamma | J_\nu) = \sum_\nu \int_{J_\nu} |\dot{\gamma}(t)| \, dt = \int_{\bigcup_\nu J_\nu} \rho(t) \, dt \\ &\le \int_J \rho(t) \, dt \, , \end{aligned}$$

as desired. □

An interesting fact about higher-dimensional Hausdorff measures—it will be required in the application we have in mind for Lemma 6.3.8—is communicated by an old theorem of Gross [**64**].

For want of a more appropriate place to put it, we insert it here.

THEOREM 6.3.9. *Suppose that $f : \mathbb{R}^n \to \mathbb{R}^k$ satisfies a Lipschitz condition with Lipschitz constant λ. If A is a subset of $\mathbb{R}^n$ and $p \ge 1$ is an integer, then the set E_p of points y in $\mathbb{R}^k$ for which*

$$\#[A \cap f^{-1}(\{y\})] \ge p$$

has

$$p \, m_k^*(E_p) \le \lambda^k \mathcal{H}^k(A).$$

In particular, if $\mathcal{H}^k(A) < \infty$, then $A \cap f^{-1}(\{y\})$ is a finite set for almost every y in $\mathbb{R}^k$, while if A has σ-finite $\mathcal{H}^k$-measure, then $A \cap f^{-1}(\{y\})$ is a countable set for almost every y in $\mathbb{R}^k$.

PROOF. In dealing with the first assertion it obviously suffices to treat the case where $\mathcal{H}^k(A) < \infty$. Fix $r > 0$ and let E_p' designate the set of points y in E_p with the following property: there are points $x_1, x_2, \ldots, x_p$ in $A \cap f^{-1}(\{y\})$ satisfying $|x_j - x_k| \ge r$ for $j \ne k$.

We start by showing that $p\,m_k^*(E_p') \le \lambda^k \mathcal{H}^k(A)$. Given $\varepsilon > 0$, we choose a countable covering $\{C_\nu\}$ of A by compact sets C_ν with $\operatorname{diam}(C_\nu) < r$ such that

$$\alpha(k)2^{-k} \sum_\nu \operatorname{diam}(C_\nu)^k \le \mathcal{H}_r^k(A) + \varepsilon \le \mathcal{H}^k(A) + \varepsilon\,.$$

Setting $C_\nu' = f(C_\nu)$ and $g = \sum_\nu \chi_{C_\nu'}$, we let $F_p = \{y \in \mathbb{R}^k : g(y) \ge p\}$ and assert that E_p' is contained in F_p. Given an element $y \in E_p'$, we can select points $x_1, x_2, \ldots, x_p$ in $A \cap f^{-1}(\{y\})$ such that $|x_j - x_k| \ge r$ for $j \ne k$. Since $\operatorname{diam}(C_\nu) < r$, at most one of these points lies in C_ν for any given ν. Because A is included in $\bigcup_\nu C_\nu$, at least p different members of $\{C_\nu\}$ must contain points of $A \cap f^{-1}(\{y\})$. This forces y into C_ν' for at least p distinct values of ν. This ensures that $g(y) \ge p$, making $y \in F_p$.

Thus E_p' is a subset of F_p. As F_p is plainly a Borel set, we get

$$\begin{aligned} p\,m_k^*(E_p') &\le p\,m_k(F_p) = \int_{\mathbb{R}^n} p\chi_{F_p}\,dm_k \le \int_{\mathbb{R}^k} g\,dm_k \le \sum_\nu m_k(C_\nu') \\ &\le \alpha(k)2^{-k}\sum_\nu \operatorname{diam}(C_\nu')^k \le \alpha(k)2^{-k}\lambda^k \sum_\nu \operatorname{diam}(C_\nu)^k \\ &\le \lambda^k\mathcal{H}^k(A) + \lambda^k\varepsilon\,. \end{aligned}$$

In the middle of this computation we have appealed to the "isodiametric inequality". This classical result is discussed in Gruber's book [**67**] and states (in our notation) that

$$m_k^*(S) \le \alpha(k)\big[\operatorname{diam}(S)/2\big]^k\,, \tag{6.29}$$

true for any set S in $\mathbb{R}^k$.

Letting $\varepsilon \to 0$, we obtain $p\,m_k^*(E_p') \le \lambda^k\mathcal{H}^k(A)$.

Next, for $\ell = 1, 2, \ldots$ we denote by $E_{p,\ell}$ the set E_p' obtained by taking $r = \ell^{-1}$ in the foregoing argument. Obviously $E_{p,1} \subset E_{p,2} \subset E_{p,3} \subset \cdots$ and $E_p = \bigcup_{\ell=1}^\infty E_{p,\ell}$. Based on the above considerations, we can say that

$$p\,m_k^*(E_p) = p\lim_{\ell\to\infty} m_k^*(E_{p,\ell}) \le \lambda^k\mathcal{H}^k(A)\,.$$

Finally, $E_\infty = \{y \in \mathbb{R}^k : A \cap f^{-1}(\{y\}) \text{ is an infinite set}\}$ is a subset of E_p for every $p \ge 1$. Assuming that $\mathcal{H}^k(A) < \infty$, we infer

$$m_k^*(E_\infty) \le \liminf_{p\to\infty} m_k^*(E_p) \le \lim_{p\to\infty} \frac{\lambda^k\mathcal{H}^k(A)}{p} = 0\;;$$

that is, $A \cap f^{-1}(\{y\})$ is a finite set for almost every y in $\mathbb{R}^k$.

Under the weaker assumption that A is a countable union of sets of finite $\mathcal{H}^k$-measure, we can conclude that for almost every y in $\mathbb{R}^k$ the set $A \cap f^{-1}(\{y\})$ is countable. □

For $p > 1$ we extend Hausdorff outer measure $\mathcal{H}^p$ on $\mathbb{R}^n$ to an outer measure on $\hat{\mathbb{R}}^n$ in the trivial fashion; namely, we define $\mathcal{H}^p(A) = \mathcal{H}^p(A \cap \mathbb{R}^n)$ for each subset A of $\hat{\mathbb{R}}^n$. In particular, the measures m_n for $n \ge 1$ and σ_{n-1} for $n \ge 2$, which can be obtained by completing the restrictions of $\mathcal{H}^n$ and $\mathcal{H}^{n-1}$, respectively, to the σ-algebra of Borel sets in $\mathbb{R}^n$, find themselves extended to measures on the obvious σ-algebras of sets in $\hat{\mathbb{R}}^n$.

6.4. The analytic characterisation of quasiconformality

In this section we give the analytic characterisation of a quasiconformal mapping—via the linear dilatation. One can take this characterisation as an alternative definition of a quasiconformal mapping, and we therefore should, and will subsequently, establish the equivalence between this definition and our earlier definition through the modulus of curve families.

6.4.1. Linear dilatation and the ACL-property. Let $f : D \to D'$ be a homeomorphism between domains D and D' in $\hat{\mathbb{R}}^n$ with $n \geq 2$. We recall from the final pages of our first chapter that the linear dilatation of f at a point x of $D_0 = \{x \in D : x \neq \infty,\ f(x) \neq \infty\}$ is the extended real number $H_f(x)$ given by

$$H_f(x) = \limsup_{r \to 0} \frac{L_f(x,r)}{\ell_f(x,r)} ,$$

in which

$$L_f(x,r) = \max_{|h|=r} |f(x+h) - f(x)|, \qquad \ell_f(x,r) = \min_{|h|=r} |f(x+h) - f(x)|$$

for $0 < r < \operatorname{dist}(x, D_0)$.

It was noted at the time that $H_f(x) = H[f'(x)]$ whenever f is differentiable at x and $f'(x)$ is nonsingular. For the sake of completeness we extend the definition of $H_f(x)$ to cover the situation where $x = \infty$ or $f(x) = \infty$ as follows: letting $R_{\mathbb{S}}$ denote the reflection in $\mathbb{S}^{n-1}$, we define

$$H_f(x) = \begin{cases} H_{f \circ R_{\mathbb{S}}}(0) & \text{if } x = \infty,\ f(x) \neq \infty\ ; \\ H_{R_{\mathbb{S}} \circ f}(x) & \text{if } x \neq \infty,\ f(x) = \infty\ ; \\ H_{R_{\mathbb{S}} \circ f \circ R_{\mathbb{S}}}(0) & \text{if } x = f(x) = \infty\,. \end{cases}$$

We shall prove shortly that the linear dilatation of a quasiconformal mapping is a bounded function. This fact is a consequence of the next lemma, which is of some interest in its own right.

LEMMA 6.4.1. *Suppose that $f : D \to D'$ is a homeomorphism between domains D and D' in $\mathbb{R}^n$ with $n \geq 2$ and that $K_O^*(f) \leq K < \infty$. Let U be a bounded open set whose closure lies in D, let x be a point of U, and let*

$$m = \min_{y \in \partial U} |y-x|, \quad M = \max_{y \in \partial U} |y-x|, \quad \ell = \min_{y \in \partial U} |f(y)-f(x)|, \quad L = \max_{y \in \partial U} |f(y)-f(x)|\,.$$

If $\overline{B}^n(f(x), L)$ is contained in D', then

$$L/\ell \leq \Psi_n(M/m)^\alpha \qquad \textit{for } \alpha = K^{1/(n-1)}.$$

Here $\Psi_n : (0, \infty] \to (1, \infty]$ is the function associated with the Teichmüller rings in $\hat{\mathbb{R}}^n$.

PROOF. We may assume that $\ell < L$, for the assertion is trivial otherwise. Let $x' = f(x)$, $R' = B^n(x', L) \setminus \overline{B}^n(x', \ell)$, and $R = f^{-1}(R')$. Then R is a ring whose closure lies in D and whose complementary components are $C_0 = f^{-1}[B^n(x', L)]^c$ and $C_1 = f^{-1}[\overline{B}^n(x', \ell)]$. Choose points y and z in ∂U for which $\ell = |f(y) - f(x)|$ and $L = |f(z) - f(x)|$. Clearly y belongs to C_1, z is a member of C_0, and R separates $\{x, y\}$ from $\{z, \infty\}$. From Theorems 5.4.2 and 5.1.9 we learn that

$$\operatorname{Mod}(R) \leq \log \Psi_n\left(\frac{|z-x|}{|y-x|}\right) \leq \log \Psi_n\left(\frac{M}{m}\right) .$$

By hypothesis $\operatorname{Cap}(R) \le K\operatorname{Cap}(R')$, so

$$\begin{aligned}\log \frac{L}{\ell} &= \operatorname{Mod}(R') = \left[\frac{\omega_{n-1}}{\operatorname{Cap}(R')}\right]^{1/(n-1)} \le \left[\frac{K\omega_{n-1}}{\operatorname{Cap}(R)}\right]^{1/(n-1)} \\ &= K^{1/(n-1)}\operatorname{Mod}(R) \le \alpha \log \Psi_n\left(\frac{M}{m}\right).\end{aligned}$$

The stated upper bound for L/ℓ follows easily from this calculation. □

From Lemma 6.4.1 we draw a significant conclusion.

THEOREM 6.4.2. *Let $f : D \to D'$ be a homeomorphism between domains D and D' in $\hat{\mathbb{R}}^n$ with $n \ge 2$. Suppose that $K_O^*(f) \le K < \infty$; then*

$$H_f(x) \le \Psi_n(1)^\alpha$$

holds for every x in D, where again $\alpha = K^{1/(n-1)}$.

PROOF. Let $D_0 = \{x \in D : x \ne \infty,\ f(x) \ne \infty\}$. Fix x in D. If x belongs to D_0, the desired estimate for $H_f(x)$ is obtained by restricting f to D_0 and applying Lemma 6.4.1: we simply take $U = B^n(x, r)$ for any r satisfying $0 < r < \operatorname{dist}(x, \partial D_0)$ and small enough that $L_f(x, r) < \operatorname{dist}[f(x), \partial f(D_0)]$. In the notation of Lemma 6.4.1 we have $m = M = 1$, $\ell = \ell_f(x, r)$, and $L = L_f(x, r)$ for each such choice of r, which allows us to conclude that

$$H_f(x) = \limsup_{r \to 0} \frac{L_f(x,r)}{\ell_f(x,r)} \le \Psi_n(1)^\alpha .$$

If either $x = \infty$ or $f(x) = \infty$, we merely compose f with the appropriate reflection (or reflections) and then invoke the case dealt with above, always bearing in mind that $K_O^*(f) = K_O^*(g \circ f \circ h)$ whenever g and h are Möbius transformations. □

Theorem 6.4.2 ensures that if a homeomorphism $f : D \to D'$ is quasiconformal, then the linear dilatation of f is unformly bounded in D by a quantity which depends solely on the outer ring distortion.

Our next objective is to establish the converse of this statement by bounding the ring distortions by the linear dilatation.

The first step we take toward this converse is to show that if a homeomorphism $f : D \to D'$ has a bounded linear dilatation, then it enjoys at least a modicum of smoothness—namely, f is absolutely continuous on lines.

We call a homeomorphism $f : D \to D'$ between domains D and D' in $\hat{\mathbb{R}}^n$ with $n \ge 2$ an *ACL-homeomorphism* (resp., ACL^p*-homeomorphism*) provided the restriction of f to the domain $D_0 = \{x \in D : x \ne \infty,\ f(x) \ne \infty\}$ is a member of the class $\mathrm{ACL}(D_0, \mathbb{R}^n)$ (resp., $\mathrm{ACL}^p(D_0, \mathbb{R}^n)$).

Recall that the partial derivatives $D_1 f(x), D_2(f), \ldots, D_n f(x)$ of such a homeomorphism exist at almost every point x of D_0.

A simple covering lemma will prove its worth during the forthcoming discussion.

LEMMA 6.4.3. *Suppose that F is a nonempty compact subset of a line in $\mathbb{R}^n$ and that $\varepsilon > 0$. There exists a $\delta > 0$ with the following property: corresponding to each $r \in (0, \delta)$ there is a finite set of points $x_1, x_2, \ldots, x_p$ in F such that*

(i) *the open balls $B^n(x_j, r)$ cover F,*
(ii) *$|x_j - x_k| \ge |j - k|r$ holds for $1 \le j, k \le p$, and*
(iii) *$pr \le m_1(F) + \varepsilon$.*

PROOF. It plainly suffices to treat the case $n = 1$; that is, we may assume that F is a compact set in the real line. Let $F_r = \{x \in \mathbb{R} : \operatorname{dist}(x, F) < r\}$. Then $F \subset F_r \subset F_s$ when $0 < r < s$ and $F = \bigcap_{r>0} F_r$, so $m_1(F) = \lim_{r\to 0} m_1(F_r)$. We can thus choose $\delta > 0$ with the property that

$$m_1(F_r) < m_1(F) + \varepsilon$$

whenever $0 < r < \delta$. Given such an r, we define $x_1, x_2, \dots$ recursively by $x_1 = \inf F$ and

$$x_{k+1} = \inf(F \setminus \bigcup_{j=1}^{k} I_j), \quad \text{where } I_j = (x_j - r, x_j + r).$$

It is understood that this process is to terminate if and when we arrive at a point x_p for which $F \setminus \bigcup_{j=1}^{p} I_j$ is empty. The compactness of F makes certain that this will indeed occur after finitely many steps. Property (i) is then evident, while the fact that $x_{j+1} \geq x_j + r$ for $j = 1, 2, \dots, p-1$ renders (ii) just as obvious. Finally, the intervals $I'_j = (x_j, x_j + r)$ are pairwise disjoint subsets of F_r, so

$$pr = \sum_{j=1}^{p} m_1(I'_j) = m_1\left(\bigcup_{j=1}^{p} I'_j\right) \leq m_1(F_r) < m_1(F) + \varepsilon ,$$

confirming (iii). □

The next lemma furnishes the mechanism which will enable us to pass from the information that the linear dilatation of a homeomorphism is bounded to the knowledge that it is absolutely continuous on lines. In a sense it is a topological characterisation of the ACL property.

LEMMA 6.4.4. *Let $f : D \to D'$ be a homeomorphism between domains D and D' in $\mathbb{R}^n$ with $n \geq 2$ and let $1 < a < \infty$. Assume that for each x in D there is a constant $d = \operatorname{dist}(x) > 0$ with the following property: whenever $0 < r < d$ there exist a Borel set E in D and a real number $s > 0$ such that*

$$B^n(x, r) \subset E \subset \overline{B}^n(x, ar)$$

and

$$B^n(x', s) \subset E' \subset \overline{B}^n(x', as) ,$$

where $x' = f(x)$ and $E' = f(E)$. Then f is an ACL*-homeomorphism.*

PROOF. Let $Q = [a_1, b_1] \times [a_2, b_2] \times \cdots \times [a_n, b_n]$ be a closed n-interval in D and let $1 \leq i \leq n$. We must establish the existence of a subset H of $p_i(Q)$ with $m_{n-1}[p_i(Q) \setminus H] = 0$ such that for each y in H the path γ_y defined on $[a_i, b_i]$ by $\gamma_y(t) = f(y + te_i)$ is absolutely continuous. As a means to identify such a set H we introduce a Borel measure μ on $\mathbb{R}^{n-1}_i$ as follows:

$$\mu(A) = |f[Q \cap p_i^{-1}(A)]|$$

for each Borel set A in $\mathbb{R}^{n-1}_i$. Then $\mu(\mathbb{R}^{n-1}_i) = |f(Q)| < \infty$.

In view of Theorem 6.3.2 we can be certain that

$$\mu'(y) = \limsup_{r\to 0} \frac{\mu[B^{n-1}(y, r)]}{m_{n-1}[B^{n-1}(y, r)]} = \limsup_{r\to 0} \frac{\mu[\overline{B}^{n-1}(y, r)]}{m_{n-1}[\overline{B}^{n-1}(y, r)]}$$

is finite for almost every y in $\mathbb{R}^{n-1}_i$.

Let H denote the set of points y in $p_i[\text{int}(Q)]$ for which $\mu'(y) < \infty$. Because the boundary of $p_i(Q)$ in $\mathbb{R}_i^{n-1}$ is of m_{n-1}-measure zero, it is clear from the preceding remarks that $m_{n-1}[p_i(Q) \setminus H] = 0$. Fix y in H and write $J = \text{int}(Q) \cap p_i^{-1}(\{y\})$. We shall demonstrate that the estimate

$$\mathcal{H}^1(F')^n \leq c\mu'(y) m_1(F)^{n-1} \tag{6.30}$$

holds with $c = 5\,(2a^2)^n \Omega_{n-1}/\Omega_n$ whenever F is a compact subset of J and $F' = f(F)$. Once this is known, Lemma 6.3.7 can be invoked to certify the absolute continuity of γ_y. However, y was an arbitrary point of H, so the proof of the lemma boils down to the verification of (6.30).

Fix a compact set F in J. For $\ell = 1, 2, \ldots$ let F_ℓ denote the set of points x in F with the ensuing feature: corresponding to each r in $(0, \ell^{-1})$ there exist a Borel set E and a number $s > 0$ such that

$$B^n(x, r) \subset E \subset \overline{B}^n(x, ar) \subset Q$$

and

$$B^n(x', s) \subset E' \subset \overline{B}^n(x', as)\,.$$

We claim that the sets F_ℓ are themselves compact. Momentarily taking this technical fact for granted, we shall prove (6.30) with F replaced by F_ℓ, a set in which the hypotheses of the lemma can be applied somewhat uniformly. Now $F = \bigcup_{\ell=1}^\infty F_\ell$, which implies that $F' = \bigcup_{\ell=1}^\infty F'_\ell$.

Being a compact set, F'_ℓ is $\mathcal{H}^1$-measurable. Also $F'_1 \subset F'_2 \subset F'_3 \subset \cdots$, for $F_1 \subset F_2 \subset F_3 \subset \cdots$. Thus, informed that (6.30) holds with F_ℓ in place of F, we can conclude that

$$\mathcal{H}^1(F')^n = \lim_{\ell\to\infty} \mathcal{H}^1(F'_\ell)^n \leq c\mu'(y) \lim_{\ell\to\infty} m_1(F_\ell)^{n-1} = c\mu'(y) m_1(F)\,.$$

In other words, (6.30) is also true for F.

We may suppose that the set F_ℓ is not empty. Let $\varepsilon > 0$ be given. We select δ in $(0, \ell^{-1})$ so that the conclusions of Lemma 6.4.3 are valid for F_ℓ and ε.

In order to estimate $\mathcal{H}^1(F'_\ell)$ we must first secure a bound for $\mathcal{H}^1_t(F'_\ell)$ when $t > 0$. Fixing t, we choose δ' in $(0, \ell^{-1})$ small enough that $|f(w) - f(z)| < t/2$ whenever z and w are points of Q satisfying $|w - z| \leq a\delta'$. Set $\eta = \min\{\delta, \delta'\}$ and consider r in $(0, \eta)$. Owing to the choice of δ, Lemma 6.4.3 entitles us to pick a finite set of points $x_1, x_2, \ldots, x_p$ in F_ℓ such that the balls $B^n(x_j, r)$ cover F_ℓ, such that $|x_j - x_k| \geq |j - k|r$ whenever $1 \leq j, k \leq p$, and such that $pr \leq m_1(F_\ell) + \varepsilon$. Our selection of δ' forces $L_f(x_j, ar) < t/2$ to hold for each j. We now choose for $j = 1, 2, \ldots, p$ a Borel set E_j and a number $s_j > 0$ for which it is true that

$$B^n(x_j, r) \subset E_j \subset \overline{B}^n(x_j, ar) \subset Q$$

and

$$B^n(x'_j, s_j) \subset E'_j \subset \overline{B}^n(x'_j, as_j)\,.$$

Since $0 < r < \ell^{-1}$, the definition of F_ℓ permits us to do this. The sets E'_j form a cover of F'_ℓ. As E'_j is a subset of $\overline{B}^n[x'_j, L_f(x_j, ar)]$, we have

$$\text{diam}(E'_j) \leq 2L_f(x_j, ar) < t\,.$$

Since E'_j lies in $\overline{B}^n(x'_j, as_j)$, we deduce with the help of Hölder's inequality for sums that

$$\mathbb{H}^1_t(F'_\ell) \leq \sum_{j=1}^p \operatorname{diam}(E'_j) \leq 2a \sum_{j=1}^p s_j \leq 2ap^{(n-1)/n}\left(\sum_{j=1}^p s_j^n\right)^{1/n} .$$

Accordingly,

$$\begin{aligned}\mathbb{H}^1_t(F'_\ell)^n &\leq 2^n a^n p^{n-1} \sum_{j=1}^n s_j^n = \frac{2^n a^{2n-1}\Omega_{n-1}}{\Omega_n} \cdot \frac{(pr)^{n-1}}{\Omega_{n-1}(ar)^{n-1}} \cdot \sum_{j=1}^p \Omega_n s_j^n \\ &\leq \frac{b\Big[\sum_{j=1}^p m_n(E'_j)\Big][m_1(F_\ell)+\varepsilon]^{n-1}}{m_{n-1}[\overline{B}^{n-1}(y, ar)]} ,\end{aligned}$$

where $b = 2^n a^{2n-1}\Omega_{n-1}/\Omega_n$. If a point x happens to be in $E_j \cap E_k$, then x must belong to both $\overline{B}^n(x_j, ar)$ and $\overline{B}^n(x_k, ar)$, so

$$2ar \geq |x_j - x| + |x_k - x| \geq |x_j - x_k| \geq |j-k|r \ ;$$

i.e., $|j-k| \leq 2a$. It follows that any point of $A = \bigcup_{j=1}^p E_j$ lies in E_j for at most $5a$ values of j and hence, that each point of $A' = f(A) = \bigcup_{j=1}^p E'_j$ lies in E'_j for at most $5a$ values of j. As a result,

$$\sum_{j=1}^p m_n(E'_j) \leq 5a\, m_n(A') .$$

Now E_j is contained in $\overline{B}^n(x_j, ar)$, which means that

$$p_i(A) = \bigcup_{j=1}^p p_i(E_j) \subset \bigcup_{j=1}^p p_i\big[\overline{B}^n(x_j, ar)\big] = \overline{B}^{n-1}(y, ar) ,$$

or, stated differently, that A is a subset of the tube $T = Q \cap p_i^{-1}[\overline{B}^{n-1}(y, ar)]$. Thus A' is contained in $T' = f(T)$ and

$$m_n(A') \leq m_n(T') = \mu\big[\overline{B}^{n-1}(y, ar)\big] .$$

These considerations permit us to refine the estimate obtained earlier for $\mathbb{H}^1_t(F')$: for any r in $(0, \eta)$ we are now assured that

$$\mathbb{H}^1_t(F'_\ell)^n \leq \frac{c\mu[\overline{B}^{n-1}(y, ar)]}{m_{n-1}[\overline{B}^{n-1}(y, ar)]}\ \big[m_1(F_\ell) + \varepsilon\big]^{n-1} ,$$

with $c = 5ab$. Taking the limit superior as $r \to 0$ we arrive at

$$\mathbb{H}^1_t(F'_\ell)^n \leq c\mu'(y)\big[m_1(F_\ell) + \varepsilon\big]^{n-1} ,$$

an inequality which is valid for every $\varepsilon > 0$ and $t > 0$. Passing to the limits as $\varepsilon \to 0$ and $t \to 0$, we see that (6.30) does indeed hold for F_ℓ and, therefore, for F.

It remains only to address the technical question of whether F_ℓ is compact. What needs to be verified is that F_ℓ is closed.

Let $\langle z_\nu \rangle$ be a sequence in F_ℓ such that $z_\nu \to z_0$. Certainly z_0 belongs to F. We want z_0 in F_ℓ. Given r in $(0, \ell^{-1})$, we select for each ν a Borel set E_ν and a number $s_\nu > 0$ such that

$$B^n(z_\nu, r) \subset E_\nu \subset \overline{B}^n(z_\nu, ar) \subset Q$$

and

$$B^n(z'_\nu, s_\nu) \subset E'_\nu \subset \overline{B}^n(z'_\nu, as_\nu).$$

Let $E_0 = \bigcap_{\lambda=1}^{\infty}(\bigcup_{\nu=\lambda}^{\infty} E_\nu)$, a Borel set, and $s_0 = \liminf_{\nu\to\infty} s_\nu$. Replacing $\langle z_\nu \rangle$ by a subsequence, if need be, we may presume that $s_\nu \to s_0$. Since $B^n(z'_\nu, s_\nu)$ lies in $f(Q)$, it is evident that $s_0 < \infty$. The uniform continuity of f^{-1} on $f(Q)$, together with the fact that $\operatorname{diam}(E_\nu) \geq 2r$, implies that $2as_\nu \geq \inf_\nu \operatorname{diam}(E'_\nu) > 0$ for every ν, so $s_0 > 0$. If $0 < \varepsilon < r$, then

$$B^n(z_0, r-\varepsilon) \subset B^n(z_\nu, r) \subset E_\nu \subset \overline{B}^n(z_\nu, ar) \subset B^n(z_0, ar+\varepsilon)$$

for all sufficiently large ν, which makes it clear that

$$B^n(z_0, r-\varepsilon) \subset E_0 \subset \overline{B}^n(z_0, ar+\varepsilon)$$

for every such ε. Therefore

$$B^n(z_0, r) \subset E_0 \subset \overline{B}^n(z_0, ar).$$

The fact that $z_\nu \to z_0$ and $\overline{B}^n(z_\nu, ar)$ lies in Q for every ν ensures that $\overline{B}^n(z_0, ar)$ is likewise contained in Q. An argument similar to the one above reveals that

$$B^n(z'_0, s_0) \subset E'_0 \subset \overline{B}^n(z'_0, as_0).$$

We have thus demonstrated that z_0 belongs to F_ℓ. The proof of the lemma is complete. □

Lemma 6.4.4 can be brought into play when certain information about the linear dilatation of f or its inverse is at hand.

THEOREM 6.4.5. *Let $f : D \to D'$ be a homeomorphism between domains D and D' in $\mathbb{R}^n$ with $n \geq 2$ and let $g = f^{-1}$. Assume the existence of a constant H such that for each x in D either $H_f(x) \leq H$ or $H_g(x') \leq H$, where $x' = f(x)$. Then f is an* ACL*-homeomorphism.*

PROOF. Fix x in D. Suppose first that $H_f(x) \leq H$. Choose $d > 0$ with the property that $L_f(x, 2r) < 2H\ell_f(x, 2r)$ whenever $0 < r < d$. For each such r take $E = B^n(x, 2r)$ and $s = \ell_f(x, 2r)/2$. Obviously

$$B^n(x, r) \subset E \subset \overline{B}^n(x, 4Hr),$$

while the fact that

$$s < \ell_f(x, 2r) \leq L_f(x, 2r) \leq 2H\ell_f(x, 2r) = 4Hs$$

implies that $E' = f(E)$ satisfies

$$B^n(x', s) \subset E' \subset \overline{B}^n(x', 4Hs).$$

Next, assume that $H_g(x') \leq H$. In this case, choose $d' > 0$ so that

$$L_g(x', 2s) \leq 2H\ell_g(x', 2s)$$

whenever $0 < s < d'$. The function $s \mapsto \ell_g(x', 2s)/2$ is continuous and positive on the interval $(0, d')$, which it maps to $(0, d)$ for some $d > 0$. Given r in $(0, d)$, we pick any s in $(0, d')$ for which $r = \ell_g(x', 2s)/2$ and set $E = g[B^n(x', 2s)]$. Because

$$r < \ell_g(x', 2s) \leq L_g(x', 2s) \leq 2H\ell_g(x', 2s) = 4Hr,$$

we find that

$$B^n(x, r) \subset E \subset \overline{B}^n(x, 4Hr),$$

while it is evident that $E' = f(E) = B^n(x', 2s)$ satisfies

$$B^n(x', s) \subset E' \subset \overline{B}^n(x', 4Hs)\,.$$

The mapping f is thus seen to satisfy the hypotheses of Lemma 6.4.4 with $a = 4H$. We infer that f is indeed an ACL-homeomorphism. □

From Theorem 6.4.5 we extract the following corollary.

COROLLARY 6.4.6. *If a homeomorphism $f : D \to D'$ between domains D and D' in $\hat{\mathbb{R}}^n$ with $n \geq 2$ has a bounded linear dilatation, then both f and f^{-1} are* ACL-*homeomorphisms.*

By making a few simple observations we can improve upon Corollary 6.4.6 in a substantial way. We absorb these observations into a lemma.

LEMMA 6.4.7. *Let $f : D \to D'$ be a homeomorphism between domains D and D' in $\mathbb{R}^n$ with $n \geq 2$ and let $g = f^{-1}$. If x_0 is a point of D for which $H_f(x_0) < \infty$, then*

$$L_f(x_0)^n \leq H_f(x_0)^n \mu_f'(x_0) \tag{6.31}$$

and

$$L_g(y_0)^n \leq H_f(x_0)^n \mu_g'(y_0)\,, \tag{6.32}$$

where $y_0 = f(x_0)$.

If H_f is finite throughout D, then f is differentiable almost everywhere in D and g is differentiable almost everywhere in D'.

If H_f is a bounded function in D, then the partial derivatives of f are locally L^n-functions in D, while the partials of g are locally L^n-functions in D'.

PROOF. In verifying (6.31) we may assume that $\mu_f'(x_0) < \infty$. For $0 < r < \operatorname{dist}(x_0, \partial D)$ write $\ell = \ell_f(x_0, r)$, $L = L_f(x_0, r)$, and $B = B^n(x_0, r)$.

Since $B^n(y_0, \ell)$ is contained in $f(B)$, we have $\Omega_n \ell^n \leq \mu_f(B)$, whence

$$\begin{aligned} L_f(x_0)^n &= \limsup_{r\to 0} \left(\frac{L}{r}\right)^n = \limsup_{r\to 0} \left[\left(\frac{L}{\ell}\right)^n \frac{\Omega_n \ell^n}{\Omega_n r^n}\right] \\ &\leq \limsup_{r\to 0} \left(\frac{L}{\ell}\right)^n \limsup_{r\to 0} \frac{\mu_f(B)}{|B|} = H_f(x_0)^n \mu_f'(x_0)\,. \end{aligned}$$

We next turn to (6.32), assuming as we may that $\mu_g'(y_0) < \infty$.

Fix $d > 0$ with $0 < d < d(y_0, \partial D)$ so that $L_g(y_0, s) < \operatorname{dist}(x_0, \partial D)$ whenever $0 < s < d$. For such s write $r = L_g(y_0, s)$ and $L = L_f(x_0, r)$. It is not difficult to see that $\ell = \ell_f(x_0, r) = s$ and, modulo the constraint $L < d(y_0, \partial D)$, that $\ell_g(y_0, L) = r$. The latter is true for all sufficiently small s, for both r and L tend to zero when s does. If s is suitably small, therefore, and if $B' = B^n(y_0, L)$, then $g(B')$ contains $B^n(x_0, r)$ and $\Omega_n r^n \leq \mu_g(B')$. It follows that

$$\begin{aligned} L_g(y_0)^n &= \limsup_{s\to 0} \left(\frac{r}{s}\right)^n = \limsup_{s\to 0} \left[\left(\frac{L}{s}\right)^n \frac{\Omega_n r^n}{\Omega_n L^n}\right] \\ &\leq \limsup_{s\to 0} \left(\frac{L}{\ell}\right)^n \limsup_{s\to 0} \frac{\mu_g(B')}{|B'|} \leq H_f(x_0)^n \mu_g'(y_0)\,, \end{aligned}$$

which confirms (6.32).

If $H_f(x) < \infty$ everywhere in D, then Theorem 6.3.2, (6.31), and (6.32) make certain that $L_f(x) < \infty$ for almost every x in D and that $L_g(y) < \infty$ for almost every y in D', and hence, by way of the Rademacher and Stepanov theorem, that f is differentiable almost everywhere in D and g is differentiable almost everywhere in D'. Moreover, if it is the case that $H_f(x) \leq H < \infty$ for every point x of D, then for any compact set F in D and for $i = 1, 2, \ldots, n$ we have

$$\begin{aligned}\int_F |D_i f|^n \, dm_n &\leq \int_F L_f^n \, dm_n \leq \int_F H_f^n \mu_f' \, dm_n \leq H^n \int_F \mu_f' \, dm_n \\ &\leq H^n |f(F)| < \infty,\end{aligned}$$

which certifies that $D_i f$ belongs to $L^n(F)$.

A similar calculation based on (6.32) would reveal that the partial derivatives of g are locally L^n-functions in the domain D', should H_f be bounded in D. □

We combine Lemma 6.4.7 with Corollary 6.4.6 to arrive at the following theorem.

THEOREM 6.4.8. *Suppose that a homeomorphism $f : D \to D'$ between domains D and D' in $\mathbb{R}^n$ with $n \geq 2$ has a bounded linear dilatation. Then both f and f^{-1} are* ACL^n*-homeomorphisms. Furthermore, f is differentiable almost everywhere in D, with the inequality*

$$\|f'(x)\|^n \leq H_f(x)^n |J_f(x)| \tag{6.33}$$

holding at any point x of D where f is differentiable; $g = f^{-1}$ is differentiable at almost every point of D', and the estimate

$$\|g'(y)\|^n \leq H_f\big[g(y)\big]^n |J_g(y)| \tag{6.34}$$

holds at each point y of D' where g is differentiable.

One conclusion which can be drawn from Theorems 6.4.2 and 6.4.8 is that any quasiconformal mapping between domains in $\hat{\mathbb{R}}^n$ is an ACL^n-homeomorphism. Actually, a much stronger statement is true: every K-quasiconformal mapping between domains in $\hat{\mathbb{R}}^n$ is an ACL^p-homeomorphism for some $p = p(n, K) > n$. Confirmation of this fact will have to wait until later in this book.

At this point we are not far removed from an analytic characterisation of quasiconformal mappings. The next result supplies much of the information needed to bridge the remaining gaps.

THEOREM 6.4.9. *Suppose that $f : D \to D'$ is a homeomorphism between domains D and D' in $\mathbb{R}^n$ with $n \geq 2$ and that $1 \leq K < \infty$. Then $K_O(f) \leq K$ if and only if f is an* ACL*-homeomorphism that is differentiable almost everywhere in D and satisfies the inequality*

$$\|f'(x)\|^n \leq K |J_f(x)| \quad \textit{at almost every point } x \textit{ of } D. \tag{6.35}$$

We remark in passing that the differential inequality (6.35) is often taken to be an alternate definition of quasiconformality once f is assumed to be an ACL-homeomorphism (or that f lies in a related Sobolev class).

PROOF. Suppose first that $K_O(f) \leq K$. Then $K_O^*(f) \leq K$, so H_f is bounded in D by Theorem 6.4.2.

Theorem 6.4.9 informs us that f is an ACL-homeomorphism—in fact, an ACL^n-homeomorphism—which is differentiable almost everywhere in D. Theorem 6.1.2

lets us know that $\|f'(x)\|^n \le K_O(f)|J_f(x)| \le K|J_f(x)|$ at any point x of D where f is differentiable.

Turning to the converse, assume that f is an ACL-homeomorphism, that f is differentiable almost everywhere in D, and that the differential inequality

$$\|f'(x)\|^n \le K|J_f(x)| = K\mu_f'(x)$$

holds for almost every x in D. Then for every compact set F in D and for $i = 1, 2, \ldots, n$ we have

$$\int_F |D_i f|^n \, dm_n \le \int_F \|f'\|^n \, dm_n \le K \int_F \mu_f' \, dm_n \le K|f(F)| < \infty \, ,$$

which shows that f is an ACL^n-homeomorphism.

We shall demonstrate that $K_O(f) \le K$. Let Γ be a family of curves in D and set $\widetilde{\Gamma} = f(\Gamma)$. We want to verify that $M(\Gamma) \le KM(\widetilde{\Gamma})$. For this we may assume that $M(\widetilde{\Gamma}) < \infty$. Theorem 4.5.5 (applied to the coordinate functions of f) informs us that Γ_0, the family of curves in D on which f fails to be locally absolutely continuous, has $M(\Gamma_0) = 0$. Therefore $M(\Gamma) = M(\Gamma \setminus \Gamma_0)$. Take an admissible density $\tilde{\rho}$ for $\widetilde{\Gamma}$, which we shall suppose to be in $L^n(\mathbb{R}^n)$, and define $\rho : \hat{\mathbb{R}}^n \to [0, \infty]$ by

$$\rho(x) = \begin{cases} \tilde{\rho}[f(x)]L_f(x) & \text{if } x \in D \, , \\ 0 & \text{if } x \notin D \, , \end{cases}$$

subject to the convention that $0 \cdot \infty = \infty \cdot 0 = 0$. If γ is a locally rectifiable member of the family $\Gamma \setminus \Gamma_0$, then $f \circ \gamma$ is a locally rectifiable curve from $\widetilde{\Gamma}$, so by Theorem 4.1.6

$$\int_\gamma \rho \, ds = \int_\gamma (\tilde{\rho} \circ f) L_f \, ds \ge \int_{f \circ \gamma} \tilde{\rho} \, ds \ge 1 \, .$$

Accordingly, ρ finds itself in $\mathrm{Adm}(\Gamma \setminus \Gamma_0)$. Now $L_f(x) = \|f'(x)\|$ at any point x of D where f is differentiable, hence we draw from (6.26) the inference that

$$\begin{aligned} M(\Gamma) &= M(\Gamma \setminus \Gamma_0) \le \int_{\mathbb{R}^n} \rho^n \, dm_n = \int_D (\tilde{\rho} \circ f)^n L_f^n \, dm_n \\ &\le K \int_D (\tilde{\rho} \circ f)^n |J_f| \, dm = K \int_D (\tilde{\rho} \circ f)^n \mu_f' \, dm_n \le K \int_{D'} \tilde{\rho}^n \, dm_n \\ &\le K \int_{\mathbb{R}^n} \tilde{\rho}^n \, dm_n \, . \end{aligned}$$

Taking the infimum over $\tilde{\rho}$ yields $M(\Gamma) \le KM(\widetilde{\Gamma})$. We have thereby established the claim that $K_O(f) \le K$. □

We are now in a position to prove a result which we suggested earlier.

THEOREM 6.4.10. *A homeomorphism $f : D \to D'$ between domains D and D' in $\hat{\mathbb{R}}^n$ with $n \ge 2$ is a quasiconformal mapping if and only if its linear dilatation H_f is bounded in D.*

PROOF. If f is a quasiconformal mapping, then Theorem 6.4.2 guarantees the boundedness of H_f in D. Conversely, if $H = \sup_D H_f < \infty$, then according to Theorem 6.4.8 the mapping f is an ACL-homeomorphism that is differentiable almost everywhere in the domain $D_0 = \{x \in D : x \ne \infty, \, f(x) \ne \infty\}$ and satisfies

$$\|f'(x)\|^n \le H_f(x)^n |J_f(x)| \le H^n |J_f(x)|$$

for almost all x in D_0. Theorem 6.4.9 ensures that $K_O(f) = K_O(f|D_0) \leq H^n$. Theorem 6.4.8 goes on to declare that $g = f^{-1}$ is an ACL-homeomorphism, that g is differentiable almost everywhere in $D_0' = f(D_0)$, and that

$$\|g'(y)\|^n \leq H_f\left[g(y)\right]^n |J_g(y)| \leq H^n |J_g(y)|$$

is true for almost every y in D_0'. This means that $K_I(f) = K_O(g) \leq H^n$. Therefore $K(f) \leq H^n$ and we conclude that f is an H^n-quasiconformal mapping. □

We can use Theorem 6.4.10 to obtain a little extra information.

COROLLARY 6.4.11. *If $f : D \to D'$ is a homeomorphism between domains D and D' in $\hat{\mathbb{R}}^n$ with $n \geq 2$, then either all of the dilatations $K_I(f)$, $K_O(f)$, $K_I^*(f)$, and $K_O^*(f)$ are finite or none of them is.*

PROOF. If any of the dilatations above are finite, then either H_f or $H_{f^{-1}}$ is bounded by Theorem 6.4.2. In either case Theorem 6.4.10 tells us that f is a quasiconformal mapping, which subsequently implies that all four dilatations are finite. □

While the following theorem is obvious from the preceding results, it is worth pointing out separately.

THEOREM 6.4.12. *Let $f : D \to D'$ be a homeomorphism between domains D and D' in $\hat{\mathbb{R}}^n$ with $n \geq 2$, and let $g = f^{-1} : D' \to D$. Then f is quasiconformal if and only if g is quasiconformal. Further,*

$$K(f) = K(g).$$

6.4.2. The analytic characterisation of quasiconformality. On the basis of Theorems 6.4.10 and 6.4.8 we can assert that a quasiconformal mapping is differentiable almost everywhere in its domain. We now strengthen this statement by showing that the derivative of such a mapping is nonsingular almost everywhere in its domain. We remark at the outset that this feature—the existence of a nonsingular derivative almost everywhere—of quasiconformal mappings is only true in dimension $n \geq 2$. It is not true of quasisymmetric homeomorphisms of the line or of the circle. This fact has remarkable consequences in higher-dimensional topology and geometry and underpins Mostow's rigidity theorem.

THEOREM 6.4.13. *If $f : D \to D'$ is a quasiconformal mapping between domains D and D' in $\mathbb{R}^n$, with $n \geq 2$, then $\mu_f'(x) > 0$ for almost every x in D. In particular, $f'(x)$ is nonsingular for almost every x in D.*

PROOF. Let $E = \{x \in D : \mu_f'(x) = 0\}$. Because μ_f' is a Lebesgue measurable function, E is a Lebesgue measurable set. We maintain that $|E| = 0$. Suppose otherwise. Then some element of E is a point of density for E. To keep the notation simple we shall assume that the origin is such a point. Let $r > 0$ be sufficiently small such that the closed n-interval

$$Q = Q_r = [-r, r] \times [-r, r] \times \cdots \times [-r, r] \subset D.$$

Consider the curve family $\Gamma = \{\gamma_y : y \in p_n(Q)\}$, where $\gamma_y(t) = y + te_n$ for $-r \leq t \leq r$. Now $M(\Gamma) = 1$ (recall the argument in Theorem 4.2.3), whence

$M[f(\Gamma)] \leq K_I(f) < \infty$. Let $\tilde{\rho}$ in $L^n(\mathbb{R}^n)$ be an admissible density for $f(\Gamma)$. We define $\rho : \hat{\mathbb{R}}^n \to [0, \infty]$, as we have on several previous occasions:

$$\rho(x) = \begin{cases} \tilde{\rho}[f(x)]L_f(x) & \text{if } x \in D \,, \\ 0 & \text{if } x \notin D \,. \end{cases}$$

Theorem 6.4.8 implies that f is an ACL^n-homeomorphism which obeys the inequality

$$L_f(x)^n = \|f'(x)\|^n \leq K_O(f)|J_f(x)| = K_O(f)\mu_f'(x)$$

for almost all x in D. Consequently, for m_{n-1}-almost every point y of $p_n(Q)$ it is true that the path $f \circ \gamma_y$ is absolutely continuous and, hence, in view of the fact that γ_y is parametrized by arclength, that f is absolutely continuous on γ_y.

For any such y we have

$$\int_{-r}^{r} \rho(y + te_n)\, dt = \int_{\gamma_y} \rho\, ds = \int_{\gamma_y} (\tilde{\rho} \circ f)L_f\, ds \geq \int_{f \circ \gamma_y} \tilde{\rho}\, ds \geq 1 \,,$$

since $f \circ \gamma_y$ is a rectifiable curve in $f(\Gamma)$. Integrating the foregoing inequality over $p_n(Q)$ and invoking Fubini's theorem, we get

$$\begin{aligned} (2r)^{n-1} &= \int_{p_n(Q)} dm_{n-1} \leq \int_{p_n(Q)} \left[\int_{-r}^{r} \rho(y + te_n)\, dt\right] dm_{n-1}(y) \\ &= \int_Q \rho\, dm_n = \int_{Q \setminus E} \rho\, dm_n \,, \end{aligned}$$

where the last step is justified simply by the fact that $L_f = 0$—and thus that $\rho = 0$—almost everywhere in E.

Hölder's inequality and (6.26) lead to

$$\begin{aligned} |Q|^{n-1} &= (2r)^{n(n-1)} \leq \left(\int_{Q \setminus E} \rho^n\, dm_n\right)\left(\int_{Q \setminus E} dm_n\right)^{n-1} \\ &\leq |Q \setminus E|^{n-1} \int_Q \rho^n\, dm_n = |Q \setminus E|^{n-1} \int_Q (\tilde{\rho} \circ f)^n L_f^n\, dm_n \\ &\leq |Q \setminus E|^{n-1} K_O(f) \int_Q (\tilde{\rho} \circ f)^n \mu_f'\, dm_n \leq |Q \setminus E|^{n-1} K_O(f) \int_{f(Q)} \tilde{\rho}^n\, dm_n \\ &\leq |Q \setminus E|^{n-1} K_O(f) \int_{\mathbb{R}^n} \tilde{\rho}^n\, dm_n \,. \end{aligned}$$

Such being the case for all functions $\tilde{\rho}$ in $\mathrm{Adm}[f(\Gamma)] \cap L^n(\mathbb{R}^n)$, we learn that

$$\begin{aligned} |Q|^{n-1} &\leq |Q \setminus E|^{n-1} K_O(f) M\big[f(\Gamma)\big] \leq |Q \setminus E|^{n-1} K_O(f) K_I(f) M(\Gamma) \\ &= |Q \setminus E|^{n-1} K_O(f) K_I(f) \,, \end{aligned}$$

from which we infer that

$$\frac{|Q \setminus E|}{|Q|} \geq c = \big[K_O(f)K_I(f)\big]^{1/(1-n)} > 0 \,.$$

Noting that $B^n(r) \subset Q_r \subset B^n(r\sqrt{n}\,)$, we deduce from the last inequality that the estimate

$$\frac{|B^n(r\sqrt{n}\,) \setminus E|}{|B^n(r\sqrt{n}\,)|} \geq n^{n/2} c > 0 \tag{6.36}$$

holds for all suitably small $r > 0$. A contradiction ensues, for the fact that the origin is a point of density for E implies that the left-hand side of (6.36) tends to 0 with r. To escape a contradiction we must have $|E| = 0$, which means that μ'_f is positive almost everywhere in D. Finally, $|J_f(x)| = \mu'_f(x)$ at any point x of D where f is differentiable, so $f'(x)$ is nonsingular for almost every x in D. □

Since the radial stretching f of $\hat{\mathbb{R}}^n$ corresponding to a real number α is not differentiable at the origin when $0 < \alpha < 1$ and has $J_f(0) = 0$ when $\alpha > 1$, we see that a quasiconformal mapping may not be differentiable at every point of its domain and need not have a nonsingular derivative at a point where it is differentiable.

We record an important consequence of the last theorem.

THEOREM 6.4.14. *If $f : D \to D'$ is a quasiconformal mapping between domains D and D' in $\mathbb{R}^n$, then both f and f^{-1} have the Lusin property. Consequently, f transforms any Lebesgue measurable set in D to a Lebesgue measurable set, and the change of variable formula*

$$\int_{D'} \rho \, dm_n = \int_D (\rho \circ f)|J_f| \, dm_n$$

is valid for every Lebesgue measurable function $\rho : D' \to [0, \infty]$.

PROOF. Let F be a compact set in D for which $|F| = 0$. We shall verify that the set $F' = f(F)$ has $|F'| = 0$. Writing $g = f^{-1}$, we invoke (6.25) to conclude that

$$\int_{F'} \mu'_g \, dm_n \leq |g(F')| = |F| = 0 \, ,$$

so $\mu'_g = 0$ almost everywhere in F'. From Theorem 6.4.13, applied to the quasiconformal mapping g, it follows that $|F'| = 0$. Therefore f has the Lusin property. Of course, this statement is equally valid for g. The remaining assertions are supported by Theorem 6.3.3. □

In a subsequent chapter we shall be able to improve Theorem 6.4.14 considerably by establishing a measure distortion inequality. We shall demonstrate that there exist constants $a > 0$ and $0 < \alpha < 1$, depending only on n and $K(f)$, such that the inequality

$$\frac{|f(E) \cap B^n(y, s)|}{|B^n(y, s)|} \leq a \left(\frac{|E \cap B^n(x, r)|}{|B^n(x, r)|} \right)^{\alpha}$$

holds whenever x is a point of D, $0 < r < \operatorname{dist}(x, \partial D)$, E is a Lebesgue measurable set in D, $y = f(x)$, and $s = \ell_f(x, r)$.

This implies, for instance, that f maps any point in D which is a point of density for E to a point of density for $f(E)$.

We move closer to an analytic characterisation of quasiconformal mappings with the next result, which is the analogue of Corollary 6.1.3 for general homeomorphisms.

THEOREM 6.4.15. *Let $f : D \to D'$ be a homeomorphism between domains D and D' in $\hat{\mathbb{R}}^n$ with $n \geq 2$. Suppose that*

(i) *f is an* ACL*-homeomorphism,*
(ii) *f is differentiable almost everywhere in D,*
(iii) *$J_f(x) \neq 0$ holds for almost every x in D.*

Then

$$K_I(f) = \mathrm{esssup}_{x\in D} H_I\big[f'(x)\big], \qquad K_O(f) = \mathrm{esssup}_{x\in D} H_O\big[f'(x)\big]\,, \tag{6.37}$$

where esssup *denotes the essential supremum. Moreover, if any of the conditions* (i), (ii) *or* (iii) *is not satisfied, then* $K_I(f) = K_O(f) = \infty$. *In particular,* (6.37) *is true if* f *is a quasiconformal mapping.*

PROOF. Should any of the conditions (i)–(iii) fail to be met, then in light of earlier results f could not be a quasiconformal mapping, so we would have $K_O(f) = K_I(f) = \infty$. Thus, we need only concern ourselves with the situation in which (i)–(iii) hold. Write $H_O = \mathrm{esssup}_D H_O(f')$. If $K_O(f) < \infty$, then Theorem 6.4.9 and condition (iii) imply that

$$\frac{\|f'(x)\|^n}{|J_f(x)|} \leq K_O(f)$$

for almost every x in D. It follows that $H_O \leq K_O(f)$, a trivial assertion when $K_O(f) = \infty$. On the other hand, given that $H_O < \infty$, we see that

$$\|f'(x)\|^n \leq H_O |J_f(x)|$$

holds almost everywhere in D. A second appeal to Theorem 6.4.9 gives $K_O(f) \leq H_O$, an inequality which is trivial when $H_O = \infty$. In all cases, therefore, $K_O(f) = H_O$.

Next, if $K_I(f) < \infty$, then f is a quasiconformal mapping by Corollary 6.4.11. So, too, is $g = f^{-1}$, which is then seen to satisfy the counterparts of conditions (i)–(iii) in D'. Accordingly,

$$K_I(f) = K_O(g) = \mathrm{esssup}_{y\in D'} H_O\big[g'(y)\big]\,.$$

If f is differentiable at x with $J_f(x) \neq 0$, then g is differentiable at $y = f(x)$ and $g'(y) = f'(x)^{-1}$, implying that $H_O[g'(y)] = H_I[f'(x)]$.

Since f and g both have the the Lusin property and since (iii) holds, we conclude that

$$K_I(f) = \mathrm{esssup}_{y\in D'} H_O\big[g'(y)\big] = \mathrm{esssup}_{x\in D} H_I\big[f'(x)\big]\,,$$

at least when $K_I(f) < \infty$. However, if $K_I(f) = \infty$, then also $K_O(f) = \infty$. Thus

$$\mathrm{esssup}_{x\in D} H_I\big[f'(x)\big] \geq \mathrm{esssup}_{x\in D} H_O\big[f'(x)\big]^{1/(n-1)} = K_O(f)^{1/(n-1)} = \infty\,,$$

and we find that (6.37) holds for $K_I(f)$ in this case as well. □

Combining Theorem 6.4.15 with (2.12) we obtain:

COROLLARY 6.4.16. *If* $f : D \to D'$ *is a homeomorphism between domains* D *and* D' *in* $\hat{\mathbb{R}}^n$ *with* $n \geq 2$, *then* $1 \leq K_O(f) \leq K_I(f)^{n-1}$ *and* $1 \leq K_I(f) \leq K_O(f)^{n-1}$.

As a second corollary of Theorem 6.4.15 we have the following which shows the equivalence between the inner and outer ring dilatations and the usual inner and outer dilatations.

COROLLARY 6.4.17. *If* $f : D \to D'$ *is a homeomorphism between domains* D *and* D' *in* $\hat{\mathbb{R}}^n$ *with* $n \geq 2$, *then*

$$K_I^*(f) = K_I(f) \quad \text{and} \quad K_O^*(f) = K_O(f).$$

PROOF. It suffices to demonstrate that $K_O^*(f) = K_O(f)$. The assertion concerning the inner dilatations follows when this fact is applied to f^{-1}. Since $K_O^*(f) \le K_O(f)$ is always true, we must prove that $K_O(f) \le K_O^*(f)$. This is trivially the case if $K_O^*(f) = \infty$. Assuming that $K_O^*(f) < \infty$, we remark that f is a quasiconformal mapping by Corollary 6.4.11. Therefore f has properties (i)–(iii) in the statement of Theorem 6.4.15. Theorem 6.1.4 assures us that $\|f'(x)\|^n < K_O^*(f)|J_f(x)|$ at every point x of D where f is differentiable, hence, almost everywhere in D. From Theorem 6.4.15 it follows that

$$K_O(f) = \operatorname{esssup}_{x\in D} H_O\big[f'(x)\big] = \operatorname{esssup}_{x\in D}\frac{\|f'(x)\|^n}{|J_f(x)|} \le K_O^*(f)\,.$$

Note that property (iii) makes it allowable to divide by $|J_f(x)|$ almost everywhere in D. □

Theorem 6.4.15 implies that the global quantities $K_I(f)$ and $K_O(f)$ are actually shaped locally and in terms of local geometric configurations. This fact assumes some importance in the process of transporting the theory of quasiconformal mappings from $\mathbb{R}^n$ to more general Riemannian manifolds.

The following corollary to Theorem 6.4.15 more precisely exposes this local to global behaviour.

COROLLARY 6.4.18. *Let $f : D \to D'$ be a homeomorphism between domains D and D' in $\hat{\mathbb{R}}^n$ with $n \ge 2$. Assume the existence of finite constants $a > 0$ and $b > 0$ for which the following statement is true: each point x of D has an open, connected neighbourhood $U = U_x$ such that $K_I(f|U) \le a$ and $K_O(f|U) \le b$. Then $K_O(f) \le a$ and $K_I(f) \le b$.*

PROOF. Through an application of Lindelöf's theorem we are able to express D as a countable union $D = \bigcup_\nu D_\nu$, where each D_ν is a domain having the property that $K_I(f|D_\nu) \le a$ and $K_O(f|D_\nu) \le b$. It follows that $f|D_\nu$ is a quasiconformal mapping for each ν. As a result, f is absolutely continuous on lines in D_ν, is differentiable almost everywhere in D_ν, and has a nonsingular derivative almost everywhere in D_ν.

Having emphasised in the process of introducing it that the ACL-property is a local one, we infer that f is an ACL-homeomorphism. Furthermore, since the collection $\{D_\nu\}$ is countable, we see from the remarks above that f is differentiable almost everywhere in D and has $J_f(x) \ne 0$ for almost every x in D.

By Theorem 6.4.15,

$$K_I(f) = \operatorname{esssup}_{x\in D} H_I\big[f'(x)\big] \le \sup_\nu \operatorname{esssup}_{x\in D_\nu} H_I\big[f'(x)\big] = \sup_\nu K_I(f|D_\nu) \le a\,.$$

In an entirely similar manner we see $K_O(f) \le b$. □

As a final corollary of Theorem 6.4.15 we register estimates for the maximal dilatation of a quasiconformal mapping in terms of its linear dilatation.

COROLLARY 6.4.19. *If $f : D \to D'$ is a quasiconformal mapping between domains D and D' in $\hat{\mathbb{R}}^n$, then*

$$\operatorname{esssup}_{x\in D} H_f(x)^{n/2} \le K(f) \le \operatorname{esssup}_{x\in D} H_f(x)^{n-1}\,. \tag{6.38}$$

Furthermore, these bounds for $K(f)$ in terms of H_f are sharp.

PROOF. The stated estimate follows immediately from Theorem 6.4.15 once it is recalled that $H_f(x) = H[f'(x)]$ for every point x of D at which f has a nonsingular derivative and that (see (2.13))

$$H(T)^{n/2} \le \max\{H_O(T),\, H_I(T)\} \le H(T)^{n-1}$$

for any nonsingular linear transformation $T : \mathbb{R}^n \to \mathbb{R}^n$.

The sharpness of the upper estimate is confirmed by considering the linear transformation $f : \mathbb{R}^n \to \mathbb{R}^n$ given by $f(x) = (\lambda x_1, x_2, \ldots, x_n)$ with $\lambda > 1$. It is easily seen that $H_f(x) = \lambda$ for every x in $\mathbb{R}^n$, while $K(f) = \lambda^{n-1}$.

Similarly, if $g(x) = (\lambda x_1, \lambda^{-1}x_2, x_3, \ldots, x_n)$ with $\lambda > 1$, we obtain $H_g(x) = \lambda^2$ for every x in $\mathbb{R}^n$ and $K(g) = \lambda^n$, so the lower bound is also best possible. □

We make the simple observation that when $n = 2$ both $n/2$ and $n-1$ are equal to 1, and so (6.38) reads as

$$K(f) = \operatorname{esssup}_{x\in D} H_f(x).$$

Corollary 6.4.19 leads, amongst other things, to a marginal improvement of Lemma 6.2.3:

if $f : D \to D'$ is a locally λ-bilipschitz homeomorphism between domains D and D' in $\mathbb{R}^n$ with $n \ge 2$, then f is a λ^{2n-2}-quasiconformal mapping.

In this situation it is clear that $H_f(x) \le \lambda^2$ holds for every point x of D.

Putting together various pieces of what we have done so far, we finally arrive at a purely analytic description of a K-quasiconformal mapping.

THEOREM 6.4.20. *Suppose that $f : D \to D'$ is a homeomorphism between domains D and D' in $\hat{\mathbb{R}}^n$ with $n \ge 2$ and that $1 \le K < \infty$. Then f is a K-quasiconformal mapping if and only if*

- *f is an ACL-homeomorphism that is differentiable almost everywhere in D and*
- *f satisfies the distortion inequality*

$$\frac{\|f'(x)\|^n}{K} \le |J_f(x)| \le K\ell\big[f'(x)\big]^n \tag{6.39}$$

at almost every point x of D.

PROOF. Assume first that f is K-quasiconformal. According to Theorem 6.4.9, f is an ACL-homeomorphism that is differentiable almost everywhere in D and obeys the inequality $\|f'(x)\|^n \le K|J_f(x)|$ for each point x of D at which f is differentiable. If $J_f(x) = 0$ for such an x, then the right-hand inequality in (6.39) is trivial. If $J_f(x) \ne 0$, then $g = f^{-1}$ is a K-quasiconformal mapping that has a nonsingular derivative at $y = f(x)$, whence

$$|J_f(x)| = |J_g(y)|^{-1} \le K\|g'(y)\|^{-n} = K\ell\big[f'(x)\big]^n.$$

Conversely, let f be an ACL-homeomorphism that is differentiable almost everywhere in D and satisfies (6.39) at almost every point x of D. Then $K_O(f) \le K$ by Theorem 6.4.9, so f is a quasiconformal mapping by Corollary 6.4.11. In particular, $|J_f(x)| > 0$ almost everywhere in D by Theorem 6.4.13.

Theorem 6.4.15 and the right-hand inequality in (6.39) give $K_I(f) = \operatorname{esssup}_D H_I(f') \le K$. Accordingly, $K(f) = \max\{K_I(f), K_O(f)\} \le K$, showing us that f is indeed a K-quasiconformal mapping. □

We make some additional comments regarding Theorem 6.4.20.

(i) The proof presented for the theorem reveals that the phrase "at almost every point x of D" can be replaced by "at every point x of D where f is differentiable" without rendering the statement invalid.

(ii) It has been shown by Väisälä [**160**] that if the given mapping f is an ACL^n-homeomorphism, then it is necessarily differentiable almost everywhere in D. As a result, such a homeomorphism is K-quasiconformal if and only if (6.39) holds at almost every point of D.

(iii) In dimension $n = 2$ a theorem due to Gehring and Lehto [**50**] asserts that an ACL-homeomorphism is differentiable almost everywhere in its domain. (It is not known whether the corresponding statement is true in higher dimensions.) Furthermore, in the plane we have $K_O(f) = K_I(f)$ for any homeomorphism. Therefore, Theorem 6.4.20 simplifies in dimension $n = 2$ to:

the homeomorphism f is K-quasiconformal if and only if f is an ACL-homeomorphism that satisfies the single inequality $\|Df(x)\|^2 \le K|J_f(x)|$ at almost every point x of D.

We note here that the symbol f' is avoided, lest it be construed to mean the complex derivative.

(iv) Theorem 6.4.20 can be reformulated in a way that makes no explicit reference to differentiability:

the homeomorphism $f : D \to D'$ is K-quasiconformal if and only if f is an ACL-homeomorphism that satisfies $L_f(x)^n/K \le \mu_f'(x) \le K\ell_f(x)^n$ for almost every x in D.

(v) The ACL-condition cannot be removed from the sufficiency part of Theorem 6.4.20; that is, it is not implied by the other conditions.

To see this, let $\varphi : [0,1] \to [0,1]$ be the classical Cantor function, let $D = \{x : 0 < x_i < 1 \text{ for } 1 \le i \le n\}$, and let $D' = \{x : 0 < x_1 < 2, 0 < x_i < 1 \text{ for } 2 \le i \le n\}$. The homeomorphism $f : D \to D'$ defined by $f(x) = x + \varphi(x_1)e_1$ is differentiable and has $f'(x)$ as the identity matrix at almost every point x in D. Thus $\|f'(x)\| = \ell[f'(x)] = |J_f(x)| = 1$ almost everywhere in D. It follows that (6.39) holds with $K = 1$—indeed, equality holds—at almost every point of D. On the other hand, f is not an ACL-homeomorphism: for every y in $p_1(D)$ the function $t \mapsto f(y + te_1) = y + [t + \varphi(t)]e_1$ is not absolutely continuous on any nondegenerate closed subinterval of $(0, 1)$.

Therefore, f is not quasiconformal, much less 1-quasiconformal (which we shall soon see implies that f is Möbius).

6.4.3. Removable singularities and sets. In this section we fine-tune many of the results of the previous section by briefly considering the problem of "removing" singularities of quasiconformal mappings.

The basic setup is as follows: D is a domain in $\hat{\mathbb{R}}^n$ with $n \ge 2$, E is a relatively closed subset of D, and f is a K-quasiconformal mapping of the open set $D \setminus E$

into $\hat{\mathbb{R}}^n$ (meaning that, for each component G of $D \setminus E$, $f|G$ is a K-quasiconformal mapping of G onto $G' = f(G)$). Under these assumptions we pose the question:

Under what conditions on E does there exist a K-quasiconformal mapping f^* of D into $\hat{\mathbb{R}}^n$ such that $f^* = f$ in $D \setminus E$?

We discuss this problem in three situations, each supposing a different level of a priori knowledge about the extendability of f.

In the first instance, we presuppose that f admits a quasiconformal extension to D; in the second, that f is known to have a homeomorphic extension to D; in the third, that no a priori information on the extendability of f is at hand.

For various technical reasons the reader will discover that we formulate the problem somewhat differently when stating the actual results.

THEOREM 6.4.21. *Suppose that $f : D \to D'$ is a quasiconformal mapping between domains D and D' in $\hat{\mathbb{R}}^n$, that E is a relatively closed subset of D, and that each point x of $D \setminus E$ has an open, connected neighbourhood $U = U_x$ in $D \setminus E$ for which $K_I(f|U) \le a$ and $K_O(f|U) \le b$, with a and b being finite constants. If $m_n(E) = 0$, then $K_I(f) \le a$ and $K_O(f) \le b$.*

PROOF. Since f is a quasiconformal mapping, it is an ACL-homeomorphism that is differentiable almost everywhere in D and has a nonsingular derivative almost everywhere in D. If $m_n(E) = 0$, we can employ the same argument used to establish Corollary 6.4.18—this time applying it in $D \setminus E$—to conclude that

$$K_I(f) = \operatorname{esssup}_{x \in D} H_I\big[f'(x)\big] = \operatorname{esssup}_{x \in D\sigma_{\mathbb{S}} \setminus E} H_I\big[f'(x)\big] \le a\,.$$

Similarly, $K_O(f) \le b$. □

When $n = 2$ there is a sense in which Theorem 6.4.21 is a sharp result. Given a domain D in $\mathbb{R}^2$, a relatively closed proper subset E of D with $m_2(E) > 0$ and a constant $c > 0$, it is possible by application of the "measurable Riemann mapping theorem" to construct a quasiconformal mapping f of D into $\hat{\mathbb{R}}^2$ with the property that

$$H_I\big[f'(x)\big] = H_O\big[f'(x)\big] = c\chi_E(x) + 1$$

for almost every x in D.

Accordingly, each point x of $D \setminus E$ has an open, connected neighbourhood $U = U_x$ in $D \setminus E$ such that $K_I(f|U) = K_O(f|U) = 1$, whereas $K_I(f) = K_O(f) = 1 + c$.

Whether Theorem 6.4.21 is sharp in such a sense when $n \ge 3$ is an open question.

In the next result we relax the assumption made about f in Theorem 6.4.21 and hence are forced to impose more severe restrictions on E in order to obtain the same conclusion.

THEOREM 6.4.22. *Suppose that*

- *$f : D \to D'$ is a homeomorphism between domains D and D' in $\hat{\mathbb{R}}^n$ with $n \ge 2$,*
- *E is a relatively closed subset of D,*

- *each point x of $D \setminus E$ has an open, connected neighbourhood $U = U_x$ in $D \setminus E$ for which*

$$K_I(f|U) \le a \quad \text{and} \quad K_O(f|U) \le b$$

for a and b finite constants,
- *and the set E has σ-finite surface area.*

Then $K_I(f) \le a$ and $K_O(f) \le b$.

PROOF. Since E has σ-finite surface area, it follows that $m_n(E) = 0$. We shall prove that f is an ACL-homeomorphism. Once this is known, an easy application of Lindelöf's theorem puts us precisely where we were at the beginning of the proof of Theorem 6.4.21: f is an ACL-homeomorphism that is differentiable, with a nonsingular derivative, at almost every point of D. The reasoning in the earlier proof remains valid here and leads to the stated conclusion. Thus, only the absolute continuity of f on lines in D persists as a question.

Let $Q = [a_1, b_1] \times [a_2, b_2] \times \cdots \times [a_n, b_n]$ be a closed n-interval in D and let $1 \le i \le n$.

Our task is to produce a subset A of $p_i(Q)$ with $m_{n-1}(A) = 0$ so that for every y in $p_i(Q) \setminus A$ the path $\gamma_y : [a_i, b_i] \to \mathbb{R}^n$ given by $\gamma_y(t) = f(y + te_i)$ is absolutely continuous. Because $p_i : \mathbb{R}^n \to \mathbb{R}_i^{n-1}$ is a nonexpansive mapping, we know by virtue of Theorem 6.3.9 that the set B of points y in $p_i(Q)$ for which the set $E \cap p_i^{-1}(\{y\})$ is uncountable has $m_{n-1}(B) = 0$. We are told by Theorem 6.1.2 that for every point x of $D \setminus E$ at which f is differentiable we have (with $U = U_x$ as in the statement of the present theorem)

$$|D_i f(x)|^n \le \|f'(x)\|^n \le K_O(f|U)|J_f(x)| \le b|J_f(x)|\,.$$

Since $m_n(E) = 0$ and since f is differentiable almost everywhere in D, the estimate $|D_i f(x)|^n \le b|J_f(x)|$ holds almost everywhere in D. In conjunction with Hölder's inequality and (6.25) we have that

$$\begin{aligned}
&\int_{p_i(Q)} \left[\int_{a_i}^{b_i} |D_i f(y + te_i)|\, dt\right] dm_{n-1}(y) \\
&\quad = \int_Q |D_i f|\, dm_n \le |Q|^{(n-1)/n} \left[\int_Q |D_i f|^n\, dm_n\right]^{1/n} \\
&\quad \le b^{1/n}|Q|^{(n-1)/n} \left[\int_Q |J_f|\, dm_n\right]^{1/n} = b^{1/n}|Q|^{(n-1)/n} \left[\int_Q \mu_f'\, dm_n\right]^{1/n} \\
&\quad \le b^{1/n}|Q|^{(n-1)/n}|f(Q)|^{1/n} < \infty\,.
\end{aligned}$$

Together with Fubini's theorem, this fact shows that for m_{n-1}-almost every y in $p_i(Q)$ it is true that the derivative $\dot{\gamma}_y(t) = D_i f(y + te_i)$ both exists for almost every t in the interval $[a_i, b_i]$ and satisfies $\int_{a_i}^{b_i} |\dot{\gamma}_y(t)|\, dt < \infty$. Let C designate the set of y in $p_i(Q)$ for which this is not so. Then $m_{n-1}(C) = 0$. Next, choose a countable collection $\{Q_\nu\}$ of closed n-intervals in $D \setminus E$ such that $\{\text{int}(Q_\nu)\}$ covers $Q \setminus E$. As f is absolutely continuous on lines in $D \setminus E$, we can choose for each ν a subset A_ν of $p_i(Q_\nu)$ with $m_{n-1}(A_\nu) = 0$ and with the feature that for each y in $p_i(Q_\nu) \setminus A_\nu$ the analogue of the path γ_y corresponding to Q_ν is absolutely continuous. Set $A = B \cup C \cup (\bigcup_\nu A'_\nu)$, where $A'_\nu = A_\nu \cap p_i(Q)$. Then $m_{n-1}(A) = 0$. We claim that the path γ_y is absolutely continuous for each y in $p_i(Q) \setminus A$.

Fix y in $p_i(Q) \setminus A$ and write $\gamma = \gamma_y$. As y does not belong to B, the set F of all t in $[a_i, b_i]$ for which $y + te_i$ lies in E is a countable, compact set. In particular, $\gamma(F)$ is countable, so $\mathcal{H}^1[\gamma(F)] = 0$. The fact that y is not in C means that $\dot{\gamma}(t)$ exists for almost every t in $[a_i, b_i]$ and that $\int_{a_i}^{b_i} |\dot{\gamma}(t)|\, dt < \infty$. If $[c, d]$ is a closed subinterval of $[a_i, b_i] \setminus F$, we can find a partition $c = t_0 < t_1 < \cdots < t_p = d$ of $[c, d]$ with the property that for $j = 1, 2, \ldots, p$ the set $\{y + te_i : t_{j-1} \leq t \leq t_j\}$ lies in the interior of Q_ν for some $\nu = \nu(j)$. Due to the fact that y is not in $\bigcup_\nu A_\nu$, γ is absolutely continuous on $[t_{j-1}, t_j]$ for each j, and hence absolutely continuous on $[c, d]$. We can now appeal to Lemma 6.3.8 and assert that the path γ is absolutely continuous. We have thereby substantiated our claim. Thus f is an ACL-homeomorphism, so the proof of the theorem is complete. □

The condition on E in Theorem 6.4.22 cannot be weakened appreciably without affecting the conclusion. The following example will make this clear.

Let F be a compact, nowhere dense, perfect subset of $\mathbb{R}$ such that $\mathcal{H}^q(F) = 0$ for every $q > 0$. (Cantor sets of this type are not hard to construct.) As in the case of the classical Cantor middle third set, there is a nonconstant, nondecreasing, continuous function $\varphi : \mathbb{R} \to \mathbb{R}$ which is constant on each component of $\mathbb{R} \setminus F$. Define $f : \mathbb{R}^n \to \mathbb{R}^n$ by $f(x) = (x_1, x_2, \ldots, x_n + \varphi(x_n))$ and set $E = \mathbb{R}^{n-1} \times F$. Then f is a homeomorphism and E is a closed subset of $\mathbb{R}^n$. It can be shown that $\mathcal{H}^p(E) = 0$ whenever p exceeds $n - 1$. Furthermore, each point x of $\mathbb{R}^n \setminus E$ has an open, connected neighbourhood $U = U_x$ in which f takes the form $f(y) = y + c$, where c is a constant vector, and hence for which $K_I(f|U) = K_O(f|U) = 1$.

Nevertheless, $K_O(f) = K_I(f) = \infty$, so f is not quasiconformal, as this mapping is manifestly not an ACL-homeomorphism.

We infer from this example that the conclusion of Theorem 6.4.22 need not be true if we assume only that the Hausdorff dimension $\dim_{\mathcal{H}}(E) \leq n - 1$.

There is another result with close ties to Theorem 6.4.22 that is at least deserving of mention.

Let $f : D \to D'$ be a homeomorphism between domains D and D' in $\hat{\mathbb{R}}^n$ with $n \geq 2$, and let $E = \{x \in D : H_f(x) = \infty\}$. The theorem in question states: *if E has at worst σ-finite surface area and if H_f is essentially bounded in $D \setminus E$, then f is a quasiconformal mapping (and $E = \emptyset$).*

Moreover, it must be the case that f is H^{n-1}-quasiconformal, where H is the essential supremum of H_f by Theorem 6.4.10. The proof entails an argument quite close to those for the proof of Lemma 6.4.4, with Theorem 6.3.9 added to the mix. We do not give the proof here.

We note in passing that Theorem 6.4.22 can be used to introduce into the theory of quasiconformal mappings a counterpart of the classical Schwarz reflection principle found in complex analysis. We present a preliminary version of the result here, and then refine it in the next section.

A domain D in $\hat{\mathbb{R}}^n$ with $n \geq 2$ is said to be *symmetric with respect to a chordal sphere Σ in $\hat{\mathbb{R}}^n$* provided $R_\Sigma(D) = D$. When this is the case, the open set $D \setminus \Sigma$ has exactly two components, call them G and G^*. It is obvious that $R_\Sigma(G) = G^*$, $R_\Sigma(G^*) = G$, and

$$\Sigma \cap \partial G = \Sigma \cap \partial G^* = \Sigma \cap \bar{D}.$$

THEOREM 6.4.23. *Suppose that Σ is a chordal sphere in $\hat{\mathbb{R}}^n$ with $n \geq 2$, that D is a domain which is symmetric with respect to Σ, that G and G^* are the components of $D \setminus \Sigma$, that $E = \Sigma \cap D$, and that $g : G \cup E \to \hat{\mathbb{R}}^n$ is a continuous injection whose restriction to G is a quasiconformal mapping. If $g(E)$ is contained in a chordal sphere Σ' and if $g(G)$ is disjoint from Σ', then the function $f : D \to \hat{\mathbb{R}}^n$ defined by*

$$f(x) = \begin{cases} g(x) & \text{if } x \in G \cup E\ , \\ R_{\Sigma'} \circ g \circ R_{\Sigma}(x) & \text{if } x \in G^* \cup E \end{cases}$$

is a quasiconformal mapping of D with $K_I(f) = K_I(g|G)$ and $K_O(f) = K_O(g|G)$.

PROOF. The function f is a continuous injection of D into $\hat{\mathbb{R}}^n$. By the invariance of domain theorem, f is a homeomorphism of D onto $D' = f(D)$. Trivially $K_I(f|G) = K_I(g|G)$ and $K_O(f|G) = K_O(g|G)$, while $K_I(f|G^*) = K_I(g|G)$ and $K_O(f|G^*) = K_O(g|G)$ because of Lemma 6.1.1 and the fact that Möbius transformations are 1-quasiconformal. The set E is relatively closed in D; as a subset of Σ, E certainly has σ-finite surface area. Theorem 6.4.22 implies that $K_I(f) \leq K_I(g|G)$ and $K_O(f) \leq K_O(g|G)$. Since the reverse inequalities are trivial, the theorem is proved. □

The problem of deciding whether a K-quasiconformal mapping f can be extended across a singular set E becomes, in the absence of any a priori knowledge that f remains at least topologically well behaved up to E, one of considerably greater delicacy than those treated by Theorems 6.4.21 and 6.4.22. We shall be content merely to quote the best result presently known in this direction, a theorem due to Väisälä, referring the reader to [**160**] for a proof.

We shall, however, establish an important special case of Väisälä's result.

THEOREM 6.4.24. *Suppose*

- *that D is a domain in $\hat{\mathbb{R}}^n$ with $n \geq 2$,*
- *that E is a relatively closed subset of D,*
- *that $f : D \setminus E \to \hat{\mathbb{R}}^n$ is a continuous injection and that each point x of $D \setminus E$ has an open, connected neighbourhood $U = U_x$ in $D \setminus E$ for which $K_I(f|U) \leq a$ and $K_O(f|U) \leq b$, with a and b being finite constants, and*
- *that $\sigma_{n-1}(E) = 0$.*

Then f admits a unique extension to a quasiconformal mapping f^* of D into $\hat{\mathbb{R}}^n$, an extension which has $K_I(f^*) \leq a$ and $K_O(f^*) \leq b$.

The special case of Theorem 6.4.24 to which we alluded above (and will prove) asserts that compact sets of capacity zero are removable singular sets for K-quasiconformal mappings.

The proof that we give for this statement puts to effective use a theorem of Gehring dealing with the oscillation of quasiconformal mappings over spheres. This theorem is a natural analogue of a classical oscillation result for plane conformal mappings that goes by the name of "Wolff's lemma".

THEOREM 6.4.25. *If f maps a spherical ring $R = B^n(x_0, b) \setminus \overline{B}^n(x_0, a)$ quasiconformally onto a bounded domain R' in $\mathbb{R}^n$, then*

$$\int_a^b (\operatorname{osc}_{S_r} f)^n \frac{dr}{r} \leq A K_O(f)|R'|\ ,$$

where $S_r = \mathbb{S}^{n-1}(x_0, r)$ *for* $a < r < b$ *and* A *is a constant that depends solely on the dimension* n.

PROOF. Let $f_1, f_2, \ldots, f_n$ be the coordinate functions of f. Each of these functions belongs to the class $\mathrm{ACL}(R)$. Moreover, for almost every x in R we have

$$|\nabla f_i(x)|^n \leq \|f'(x)\|^n \leq K_O(f)|J_f(x)|\,.$$

Hölder's inequality yields the bound

$$(\operatorname{osc}_{S_r} f)^n \leq n^{n-1} \sum_{i=1}^{n} (\operatorname{osc}_{S_r} f_i)^n$$

for $a < r < b$. Theorem 4.5.6 then allows us to conclude that

$$\int_a^b (\operatorname{osc}_{S_r} f)^n \frac{dr}{r} \leq n^{n-1} \sum_{i=1}^{n} \int_a^b (\operatorname{osc}_{S_r} g_i)^n \frac{dr}{r} \leq n^n a_n \int_R \|f'\|^n \, dm_n$$

$$\leq AK_O(f) \int_R |J_f| \, dm_n = AK_O(f)|R'|$$

with $A = n^n a_n$. □

Now suppose that E is a compact set of capacity zero in $\hat{\mathbb{R}}^n$. The reader is reminded that E has Hausdorff dimension zero (which implies, in particular, that $\sigma_{n-1}(E) = 0$). Because it is a totally disconnected set, E also has topological dimension zero and cannot locally separate. That is, if D is a domain in $\hat{\mathbb{R}}^n$, then $D \setminus E$ is still a domain.

THEOREM 6.4.26. *Let* D *be a domain in* $\hat{\mathbb{R}}^n$, *let* E *be a compact set in* D *with* $\mathrm{Cap}(E) = 0$, *and let* $G = D \setminus E$. *If* f *maps* G *quasiconformally onto a domain* G' *in* $\hat{\mathbb{R}}^n$, *and if* $E' = C_f(E)$, *then* f *has an extension to a quasiconformal mapping* f^* *from* D *onto* $D' = G' \cup E'$ *and the dilatations of* f^* *are the same as those of* f.

PROOF. We may assume that ∞ is a point of G, that $f(\infty) = \infty$ and that $B^n(x, 1)$ lies in D for each x in E. If not, we consider in place of f the mapping $h \circ f \circ g$ of the domain $g^{-1}(D) \setminus g^{-1}(E)$ for suitably chosen Möbius transformations g and h.

Observe that $\partial G = E \cup \partial D$. Given an arbitrary point x of E, we shall verify that the cluster set $C_f(x)$ consists of a single point; label it x'. In other words, $\lim_{y \to x} f(y) = x'$. Because the set $\{x\}$ is a boundary component of G and because $B \mapsto B' = C_f(B)$ describes a bijection between the collection of boundary components of G and the collection of boundary components of G', we conclude that $\{x'\}$ is a boundary component of G' and that $x' \neq y'$ when x and y are different points of E. If we now define $f^* : D \to \hat{\mathbb{R}}^n$ by

$$f^*(x) = \begin{cases} f(x) & \text{if } x \in G\,, \\ \lim\limits_{y \to x} f(y) & \text{if } x \in E\,, \end{cases}$$

we obtain an extension of f to a continuous injection of D onto $G' \cup E' = D'$. By the invariance of domain theorem, D' is a domain and f^* is a homeomorphism of D onto D'. As $\sigma_{n-1}(E) = 0$, Theorem 3.21 confirms that f^* is a quasiconfromal mapping whose dilatations agree with the dilatations of f. Accordingly, the crux of the argument lies in showing that $C_f(x)$ reduces to a point whenever x belongs to E.

Fix an arbitrary point x_0 of E. Let $S_r = \mathbb{S}^{n-1}(x_0, r)$, $I = (0,1)$, and $J = \{r \in I : E \cap S_r = \emptyset\}$. We shall learn presently that $\int_J r^{-1}\,dr = \infty$, so J is not empty. Since E is compact, it is clear that J is an open subset of I. As such, J can be expressed as the disjoint union of countably many open intervals (a_ν, b_ν). The domain G thus contains the ring $R_\nu = B^n(x_0, b_\nu) \setminus \overline{B}^n(x_0, a_\nu)$ for each ν. We set $R'_\nu = f(R_\nu)$. If $U = B^n(x_0, 1)$, then $U' = f(G \cap U)$ is a bounded set, for by assumption, $f(G \setminus U)$ includes a neighbourhood of ∞. We deduce from Theorem 6.4.25 that

$$\begin{aligned}\int_J (\operatorname{osc}_{S_r} f)^n \frac{dr}{r} &= \sum_\nu \int_{a_\nu}^{b_\nu} (\operatorname{osc}_{S_r} f)^n \frac{dr}{r} \\ &\le AK_O(f) \sum_\nu |R'_\nu| \le AK_O(f)|U'| < \infty .\end{aligned}$$

Next, because $G \cap B^n(x_0, r) = B(x_0, r) \setminus E$ is a domain for each $r > 0$, it is straightforward to establish the existence of an arc A that, except for having one of its endpoints at x_0, lies in G and that finds its second endpoint on the sphere S_1. The half-open arc $F = A \setminus \{x_0\}$ is contained in G and $F \cap S_r \neq \emptyset$ for every r in I. We claim that the curve family $\Gamma = \Delta(E, F : \hat{\mathbb{R}}^n)$ has $M(\Gamma) = 0$. To see this, represent F in the fashion $F = \bigcup_\nu A_\nu$, where $A_1 \subset A_2 \subset \cdots$ is a sequence of arcs. In light of Lemma 4.3.9 and the definition of a set of capacity zero we obtain

$$M(\Gamma) = \lim_{\nu \to \infty} M\big[\Delta(E, A_\nu : \hat{\mathbb{R}}^n)\big] = 0 ,$$

as maintained. Now consider an arbitrary admissible density ρ for Γ. If r belongs to $I \setminus J$, then the sets $E \cap S_r$ and $F \cap S_r$ are nonempty. Also, ρ is an admissible density for the family $\Delta(E \cap S_r,\ F \cap S_r : S_r)$. Theorem 4.2.20 implies that

$$\int_{S_r} \rho^n \, d\sigma_{n-1} \ge \frac{c_n}{r}$$

for any such r, where $c_n = 2^n b(n,n)$. As a result,

$$c_n \int_{I\setminus J} \frac{dr}{r} \le \int_{I \setminus J} \left[\int_{S_r} \rho^n \, d\sigma_{n-1}\right] dr \le \int_{\mathbb{R}^n} \rho^n \, dm_n$$

whenever ρ is a member of $\operatorname{Adm}(\Gamma)$. It follows that

$$c_n \int_{I\setminus J} \frac{dr}{r} \le M(\Gamma) = 0 ,$$

from which we infer that

$$\int_{I\setminus J} \frac{dr}{r} = 0 .$$

Since $\int_0^t r^{-1}\,dr = \infty$ for every $t > 0$, we have thus arrived at the following situation:

$$\int_J (\operatorname{osc}_{S_r} f)^n \frac{dr}{r} < \infty, \qquad \int_{J_t} \frac{dr}{r} = \infty$$

whenever $0 < t < 1$, where $J_t = J \cap (0,t)$. For this to be possible it is certainly necessary that

$$\liminf_{J \ni r \to 0} d\big[f(S_r)\big] = \liminf_{J \ni r \to 0} \operatorname{osc}_{S_r} f = 0 .$$

Stated differently, J must contain a sequence $\langle r_\nu \rangle$ such that $r_\nu \to 0$ and $\operatorname{diam}(S'_\nu) \to 0$, in which we write $S_\nu = S_{r_\nu}$ and $S'_\nu = f(S_\nu)$. Let $B_\nu = B^n(x_0, r_\nu)$. The sphere

S_ν separates the connected set $G \cap B_\nu$ from ∞. It follows without difficulty that S'_ν separates $f(G \cap B_\nu)$ from $f(\infty) = \infty$; i.e., $f(G \cap B_\nu)$ lies in the bounded component of $\mathbb{R}^n \setminus S'_\nu$. This leads to the conclusion that

$$d\big[\,\overline{f(G \cap B_\nu)}\,\big] \leq \operatorname{diam}(S'_\nu) \to 0$$

as $\nu \to \infty$. But $C_f(x_0)$ lies in the closure of $f(G \cap B_\nu)$ for each ν, hence $\operatorname{diam}[C_f(x_0)] = 0$. Now $C_f(x_0) \neq \emptyset$, so this set consists of exactly one point, as we had hoped. □

An immediate corollary of Theorem 6.4.26 addresses the case of an isolated singularity.

COROLLARY 6.4.27. *Suppose that $f : G \to G'$ is a quasiconformal mapping between domains G and G' in $\hat{\mathbb{R}}^n$ and that x_0 is an isolated boundary point of G. Then $x'_0 = \lim_{x \to x_0} f(x)$ exists, x'_0 is an isolated point of $\partial G'$, and the extension f^* of f to the domain $D = G \cup \{x_0\}$ defined by $f^*(x_0) = x'_0$ is a quasiconformal mapping of D onto the domain $D' = G'_0 \cup \{x'_0\}$ which has the same dilatations as f.*

Corollary 6.4.27 implies that the image of a domain D of the type $D = \hat{\mathbb{R}}^n \setminus \{x_0\}$ under a quasiconformal mapping must be a domain D' of the same sort, $D' = \hat{\mathbb{R}}^n \setminus \{x'_0\}$, as after extending over the point x_0 the image must be both open and closed, and therefore equal to $\hat{\mathbb{R}}^n$. In particular, it shows the impossibility of mapping $\mathbb{R}^n$ quasiconformally to any proper subdomain of itself.

6.5. The boundary behavior of quasiconformal mappings

Suppose that $f : D \to D'$ is a quasiconformal mapping between domains D and D' in $\hat{\mathbb{R}}^n$. In this section we take up the following question, which is prompted by analogy with well-known results on the boundary behavior of conformal mappings in the plane:

What conditions on D and D' ensure that f can be extended to a continuous mapping, perhaps even to a homeomorphism, of $\bar{D}$ onto $\bar{D}'$?

In two dimensions, if D and D' are Jordan domains, that is, domains bounded by the homeomorphic image of a circle, then the classical Carathéodory theorem asserts that any conformal mapping between D and D' extends to a homeomorphism between D and D'. We will see in a moment that this is also true for quasiconformal mappings—in two dimensions one can reduce the problem to the extension problem for quasiconformal self-mappings of the disk via the Riemann mapping theorem.

We would certainly like such a theorem to be true in higher dimensions, but these hopes must go unrealised as there are considerable topological complications. For instance, it is not even true that a domain in $\mathbb{R}^n$ bounded by the homeomorphic image of the $(n-1)$-sphere $\mathbb{S}^{n-1}$ needs to be simply connected—we will construct such a domain bounded by the bilipschitz image of the $(n-1)$–sphere in the next chapter. This certainly presents an obstruction to such a domain being homeomorphic to the ball B^n, let alone quasiconformally equivalent to it.

Thus we must settle for rather less. In fact we treat this problem only under the added constraint that either D or D' is "smoothly bounded", in a sense shortly to be made explicit. Modulo this restriction, we give necessary and sufficient geometric conditions for the existence of a continuous (respectively, homeomorphic) extension of f to the boundary.

The lion's share of what we say here is drawn from work of J. Väisälä [**161, 161**] and R. Näkki [**128, 129**]. We especially wish to direct the reader's attention to Näkki's studies of the boundary behavior of mappings between "non-smooth" domains by transporting Carathéodory's theory of prime ends from the context of conformal mappings in the plane to the setting of quasiconformal mappings in n-space. We do not have the space to go into that here.

6.5.1. Quasiconformally collared domains. A domain D in $\hat{\mathbb{R}}^n$ with $n \geq 2$ is said to be *locally quasiconformally collared* at a point x_0 of ∂D provided x_0 has an open neighbourhood U with the following properties: there exists a homeomorphism φ of $U \cap \bar{D}$ onto $B^n_+ = B^n \cap \bar{\mathbb{H}}^n$ whose restriction to $U \cap D$ is a quasiconformal mapping. Naturally

$$\varphi(U \cap D) = B^n \cap \mathbb{H}^n \quad \text{and} \quad \varphi(U \cap \partial D) = B^{n-1} \subset \mathbb{R}^{n-1} \hookrightarrow \mathbb{R}^n.$$

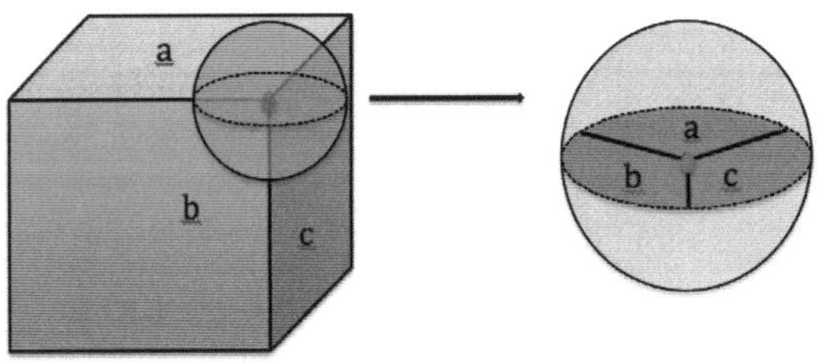

The corner of a cube is quasiconformally collared.

We invariably assume that $\varphi(x_0) = 0$, for if $\varphi(x_0) \neq 0$ we can always replace φ with $h \circ \varphi$, where h is a Möbius transformation that leaves B^n_+ invariant and moves $\varphi(x_0)$ to the origin.

Suppose that W is an arbitrary neighbourhood of x_0. Letting U and φ be as above, including the requirement that $\varphi(x_0) = 0$, we choose r in $(0, 1)$ so that $\varphi^{-1}[B^n_+(r)]$ is a subset of W. Then V and ψ, where $V = \varphi^{-1}[B^n_+(r)] \cup (W \setminus \bar{D}\,)$ and $\psi : V \cap \bar{D} \to \mathbb{R}^n$ is given by $\psi(x) = r^{-1}\varphi(x)$, furnish an open neighbourhood V of x_0 that is contained in W and a homeomorphism ψ of $V \cap \bar{D}$ onto B^n_+ such that $\psi|V \cap D$ is a quasiconformal mapping. Through the choice of W we can arrange to make V as small as we like. In other words, the neighbourhood U that appears in the definition of local quasiconformal collardness at x_0 can be taken arbitrarily small.

A proper subdomain D of $\hat{\mathbb{R}}^n$ that is locally quasiconformally collared at each of its boundary points is called, for brevity's sake, a *quasiconformally collared domain.*

While these domains can be relatively complicated and certainly are not necessarily bounded by smooth codimension 1 submanifolds, these are the domains we had in mind when we spoke earlier of "smoothly bounded" domains.

It is easy to check that a quasiconformally collared domain has only finitely many boundary components, each of which is a compact $(n-1)$-dimensional topological submanifold of $\hat{\mathbb{R}}^n$ without boundary. This is because at each point $x_0 \in \partial D$ with local collar (U, φ) we are provided with local coordinates giving the manifold structure by $(U \cap \bar{D}, \varphi|U \cap \bar{D})$.

The boundary of a quasiconformally collared domain in $\hat{\mathbb{R}}^2$, for instance, consists of finitely many disjoint Jordan curves in $\hat{\mathbb{R}}^2$.

It follows straight from its definition that this notion of being smoothly bounded exhibits a certain degree of invariance under a quasiconformal mapping: if D and D' are domains in $\hat{\mathbb{R}}^n$ and if f is a homeomorphism of $\bar{D}$ onto $\bar{D}'$ that maps D quasiconformally to D', then D is locally quasiconformally collared at a point x_0 of ∂D precisely when D' enjoys this property at $f(x_0)$; hence, the domain D is quasiconformally collared precisely when D' is quasiconformally collared.

The simplest examples of quasiconformally collared domains are Euclidean balls and half-spaces in $\mathbb{R}^n$. By what we have just said, if f is a homeomorphism of $\bar{B}^n$ which is quasiconformal on B^n, then $D = f(B^n)$ is quasiconformally collared.

More generally, a domain D in $\hat{\mathbb{R}}^n$ is quasiconformally collared if ∂D is a nonempty set with at most finitely many components and if each of these components is a chordal sphere. A broader class of quasiconformally collared domains is the class of domains with finitely many boundary components each of which is a quasi-sphere:

A subset S of $\hat{\mathbb{R}}^n$ is termed a *quasi-sphere*—short for *quasiconformal sphere*—if there is a quasiconformal self-mapping f of $\hat{\mathbb{R}}^n$ such that $S = f(\mathbb{S}^{n-1})$; if f is K-quasiconformal, we call S a *K-quasi-sphere*.

The wedges $W(n, \alpha)$ with $0 < \alpha < 2\pi$ and cones $C(n, \alpha)$ with $0 < \alpha < \pi$, which we met in earlier as mappings between wedges and cones (see (6.17)), are quasiconformally collared domains, for it was pointed out at the time that the natural foldings of an open half-space to these domains are, in fact, homeomorphisms of the closed half-space. Indeed the boundaries of such wedges and cones are actually quasi-spheres, though this fact is not immediately obvious.

6.5.2. Lipschitz domains. Suppose that D is a domain in $\hat{\mathbb{R}}^n$ and that x_0 is a finite point of $\partial\bar{D}$. Let there be an open neighbourhood V of x_0 such that $V \cap \partial D$ is up to a Euclidean isometry the graph of a Lipschitz function of the variables $x_1, x_2, \ldots, x_{n-1}$, a statement we interpret to mean that $V \cap \partial D$ can be transformed by a Euclidean isometry σ to a set S that for some $r > 0$ fits the description

$$S = \{x \in \mathbb{R}^n : (x_1, x_2, \ldots, x_{n-1}) \in B^{n-1}(r), \quad x_n = u(x_1, x_2, \ldots, x_{n-1})\},$$

in which $u : B^{n-1}(r) \to \mathbb{R}$ is a Lipschitz function.

The implicit function theorem implies, for example, that these conditions will be met for some V if x_0 has an open neighbourhood W with the property that $W \cap \partial D = \{x \in W : g(x) = 0\}$, where g is a function from the class $C^1(W)$ for which $\nabla g(x_0) \neq 0$.

We can define a homeomorphism f of the cylinder $C = B^{n-1}(r) \times (-\infty, \infty)$ onto itself by $f(x) = x - u(x_1, x_2, \ldots, x_{n-1})e_n$. Then f is plainly an ACL-homeomorphism. Also, since the Rademacher-Stepanov theorem asserts that u is differentiable almost everywhere in $B^{n-1}(r)$ with respect to the measure m_{n-1}, f is seen to be differentiable m_n-almost everywhere in C. Finally, $J_f(x) = 1$ and $\|f'(x)\| \leq \sqrt{2 + \lambda^2}$ for almost every x in C, where λ is a Lipschitz constant for u; see for instance Lemma 6.2.5. Theorem 6.4.9 tells us that $K_O(f) \leq (2 + \lambda^2)^{n/2} < \infty$; that is, f is a quasiconformal mapping of C onto itself, one which maps S to $B^{n-1}(r)$. Let $\psi = f \circ \sigma$ in $\sigma^{-1}(C)$, and let $y_0 = \psi(x_0)$. Choose $s > 0$ so that

$B = B^n(y_0, s)$ lies in C and so that $U = \psi^{-1}(B)$ is contained in V. Since $\psi(U \cap \partial D) = B^{n-1}(y_0, s)$, $\psi^{-1}(B \cap H^n)$ is either a subset of D or a subset of D^c. The same is true of the set $\psi^{-1}(B \setminus \bar{H}^n)$. The fact that x_0 is a point of $\partial \bar{D}$ implies that one of these sets lies in D, and the other in D^c.

Thus ψ maps $U \cap \bar{D}$ to either $B_+ = B \cap \bar{\mathbb{H}}^n$ or $B_- = B \setminus \mathbb{H}^n$.

Defining $\varphi : U \to \mathbb{R}^n$ by $\varphi(x) = s^{-1}[\psi(x) - y_0]$ we obtain a quasiconformal mapping of U onto B^n that transforms $U \cap \bar{D}$ to either B^n_+ or B^n_-. Replacing φ by $-\varphi$, if necessary, we may assume that $\varphi(U \cap \bar{D}) = B^n_+$.

We conclude that D is locally quasiconformally collared at x_0—indeed, it is locally quasiconformally "bi-collared" at x_0, for φ is a quasiconformal mapping of U, not just a homeomorphism of $U \cap \bar{D}$.

A bounded domain D in $\mathbb{R}^n$ with $n \geq 2$ is known as a *Lipschitz domain* if $\partial D = \partial \bar{D}$ and if ∂D can be covered by a finite collection of open sets $V_1, V_2, \ldots, V_p$ with the property that each of the sets $V_j \cap \partial D$ is modulo a Euclidean isometry the graph of a Lipschitz function of $x_1, x_2, \ldots, x_{n-1}$.

The considerations above reveal that every domain of this sort is a quasiconformally collared domain.

THEOREM 6.5.1. *A bounded Lipschitz domain D is quasiconformally collared.*

In particular, the class of quasiconformally collared domains includes any bounded domain D in $\mathbb{R}^n$ for which $\partial D = \partial \bar{D}$ and for which a finite open covering W_1,
$W_2, \ldots, W_q$ of ∂D exists such that $W_j \cap \partial D = \{x \in W_j : g_j(x) = 0\}$, where g_j is a member of $C^1(W_j)$ and $\nabla g_j(x) \neq 0$ holds for every x in $W_j \cap \partial D$.

Thus the class of quasiconformally collared domains encompasses any domain that would by conventional standards be considered smoothly bounded.

If D is a quasiconformally collared domain in $\hat{\mathbb{R}}^n$, then the modulus of a curve family of the type $\Delta(E, F : D)$, where E and F are connected sets in D, behaves qualitatively very much like the modulus of $\Delta(E, F : \hat{\mathbb{R}}^n)$. We document this statement by deriving two general modulus principles for collared domains. The first is captured in the next theorem.

THEOREM 6.5.2. *Suppose that D is a quasiconformally collared domain in $\hat{\mathbb{R}}^n$ and that $0 < r \leq 2$. For each $m > 0$ there exists a constant $s = s(m, r, D) > 0$ with the property that $M[\Delta(E, F : D)] > m$ whenever E and F are connected sets in D satisfying the conditions $q(E) \geq r$, $q(F) \geq r$, and $q(E, F) < s$. In particular, $M[\Delta(E, F : D)] = \infty$ whenever E and F are nondegenerate connected sets in D for which $\bar{E} \cap \bar{F} \neq \emptyset$.*

PROOF. We may assume by subjecting D to a preliminary chordal isometry that ∞ lies in D^c. Let x be an arbitrary point of $\bar{D}$. Given $m > 0$, we shall produce a $\delta_x > 0$ so that the chordal ball $B_x = \{y : q(y, x) < \delta_x\}$ has the following property: $M[\Delta(E, F : D) > m$ for each pair of connected sets E and F in D such that $q(E) \geq r$ and $q(F) \geq r$ and such that both sets have nonempty intersection with B_x.

Naturally, we need only concern ourselves with sets E and F that are disjoint. Suppose first that x belongs to D. Choose a Euclidean ball $B^n(x, b)$ whose closure lies in D and whose chordal diameter is less than r, fix a in $(0, b)$ such that

$c_n \log(b/a) > m$, where c_n is the constant identified in the cap inequality Theorem 4.2.22, and pick $\delta_x > 0$ small enough that B_x lies in $B^n(x,a)$. Let E and F be disjoint connected sets in D with

$$q(E) \geq r, \quad q(F) \geq r, \quad E \cap B_x \neq \emptyset, \quad \text{and} \quad F \cap B_x \neq \emptyset.$$

Then both E and F intersect $\mathbb{S}^{n-1}(x,t)$ for every t in (a,b), so the cap inequality implies that

$$M\bigl[\Delta(E,F:D)\bigr] \geq c_n \log\frac{b}{a} > m\,.$$

Assume next that x belongs to ∂D. Take an open neighbourhood U of x with $q(U) < r$ for which there exists a homeomorphism φ of $U \cap \bar{D}$ onto B^n_+ that maps x to 0 and has $K = K(\varphi|U \cap D) < \infty$. Let a in $(0,1)$ be such that $b_n \log(1/a) > Km$, where b_n is the other constant that surfaced in Theorem 4.2.22, and let $\delta_x > 0$ be such that $B_x \cap \bar{D}$ is a subset of $\varphi^{-1}[B^n_+(a)]$. Consider a disjoint pair of connected sets E and F in D, each of which meets B_x and has chordal diameter at least r. It is clear that both $E' = \varphi(U \cap E)$ and $F' = \varphi(U \cap F)$ have nonempty intersection with the cap

$$C_t = \mathbb{S}^{n-1}(t) \cap \mathbb{H}^n$$

for every t in the interval $(a,1)$. Theorem 4.2.22 assures us that

$$\begin{aligned} M\bigl[\Delta(E,F:D)\bigr] &\geq M\bigl[\Delta(U\cap E, U\cap F : U \cap D)\bigr] \\ &\geq \frac{M[\Delta(E',F':B^n \cap H^n)]}{K} \geq \frac{b_n}{K}\log\frac{1}{a} > m \end{aligned}$$

for any sets E and F of the sort under discussion.

Finally, for each x in $\bar{D}$ set $B'_x = \{y : q(y,x) < \delta_x/2\}$. Extract from the open covering $\{B'_x : x \in \bar{D}\}$ of $\bar{D}$ a finite subcovering—say $B'_1 = B'_{x_1}$, $B'_2 = B'_{x_2}, \dots, B'_p = B'_{x_p}$—and let $s = 2^{-1}\min\{\delta_{x_j} : 1 \leq j \leq p\}$. Given connected subsets E and F of D with $q(E) \geq r$, $q(F) \geq r$, and $q(E,F) < s$, we fix z in E and w in F for which $q(z,w) < s$. Now z belongs to B'_j for some j, which implies that both z and w lie in B_{x_j}. By the manner in which B_{x_j} was selected we have $M[\Delta(E,F:D)] > m$. □

Theorem 6.5.2 generalizes what we already knew to be true about $D = \hat{\mathbb{R}}^n$ (recall the comment following Theorem 4.2.24 and the estimate (5.33) from Corollary 5.4.4). The second modulus principle for quasiconformally collared domains that we wish to cite is more subtle than Theorem 6.5.2. Its treatment requires preparation in the form of several preliminary results. We begin with one that is of independent interest.

THEOREM 6.5.3. *Let D be a domain in $\hat{\mathbb{R}}^n$ with $n \geq 2$. If x_0 and y_0 are points of D, then there exists a quasiconformal mapping f of $\hat{\mathbb{R}}^n$ onto itself such that $f(D) = D$, $f(x_0) = y_0$, and $f(x) = x$ for every x in D^c.*

PROOF. If $D = \hat{\mathbb{R}}^n$, the statement is trivial: there is a Möbius transformation of $\hat{\mathbb{R}}^n$ that maps x_0 to y_0. We proceed assuming that D^c is nonempty. In fact, since the conclusion of the theorem is obviously Möbius invariant, we are free to suppose that ∞ lies in D^c. We first deal with the special case $D = \mathbb{H}^n$. It suffices to show that for each point x_0 of $\mathbb{H}^n$ there is a quasiconformal self-mapping f of $\hat{\mathbb{R}}^n$ which leaves $\mathbb{H}^n$ invariant, transforms e_n to x_0, and fixes every point not in $\mathbb{H}^n$. For this, let $T : \mathbb{R}^n \to \mathbb{R}^n$ be the nonsingular linear transformation that maps e_n to x_0 and fixes e_i when $1 \le i \le n-1$. Clearly $T(\mathbb{H}^n) = \mathbb{H}^n$ and $T(x) = x$ for x belonging to $\mathbb{R}^{n-1}$. Define $f : \hat{\mathbb{R}}^n \to \hat{\mathbb{R}}^n$ by

$$f(x) = \begin{cases} T(x) & \text{if } x \in H^n, \\ x & \text{if } x \notin H^n. \end{cases}$$

Then f is a homeomorphism of $\hat{\mathbb{R}}^n$, $f(\mathbb{H}^n) = \mathbb{H}^n$, $f(e_n) = x_0$, and $f(x) = x$ outside $\mathbb{H}^n$. Because the restrictions of f to both $\mathbb{H}^n$ and $\mathbb{R}^n \setminus \bar{\mathbb{H}}^n$ are quasiconformal, Theorem 6.4.22 shows that f itself is quasiconformal. Accordingly, the theorem is true when $D = \mathbb{H}^n$. In view of the indicated Möbius invariance, the theorem is also true if D is a Euclidean ball in $\mathbb{R}^n$.

Now let D be an arbitrary subdomain of $\mathbb{R}^n$. Given x_0 and y_0 in D, we can construct a finite chain of open Euclidean balls $B_0, B_1, \ldots, B_p$ whose closures lie in D such that x_0 belongs to B_0, y_0 belongs to B_p, and $B_{j-1} \cap B_j$ is nonempty when $1 \le j \le p$. For each j in this range choose a point x_j in $B_{j-1} \cap B_j$, and set $x_{p+1} = y_0$. We know from what was said above that there exist quasiconformal self-mappings $f_0, f_1, \ldots, f_p$ of $\hat{\mathbb{R}}^n$ such that $f_j(B_j) = B_j$, $f_j(x_j) = x_{j+1}$, and $f_j(x) = x$ for every x in B_j^c. These conditions imply that $f_j(D) = D$ and $f_j(x) = x$ for every x in D^c. It follows that the mapping $f = f_p \circ f_{p-1} \circ \cdots \circ f_0$ meets all the demands of the theorem. □

On the strength of Corollary 5.4.4 we can assert that $M[\Delta(E, F : \hat{\mathbb{R}}^n)] > 0$ whenever E and F are continua in $\hat{\mathbb{R}}^n$. Theorem 6.5.3 leads to a generalisation of this fact.

THEOREM 6.5.4. *Suppose that E and F are continua in a subdomain D of $\hat{\mathbb{R}}^n$ with $n \ge 2$. Then $M[\Delta(E, F : D)] > 0$.*

PROOF. By what was just remarked, we only have to worry about a proper subdomain D of $\hat{\mathbb{R}}^n$. In fact, due to the Möbius invariance of the assertion we may assume that D lies in $\mathbb{R}^n$. We may further suppose that E and F are disjoint, for $M[\Delta(E, F : D)] = \infty$ otherwise. Let G denote the component of $D \setminus E$ that contains F. Then $E \cap \partial G$ is nonempty. We pick a point z_0 of $E \cap \partial G$ and then fix y_0 in G for which $d = |z_0 - y_0|$ is small enough such that $\overline{B}^n(z_0, 2d)$ is contained in D and $4d < \operatorname{diam}(E)$. We also choose a point x_0 of F. Appealing to Theorem 6.5.3, we obtain a quasiconformal self-mapping f of $\hat{\mathbb{R}}^n$ such that $f(G) = G$, $f(x_0) = y_0$, and $f(x) = x$ for every x in G^c. These conditions make certain that $f(D) = D$ and $f(E) = E$. Writing F' for $f(F)$, we conclude that

$$M\big[\Delta(E, F : D)\big] \ge K(f)^{-1} M\big[\Delta(E, F' : D)\big].$$

To complete the proof we need only demonstrate that $M[\Delta(E, F' : D)] > 0$. If F' is not a subset of $\mathbb{S}^{n-1}(z_0, d)$, then the interval $(0, 2d)$ has a subinterval (a, b) such that $a = d$ or $b = d$ and such that the sphere $\mathbb{S}^{n-1}(z_0, r)$ intersects both of the sets E and F' for each r satisfying $a < r < b$. In this case Theorem 4.2.22 surrenders the information that

$$M\big[\Delta(E, F' : D)\big] \ge c_n \log \frac{b}{a} > 0.$$

If, on the other hand, F' happens to be a subset of $\mathbb{S}^{n-1}(z_0, d)$, then there is a b in $(d/2, d)$ such that the sphere $\mathbb{S}^{n-1}(w_0, r)$, where $w_0 = (z_0 + y_0)/2$, has a nonempty intersection with both E and F' whenever $d/2 < r < b$. In the latter situation, Theorem 4.2.22 gives

$$M\big[\Delta(E, F' : D)\big] \geq c_n \log \frac{2b}{d} > 0\ ,$$

so again the desired conclusion follows. □

In a quasiconformally collared domain Theorem 6.5.4 can be strengthened considerably, a statement to which Theorem 6.5.7 will attest. To set the stage for that result we must establish two technical lemmas.

LEMMA 6.5.5. *Suppose that D is a quasiconformally collared domain in $\mathbb{R}^n$, that C is a continuum in D, that x is a point of $\bar{D}$, and that $0 < r \leq 2$. Then there exist a chordal ball $B = \{y : q(y, x) < \delta\}$ and a constant $m > 0$ for which the following is true: $M[\Delta(C, E : D)] \geq m$ whenever E is a connected set in D satisfying the conditions $q(E) \geq r$ and $E \cap B \neq \emptyset$. Both δ and m depend, in general, on n, x, r, C and D.*

PROOF. Consider first the case where x is a point of C. Let d be such that $0 < 4d < \operatorname{diam}(C)$, that $U = B^n(x, 2d)$ is contained in D, and that $q(U) < r$. Choose $\delta > 0$ so that $B = \{y : q(y, x) < \delta\}$ lies in $B^n(x, d)$. If E is a connected set in D satisfying the conditions $q(E) \geq r$ and $E \cap B \neq \emptyset$, then by Theorem 4.2.22 we have $M[\Delta(C, E : D)] \geq c_n \log 2$, so the conclusion of the theorem holds in this case with B as indicated and $m = c_n \log 2$.

For the remainder of the proof we assume that x lies in $\bar{D} \setminus C$. We select a neighbourhood U of x, a number $K \geq 1$, a continuum A in U, and a chordal ball $B = \{y : q(y, x) < \delta\}$ according to the following prescriptions. If x belongs to D: we let $U = B^n(x, 3d)$ with $d > 0$ sufficiently small such that U is contained in D, $U \cap C = \emptyset$, and $q(U) < r$; we take $K = 1$; we set $A = \{x + te_n : d \leq t \leq 2d\}$; we insist that B lie in $B^n(x, d)$. If x is a point of ∂D: we pick an open neighbourhood U of x with $U \cap C = \emptyset$ and $q(U) < r$ for which there is a homeomorphism φ of $U \cap \bar{D}$ onto B^n_+ satisfying the conditions $\varphi(x) = 0$ and $K = K(\varphi | U \cap D) < \infty$; we let $A = \varphi^{-1}(I)$ for $I = \{te_n : 1/3 \leq t \leq 2/3\}$; we choose B so that $\varphi(B \cap \bar{D})$ is a subset of $B^n_+(1/3)$. By Theorems 4.2.15 and 6.5.4,

$$M = M[\bar{\Delta}(C, A : D)] = M[\Delta(C, A : D)] > 0.$$

Moreover, since $q(C, A) > 0$, M is finite by Lemma 4.2.8.

We now assert that, with A, K, B and M determined as above,

$$M\big[\Delta(C, E : D)\big] \geq m = \frac{1}{2^n} \min\left\{M, \frac{b_n}{K} \log \frac{3}{2}\right\}$$

for every connected subset E of D such that $q(E) \geq r$ and $E \cap B \neq \emptyset$. Here b_n is the constant from Theorem 4.2.22.

Fix a set E of the type described. In view of Theorem 4.2.15, it suffices to verify that $M(\Gamma) \geq m$ for $\Gamma = \bar{\Delta}(C, E : D)$. This would trivially be the case were $M(\Gamma) = \infty$, so we assume that $M(\Gamma) < \infty$, which mandates that C and E be disjoint. Let ρ be an admissible density for Γ. It may happen that $\int_\alpha \rho\, ds \geq 1/2$ for every rectifiable member α of the family $\Gamma_1 = \bar{\Delta}(C, A : D)$. If so, 2ρ belongs to $\operatorname{Adm}(\Gamma_1)$ and we obtain the lower bound

$$\int_{\mathbb{R}^n} \rho^n \, dm_n \geq 2^{-n} M(\Gamma_1) = 2^{-n} M\,.$$

Otherwise we can fix a rectifiable path α in Γ_1—say, $\alpha : [0,1] \to D$ with $\alpha(0)$ in C and $\alpha(1)$ in A—for which $\int_\alpha \rho\, ds < 1/2$. It follows that $\int_\beta \rho\, ds \geq 1/2$ for every rectifiable path β belonging to the family $\Gamma_2 = \bar{\Delta}(|\alpha|, E : D)$. Indeed, let $\beta : [a,b] \to D$ be such a path. We may suppose that $\beta(a) = \alpha(c)$ and that $\beta(b)$ is in E. Then $\gamma = \alpha|[0,c] + \beta$ is a rectifiable path from Γ, so

$$1 \leq \int_\gamma \rho\, ds \leq \int_\alpha \rho\, ds + \int_\beta \rho\, ds < \frac{1}{2} + \int_\beta \rho\, ds\,,$$

which indeed shows $\int_\beta \rho\, ds \geq 1/2$.

We infer in this situation that 2ρ is an admissible density for Γ_2 and hence that

$$\int_{\mathbb{R}^n} \rho^n\, dm_n \geq 2^{-n} M(\Gamma_2)\,.$$

Again recalling Theorems 4.2.15 and 4.2.22 we note that if x is in $D \setminus C$, then

$$M(\Gamma_2) = M\big[\Delta(|\alpha|, E : D)\big] \geq c_n \log \frac{3}{2} \geq b_n \log \frac{3}{2} \geq m\,.$$

In this case both E and $|\alpha|$ meet $\mathbb{S}^{n-1}(x,t)$ for $2d \leq t \leq 3d$.

If x belongs to ∂D, on the other hand, we can set $F_0 = U \cap |\alpha|$, $E_0 = U \cap E$, $F_0' = \varphi(F_0)$, and $E_0' = \varphi(F_0)$ and conclude on the basis of the same two theorems just cited that

$$M(\Gamma_2) \geq M\big[\Delta(F_0, E_0 : U \cap D)\big] \geq K^{-1} M\big[\Delta(E_0', F_0' : B^n \cap H^n)\big] \geq \frac{b_n}{K} \log \frac{3}{2} \geq m\,,$$

where here both E_0' and F_0' intersect the cap $C_t = \mathbb{S}^{n-1}(t) \cap \mathbb{H}^n$ whenever $2/3 < t < 1$. We thus arrive at an inequality

$$\int_{\mathbb{R}^n} \rho^n\, dm_n \geq m$$

which is valid in all cases that can occur, and be x in $D \setminus C$ or on ∂D. Take note, however, that the numbers M and K do vary with x. Taking the infimum over ρ we get

$$M[\Delta(C, E : D)] = M(\Gamma) \geq m$$

for every set E of the kind being considered. □

The argument we have just presented includes a proof of the following fact: *if a domain D is locally quasiconformally collared at a point x_0 of ∂D, if C is a continuum in D, and if $0 < r \leq 2$, then there exist a chordal ball $B = \{x : q(x, x_0) < \delta\}$ and a constant $m > 0$ such that $M[\Delta(C, E : D)] \geq m$ whenever E is a connected set in D satisfying the conditions $q(E) \geq r$ and $E \cap B \neq \emptyset$.* The second lemma that we require builds upon the first.

LEMMA 6.5.6. *Suppose that D is a quasiconformally collared domain in $\mathbb{R}^n$, that x_1 and x_2 are points of $\bar{D}$ (possibly $x_1 = x_2$), and that $0 < r \leq 2$. Then there exist chordal balls $B_1 = \{x : q(x, x_1) < \delta\}$, $B_2 = \{x : q(x, x_2) < \delta\}$ and a constant $m > 0$ for which the following is true: $M[\Delta(E_1, E_2 : D)] \geq m$ for every pair of connected sets E_1 and E_2 in D such that $q(E_1) \geq r$, $q(E_2) \geq r$, $E_1 \cap B_1 \neq \emptyset$, and $E_2 \cap B_2 \neq \emptyset$. Both δ and m depend on n, x_1, x_2, r, and D.*

PROOF. To start with, assume that $x_1 = x_2$. The proof of Theorem 6.5.2 establishes the existence of a chordal ball $B = \{x : q(x, x_1) < \delta\}$ with the property that $M[\Delta(E_1, E_2 : D)] \geq 1$ whenever E_1 and E_2 are connected sets in D that intersect B and have chordal diameter no smaller than r.

Thus only the case $x_1 \neq x_2$ demands further attention. Fix a point x_0 in $D \setminus \{x_1, x_2\}$, choose $d > 0$ so that $U = B^n(x_0, 4d)$ has its closure in $D \setminus \{x_1, x_2\}$, and set $C = \overline{B}^n(x_0, d)$. Lemma 6.5.5 ensures the existence of chordal balls $B_j = \{x : q(x, x_j) < \delta\}$ and a constant $m_0 > 0$ for which it is the case that $M[\Delta(C, E : D)] \geq m_0$ whenever E is a connected set in D satisfying $q(E) \geq r$ and $E \cap B_j \neq \emptyset$ for $j = 1$ or $j = 2$. We may assume that δ is small enough to make U, B_1, and B_2 disjoint. Consider a pair of connected sets E_1 and E_2 in D that satisfy the conditions $q(E_j) \geq r$ and $E_j \cap B_j \neq \emptyset$. Write

$$\Gamma = \Delta(E_1, E_2 : D),\ \bar{\gamma} = \bar{\Delta}(E_1, E_2 : D),\ \Gamma_j = \Delta(C, E_j : D),\ \bar{\gamma}_j = \bar{\Delta}(C, E_j : D).$$

If neither E_1 nor E_2 intrudes into the ball $B^n(x_0, 2d)$, then Theorem 4.2.22 implies that

$$M\big[\bar{\Delta}(|\gamma_1|, |\gamma_2| : D)\big] \geq c_n \log 2$$

whenever γ_1 is in Γ_1 and γ_2 in Γ_2, whence Theorems 4.2.15 and 5.5.1 yield the lower bound

$$M(\Gamma) = M(\bar{\gamma}) \geq 3^{-n} \min\{m_0, c_n \log 2\}\,. \tag{6.40}$$

If both E_1 and E_2 intersect the ball $B^n(x_0, 3d)$, then Theorem 4.2.22 shows that

$$M(\Gamma) \geq c_n \log \frac{4}{3}\,. \tag{6.41}$$

There remains the case where one of the sets E_1 or E_2 meets $B^n(x_0, 2d)$, while the other is disjoint from $B^n(x_0, 3d)$. We suppose that E_1 enjoys the former property and E_2 the latter. Assuming that $M(\bar{\gamma}) < \infty$, we let ρ be a member of $\mathrm{Adm}(\bar{\gamma})$. If 2ρ is an admissible density for $\bar{\gamma}_2$, then we have

$$\int_{\mathbb{R}^n} \rho^n \, dm_n \geq 2^{-n} M(\bar{\gamma}_2) = 2^{-n} M(\Gamma_2) \geq 2^{-n} m_0\ ; \tag{6.42}$$

if not, there is a rectifiable path α in $\bar{\gamma}_2$ for which $\int_\alpha \rho\, ds < 1/2$. We argue as we did in the proof of Lemma 6.5.5 to conclude that, in the second case, 2ρ is an admissible density for the family $\bar{\Delta}(|\alpha|, E_1 : D)$, which leads by way of Theorem 4.2.22 to

$$\int_{\mathbb{R}^n} \rho^n \, dm_n \geq 2^{-n} M\big[\bar{\Delta}(|\alpha|, E_1 : D)\big] \geq 2^{-n} c_n \log \frac{3}{2}\,. \tag{6.43}$$

The implication of (6.42) and (6.43) is that

$$M(\Gamma) = M(\bar{\gamma}) \geq 2^{-n} \min\Big\{m_0, c_n \log \frac{3}{2}\Big\} \tag{6.44}$$

when one of the sets E_1 or E_2 hits $B^n(x_0, 2d)$ and the other misses $B^n(x_0, 3d)$. Reviewing (6.40), (6.41), and (6.44), we discover that

$$M(\Gamma) \geq m = 3^{-n} \min\{m_0, c_n \log(3/2)\}$$

in every instance. □

Let $0 < r \leq 2$. Corollary 5.4.4 and the analogue of Theorem 4.2.24 in $\hat{\mathbb{R}}^n$ deliver the inequality

$$M\big[\Delta(E, F : \hat{\mathbb{R}}^n)\big] = M\big[\Delta(\bar{E}, \bar{F} : \hat{\mathbb{R}}^n)\big] \geq \omega_{n-1}\big[\log \Psi_n(4r^{-2})\big]^{1-n} > 0$$

for each pair of connected sets E and F in $\hat{\mathbb{R}}^n$ having $q(E) \geq r$ and $q(F) \geq r$. Our second modulus principle for quasiconformally collared domains is a refinement of Theorem 6.5.4 which mimics the estimate above.

THEOREM 6.5.7. *Suppose that D is a quasiconformally collared domain in $\hat{\mathbb{R}}^n$ and that $0 < r \leq 2$. There exists a constant $m = m(r, D) > 0$ such that $M[\Delta(E, F : D)] \geq m$ whenever E and F are connected sets in D satisfying the conditions $q(E) \geq r$ and $q(F) \geq r$.*

PROOF. After a preliminary chordal isometry we may assume that D lies in $\mathbb{R}^n$. If the theorem is false, then there exist sequences $\langle E_\nu \rangle$ and $\langle F_\nu \rangle$ of connected sets in D such that $q(E_\nu) \geq r$ and $q(F_\nu) \geq r$ for every ν, yet $M[\Delta(E_\nu, F_\nu : D)] \to 0$. Choose for each ν a point x_ν in E_ν and a point y_ν in F_ν. By passing to subsequences and relabeling we may suppose that $x_\nu \to x_0$ and $y_\nu \to y_0$, where x_0 and y_0 are points of $\bar{D}$. According to Lemma 6.5.6 we can produce chordal balls $B = \{x : q(x, x_0) < \delta\}$ and $B' = \{y : q(y, y_0) < \delta\}$, together with a positive constant m_0, for which it is true that $M[\Delta(E, F : D)] \geq m_0$ whenever E and F are connected subsets of D with $q(E) \geq r$, $q(F) \geq r$, $E \cap B \neq \emptyset$, and $F \cap B' \neq \emptyset$. Taking $E = E_\nu$ and $F = F_\nu$ for sufficiently large ν, we run up against a contradiction. Therefore the theorem must be true as stated. □

6.5.3. A higher-dimensional Carathéodory theorem. Here we establish results which will yield a reasonable condition to assure that a quasiconformal mapping between domains extends continuously and homeomorphically to the boundary. Again we start with some topological preliminaries.

A domain D in $\hat{\mathbb{R}}^n$ with $n \geq 2$ is said to be *locally connected* at a point x_0 of ∂D if x_0 has arbitrarily small open neighbourhoods U for which $U \cap D$ is connected.

Observe that D enjoys this property if (and, trivially, only if) each open neighbourhood W of x_0 contains an open neighbourhood V of x_0 such that $V \cap D$ lies in a component of $W \cap D$. Indeed, if G denotes the component in question, then $U = V \cup G$ is an open neighbourhood of x_0, U lies in W, and $U \cap W = G$ is connected.

A domain D that is locally quasiconformally collared at x_0 is plainly locally connected at x_0. A proper subdomain D of $\hat{\mathbb{R}}^n$ is *locally connected along its boundary* if D is locally connected at each point of ∂D.

If, for example, ∂D is homeomorphic to $\mathbb{S}^{n-1}$ (a domain with this property is often referred to as a *Jordan domain*), then it can be shown, although the proof is not elementary, that D is locally connected along its boundary. This implies that a domain D in $\hat{\mathbb{R}}^n$ with only finitely many boundary components each of which is homeomorphic to $\mathbb{S}^{n-1}$ is, for instance, locally connected along its boundary.

A domain D is dubbed *finitely connected* at a point x_0 of ∂D provided x_0 has arbitrarily small open neighbourhoods U such that $U \cap D$ has only finitely many components. The actual number of such components is allowed to vary and even to increase without bound as U shrinks to x_0. This property is in evidence if and only if each open neighbourhood W of x_0 contains an open neighbourhood V of x_0 such that $V \cap D$ intersects at most a finite number of components of $W \cap D$. A proper subdomain D of $\hat{\mathbb{R}}^n$ that is finitely connected at each point of its boundary is said to be *finitely connected along its boundary*.

The fact that the above notions of local connectedness are pertinent to the boundary behavior of quasiconformal mappings will soon become apparent. Before stating the first extension theorem, we remind the reader that a homeomorphism $f : D \to D'$ between domains D and D' in $\hat{\mathbb{R}}^n$ admits an extension to a continuous mapping f^* of $\bar{D}$ onto $\bar{D}'$ if and only if for each point x of ∂D the cluster set $C_f(x)$ consists of a single point. Indeed, the assertion that $C_f(x)$ reduces to a point is equivalent to the statement that $\lim_{y\to x} f(y)$ exists. If this is the case at every point of ∂D, it is a simple matter to verify that

$$f^*(x) = \begin{cases} f(x) & \text{if } x \in D \ , \\ \lim\limits_{y\to x} f(y) & \text{if } x \in \partial D \end{cases}$$

defines a continuous mapping of $\bar{D}$ onto $\bar{D}'$.

THEOREM 6.5.8. *Let D be a quasiconformally collared domain in $\hat{\mathbb{R}}^n$, and let f be a quasiconformal mapping of D onto a domain D'. Then f admits a continuous extension to $\bar{D}$ if and only if D' is finitely connected along its boundary.*

PROOF. Suppose first that D' is finitely connected along its boundary. Consider an arbitrary point x of ∂D. We must demonstrate that $C_f(x)$ consists of a single point. If not, there are sequences $\langle x_\nu\rangle$ and $\langle y_\nu\rangle$ in D such that $x_\nu \to x$, $y_\nu \to x$, $x'_\nu = f(x_\nu) \to x'$, and $y'_\nu = f(y_\nu) \to y'$, where x' and y' are different points of $\partial D'$. We fix open neighbourhoods U and V of x' and y', respectively, such that $\bar{U} \cap \bar{V} = \emptyset$ and such that each of the sets $U \cap D'$ and $V \cap D'$ has only a finite number of components. We can then choose a component E' of $U \cap D'$ with the property that x'_ν lies in E for infinitely many values of ν and a component F' of $V \cap D'$ that contains y'_ν for infinitely many values of ν. It follows that $E = f^{-1}(E')$ and $F = f^{-1}(F')$ are connected sets in D and that x belongs to $\bar{E} \cap \bar{F}$. By Theorem 6.5.2, $M[\Delta(E, F : D)] = \infty$. On the other hand,

$$M\big[\Delta(E,F:D)\big] \le K(f)M\big[\Delta(E',F':D')\big] \le K(f)M\big[\Delta(E',F':\hat{\mathbb{R}}^n)\big] < \infty \ ,$$

for $q(E', F') > 0$, where we have used Lemma 4.2.8. This contradiction shows that $C_f(x)$ has only one element, as desired.

Turning to the converse, let f^* be the (obviously unique) continuous extension of f to $\bar{D}$. Suppose that D' fails to be finitely connected at some point x'_0 of $\partial D'$. Then x'_0 must have an open neighbourhood W' with the following characteristic: every open neighbourhood V' of x'_0 that is contained in W' intersects infinitely many components of $W' \cap D'$. This state of affairs makes it possible to select a sequence $\langle x'_\nu\rangle$ in D' such that $x'_\nu \to x'_0$, but such that all the points x'_ν lie in different components of $W' \cap D$.

By passing to a subsequence we may assume that $x_\nu = f^{-1}(x'_\nu) \to x_0$, a point of ∂D. Since f^* is continuous at x_0, $f^*(x_0) = x'_0$. Also, as D is locally connected at x_0, we can pick an open neighbourhood U of x_0 such that $U \cap D$ is connected and $f^*(U \cap \bar{D})$ is contained in W'. Then x'_ν belongs to $f(U \cap D)$ for all large ν. But $f(U \cap D)$, a connected set in $W' \cap D$, lies in a single component of $W' \cap D$, contrary to the way in which the sequence $\langle x'_\nu\rangle$ was chosen. The contradiction shows that no such x'_0 can exist; that is, D' is locally connected along its boundary. □

We note the following corollary of the first part of the preceding proof:

COROLLARY 6.5.9. *If f maps a domain D in $\hat{\mathbb{R}}^n$ quasiconformally onto a domain D', if x_0 is a point of ∂D having the feature that $M[\Delta(E, F : D)] = \infty$ whenever E and F are connected sets in D with $x_0 \in \bar{E} \cap \bar{F}$, and if D' is finitely connected at each point of $C_f(x_0)$, then $\lim_{x \to x_0} f(x)$ exists.*

The stated modulus condition of the corollary is certainly satisfied when D is locally quasiconformally collared at x_0.

More can be said about the nature of the extension in Theorem 6.5.8.

THEOREM 6.5.10. *Suppose that D is a quasiconformally collared domain in $\hat{\mathbb{R}}^n$, that f is a continuous mapping of $\bar{D}$ into $\hat{\mathbb{R}}^n$ which maps D quasiconformally onto a domain D', and that $A = f^{-1}(\{x'\})$, with x' being a point of $\partial D'$. Then A is a totally disconnected set.*

Furthermore, A consists of exactly p points, where p is a positive integer, if and only if x' has arbitrarily small open neighbourhoods U' with the following two properties:

(i) *$U' \cap D'$ has exactly p components;*
(ii) *x' lies on the boundary of each component of $U' \cap D'$.*

PROOF. Fix a continuum E in D and let C be an arbitrary component of A. We first show that C is degenerate. Assuming to the contrary that C is nondegenerate, we set $r = \min\{q(E), q(C)\} > 0$ and let $m = m(r, D) > 0$ be a constant with the property described in Theorem 6.5.7. We also select a chordal ball $B' = \{y : q(y, x') < \delta\}$ for which

$$M\big[\Delta(E', B' : \hat{\mathbb{R}}^n)\big] < m/K\ ,$$

where $E' = f(E)$ and $K = K(f)$. Since x' is not a point of E', the modulus on the left tends to zero as $\delta \to 0$, so the choice of B' poses no problem.

Because f is a continuous function on $\bar{D}$ that maps the set C to x' and because D is locally connected at every point of C, it is not difficult to establish the existence of a domain G containing C such that $G \cap D$ is connected and $f(G \cap D)$ is a subset of B' as follows. Fixing x_0 in C and denoting by S the set of x in C for which there exists a domain G_x with x_0 and x in G_x, $G_x \cap D$ connected, and $f(G_x \cap D)$ contained in B', one checks that S is a nonempty set that is both open and closed in C. Thus $S = C$ and $G = \bigcup_{x \in C} G_x$ is a domain with the desired features.

Clearly $q(G \cap D) \geq q(C) \geq r$. By Theorem 6.5.7 and our choice of B',

$$\begin{aligned} m &\leq M\big[\Delta(E, G \cap D : D)\big] \leq KM\big[\Delta(E', f(G \cap D) : D')\big] \\ &\leq KM\big[\Delta(E', B' : \hat{\mathbb{R}}^n)\big] < m. \end{aligned}$$

This contradiction is unavoidable unless the suggestion that C is nondegenerate is rejected. Therefore, every component of A is a singleton, making A a totally disconnected set.

Suppose next that p is a positive integer and that $U_1' \supset U_2' \supset U_3' \supset \cdots$ is a sequence of open neighbourhoods of x' such that $q(U_\nu') \to 0$ and such that each U_ν' has properties (i) and (ii). We can list the components of U_ν' as $G_\nu^1, G_\nu^2, \ldots, G_\nu^p$ in such a way that $G_1^k \supset G_2^k \supset G_3^k \cdots$ for $k = 1, 2, \ldots, p$ (property (ii) is instrumental here). Since $\bar{G}_\nu^k \to \{x'\}$ in the Hausdorff metric as $\nu \to \infty$, we can assert on the strength of Theorem 4.3.15 that

$$M\big[\Delta(E', G_\nu^k : D')\big] \leq M\big[\Delta(E', \bar{G}_\nu^k : \hat{\mathbb{R}}^n)\big] \to M\big[\Delta(E', \{x\} : \hat{\mathbb{R}}^n)\big] = 0\,.$$

As a consequence, $M[\Delta(E, F_\nu^k : D)] \to 0$ for each k, where $F_\nu^k = f^{-1}(G_\nu^k)$. Theorem 6.5.7 implies that $q(F_\nu^k) \to 0$, which means that the set $\bigcap_{\nu=1}^\infty \bar{F}_\nu^k$ consists of a single point, call it x_k. Obviously

$$f(x_k) \in \bigcap_{\nu=1}^{\infty} \bar{G}_\nu^k = \{x'\} ,$$

which places x_k in the set A. Let U_k be an open neighbourhood of x_k for which $U_k \cap D$ is connected and $f(U_k \cap D)$ is a subset of U_1'. The connected set $f(U_k \cap D)$ lies in some component of $U_1' \cap D'$ and, as $U_k \cap f^{-1}(G_1^k) \neq \emptyset$ by the definition of x_k, said component must be G_1^k. In particular, the sets $U_1 \cap D, U_2 \cap D, \ldots, U_p \cap D$ must be pairwise disjoint, which demands that $x_1, x_2, \ldots, x_p$ be distinct. If x were a point of $A \setminus \{x_1, x_2, \ldots, x_p\}$, then x would lie in $\bar{D} \setminus \bigcup_{k=1}^p \bar{F}_\nu^k$ for some ν. However, x would have an open neighbourhood U with $U \cap D$ connected and $f(U \cap D)$ contained in U_ν', implying that $f(U \cap D)$ is contained in G_ν^k for some k and forcing x into $\bar{F}_\nu^k$, a contradiction. Therefore, $A = \{x_1, x_2, \ldots, x_p\}$ consists of precisely p points.

Conversely, informed that $A = \{x_1, x_2, \ldots, x_p\}$ is a p-point set and presented with an open neighbourhood V' of x', we exhibit an open neighbourhood U' of x' such that U' is contained in V' and is endowed with properties (i) and (ii). We do so by picking pairwise disjoint open neighbourhoods $U_1, U_2, \ldots, U_p$ of $x_1, x_2, \ldots, x_p$ for which $U_k \cap D$ is connected and $G_k' = f(U_k \cap D)$ lies in V'. We take $U' = \operatorname{int}[G_1' \cup G_2' \cup \cdots \cup G_p' \cup (V' \setminus D')]$. It is obvious that the open set U' is a subset of V', that $G_1', G_2', \ldots, G_p'$ are the components of $U' \cap D'$, and that $x' = f(x_k)$ belongs to $\partial G_k'$ for each k. Finally, x' lies in U'. Otherwise there would be a sequence $\langle y_\nu' \rangle$ in $D' \setminus (G_1' \cup G_2' \cup \cdots \cup G_p')$ such that $y_\nu' \to x'$. Then $y_\nu = f^{-1}(y_\nu')$ would belong to $D \setminus (U_1 \cup U_2 \cup \cdots \cup U_p)$. Passing to a subsequence if necessary, we could suppose that $y_\nu \to x$, a point of ∂D different from $x_1, x_2, \ldots, x_p$. The continuity of f, however, would force $f(x) = x'$ and hence would force x to be a point of A, thereby producing a contradiction. We conclude that U' is an open neighbourhood of x' with properties (i) and (ii). □

It may happen in the context of Theorem 6.4.8 that A is an infinite set, perhaps even an uncountable one. This is well known to be the case when $n = 2$, for the Riemann mapping theorem makes conformal examples in which A is a Cantor set easy to construct. We describe a simple example in dimension $n \geq 3$ for which the set A is infinite. Examples in which the set A is uncountable are more complicated to construct in dimensions three and above, but they do exist.

For $\nu = 0, 1, 2, \ldots$ let $B_\nu = B^{n-1}(3\nu e_1, 1)$, let $D_\nu = B_\nu \times (-\infty, 0)$, and let $D_\nu' = B_\nu \times (-\infty, \infty)$. By conjugating the mapping in Lemma 6.2.1 with a translation we can exhibit a continuous mapping f_ν of $\bar{D}_\nu$ onto $\bar{D}_\nu'$ such that $f_\nu(3\nu e_1) = \infty$, such that $f_\nu(x) = x$ for every x in $\partial D_\nu \cap \partial D_\nu'$, and such that $K = K(f_\nu | D_\nu)$ is a finite number independent of ν. Now take $D = \mathbb{R}^n \setminus \bar{\mathbb{H}}^n$ and $D' = D \cup (\bigcup_{\nu=0}^\infty D_\nu')$. If $f : \bar{D} \to \bar{D}'$ is defined by

$$f(x) = \begin{cases} f_\nu(x) & \text{if } x \in \bar{D}_\nu \ , \\ x & \text{if } x \in \bar{D} \setminus \bigcup_{\nu=0}^\infty \bar{D}_\nu \ , \end{cases}$$

then f is a continuous mapping of $\bar{D}$ onto $\bar{D}'$ which maps D homeomorphically to D'. Theorem 6.4.22 implies that f maps D in a K-quasiconformal fashion onto D'. Finally, it is easy to see that $f^{-1}(\{\infty\}) = \{3\nu e_1 : \nu = 0, 1, 2, \ldots\} \cup \{\infty\}$, an infinite set.

If the boundary components of the domain D in Theorem 6.5.10 are known to be quasiconformal spheres, then the first statement about A can be improved; namely, it is true in this case that $\mathrm{Cap}(A) = 0$. To see this, let S denote the boundary component of D that contains A. Then $f(S) = S'$, the component of $\partial D'$ that contains x'. Fix a continuum F in D and a quasiconformal mapping g of $\hat{\mathbb{R}}^n$ onto itself that maps $\mathbb{S}^{n-1}$ to S and B^n to G, the component of S^c which contains D. Since f transforms $\Delta(A, F : D)$ to a subfamily of $\Delta(\{x'\}, f(F) : D')$, Corollary 4.2.14 gives

$$M\big[\Delta(A, F : D)\big] \le K(f) M\big[\Delta(\{x'\}, f(F) : D')\big] = 0\,.$$

Similarly, $M[\Delta(A, \partial D \setminus S : D)] = 0$. The family $\Delta(A, F : G)$ is minorized by the union of $\Delta(A, F : D)$ and $\Delta(A, \partial D \setminus S : D)$, so $M[\Delta(A, F : G)] = 0$. Writing $A_0 = g^{-1}(A)$, $F_0 = g^{-1}(F)$, and $K = K(g)$, we invoke Corollary 4.3.4 to conclude that

$$\begin{aligned} M\big[\Delta(A, F : \hat{\mathbb{R}}^n)\big] &\le KM\big[\Delta(A_0, F_0 : \hat{\mathbb{R}}^n)\big] \le KM\big[\Delta(A_0, F_0 \cup F_0^* : \hat{\mathbb{R}}^n)\big] \\ &= 2KM\big[\Delta(A_0, F_0 : B^n)\big] \le 2K^2 M\big[\Delta(A, F : G)\big] = 0\,. \end{aligned}$$

Corollary 5.5.2 informs us that $\mathrm{Cap}(A) = 0$.

A continuous bijection $\varphi : X \to Y$, where X is a compact space and Y is a Hausdorff space, is necessarily a homeomorphism. In concert with Theorems 6.5.8 and 6.5.10 this fact makes the passage from continuous to homeomorphic extensions a simple one to negotiate.

THEOREM 6.5.11. *Let D be a quasiconformally collared domain in $\hat{\mathbb{R}}^n$, and let f be a quasiconformal mapping of D onto a domain D'. Then f admits an extension to a homeomorphism of $\bar{D}$ onto $\bar{D}'$ if and only if D' is locally connected along its boundary.*

PROOF. If f extends to a homeomorphism of $\bar{D}$ onto $\bar{D}'$, then D' is itself quasiconformally collared and hence is locally connected along its boundary. Conversely, if D' is locally connected along its boundary, then Theorem 6.5.8 guarantees the existence of an extension of f to a continuous mapping f^* of $\bar{D}$ onto $\bar{D}'$. Furthermore, each point x' of $\partial D'$ has arbitrarily small open neighbourhoods U' for which conditions (i) and (ii) in Theorem 6.5.10 are satisfied with $p = 1$, so that theorem certifies that f^* is injective. Thus f^* is a homeomorphism. □

Embedded in Theorem 6.5.11 is a generalization of the Carathéodory-Osgood theorem from the classical theory of conformal mappings.

COROLLARY 6.5.12. *Let D be an open Euclidean ball or open half-space in $\mathbb{R}^n$, and let f be a quasiconformal mapping of D onto a domain D' in $\hat{\mathbb{R}}^n$. Then f extends to a homeomorphism of $\bar{D}$ onto $\bar{D}'$ if and only if D' is a Jordan domain. In particular, such an extension always exists if D' is itself an open Euclidean ball or open half-space in $\mathbb{R}^n$.*

PROOF. If a homeomorphic extension of f exists, then certainly $\partial D'$ is homeomorphic to ∂D (and thus to $\mathbb{S}^{n-1}$), so D is a Jordan domain. On the other hand, it has already been pointed out that a Jordan domain in $\hat{\mathbb{R}}^n$ is locally connected along its boundary. Accordingly, if D' is a Jordan domain, then Theorem 6.5.11 ensures that f extends to a homeomorphism of $\bar{D}$ onto $\bar{D}'$. □

Another immediate corollary of Theorem 6.5.11 worth noting is:

COROLLARY 6.5.13. *Suppose that a domain D in $\hat{\mathbb{R}}^n$ is the image under a quasiconformal mapping of some quasiconformally collared domain. Then D is quasiconformally collared if and only if it is locally connected along its boundary.*

Let D be a domain in $\hat{\mathbb{R}}^2$ that has only finitely many boundary components, each of these nondegenerate. With the aid of Corollary 6.5.13 we can confirm that D is quasiconformally collared if and only if each component of ∂D is a Jordan curve in $\hat{\mathbb{R}}^2$. We appeal to the known fact that any plane domain D of the given description is conformally equivalent to a domain D' whose finitely many boundary components are all Euclidean circles. Such a domain D' is quasiconformally collared.

It is of course natural to ask how crucial the role of quasiconformal collardness is in Theorem 6.5.11. If, for instance, $n \geq 3$ and f is a quasiconformal mapping of a quasiconformally collared Jordan domain D in $\hat{\mathbb{R}}^n$ onto a Jordan domain D', then we are assured by Theorem 6.5.11 that f has a homeomorphic extension to $\bar{D}$.

But might not the same be true if D were an arbitrary Jordan domain rather than a collared one? This question was answered by T. Kuusalo in 1983 [**94**] when he constructed a Jordan domain D in $\hat{\mathbb{R}}^3$ and a quasiconformal self-mapping f of D that does not extend to a homeomorphism of $\bar{D}$ onto itself. Thus the property of having a quasiconformal collar—or some substitute for it—is vital to Theorem 6.5.11.

Theorem 6.5.11 can be employed to demonstrate that certain domains are not quasiconformally equivalent. Consider, for example, the domain D in $\mathbb{R}^n$ with $n \geq 3$ that is bounded by the hyperplanes with equations $x_n = 0$ and $x_n = 1$. Could D be the image of B^n under a quasiconformal mapping f? First we note that it is not difficult to construct a C^∞-diffeomorphism that maps B^n to D. However if there were a quasiconformal mapping $B^n \to D$, then f could be extended to a homeomorphism of $\overline{B}^n$ onto $\bar{D}$, for D is locally connected along its boundary.

However, ∂D is definitely not homeomorphic to $\mathbb{S}^{n-1}$: the set $\partial D \setminus \{\infty\}$ is disconnected, whereas removing a point from a topological $(n-1)$-sphere S does not disconnect S. We conclude that no such mapping f exists.

Although they are diffeomorphic, B^n and D are not quasiconformally equivalent.

Yet to be determined are conditions on a domain D under which a quasiconformal mapping $f : D \to D'$ will admit a continuous extension to $\bar{D}$, given that D' is smoothly bounded.

In order to establish extension criteria that cover this situation we shall need the following variant of Theorem 6.4.25.

THEOREM 6.5.14. *Suppose that g maps $G = B^n \cap \mathbb{H}^n$ quasiconformally onto a bounded domain G' in $\mathbb{R}^n$. Then*

$$\int_a^b \left(\operatorname{osc}_{C_r} g\right)^n \frac{dr}{r} \leq A K_O(g) |G'(a,b)|$$

whenever $0 < a < b < 1$. Here $C_r = \mathbb{S}^{n-1}(r) \cap \mathbb{H}^n$, $G'(a,b)$ is the image under g of $G(a,b) = \{x \in G : a < |x| < b\}$, and A is a constant that depends only on n.

In particular, for each ρ *in* $(0,1)$ *there exists an* r *in* $[\rho, \sqrt{\rho}]$ *such that*

$$d\big[g(C_r)\big] \le \left[\frac{2AK_O(g)|G'|}{\log \frac{1}{\rho}}\right]^{1/n}.$$

PROOF. The proof of the first inequality retraces essentially step by step the proof of Theorem 6.4.25, but appeals to Theorem 4.5.7 at the point where the earlier argument invoked Theorem 4.5.6. Given ρ in $(0,1)$, we infer that

$$\Big(\frac{1}{2}\log\frac{1}{\rho}\Big) \min_{\rho \le r \le \sqrt{\rho}} \Big(\operatorname{osc}_{C_r} g\Big)^n \le AK_O(g)|G'(\rho, \sqrt{\rho})| \le AK_O(g)|G'|\,,$$

which implies the final assertion of the theorem. Note that $r \mapsto \operatorname{osc}_{C_r} g = \operatorname{diam}[g(C_r)]$ defines a continuous function on $(0,1)$, so the minimum that appears in the above expression is well defined. □

Let D be a domain in $\hat{\mathbb{R}}^n$. We define the *relative chordal distance* $q_{\sigma_D}(x,y)$ between points x and y of D by

$$q_{\sigma_D}(x,y) = \inf_{\gamma} q(|\gamma|)\,,$$

the infimum being taken over all paths γ in D having x and y as endpoints. For nonempty subsets E and F of D we set

$$q_{\sigma_D}(E,F) = \inf\big\{q_{\sigma_D}(x,y) : x \in E\,,\ y \in F\big\}.$$

Equivalently,

$$q_{\sigma_D}(E,F) = \inf\big\{q(|\gamma|) : \gamma \in \bar{\Delta}(E, F : D)\big\}.$$

Notice that $q_{\sigma_D}(E,F)$ may be relatively large in comparison with $q(E,F)$. If $\bar{E} \cap \bar{F} \cap D \ne \emptyset$, then obviously $q_{\sigma_D}(E,F) = q(E,F) = 0$; if $\bar{E} \cap \bar{F}$ contains a point x of ∂D at which D is locally connected, the same is true. In particular, if D is locally connected along its boundary, then $q_{\sigma_D}(E,F) = 0$ whenever E and F are connected sets in D for which $\bar{E} \cap \bar{F} \ne \emptyset$.

THEOREM 6.5.15. *Let D be a domain in $\hat{\mathbb{R}}^n$, and let f be a quasiconformal mapping of D onto a quasiconformally collared domain D'. Then f admits a continuous extension to $\bar{D}$ if and only if $q_{\sigma_D}(E,F) = 0$ whenever E and F are connected sets in D with $\bar{E} \cap \bar{F} \ne \emptyset$.*

PROOF. Suppose first that the relative distance condition is met. Consider an arbitrary point x of ∂D. We must show that the cluster set $C_f(x)$ consists of a single point. Assume otherwise and choose a pair of distinct points x' and y' belonging to $C_f(x)$. We fix a continuum C in D and select open neighbourhoods U' and V' of x' and y', respectively, such that the sets $E' = U' \cap D'$ and $F' = V' \cap D'$ are connected and such that $\bar{U}', \bar{V}'$, and $C' = f(C)$ are pairwise disjoint. Writing $E = f^{-1}(E')$ and $F = f^{-1}(F')$, we observe that x is a point of $\bar{E} \cap \bar{F}$. By hypothesis this forces $q_{\sigma_D}(E,F) = 0$. We can thus find a sequence $\langle \gamma_\nu \rangle$ in $\bar{\Delta}(E, F : D)$ for which $q(|\gamma_\nu|) \to 0$. Now $0 < d = q(C, E \cup F) < 2$, and we may presume that $q(|\gamma_\nu|) < d/4$ for every ν. Select a point x_ν of $|\gamma_\nu| \cap E$ and consider the ring $R_\nu = \{x \in \hat{\mathbb{R}}^n : 2q(|\gamma_\nu|) < q(x, x_\nu) < 3d/4\}$. Since R_ν separates C and $|\gamma_\nu|$ in $\hat{\mathbb{R}}^n$,

$$M\big[\Delta(C, |\gamma_\nu| : D)\big] \le M\big[\Delta(C, |\gamma_\nu| : \hat{\mathbb{R}}^n)\big] \le \operatorname{Cap}(R_\nu) \le \frac{\omega_{n-1}}{\big[\log \frac{3d}{8q(|\gamma_\nu|)}\big]^{n-1}} \to 0$$

as $\nu \to \infty$, where we recall (5.6). As a result,

$$M\big[\Delta(C', f(|\gamma_\nu|) : D')\big] \le K(f) M\big[\Delta(C, |\gamma_\nu| : D)\big] \to 0\,.$$

On the other hand, the chordal diameters of C' and $f(|\gamma_\nu|)$ are no smaller than $r = \min\{q(C'), q(U', V')\} > 0$, so Theorem 6.5.7 implies that

$$\inf_{\nu} M\big[\Delta(C', f(|\gamma_\nu|) : D')\big] > 0\ ,$$

a contradiction. Hence $C_f(x)$ must comprise but a single point, as we had anticipated.

As for the converse, suppose that f has a continuous extension f^* to $\bar{D}$. By subjecting D to a chordal isometry g and then considering $f \circ g$ in place of f, we are at liberty to assume that D contains the point ∞. (It is evident that D satisfies the stated relative distance condition if and only if $g^{-1}(D)$ does.) We fix $\delta > 0$ for which $B = \overline{B}^n(\infty, \delta)$ lies in D. Let E and F be connected sets in D for which $\bar{E} \cap \bar{F} \neq \emptyset$. If $\bar{E} \cap \bar{F} \cap D \neq \emptyset$, then trivially $q_{\sigma_D}(E, F) = 0$. Assuming that $\bar{E} \cap \bar{F} \cap D = \emptyset$, we choose a point x_0 of ∂D belonging to $\bar{E} \cap \bar{F}$. In this case $E' = f(E)$ and $F' = f(E)$ are nondegenerate connected sets. Moreover, due to the continuity of f^*, $x_0' = f^*(x_0)$ is a point of $\bar{E}' \cap \bar{F}' \cap \partial D'$. Because D' is known to be quasiconformally collared, we can find an open neighbourhood U' of x_0' for which there exists a homeomorphism φ of $\bar{D}' \cap U'$ onto B^n_+ with $\varphi(x_0') = 0$ whose restriction to $U' \cap D'$ is quasiconformal. Furthermore, we may take U' sufficiently small that $f^{-1}(U') \cap B = \emptyset$, $q(U') < q(E')$, and $q(U') < q(F')$. The last conditions ensure that both $\varphi(E' \cap U')$ and $\varphi(F' \cap U')$ intersect the hemisphere $C_r = \mathbb{S}^{n-1}(r) \cap H^n$ whenever $0 < r < 1$. Write $G = B^n \cap \mathbb{H}^n$ and let $g = f^{-1} \circ (\varphi | U' \cap D')^{-1}$. Then g is a quasiconformal mapping of G onto a bounded domain G' in $\mathbb{R}^n$ since by construction, G' is contained in B^c.

Given $\varepsilon > 0$ we can pick ρ in $(0, 1)$ such that

$$\frac{2AK_O(g)|G'|}{\log \frac{1}{\rho}} < \frac{\varepsilon}{2}\ ,$$

where A is the constant that arose in Theorem 6.5.14. We conclude on the basis of this theorem that there is an r in $(\rho, \sqrt{\rho})$ for which

$$q\big[g(C_r)\big] \le 2d\big[g(C_r)\big] < \varepsilon\,.$$

As both $\varphi(E' \cap U')$ and $\varphi(F' \cap U')$ meet C_r, we can select a path β that belongs to the family $\bar{\Delta}[\varphi(E' \cap U'), \varphi(F' \cap U') : C_r]$. Then $\gamma = g \circ \beta$ is a member of $\bar{\Delta}(E, F : D)$ and

$$q_{\sigma_D}(E, F) \le q(|\gamma|) \le q\big[g(C_r)\big] < \varepsilon\,.$$

As $\varepsilon > 0$ was arbitrary, it is still true when $\bar{E} \cap \bar{F} \cap D = \emptyset$ that $q_{\sigma_D}(E, F) = 0$. □

There is a local version of Theorem 6.5.15 that is arrived at by combining the remark after the proof of Lemma 6.5.5 with the first part of the proof of Theorem 6.5.15:

THEOREM 6.5.16. *If f maps a domain D in $\hat{\mathbb{R}}^n$ quasiconformally onto a domain D', if x_0 is a point of ∂D at which D is locally connected, and if $C_f(x_0)$ contains a point x_0' at which D' is locally quasiconformally collared, then it must be the case that* $\lim_{x \to x_0} f(x) = x_0'$.

Theorem 6.5.10 has a counterpart in the setting of Theorem 6.5.15. Before stating it we remind the reader that a boundary point x_0 of a domain D in $\hat{\mathbb{R}}^n$ is *accessible from* D provided there exists a curve $\gamma : [0,1) \to D$ such that $\lim_{t\to 1^-} \gamma(t) = x_0$. If so, then it is always possible to produce an injective curve γ with this property, in which event the half-open arc $|\gamma|$ is termed an *endcut from* D *to* x_0.

If, for example, D is locally connected at a point x_0 of ∂D, then x_0 is accessible from D.

THEOREM 6.5.17. *Suppose that* D *is a domain in* $\hat{\mathbb{R}}^n$, *that* f *is a continuous mapping of* $\bar{D}$ *into* $\hat{\mathbb{R}}^n$ *which maps* D *quasiconformally onto a quasiconformally collared domain* D', *and that* $A = f^{-1}(\{x'\})$, *with* x' *being a point of* $\partial D'$. *Then either* A *consists of a single point* x *of* ∂D *or* A *is a continuum with the property that* $M[\Delta(A, C : D)] = 0$ *for every continuum* C *in* D. *In the former case* D *is locally connected at* x; *in either case there is at most one point of* A *that is accessible from* D.

PROOF. Let $U_1' \supset U_2' \supset U_3' \supset \cdots$ be a sequence of open neighbourhoods of x' such that $U_\nu' \cap D'$ is connected and such that $q(U_\nu') \to 0$. It is easily verified that

$$A = \bigcap_{\nu=1}^{\infty} \overline{f^{-1}(U_\nu' \cap D')}\,.$$

As the intersection of a nonincreasing sequence of continua the set A either reduces to a single point x of ∂D (for this to happen it must be true that $q[f^{-1}(U_\nu' \cap D')] \to 0$ as $\nu \to \infty$) or is a continuum in ∂D. For any continuum C in D the image of $\Delta(A, C : D)$ under f is a subfamily of $\Delta(\{x'\}, f(C) : D')$, so by Corollary 4.2.14

$$M\big[\Delta(A, C : D)\big] \le K(f) M\big[\Delta(\{x'\}, f(C) : D')\big] = 0\,.$$

If $A = \{x\}$ and if V is an arbitrary open neighbourhood of x, we can fix ν for which $f^{-1}(U_\nu' \cap D')$ is a subset of V. Then $U = \operatorname{int}[f^{-1}(U_\nu' \cap D') \cup (V \setminus D)]$ is readily seen to be an open neighbourhood of x which is contained in V and for which $U \cap D = f^{-1}(U_\nu' \cap D')$ is connected; thus D is locally connected at x.

Finally, if A contained distinct points x and y that were accessible from D, we could find disjoint endcuts E and F from D to x and from D to y, respectively. Then $q_{\sigma_D}(E, F) \ge q(E, F) > 0$. On the other hand, if $E' = f(E)$ and $F' = f(F)$, then x' would belong to $\bar{E}' \cap \bar{F}'$. The argument used to establish the necessity of the relative distance condition in Theorem 6.5.15—it relied solely on the information that D' was quasiconformally collared and that $\bar{E}' \cap \bar{F}' \cap \partial D' \ne \emptyset$—could be repeated here to show that $q_{\sigma_D}(E, F) = 0$. This contradiction forces us to conclude that at most one point of A can be accessible from D. □

We sketch an example in which the set A occurring in Theorem 6.5.17 is a continuum. Since examples of this type abound in the context of plane conformal mappings, we present one for dimension $n \ge 3$. For $\nu = 0, 1, 2, \ldots$ let $B_\nu = B^{n-1}(2^{-\nu} e_1, 2^{-\nu-3})$, let $D_\nu = B_\nu \times (-\infty, 0)$, and let $D_\nu' = B_\nu \times (-\infty, 1)$. By taking the mapping h that arose as an afterthought to Lemma 6.2.1 and conjugating it with a similarity transformation, we can exhibit a homeomorphism h_ν of $\bar{D}_\nu$ onto $\bar{D}_\nu'$ such that $h_\nu(x) = x$ for every x in $\partial D_\nu \cap \partial D_\nu'$ and such that $K = K(h_\nu | D_\nu)$ is the same finite number for every ν. Let $D' = \{\mathbb{R}^n \setminus \bar{\mathbb{H}}^n\}$, let $D = D' \cup (\bigcup_{\nu=0}^{\infty} D_\nu')$, define $g : D' \to D$ by

$$g(x) = \begin{cases} h_\nu(x) & \text{if } x \in D_\nu \ , \\ x & \text{if } x \in D' \setminus \bigcup_{\nu=0}^\infty D_\nu \ , \end{cases}$$

and set $f = g^{-1}$. Then g is a homeomorphism of D' onto D. Theorem 6.4.22 can be used to show that g is a K-quasiconformal mapping, which therefore makes f a K-quasiconformal mapping as well.

The domain D is easily seen to satisfy the relative distance condition in Theorem 6.5.15, so f has an extension to a continuous mapping of $\bar{D}$ onto $\bar{D}'$, an extension we continue to call f. The line segment $A = \{te_n : 0 \le t \le 1\}$ lies on ∂D. Write $E = \{te_n : -2 \le t \le -1\}$. For $\nu = 0, 1, 2, \ldots$ let $r_\nu = 7 \cdot 2^{-\nu-3}$, let $C_\nu = \{x \in \mathbb{S}^{n-1}(r_\nu) : x_n < 0\}$, and let F_ν be the component of $D \setminus C_\nu$ that does not contain E. Then $F_1 \supset F_2 \supset F_3 \supset \cdots$. Also, A lies on ∂F_ν, so $A' = f(A)$ is contained in the closure of $F'_\nu = f(F_\nu)$. The curve family $\Delta(E, F_\nu : D)$ is minorized by $\Delta(\mathbb{S}^{n-1}, \mathbb{S}^{n-1}(r_\nu) : \mathbb{R}^n)$, which implies that $M[\Delta(E, F_\nu : D)] \to 0$ as $\nu \to \infty$. Noting that $f(E) = E$, we infer that $M[\Delta(E, F'_\nu : D')] \to 0$ and then conclude on the basis of Theorem 6.5.7 that $q(F'_\nu) \to 0$.

It follows that A' consists of a single point. As a matter of fact, $A' = \{0\}$, for clearly $f(0) = 0$. It is easy to see that $f(x) \neq 0$ when x is not a point of A, so $A = f^{-1}(\{0\})$. In this example the origin is the only point of A that is accessible from D.

Combining Theorems 6.5.15 and 6.5.17 we are able to generate a companion to Theorem 6.5.11.

THEOREM 6.5.18. *Let D be a domain in $\hat{\mathbb{R}}^n$ and let f be a quasiconformal mapping of D onto a quasiconformally collared domain D'. Then f admits an extension to a homeomorphism of $\bar{D}$ onto $\bar{D}'$ if and only if*

(i) *the relative chordal distance $q_D(E, F) = 0$ whenever both E and F are connected sets in D such that $\bar{E} \cap \bar{F} \neq \emptyset$ and*
(ii) *each point of ∂D is accessible from D.*

PROOF. Assume that f has a homeomorphism extension to $\bar{D}$. Theorem 6.5.15 shows that condition (i) is satisfied. Moreover, since f^{-1} extends to a homeomorphism of $\bar{D}'$ onto $\bar{D}$, Theorem 6.5.11 tells us that D has to be locally connected at each of its boundary points. This implies that each such point is accessible from D. As for the converse, Theorem 6.5.15 guarantees that f extends to a continuous mapping f^* of $\bar{D}$ onto $\bar{D}'$. In light of condition (ii), Theorem 6.5.17 confirms that f^* is injective. As a continuous injection from a compact space onto a Hausdorff space, f^* is a homeomorphism. □

Sometimes when it is impossible to extend a quasiconformal mapping $f : D \to D'$ continuously to $\bar{D}$, it may still be possible to get a continuous extension of f to a portion of ∂D. We record one theorem that illustrates this state of affairs.

THEOREM 6.5.19. *Suppose that $f : D \to D'$ is a quasiconformal mapping between domains D and D' in $\hat{\mathbb{R}}^n$, that D is locally quasiconformally collared at every point of a nonempty set E in ∂D, and that D' is locally connected at every point of $E' = C_f(E)$. Then f extends to a homeomorphism of $D \cup E$ onto $D' \cup E'$.*

PROOF. The comment following the proof of Theorem 6.5.8 reveals that

$$f^*(x) = \begin{cases} f(x) & \text{if } x \in D\ , \\ \lim_{y \to x} f(y) & \text{if } x \in E \end{cases}$$

delivers a well-defined extension of f to a continuous function on $D \cup E$ which maps E to E'. Write $g = f^{-1}$. The observation subsequent to the proof of Theorem 6.5.15 shows that the function $g^* : D' \cup E' \to \bar{D}$ given by

$$g^*(x') = \begin{cases} g(x') & \text{if } x \in D'\ , \\ \lim_{y' \to x'} g(y') & \text{if } x' \in E' \end{cases}$$

is likewise well defined and continuous, with $g^*(E') = E$. Obviously f^* and g^* are inverse to one another, so f^* is a homeomorphism. □

Building on Theorem 6.5.19, we can improve upon our initial version of the reflection principle for quasiconformal mappings previously mentioned in Theorem 6.4.23.

THEOREM 6.5.20. *Suppose that Σ and Σ' are chordal spheres in $\hat{\mathbb{R}}^n$ with $n \geq 2$, that D is a domain which is symmetric with respect to Σ, that G is a component of $D \setminus \Sigma$ and $E = \Sigma \cap D$, and that g is a quasiconformal mapping of G onto a domain G' disjoint from Σ'. If $E' = C_g(E)$ is contained in Σ' and if $C_g(\partial G \setminus E)$ is disjoint from E', then g has a unique extension to a quasiconformal mapping f of D onto $D' = G' \cup E' \cup R_{\Sigma'}(G')$ with the feature that $f \circ R_\Sigma = R_{\Sigma'} \circ f$ in D. Furthermore, f has the same dilatations as g.*

PROOF. If x is a point of E and $B = B^n(x, r)$ is an open ball that lies in D, then $B \cap \partial G = B \cap E = B \cap \Sigma$ and $B \cap G$ is one of the two components of $B \setminus \Sigma$. It follows without difficulty from this comment that E is a relatively open subset of ∂G and that G is locally quasiconformally collared at every point of E. The set $C_g(\partial G \setminus E)$ is a compact set. We infer that $E' = \partial G' \setminus C_g(\partial G \setminus E)$ is relatively open in $\partial G'$. For x' in E' it is thus true of the open ball $B' = B^n(x', s)$ with s sufficiently small that $B' \cap \partial G' = B' \cap E'$. The last condition ensures that $B' \cap \partial G' = B' \cap \Sigma'$ and that $B' \cap G'$ is one of the components of $B' \setminus \Sigma'$, which enables us to conclude that G' is locally connected at x'. Theorem 6.5.19 shows that g can be extended to a homeomorphism g^* of $G \cup E$ onto $G' \cup E'$. To complete the proof we apply Theorem 3.22 to g^*. □

We close this section with an observation concerning the boundary correspondence induced by a quasiconformal self-mapping of $\mathbb{H}^{n+1}$ with $n \geq 2$.

THEOREM 6.5.21. *Let $n \geq 2$, let F be a quasiconformal mapping of $\mathbb{H}^{n+1}$ onto itself, and let f be the restriction of F^* to $\hat{\mathbb{R}}^n$, where F^* is the homeomorphic extension of F to $\hat{\mathbb{H}}^{n+1}$. Then f is a quasiconformal self-mapping of $\hat{\mathbb{R}}^n$.*

PROOF. No loss of generality is suffered by assuming that $F^*(\infty) = \infty$. We extend F^* by reflection to a quasiconformal mapping of $\hat{\mathbb{R}}^{n+1}$ onto itself. Label this extension $\tilde{F}$. If x belongs to $\mathbb{R}^n$ and if $r > 0$, it is clear that

$$\ell_f(x, r) \geq \ell_{\tilde{F}}(x, r), \qquad L_f(x, r) \leq L_{\tilde{F}}(x, r)\ ,$$

so by Theorem 6.4.2 we have

$$H_f(x) = \limsup_{r\to 0} \frac{L_f(x,r)}{\ell_f(x,r)} \le \limsup_{r\to\infty} \frac{L_{\tilde{F}}(x,r)}{\ell_{\tilde{F}}(x,r)} = H_{\tilde{F}}(x) \le \sup_{x\in\hat{\mathbb{R}}^{n+1}} H_{\tilde{F}}(x) < \infty\,.$$

According to Theorem 6.4.10, f maps $\mathbb{R}^n$ quasiconformally onto itself and hence maps $\hat{\mathbb{R}}^n$ quasiconformally onto itself. □

With some nontrivial extra work it can be shown that the mapping f in Theorem 6.5.21 obeys the inequalities

$$H_I\big[f'(x)\big] \le K_I(F), \qquad H_O\big[f'(x)\big] \le K_O(F)$$

whenever x is a point of $\hat{\mathbb{R}}^n$ at which f possesses a nonsingular derivative. As a result, we learn from Theorem 6.4.15 that $K_I(f) \le K_I(F)$ and $K_O(f) \le K_O(F)$.

Let F be a quasiconformal mapping of $\mathbb{H}^2$ onto itself for which $F^*(\infty) = \infty$. Lemma 6.4.1 and the proof offered for Theorem 6.5.21 reveal that the associated boundary map $f = F^*|\mathbb{R}$, which is a homeomorphism of $\mathbb{R}$ onto itself, satisfies

$$\frac{1}{k} \le \frac{f(x+r) - f(x)}{f(x) - f(x-r)} \le k \tag{6.45}$$

for every x in $\mathbb{R}$ and every $r > 0$, where $k \ge 1$ is a constant. In fact, the k involved here depends only on n and $K(F)$. Any homeomorphism of $\mathbb{R}$ onto itself which satisfies condition (6.45) is called a *k-quasisymmetric homeomorphism* of the real line.

The quasisymmetric homeomorphisms of $\mathbb{R}$ can be thought of as "one-dimensional quasiconformal mappings", even though they do not have all the properties enjoyed by their higher-dimensional counterparts (for instance, a quasisymmetric homeomorphism $f : \mathbb{R} \to \mathbb{R}$ may fail to satisfy Lusin's condition).

Given a quasiconformal mapping f of $\mathbb{R}^n$ onto itself—when $n = 1$ interpret this to mean a quasisymmetric homeomorphism of the real line—it is natural to ask whether there exists a quasiconformal self-mapping F of $\mathbb{H}^{n+1}$ such that, in the notation of Theorem 6.5.21, $f = F^*|\mathbb{R}^n$. An affirmative answer to this question was provided in 1982 by Pekka Tukia and Jussi Väisälä [**158**]. Prior to their paper the answer was known to be "yes" for $n = 1, 2,$ and 3. We shall return to this topic in a later chapter at which time we shall say more about the history of the problem and present the proof of this result.

6.6. The distortion, compactness and convergence properties of quasiconformal families

Earlier we looked at distortion, compactness and convergence properties of Möbius transformations. This section aims to show that most of the results obtained for conformal mappings have analogues for quasiconformal mappings. In particular we will establish quite general equicontinuity results and therefore the normal families property. These equicontinuity bounds arise from distortion estimates such Mori's distortion theorem which will be discussed first. One should also expect a quasi-invariance of cross ratio, since these quantities have already appeared, though disguised, in our formulas for the modulus of the extremal Teichmüller ring. This leads directly to the notion of quasisymmetry which we have already discussed.

The normal families properties of quasiconformal mappings are extremely important for the theory. In high dimensions there are few mappings which can be explicitly written—although we saw a few such earlier, this almost exhausts the list. In most applications (indeed in much of analysis) the construction of mappings is through an approximation and limiting process. We therefore need results to assure ourselves that these limits are quasiconformal.

6.6.1. Distortion under quasiconformal mappings. We commence the discussion with an n-dimensional version of a classical theorem due to Mori [**120, 121**].

THEOREM 6.6.1. *If f maps B^n quasiconformally onto itself and fixes the origin, then*

$$|f(y) - f(x)| \leq 4\lambda_n^2 |y-x|^\alpha \tag{6.46}$$

for all x and y in B^n, where $\alpha = K_I(f)^{1/(1-n)}$ and λ_n is the constant associated with the Grötzsch rings in $\hat{\mathbb{R}}^n$.

Before proving this we note that $g = f^{-1} : B^n \to B^n$ has $g(0) = 0$ and therefore satisfies the hypotheses of the theorem. Since $K_I(g) = K_O(f)$ Theorem 6.6.1 directly implies the estimate

$$\frac{1}{4\lambda_n^2}|y-x|^{K^{1/(n-1)}} \leq |f(y) - f(x)| \leq 4\lambda_n^2 |y-x|^{1/K^{1/(n-1)}} \tag{6.47}$$

where $K = K(f)$.

PROOF. Theorem 6.5.21 assures us that f can be extended by reflection to a quasiconformal self-mapping of $\hat{\mathbb{R}}^n$, call it F. Furthermore, $K_I(F) = K_I(f)$.

Since $f(0) = 0$, we see that $F(0) = 0$ and $F(\infty) = \infty$. Fix a pair of distinct points x and y in B^n, say with $x \neq 0$. Then

$$|y - x| \leq |y| + |x| < 1 + |x| .$$

The ring $R = \{z : |y-x| < |z-x| < 1+|x|\}$ has

$$\mathrm{Mod}(R) = \log\frac{1+|x|}{|y-x|} > \log\frac{1}{|y-x|} .$$

Setting $z = -|x|^{-1}x$, we note that $|z-x| = 1+|x|$. Thus R separates x and y from z and ∞, from which it follows that $R' = F(R)$ separates $F(x)$ and $F(y)$ from $F(z)$ and $F(\infty) = \infty$.

Because $F(x) = f(x)$ and $F(y) = f(y)$ lie in B^n, while $|F(z)| = 1$, we obtain from Theorems 5.4.2 and 5.1.9 that

$$\begin{aligned} \mathrm{Mod}(R') &\leq \log\Psi_n\left(\frac{|F(z)-F(x)|}{|F(y)-F(x)|}\right) \leq \log\Psi_n\left(\frac{2}{|f(y)-f(x)|}\right) \\ &\leq \log\left(\lambda_n^2\left[\frac{2}{|f(y)-f(x)|}+1\right]\right) = \log\left(\frac{\lambda_n^2[2+|f(y)-f(x)|}{|f(y)-f(x)|}\right) \\ &\leq \log\frac{4\lambda_n^2}{|f(y)-f(x)|} . \end{aligned}$$

Now

$$\mathrm{Cap}(R') \leq K_I(F)\mathrm{Cap}(R) = K_I(f)\mathrm{Cap}(R),$$

which yields the estimate

$$\log \frac{1}{|y-x|} \le \operatorname{Mod}(R) \le K_I(f)^{1/(n-1)} \operatorname{Mod}(R') \le \frac{1}{\alpha} \log \frac{4\lambda n^2}{|f(y)-f(x)|}.$$

This leads to

$$|f(y)-f(x)| \le 4\lambda_n^2 |y-x|^\alpha ,$$

as asserted. □

It was noted in Theorem 5.1.8 that $\lambda_2 = 4$.

Since $K_I(f) = K_O(f) = K(f)$ when $n = 2$, Theorem 6.6.1 declares that any K-quasiconformal self-mapping f of B^2 which fixes the origin must satisfy

$$|f(w)-f(z)| \le 64|w-z|^{1/K}$$

for all z and w in B^2.

Mori showed a better estimate—namely,

THEOREM 6.6.2. *If f maps B^2 quasiconformally onto itself and fixes the origin, then*

$$|f(y)-f(x)| \le 16|y-x|^K$$

for all x and y in B^2 and $K = K(f)$. The number 16 is the smallest possible absolute constant here.

One can do a little better here by replacing 16 with a constant that depends on K. The Mori Conjecture places the number $16^{1-1/K}$ here as best possible.

What the corresponding sharp result might be in dimension three or above is anybody's guess, although recently there have been some promising developments on the subject. Recall one has the bound (6.46) $|f(y)-f(x)| \le 4\lambda_n^{2(1-1/\alpha)} |y-x|^\alpha$ with $\alpha = K_I(f)^{1/(1-n)}$. These refinements of Theorem 6.6.1 become quite technical, and we refer the reader to the papers of M. Vuorinen and coauthors, in particular see [**165, 18**] and the references therein.

From Theorem 6.6.1 it is straightforward to derive a distortion theorem for a general quasiconformal self-mapping of B^n.

THEOREM 6.6.3. *If f maps B^n quasiconformally onto itself, then*

$$|f(y)-f(x)| \le 4\lambda_n^2 e^{d_{\mathbb{H}}[0,f(0)]} |y-x|^\alpha$$

for all x and y in B^n, where $\alpha = K_I(f)^{1/(1-n)}$.

PROOF. Write $x_0 = f(0)$. We must deal with the case where $x_0 \neq 0$. Let x_0^* be the point that is symmetric to x_0 with respect to $\mathbb{S}^{n-1}$, and let R be the reflection in the sphere $\Sigma = \mathbb{S}^{n-1}(x_0^*, r)$, where $1 + r^2 = |x_0^*|^2$. Since Σ is orthogonal to $\mathbb{S}^{n-1}$, R maps B^n onto itself. Also, $R(x_0) = 0$. Thus $g = R \circ f$ is a quasiconformal mapping of B^n onto itself, $g(0) = 0$, and $K_I(g) = K_I(f)$. Moreover, the fact that $R = R^{-1}$ means that $f = R \circ g$.

Applying (3.21), Theorem 6.6.1 and (3.47), we obtain

$$\begin{aligned} |f(y)-f(x)| &= \big|R[g(y)]-R[g(x)]\big| = \frac{r^2|g(y)-g(x)|}{|g(y)-x_0^*|\,|g(x)-x_0^*|} \\ &\le \frac{|x_0^*|^2-1}{(|x_0^*|-1)^2}\cdot 4\lambda_n^2|y-x|^\alpha = \frac{|x_0^*|+1}{|x_0^*|-1}\cdot 4\lambda_n^2|y-x|^\alpha \\ &= \frac{1+|x_0|}{1-|x_0|}\cdot 4\lambda_n^2|y-x|^\alpha = 4\lambda_n^2 e^{d_{\mathbb{H}}(0,x_0)}|y-x|^\alpha \end{aligned}$$

for all x and y in B^n. □

There is a local version of Theorems 6.5.2 and 6.5.3 which holds in a much more general setting. It involves a Hölder estimate which, derived as it is from a certain distortion function, is not quite as explicit as the ones we have obtained above.

For $n \ge 2$ and $1 \le K < \infty$ we define the *distortion function* Θ_K^n on the interval $(0,1)$ as follows:

$$\Theta_K^n(t) = \frac{1}{\Psi_n^{-1}[\Phi_n(\frac{1}{n})^\beta]}\,, \tag{6.48}$$

where

- $\beta = K^{1/(1-n)}$.
- $\Phi_n : (1,\infty) \to (0,\infty)$ is the function that is associated with the Grötzsch ring.
- $\Psi_n^{-1} : (1,\infty) \to (0,\infty)$ is the function that is inverse to Ψ_n, the function associated with the Teichmüller ring.

Then Θ_K^n is an increasing, continuous function satisfying $\lim_{t\to 0}\Theta_K^n(t) = 0$ and $\lim_{t\to 1}\Theta_K^n(t) = \infty$. Moreover, since $\Phi_n(r)/r \to \lambda_n$ and $\Psi_n(r)/r \to \lambda_n^2$ as $r \to \infty$, we observe the important asymptotic information

$$\Theta_K^n(t) \sim \lambda_n^{2-\beta}t^\beta, \qquad \text{as } t \to 0. \tag{6.49}$$

Much more precise information is available about the distortion function Θ_K^n in the monograph [**12**]. We will not really need any of this.

We now establish the following distortion theorem.

THEOREM 6.6.4. *If D is a proper subdomain of $\mathbb{R}^n$ and f is a quasiconformal mapping of D onto a domain D' in $\mathbb{R}^n$, then the inequality*

$$\frac{|f(y)-f(x)|}{\operatorname{dist}[f(x),\partial D']} \le \Theta_K^n\left(\frac{|y-x|}{\operatorname{dist}(x,\partial D)}\right)$$

holds for $K = K_I(f)$ whenever x belongs to D and $|y-x| < \operatorname{dist}(x,\partial D)$.

PROOF. Corollary 6.4.27 implies that D' is also a proper subdomain of $\mathbb{R}^n$. Consider a pair of distinct points x and y of D for which $|y-x| < d$, where we write $d = \operatorname{dist}(x,\partial D)$. Fixing r with $|y-x| < r < d$, we denote by R the ring whose boundary components are the sphere $B_0 = \mathbb{S}^{n-1}(x,r)$ and the line segment B_1 having endpoints x and y. Then $\bar{R}$ lies in D. Since R is Möbius equivalent to the Grötzsch ring $R_G(n, r/|y-x|)$, its modulus is given by

$$\operatorname{Mod}(R) = \log\Phi_n\left(\frac{r}{|y-x|}\right).$$

Write $B_0' = f(B_0)$, let $s = \text{dist}[f(x), B_0']$, and choose z in B_0' for which $|z - f(x)| = s$. The ring $R' = f(R)$ separates the points $f(x)$ and $f(y)$ from the points z and ∞. Plainly, $s \le d' = \text{dist}[f(x), \partial D]$, so by Theorem 5.4.2

$$\text{Mod}(R') \le \log \Psi_n \left(\frac{s}{|f(y) - f(x)|} \right) \le \log \Psi_n \left(\frac{d'}{|f(y) - f(x)|} \right) .$$

Because f is quasiconformal, we know that

$$\text{Mod}(R) \le K_I(f)^{1/(n-1)} \text{Mod}(R') .$$

In view of the remarks above, therefore,

$$\log \Phi_n \left(\frac{r}{|y - x|} \right) \le K_I(f)^{1/(n-1)} \log \Psi_n \left(\frac{d'}{|f(y) - f(x)|} \right)$$

which translates to

$$\frac{|f(y) - f(x)|}{d'} \le \Theta_K^n \left(\frac{|y - x|}{r} \right)$$

with $K = K_I(f)$. Letting $r \to d$, we obtain the estimate announced by the theorem. □

Given the asymptotic behavior of $\Theta_K^n(t)$ as $t \to 0$ described in (6.49), it is implicit in Theorem 6.6.4 that f satisfies a uniform Hölder condition on each compact set in D.

COROLLARY 6.6.5. *If D is a proper subdomain of $\mathbb{R}^n$ and f is a quasiconformal mapping of D onto a domain D' in $\mathbb{R}^n$, then for each compact set A in D there exists a constant $a > 0$, which depends on f, such that*

$$|f(y) - f(x)| \le a |y - x|^\alpha$$

holds for all points x and y of A, where $\alpha = K_I(f)^{1/(1-n)}$.

PROOF. Fix a nonempty compact subset A of D. Set $d = \text{dist}(A, \partial D)$, $b = \max\{\text{dist}(x', \partial D') : x' \in f(A)\}$, and

$$c = \sup_{0 < t \le 1/2} \frac{\Theta_K^n(t)}{t^\alpha} ,$$

where $K = K_I(f)$ and $\alpha = K^{1/(1-n)}$. Because $\Theta_K^n(t) \sim \lambda_n^{2-\alpha} t^\alpha$ as $t \to 0$, we have $0 < c < \infty$.

Let x and y be points of A. If $|y - x| < d/2$, then $|y - x| < \text{dist}(x, \partial D)/2$. Thus Theorem 6.6.4 leads to the inequality

$$|f(y) - f(x)| \le d\big[f(x), \partial D'\big] \Theta_K^n \left(\frac{|y - x|}{\text{dist}(x, \partial D)} \right) \le bcd^{-\alpha} |y - x|^\alpha .$$

On the other hand, when $|y - x| \ge d/2$ it is trivially the case that

$$|f(y) - f(x)| \le d\big[f(A)\big] \le 2^\alpha d^{-\alpha} d\big[f(A)\big] |y - x|^\alpha .$$

The estimate asserted by the corollary follows, with

$$a = \max\{bcd^{-\alpha}, 2^\alpha b^{-\alpha} \text{diam}[f(A)]\},$$

which completes the proof. □

It is because of these sorts of results, Theorem 6.6.4, and Corollary 6.6.5 that the function Θ_K^n has been fairly intensively studied; see [**165**].

6.6.2. Quasiconformality and quasisymmetry. We can now use the distortion estimate of Theorem 6.6.4 together with our knowledge that the distortion function Θ_K^n is increasing, and hence invertible, to give important local and global properties of quasiconformal mappings. We begin with a definition of a crucial concept—quasisymmetry—that will be used later in the book. It is this concept that has allowed the generalisation of much of the theory of quasiconformal mappings to more general metric measure spaces and related spaces.

Let D be a domain in $\mathbb{R}^n$ and $f : D \to \mathbb{R}^n$ be a homeomorphism. Then f is said to be *quasiymmetric* if there is a homeomorphism $\eta : (0,\infty) \to (0,\infty)$ such that for all triples of distinct points $x, y, z \in D$ we have

$$\frac{|f(x) - f(y)|}{|f(x) - f(z)|} \leq \eta\Big(\frac{|x-y|}{|x-z|}\Big). \tag{6.50}$$

There is a related concept here. The homeomorphism f is said to be *weakly quasisymmetric* if there is a constant H such that for all triples of distinct points $x, y, z \in D$ such that $|x-y| = |x-z|$, then

$$|f(x) - f(y)| \leq H|f(x) - f(z)|. \tag{6.51}$$

It is immediate from the definitions that a quasisymmetric homeomorphism is weakly quasisymmetric with $H = \eta(1)$. The relationship with quasiconformality is also straightforward.

THEOREM 6.6.6. *Let $f : D \to \mathbb{R}^n$ be weakly quasisymmetric with constant H. Then f is K quasiconformal with $K \leq H^{n-1}$.*

PROOF. We simply estimate the linear dilatation of the mapping f to be

$$\limsup_{|h|\to 0} \frac{\max\, |f(x+h) - f(x)|}{\min\, |f(x+h) - f(x)|} = \limsup_{r\to 0} \frac{\max_{y\in\mathbb{S}^{n-1}(x,r)}\, |f(x) - f(y)|}{\min_{z\in\mathbb{S}^{n-1}(x,r)}\, |f(x) - f(z)|} \leq H.$$

The result follows as the maximal distortion K is dominated by the $(n-1)^{th}$ power of the linear distortion. □

The converse is not true in complete generality. However it is true for entire mappings.

THEOREM 6.6.7. *Let $f : \mathbb{R}^n \to \mathbb{R}^n$ be a K-quasiconformal homeomorphism. Then f is quasisymmetric. Indeed, let $\eta(t)$ be defined by*

$$\eta(t) = \begin{cases} \Theta_K^n\,(t)\,, & t \geq 1, \\ 1/\Theta_K^n\,(1/t)\,, & t < 1. \end{cases} \tag{6.52}$$

Then η is increasing on $[0,1]$ and on $[1,\infty]$ with $\eta(t) \to 0$ as $t \to 0$ and $\eta(t) \to \infty$ as $t \to \infty$ and we have the inequality

$$\frac{|f(x) - f(y)|}{|f(x) - f(z)|} \leq \eta\left(\frac{|x-y|}{|x-z|}\right).$$

Then f is quasisymmetric for any homeomorphism which dominates η.

PROOF. Let x, y, z be points of $\mathbb{R}^n$. First suppose that $|f(x) - f(y)| > |f(x) - f(z)|$. Let R be the Grötzsch ring

$$R_G = R(B^n[f(x), |f(x) - f(z)|], [f(y), \infty); \mathbb{R}^n)$$

and set $T = f^{-1}(R)$. As f^{-1} is K-quasiconformal,

$$\Phi_n\Big(\frac{|f(x)-f(y)|}{|f(x)-f(z)|}\Big) = \mathrm{Mod}(R) \leq K\ \mathrm{Mod}(T) \leq K\ \Psi_n\left(\frac{|x-y|}{|x-z|}\right)$$

as the ring T separates x and z from y and ∞, so Theorem 5.4.2 applies to give the Teichmüller estimate by the function Ψ_n. This gives

$$\frac{|f(x)-f(y)|}{|f(x)-f(z)|} \leq \Theta_K^n\left(\frac{|x-y|}{|x-z|}\right). \tag{6.53}$$

Next, suppose that $|f(x)-f(y)| < |f(x)-f(z)|$. We note that $\Theta_K^n(1) > 1$ and Θ_K^n is increasing, and thus (6.53) holds trivially unless $|x-y| < |x-z|$. In this case we consider the Grötzsch ring

$$R_G = R[B^n(|x-y|), [z,\infty); \mathbb{R}^n]$$

and following the above argument we arrive at the estimate

$$\frac{|x-z|}{|x-y|} \leq \Theta_K^n\left(\frac{|f(x)-f(z)|}{|f(x)-f(y)|}\right)$$

which we may write as

$$\frac{|f(x)-f(y)|}{|f(x)-f(z)|} \leq 1/(\Theta_K^n)^{-1}\left(\frac{1}{\frac{|x-y|}{|x-z|}}\right). \tag{6.54}$$

Thus the desired inequality holds for η as defined at (6.52). It is an easy matter to check that η is increasing on the indicated intervals with the ascribed limits. Thus η is dominated by a homeomorphism. Then continuity accounts for the situation where $|f(x)-f(y)| = |f(x)-f(z)|$. This completes the proof. □

It would be good to have a local version of Theorem 6.6.7. A moment's thought shows that this cannot be true. In two dimensions the Riemann mapping $\varphi : B^2 \to \{|\Im m(z)| < 1\}$, $\varphi(0) = 0$, between the disk and the strip cannot satisfy such an estimate. To see this consider $r \to 1$ and points z_1 and z_2, $|z_1| = |z_2| = r$ with z_1 chosen so $\varphi(z_1)$ tends to the boundary point i and $\varphi(z_2) \to \infty$. Then we would have

$$\frac{|\varphi(z_2)-\varphi(0)|}{|\varphi(z_1)-\varphi(0)|} \leq \eta(1)$$

which is impossible as the left-hand side tends to ∞. Examples such as this persist in higher dimensions, as the quasiconformal mapping between the infinite cylinder and the ball show.

The best one can expect without additional assumptions on the geometry of the domain is that quasiconformality implies some form of local quasisymmetry. Next we prove an elementary version of this; however, a more general result which we establish later in Theorem 7.5.10 shows that U in the theorem below can be taken to be $B^n(x,\delta)$ for any $\delta < \mathrm{dist}(x, \partial D)$.

THEOREM 6.6.8. *Let D be a domain in $\mathbb{R}^n$ and $f : D \to D'$ a homeomorphism. Then f is locally quasisymmetric: given $x \in D$ there is a neighbourhood U of x such that $f|U$ is quasisymmetric.*

PROOF. We first prove a local estimate. Let $x \in D$ and put $d' = \text{dist}[f(x), \partial D']$. Let

$$d_x = \text{dist}\big(f(x), \partial f^{-1}[B^n(f(x), d')]\big) > 0.$$

Let $y, z \in B^n(x, d_x)$. If $|f(x) - f(z)| < |f(x) - f(y)|$ consider the ring $R = B^n(f(x), |f(x) - f(y)|) \setminus \bar{B}^n(f(x), |f(x) - f(z)|)$. Then $f^{-1}(R)$ separates x, z from y, ∞, and so

$$\log \frac{|f(x) - f(y)|}{|f(x) - f(z)|} = \text{Mod}(R) \leq K\text{Mod}[f^{-1}(R)] \leq K\Psi_n\Big(\frac{|x - y|}{|x - z|}\Big).$$

On the other hand, if $|f(x) - f(z)| > |f(x) - f(y)|$, we have the above estimate holding unless $|x - y| < |x - z|$. So supposing otherwise, that is, supposing $|x - y| > |x - z|$, we argue as above with the ring $R = B^n(x, |x - z|) \setminus \bar{B}^n(x, |x - y|)$ to get

$$\log \frac{|x - z|}{|x - y|} \leq K\Psi_n\Big(\frac{|f(x) - f(y)|}{|f(x) - f(z)|}\Big).$$

Thus we have

$$\frac{|f(x) - f(y)|}{|f(x) - f(z)|} \leq \eta\Big(\frac{|x - y|}{|x - z|}\Big), \tag{6.55}$$

where we have defined

$$\eta(t) = \begin{cases} \exp[K\Psi_n(t)], & t \geq 1, \\ \exp[-K\Psi_n(1/t)], & t < 1. \end{cases} \tag{6.56}$$

The function η is independent of x, but the estimate (6.55) holds only in $B^n(x, d_x)$, and we cannot yet claim quasisymmetry since we do not have three arbitrary points in $B^n(x, d_x)$—our estimate chooses one of them to be x. We can resolve this situation by putting

$$U = \bigcap \Big\{B^n(u, d_u) : u \in B^n\Big(x, \frac{d_x}{3}\Big)\Big\}.$$

Thus if $u, v, w \in U$, then $u, v, w \in B^n(u, d_u)$, and our estimates hold. The triangle inequality shows us that U is a neighbourhood of x. This completes the proof of the theorem since the function η defined by (6.56) is dominated by a homeomorphism. $\square$

Given these local estimates, one can iterate them to obtain quasisymmetry on any relatively compact $U \subset D$. In fact when D has reasonable geometry—for instance if D is a so-called uniform domain—one can get global estimates from local ones. We will see this in part later in Chapter 7.

6.6.3. Hölder estimates in the spherical metric. If D is a proper subdomain of $\mathbb{R}^n$ and f is a quasiconformal mapping of D onto a domain D' in $\mathbb{R}^n$, then the inequality

$$\frac{|f(y) - f(x)|}{\text{dist}[f(x), \partial D']} \leq \Theta_K^n\left(\frac{|y - x|}{\text{dist}(x, \partial D)}\right)$$

holds for $K = K_I(f)$ whenever x belongs to D and $|y - x| < \text{dist}(x, \partial D)$.

The preceding results give reasonably sharp bounds for $|f(y) - f(x)|$ in terms of $|y - x|$. The next theorem, which is motivated in part by Theorem 3.6.3, delivers an upper bound for $q[f(x), f(y)]$ in terms of $q(x, y)$. Although this bound is not quite as precise as those above, it compensates with its generality.

THEOREM 6.6.9. *If D is a domain in $\hat{\mathbb{R}}^n$ with at least two boundary points and f is a quasiconformal mapping of D into $\hat{\mathbb{R}}^n$, then*

$$q\big[f(x), f(y)\big]\, q\big[f(D)^c\big] \le 8\lambda_n^2 \left[\frac{q(x,y)}{\max\{q(x,D^c), q(y,D^c)\}}\right]^\alpha$$

for all x and y in D, where $\alpha = K_I(f)^{1/(1-n)}$.

PROOF. We infer from the comment following Corollary 6.4.27 that $f(D)^c$ contains at least two points, and hence that $q[f(D)^c] > 0$. We shall establish the inequality

$$q\big[f(x), f(y)\big]\, q\big[f(D)^c\big] \le 8\lambda_n^2 \left[\frac{q(x,y)}{q(x,D^c)}\right]^\alpha \tag{6.57}$$

for all x and y in D, with α as indicated in the statement of the theorem. By symmetry

$$q\big[f(x), f(y)\big]\, q\big[f(D)^c\big] \le 8\lambda_n^2 \left[\frac{q(x,y)}{q(y,D^c)}\right]^\alpha$$

also holds for all x and y in D. In combination these two inequalities produce the desired conclusion.

Fix x and y in D, $x \ne y$. Set $r = q(x,y)$ and $s = q(x,D^c)$. Since D^c contains more than one point, $s < 2$. Assume initially that $r < s$, and consider the chordal spherical ring $R = \{z \in \hat{\mathbb{R}}^n : r < q(z,x) < s\}$. According to (5.6),

$$\operatorname{Mod}(R) = \log \frac{s(4-r^2)^{1/2}}{r(4-s^2)^{1/2}} > \log \frac{s}{r}\,.$$

The ring R separates x and y from D^c, so its image $R' = f(R)$ separates $f(x)$ and $f(y)$ from $f(D)^c$. Let C_0 and C_1 denote the complementary components of R', labeled so that $f(x)$ and $f(y)$ lie in C_0, while C_1 contains $f(D)^c$. By virtue of Corollary 5.4.4 and Theorem 5.1.9 we can assert that

$$\begin{aligned}
\operatorname{Mod}(R') &\le \log \Psi_n \left[\frac{4}{q(C_0)q(C_1)}\right] \le \log\left(\lambda_n^2\left[\frac{4}{q(C_0)q(C_1)}+1\right]\right) \\
&= \log \frac{\lambda_n^2[4+q(C_0)q(C_1)]}{q(C_0)q(C_1)} \le \log \frac{8\lambda_n^2}{q(C_0)q(C_1)} \\
&\le \log \frac{8\lambda_n^2}{q[f(x),f(y)]\, q[f(D)^c]}\,.
\end{aligned}$$

Since

$$\operatorname{Mod}(R) \le K_I(f)^{1/(n-1)} \operatorname{Mod}(R') = \alpha^{-1}\operatorname{Mod}(R')\,,$$

we conclude that

$$\alpha \log \frac{q(x,D^c)}{q(x,y)} = \alpha \log \frac{s}{r} \le \alpha \operatorname{Mod}(R) \le \operatorname{Mod}(R') \le \log \frac{8\lambda_n^2}{q[f(x),f(y)]\, q[f(D)^c]}\,.$$

This leads directly to (6.57), at least when $q(x,y) < q(x,D^c)$. On the other hand, if $q(x,y) \ge q(x,D^c)$, then it is essentially trivial that

$$q\big[f(x), f(y)\big]\, q\big[f(D)^c\big] \le 4 \le 4\left[\frac{q(x,y)}{q(x,D^c)}\right]^\alpha \le 8\lambda_n^2 \left[\frac{q(x,y)}{q(x,D^c)}\right]^\alpha,$$

for $8\lambda_n^2 \ge 4$. Thus inequality (6.57) is valid in all cases. □

Theorem 6.6.9 ensures that quasiconformal mappings are locally Hölder continuous with respect to the chordal metric.

COROLLARY 6.6.10. *Suppose that f is a quasiconformal mapping of a subdomain D of $\hat{\mathbb{R}}^n$ into $\hat{\mathbb{R}}^n$. If $q(D^c) > 0$ and if A is a compact set in D, then*

$$q\big[f(x), f(y)\big] \le a q(x,y)^\alpha \tag{6.58}$$

for all x and y in A, where $\alpha = K_I(f)^{1/(1-n)}$ and

$$a = \frac{8\lambda_n^2}{q[f(D)^c]\, q(A, D^c)^\alpha}\,.$$

If $q(D^c) = 0$, then there is a constant $a > 0$, which depends on f, such that (6.58) holds with α as above for all x and y in D.

PROOF. In the case where D^c contains at least two points, (6.58) follows, with α and a as specified, from Theorem 6.6.9. Turning to the situation in which $q(D^c) = 0$, we note that either $D = \hat{\mathbb{R}}^n$ or $D = \hat{\mathbb{R}}^n \setminus \{x_0\}$ for some x_0. In the latter instance Corollary 6.4.27 allows us to extend f to a quasiconformal self-mapping of $\hat{\mathbb{R}}^n$ whose dilatations coincide with those of f, so we may as well assume from the outset that $D = \hat{\mathbb{R}}^n$.

Consider the domains

$$D_1 = \big\{x \in \hat{\mathbb{R}}^n : q(x,0) < 7/4\big\}, \qquad D_2 = \big\{x \in \hat{\mathbb{R}}^n : q(x,0) > 1/4\big\}$$

and the compact sets

$$A_1 = \big\{x \in \hat{\mathbb{R}}^n : q(x,0) \le 3/2\big\}, \qquad A_2 = \big\{x \in \hat{\mathbb{R}}^n : q(x,0) \le 1/2\big\}\,.$$

Applying what we learned above to $f|D_j$, we come up with constants a_1 and a_2, which depend on $q[f(D_1)^c]$ and $q[f(D_2)^c]$, such that $q[f(x), f(y)] \le a_j q(x,y)^\alpha$ holds with $\alpha = K_I(f)^{1/(1-n)}$ whenever both x and y belong to A_j. Now let x and y be arbitrary points of $\hat{\mathbb{R}}^n$. Assuming that neither A_1 nor A_2 contains both x and y, one of these points must lie in $A_1 \setminus A_2$ and the other in $A_2 \setminus A_1$. In this event $q(x,y) \ge 1$, whence

$$q\big[f(x), f(y)\big] \le 2 \le 2q(x,y)^\alpha\,.$$

We have now shown that (6.58) holds with $a = \max\{a_1, a_2, 2\}$ for all x and y in $\hat{\mathbb{R}}^n$. ☐

From Corollary 6.6.10 we derive an analogue of Theorem 3.6.2.

COROLLARY 6.6.11. *Let f be a K-quasiconformal mapping of $\hat{\mathbb{R}}^n$ onto itself. There is a constant b, which depends on f, such that*

$$b^{-1} q(x,y)^{1/\beta} \le q\big[f(x), f(y)\big] \le b\, q(x,y)^\beta$$

for all x and y in $\hat{\mathbb{R}}^n$, where $\beta = K^{1/(1-n)}$.

PROOF. By Corollary 6.6.10 there exists a constant $a_1 = a_1(f) > 0$ such that

$$q\big[f(x), f(y)\big] \le a_1 q(x,y)^\alpha$$

for all x and y in $\hat{\mathbb{R}}^n$, where $\alpha = K_I(f)^{1/(1-n)}$.

Now $1 \leq K_I(f) \leq K$, so $\beta \leq \alpha \leq 1$. Therefore

$$\begin{aligned} q\big[f(x), f(y)\big] &\leq a_1 q(x,y)^\alpha = 2^\alpha a_1 \left[\frac{q(x,y)}{2}\right]^\alpha \\ &\leq 2^\alpha a_1 \left[\frac{q(x,y)}{2}\right]^\beta = 2^{\alpha-\beta} a_1 q(x,y)^\beta \leq 2a_1 q(x,y)^\beta \end{aligned}$$

is also true for all x and y in $\hat{\mathbb{R}}^n$. The same reasoning applied to f^{-1} delivers a constant $a_2 > 0$ with the property that

$$q\big[f^{-1}(x), f^{-1}(y)\big] \leq 2a_2 q(x,y)^\beta$$

or, equivalently,

$$q\big[f(x), f(y)\big] \geq (2a_2)^{-1/\beta} q(x,y)^{1/\beta}$$

for all x and y in $\hat{\mathbb{R}}^n$. The conclusion follows, with $b = \max\{2a_1, (2a_2)^{1/\beta}\}$. □

The Hölder exponent that arises in Corollaries 6.6.5 and 6.6.10 is optimal. To convince oneself of this, one need look no further than the radial stretching f of $\hat{\mathbb{R}}^n$ associated with a number α in $(0,1]$ that we considered in Section 6.2.1:

$$f(x) = |x|^{\alpha-1} x.$$

As we saw, this mapping has $K_I(f) = \alpha^{1-n}$—which means that $K_I(f)^{1/(1-n)} = \alpha$—and satisfies $|f(x)| = |x|^\alpha$ for every x in $\mathbb{R}^n$, while

$$q(x,0)^\alpha \leq q\big[f(x), f(0)\big] \leq 4q(x,0)^\alpha$$

holds for every x in $A = \overline{B}^n$.

Notice that the constant a in the first part of Corollary 6.6.10 tends to ∞ as A approaches ∂D. This suggests that, in the case $q(D^c) > 0$, the mapping f might not obey a global Hölder condition in D, at least not one whose order is $\alpha = K_I(f)^{1/(1-n)}$. Examples can be found in [**52**] which confirm that this suspicion is correct: for a quasiconformal mapping $f : D \to D'$ with $q(D^c) > 0$ there need be no α in $(0,1]$ such that f satisfies a uniform Hölder condition of order α throughout D.

The literature on this subject gives a number of results on the topic of Hölder continuity of quasiconformal mappings. Some of these describe conditions under which Hölder continuity is manifested on a global scale—these basically depend on the regularity of the boundary of the domain in various technical senses. The case of quasiballs is considered in [**52**].

We leave the reader to discover other such results for themselves; the paper [**14**] and the references therein would be a reasonable place to start.

6.6.4. Normal families of quasiconformal mappings. Given the material in the preceding section and the knowledge of the importance of the Arzela-Ascoli theorem in analysis, the following generalization of Theorem 3.6.5 should come as no great surprise.

THEOREM 6.6.12. *Suppose that D is a domain in $\hat{\mathbb{R}}^n$ and that $\mathcal{F}$ is a non-empty family of K-quasiconformal mappings of D into $\hat{\mathbb{R}}^n$. If each point x of D has an open neighbourhood $U = U_x$ in D with the property that* $\inf\{q[f(U)^c] : f \in \mathcal{F}\} > 0$, *then $\mathcal{F}$ is a normal family.*

PROOF. In light of Theorem 3.6.4 it suffices to verify that $\mathcal{F}$ is equicontinuous at each point of D. Fixing x_0 in D, we pick r in $(0,1)$ for which $U = \{x : q(x,x_0) < 2r\}$ has closure in D and for which $d = \inf\{q[f(U)^c] : f \in \mathcal{F}\} > 0$. Given f in $\mathcal{F}$, we apply Corollary 6.6.10 to $f|U$ and $A = \{x : q(x,x_0) \leq r\}$ to conclude that

$$q\big[f(x), f(x_0)\big] \leq aq(x,x_0)^\alpha \leq bq(x,x_0)^\beta$$

for every x in A, where $\alpha = K_I(f)^{1/(1-n)}$, $a = 8\lambda_n^2 d^{-1} r^{-\alpha}$, $\beta = K^{1/(1-n)}$, and $b = 8\lambda_n^2 d^{-1} r^{-1}$. Since b and β are plainly independent of f, the equicontinuity of $\mathcal{F}$ at x_0 becomes apparent. □

We list in the following corollary some specific circumstances in which Theorem 6.6.12 tells us directly that a family of K-quasiconformal mappings is normal.

COROLLARY 6.6.13. *Suppose that D is a domain in $\hat{\mathbb{R}}^n$ and that $\mathcal{F}$ is a nonempty family of K-quasiconformal mappings of D into $\hat{\mathbb{R}}^n$. Under any of the following conditions $\mathcal{F}$ is a normal family:*

(i) $r = \inf\{q[f(D)^c] : f \in \mathcal{F}\} > 0$;
(ii) *there exist a constant $r > 0$ and a pair of distinct points a and b in D such that for each member f of $\mathcal{F}$ the set $f(D)^c$ contains a point y_f with the property that $q[f(a), y_f] \geq r$ and $q[f(b), y_f] \geq r$;*
(iii) *there exist a constant $r > 0$ and distinct points a, b and c of D such that*

$$q\big[f(a), f(b)\big] \geq r, \qquad q\big[f(a), f(c)\big] \geq r, \qquad q\big[f(b), f(c)\big] \geq r$$

for every f in $\mathcal{F}$.

PROOF. We apply Theorem 6.6.12. Given x in D we choose $U = U_x$ as follows: in case (i), we take $U = D$; in case (ii), we use $U = D \setminus \{a\}$ if $x \neq a$, and $U = D \setminus \{b\}$ if $x = a$; in case (iii), take $U = D \setminus \{a,b\}$ if x is neither a nor b, $U = D \setminus \{b,c\}$ if $x = a$, and $U = D \setminus \{a,c\}$ if $x = b$. In each case $q[f(U)^c] \geq r$ for every f in $\mathcal{F}$. □

We cite another corollary of Theorem 6.6.12.

COROLLARY 6.6.14. *Suppose that D is a domain in $\hat{\mathbb{R}}^n$ and that $\mathcal{F}$ is a nonempty family of K-quasiconformal mappings of D into $\hat{\mathbb{R}}^n$ which is not a normal family. Then there exist a point x_0 of D and an infinite subfamily $\mathcal{F}_0$ of $\mathcal{F}$ such that $\mathcal{F}_0|D_0$ is a normal family, where $D_0 = D \setminus \{x_0\}$.*

PROOF. Because of Theorem 3.6.4 there must be a point of D at which $\mathcal{F}$ fails to be equicontinuous. Choose such a point and label it x_0.

The failure of $\mathcal{F}$ to be equicontinuous at x_0 means that there is an $\varepsilon > 0$ for which the following is true: for each $\delta > 0$ the family $\mathcal{F}$ contains infinitely many functions f with the property that $q[f(x), f(x_0)] \geq \varepsilon$ holds for some point $x = x_f$ satisfying $q(x,x_0) < \delta$. This makes it possible to produce a sequence $\langle f_\nu \rangle$ of distinct mappings from $\mathcal{F}$ and an associated sequence $\langle x_\nu \rangle$ of points in D such that $x_\nu \to x_0$, while

$$q[f_\nu(x_\nu), f_\nu(x_0)] \geq \varepsilon.$$

We take $\mathcal{F}_0 = \{f_\nu : \nu = 1, 2, \ldots\}$. For $\nu \geq 1$ we write

$$D_\nu = D \setminus \{x_0, x_\nu, x_{\nu+1}, \ldots\}.$$

Given x in $D_0 = D \setminus \{x_0\}$, we can select μ such that D_μ contains x and set $U = U_x = D_\mu$. Then $q[f_\nu(U)^c] \geq q[f_\nu(x_\nu), f_\nu(x_0)] \geq \varepsilon$ for $\nu \geq \mu$, whence

$$q\big[f(U)^c\big] \geq \min\Big\{\varepsilon, \min_{1\leq\nu\leq\mu} q\big[f_\nu(x_\mu), f_\nu(x_0)\big]\Big\} > 0$$

for every f in $\mathcal{F}_0$. We can now appeal to Theorem 6.6.12 to persuade ourselves that $\mathcal{F}_0|D_0$ is a normal family. □

Sometimes it is important to have at one's disposal a normal family criterion for the Euclidean metric. Such a criterion can be extracted from Theorem 6.6.12.

THEOREM 6.6.15. *Suppose that D is a domain in $\mathbb{R}^n$ and that $\mathcal{F}$ is a nonempty family of K-quasiconformal mappings of D into $\mathbb{R}^n$. Then $\mathcal{F}$ is a normal family with respect to the Euclidean metric if and only if $\mathcal{F}$ is pointwise bounded in D.*

PROOF. The necessity follows directly from Theorem 3.6.4. For the sufficiency assume that $\mathcal{F}$ is pointwise bounded in D. To confirm the normality of $\mathcal{F}$ relative to the Euclidean metric it suffices to demonstrate that $\mathcal{F}$ is equicontinuous with respect to the Euclidean metric at an arbitrary point x_0 of D as Theorem 3.6.4 applies here.

For each x in D it is true that $\inf\{q[f(x), \infty] : f \in \mathcal{F}\} > 0$. By Corollary 6.6.13(ii), therefore, $\mathcal{F}$ is at least a normal family with respect to the chordal metric. Let $r = \inf\{q[f(x_0), \infty] : f \in \mathcal{F}\}$. We exploit the chordal normality of $\mathcal{F}$ to select a number $s > 0$ with the property that $q[f(x), f(x_0)] < r/2$ whenever f is in $\mathcal{F}$ and $q(x, x_0) < 2s$. Then $q[f(x), \infty] \geq r/2$ for such x and f, from which we infer that $c = \sup\{|f(x)| : f \in \mathcal{F},\ q(x, x_0) < 2s\} < \infty$. Finally, given $\varepsilon > 0$ we choose δ in $(0, s)$ so that $B^n(x_0, \delta)$ lies in D and so that $q[f(x), f(x_0)] < \varepsilon$ holds for every member f of $\mathcal{F}$ whenever $q(x, x_0) < 2\delta$. Since $q(x, x_0) \leq 2|x - x_0|$, we see that

$$\begin{aligned} |f(x) - f(x_0)| &= 2\big(1 + |f(x)|^2\big)^{1/2}\big(1 + |f(x_0)|^2\big)^{1/2} q\big[f(x), f(x_0)\big] \\ &\leq 2(1 + c^2) q\big[f(x), f(x_0)\big] \leq 2(1 + c^2)\varepsilon \end{aligned}$$

if f comes from $\mathcal{F}$ and if $|x - x_0| < \delta$. We have thus established the desired Euclidean equicontinuity of $\mathcal{F}$ at x_0. □

6.6.5. Convergence of quasiconformal mappings. We start this section by recalling a general situation in which locally uniform convergence is implied by pointwise convergence.

LEMMA 6.6.16. *Let $\mathcal{F}$ be an equicontinuous family of functions from a locally compact metric space X to a complete metric space Y. If a sequence $\langle f_\nu \rangle$ from $\mathcal{F}$ converges pointwise on a dense subset E of X, then $\langle f_\nu \rangle$ is locally uniformly convergent in X.*

PROOF. We write d and d' for the metrics in X and Y, respectively. Let K be a compact subset of X. We shall show that $\langle f_\nu \rangle$ is a uniform Cauchy sequence on K. Given $\varepsilon > 0$ we exploit the equicontinuity of $\mathcal{F}$ to choose for each p in K a number $\delta_p > 0$ so that $d'[f(q), f(p)] < \varepsilon$ whenever f belongs to $\mathcal{F}$ and $d(q, p) < \delta_p$. The collection of balls $B_p = \{q \in X : d(q, p) < \delta_p\}$ with p in K forms an open covering of K. We extract from this cover a finite subcovering, say $B_1 = B_{p_1}$, $B_2 = B_{p_2}, \dots, B_r = B_{p_r}$, and then choose for each j a point e_j in $E \cap B_j$. If p is an arbitrary point of K, then p lies in B_j for some j, so

$$d'\big[f(p), f(e_j)\big] \leq d'\big[f(p), f(p_j)\big] + d'\big[f(p_j), f(e_j)\big] < 2\varepsilon$$

for every f in $\mathcal{F}$. The pointwise convergence of $\langle f_\nu \rangle$ on E implies the existence of an index N with the feature that $d'[f_\nu(e_j), f_\mu(e_j)] < \varepsilon$ for $j = 1, 2, \ldots, r$ whenever $\mu \geq \nu \geq N$. Combining these two observations we discover that $d'[f_\nu(p), f_\mu(p)] < 5\varepsilon$ for every p in K whenever $\mu \geq \nu \geq N$, which confirms that $\langle f_\nu \rangle$ is a uniform Cauchy sequence on K. The completeness of Y guarantees that $\langle f_\nu \rangle$ converges uniformly on K, an arbitrary compact subset of X. Since each point of X has a compact neighbourhood, the local uniform convergence of $\langle f_\nu \rangle$ in X is assured. □

Let $\langle A_\nu \rangle$ be a sequence of sets in $\hat{\mathbb{R}}^n$ with $n \geq 2$. The *kernel of* $\langle A_\nu \rangle$, denoted by $\ker_{\nu\to\infty} A_\nu$, is the set defined as follows:

$$\ker_{\nu\to\infty} A_\nu = \bigcup_{\nu=1}^{\infty} \operatorname{int}\left(\bigcap_{\mu=\nu}^{\infty} A_\mu\right) .$$

In other words, a point x belongs to $\ker_{\nu\to\infty} A_\nu$ if and only if x has a neighbourhood U that is included in A_ν for all sufficiently large ν. We conclude that each compact subset of $\ker_{\nu\to\infty} A_\nu$ is contained in A_ν once ν is suitably large.

Clearly $\ker_{\nu\to\infty} A_\nu$ is an open set. It is not difficult to give examples where this set need not be connected, however, even if each of the sets A_ν is a domain.

Suppose now that $f_\nu : A_\nu \to \hat{\mathbb{R}}^n$ and that D is a domain which lies in $\ker_{\nu\to\infty} A_\nu$.

Under these conditions it makes perfectly good sense to speak of the sequence $\langle f_\nu \rangle$ converging in D, whether the convergence is pointwise or locally uniform, to a function $f : D \to \hat{\mathbb{R}}^n$. We simply require that for each point x of D there exist an index N and a neighbourhood U of x in D such that U lies in A_ν for all $\nu \geq N$ and such that the truncated sequence $\langle f_\nu \rangle_{\nu \geq N}$ converges to f pointwise in U (respectively, uniformly on U).

Our first convergence result dealing specifically with quasiconformal mappings is reminiscent of Theorem 3.6.7.

THEOREM 6.6.17. *Let $\langle D_\nu \rangle$ be a sequence of domains in $\hat{\mathbb{R}}^n$ with $n \geq 2$, let f_ν be a K-quasiconformal mapping of D_ν onto a domain D'_ν, and let D be a subdomain of $\ker_{\nu\to\infty} D_\nu$. Assume that $f_\nu \to f$ pointwise in D. There are the following three possibilities for the limit mapping f:*

(i) *f might take exactly two values in D—one of these at precisely one point of D—in which event the convergence of $\langle f_\nu \rangle$ is not locally uniform in D;*
(ii) *f might be a homeomorphism of D onto a subdomain of $\ker_{\nu\to\infty} D'_\nu$, in which event the convergence of $\langle f_\nu \rangle$ is locally uniform in D;*
(iii) *f might be constant in D, in which event the convergence of $\langle f_\nu \rangle$ may or may not be locally uniform in D.*

PROOF. It obviously suffices to prove that, for each domain G whose closure lies in D, one of the statements (i), (ii) or (iii) is true with G in place of D. Fix such a domain G.

There is no loss of generality incurred by assuming that G is contained in D_ν for every ν. Suppose first that f takes exactly two values in G, call them a and b. Choose points x_0 and x'_0 of G for which $f(x_0) = a$ and $f(x'_0) = b$. Since

$$\lim_{\nu\to\infty} q\big[f_\nu(x_0), f_\nu(x'_0)\big] = q\big[f(x_0), f(x'_0)\big] = q(a, b) > 0 ,$$

it is clear that with $G_0 = G \setminus \{x_0, x_0'\}$ we have

$$\inf_\nu q\big[f_\nu(G_0)^c\big] \geq \inf_\nu q\big[f_\nu(x_0), f_\nu(x_0')\big] > 0\,.$$

By Corollary 6.6.13(i) the family $\mathcal{F}_0 = \{f_\nu|G_0 : \nu = 1, 2, \ldots\}$ is a normal family. In particular, $\mathcal{F}_0$ is equicontinuous in G_0. In conjunction with the pointwise convergence of $\langle f_\nu \rangle$ in G_0, equicontinuity forces the convergence of $\langle f_\nu \rangle$ to be locally uniform there (Lemma 6.6.16). It follows that $f|G_0$ is continuous. Now the set $f(G_0)$ is connected and contained in $\{a, b\}$, so either $f(G_0) = \{a\}$ or $f(G_0) = \{b\}$; i.e., either $f \equiv a$ in $G \setminus \{x_0'\}$ or $f \equiv b$ in $G \setminus \{x_0\}$. Because f is necessarily discontinuous at either x_0 or x_0', the convergence of $\langle f_\nu \rangle$ could not be locally uniform in G. In other words, we find ourselves in case (i).

We next assume that there exist points x_1, x_2, and x_3 of G for which the values $f(x_1), f(x_2)$ and $f(x_3)$ are distinct. We shall show that in this case the convergence of $\langle f_\nu \rangle$ is locally uniform in G and that f is a continuous, one-to-one mapping of G onto a subdomain of $\bigcap_{\nu=1}^\infty D_\nu'$, putting us into case (ii). Incidentally, by doing so we complete the proof, for (iii) plainly offers the only alternative to (i) and (ii).

It is evident that

$$\inf_\nu \min_{k \neq \ell} q\big[f_\nu(x_k), f_\nu(x_\ell)\big] > 0\,.$$

Theorem 3.6.4 and Corollary 6.6.13(iii) imply that the family $\mathcal{F} = \{f_\nu|G : \nu = 1, 2, \ldots\}$ is equicontinuous in D, so $f_\nu \to f$ locally uniformly in G. We again cite Lemma 6.6.16. Accordingly, f is continuous in G. Consider an arbitrary point x_0 in G. We select a chordal ball $B = \{x : q(x, x_0) < \delta\}$ such that $\overline{B}$ is contained in G and such that $q[f_\nu(B)] < 1$ for every ν.

The equicontinuity of $\mathcal{F}$ makes certain that we can achieve the latter. We claim that either f is injective in B or it is constant in B. If the claim were false, we could find distinct points x, y and z in B satisfying $f(x) = f(y) \neq f(z)$. Assuming this to be the situation, let $R = \mathcal{R}(C_0, C_1)$ be a ring that meets the following specifications: C_0 is an arc in $B \setminus \{y\}$ with endpoints x and z; $C_1 = B^c \cup A$, where A is an arc in C_0^c that joins y to ∂B. Then R lies in B.

We now have $R_\nu' = f_\nu(R) = \mathcal{R}(C_0^\nu, C_1^\nu)$, where $C_0^\nu = f_\nu(C_0)$ and C_1^ν is a continuum that contains both $f_\nu(B)^c$ and $f_\nu(y)$. Clearly $\operatorname{Cap}(R) < \infty$, so $\operatorname{Mod}(R) > 0$. In view of the choices of x and z we must have

$$c = \inf_\nu q(C_0^\nu) \geq \inf_\nu q\big[f_\nu(x), f_\nu(z)\big] > 0\,,$$

for $f_\nu(x) \to f(x)$ and $f_\nu(z) \to f(z)$. The fact that $q[f_\nu(B)] < 1$ for every ν means that

$$\inf_\nu q(C_1^\nu) \geq \inf_\nu q\big[f_\nu(B)^c\big] \geq 1\,.$$

Because $f(x) = f(y)$ we also discover that

$$d_\nu = q(C_0^\nu, C_1^\nu) \leq q\big[f_\nu(x), f_\nu(y)\big] \to q\big[f(x), f(y)\big] = 0$$

as $\nu \to \infty$. Corollary 5.4.4 provides the estimate

$$\operatorname{Mod}(R_\nu') \leq \log \Psi_n \left[\frac{8q(C_0^\nu, C_1^\nu)}{q(C_0^\nu)q(C_1^\nu)}\right] \leq \log \Psi_n \left(\frac{8d_\nu}{c}\right)\,.$$

Now, if we let $\nu \to \infty$, we run into a contradiction:

$$0 < \operatorname{Mod}(R) \leq K^{1/(n-1)} \operatorname{Mod}(R_\nu') \leq K^{1/(n-1)} \log \Psi_n \left(\frac{8d_\nu}{c}\right) \to 0\,,$$

since $\Psi_n(t) \to 1$ as $t \to 0$. This contradiction is avoided only if f is either injective in B or constant there.

Let U denote the set of x in G such that f is injective in some open neighbourhood of x, and let V be the set of x in D such that f is constant in some open neighbourhood of x. The sets U and V are open by definition and $U \cap V = \emptyset$. The argument presented above shows that $G = U \cup V$. It follows that $G = U$ or $G = V$. A locally constant function in G would be constant throughout G, which f by assumption is not. As a consequence, $G = U$; thus f is at least locally injective in G.

Finally, suppose that $f(x) = f(y)$, where x and y are distinct points of G. We may assume $x \neq \infty$. Fix $r > 0$ so that $\overline{B}^n(x,r)$ is contained in G, so that f is injective in $\overline{B}^n(x,r)$, and so that y is not a point of $\overline{B}^n(x,r)$. Then $S = \mathbb{S}^{n-1}(x,r)$ separates x and y in $\hat{\mathbb{R}}^n$, and $f_\nu(S)$ separates $f_\nu(x)$ and $f_\nu(y)$ in $\hat{\mathbb{R}}^n$. We deduce the existence of a point z_ν of S for which $q[f_\nu(z_\nu), f_\nu(x)] \leq q[f_\nu(y), f_\nu(x)]$. Otherwise the chordal ball $\{w : q[w, f_\nu(x)] \leq q[f_\nu(y), f_\nu(x)]\}$ would be a connected set in $f_\nu(S)^c$ containing both $f_\nu(x)$ and $f_\nu(y)$. It now follows that

$$q\big[f_\nu(z_\nu), f_\nu(x)\big] \leq q\big[f_\nu(y), f_\nu(x)\big] \to q\big[f(y), f(x)\big] = 0$$

as $\nu \to \infty$. Passing to a subsequence if need be, we may assume that $z_\nu \to z$, a point of S.

The fact that $f_\nu \to f$ uniformly on S implies that $f_\nu(z_\nu) \to f(z)$. However, this leads to

$$q\big[f(z), f(x)\big] = \lim_{\nu\to\infty} q\big[f_\nu(z_\nu), f_\nu(x)\big] = 0 \ ,$$

so $f(z) = f(x)$, contradicting the one-to-oneness of f in $\overline{B}^n(x,r)$. We conclude that f must be injective in G.

Finally, since G lies in D_ν for all ν, it is clear that $G' = f(G)$ is a subdomain of $\ker_{\nu\to\infty} D'_\nu$. □

Each of the possibilities permitted by Theorem 6.6.17 for the limit mapping f can actually occur: To see this we provide the following examples:

(i) if $f_\nu : \mathbb{R}^n \to \mathbb{R}^n$ is given by $f_\nu(x) = (1+2^{-\nu})x$, then f_ν is 1-quasiconformal and $f_\nu \to id$, the identity mapping of $\hat{\mathbb{R}}^n$, uniformly on $\hat{\mathbb{R}}^n$;

(ii) if $f_\nu : \hat{\mathbb{R}}^n \to \hat{\mathbb{R}}^n$ is given by $f_\nu(x) = 2^{-\nu}x$, then f_ν is 1-quasiconformal and $f_\nu \to f$ pointwise in $\hat{\mathbb{R}}^n$, where $f(x) = 0$ for x in $\mathbb{R}^n$ and $f(\infty) = \infty$;

(iii) if $f_\nu : \hat{\mathbb{R}}^n \to \hat{\mathbb{R}}^n$ is given by $f_\nu(x) = x + \nu e_1$, then f_ν is 1-quasiconformal and $f_\nu \to \infty$ pointwise in $\hat{\mathbb{R}}^n$, the convergence being locally uniform in $\mathbb{R}^n$ but not locally uniform (which is to say uniform) in $\hat{\mathbb{R}}^n$.

Suppose that D is a domain in $\mathbb{R}^n$ and that $\langle f_\nu \rangle$, a sequence of K-quasiconformal mappings of D into $\hat{\mathbb{R}}^n$, is known to converge pointwise in D to a homeomorphism f. In view of Theorem 6.6.17, $f_\nu \to f$ locally uniformly in D. There are situations in which one can draw an even stronger conclusion, namely, that $f_\nu \to f$ uniformly on D. For instance, if each term f_ν in the sequence admits a continuous extension f_ν^* to $\bar{D}$ and if the sequence $\langle f_\nu^* \rangle$ is uniformly convergent on ∂D, then it is not difficult to show that $f_\nu \to f$ uniformly on D.

We now proceed to refine the information obtained in case (ii) of Theorem 6.6.17.

6.6.6. Lower semicontinuity of distortions. In this subsection it is our intention to prove the lower semicontinuity of the inner, outer, and maximal distortion functions. These facts are crucial to results concerning the existence of extremal quasiconformal mappings for a number of interesting geometric problems. These results are encapsulated in the following theorem.

THEOREM 6.6.18. *Let $\langle D_\nu \rangle$ be a sequence of domains in $\hat{\mathbb{R}}^n$ with $n \geq 2$, let f_ν be a homeomorphism of D_ν into $\hat{\mathbb{R}}^n$, and let D be a subdomain of $\ker_{\nu\to\infty} D_\nu$. Suppose that $f_\nu \to f$ locally uniformly in D, where f is a homeomorphism. Then*

$$K_I(f) \leq \liminf_{\nu\to\infty} K_I(f_\nu), \qquad K_O(f) \leq \liminf_{\nu\to\infty} K_O(f_\nu).$$

In particular, if each of the mappings f_ν is K-quasiconformal, then f is K-quasiconformal as well.

PROOF. Write $K = \liminf_{\nu\to\infty} K_I(f)$. If $K = \infty$, then $K_I(f) \leq K$ holds trivially. We proceed assuming that $K < \infty$. Consider an arbitrary ring R whose closure lies in D. We shall verify that $R' = f(R)$ has $\operatorname{Cap}(R') \leq K\operatorname{Cap}(R)$. From this we infer that $K_I^*(f) \leq K$, which by reason of Corollary 6.4.17 is equivalent to the assertion that $K_I(f) \leq K$. It obviously suffices to deal with the case where R is a nondegenerate ring. We know that $\bar{R}$ is a subset of D_ν for all large ν—to avoid notational complications we presume this to be the case for every ν. Let B_0 and B_1 denote the boundary components of R. The uniform convergence of $\langle f_\nu \rangle$ on $\bar{R}$ makes it apparent that $f_\nu(B_0) \to f(B_0)$ and $f_\nu(B_1) \to f(B_1)$ in the Hausdorff metric. If we write $R'_\nu = f(R_\nu)$, then Theorem 4.3.15 informs us that

$$\begin{aligned}\lim_{\nu\to\infty} \operatorname{Cap}(R'_\nu) &= \lim_{\nu\to\infty} M\big(\Delta\big[f_\nu(B_0), f_\nu(B_1) : \hat{\mathbb{R}}^n\big]\big) \\ &= M\Big(\Delta\big[f(B_0), f(B_1) : \hat{\mathbb{R}}^n\big]\Big) = \operatorname{Cap}(R').\end{aligned}$$

The fact that $0 < \operatorname{Cap}(R) < \infty$ means that $\operatorname{Cap}(R'_\nu) \leq K_I(f_\nu)\operatorname{Cap}(R)$ for all ν (even if $K_I(f_\nu) = \infty$), so

$$\operatorname{Cap}(R') = \lim_{\nu\to\infty} \operatorname{Cap}(R'_\nu) \leq \liminf_{\nu\to\infty} K_I(f_\nu)\operatorname{Cap}(R) = K\operatorname{Cap}(R).$$

The statement concerning $K_O(f)$ is handled similarly. □

This result can be rephrased in terms of the continuity properties of these distortion functions as follows.

COROLLARY 6.6.19. *Let D be a domain in $\mathbb{R}^n$. The functions $K_I(f)$, $K_O(f)$, and $K(f)$ defined on the space of quasiconformal mappings of D into $\mathbb{R}^n$ with the topology of local uniform convergence are lower semicontinuous.*

Of course this result applies for domains in $\hat{\mathbb{R}}^n$ with the obvious modifications. However it does not hold for the linear distortion in higher dimensions $n \geq 3$ as we will see in subsection 6.6.9.

6.6.7. Convergence and kernels of domains. We now wish to examine more closely the implications of Theorem 6.6.18 for situations in which the domain D is a component of $\ker_{\nu\to\infty} D_\nu$.

In the first instance we assume that $D = \hat{\mathbb{R}}^n$. This, of course, ensures that $D_\nu = D'_\nu = \hat{\mathbb{R}}^n$ for all large ν, so we may as well assume this to be so for all ν.

We recall that any homeomorphism φ of $\hat{\mathbb{R}}^n$ into $\hat{\mathbb{R}}^n$ must have $\varphi(\hat{\mathbb{R}}^n) = \hat{\mathbb{R}}^n$, for $\varphi(\hat{\mathbb{R}}^n)$ is both open and closed in $\hat{\mathbb{R}}^n$.

THEOREM 6.6.20. *Suppose that $\langle f_\nu \rangle$ is a sequence of K-quasiconformal mappings of $\hat{\mathbb{R}}^n$ onto itself and that $f_\nu \to f$ uniformly on $\hat{\mathbb{R}}^n$. Then f is a K-quasiconformal mapping of $\hat{\mathbb{R}}^n$ onto itself.*

Moreover, $f_\nu^{-1} \to f^{-1}$ uniformly on $\hat{\mathbb{R}}^n$ as well.

PROOF. Theorems 6.6.17 and 6.6.18 show that the limit mapping f is either a K-quasiconformal self-mapping of $\hat{\mathbb{R}}^n$ or a constant mapping. The latter possibility is ruled out by the fact that convergence is uniform: we have $q[f(\hat{\mathbb{R}}^n)] = \lim_{\nu\to\infty} q[f_\nu(\hat{\mathbb{R}}^n)] = 2$. Finally,

$$\begin{aligned} \max_{x\in\hat{\mathbb{R}}^n} q\big[f_\nu^{-1}(x), f^{-1}(x)\big] &= \max_{y\in\hat{\mathbb{R}}^n} q\big[y, f^{-1}\circ f_\nu(y)\big] \\ &= \max_{y\in\hat{\mathbb{R}}^n} q\big[f^{-1}\circ f(y), f^{-1}\circ f_\nu(y)\big] \to 0 \end{aligned}$$

as $\nu\to\infty$, since $f_\nu \to f$ uniformly on $\hat{\mathbb{R}}^n$ and f^{-1} is uniformly continuous there. Thus $f_\nu^{-1} \to f^{-1}$ uniformly on $\mathbb{R}^n$. □

As the proof above shows, the last statement of Theorem 6.6.20 is a special case of a more general fact:

LEMMA 6.6.21. *If $\langle f_\nu \rangle$ is a sequence of homeomorphisms of $\hat{\mathbb{R}}^n$ and if $\langle f_\nu \rangle$ converges uniformly on $\hat{\mathbb{R}}^n$ to a homeomorphism f, then $f_\nu^{-1} \to f^{-1}$ uniformly on $\hat{\mathbb{R}}^n$.*

As a consequence of the lemma, the family of all self-homeomorphisms of $\hat{\mathbb{R}}^n$, when endowed with the topology of uniform convergence, becomes a topological group.

We say that a sequence $\langle D_\nu \rangle$ of domains in $\hat{\mathbb{R}}^n$ *converges to its kernel* if every subsequence $\langle D_{\nu_k} \rangle$ of $\langle D_\nu \rangle$ has the same kernel as $\langle D_\nu \rangle$.

For example, let $\langle x_\nu \rangle$ be a sequence in $\hat{\mathbb{R}}^n$ and let $D_\nu = \hat{\mathbb{R}}^n \setminus \{x_\nu\}$. If x is an accumulation point of $\langle x_\nu \rangle$, then there is a subsequence $\langle x_{\nu_k} \rangle$ of $\langle x_\nu \rangle$ such that $x_{\nu_k} \to x$. The kernel of the subsequence $\langle D_{\nu_k} \rangle$ is readily seen to be $\hat{\mathbb{R}}^n \setminus \{x\}$. It follows that $\langle D_\nu \rangle$ converges to its kernel if and only if the sequence $\langle x_\nu \rangle$ has a limit, say x_0, in which event $\ker_{\nu\to\infty} D_\nu = \hat{\mathbb{R}}^n \setminus \{x_0\}$.

This observation is germane to the next theorem, as is the statement following Corollary 6.4.27.

THEOREM 6.6.22. *Let $\langle x_\nu \rangle$ be a convergent sequence in $\hat{\mathbb{R}}^n$ with $n \geq 2$, let x_0 be its limit, and let $D = \hat{\mathbb{R}}^n \setminus \{x_0\}$. If $D_\nu = \hat{\mathbb{R}}^n \setminus \{x_\nu\}$, if f_ν is a K-quasiconformal mapping of D_ν onto $D'_\nu = \hat{\mathbb{R}}^n \setminus \{x'_\nu\}$, and if $f_\nu \to f$ locally uniformly in D, then either f is a constant mapping or it is a K-quasiconformal mapping of D onto $D' = \hat{\mathbb{R}}^n \setminus \{x'_0\}$, where $x'_0 = \lim_{\nu\to\infty} x'_\nu$. In the latter case $f_\nu^{-1} \to f^{-1}$ locally uniformly in D'.*

PROOF. Corollary 6.4.27 tells us that f_ν can be extended to a K-quasiconformal self-mapping f_ν^* of $\hat{\mathbb{R}}^n$; this extension is achieved by setting $f^*(x_\nu) = x'_\nu$. Assume that the limit mapping f is nonconstant. Theorems 6.6.17 and 6.6.18 imply that f is a K-quasiconformal mapping of D into $\hat{\mathbb{R}}^n$. Its range D' must have the form $D' = \hat{\mathbb{R}}^n \setminus \{x'_0\}$ for some x'_0 in $\hat{\mathbb{R}}^n$.

In accordance with the situation for f_ν, the extension f^* has $f^*(x_0) = x_0'$. We show that $x_\nu' \to x_0'$. For this, let y_1, y_2, and y_3 be three distinct points of D. Since $f_\nu^*(y_j) \to f^*(y_j)$ for $1 \le j \le 3$, it is evident that

$$\inf_\nu \min_{j \neq k} q\big[f_\nu^*(y_j), f_\nu^*(y_k)\big] > 0\,.$$

Corollary 6.6.13 shows that $\mathcal{F}^* = \{f_\nu^* : \nu = 1, 2, \ldots\}$ is a normal family of self-mappings of $\hat{\mathbb{R}}^n$.

The equicontinuity of $\mathcal{F}^*$ in $\hat{\mathbb{R}}^n$, in tandem with the pointwise convergence of $\langle f_\nu^* \rangle$ in the dense subset D of $\hat{\mathbb{R}}^n$, guarantees that $\langle f_\nu^* \rangle$ converges uniformly on $\hat{\mathbb{R}}^n$ to a continuous limit function, call it F. Here we have once more appealed to Lemma 6.6.16.

As $F = f^*$ in D and f^* is continuous in $\hat{\mathbb{R}}^n$, we conclude that $F = f^*$. In particular $x_\nu' = f_\nu^*(x_0) \to f^*(x_0) = x_0'$. This makes certain that $D' = \ker_{\nu\to\infty} D_\nu'$. According to Theorem 6.6.20, $(f_\nu^*)^{-1} \to (f^*)^{-1}$ uniformly on $\hat{\mathbb{R}}^n$, so $f_\nu^{-1} \to f^{-1}$ locally uniformly in D'. □

If in Theorem 6.6.22 we have $D_\nu = D$ for all ν and if the limit mapping f is nonconstant, then the above proof reveals that $f_\nu \to f$ uniformly in D.

Having now dispensed with a few exceptional cases in the results above, we turn to a theorem that gives precise information about the convergence of a sequence of K-quasiconformal mappings under conditions which are quite general.

THEOREM 6.6.23. *Suppose that $\langle D_\nu \rangle$ is a sequence of domains in $\hat{\mathbb{R}}^n$ which converges to a kernel whose complement contains at least two points, that f_ν is a K-quasiconformal mapping of D_ν onto a domain D_ν', and that D is a component of* $\ker_{\nu\to\infty} D_\nu$.

If $f_\nu \to f$ locally uniformly in D, then either f is a constant mapping of D that takes its value in the set

$$f(D) \in [\ker_{\nu\to\infty} D_\nu' \cup \ker_{\nu\to\infty}(D_\nu')^c]^c$$

or $f : D \to D'$ is a K-quasiconformal mapping onto a component D' of $\ker_{\nu\to\infty} D_\nu'$.

In the latter case $f_\nu^{-1} \to f^{-1}$ locally uniformly in D' as well.

PROOF. We first make the observation that $\liminf_{\nu\to\infty} q(D_\nu^c) > 0$. Suppose to the contrary that there were a subsequence $\langle D_{\nu_k} \rangle$ of $\langle D_\nu \rangle$ for which $q(D_{\nu_k}^c) \to 0$. Should $D_{\nu_k} = \hat{\mathbb{R}}^n$ for all large k, we would have $\ker_{k\to\infty} D_{\nu_k} = \hat{\mathbb{R}}^n$; should $D_{\nu_k} \neq \hat{\mathbb{R}}^n$ hold for infinitely many k, we could assume (via passage to a further subsequence) the existence of x_{ν_k} in $D_{\nu_k}^c$ such that $\langle x_{\nu_k} \rangle$ converged to some point x_0 of $\hat{\mathbb{R}}^n$, in which event we would clearly have $\ker_{k\to\infty} D_{\nu_k} = \hat{\mathbb{R}}^n \setminus \{x_0\}$.

On the other hand, by hypothesis we have $\ker_{k\to\infty} D_{\nu_k} = \ker_{\nu\to\infty} D_\nu$, a set whose complement is known to contain at least two points. The only way this contradiction can be avoided is to have $\liminf_{\nu\to\infty} q(D_\nu^c) > 0$. After possibly deleting a finite number of domains from the original sequence we are free to assume that $q(D_\nu^c) \ge a > 0$ for all ν.

According to Theorems 6.6.17 and 6.6.18 the mapping f must be either a constant mapping or a K-quasiconformal mapping of D onto a subdomain D' of $\ker_{\nu\to\infty} D_\nu'$. Suppose that this limit is constant, say $f(x) = c$ for every x in D. Fixing x in D, we observe that any neighbourhood U of c contains $f_\nu(x)$, a point of D'_ν, for all suitably large ν. This fact makes it plain that c does not lie in $\ker_{\nu\to\infty}(D'_\nu)^c$.

We now consider if c in fact belongs to $\ker_{\nu\to\infty} D'_\nu$. If it does, we can choose $r > 0$ so that $G' = \{y : q(y, c) < r\}$ is contained in D'_ν as soon as ν becomes sufficiently large. To keep matters simple we assume this to be so for all ν, and we write $g_\nu = f_\nu^{-1}|G'$. Since

$$\inf_\nu q\big[g_\nu(G')^c\big] \geq \inf_\nu q(D_\nu^c) \geq a > 0 \ ,$$

the family $\mathcal{G} = \{g_\nu : \nu = 1, 2, \ldots\}$ is a normal family. We are thus at liberty to choose a subsequence $\langle g_{\nu_k}\rangle$ of $\langle g_\nu\rangle$ that converges locally uniformly in G' to a function g. Given x in D, we have $f_{\nu_k}(x) \to c$. We may presume that $f_{\nu_k}(x)$ is in G' for all k. In conjunction with the local uniform convergence of $\langle g_{\nu_k}\rangle$ in G' these circumstances force

$$g(c) = \lim_{k\to\infty} g_{\nu_k}\big[f_{\nu_k}(x)\big] = \lim_{k\to\infty} x = x$$

to hold for every x in D, which is certainly not possible. We conclude that $c \notin \ker_{\nu\to\infty} D'_\nu$ either. Therefore $c = f(D)$ belongs to $[\ker_{\nu\to\infty} D'_\nu \cup \ker_{\nu\to\infty}(D'_\nu)^c]^c$ as claimed.

We must still deal with the case where f maps D in a K-quasiconformal fashion onto a domain D' that lies in $\ker_{\nu\to\infty} D'_\nu$.

Let D'_0 denote the component of $\ker_{\nu\to\infty} D'_\nu$ that contains D'. If $D' \neq D'_0$, then $D'_0 \cap \partial D'$ is not empty. Consider a point y_0 of $D'_0 \cap \partial D'$. Fix a chordal ball $G' = \{y : q(y, y_0) < r\}$ that lies in D'_ν for all large ν. Once more it entails no loss of generality to assume G' is a subset of D'_ν for all ν. We again set $g_\nu = f_\nu^{-1}|G'$. As earlier, we can extract from $\langle g_\nu\rangle$ a subsequence $\langle g_{\nu_k}\rangle$ that converges locally uniformly in G' to a limit g. For any x in $f^{-1}(G' \cap D')$ we have $f_\nu(x) \to f(x)$, whence $f_{\nu_k}(x)$ belongs to G' for large k and

$$g\big[f(x)\big] = \lim_{k\to\infty} g_{\nu_k}\big[f_{\nu_k}(x)\big] = x \,.$$

This implies, among other things, that g is not a constant mapping of G' and hence, in view of Theorem 6.6.17, that g is a homeomorphism which maps G' onto a subdomain G of $\ker_{k\to\infty} g_{\nu_k}(G')$. Certainly G is contained in $\ker_{k\to\infty} D_{\nu_k} = \ker_{\nu\to\infty} D_\nu$, so G lies in some component of $\ker_{\nu\to\infty} D_\nu$. The argument presented above shows that $g(G' \cap D')$ is a subset of D. Accordingly, G is contained in D. In particular, $x_0 = g(y_0)$ is a point of D. It follows that

$$f(x_0) = f\big[g(y_0)\big] = \lim_{k\to\infty} f_{\nu_k}\big[g_{\nu_k}(y_0)\big] = y_0 \ ,$$

which clearly places y_0 in D', contrary to our supposition. To escape this contradiction we must have $D' = D'_0$; i.e., D' has to be a component of $\ker_{\nu\to\infty} D'_\nu$.

The only thing left to prove is that, assuming f to be nonconstant, $f_\nu^{-1} \to f^{-1}$ locally uniformly in D'. By virtue of Theorem 6.6.17 it is enough to check that $f_\nu^{-1} \to f^{-1}$ pointwise in D'. Fix y_0 in D' and let x_0 be an arbitrary accumulation point of the sequence $\langle f_\nu^{-1}(y_0)\rangle$ (we assume that y_0 is in D'_ν for every ν). We maintain that x_0 belongs to D and that $f(x_0) = y_0$. Choose a subsequence $\langle f_{\nu_k}\rangle$ of $\langle f_\nu\rangle$ such that $f_{\nu_k}^{-1}(y_0) \to x_0$. Reasoning as we did above, we may suppose—upon passage to a further subsequence, if necessary—that the sequence $\langle f_{\nu_k}^{-1}\rangle$ converges locally uniformly in some chordal ball $G' = \{y : q(y, y_0) < r\}$ contained in D' to a homeomorphism g of G' onto a subdomain G of D and that $g[f(x)] = x$ holds for every x in G. As a result, $x_0 = g(y_0)$ lies in G and $g[f(x_0)] = x_0 = g(y_0)$, so $f(x_0) = y_0$. We deduce that $f_\nu^{-1}(y_0) \to f^{-1}(y_0)$ for each y_0 in D', as desired. □

The most common applications of Theorem 6.6.23 are to situations where the mappings f_ν are defined in the same domain. In this case the theorem becomes:

COROLLARY 6.6.24. *Let D be a domain in $\hat{\mathbb{R}}^n$ with at least two boundary points, and for $\nu = 1, 2, \ldots$ let f_ν be a K-quasiconformal mapping of D onto a domain D'_ν. If $f_\nu \to f$ locally uniformly in D, then either f is a constant mapping taking its value in the set $[\ker_{\nu\to\infty} D'_\nu \cup \ker_{\nu\to\infty}(D'_\nu)^c]^c$ or it is a K-quasiconformal mapping of D onto a component D' of $\ker_{\nu\to\infty} D'_\nu$. In the latter case $f_\nu^{-1} \to f^{-1}$ locally uniformly in D'.*

In the case of a locally uniformly convergent sequence of K-quasiconformal mappings between two fixed domains, the prospects of getting a constant limit are substantially reduced. The next result spells out the possibilities completely.

THEOREM 6.6.25. *Suppose that $\langle f_\nu \rangle$ is a sequence of K-quasiconformal mappings of a domain D in $\hat{\mathbb{R}}^n$ onto a domain D' and that $f_\nu \to f$ locally uniformly in D.*

Then either f is a K-quasiconformal mapping of D onto D', in which event $f_\nu^{-1} \to f^{-1}$ locally uniformly in D', or it is a constant mapping.

In the latter case the value taken by f lies in $\partial D'$, except possibly when D^c reduces to a single point.

A constant limit is a possibility only in the following situations:

(i) *each of the sets D^c and $(D')^c$ has at most two points;*
(ii) *both D^c and $(D')^c$ are continua;*
(iii) *each of the sets D^c and $(D')^c$ has infinitely many components, in which case the value taken by f is a point whose every neighbourhood contains infinitely many components of $(D')^c$.*

PROOF. It is only the case of a constant limit that requires further attention. Assume that $f_\nu \to c'$, a constant, locally uniformly in D. With allowance for an exception when D^c consists of a single point, Theorem 6.6.23 tells us that c' lies in $[D' \cup \text{ext}(D')]^c = \partial D'$. Remembering that the components of D^c are in one-to-one correspondence with those of $(D')^c$, we shall make the assumption that D^c is a disconnected set with three or more elements (then $(D')^c$ is likewise disconnected and, owing to Corollary 6.4.27, contains at least three points) and then prove that statement (iii) correctly explains the situation.

We first observe that c' cannot be an isolated point of $\partial D'$. If it were, then $g_\nu = f_\nu^{-1}$ would have an extension to a K-quasiconformal mapping g_ν^* of $G' = D' \cup \{c'\}$ onto a domain G_ν that is a subset of $\bar{D}$. By assumption D^c has, at the very minimum, three points. Since G_ν would contain only one point of D^c, it is evident that $\inf\{q(G_\nu^c) : \nu = 1, 2, \ldots\} > 0$ would be true. Thus $\mathcal{G}^* = \{g_\nu^* : \nu = 1, 2, \ldots\}$ would be an equicontinuous family. This fact would enable us to choose a neighbourhood U of c' such that $q[g_\nu^*(U)] < q(D)/2$ for all ν. Given x and x' in D, we would find $y = f_\nu(x)$ and $y' = f(x')$ in U if ν were taken suitably large, whence

$$q(x, x') = q\big[g_\nu(y), g_\nu(y')\big] \leq q\big[g_\nu^*(U)\big] < q(D)/2$$

would hold for all x and x' in D. This would imply that

$$q(D) = \sup_{x,x' \in D} q(x,x') \le q(D)/2 ,$$

an obvious contradiction. Accordingly, c' is not isolated in $\partial D'$.

Next, let V be an arbitrary open neighbourhood of c'. We assert that $(V \cap D')^c$ is not connected. To see this, write $A' = V \cap D'$ and $A_\nu = g_\nu(A')$. The fact that $f_\nu \to c'$ locally uniformly in D is easily seen to imply that

$$D = \ker_{\nu \to \infty} A_\nu = \bigcup_{\nu=1}^{\infty} \operatorname{int}\left(\bigcap_{\mu=\nu}^{\infty} A_\nu\right) .$$

This allows us to express D^c in the manner

$$D^c = \bigcap_{\nu=1}^{\infty}\left[\operatorname{int}\left(\bigcap_{\mu=\nu}^{\infty} A_\mu\right)\right]^c = \bigcap_{\nu=1}^{\infty}\overline{\left(\bigcap_{\mu=\nu} A_\mu\right)^c} = \bigcap_{\nu=1}^{\infty}\overline{\bigcup_{\mu=\nu} A_\mu^c} = \bigcap_{\nu=1} \bar{F}_\nu ,$$

where for $\nu = 1, 2, \ldots$ we set

$$F_\nu = \bigcup_{\mu=\nu}^{\infty} A_\mu^c .$$

If $(A')^c$ happened to be connected, it would follow from elementary topological deliberations that A_ν^c was connected for each ν.

Since D^c is contained in A_ν^c for every ν, F_ν would be connected for every ν, as would $\bar{F}_\nu$. Now $\bar{F}_1 \supset \bar{F}_2 \supset \bar{F}_3 \supset \cdots$ and the intersection of a decreasing sequence of compact, connected sets in $\hat{\mathbb{R}}^n$ is connected. The upshot of the preceding remarks is that if $(V \cap D')^c$ were connected, D^c would be connected, which runs counter to our assumption. Therefore $(V \cap D')^c$ is disconnected.

Let F be the component of $(D')^c$ which contains c'. Suppose that c' had an open neighbourhood V which encompassed at most finitely many components of $(D')^c$. By replacing V with a smaller neighbourhood we could assume that V^c was a connected set, that V^c had nonempty intersection with each component of $(D')^c$ other than F, and that V^c met F, too, should F be nondegenerate. If $F \ne \{c'\}$, then $(V \cap D')^c = V^c \cup (D')^c$ would clearly be a connected set; if $F = \{c'\}$, then the fact that c' is not an isolated point of $(D')^c$ would mean that

$$(V \cap D')^c = V' \cup (D')^c = V^c \cup \overline{\left[(D')^c \setminus \{c'\}\right]} = \overline{V^c \cup \left[(D')^c \setminus \{c'\}\right]} ,$$

so again $(V \cap D')^c$ would be connected. The connectedness of $(V \cap D')^c$, however, would stand in direct conflict with an earlier observation. We conclude that no such neighbourhood V of c' can exist. Thus each neighbourhood of c' contains infinitely many components of $(D')^c$, as specified in (iii). □

An example illustrating case (iii) of Theorem 6.6.25 is quickly constructed as follows.

Take $D = D' = H^n \setminus \bigcup_{k=-\infty}^{\infty} \overline{B}^n(2^k e_n, 2^{k-2})$ and for $\nu = 1, 2, \ldots$ let $f_\nu : D \to D'$ be defined by $f_\nu(x) = 2^{-\nu} x$. Then f_ν is a 1-quasiconformal mapping of D onto itself, and $f_\nu \to 0$ locally uniformly in D. Every neighbourhood of the origin contains infinitely many of the balls $\overline{B}^n(2^k e_n, 2^{k-2})$, each of which is a component of D^c.

The next theorem exposes one of the key ideas underlying the concept of a "convergence group" (see [**51, 104, 107**]). These are basically discrete groups of homeomorphisms acting on relatively nice spaces for which every infinite sequence must have (ii) below holding.

In recent years the study of these groups has fostered an intriguing interplay between the theory of quasiconformal mappings and certain areas of topology and low-dimensional topology and geometry. It is quite remarkable that the convergence property described in Theorem 6.6.26 below actually holds for groups of isometries of a Hadamard space—a simply connected and nonpositively curved manifold—when acting on the ideal boundary; see [**105**]. In fact in dimension $n = 1$ D. Gabai completed a project initiated by P. Tukia to show that this convergence property characterises the topological conjugates of Fuchsian groups acting on the circle [**43**], while in two dimensions the notion underpins a central problem in low-dimensional topology—the Cannon conjecture.

THEOREM 6.6.26. *Let $\mathcal{F}$ be an infinite family of K-quasiconformal self-mappings of $\hat{\mathbb{R}}^n$. There exists in $\mathcal{F}$ a sequence $\langle f_\nu \rangle$ whose terms are distinct and for which one of the following two conditions holds:*

(i) *$f_\nu \to f$ and $f_\nu^{-1} \to f^{-1}$ uniformly on $\hat{\mathbb{R}}^n$, where f is a K-quasiconformal self-mapping of $\hat{\mathbb{R}}^n$;*

(ii) *there are points x_0 and y_0 in $\hat{\mathbb{R}}^n$—possibly $x_0 = y_0$—such that $f_\nu \to y_0$ locally uniformly in $\hat{\mathbb{R}}^n \setminus \{x_0\}$ and $f_\nu^{-1} \to x_0$ locally uniformly in $\hat{\mathbb{R}}^n \setminus \{y_0\}$.*

PROOF. Should $\mathcal{F}$ contain an infinite subfamily that is normal, we could certainly produce a uniformly convergent sequence $\langle f_\nu \rangle$ in $\mathcal{F}$ whose terms were distinct. Theorem 6.6.20 would then attest to the fact that $\langle f_\nu \rangle$ was of type (i).

We may therefore assume that $\mathcal{F}$ contains no infinite normal subfamily. Corollary 6.6.14 announces the existence of a point x_0 in $\hat{\mathbb{R}}^n$ and an infinite subfamily $\mathcal{F}_0$ of $\mathcal{F}$ such that $\mathcal{F}_0 | D_0$ is a normal family, where $D_0 = \hat{\mathbb{R}}^n \setminus \{x_0\}$.

Normality on the part of $\mathcal{F}_0^{-1} = \{f^{-1} : f \in \mathcal{F}_0\}$ would, via Theorem 5.15, imply that $\mathcal{F}_0$ was normal. We conclude that $\mathcal{F}_0^{-1}$ is not a normal family. A second appeal to Corollary 6.6.14 confirms the existence of a point y_0 in $\hat{\mathbb{R}}^n$ and an infinite subfamily $\mathcal{G}_0$ of $\mathcal{F}_0^{-1}$ such that $\mathcal{G}_0 | D_0'$ is normal, where $D_0' = \hat{\mathbb{R}}^n \setminus \{y_0\}$. The possibilty that $x_0 = y_0$ cannot be ruled out. Let $\langle f_\nu \rangle$ be a sequence of distinct mappings drawn from $\mathcal{G}_0^{-1}$. By passing to subsequences and relabeling, we may presume that $f_\nu \to f$ pointwise in $\hat{\mathbb{R}}^n$ and locally uniformly in $\hat{\mathbb{R}}^n \setminus \{x_0\}$ and that $g_\nu \to g$ pointwise in $\hat{\mathbb{R}}^n$ and locally uniformly in $\hat{\mathbb{R}}^n \setminus \{y_0\}$, where $g_\nu = f_\nu^{-1}$. In neither case can convergence be uniform on $\hat{\mathbb{R}}^n$, for uniform convergence in either case would imply that $\{f_\nu : \nu = 1, 2, \ldots\}$ was an infinite normal subfamily of $\mathcal{F}$. If, however, either f were nonconstant in $\hat{\mathbb{R}}^n \setminus \{x_0\}$ or g were nonconstant in $\hat{\mathbb{R}}^n \setminus \{y_0\}$, the convergence of the associated sequence would be uniform on $\hat{\mathbb{R}}^n$, as the proof of Theorem 6.6.22 makes evident.

We deduce that $f_\nu \to c$ locally uniformly in $\hat{\mathbb{R}}^n \setminus \{x_0\}$ and that $g_\nu \to c'$ locally uniformly in $\hat{\mathbb{R}}^n \setminus \{y_0\}$, where c and c' are constants. We claim that $c = y_0$ and $c' = x_0$. If $c \neq y_0$, we know that $g_\nu \to c'$ uniformly in some neighbourhood of c. Fix a point x different from x_0 and c'. Then $f_\nu(x) \to c$, so $c' = \lim_{\nu \to \infty} g_\nu[f_\nu(x)] = x$, a contradiction. Thus $c = y_0$. Similarly, $c' = x_0$. We conclude that the sequence $\langle f_\nu \rangle$ is of type (ii). □

6.6.8. Normal families in the Euclidean metric. The results of the previous section presented a fairly comprehensive view of the convergence of sequences of K-quasiconformal mappings relative to the chordal metric. Here we now seek to say a little more about the normality of families of K-quasiconformal mappings with respect to the Euclidean metric in domains of $\mathbb{R}^n$.

THEOREM 6.6.27. *Suppose that D is a domain in $\mathbb{R}^n$ and that $\mathcal{F}$ is a normal family of K-quasiconformal mappings of D into $\mathbb{R}^n$. Either $\mathcal{F}$ is a normal family with respect to the Euclidean metric or $\mathcal{F}$ contains a sequence $\langle f_\nu \rangle$ such that $f_\nu \to \infty$ locally uniformly in D.*

PROOF. If $\mathcal{F}$ is not a normal family with respect to the Euclidean metric, then by Theorem 6.6.15 there must be a point x_0 of D and a sequence $\langle f_\nu \rangle$ in $\mathcal{F}$ such that $f_\nu(x_0) \to \infty$. Passing to a subsequence we may assume that $f_\nu \to f$ locally uniformly in D with respect to the chordal metric. Of course, $f(x_0) = \infty$. Either $f \equiv \infty$ in D or f is a K-quasiconformal mapping of D onto a domain D'. In the latter case, Theorem 6.6.17 would place D' in $\ker_{\nu\to\infty} f_\nu(D)$, a subset of $\mathbb{R}^n$. As a result, $f_\nu \to \infty$ locally uniformly in D. □

We can strengthen Theorem 6.6.15 as follows.

COROLLARY 6.6.28. *Suppose that D is a domain in $\mathbb{R}^n$ and that $\mathcal{F}$ is a non-empty family of K-quasiconformal mappings of D into $\mathbb{R}^n$. If there exist distinct points a and b of D for which both $\mathcal{F}(a)$ and $\mathcal{F}(b)$ are bounded sets, then $\mathcal{F}$ is a normal family with respect to the Euclidean metric.*

PROOF. According to Corollary 6.6.13(ii), $\mathcal{F}$ is a normal family relative to the chordal metric (take $y_f = \infty$). Since $\mathcal{F}$ contains no sequence tending to ∞ pointwise in D, $\mathcal{F}$ is normal with respect to the Euclidean metric. □

The family $\mathcal{F} = \{f_\nu : \nu = 1, 2, \ldots\}$, where $f_\nu : \mathbb{R}^n \to \mathbb{R}^n$ is given by $f_\nu(x) = \nu x$, shows that two points a and b are needed in Corollary 6.6.28.

6.6.9. Lower semicontinuity of the linear distortion fails. From the beginning of the higher-dimensional theory of quasiregular mappings it was widely believed that the class of H-quasiconformal self-homeomorphisms of $\mathbb{R}^n$ was closed with respect to uniform convergence when H had the linear dilatation,

$$H(x, f) = \frac{\max_{h\in\mathbb{R}^n, |h|=1} |Df(x)h|}{\min_{h\in\mathbb{R}^n, |h|=1} |Df(x)h|} \tag{6.59}$$

and

$$H = H(f) = \|H(x, f)\|_{L^\infty(D)} \tag{6.60}$$

or identically, $H = \operatorname{essup}_{x\in D} H(x, f)$.

We already established the lower semicontinuity of the maximal distortion function $K(f)$ derived from the inner and outer distortions of f at (6.6.19), so such a belief seemed perfectly reasonable. However in response to a direct question from Curt McMullen, Tadeusz Iwaniec gave a striking example which refutes this belief in his paper [**83**]. The key idea for the construction is that the linear dilatation function fails to be rank-one convex in dimensions $n \geq 3$.

We will not give a complete proof for this result as it is already covered in other places, for instance [**84**], but we will outline some ideas as they fit in with the scope of this book.

THEOREM 6.6.29. *For each $n \geq 3$ and $K > 1$, there exists a $C > 1$ and a sequence $\{f_\nu\}_{\nu=1}^{\infty}$ of uniformly bilipschitz, piecewise linear, self-maps of $\mathbb{R}^n$,*

$$\frac{1}{C}\,|x-y| \leq |f_\nu(x) - f_\nu(y)| \leq C|x-y|, \qquad \textit{all } x, y \in \mathbb{R}^n \textit{ and } \nu = 1, 2, \ldots,$$

with the following properties:

(i) *The linear dilatation function of each mapping is the same constant H and almost everywhere $H(z, f_\nu) = H$, for every $\nu = 1, 2, \ldots$ and almost every $x \in \mathbb{R}^n$.*
(ii) *The sequence f_ν converges uniformly in the spherical metric of $\hat{\mathbb{R}}^n$ to a linear map $f_0 : \mathbb{R}^n \to \mathbb{R}^n$ whose dilatation is equal to $H_0 > H$.*

In particular the linear dilatation function is not lower semi-continuous with respect to uniform convergence in dimensions greater than two.

The key idea that underpins these examples is the lack of rank-one convexity for the linear distortion function. We explain this. Recall that $\mathbb{R}^{n\times n}$ denotes the real vector space of all $n \times n$ matrices endowed with the norm

$$|A| = \max_{|h|=1} |Ah|.$$

A matrix $X \in \mathbb{R}^{n\times n}$ is *rank one* if it can be written as the tensor product of two vectors. Thus there are $x = (x_1, x_2, \ldots, x_n)$ and $y = (y_1, y_2, \ldots, y_n)$ in $\mathbb{R}^n$ and

$$X = [w_{ij}]_{i,j=1,\ldots,n}, \qquad w_{ij} = x_i\, y_j.$$

A function $\mathcal{L} : U \to \mathbb{R}^m$, $m \geq 1$, defined on an open set $U \subset \mathbb{R}^{n\times n}$ is said to be *rank-one convex* at a given matrix $A \in U$ if for every rank-one matrix $X \in \mathbb{R}^{n\times n}$, the function

$$X \mapsto \mathcal{L}(A + tX)$$

is convex near $t = 0$. It transpires that the linear dilatation function

$$\mathcal{H}(A) = \frac{\max_{|h|=1}\ |Ah|}{\min_{|h|=1}\ |Ah|}$$

is not rank-one convex. This is the content of the next lemma which we will not prove, despite the fact that there is an interesting computation necessary to establish it.

LEMMA 6.6.30. *Given $n \geq 3$ and $H > 1$, there is a matrix A and a rank-one matrix X and numbers $t, s > 0$ such that*

$$\mathcal{H}(A - sX) = \mathcal{H}(A + tX) = H < \mathcal{H}(A).$$

PROOF FOR THEOREM 6.6.29. This follows quite quickly with the aid of the lemma. We have $X = p \otimes q$ for vectors $p, q \in \mathbb{R}^n$. We set $f_0(x) = Ax$, and for $\nu = 1, 2, \ldots$ define

$$f_\nu = Ax + \frac{1}{\nu}\, u(\nu p \cdot x)\, q,$$

where u is a periodic piecewise linear function on $\mathbb{R}$ defined as follows: Given s and t of the lemma

$$u(r) = \begin{cases} rs & -s^{-1} \leq t \leq 0, \\ rt & 0 \leq t \leq t^{\ 1}. \end{cases}$$

Then we extend u periodically to $\mathbb{R}$ as a bounded Lipschitz function whose derivative assumes only the two values $-s$ and t. The sequence $\{f_\nu\}$ is easily seen to converge uniformly to f_0. Indeed the differentials of f_ν also assume only two values—independent of ν. These are

$$Df_\nu = A + u'(\nu p \cdot x) p \otimes q \in \{A - sX, A + tX\}.$$

In either case, the linear dilatation of $Df_\nu(x)$ is equal to H, and this proves the theorem once we note the elementary facts that each f_ν is piecewise linear and the family $\{f_\nu\}$ is uniformly bilipschitz due to the bounds on the differential. □

Of course if $\{f_\nu\}$ is a sequence of quasiconformal mappings with uniformly bounded linear dilatation, then the maximal dilatation also has a finite uniform bound. Thus if $f\nu \to f$ uniformly, f will be quasiconformal and hence have bounded linear dilatation. Theorem 6.6.29 shows that the linear dilatation can jump up, and the questions remains "by how much?".

The best known estimate for the linear dilatation of the mapping $\lim f_\nu$ was given by F. Gehring and T. Iwaniec as

$$H(x, f) \leq 2^{-n/2} \big(H^{n-1} + H\big)^{n/2}$$

which is, of course, sharp in two dimensions. It is not known in general what the sharp bound might be.

6.6.10. Beurling's compactness criterion. We next show that, when it comes to distinguishing general self-homeomorphisms of $\mathbb{R}^n$ (or $\hat{\mathbb{R}}^n$) from quasiconformal ones, the compactness and convergence properties discussed in the preceding subsections are close to being the heart of the matter.

Suppose that $n \geq 2$ and that $\mathcal{F}$ is a nonempty family of homeomorphisms of $\mathbb{R}^n$ onto itself. We say that $\mathcal{F}$ is *stable with respect to similarity transformation* provided $\psi \circ f \circ \varphi$ belongs to $\mathcal{F}$ whenever f is a member of $\mathcal{F}$ and the mappings φ and ψ come from $\mathbf{S}(n)$, the group of similarity transformations of $\mathbb{R}^n$.

We term such a family a *precompact modulo two-point normalization* if every sequence $\langle f_\nu \rangle$ from $\mathcal{F}$ whose terms satisfy $f_\nu(0) = 0$ and $f_\nu(e_1) = e_1$ has a subsequence that converges locally uniformly in $\mathbb{R}^n$ to a homeomorphism.

Our choice of normalization here might seem somewhat arbitrary. However, since $\mathcal{F}$ is stable with respect to the similarity group, any sequence $\langle g_\nu \rangle$ from $\mathcal{F}$ subject to a normalization by $g_\nu(a) = a'$ and $g_\nu(b) = b'$, where a, b, a' and b' are fixed points of $\mathbb{R}^n$ and $a \neq b$, will have a subsequence that is locally uniformly convergent in $\mathbb{R}^n$ to a homeomorphism. Indeed, just consider $f_\nu = \psi \circ g_\nu \circ \varphi$, where φ is a similarity transformation taking 0 to a and e_1 to b, and ψ is a similarity transformation that carries a' to 0 and b' to e_1. Then $\langle f_\nu \rangle$ is a sequence from $\mathcal{F}$, $f_\nu(0) = 0$, and $f_\nu(e_1) = e_1$. If $f_{\nu_k} \to f$ locally uniformly in $\mathbb{R}^n$, where f is a homeomorphism, then $g_{\nu_k} \to g$ locally uniformly in $\mathbb{R}^n$, where $g = \psi^{-1} \circ f \circ \varphi^{-1}$.

We now establish a result that is most often credited to A. Beurling.

THEOREM 6.6.31. *Let $n \geq 2$ and let $\mathcal{F}$ be a nonempty family of homeomorphisms of $\mathbb{R}^n$ onto itself that is stable with respect to similarity transformation. Then $\mathcal{F}$ is a pre-compact modulo two-point normalization if and only if*

$$\sup\{K(f) : f \in \mathcal{F}\} < \infty .$$

That is, if and only if each $f \in \mathcal{F}$ is K-quasiconformal for some fixed and finite K.

PROOF. Assume first that $K = \sup\{K(f) : f \in \mathcal{F}\} < \infty$. Then each member of $\mathcal{F}$ is a K-quasiconformal mapping of $\mathbb{R}^n$ onto itself. We thus know that each f in $\mathcal{F}$ extends to a K-quasiconformal self-mapping f^* of $\hat{\mathbb{R}}^n$ satisfying $f^*(\infty) = \infty$. Corollary 6.6.13(iii) shows that $\{f^* : f \in \mathcal{F},\ f(0) = 0,\ f(e_1) = e_1\}$ is a normal family. Given a sequence $\langle f_\nu \rangle$ from $\mathcal{F}$ normalized by $f_\nu(0) = 0$ and $f_\nu(e_1) = e_1$, we can extract from $\langle f_\nu^* \rangle$ a subsequence $\langle f_{\nu_k}^* \rangle$ that converges uniformly on $\hat{\mathbb{R}}^n$, say to a function f^*. Theorem 6.6.20 identifies f^* as a K-quasiconformal self-mapping of $\hat{\mathbb{R}}^n$, one which satisfies $f^*(\infty) = \infty$. The mapping $f = f^*|\mathbb{R}^n$ is a K-quasiconformal homeomorphism of $\mathbb{R}^n$, and $f_{\nu_k} \to f$ locally uniformly in $\mathbb{R}^n$—in fact, uniformly on $\mathbb{R}^n$. It follows that $\mathcal{F}$ is a precompact modulo two-point normalization.

In addressing the converse, we assume that $\mathcal{F}$ is a precompact modulo two-point normalization. Let $\mathcal{F}_0 = \{f \in \mathcal{F} : f(0) = 0,\ f(e_1) = e_1\}$. Define H in $(0, \infty]$ by

$$H = \sup_{f \in \mathcal{F}_0} L_f(0,1) = \sup_{f \in \mathcal{F}_0} \max_{|x|=1} |f(x)| ,$$

and select a sequence $\langle f_\nu \rangle$ in $\mathcal{F}_0$ such that $L_{f_\nu}(0,1) \to H$. After possibly passing to a subsequence, we may suppose that $f_\nu \to f_0$ locally uniformly in $\mathbb{R}^n$, where f_0 is a homeomorphism. In particular, $f_\nu \to f_0$ uniformly on $\mathbb{S}^{n-1}$, which implies that

$$H = \lim_{\nu \to \infty} \max_{|x|=1} |f_\nu(x)| = \max_{|x|=1} |f_0(x)| < \infty .$$

Fix f in $\mathcal{F}$. Given x in $\mathbb{R}^n$ and $r > 0$, we pick y on $\mathbb{S}^{n-1}(x, r)$ for which $s = \ell_f(x, r) = |f(y) - f(x)|$. We then choose φ and ψ in $\mathbf{S}(n)$ so that $\varphi(0) = x$, $\varphi(e_1) = y$, $\psi(0) = f(x)$, and $\psi(e_1) = f(y)$. Clearly φ maps $\mathbb{S}^{n-1}$ to $\mathbb{S}^{n-1}(x, r)$, ψ transforms $\mathbb{S}^{n-1}$ to $\mathbb{S}^{n-1}[f(x), s]$, and $g = \psi^{-1} \circ f \circ \varphi$ belongs to $\mathcal{F}_0$. Accordingly,

$$\begin{aligned} L_f(x,r) &= \max_{|z-x|=r} |f(z) - f(x)| = \max_{|z|=1} |f \circ \varphi(z) - f \circ \varphi(0)| \\ &= \max_{|z|=1} |\psi \circ g(z) - \psi \circ g(0)| = s \max_{|z|=1} |g(z)| \leq Hs = H\ell_f(x,r) . \end{aligned}$$

Accordingly, we have $L_f(x, r)/\ell_f(x, r) \leq H$ for every x in $\mathbb{R}^n$ and every $r > 0$. We conclude that $H_f(x) \leq H$ for each x in $\mathbb{R}^n$. Corollary 6.4.19 implies that $K(f) \leq H^{n-1}$. Since f was an arbitrary member of $\mathcal{F}$, $\sup\{K(f) : f \in \mathcal{F}\} \leq H^{n-1} < \infty$, as proclaimed. □

As a spin-off from Theorem 6.6.31 we obtain:

COROLLARY 6.6.32. *Let $n \geq 2$. A homeomorphism f of $\mathbb{R}^n$ onto itself is quasiconformal if and only if the family $\mathcal{F}_f = \{\psi \circ f \circ \varphi : \psi, \varphi \in \mathbf{S}(n)\}$ is a precompact modulo two-point normalization.*

Something completely analogous to the above can be carried out in $\hat{\mathbb{R}}^n$. It is clear what it should mean for a nonempty family $\mathcal{F}$ of a homeomorphism of $\hat{\mathbb{R}}^n$ onto itself to be *stable with respect to Möbius transformation*: $\psi \circ f \circ \varphi$ finds itself in $\mathcal{F}$ whenever f is in $\mathcal{F}$ and the functions φ and ψ are members of $\mathrm{M\ddot{o}b}(n)$. A family of this type is a *pre-compact modulo three-point normalization* if each sequence $\langle f_\nu \rangle$ from $\mathcal{F}$ whose terms are normalized by $f_\nu(0) = 0$, $f_\nu(e_1) = e_1$, and $f_\nu(\infty) = \infty$ has a subsequence that converges uniformly on $\hat{\mathbb{R}}^n$ to a homeomorphism. We easily extract from Theorem 6.6.31 its counterpart in Möbius space.

THEOREM 6.6.33. *Let $n \geq 2$ and let $\mathcal{F}$ be a nonempty family of homeomorphisms of $\hat{\mathbb{R}}^n$ onto itself that is stable with respect to Möbius transformation. Then $\mathcal{F}$ is a precompact modulo three-point normalization if and only if* $\sup\{K(f) : f \in \mathcal{F}\} < \infty$.

PROOF. The sufficiency follows from Corollary 6.6.13(iii) and Theorem 6.6.22. As for the necessity, assume that $\mathcal{F}$ is a precompact modulo three-point normalization. If $\mathcal{F}_\infty = \{f \in \mathcal{F} : f(\infty) = \infty\}$, then $\mathcal{F}_\infty | \mathbb{R}^n$ is stable with respect to similarity transformation and is a precompact modulo two-point normalization. Since $K(f) = K(f|\mathbb{R}^n)$, Theorem 6.6.31 affirms that $K = \sup\{K(f) : f \in \mathcal{F}_\infty\} < \infty$. However, every f in $\mathcal{F}$ can be written in the form $f = \varphi \circ g$ with g a member of $\mathcal{F}_\infty$ and φ belonging to $\text{Möb}(n)$. Then $K(f) = K(g) \leq K$, so we see that $\sup\{K(f) : f \in \mathcal{F}\} \leq K$. □

Theorem 6.6.33 has as an obvious corollary, a result that parallels Corollary 6.6.32 but applies to homeomorphisms of $\hat{\mathbb{R}}^n$.

6.7. Quasiconformal mappings of $\mathbb{H}^n$ with the same boundary values

In this section we show how compactness results such as Theorem 6.6.26 and its presursors can yield useful geometric information. For instance here we shall prove that K-quasiconformal mappings of the unit ball B^n with the same boundary values—and Theorem 6.5.20 assures us that these boundary values exist as a quasiconformal mapping homeomorphism of $\mathbb{S}^{n-1}$—are a bounded distance apart in the hyperbolic metric and that this bound depends only on the distortion K and the dimension n. This will in part support and motivate a more general discussion in the case of boundary values of quasi-isometries in our discussion of the Mostow rigidity theorem.

We begin with a result of independent interest.

THEOREM 6.7.1. *For every $K \geq 1$ and $n \geq 2$ there is a number $k_0 < 1$ depending only on these quantities and with the following property. If $f : \bar{B}^n \to \bar{B}^n$ is a homeomorphism which is K-quasiconformal on B^n and if $f|\partial B^n =$ identity, then $|f(0)| \leq k_n$. Moreover $k_n \to 0$ as $K \to 1$.*

PROOF. It is an elementary consequence of Theorem 6.6.26 that the family of mappings described by the hypotheses of the theorem is compact, as they can all be extended to quasiconformal maps of $\mathbb{R}^n$ by either reflection or by the identity. Therefore the infimum

$$k_0 = \inf\{|f(0)| : f : B^n \to B^n \text{ is } K\text{-quasiconformal with } f|\mathbb{S}^{n-1} = \textit{identity}\}$$

is achieved for some K-quasiconformal mapping by the lower semicontiniuity of the distortion. This last observation also easily establishes that $k_0 \to 1$ as $K \to 1$, since any orientation-preserving Möbius transformation which fixes every point of $\mathbb{S}^{n-1}$ is equal to the identity. □

One can achieve explicit estimates for the number k_0 using moduli estimates. For instance one could estimate the moduli of $R = (\mathbb{S}^{-1}_{+}, [0, -e_3 \cdot \infty); \mathbb{R}^n)$ and its image $R' = (\mathbb{S}^{-1}_{+}, \alpha; \mathbb{R}^n)$, where α is an arc from $f(0)$ to ∞, and then use Teichmüller's estimate, and although there are better—if more complicated—approaches, these do not seem to produce optimal results. In two dimensions where there are rather more sophisticated techniques available—such as holomorphic motions, [**15**]—one can establish the sharp relationship between k_0 and K as

$$K = \log \frac{1+k_0}{1-k_0} = \rho_{\mathbb{H}}(0, k_0). \tag{6.61}$$

This is known as Teichmüller's problem. In fact (6.61) can be used to provide a lower bound on k_0 in all dimensions. To see this, map B^n to the half-space $\mathbb{H}^n$ with $0 \mapsto \mathbf{j} = (0, 0, \ldots, 0, 1) \in \mathbb{H}^n$. Then consider the map $\mathbb{H}^n \to \mathbb{H}^n$ which is the identity on $\mathbb{R}^{n-1} = \partial \mathbb{H}^n$ defined for $\lambda \geq 1$ by the formula

$$h(x_1, x_2, \ldots, x_n) = (x_1, x_2, \ldots, \lambda x_n). \tag{6.62}$$

This map has $|Dh| = \lambda$ and $J(x, h) = \lambda$ so that $K(x, h) = \lambda^{n-1}$ while $h(\mathbf{j}) = \lambda \mathbf{j}$, and so

$$\rho_{\mathbb{H}}(\mathbf{j}, h(\mathbf{j})) = \log \lambda.$$

Once we remove our normalisations by mapping back to the ball we find that

$$\rho_{\mathbb{H}}(0, \tilde{h}(0)) = \log \frac{1+|\tilde{h}(0)|}{1+|\tilde{h}(0)|} = \log K^{1/(n-1)}, \quad |\tilde{h}(0)| = \frac{K^{1/(n-1)} - 1}{K^{1/(n-1)} + 1}.$$

We now use Theorem 6.7.1 to produce a more general result which is already suggested by the above calculations. It shows that a quasiconformal mapping which is the identity on the boundary of $\mathbb{H}^n$ or B^n is a bounded distance in the hyperbolic metric from the identity in that domain.

THEOREM 6.7.2. *For every $K \geq 1$ and $n \geq 2$ the number*

$$A(n, K) = \log \frac{1+k_0}{1-k_0},$$

where k_0 comes from Theorem 6.7.1, has the following property. If $f : \bar{\mathbb{H}}^n \to \bar{\mathbb{H}}^n$ is a homeomorphism which is K-quasiconformal on $\mathbb{H}^n$ and if $f|\partial \mathbb{H}^n =$ identity, then for all $x \in \mathbb{H}^n$

$$\rho_{\mathbb{H}}[x, f(x)] \leq A(n, K). \tag{6.63}$$

PROOF. Let $x \in \mathbb{H}^n$ and choose a Möbius transformation ϕ such that $\phi(\mathbf{j}) = x$. Let $h = \phi^{-1} \circ f \circ \phi$. Then $h : \mathbb{H}^n \to \mathbb{H}^n$ is K-quasiconformal and so by using Theorem 6.7.1 we are able to make the following calculation:

$$\begin{aligned} \rho_{\mathbb{H}}[x, f(x)] &= \rho_{\mathbb{H}}(\phi(\mathbf{j}), f[\phi(\mathbf{j})]) = \rho_{\mathbb{H}}[\mathbf{j}, \phi^{-1}(f[\phi(\mathbf{j})])] \\ &= \rho_{\mathbb{H}}[\mathbf{j}, h(\mathbf{j})] \leq \log \frac{1+k_0}{1-k_0}, \end{aligned}$$

which in view of (6.63) is what we wanted to prove. □

This now quickly leads to the following corollary which was the point of this section. It shows that quasiconformal mappings of $\mathbb{H}^n$ with the same boundary values are only a bounded distance apart in the hyperbolic metric—the bound depending only on the distortion and the dimension.

COROLLARY 6.7.3. *Let f and g be K-quasiconformal self-homeomorphisms of $\mathbb{H}^n$. If the two homeomorphic extensions of f and g to the boundary $\mathbb{H}^n$ agree, then we have the uniform estimate*

$$\sup_{x\in\mathbb{H}^n} \rho_{\mathbb{H}}[f(x), g(x)] \leq A(n, K^2). \tag{6.64}$$

PROOF. Write $h = f \circ g^{-1} : \mathbb{H}^n \to \mathbb{H}^n$. Then h is K^2-quasiconformal and the extension of h as a boundary mapping is by the identity. Therefore Theorem 6.7.2 applies. □

In two dimensions we know that $A(2, K^2) = 2A(2, K)$. It would be interesting if a similar result could hold in all dimensions, however it appears that nothing is known concerning this.

6.8. The 1-quasiconformal mappings

We are now in a position to extend our discussion of the Liouville theorem. We already know from our earlier calculations that conformal mappings $f : D \to D'$ are 1-quasiconformal. We naturally seek the converse, and expect in higher dimensions $n \geq 3$ that if f is a 1-quasiconformal mapping, then $f = \varphi|D$ and φ is a Möbius transformation of $\hat{\mathbb{R}}^n$.

6.8.1. Two dimensions. We start by examining the situation in the classical setting of $\hat{\mathbb{R}}^2$, which we view as the extended complex plane $\hat{\mathbb{C}}$.

THEOREM 6.8.1. *Let D and D' be domains in $\hat{\mathbb{C}}$, and let f be a homeomorphism of D onto D'. Then $K(f) = 1$ if and only if f is a meromorphic function (f sense-preserving) or the complex conjugate of a meromorphic function (f sense-reversing).*

PROOF. If f is sense-reversing, then $\bar{f}$ is a sense-preserving homeomorphism with $K(\bar{f}) = K(f)$. Thus it suffices to deal with the sense-preserving case.

Let $D_0 = \{z \in D : z \neq \infty,\ f(z) \neq \infty\}$. If f is a meromorphic function, then its restriction to D_0 is a (complex) analytic diffeomorphism of D_0 into $\mathbb{C}$. We have already observed that such a function maps D_0 conformally into $\mathbb{C}$, whence $K(f) = K(f|D_0) = 1$.

Conversely, assume that $K(f) = 1$. We shall prove that f is analytic in D_0. Assuming this to be the case, we observe that the function f is analytic in D except for two possible isolated singularities—namely, the one at ∞ when ∞ belongs to D and the one at $f^{-1}(\infty)$ when ∞ is a point of D'. Because f is continuous in the chordal metric, neither potential singularity could be essential; f has no singularity worse than a pole in D. Since f is one-to-one, it is easily seen that these singularities must be simple poles. Therefore, f is a meromorphic function.

Let z be a point of D_0 at which f is differentiable in the real sense. Lest there be confusion with complex derivatives, we use $Df(z)$ to denote the Fréchet derivative of f at such a point.

Recall that, in notation proper to $\mathbb{C}$,

$$\|Df(z)\| = |\partial f(z)| + |\overline{\partial} f(z)|, \qquad J_f(z) = |\partial f(z)|^2 - |\overline{\partial} f(z)|^2 .$$

The information that $K(f) = 1$ thus tells us that

$$\left(|\partial f(z)| + |\overline{\partial} f(z)|\right)^2 \leq |\partial f(z)|^2 - |\overline{\partial} f(z)|^2 .$$

Due to our assumption that f is sense-preserving, $J_f(z) \geq 0$. If $J_f(z)$ is actually positive, then $|\partial f(z)| + |\partial \bar{f}(z)|$ is also positive, in which event the inequality above simplifies to

$$|\partial f(z)| + |\overline{\partial} f(z)| \leq |\partial f(z)| - |\overline{\partial} f(z)|\,.$$

This clearly dictates that $\overline{\partial} f(z) = 0$.

Now, as a quasiconformal mapping, f has a nonsingular derivative at almost every point z of D_0. It follows that $\overline{\partial} f = 0$—or, equivalently, that the real and imaginary parts of f obey the Cauchy-Riemann relations—almost everywhere in D_0. If f were known to be a C^1-function in D_0, we could first infer that $\overline{\partial} f$ vanished everywhere in D_0 and then conclude that f was analytic there. Since we possess no such a priori knowledge, a further argument is required.

It is enough to verify that f is analytic in each open disk U whose closure lies in D_0. Fix such a disk U, along with a second concentric open disk V such that V contains $\bar{U}$ and D_0 contains $\bar{V}$. Choose a C^∞-function $\psi : \mathbb{C} \to \mathbb{R}$ that has support in V and is identically 1 in a neighbourhood of $\bar{U}$. The function $g : \mathbb{C} \to \mathbb{C}$ that coincides with ψf in D_0 and vanishes in $\mathbb{C} \setminus D_0$ is a compactly supported ACL^2-function with $\overline{\partial} g = \psi \overline{\partial} f + f \overline{\partial} \psi$ almost everywhere in D_0. In particular, $\overline{\partial} g = 0$ almost everywhere in a neighbourhood of $\bar{U}$. Set $f_\nu = g * \varphi_\nu$, where $\langle \varphi_\nu \rangle$ is a standard sequence of C_0^∞-mollifiers. Then $\langle f_\nu \rangle$ is a sequence of C^∞-functions. Because g is an ACL^2-function, we can write $\overline{\partial} f_\nu = (\overline{\partial} g) * \varphi_\nu$. This implies that $\overline{\partial} f_\nu$ vanishes in U—hence, that f_ν is analytic in U—once ν is sufficiently large. Now $f_\nu \to g$ uniformly in U and $g \equiv f$ in that set, so standard results from complex analysis show f to be analytic in U, as we had foreseen. □

The last part of the preceding argument owes nothing to the univalence of f. Indeed, it establishes the following result:

THEOREM 6.8.2. *If D is a domain in the complex plane and $f : D \to \mathbb{C}$ is an* ACL^2*-function with the property that $\overline{\partial} f = 0$ almost everywhere in D, then f is an analytic function.*

This fact belongs to the circle of ideas surrounding "Weyl's lemma" and the Looman-Menchoff theorem. Indeed, the following is true.

THEOREM 6.8.3. *Suppose f is locally integrable on D and, as a distribution, satisfies the Cauchy-Riemann equations. Then f agrees almost everywhere with an analytic function.*

One should compare this result with the following example:

$$f(z) = \begin{cases} \exp(-z^{-4}) & z \neq 0, \\ 0 & z = 0. \end{cases} \tag{6.65}$$

As Looman noticed, it is a simple matter to show that f satisfies the Cauchy-Riemann equations *everywhere*. But $f(z)/z \to \infty$ as $z \to 0$ with $\arg(z) = \pi/4$, and so f cannot be analytic. This, with some history and more, is discussed in an expository paper of J. Gray and S. Morris [**59**].

6.8.2. Liouville's theorem. We turn next to the identification of the 1-quasiconformal mappings in dimensions three and above.

The centerpiece of the discussion is a theorem discovered independently by F.W. Gehring [**46**] and Yu. G. Reshetnyak [**141**].

THEOREM 6.8.4. *Let D and D' be domains in $\hat{\mathbb{R}}^n$ with $n \geq 3$, and let f be a homeomorphism of D onto D'. Then $K(f) = 1$ if and only if f is the restriction to D of a Möbius transformation.*

As we have already noted, we are aware of the fact that the members of Möb(n) are 1-quasiconformal self-mappings of $\hat{\mathbb{R}}^n$; the implication in one direction is obvious.

In the other direction, we may assume by restricting f to $D_0 = \{x \in D : x \neq \infty, f(x) \neq \infty\}$ that both D and D' lie in $\mathbb{R}^n$, for $K(f) = K(f|D_0)$ and a Möbius transformation will coincide with f in D precisely when it does so in D_0.

At any point x of D where f is differentiable the inequality

$$\|f'(x)\|^n \leq |J_f(x)|$$

follows from the condition $K(f) = 1$. Since the reverse inequality is always true—recall (2.8)—we learn that $\|f'(x)\|^n = |J_f(x)|$ for each such x. Corollary 6.6.5, applied to f and f^{-1}, implies that f is locally bilipschitz in D. This requires, among other things, that

$$\|f'(x)\| = L_f(x) \geq \ell_f(x) > 0$$

holds for any point x at which f has a derivative.

On the strength of Theorem 3.1.1 we draw from these considerations the following conclusion: if f is differentiable at x, then $f'(x)$ is a conformal linear transformation. Suppose we were somehow presented with the information that the mapping f was quite smooth—a C^∞-mapping, say. If so, f would be differentiable throughout D and thus would be a conformal mapping of D onto D'. Theorem 3.8.7 would then justify the statement that f was the restriction to D of a Möbius transformation.

The key to proving Theorem 6.8.4, therefore, is to establish a sufficient degree of smoothness in f (actually, it is enough that f be a C^4-mapping).

There are several ways to achieve this end. We shall do it as it was done in Gehring's original proof, by exploiting extremal functions for condensers. The main element in the proof is supplied by the following lemma.

LEMMA 6.8.5. *Let D and D' be domains in $\mathbb{R}^n$ with $n \geq 2$, and let f be a homeomorphism of D onto D' for which $K(f) = 1$.*

Then there exists a constant $\tau = \tau(n)$ in $(0,1)$ with the following property: for any x_0 in D the function u defined in D by $u(x) = |f(x) - f(x_0)|$ is real-analytic in the punctured ball $B^* = B^n(x_0, \delta) \setminus \{x_0\}$, where $\delta = \tau \operatorname{dist}(x_0, \partial D)$ if $D \neq \mathbb{R}^n$ and $\delta = 1$ if $D = \mathbb{R}^n$.

PROOF. Let $\Theta = \Theta_1^n$ be the distortion function that was introduced for Theorem 6.6.4, and let $\tau = \tau(n)$ be any number in $(0,1)$ for which $\Theta(\tau) < 1$. We consider a point x_0 of D. To simplify notation we shall assume that $f(x_0) = 0$. Let $b = L_f(x_0, \delta)$. We make note that $b < d(0, \partial D')$. This is trivially the case if $D = \mathbb{R}^n$ and is a consequence of Theorem 6.6.4 otherwise: fixing x on ∂B^* for which $|f(x)| = b$, we see that

$$b = |f(x)| \leq d(0, \partial D')\Theta\left[\frac{|x - x_0|}{\operatorname{dist}(x_0, \partial D)}\right] = d(0, \partial D')\Theta(\tau) < d(0, \partial D').$$

Thus the spherical ring $R' = B^n(b) \setminus \overline{B}^n(a)$ has closure in D' whenever $0 < a < b$. We fix such a ring R' and let $R = f^{-1}(R')$. Because $K(f) = 1$, $\operatorname{Cap}(R) = \operatorname{Cap}(R')$.

The extremal function w for the conformal capacity of the condenser $C = (B^n(b), \overline{B}^n(a))$ was identified earlier as

$$w(y) = \frac{\log|y| - \log a}{\log b - \log a},$$

for y in $\bar{R}'$. Define a function $v : R \to [0,1]$ by $v(x) = w[f(x)]$. Corollary 5.4 ensures that f is bilipschitz on $\bar{R}$—say

$$\lambda^{-1}|x' - x| \le |f(x') - f(x)| \le \lambda|x' - x|$$

for all x and x' in $\bar{R}$, where $1 \le \lambda < \infty$. Since w is a C^∞-function in R', it follows that v satisfies a Lipschitz condition on each compact set in R and hence that v is an ACL-function in R. At each point x of R where v is differentiable we have $|\nabla v(x)| = |\nabla w[f(x)]|\,\|f'(x)\|$, which leads to the estimate

$$\frac{1}{\lambda b}\left(\log\frac{b}{a}\right)^{-1} \le |\nabla v(x)| \le \frac{\lambda}{a}\left(\log\frac{b}{a}\right)^{-1}. \tag{6.66}$$

By the Rademacher-Stepanov theorem, (6.66) holds for almost every x in R.

Now $C = f^{-1}(C') = (U, F)$, where $U = f^{-1}[B^n(b)]$ and $F = f^{-1}[\overline{B}^n(a)]$. Clearly $v(x) \to 0$ as $x \to \partial U \cap \partial R = f^{-1}[S^{n-1}(b)]$, while $v(x) \to 1$ as $x \to F \cap \partial U = f^{-1}[S^{n-1}(a)]$. We conclude that v is a member of the class $\mathcal{A}_3(C)$. From this we infer that

$$\begin{aligned}\operatorname{Cap}(C) &\le \int_R |\nabla v|^n\, dm_n = \int_R |\nabla w \circ f|^n \|f'\|^n\, dm_n \\ &= \int_R |\nabla w \circ f|^n |J_f|\, dm_n = \int_{R'} |\nabla w|^n\, dm_n = \operatorname{Cap}(C') = \operatorname{Cap}(C).\end{aligned}$$

Thus, if we extend v to $\hat{\mathbb{R}}^n$ by defining $v(x) = 1$ for x in F and $v(x) = 0$ for x in U^c, we obtain the extremal function for the conformal capacity of C. Moreover, (6.66) shows that v is essentially nonsingular. Theorem 5.6.14 affirms that v is real-analytic in R. But $u = a(b/a)^v$ in R', so u is also real-analytic there. As this is true whenever $0 < a < b$, we can be certain that u is real-analytic in $f^{-1}[B^n(b) \setminus \{0\}]$, a set which includes the punctured ball B^*. □

With Lemma 6.8.5 in our grasp, the proof of Theorem 6.8.4 becomes straightforward.

Proof of Theorem 6.8.4. As indicated in the opening remarks of this section, following the statement of the theorem, the only point that remains in doubt is whether f is sufficiently smooth to permit us to use Theorem 3.8.7.

We assume, as we may, that D and D' lie in $\mathbb{R}^n$. We shall verify that the coordinate functions $f_1, f_2, \dots, f_n$ of f are real-analytic in D. This instantly places f in the class $C^\infty(D, \mathbb{R}^n)$, which is more than enough smoothness for our purposes.

Fix i with $1 \le i \le n$. It suffices to show that each point of D has a neighbourhood in which f_i is real-analytic. Consider, therefore, an arbitrary point x_0 of D and the ball $B = B^n(x_0, \delta/4)$, in which $\delta = \tau \operatorname{dist}(x_0, \partial D)$ if $D \ne \mathbb{R}^n$—τ being the constant from Lemma 6.8.5—and $\delta = 1$ if $D = \mathbb{R}^n$. Without loss of generality we may assume that $f(x_0) = 0$. The proof of Lemma 6.8.5 included the observation that $b = L_f(x_0, \delta) < d(0, \partial D')$, so D' encompasses the closed ball $\overline{B}^n(b)$. Set $S = \mathbb{S}^{n-1}(x_0, \delta/4)$. For each y on S it is clear that $\operatorname{dist}(y, D) \ge \operatorname{dist}(x_0, D)/2$ and hence that the punctured ball $B_y^* = B^n(y, \delta_y) \setminus \{y\}$, where $\delta_y = \tau \mathrm{dist}(y, D)$ when $D \neq \mathbb{R}^n$ and $\delta_y = 1$ when $D = \mathbb{R}^n$, contains B.

Because $f(S)$ separates $\mathbb{S}^{n-1}(b)$ from the origin in D', both the positive x_i-axis and the negative x_i-axis must intersect $f(S)$. We can thus choose points y and z of S such that $f(y) = se_i$ and $f(z) = -te_i$ for some s and t in $(0, b)$. According to Lemma 6.8.5, the functions u and v defined in D by $u(x) = |f(x) - se_i|$ and $v(x) = |f(x)+te_i|$ are real-analytic in B. However, $f_i = (v^2-u^2+s^2-t^2)/(2s+2t)$ in B, which plainly demands that f_i be real-analytic there as well. We have thereby demonstrated that f_i is a real-analytic function in D, the last item needed to complete the proof. □

The argument just presented also demonstrates that a 1-quasiconformal mapping f between domains in the complex plane $\mathbb{C}$ is a C^∞-function.

In the proof of Theorem 6.8.1 we opted to side step the smoothness issue, preferring to arrive at the analyticity of f in the domain D_0 by a more direct and elementary route that circumvented any use of extremal functions for condensers.

We list several corollaries of Theorem 6.8.4. The first recovers Liouville's theorem in its nicest form.

COROLLARY 6.8.6. *If D is a domain in $\mathbb{R}^n$ with $n \geq 3$ and $f : D \to \mathbb{R}^n$ is a conformal mapping, then f is the restriction to D of a Möbius transformation.*

PROOF. We have $H_f(x) = H[f'(x)] = 1$ for every x in D. Theorem 6.4.10 and Corollary 6.4.19 inform us that $K(f) = 1$, creating a situation to which Theorem 6.8.4 applies. □

Theorem 6.8.4 affords considerable leeway for weakening the a priori smoothness assumptions in Liouville's theorem, even allowing for the presence of a singular set where no conformality need be assumed.

COROLLARY 6.8.7. *Let D be a domain in $\mathbb{R}^n$ with $n \geq 3$, and let $f : D \to \mathbb{R}^n$ be a continuous injection. If there is a relatively closed subset E of D such that E has σ-finite surface area and such that $H_f(x) = 1$ for every point x of $D \setminus E$, then f is the restriction to D of a Möbius transformation.*

PROOF. Theorem 6.4.10, Corollary 6.4.19 and Theorem 6.4.22 lead in combination to the conclusion that $K(f) = 1$. Theorem 6.8.4 does the rest. □

As a matter of fact, the hypotheses in Corollary 6.8.7 can be weakened still further: the conclusion holds if the set $E = \{x \in D : H_f(x) = \infty\}$ has at worst σ-finite surface area and if $H_f(x) = 1$ for almost every x in $D \setminus E$. The example after Theorem 6.4.22 shows, however, that the requirement in Corollary 6.8.7 that E have σ-finite surface area cannot be replaced by the condition that the Hausdorff dimension of E is at most $n - 1$ if the conclusion is to stand.

A third corollary of Theorem 6.8.4 dispenses, to a certain extent, with the injectivity assumptions in Liouville's theorem.

COROLLARY 6.8.8. *Let D be a domain in $\mathbb{R}^n$ with $n \geq 3$, let $f : D \to \mathbb{R}^n$ be a continuous mapping, and let E be the set of points x in D such that x has no neighbourhood in which f is injective. Assume that E is nowhere dense in D and that $D \setminus E$ is connected. If $H_f(x) = 1$ for each point x of $D \setminus E$, then f is the restriction to D of a Möbius transformation.*

PROOF. The set E is easily seen to be closed in D, and therefore $D \setminus E$ is a domain. Each point x of $D \setminus E$ is the center of an open ball B_x in which f is injective. Since H_f is identically 1 in B_x, $K(f|B_x) = 1$ by Theorem 6.4.10 and Corollary 6.4.19. By Theorem 6.8.4 there exists for each x in $D \setminus E$ a Möbius transformation g_x that coincides with f in B_x. A standard connectedness argument (recall the proof of Lemma 3.8.6) reveals that $g_x = g_y$ for all x and y in $D \setminus E$. If g denotes the Möbius transformation that arises in this way, then f and g coincide in $D \setminus E$. Because E is nowhere dense in D, the continuity of f and g then forces f and g to agree throughout D. □

The function $f : \mathbb{R}^n \to \mathbb{R}^n$ given by $f(x_1, x_2, \ldots, x_n) = (x_1, x_2, \ldots, |x_n|)$ shows that the validity of Corollary 6.8.8 depends in a critical way on the assumption that E not separate D. In this example $E = \mathbb{R}^{n-1}$. Thus $\mathbb{R}^n \setminus E$ has two components, in each of which f is the restriction of a Möbius transformation. This obviously cannot be said about f in all of $\mathbb{R}^n$.

In 1967 Reshetnyak published a far-reaching generalization of Corollary 6.8.8, in the process giving birth to the theory of what are nowadays called "quasiregular mappings" [**140**].

In order to make sense of the statement of Reshetnyak's result, we remind the reader that an ACL-mapping $f : D \to \mathbb{R}^n$, where D is a domain in $\mathbb{R}^n$, has a formal derivative matrix $f'(x)$ for almost every x in D. For each x with this property we obtain a linear transformation $T = T_x$ of $\mathbb{R}^n : T(h) = f'(x)h$. Of course, f may fail to be differentiable at such a point x; i.e., it may happen that $\|f(x+h) - f(x) - T(h)\|/\|h\|$ fails to tend to zero with h. We are free, nevertheless, to retain the notation $\|f'(x)\|$ and $J_f(x)$ for the norm of T and determinant of T, respectively.

THEOREM 6.8.9. *If D is a domain in $\mathbb{R}^n$ with $n \geq 3$ and $f : D \to \mathbb{R}^n$ is a nonconstant ACL^n-mapping with the property that $\|f'(x)\|^n = J_f(x)$ for almost every x in D, then f is the restriction to D of a (sense-preserving) Möbius transformation.*

The original proof of Theorem 6.8.9 was long and complicated. What might reasonably be termed an "elementary" proof of the result—the most sophisticated tool used in it is the Sobolev embedding theorem—was discovered by Bojarski and Iwaniec. It appeared in 1982 [**19**]. Even their "elementary" argument contains several daunting computations. The latest contribution to the literature on Liouville's theorem, a 1993 paper of Iwaniec and Martin [**85**], includes the following result for even dimensions $2n$.

THEOREM 6.8.10. *Let D be a domain in $\mathbb{R}^n$ with $n = 2m \geq 2$, and let $f : D \to \mathbb{R}^n$ be a nonconstant ACL^p-mapping with the property that $\|f'(x)\|^m = J_f(x)$ for almost every x in D. If $p \geq m$, then f is the restriction to D of a (sense-preserving) Möbius transformation. The same need not be true when $1 \leq p < m$.*

In the framework of partial differential equations, the higher-dimensional Cauchy–Riemann equations would read

$$Df^t(x)\, Df(x) = J(x, f)^{2/n}, \mathit{identity}, \tag{6.67}$$

and so the Iwaniec-Martin theorem delivers in even dimensions the best possible version of Liouville's theorem, meaning that it derives the conclusion of Liouville's theorem from the weakest possible differentiability assumptions. This is analogous to the distributional version of the solutions to the Cauchy-Riemann equations we discussed above in the classical complex analytic setting.

What the corresponding "sharp" Liouville theorem is in a Euclidean space of dimension $2m+1$ with $m \geq 1$ is still open to conjecture.

CHAPTER 7

Mapping Problems

One of the seminal results of complex analysis, and certainly mathematics of the 19th century, is the Riemann mapping theorem.

THEOREM 7.0.11 (Riemann mapping theorem). *A simply connected proper subdomain Ω of the complex plane $\mathbb{C}$ is conformally equivalent to the unit disk $\mathbb{D}$. That is, there is a bi-holomorphic homeomorphism $\varphi : \mathbb{D} \to \Omega$.*

Here bi-holomorphic means that both φ and $\varphi^{-1} : \varphi(D) \to \mathbb{C}$ are holomorphic, that is, complex analytic. As noted earlier in our discussion of the Liouville theorem this result was stated by Bernhard Riemann, for domains with piecewise smooth boundary, in his PhD thesis of 1851. Riemann's proof depended on the use of the Dirichlet principle (coined by Riemann), which asserts the existence of a minimiser to the Dirichlet energy functional

$$\mathcal{E}[u] = \int_\Omega \Big(|\nabla u(z)|^2 - u(z)f(z) \Big)\, dz \tag{7.1}$$

with given boundary values $u(z) = g(z)$ for $z \in \partial\Omega$, and that this minimiser satisfies the Laplace equation $\Delta u = f$, $u|\partial\Omega = g$. Ultimately David Hilbert showed that Riemann's use of this principle was sound in the setting in which he was using it, after Karl Weierstrass had showed that the Dirichlet principle does not hold in complete generality, even for simply connected domains. The first complete proof of the theorem as stated was given by Carathéodory in 1912. This proof was simplified to what is now the most common approach by Paul Koebe and can be found in almost any graduate text on complex analysis.

Carathéodory's extension of the Riemann mapping theorem goes on to assert that if Ω is a *Jordan domain*, that is, if $\partial\Omega$ is homeomorphic to the circle $\mathbb{S}$, then any Riemann mapping $f : \Omega \to \mathbb{D}$ extends homeomorphically as a map $f : \overline{\Omega} \to \overline{\mathbb{D}}$—we have already discussed higher-dimensional analogues of this in Theorem 6.5.11.

The applications of the Riemann mapping theorem are many and varied, and it would be wonderful to have a useful higher-dimensional version of it. This is in part due to the fact that many functionals of physical interest—and their associated partial differential equations—transform in natural ways under conformal mappings. For instance, if u is harmonic on a domain Ω and $\varphi : \Omega' \to \Omega$ is conformal, then $u \circ \varphi$ is harmonic on Ω'. Thus, as a consequence of the Riemann mapping theorem, we are able to solve the Laplace equation on a simply connected domain (not $\mathbb{C}$) as soon as we can solve it on the disk and have paid due attention to the question of boundary values. In higher dimensions there are not enough conformal mappings to make this observation worthwhile. However, quasiconformal mappings can play a role here as they may be used—in all dimensions—to transform second order equations of divergence type, such as the Laplacian, on one (possibly irregular) domain to another nicer domain while preserving the key property of ellipticity needed in the general theory of existence and regularity. This is exploited to great extent in [**15**] in two dimensions and is discussed in [**84**] in higher dimensions.

This brings us to what is described as the main problem of F.W. Gehring and J. Väisälä in their important Acta paper of 1965, [**54**]. They ask "*what are the domains D in $\mathbb{R}^n$ that are quasiconformally equivalent to the unit ball*".

No doubt unbeknownst to the reader, many of the results and ideas from that paper have already made their way into this book. In this section we will present other (now) classical results from that paper—in particular the geometric obstructions of an inward directed cusp (they used the term spire) and an outward directed ridge.

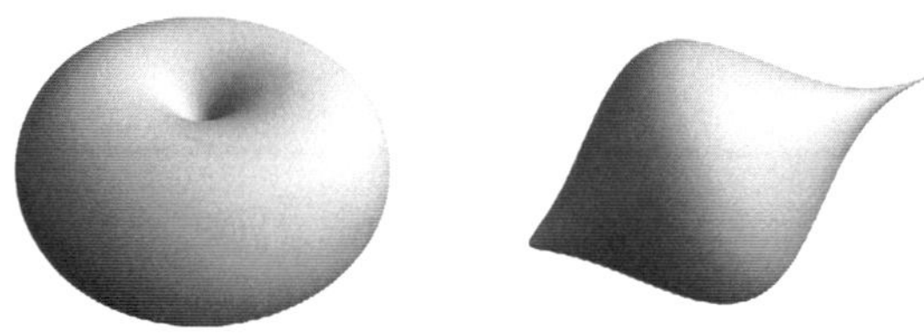

Domains with cusps.
The inward directed cusp on the left is not a quasiball, while the outward directed cusp on the right is a quasiball. Neither is a quasisphere.

We will also present positive results through the solution of the quasiconformal Schoenflies problem along with another interesting result of Väisälä. In the paper mentioned above, the sharp outer distortion estimate for a quasiconformal mapping $f : B^3 \to B^2 \times \mathbb{R}$, the infinite cylinder, was determined and uniqueness was established. In fact if $f : B^3 \to B^2 \times \mathbb{R}$ is quasiconformal, then

$$K_O(f) \geq q_0 = \frac{1}{2}\int_0^{\pi/2} \frac{dt}{\sqrt{\sin t}} \approx 1.3110\dots. \tag{7.2}$$

This bound is achieved in the example we gave earlier. This example motivates Jussi Väisälä's complete description of all cylindrical domains $\Omega \times \mathbb{R}$, Ω a planar domain, which are quasiconformally equivalent to the unit ball. We present this later in Section 7.5.

There are in fact two problems here. We mentioned above the question of domains quasiconformally equivalent to the ball. There is an associated problem: "*What are the domains D in $\mathbb{R}^n$ that are equivalent to the unit ball, under a quasiconformal homeomorphism of $\hat{\mathbb{R}}^n$*". More precisely we ask when there is a quasiconformal homeomophism $f : \hat{\mathbb{R}}^n \to \hat{\mathbb{R}}^n$ such that $f(B^n) = D$. In this case the boundary of D is called a quasisphere.

In two dimensions complete answers to these questions are known with the answer to the last question effectively controlled by the geometry of the boundary of Ω through Ahlfors' three-point, or bounded turning condition, which we will discuss later. There seems no hope for such a simple answer in three or more dimensions.

We have already seen in subsections 6.2.1–6.2.7 many examples of domains quasiconformally equivalent to a ball. These included cones, dihedral wedges, the infinite cylindrical domain and related examples. A natural question is to ask when a domain is quasiconformally equivalent to the ball is "*is there a best possible mapping*". Such a mapping will be called an *extremal mapping*. We dispense with the existence of such a mapping in a moment. Virtually nothing is known in dimension $n \geq 3$ about the problem of uniqueness for extremal mappings. In dimension two the question is of course trivial; there is a conformal mapping unique up to a normalisation.

7.1. Existence of extremal mappings

The existence of extremal mappings follows from the lower-semicontinuity of the distortion. Let us define the *coefficients of quasiconformality* of a domain as follows:

$$K_I(D) = \inf\{K : \text{there is a quasiconformal } f : D \to B^n \text{ and } K_I(f) \leq K\}.$$

We define the numbers $K_O(D)$ and $K(D)$ similarly.

We already know that these three coefficients are simultaneously finite or infinite, and when they are finite we say that D is quasiconformally equivalent to a ball.

THEOREM 7.1.1. *Let D be a domain in $\mathbb{R}^n$ which is quasiconformally equivalent to the ball B^n. Then there exist extremal quasiconformal homeomorphisms $f_I, f_O, f : D \to B^n$ for which $K_I(f_I) = K_I(D)$, $K_O(f_O) = K_O(D)$, and $K(f) = K(D)$.*

PROOF. Fix K so that $K_I(D) < K < \infty$. Then we may choose a sequence of homeomorphisms $\{f_m\}$ of D onto B^n such that

$$K_I(D) = \lim_{m\to\infty} K_I(f_m) \tag{7.3}$$

and such that $K_I(x, f_m) \leq K$ for all m. By composing f_m with a suitable Möbius transformation of B^n onto itself, we may further assume that $f_m(x_0) = x_1$, where x_0 and x_1 are fixed points in D and B^n respectively. Now the f_m are all K-quasiconformal, and hence by Corollary 6.6.19 and its precursors, there exists a subsequence which converges to a homeomorphism $f : D \to B^n$ locally uniformly on D. We then have

$$K_I(D) \leq K_I(f) \leq \liminf K_I(f_m). \tag{7.4}$$

This establishes the existence of an extremal mapping for $K_I(f)$. The proof for the existence of extrema for K_O and the maximal distortion follow the same lines. □

At this point it is worthwhile recalling the relationships between these distortion functions for domains. The following lemma is a direct consequence of what we established in Corollary 6.4.16.

LEMMA 7.1.2. *Let D be a domain in $\mathbb{R}^n$; then*

$$1 \leq K_O(D) \leq K_I(D)^{n-1}, \qquad 1 \leq K_I(D) \leq K_O(D)^{n-1}$$

and

$$\min\{K_I(D), K_O(D)\} \leq K(D)^{n/2}, \qquad \max\{K_I(D), K_O(D)\} \leq K(D)^{n-1}.$$

From the very definition of a quasiconformal mapping, the coefficients of quasiconformality are determined by the moduli of curve families within the domain and their images in the ball.

LEMMA 7.1.3. *Let D be a domain in $\mathbb{R}^n$ and $f : D \to \mathbb{R}^n$ a homeomorphism. Then*

$$K_I(f)^{n-1} = \sup_{\Gamma} \frac{M(\Gamma')}{M(\Gamma)}, \qquad K_O(f)^{n-1} = \sup_{\Gamma} \frac{M(\Gamma)}{M(\Gamma')},$$

where the supremum is taken over all curve families Γ in D, $\Gamma' = f(\Gamma)$, and for which $M(\Gamma), M(\Gamma')$ are not equal to 0 and ∞.

In light of subsection 6.6.9 concerning the limits of quasiconformal mappings when one uses the linear dilatation in higher dimensions, $n \geq 3$, there will in general be no extremal mapping for the linear dilatation, though as noted there every normalised sequence of quasiconformal mappings with uniformly bounded linear dilatation will have a locally uniformly convergent subsequence. The point is that in the limit the linear dilatation may jump up.

7.2. Topological obstructions: Wild bilipschitz spheres

In this section we outline a fairly standard topological construction due to R.H. Fox and E. Artin (see [**39**]) to prove the following theorem which asserts the existence of a wild sphere in the bilipschitz category.

THEOREM 7.2.1. *In three dimensions there are*

(i) *a bilipschitz homeomorphism $f_1 : \bar{B}^3 \to \mathbb{R}^3$ such that both $D = f(B^3)$ and $D^c = \hat{\mathbb{R}}^3 \setminus \bar{D}$ are simply connected Jordan domains, but D^c is not topologically equivalent to the ball,*
(ii) *a bilipschitz homeomorphism $f_2 : \mathbb{S}^2 \to \mathbb{R}^3$ such that both of the components of $\hat{\mathbb{R}}^3 \setminus f_2(\mathbb{S}^2)$ are simply connected Jordan domains, neither of which is topologically equivalent to the ball,*
(iii) *a bilipschitz homeomorphism $f_3 : \mathbb{S}^2 \to \mathbb{R}^3$ such that both of the components of $\hat{\mathbb{R}}^3 \setminus f_2(\mathbb{S}^2)$ are Jordan domains neither is of which is simply connected.*

PROOF. Once we have constructed an example to show that (i) is true, the reader will have no problems with (ii) and (iii). The first example is a "thickened" version of the wild Fox-Artin knot illustrated below.

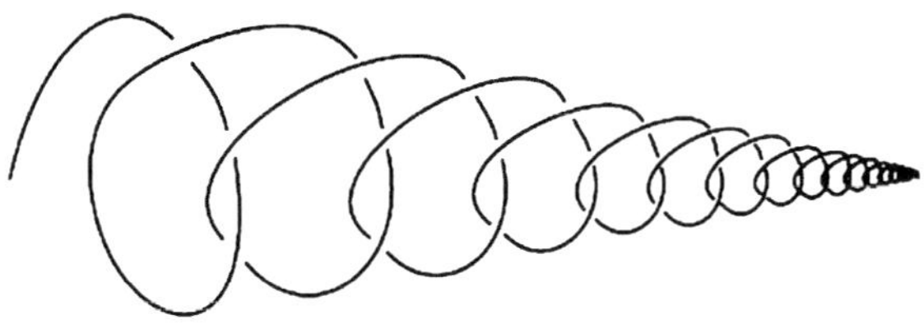

The Fox-Artin knot $\mathcal{K}$.

We can parameterise this knot by a homeomorphism $\alpha : [-1, 0] \to \mathbb{R}^3$, C^∞-smooth on the interval $(-1, 0)$, $\alpha(0) = 0$, $\alpha(-1) = -e_1$, and with the periodic relationship

$$\alpha(2t) = 2\alpha(t), \qquad 0 \geq t \geq -\frac{1}{2}. \tag{7.5}$$

Using this periodicity we can further arrange the following separation property. Write $\alpha(t) = (\alpha_1(t), \alpha_2(t), \alpha_3(t))$. Then there is $\delta_1 > 0$ such that

$$|\alpha_1(t) - \alpha_1(s)| > \delta, \qquad \text{whenever } t \in [-1, -\tfrac{1}{2}] \text{ and } s \in [-\tfrac{1}{4}, \tfrac{1}{8}]. \tag{7.6}$$

Then, because of the periodicity of α, we see that

$$|\alpha_1(t) - \alpha_1(s)| > 2^{-k}\delta_1 \tag{7.7}$$

for $t \in [-2^{-k}, -2^{-k-1}]$ and $s \in [-2^{-k-2}, -2^{-k-3}]$.

Next, the image of $[-1, -\frac{1}{4}]$ is a C^∞-smooth arc in $\mathbb{R}^3$ and as such has curvature bounded below. This means that there is $\delta_2 > 0$ such that if Π_t and Π_s are orthogonal planes to $\alpha'(t)$ and $\alpha'(s)$ respectively, then

$$\left[\alpha(t) + B^3(\delta_2) \cap \Pi_t\right] \cap \left[\alpha(s) + B^3(\delta_2) \cap \Pi_s\right] = \emptyset. \tag{7.8}$$

In order to achieve both separation and curvature boundedness, we put

$$\delta = \frac{1}{8} \min\{\delta_1, \delta_2, 1\}. \tag{7.9}$$

Let $\mathcal{C}$ denote the closed cone

$$\mathcal{C} = \{(x_1, x_2, x_3) \in \mathbb{R}^3 : -1 \leq x_1 \leq 0, x_2^2 + x_3^2 \leq \delta\,|x_1|\}$$

and $\mathcal{C}_+ = \{x \in \mathcal{C} : -1 \leq x_1 < \frac{1}{2}\}$. Then

$$\mathcal{C} \setminus \{0\} = \bigcup_{k=0}^{\infty} 2^{-n}\mathcal{C}_+.$$

On $\mathcal{C}_+$ we can introduce cylindrical coodinates (t, z), $t \in [-1, -\frac{1}{2}]$, $(x_2, x_3) \in B^2(0, \delta|x_1|)$. Let $O(t)$ be a smoothly varying family of orthogonal transformations for which

$$O(t)e_1 = \frac{\alpha'(t)}{\|\alpha'(t)\|}.$$

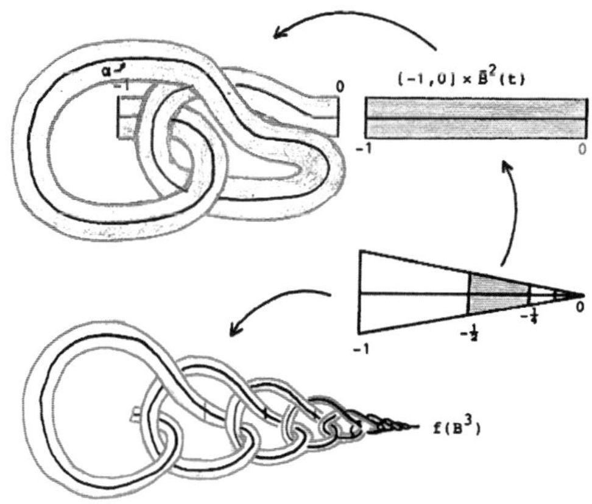

The wild Fox-Artin sphere.

We now define a diffeomorphism $h_1 : \mathcal{C}_+ \to \mathbb{R}^3$ by

$$h_1(x_1, x_2, x_3) = \alpha(x_1) + |x_1|\, O(x_1) \cdot (0, x_2, x_3).$$

As h_1 is seen to be a diffeomorphism of a slightly larger neighbourhood, it is bilipschitz on its compact domain of definition—the map h_1 is bilipschitz. In fact the reader should see that $h_1(\mathcal{C}_+)$ is a "conical" neighbourhood of the initial segment $\alpha([\frac{1}{2}, 1])$ of α.

We now use the periodicity of α to extend h_1.

For $x \in \mathcal{C} \setminus \{0\}$, there is a unique integer k_x so that $2^k x \in \mathcal{C}_+$. Then define

$$f(x) = 2^{-k_x}\, h_1(2^{k_x} x), \quad f(0) = 0. \tag{7.10}$$

We have to prove that f is a well-defined bilipschitz homeomorphism of $\mathcal{C}$. On the core, $x_1 \in [0,1]$, $x_2 = x_3 = 0$, and we have $h_1(x_1, x_2, x_3) = \alpha(x_1)$ by the periodicity of α at (7.5). The gluing at the ends of each $2^{-k}\mathcal{C}_+$ also matches from this periodicity, and because the map is defined only in terms of the local data, it is clear that f is smooth over each such end.

Next it is also clear that

$$f : \mathcal{C}_+ \cup \frac{1}{2}\, \mathcal{C}_+ \to \mathbb{R}^3 \tag{7.11}$$

is an L-bilipschitz diffeomorphism, for some $L > 1$, as this is in fact the set in which we obtained our curvature bound at (7.8).

These facts together imply that $f : \mathcal{C} \to \mathbb{R}^3$ is locally L-bilipschitz, and so as $\mathcal{C}$ is convex, the map f is locally L_1-Lipschitz.

To complete the proof of (i) we first need to get an appropriate lower bound on $|f(x) - f(y)|$, $x, y \in \mathcal{C}$. If $x, y \in 2^{-k}\mathcal{C}_+ \cup 2^{-k-1}\mathcal{C}_+$, then periodicity gives

$$|f(x) - f(y)| = 2^{-k}|f(2^k x) - f(2^k y)| \geq 2^{-k}\frac{1}{L}\, |(2^k x - 2^k y)| = \frac{1}{L_1}\, |x - y| \tag{7.12}$$

since $2^k x, 2^k y \in \mathcal{C}_+ \cup \frac{1}{2}\mathcal{C}_+$, where f is L_1-bilipschitz. Next, if $x \in 2^{-k}\mathcal{C}_+$ and $y \in 2^{-j}\mathcal{C}_+$, with $j - k > 2$, we proceed as follows. Let $x = (x_1, x_2, x_3)$ and

$y = (y_1, y_2, y_3)$. Then applications of the triangle inequality, periodicity, and the definition of the map f gives

$$
\begin{array}{l}
|f(x) - f(y)| \\
\quad = \quad 2^{-k}|f(2^k x) - f(2^k y)| \\
\quad = \quad 2^{-k}|f(2^k x) - f(2^k x_1) + f(2^k x_1) - f(2^k y_1) + f(2^k y_1) - f(2^k y)| \\
\quad \geq \quad -2^{-k}|f(2^k x) - f(2^k x_1)| + 2^{-k}|f(2^k x_1) - f(2^k y_1)| - 2^{-k}|f(2^k y_1) - f(2^k y)| \\
\quad \geq \quad -2^{-k}\delta + 2^{-k}|f(2^k x_1) - f(2^k y_1)| - 2^{-k}|f(2^k y_1) - f(2^k y)| \\
\quad \geq \quad -2^{-k}\delta + 2^{-k}|f(2^k x_1) - f(2^k y_1)| - 2^{-j}|f(2^{-j} y_1) - f(2^{-j} y)| \\
\quad \geq \quad -2^{-k}\delta - 2^{-j}\delta + 2^{-k}|f(2^k x_1) - f(2^k y_1)| \\
\quad \geq \quad -2^{-k}\delta - 2^{-j}\delta + 2^{-k}|\alpha(x_1) - \alpha(y_1)|^{-k} \\
\quad \geq \quad -2^{-k}\delta - 2^{-j}\delta + 2^{-k}\delta_1 \geq \left(2^{3-k} - 2^{-k} - 2^{-k-2}\right)\delta \geq 6 \times 2^{-k}\delta.
\end{array}
$$

The last estimate follows from the separation property (7.6) and our choice at (7.9) giving $\delta < \delta_1/8$. Next,

$$
\begin{aligned}
|x - y| &\leq |x_1 - y_1| + |x - x_1| + |y - y_1| \\
&\leq |x_1 - y_1| + 2^{-k}\delta + 2^{-j}\delta \leq 2^{-k-1} + 2^{-k}\delta + 2^{-j}\delta \leq 2^{-k}
\end{aligned}
$$

since $\delta \leq \frac{1}{8}$ by our choice at (7.9). These two estimates show that in this case

$$\frac{|f(x) - f(y)|}{|x - y|} \geq \frac{6\delta 2^{-k}}{2^{-k}} = 6\delta.$$

We have therefore found that f is L-bilipschitz on $\mathcal{C}$ as soon as we put

$$L = \max\{L_1, \frac{1}{6\delta}\},$$

and observe that continuity extends all our estimates to the point 0.

We put

$$D = f[\operatorname{int}(\mathcal{C})]$$

and refer to ∂D as a "wild" Fox-Artin sphere after their 1948 paper [**39**] which introduced such things in the topological category.

Next we observe that $\mathcal{C}$ itself is the bilipschitz image of the closed unit ball B^n, and therefore there is a bilipschitz homeomorphism $f : \bar{B}^3 \to \bar{D}$ justifying the term "sphere" used above. Of course D is a *quasiball*—the quasiconformal image of a ball. Also, D^c (in $\hat{\mathbb{R}}^n$) is simply connected. There is an obvious retract of $\bar{D}$ to the central core α, and then retract α by

$$H(t, \alpha(s)) = \alpha(st), \quad t \in [0, 1], \quad s \in [-1, 0].$$

Then $H(1, \alpha(s)) = \alpha(s)$ and $H(0, \alpha(s)) = 0$. This shows D^c has the homotopy type of a point. However D^c is not topologically a closed ball. It is not locally simply connected at the boundary point 0.

An open set U in $\hat{\mathbb{R}}^3$ is *locally simply connected at its boundary* if for every $x \in \partial U$ and $\epsilon > 0$ there is $\delta > 0$ such that any loop in $B^3(x, \delta) \cap U$ can be homotoped to a point in $B^3(x, \epsilon) \cap U$. D^c obviously does not have this property at 0—we can't move continuously through the infinitely tangled part of the knot and so must always go over the "end" at $-e_1$. This local simple connectedness is clearly a property preserved by homeomorphisms.

This completes part (i) of the theorem.

Part (ii) of the theorem follows in a similar fashion, but this time we use the bi-infinite Fox-Artin knot $\alpha \cup \{-1-\alpha\}$. Then part (iii) follows by attaching an inward and an outward directed wild sphere to $\mathbb{S}^n$. We leave it to the reader to think about this. □

7.3. Geometric obstructions to existence

As we noted in the introduction to this chapter, an elementary geometric criterion to determine whether a topological 2-sphere in $\mathbb{R}^3$ is a quasi-sphere has long been sought. Ahlfors' bounded turning condition fulfils this wish in an elegant and precise way in two dimensions. This section shows that such hopes in three or more dimensions are probably futile.

THEOREM 7.3.1. *Suppose that $0 < a < b$, that D is a domain in $\mathbb{R}^3$, and that $D^c \cap B^3(b)$ has at least two components which meet $S^2(a)$. Then*

$$K_I(D) \geq \sqrt{\frac{\Psi(\frac{1}{2})}{4\pi}} \log \frac{b}{a}. \tag{7.13}$$

PROOF. Let f be any homeomorphism of D onto B^3. We seek to establish (7.13). Since the right-hand side of (7.13) is continuous in b, it is sufficient to establish (7.13) under the slightly stronger hypothesis that the closed set $E = D^c \cap B^3(b)$ has at least two components which meet $S^2(a)$. We in fact need only discuss the special case where f can be extended to be a homeomorphism of D onto B^3, for we may take an integer $n > 0$, and then let D_n denote the image of $B^3(n/(n+1))$ under f^{-1}. The continuity of moduli and the lower semicontinuity of the distortion of a quasiconformal mapping will now imply the result we want as we let $n \to \infty$.

By hypothesis, there exist points $p_1, p_2 \in \mathbb{S}^2(a)$ which belong to different components of E, and hence we can find disjoint compact sets E_1 and E_2 such that $E = E_1 \cup E_2$ with $p_1 \in E_1$ and $p_2 \in E_2$. Let P_1 be the last point of $[p_1, p_2] \cap E_1$ and let P_2 be the first point in $[p_1, p_2] \cap E_2$ as we move from P_1 toward p_2 along $[p_1, p_2]$. Next let $E_0 = [P_1, P_2]$ be the closed segment. Then $E_0 \subset \bar{D}$ and P_1 and P_2 are points of ∂D which lie in different components of E.

Next let $F_1 = \partial D \cap (B^3(b))^c$ and let C be any connected set in ∂D which contains both P_1 and P_2. Since P_1 and P_2 belong to different components of E, we must have $F_1 \cap C \neq \emptyset$. Hence F_1 separates P_1 and P_2 in ∂D, and as ∂D is homeomorphic to a 2-sphere we can find a continuum $F_0 \subset F_1$ which separates P_1 and P_2 in ∂D.

We now define Γ to be the family of arcs which join the segment E_0 and the separating continuum F_0 in D. Since $E_0 \subset B^3(a)$ and $F_0 \subset B^3(b)^c$ the curve family Γ is minorized by the family of curves joining $S^2(a)$ and $S^2(b)$ in $\mathbb{R}^3$. Hence

$$M(\Gamma) \leq 4\pi\Big(\log \frac{b}{a}\Big)^{-2}. \tag{7.14}$$

Now we need to consider the images of these sets. Set $E_0' = f(E_0)$, $F_0' = f(F_0)$, and $P_i' = f(P_i)$, $i = 1, 2$. Now E_0' joins the points P_1' and P_2' in B^3. Also F_1' separates P_1' and P_2' in $\mathbb{S}^2$. Let us map conformally to the half-space $\mathbb{H}^3$ by Möbius φ so that $\varphi(P_1') = 0$ and $\varphi(P_2') = \infty$. Then the image $\varphi(E_0')$ joins 0 and ∞ in $\mathbb{H}^3$

and $X = \varphi(E_1')$ separates 0 and ∞ in ∂D. Let x_0 be the point with the smallest absolute value in X. Then the diametrically opposite point x_0, call it $x_1 \in X$, has

$$0 < |x_1| < \frac{1}{2}|x_1 - x_2|,$$

so the extremal Teichmüller modulus estimate gives

$$M[\varphi(\Gamma)] = M[f(\Gamma)] \geq \Psi\Big(\frac{1}{2}\Big),$$

since Ψ is decreasing. We put this together with (7.14) to see

$$\frac{M[f(\Gamma)]}{M(\Gamma)} \geq \frac{\Psi\left(\frac{1}{2}\right)}{4\pi}\Big(\log\frac{b}{a}\Big)^2.$$

We finally apply Lemma 7.1.3, with $n = 3$, to obtain the result we want. □

The observant reader will have seen that buried in the last part of the proof of the theorem is an estimate on the moduli of curve families of linking continua. Say E and F continua in $\mathbb{S}^3$ are linked if there is no topological ball B which contains one of E or F and does not meet the other. If Γ is the family of curves joining E to F, then the above argument shows there is a universal lower bound $M(\Gamma) \geq M_0$ and it seems reasonable to conjecture the extremal is achieved in the Hopf link. There is an interesting survey around this problem in [**113**].

One of the geometric obstructions to a domain D homeomorphic to a ball being quasiconformally equivalent to a ball are these sorts of "linking" phenomena which force a lower bound on the moduli of a topologically configured curve family in the ball, which may have had a small modulus in D because of its geometry.

Here is an alternative formulation of basically the same result which the reader should have no trouble in proving for themselves.

THEOREM 7.3.2. *Suppose that* $0 < a < b$, *that* D *is a domain in* $\mathbb{R}^3$, *and that* $D^c \cap B^3(a)^c$ *has at least two components which meet* $S^2(b)$. *Then*

$$K_I(D) \geq \sqrt{\frac{\Psi(\frac{1}{2})}{4\pi}}\,\log\frac{b}{a}.$$

We will use this result to study ridges in the boundary of a domain.

7.3.1. Cusps in the boundary. A set in $\mathbb{R}^3$ is called a cusp if it can be transformed by a similarity onto

$$S = \{x = (r, \theta, t) \in \mathbb{R}^3 : r = g(t), 0 < t \leq a\}$$

for some finite a and where g is subject to the restrictions

(i) g is continuous in $[0, a]$ with $g(a) = 0$,
(ii) g' is continuous and increasing on $(0, a)$, and
(iii) $\lim_{t\to a} g'(t) = 0$.

These conditions imply that $g > 0$ in $[0, a)$ and that

$$\int_0^a \frac{dt}{g(t)} = +\infty. \tag{7.15}$$

A domain D is said to have a boundary cusp at $p \in \partial D$ if a neighbourhood of p can be parameterised as above.

We leave it to the reader to decide for themselves what it means for a cusp to be inward directed or outward directed, but how this idea is used appears in the proof of the next theorem.

We have the following obstruction to a domain D, which we may as well assume is a topological ball in the following, being the quasiconformal image of a ball.

THEOREM 7.3.3. *If D is a domain in $\mathbb{R}^3$ whose boundary contains an inward directed cusp, then $K(D) = +\infty$.*

PROOF. After a similarity transformation, we may assume that the tip of the cusp S is the origin and that it is parametrised as above on the segment $[-e_3, 0]$. There is some $a > 0$ such that $S \cap B^3(a) = \partial D \cap B^3(a)$.

Now S splits $B^3(a)$ into two domains, and since S is inward directed, we see $\bar{D}^c \cap B^3(a)$ is the component of $B^3(a) \backslash S$ which contains the segment $\{r = 0, 0 < x_3 < a\}$. Fix $c \in (0, 1)$. Since S is a cusp, we can find $b \in (0, a/2)$ such that $\mathbb{S}^2(be_3, b)$ separates 0 from ∞ in D^c and hence $D^c \cap [B^3(be_3, bc)]^c$ has two components which meet $\mathbb{S}^2(be_3, b)$. From Theorem 7.3.1 we have

$$K_I(D) \geq \sqrt{\frac{\Psi(\frac{1}{2})}{4\pi}} \log \frac{1}{c}.$$

We can now let $c \to 0$ to obtain the proof of the theorem. □

We have already shown that the domain bounded by the right circular cylinder

$$D = \{(x, y, z) : z^2 + y^2 < \frac{1}{2}, x > \frac{\sqrt{3}}{2}\}$$

is the quasiconformal image of a ball. Consider the image of D, say D^*, under inversion in the unit sphere. A brief calculation shows that the part of the boundary of D^* inside the unit ball is parameterised by

$$S = \{(t, g, g) \in \mathbb{R}^3 : 2g = t^2 + g^2, 0 \leq t \leq \sqrt{3}/2\}.$$

Then $g = 1 - \sqrt{1 - t^2}$ and $g'(t) = \frac{t}{\sqrt{1-t^2}} \to 0$ as $t \to 0$, meeting our definition of a cusp. As the conformal image of D, the domain D^* is the quasiconformal image of the ball and D^* has an outward directed cusp.

In summary: *outward directed cusps are allowed for quasiballs, but inward directed cusps are not.*

Note that in the proof given for Theorem 7.3.3 we never actually used the precise parametrisation of the cusp. It was simply enough to find a sequence of $c \to 0$ with the separation properties needed for the application of the modulus estimates of Theorem 7.3.1. This is clearly a far wider class of domains.

7.3.2. Ridges in the boundary. A set in $R \subset \mathbb{R}^3$ is called a *boundary ridge* if it can be transformed by a similarity transformation onto

$$S = \{(x_1, x_2, x_3) \in \mathbb{R}^3 : |x_2| = g(x_1), 0 < x_1 \leq a, |x_3| < b\},$$

where $a < \infty$, $b \leq \infty$, and g also satisfies the conditions above for a cusp. The image of the line segment

$$E = \{(x, y, z) \in \mathbb{R}^3 : x = a, y = 0, |z| < b\}$$

under a similarity is called the edge of the ridge R.

Again we leave it to the reader to figure out what we mean by "inward" and "outward" ridges for the boundary of a domain—it is the same as for a cusp of course.

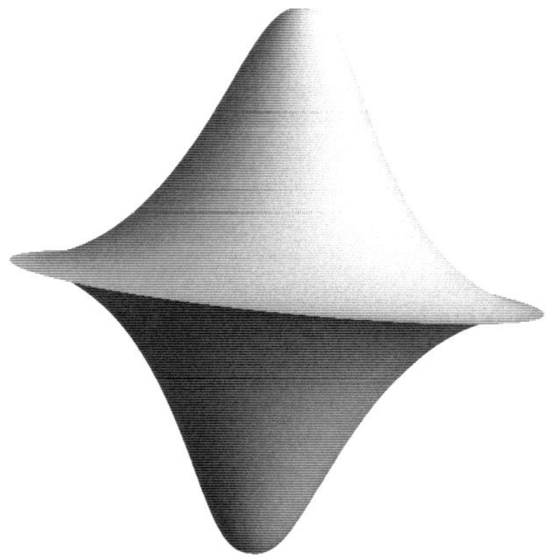

A domain with an outward directed ridge.

We have the following analogue of Theorem 7.3.3 for ridges.

THEOREM 7.3.4. *Let D be a domain in $\mathbb{R}^3$ whose boundary contains an outward directed ridge; then $K(D) = +\infty$.*

PROOF. We may assume that the edge of the ridge S is the line segment $x = y = 0$ and $|z| < 1$, and that for some $a > 0$ (since the ridge is outward directed)

$$S \cap B^3(a) = \partial D \cap B^3(a).$$

Then S divides $B^3(a)$ into two components, and again since S is outward directed, $D \cap B^3(a)$ is the component of $B^3(a) \setminus S$ containing the interval $0 < x < a$, $y = z = 0$. Because S is a ridge, given $0 < c < 1$, we can choose $0 < b < a/2$ so that D separates $(b, bc, 0)$ and $(b, -bc, 0)$ in $B^3(be_1, b)$. Thus $D^c \cap B^3(be_1, b)$ has two components which meet $\mathbb{S}^2(be_1, bc)$, and we achieve from Theorem 7.3.2 the estimate

$$K_I(D) \geq \sqrt{\frac{\Psi(\frac{1}{2})}{4\pi}} \log \frac{1}{c}.$$

We can now let $c \to 0$ to obtain the proof of the theorem. □

As with the case of cusps and in contrast to the above situation, there exist domains with inward directed ridges in their boundaries which are quasiballs. For example, given $0 < a < \infty$, set $g(u) = \min\{u^2, a^2\}$ and let

$$T = \{x = (x_1, x_2, x_3) : |x_3| = g(x_1),\ x_1 \geq 0\}.$$

Then T bounds a domain D which has an inward directed ridge in its boundary. We can define a quasiconformal mapping f of D onto a Dihedral wedge $D^* \subset \mathbb{R}^3$ of angle 2π

$$D^* = \{(r, \theta, x_3) : 0 < \theta < 2\pi\}$$

by

$$f(x) = \begin{cases} x - g(x_1)\, sgn(x_2),\ e_2 & x_1 \geq 0, \\ x, & x_1 < 0, \end{cases} \tag{7.16}$$

where the sign function sgn has $sgn(0) = 0$. We leave to the reader the verification of the distortion estimate

$$K_I(f) \leq (2a+1)^{3/2}.$$

Now the dihedral wedge maps to the half-space by

$$(r, \theta, x_3) \mapsto (r, \theta/2, x_3/\sqrt{2})$$

of maximal distortion less than $2^{3/4}$.

In summary: *inward directed ridges are allowed for quasiballs, but outward directed ridges are not.*

7.3.3. Rickman's rugs. There are other natural geometric obstructions on the boundary of a domain D that is at least homeomorphic to B^n which prevent D from being quasiconformally equivalent to the ball B^n.

Among other things, the linear distortion measures the anisotropic nature of the stretching of a mapping. It might be supposed, therefore, that if the boundary of a domain D is highly anisotropic, then as we approach the boundary of D from the interior by a quasiconformal mapping from a highly isotropic body (the ball B^n), the level of anisotropy should be reflected at ever higher levels and so no uniform bound on the distortion can be obtained. Therefore that domain should not be a quasiball.

This situation might arise if for instance the boundary of D were totally unrectifiable in one direction and smooth in another. This would need to occur in at least three dimensions and thus be another higher-dimensional obstruction. We next show that these guesses—when suitably framed—lead to theorems.

The argument we present here is due to J. Väisälä and generalises an observation of S. Rickman. It is presented by P. Tukia in his paper on quasiconformal groups not isomorphic to Lie groups [**156**].

A metric arc A is a metric space homeomorphic to a closed interval. The distance of two points $x, y \in A$ is denoted by $|x - y|$. Note that this might not be the Euclidean distance, even should A come as a subset of $\mathbb{R}^n$. We fix a metric arc A in what follows.

LEMMA 7.3.5. *Let $\ell(A)$ be the length of A. Let $M, \delta > 0$. Then there is a subdivision of A into subarcs $A_1, A_2, \ldots, A_r$ by successive points $x_0, x_1, \ldots, x_r \in A$ such that with $\lambda_i = |x_i - x_{i-1}|$, $i = 1, 2, \ldots, r$,*

(a) $\lambda_1 + \lambda_2 + \cdots + \lambda_r \geq \min\{M, \ell(A) - M^{-1}\}$,
(b) $\frac{1}{2} \leq \frac{\lambda_i}{\lambda_j} \leq 2$, *$i, j \leq r$, and*
(c) $\lambda_i \leq \delta$, $i = 1, 2, \ldots, r$.

PROOF. There is obviously a subdivision $A_1, \ldots, A_r$ for which (a) is true. Set

$$\alpha = \frac{1}{2} \min\{\delta, \lambda_1, \ldots, \lambda_r\},$$

so each $\lambda_i \geq 2\alpha$. We proceed to subdivide each A_i as follows. Let y_1 be the last point on A_i from x_{i-1} which is on $B^n(x_{i-1}, \alpha)$. If $|y_1 - x_1| < 2\alpha$ we do not do anything more. Otherwise, let y_2 be the last point of A_i on $B^n(y_1, \alpha)$. Continuing in this way we get a new subdivision $x_{i-1} = y_0 < y_1 < \cdots < y_k = x_i$ of A_i such that each distance $|y_i - y_{i-1}| = \alpha$ except possibly the last one, for which $\alpha \leq |y_k - y_{k-1}| < 2\alpha$.

The new subdivision again satisfies (a), as well as (b) and (c). □

We now need the reader to review our earlier discussion of quasisymmetry in Theorems 6.6.7 and 6.6.8 so that we can prove the following result.

LEMMA 7.3.6. *Let A be a nonrectifiable metric arc and let $k > 1$. Then $A \times [0,1]^k$ cannot be embedded into $\mathbb{R}^{k+1}$ by a quasisymmetric map when the metric is given by*

$$|(x,y) - (x',y')| = \sqrt{|x-x'|^2 + |y-y'|^2}, \quad x \in A, y \in \mathbb{R}^k. \tag{7.17}$$

PROOF. Set $I^k = [0,1]^k$ and let $h : A \times I^k \to \mathbb{R}^{k+1}$ be η-quasisymmetric. Recall this means that

$$|h(y) - h(x_0)| \leq \eta(\rho)|h(w) - h(x_0)|$$

for all $x_0, y, w \in A \times I^k$ satisfying $|y - x_0| = \rho\,|w - x_0|$.

We show that this leads to a contradiction.

For $a \in A$ let $I_a = \{a\} \times I^k$. It is a moment's thought to see that then every $h|I_a$ is also η-quasisymmetric.

Now choose $M > 0$ and with $\delta = \frac{1}{2}$ find the subdivision $A_1, \ldots, A_r$ of A by points $x_0, \ldots, x_r$ promised to us by Lemma 7.3.5. Put

$$\lambda = \max_i \lambda_i, \qquad \text{so} \quad \frac{\lambda}{2} \leq \lambda_i \leq \lambda \leq \frac{1}{2}, \quad i = 1, 2, \ldots, r.$$

Next we subdivide $[0,1]$ by points $y_0 = 0 < y_1 < \cdots < y_s = 1$ into intervals $B_j = [y_{j-1}, y_j]$ all of equal length $\mu = \frac{1}{s}$ such that $\lambda \leq \mu \leq 2\lambda \leq 1$.

If $\nu = (j_1, \ldots, j_k)$, $j_i \in \{1, 2, \ldots, s\}$, we put

$$B_\nu = B_{j_1} \times \cdots \times B_{j_k}.$$

Let $z_j = (y_j + y_{j-1})/2$ and $z_\nu = (z_{j_1}, \ldots, z_{j_k})$. Then z_ν is the center of B_ν. Define $Q_{i\nu} = \operatorname{int}(A_i \times B_\nu)$. Choose $z_i' \in A_i$ such that

$$|z_i' - x_i| = \frac{\lambda_i}{2} = \frac{|x_i - x_{i-1}|}{2}$$

and set $z_{i\nu} = (z_i', z_\nu)$. Then $z \in \partial Q_{i\nu}$ implies the estimate

$$|z - z_{i\nu}| \geq \min\{\mu/2, \lambda_i/2\} = \lambda_i/2 \geq \lambda/4.$$

If we put $a_{i\nu} = (x_i, z_\nu)$, then $|a_{i\nu} - z_{i\nu}| = |x_i - z_i'| = \lambda_i/2$, and so $|z - z_{i\nu}| \geq |a_{i\nu} - z_{i\nu}|$ for all $z \in \partial Q_{i\nu}$. Hence

$$|h(z) - h(z_{i\nu})| \geq \frac{1}{H}\,|h(z_{i\nu}) - h(a_{i\nu})|.$$

Now choose $y_{i\nu}' \in [z_{j_1}, y_{j_1}]$ such that $|y_{i\nu}' - z_{j_1}| = \lambda_i/2 = |z_{i\nu} - a_{i\nu}|$ and let

$$\begin{aligned} b_{i\nu}' &= (x_i, y_{i\nu}', y_{j_2}, \ldots, y_{j_k}), \\ b_{i\nu} &= (x_i, y_{j_1}, y_{j_2}, \ldots, y_{j_k}), \\ b_{i\nu}'' &= (x_i, y_{j_1-1}, y_{j_2}, \ldots, y_{j_k}). \end{aligned}$$

Then

$$|h(z) - h(z_{i\nu})| \geq \frac{1}{H}\,|h(z_{i\nu}) - h(a_{i\nu})| = \frac{1}{H^2}\,|h(b_{i\nu}') - h(a_{i\nu})| \tag{7.18}$$

for all $z \in \partial Q_i$ and all i and ν. Since

$$|b'_{i\nu} - a_{i\nu}| = \frac{\lambda_i}{\mu}\,|b_{i\nu} - a_{i\nu}| \geq \frac{1}{4}\,|b_{i\nu} - a_{i\nu}|,$$

and since $|a_{i\nu} - b_{i\nu}| = \frac{1}{2}\,|b''_{i\nu} - b_{i\nu}|$, (??) implies that

$$\begin{aligned} |h(b'_{i\nu}) - h(a_{i\nu})| &\geq \eta\Big(\frac{1}{4}\Big)|h(b_{i\nu}) - h(a_{i\nu})| \\ &\geq \eta\Big(\frac{1}{4}\Big)\eta\Big(\frac{1}{2}\Big)\beta_{i\nu}, \quad \beta_{i\nu} = |h(b''_{i\nu}) - h(b_{i\nu})|. \end{aligned} \tag{7.19}$$

It follows by (7.18) and (7.19) that for every $z \in \partial Q_{i\nu}$ we have $|h(z) - h(z_{i\nu}| \geq c\beta_{i\nu}$ where the constant

$$c = \frac{1}{H^2}\,\eta\Big(\frac{1}{4}\Big)\eta\Big(\frac{1}{2}\Big).$$

This means that

$$\operatorname{int}[B^{k+1}(h(z_{i\nu}), c\beta_{i\nu})] \subset h(Q_{i\nu}),$$

which certainly shows that the $(k+1)$-measure $|h(Q_{i\nu})| \geq c'\beta_{i\nu}^{k+1}$.

We now put

$$\beta = \inf\{|h(x,0,y) - h(x,1,y)| : x \in A, y \in I^{k-1}\} > 0.$$

Then for every i and $\nu' = (j_2, \ldots, j_k)$ we have

$$\beta \leq \sum_{j=1}^{s} \beta_{ij\nu'}.$$

Now Hölder's inequality, together with our lower bound on the measure of $Q_{i\nu}$, implies that

$$\beta^{k+1} \leq s^k \sum_j \beta_{ij\nu'}^{k+1} \leq \frac{s^k}{c'} \sum_j |h(Q_{ij\nu'})|.$$

We may now sum over all i and ν' to obtain

$$rs^{k-1}\beta^{k+1} \leq \frac{s^k}{c'} \sum_{i,\nu} |h(Q_{i\nu})| \leq \frac{s^k}{c'}\,|h(A \times I^k)|.$$

Thus $r/s \leq c''$, which is a constant depending only on h but not on M. On the other hand,

$$\frac{1}{s} = \mu \geq \lambda \geq \lambda_i,$$

which implies that

$$\frac{r}{s} \geq \sum_i \lambda_i \geq M,$$

when M is chosen sufficiently large. This leads to a contradiction since we are certainly at liberty to choose $M > c''$. □

We now come to the point of this lemma.

THEOREM 7.3.7. *Let $\alpha : (-\infty, \infty) \to \mathbb{C}$ be an arc in the complex plane with image $X = \alpha(-\infty, +\infty)$, with $\Re(\alpha(t)) \to -\infty$ as $t \to -\infty$ and with $\Re(\alpha(t)) \to +\infty$ as $t \to +\infty$. Suppose that α is not rectifiable on some interval $[a, b]$. Then the topological hyperplane*

$$S = X \times \mathbb{R}^{n-2} \subset \mathbb{R}^n$$

cannot be the image of $\mathbb{R}^{n-1}$ *under a quasiconformal mapping of* $\mathbb{R}^n$. *In particular* $S \cup \{\infty\}$ *is not a quasisphere in* $\hat{\mathbb{R}}^n$.

PROOF. We have already seen in Theorem 6.6.7 that the distortion estimates imply that if f is a quasiconformal mapping of $\mathbb{R}^n$ to itself, then f is quasisymmetric and so surely quasisymmetric in a large ball containing $\alpha([a,b]) \times [-1,2]^{k-2}$. As a consequence, supposing there is a quasiconformal mapping $f : \mathbb{R}^n \to \mathbb{R}^n$ with $f(\mathbb{R}^{n-1}) = S$, then $f^{-1}|S : S \to \mathbb{R}^{n-1}$ will be a quasisymmetric embedding of S into $\mathbb{R}^{n-1}$. Then Lemma 7.3.6 gives a contradiction as α is not rectifiable on $[a,b]$. □

A particular example of such a set X is the infinite von Koch snowflake. This curve first appeared in a 1904 paper entitled "On a continuous curve without tangents, constructible from elementary geometry" by the Swedish mathematician Helge von Koch. It was S. Rickman who first observed that this set crossed with the real line is not a two-dimensional quasisphere, and for this reason the topological hyperplane

$$\Pi_R = \mathcal{L} \times \mathbb{R}^{n-2}$$

is called *Rickman's rug.*

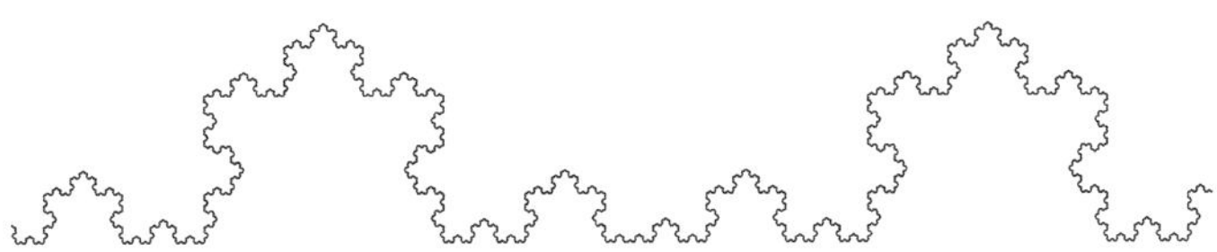

The snowflake line $\mathcal{L}$.

There is another remarkable feature of this curve. First it satisfies Ahlfors' bounded turning condition and is therefore the quasiconformal image of $\mathbb{R} \subset \mathbb{C}$—a quasiline. Ahlfors has proven [**7**] that each quasiline admits a bilipschitz reflection. This is a bilipschitz homeomorphism $h : \mathbb{C} \to \mathbb{C}$, $h|\mathcal{L} = identity$, so $h(\mathcal{L}) = \mathcal{L}$, and also h interchanges the components of $\mathbb{C} \setminus \mathcal{L}$. Indeed the existence of a such a quasireflection characterises quasilines in $\mathbb{C}$. Therefore quasicircles in the plane are characterised by the existence of a quasiconformal reflection—these reflections cannot be bilipschitz, as the point at ∞ will have to be moved to a finite point by any reflection in a finite circle.

Evidently the map $h \times identity : \mathbb{R}^2 \times \mathbb{R}^{n-2} \to \mathbb{R}^n$ is a bilipschitz reflection (in the spherical metric). However its fixed set $h(\mathcal{L} \times \mathbb{R}^{n-1})$ is a topological hyperplane, but it is not a quasihyperplane, by which we mean the quasiconformal image of $\mathbb{R}^{n-1} \subset \mathbb{R}^n$.

If we move ∞ to a finite point by a Möbius transformation for which the preimage of ∞ is not a point on $\mathcal{L}$, then

$$S = \varphi(\Pi_R) = \varphi(\mathcal{L} \times \mathbb{R}^{n-2}) \cup \varphi(\infty) \tag{7.20}$$

is a topological sphere which cannot be a quasisphere. We state these observations above as a theorem.

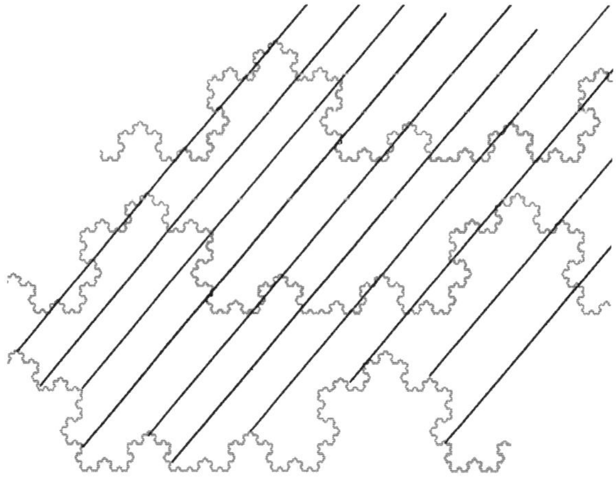

Rickman's rug $\mathcal{L} \times \mathbb{R} \subset \mathbb{R}^3$.

THEOREM 7.3.8. *In all dimensions $n \geq 3$ there is a topological $(n-1)$-sphere $S \subset \hat{\mathbb{R}}^n$ and a quasiconformal homeomorphism $h : \hat{\mathbb{R}}^n \to \hat{\mathbb{R}}^n$ such that*

- *$h|S =$ identity, so in particular $h(S) = S$,*
- *h interchanges the two components of $\hat{\mathbb{R}}^n \setminus S$,*

and yet S is not the quasiconformal image of the round $(n-1)$-sphere $\mathbb{S}^{n-1}$.

As a corollary, we find that the set S, satisfying the hypotheses of Theorem 7.3.8, allows us to claim the following.

COROLLARY 7.3.9. *In all dimensions $n \geq 3$ there is a topological $(n-1)$-sphere $S \subset \hat{\mathbb{R}}^3$ with the following properties:*

- *There is a quasiconformal homeomorphism $h : \hat{\mathbb{R}}^n \to \hat{\mathbb{R}}^n$ such that $h(S) = S$.*
- *The quasiconformal mapping h interchanges the two components of $\hat{\mathbb{R}}^n \setminus S$.*
- *Neither component of $\hat{\mathbb{R}}^n \setminus S$ is quasiconformally equivalent to the ball B^n.*

PROOF. Let S be the image of Rickman's rug as per (7.20) and let f denote the reflection interchanging the components of $\hat{\mathbb{R}}^n \setminus S$. We only need to show that one component, say D, of $\hat{\mathbb{R}}^n \setminus S$ is not a quasiball. Let $g : B^n \to D$ be quasiconformal. As D is a Jordan domain, g extends homeomorphically to a map $\bar{B}^n \to \bar{D}$. Since $f(x) = x$ on S the mapping defined by

$$h(x) = \begin{cases} (f \circ g \circ f)(x) & x \in \hat{\mathbb{R}}^n \setminus D, \\ g(x) & x \in \bar{D} \end{cases}$$

is a homeomorphism of $\hat{\mathbb{R}}^n$ which is quasiconformal away from the Jordan sphere S of Lebesgue n-measure 0. Thus h is quasiconformal, and hence $S = h(\mathbb{S}^{n-1})$ is a quasisphere. This contradicts Lemma 7.3.7. Thus D is not a quasiball. □

P. Tukia, in his paper on quasiconformal groups [**156**] from which the material in this section was largely taken, goes on to discuss other properties of Rickman's rug, such as its quasiconformal homogeneity. We will briefly discuss this property at the end of this chapter.

We now turn to discuss some positive results.

7.4. Existence: The Schoenflies theorem

We begin by recalling our higher-dimensional versions of the Carathéodory extension theorem and explaining when a quasiconformal mapping $f : D \to B^n$ extends to a homeomorphism $\hat{f} : \bar{D} \to \bar{B}^n$. The condition we required of D was that each point x_0 of the boundary of D was quasiconformally collared. This section explores a converse to this result in that we suppose the existence of a homeomorphic mapping of the boundaries of the two domains D and B^n and ask if there is an extension of this boundary value mapping to the interiors of D and B^n.

The first obvious restriction is that the boundary of D should be homeomorphic to the $(n-1)$-sphere $\mathbb{S}^{n-1}$ of $\mathbb{R}^n$. If this is true we call the domain D a *Jordan domain.* To motivate aspects of the developments we present here, we sketch a proof of the classical Jordan curve theorem from low-dimensional topology in the quasiconformal setting.

THEOREM 7.4.1. *Let $\mathbb{S}$ denote the unit circle and let $h_0 : \mathbb{S} \to \mathbb{C}$ be a homeomorphism. Then there are homeomorphisms $f : \bar{B}^2 \to \mathbb{C}$ and $g : \hat{\mathbb{C}} \setminus B^2 \to \hat{\mathbb{C}}$ such that*

$$f|\mathbb{S} = g|\mathbb{S} = h_0.$$

Moreover, if h_0 is quasisymmetric, then both f and g can be chosen to be quasiconformal.

PROOF. We only construct the mapping g, as f can be constructed in an entirely similar fashion.

Our first observation is that the finite region D bounded by $h_0(\mathbb{S})$ is simply connected and therefore by the Riemann mapping theorem there is a conformal mapping $\varphi : B^2 \to D$. The Carathéodory extension theorem tells us that φ can be extended to a homeomorphism $\hat{\varphi} : \bar{B}^2 \to \bar{D}$. Consider the homeomorphism $g_0 = \hat{\varphi}^{-1} \circ h_0 : \mathbb{S} \to \mathbb{S}$. We extend g_0 to the closed disk by the radial extension

$$g(re^{i\theta}) = rg_0(e^{i\theta}), \qquad r \in [0,1],\ \theta \in [0, 2\pi].$$

Then we put

$$h = \hat{\varphi} \circ g : B^2 \to D$$

and it is a moment's work to check that $h : \bar{B}^2 \to \bar{D}$ is a homeomorphism extending h_0. This completes the proof if we are only interested in the topological setting.

In the quasiconformal setting there are two obstacles to overcome if we want to use this route. First, we must show that the mapping $g_0 : \mathbb{S} \to \mathbb{S}$ is quasisymmetric. Second, we must show that this quasisymmetric mapping $g_0 : \mathbb{S} \to \mathbb{S}$ extends to a quasiconformal mapping $g : B^2 \to B^2$. Then as φ is conformal the result will follow.

Fortunately, in two dimensions the second problem here is not too difficult to resolve since the hard work has already been done. Although the radial extension we used will, in general, not be quasiconformal unless the boundary map g_0 is bilipschitz, the Beurling-Ahlfors extension at (8.1) provides the mapping g (Chapter 8 has a discussion of this as well as a proof for the highly nontrivial higher-dimensional result).

Unfortunately, the first part of the problem is not so straightforward either, and to give a complete proof would require us to develop more two-dimensional theory than space allows. It suffices to say that this problem has been considered by P. Tukia [**153**]. The basic idea is as follows: Use Möbius transformations to normalise the situation so as to replace h_0 by a quasisymmetric mapping of $\mathbb{R} \hookrightarrow \mathbb{C}$, again called h_0, with $h_0(\infty) = \infty$. The quasisymmetry condition now implies that $J = h_0(\mathbb{R})$ has *bounded turning*. This means that there is a finite constant c such that if $x, y \in h_0(\mathbb{R})$, then

$$\operatorname{diam}(J') \leq c|x - y|,$$

where J' is the finite arc of $J \setminus \{x, y\}$. In a famous paper from 1963 Ahlfors showed that this geometric condition implies that J is the image under the real line of a quasiconformal mapping of $\mathbb{C}$, [**7**]. With this in hand the proof is easy. Actually Ahlfors' proof itself is not too hard, and the reader has enough tools at hand from our discussion of moduli to establish it for themselves. The key idea is to consider for $x \in \mathbb{R}$ and $\delta > 0$ the moduli of the curve families connecting the arcs $h_0(-\infty, x - \delta)$ and $h_0(x, x + \delta)$ in the component of $\mathbb{C} \setminus J$ and note that these moduli are preserved by the (inverse of) the Riemann map $\varphi : \bar{\mathbb{H}}^2 \to \bar{D}$, D a component of $\mathbb{C} \setminus J$. This leads to the quasisymmetry condition for $\varphi^{-1} \circ h_0 : \mathbb{R} \to \mathbb{R}$. □

The key ingredients of this two-dimensional proof are:

- The Riemann mapping theorem.
- The Carathéodory extension theorem.
- The extension problem: given a quasisymmetric $g_0 : \mathbb{R}^{n-1} \to \mathbb{R}^{n-1}$, construct an extension $g : \mathbb{H}^n \to \mathbb{H}^n$.

In this section we prove a higher-dimensional version of Theorem 7.4.1, a result which was alluded to when we discussed domains which are quasiconformally collared in our higher-dimensional versions of the Carathéodory extension theorem; see Theorems 6.5.18 and the local version, Theorem 6.5.19.

The example of a wild Fox-Artin sphere, bilipschitz equivalent to the $(n-1)$-sphere, shows that there are topological issues arising in higher dimensions that are not present in two dimensions. In particular there will be no Riemann map, and even if there were a reasonable replacement for it, it is not clear that it would have an extension to the boundary.

Our approach will be to deal with both of these issues simultaneously through the notion of quasiconformally locally flat domains, a slight refinement of the notion of quasiconformally collared domains that we have already introduced.

As the first step we prove the following version of the Schoenflies theorem. It was originally due to Gehring in 1967 who adapted ideas of B. Mazur, but the proof we give here follows D. Gauld and M. Vamanamurthy [**44, 45**] who adapted ideas of M. Brown [**25**].

7.4.1. Locally flat implies flat. We first just recall that in $\mathbb{R}^n$, $n \geq 2$, a *Jordan domain* is a domain bounded by a topological $(n-1)$-sphere—the homeomorphic image of $\mathbb{S}^{n-1}$.

THEOREM 7.4.2. *Let D be a Jordan domain in $\mathbb{R}^n$. Then a homeomorphism $h_0 : \partial D \to \mathbb{S}^{n-1}$ extends to a homeomorphism $h : D \to B^n$ with $h|D$ quasiconformal, if and only if each point $x_0 \in \partial D$ has a neighbourhood in D over which h_0 extends quasiconformally.*

The proof for this theorem will occupy the next several pages. However we wish first to observe that if $h_0 : \partial D \to \mathbb{S}^{n-1}$ extends to a homeomorphism $h : D \to B^n$ with $h|D$ quasiconformal, then ∂D is certainly quasiconformally collared. Also if each point $x_0 \in \partial D$ has a neighbourhood in D over which h_0 extends quasiconformally, then ∂D is locally collared.

We begin with the following definition and then a lemma which makes assumptions about the local extension properties of the boundary mapping.

Suppose D is a Jordan domain in $\mathbb{R}^n$ and suppose $h : \partial D \to \mathbb{S}^{n-1}$. We say h is *locally quasiconformally flat* if there is a real number $0 < a < 1$ and a covering $\mathcal{U}$ of ∂D by open sets of $\bar{D}$ such that for each $U \in \mathcal{U}$ the map $h|U \cap \partial D$ extends to a homeomorphism

$$h_U : \bar{U} \to \{rz \in \mathbb{R}^n : a \leq r \leq 1, \ \ z \in h(\bar{U} \cap \partial D) \subset \mathbb{S}^{n-1}\}$$

and $h_U : U \cap D \to B^n$ is quasiconformal.

LEMMA 7.4.3. *Suppose D is a Jordan domain in $\mathbb{R}^n$ and suppose $h : \partial D \to \mathbb{S}^{n-1}$ is locally quasiconformally flat. Then h is quasiconformally flat, meaning there is a neighbourhood V of ∂D and $h : V \cap D \to B^n$ is quasiconformal.*

PROOF. The compactness of ∂D allows us to identify a finite cover of ∂D by sets $U_1, U_2, \ldots, U_N$ on which the map h locally admits an extension. We can also dilate each local extension so as to assume $a \leq 1/3$.

To alleviate notation, we abbreviate the mapping h_{U_i} by h_i, $i = 1, 2, \ldots, N$. Let D^* be the topological space obtained from the disjoint union of $\bar{D}$ and $\partial D \times [1, 2]$ by identifying $x \in \partial D$ with $(x, 1)$. We want to find a collar on ∂D lying inside D. Now D^* is D together with the collar $\partial D \times [1, 2]$, and the idea is to use the local information from h to push this collar inside D.

To do this we will define a homeomorphism $e_m : \bar{D} \to D^*$. The extension of h to a neighbourhood V of ∂D in D as required by the statement of the lemma will use the product structure of $\partial D \times (1, 2)$ with $V = e_m^{-1}[\partial D \times (1, 2)]$. Let

$$U_i^* = U_i \cup (U_i \cap \partial D) \times [1, 2]$$

and extend h_i to

$$h_i^* : U_I^* \to \left\{ rz \in \mathbb{R}^n : \frac{1}{3} < r \leq 2, \ \ z \in h(\bar{U}_i \cap \partial D) \subset \mathbb{S}^{n-1} \right\}$$

by letting $h_i(x, t) = th(x)$ when $x \in U_i \cap \partial D$ and $t \in [1, 2]$. Define projections

$$p_i : U_i^* \to U_i \cap \partial D \quad \text{and} \quad q_i : U_i^* \to \left[\frac{1}{3}, 2\right]$$

by $p_i(x) = h^{-1}(z)$ and $q_i(x) = r$ if $x \in U_i$ with $h_i(x) = rz$ for $1/3 < r \leq 1$ and $z \in h(U_i \cap \partial D)$. If $(x, t) \in (U_i \cap \partial D) \times [1, 2]$ we put

$$p_i(x, t) = x \quad \text{and} \quad q_i(x, t) = t.$$

Let $\{\alpha_i : \mathbb{S}^{n-1} \to [0, 1], i = 1, 2, \ldots, N\}$ be a smooth partition of unity subordinate to the cover $\{h(U_i) \cap \partial D\}$ of $\mathbb{S}^{n-1}$ and define $\beta_i : \partial D \to [0, 1/2]$ for $x \in \partial D$ by

$$\beta_i(x) = \frac{1}{2}\alpha_i[h(x)].$$

Next for $i = 1, 2, \dots, N$ define $\gamma_i : \partial D \to [0, 1/2]$ for $x \in \partial D$ by

$$\gamma_i(x) = \sum_{j=1}^{i} \beta_i(x).$$

Then $\gamma_0(\partial D) = 0$ and $\gamma_N(\partial D) = 1/2$. Using induction on k we now construct embeddings $e_k : \bar{D} \to D^*$, $k = 0, 1, \dots, N$, to satisfy the following properties:

(i) $e_k(\bar{D}) = \bar{D} \cup \{(x, t) \in \partial D \times [1, 2] : t \leq 1 + \gamma_k(x)\}$.

(ii) For every $j = 1, 2, \dots, N$ the embedding $h_j^* e_k$ is quasiconformal on the interior of $e_k^{-1}(U_j^*)$—its domain of definition.

The induction begins by letting e_0 be the inclusion and $e_0(\bar{D}) = \bar{D}$, so (i) is satisfied, while the domain of $h_j^* \circ e_0$ is U_j, on the interior of which $h_j^* \circ e_0 = h_j$ is quasiconformal. Suppose now that e_{k-1} has been constructed satisfying conditions (i) and (ii). Define e_k as follows:

(a) If $e_{k-1}(x) \in D^* \setminus U_k^*$ or if $e_{k-1}(x) \in U_k$ and $1/3 < q_k[e_{k-1}(x)] \leq 1/2$, let

$$e_k(x) = e_{k-1}(x).$$

(b) If $e_{k-1}(x) \in U_k$ and $1/2 < q_k[e_{k-1}(x)] \leq 2$, let

$$e_k(x) = \left(h_k^*\right)^{-1}\left[\left(q_k[e_{k-1}(x)] + \frac{\beta(p_k[e_{k-1}(x)])(2q_k[e_{k-1}(x)] - 1)}{1 + 2\gamma_{k-1}(p_k[e_{k-1}(x)])}\right) h[p_k(e_{k-1}(x))]\right].$$

While this is a somewhat complicated formula, the effect of (b) is to map x to the point $(h_k^* \circ e_{k-1})(x)$ in the "nicer" region $\{rz \in \mathbb{R}^n : 1/2 \leq r \leq 2 \text{ and } z \in h(U_k \cap \partial D)\}$, then apply a radial diffeomorphism to the image which stretches (for $z \in h(U_k \cap \partial D)$) the radial line $\{rz : 1/2 \leq r \leq 1 + \gamma_{k-1}[h^{-1}(z)]\}$ onto the radial line $\{rz : 1/2 \leq r \leq 1 + \gamma_k[h^{-1}(z)]\}$, and then to apply $\left(h_k^*\right)^{-1}$.

We have to verify that each e_k is a well-defined embedding. If $e_{k-1}(x) \in U_k^*$ and if $1/2 \leq q_k[e_{k-1}(x)]$, then by the inductive assumption (i),

$$\frac{1}{2} \leq q_k[e_{k-1}(x)] \leq 1 + \gamma_{k-1}(p_k[e_{k-1}(x)])$$

and also

$$0 \leq \beta_k(p_k[e_{k-1}(x)]) \leq \frac{1}{2} - \gamma_{k-1}(p_k[e_{k-1}(x)]),$$

we see that

$$\frac{1}{2} \leq \left(q_k[e_{k-1}(x)] + \frac{\beta(p_k[e_{k-1}(x)])(2q_k[e_{k-1}(x)] - 1)}{1 + 2\gamma_{k-1}(p_k[e_{k-1}(x)])}\right) h[p_k(e_{k-1}(x))]$$

$$\leq 1 + \gamma_k(p_k[e_{k-1}(x)]) \leq 2 \tag{7.21}$$

so

$$\left(q_k[e_{k-1}(x)] + \frac{\beta(p_k[e_{k-1}(x)])(2q_k[e_{k-1}(x)] - 1)}{1 + 2\gamma_{k-1}(p_k[e_{k-1}(x)])}\right) h[p_k(e_{k-1}(x))]$$

is in the image of the embedding h_k^*, so at least (b) makes sense.

If $e_{k-1}(x) \in U_k^*$ and $q_k[e_{k-1}(x)] = \frac{1}{2}$, then the inequality above becomes an equality, and definition (b) of $e_k(x)$ reduces to

$$\left(h_k^*\right)^{-1}\left(\frac{1}{2} h(p_k[e_{k-1}(x)]\right) = (h_k)^{-1}(q_k[e_{k-1}(x)]), \cdot h(p_k[e_{k-1}(x)]) = e_{k-1}(x)$$

so the definitions (a) and (b) are compatible.

Next if $e_{k-1}(x) \in U_k^*$ and $1/2 \leq q_k[e_{k-1}(x)]$, then, provided $p_k[e_{k-1}(x)]$ is sufficiently close to the boundary of $U_k \cap \partial D$, we have $\beta_k(p_k[e_{k-1}(x)]) = 0$ since the partition of unity is subordinate to the cover $\{h(U_i \cap \partial D) : i = 1, 2, \ldots, N\}$ by construction. Then (b) gives $e_k(x) = e_{k-1}(x)$.

This shows that e_k is continuous. Furthermore, e_k is an embedding—being a continuous injective map from a compact space to a Hausdorff space.

We now have to show that e_k satisfies both properties (i) and (ii).

For (i) we need only verify the equality on U_k^*, since it is only here that e_{k-1} and e_k differ (along with γ_{k-1} and γ_k).

By the inductive hypothesis (i), $q_k[e_{k-1}(x)]$ can be as large as

$$1 + \gamma_{k-1}(p_k[e_{k-1}(x)]),$$

and in this case the right inequality of (7.21) is an equality. Thus the frontier of $e_{k-1}(\bar{D})$ has been pushed (within U_k^*) out as far as

$$\{(x,t) \in \partial D \times [1,2] : t - 1 + \gamma_k(x)\}$$

in order to construct $e_k(\bar{D})$.

Of course $e_k(\bar{D})$ includes everything out to this frontier, and thus property (i) is satisfied.

Again we need only verify (ii) inside $(e_{k-1})^{-1}(U_k^*)$ as e_k and e_{k-1} agree from this set and quasiconformality is the only issue. Let $j \in \{1, 2, \ldots, N\}$. From (b) for e_k we see that $h_j^* \circ e_k$ has the form of a composition of three mappings,

$$h_j^* \circ e_k = (h_j^* \circ (h_k^*)^{-1}) \circ (\text{stretch map}) \circ (h_k^* \circ e_{k-1}).$$

Each of these is quasiconformal by the inductive hypothesis. The only two possible issues are with the stretching map (and a moment's thought shows this to be fine) and to observe that on $h_k^*[(U_k^* \cap U_j^*) \setminus (U_K \cap U_j)]$ the map $h_j^* \circ (h_k^*)^{-1}$ is the identity.

We have now established the inductive construction of the embeddings e_k. Further, we have noted $\gamma_N(\partial D) = 1/2$, and so (i) implies $e_N(\bar{D}) = D^*$, so e_N is a surjection and therefore a homeomorphism.

Let $U = e_N^{-1}(\partial D \times [1,2])$. Then U is a neighbourhood of ∂D in D. If we define $H : U \to [1/2, 1]$ by $H(x) = (q[e_m(x)] - \frac{1}{2})h(p[e_m(x)])$, where p and q are the usual projections to ∂D and $[1,2]$, then:

- H is a homeomorphism with the above range. In fact, H is the composition of the three homeomorphisms: $e_m : U \to \partial D \times [1,2]$; the obvious radial extension of h to a homeomorphism from $\partial D \times [1,2]$ onto the annulus $\{1 < |x| < 2 : x \in \mathbb{R}^n\}$; and a radial shrinking of the annulus $\{1 < |x| < 2\}$ to $\{\frac{1}{2} < |x| < 1\}$.
- H extends h. If $x \in \partial D$, then from (b) $e_k(x) = (x, l + \gamma_k(x))$, so $e_N(x) = (x, 2)$ and hence $pe_N(x) = x$ and $qe_N(x) = 2$. Thus $H(x) = h(x)$.
- H is quasiconformal on $U \cap D$. This follows from (ii) as we have discussed above, though note that the finiteness of the cover is necessary here.

Finally, these last three observations complete the proof as H is the flat mapping we are seeking. □

7.4.2. Flatness of non-Jordan domains. Suppose that D is some domain in $\mathbb{R}^n$ and ∂D is a topological $(n-1)$-manifold without boundary. Provided we can find a model for ∂D which is nicely embedded in $\mathbb{R}^n$, the techniques above generalise to domains such as D. The main criterion here for "niceness" of the model, say M, is that there is an embedding of $M \times [0,1)$ into $\mathbb{R}^n$ whose restriction to $M \times \{0\}$ is the projection on M. The $[0,1)$ factor of this embedding plays the same role in this case as the radial structure of $\mathbb{R}^n$ in the previous case.

As an example, the 2-torus $\mathbb{S} \times \mathbb{S}$ has a nice embedding in $\mathbb{R}^3$ obtained by revolving the circle $\{(x,y,z) : (x-2)^2 + z^2 = 1, y = 0\}$ around the z-axis.

Given a domain D in $\mathbb{R}^n$ whose boundary is homeomorphic to the torus, we can speak of local quasiconformal flatness of the boundary in D and that the techniques used to prove the above Lemma 7.4.3 carry over to this case to prove that if ∂D is locally quasiconformally flat in D, then ∂D is quasiconformally flat in D.

In certain cases where ∂D is not an $(n-1)$-manifold, the above techniques can still be used to go from local flatness to global flatness. Here is an example of this that we will use later.

THEOREM 7.4.4. *Suppose X is a compact subset of $\mathbb{R}^n$ and $h : X \to \bar{B}^{n-1}$ is a homeomorphism. Let $\mathcal{U}$ be an open cover of X so that for each $U \in \mathcal{U}$, $h|U \cap X$ extends to a quasiconformal embedding $h_U : U \to \mathbb{R}^n$. Then h extends to a quasiconformal embedding of a neighbourhood of X.*

PROOF. Let

$$Y = (\mathbb{R}^n \setminus X) \cup (X \times \{-1,1\})/\{(x,-1) \sim (x,1)\}, \qquad x \in h^{-1}(\mathbb{S}^{n-2}),$$

so Y is roughly $\mathbb{R}^n$ opened up along X. Define $\pi : Y \to \mathbb{R}^n$ by $\pi(y) = y$ if $y \in \mathbb{R}^n \setminus X$ and $\pi(x, \pm 1) = x$ if $x \in X$. The map π closes up the slit.

Now $\hat{\mathbb{R}}^n \setminus \bar{B}^n$ is conformally equivalent to $\mathbb{H}^n$ and $\hat{\mathbb{R}}^n \setminus \bar{B}^{n-1}$ is conformally (Möbius) equivalent to $W = \hat{\mathbb{R}}^n \setminus \{(x_1, x_2, \ldots, x_n) : x_n = 0, x_{n-1} \geq 0\}$, a "slit" space. There is an obvious quasiconformal folding $\hat{\mathbb{R}}^n \setminus \bar{B}^n \to \hat{\mathbb{R}}^n \setminus \bar{B}^{n-1}$ induced by the fold $\mathbb{H}^n \to W$ given by doubling the angle along the copy of $\mathbb{R}^{n-2} \subset \partial\mathbb{H} = \mathbb{R}^{n-1}$. Now we are in the situation of Lemma 7.4.3 as the map $h^* : Y \to \mathbb{R}^n \setminus \bar{B}^n$ is locally flat. The lemma tells us that this map is flat, and then we push this flat map back down to $\mathbb{R}^n \setminus X \to \mathbb{R}^n \setminus \bar{B}^{n-1}$ by using π. We leave a few minor details to the reader to confirm that this is a flattening of h. The proof is complete. □

7.4.3. Quasiconformally flat spheres bound quasiballs. We now take a further step towards the main result of this section with the following theorem which usually goes by the name of the higher-dimensional *Schoenflies theorem.*

THEOREM 7.4.5. *Let $e : N \to \mathbb{R}^n$ be an embedding where N is a neighbourhood of $\mathbb{S}^{n-1}$ in $\bar{B}^n$. Then $e|\mathbb{S}^{n-1}$ extends to an embedding $\hat{e} : \bar{B}^n \to \mathbb{R}^n$. If $e|N \cap B^n$ is quasiconformal, then so is $\hat{e}|B^n$.*

PROOF. We may suppose, by precomposing with a radial stretching should it be necessary, that

$$N = \Big\{x \in \mathbb{R}^n : \frac{1}{2} \leq |x| \leq 1\Big\}.$$

Let C denote the component of $\mathbb{R}^n \setminus f(\mathbb{S}^{n-1})$ for which $f(N) \subset \bar{C}$.

As a first step we find an embedding $f : \bar{B}^n \to \mathbb{R}^n$, f quasiconformal on B^n, such that

$$f(\bar{B}^n) \cup e[\mathrm{int}(N)] = C. \tag{7.22}$$

We may as well work in $\hat{\mathbb{R}}^n$, and in doing so we can assume $\infty \in e[\mathrm{int}(N)]$, $\partial e(N) \subset B^n$ and $0 \in C_2$, the bounded component of $\hat{\mathbb{R}}^n \setminus e(\mathbb{S}^{n-1}(\frac{1}{2}))$. Let C_1 be the bounded component of $\hat{\mathbb{R}}^n \setminus e(\mathbb{S}^{n-1})$. Then $C_1 \cup C_2 \subset B^n$.

Let $\alpha : N \to \bar{B}^n$ be the radial stretch $\alpha(tx) = (2t-1)x$ for $t \in [\frac{1}{2}, 1]$ and $|x| = 1$. Then $\alpha \circ e^{-1}$ has a continuous extension over C_2 which sends C_2 to the origin. Next, let $\beta : \hat{\mathbb{R}}^n \to \hat{\mathbb{R}}^n$ be a quasiconformal mapping which is the identity outside B^n and so that $\beta[\alpha(e^{-1}(\infty)] = 0$. Define $g : \mathbb{R}^n \to \hat{\mathbb{R}}^n$ by

$$g(x) = \begin{cases} (e \circ \alpha^{-1} \circ \beta \circ \alpha \circ e^{-1})(x), & x \notin C_1, \\ x, & x \in C_1. \end{cases} \tag{7.23}$$

If $x \in \partial C_1$, then $\alpha[e^{-1}(x)] \in \mathbb{S}^{n-1}$ where the map β is the identity, and so for these x, $g(x) = x$, and therefore g is continuous. Now g is an embedding only on $\mathbb{R}^n \setminus \bar{C}_2$ and is not quasiconformal on this set even when e is quasiconformal. However, when e is quasiconformal the map g will be locally quasiconformal, which is all we will need.

Let $p_0 = g(\bar{C}_2)$. Choose $r > 0$ so that $V = B^n(p_0, 2r)$ has $\bar{V} \subset e[\mathrm{int}(N)]$. Set $U = B^n(p_0, r)$ and $W = B^n(0, s)$ with s large enough that $\bar{C}_2 \subset W$. Let $\gamma : \hat{\mathbb{R}}^n \to \hat{\mathbb{R}}^n$ be a quasiconformal mapping which takes the annulus $\hat{\mathbb{R}}^n \setminus (U \cup W)$ onto the annulus $\bar{V} \setminus U$ leaving points of $\bar{U}$ and the origin fixed.

We can now define the required map f as follows:

$$f(x) = \begin{cases} (g^{-1} \circ \gamma \circ g)(x), & x \in \overline{e[\mathrm{int}(N)]}, \\ x, & x \in g^{-1}(U). \end{cases} \tag{7.24}$$

Then the map f is an embedding with

$$f(\bar{B}^n) \cup e[\mathrm{int}(N)] = C,$$

and if e is quasiconformal on $\mathrm{int}(N)$, then f is quasiconformal on B^n.

Now we have a quasiconformal mapping f covering the "hole"— the bounded component of $\mathbb{R}^n \setminus e(N)$. Brown's innovation was to show how to use the map f to draw in $e(N)$ to fill this hole homeomorphically, thereby constructing an extension of e to the ball B^n. Gauld and Vamanamurthy showed how to make Brown's construction respect quasiconformality.

Now, given our initial embedding e and the embedding f we have just constructed, we need to further refine f so as to match up a bit better with e.

Let

$$A(s,t) = \{x \in \mathbb{R}^n : s < |x| < t\}$$

denote the annulus. We claim that there are numbers a, b, c and an embedding $F : \bar{B}^n \to \mathbb{R}^n$ such that

(i) $0 < a < b < 1$ and $\frac{1}{2} < c < 1$,
(ii) $F[\bar{A}(b,1)] \subset e[A(c,1)]$,
(iii) $F[\mathbb{S}^{n-1}(a)] \subset e[A(\frac{1}{2}, c)]$,
(iv) for every $x \in \mathbb{S}^{n-1}$ we have $F(x) = f(x)$, and
(v) if e is quasiconformal, then so is F.

We continue with C and C_2 as above and assume further, as we may, that $\bar{C}$ is bounded and $f(0) \in C_2$. Since $f(\mathbb{S}^{n-1}) \subset e[A(\frac{1}{2}, 1)]$ there is a $b < 1$ such that $f[A(b, 1)] \subset e[A(\frac{1}{2}, 1)]$. This means that $e[\mathbb{S}^{n-1}(\frac{1}{2})] \subset f[A(0, b)]$, so there certainly is a $c > 1/2$ such that $e[\bar{A}(\frac{1}{2}, 1)] \subset f[A(0, b)]$. As $f(0) \in C_2$, there is $a > 0$ such that $f[\bar{A}(0, a)] \subset C_2$.

Define a quasiconformal radial homeomorphism $\alpha_0 : A(0, 1) \to A(0, c)$ which fixes $A(0, \frac{1}{2})$ and radially shrinks $\bar{A}(\frac{1}{2}, 1)$ to $\bar{A}(\frac{1}{2}, c)$. Also define the quasiconformal $\beta_0 : \bar{B}^n \to \bar{B}^n$ to be a radial homeomorphism with $\beta[\mathbb{S}^{n-1}(a)] = \mathbb{S}^{n-1}(b)$. Using these auxiliary maps we construct the following intermediary maps:

$$\alpha(x) = \begin{cases} (e \circ \alpha_0 \circ e^{-1})(x), & x \in e[\bar{A}(\frac{1}{2}, 1)], \\ x, & \text{otherwise}. \end{cases}$$

$$\beta(x) = \begin{cases} (f \circ \beta_0 \circ f^{-1})(x), & x \in f(\bar{B}^n), \\ x, & \text{otherwise}. \end{cases}$$

Clearly when e and f are quasiconformal, the same can be said of α and β. We are therefore able to define the embedding F required above by the following formula:

$$F(x) = \begin{cases} f(x), & x \in e[A(c, 1)], \\ (\alpha \circ \beta \circ \alpha^{-1})(x), & \text{otherwise}. \end{cases}$$

One can easily check that F has all the requirements (i)–(v) that we asked for.

Continuing our proof, we next ask for $d \in (\frac{1}{2}, c)$ so that $\mathbb{S}^{n-1}(a) \subset e[A(\frac{1}{2}, d)]$—a number which is easily seen to exist. We require some more radial (and quasiconformal) homeomorphisms. Namely

$$\begin{aligned} &\chi : \bar{A}(0, c) \to \bar{A}(0, 1), && \chi\Big[\mathbb{S}^{n-1}\Big(\frac{1}{2}\Big)\Big] = \mathbb{S}^{n-1}(c), \\ &\lambda : \bar{A}(0, 1) \to \bar{A}(0, 1), && \lambda[\mathbb{S}^{n-1}(b)] = \mathbb{S}^{n-1}(a), \lambda(x) = x, \ \text{for} |x| \geq 1, \\ &\mu : \bar{A}(0, d) \to \bar{A}(0, c), && \mu[\mathbb{S}^{n-1}(d)] = \mathbb{S}^{n-1}(d). \end{aligned}$$

We then define an embedding $e_0 : \bar{A}\Big(\frac{1}{2}, 1\Big) \to \mathbb{R}^n$ as follows:

$$e_0(x) = \begin{cases} e(x), & |x| \geq c, \\ (e \circ \mu \circ e^{-1} \circ \lambda \circ e \circ \chi)(x), & |x| \leq c \ \text{and} \ (\lambda \circ e \circ \chi)(x) \in e[\bar{A}(d, 1)], \\ (\lambda \circ e \circ \chi)(x), & (\lambda \circ e \circ \chi)(x) \in e[\bar{A}(\frac{1}{2}, d)] \cup \bar{A}(0, a). \end{cases}$$

We have to show that the map e_0 is well defined and quasiconformal when desired. We do this in the following three steps:

(1) $1/2 \leq |x| \leq c$.

In this case $\chi(x) \in \bar{A}(0, 1)$ which lies in the domain of e so that $e \circ \chi$ is defined, as is $\lambda \circ e \circ \chi$. If we also have $(\lambda \circ e \circ \chi)(x) \in e[\bar{A}(d, 1)]$, then $(e^{-1} \circ \lambda \circ e \circ \chi)(x) \in \bar{A}(d, 1)$ so that $(\mu \circ e^{-1} \circ \lambda \circ e \circ \chi)(x) \in \mu[\bar{A}(d, 1)] = \bar{A}(d, c)$ which also lies in the domain of e, so the long map $e \circ \mu \circ e^{-1} \circ \lambda \circ e \circ \chi$ is defined.

(2) $(\lambda \circ e \circ \chi)(x) \in e[\mathbb{S}^{n-1}(d)]$.

In this case $(e^{-1} \circ \lambda \circ e \circ \chi)(x) \in \mathbb{S}^{n-1}(d)$, and so

$$e \circ \mu \circ e^{-1} \circ \lambda \circ e \circ \chi = e \circ e^{-1} \circ \lambda \circ e \circ \chi = \lambda \circ e \circ \chi.$$

So e_0 is well defined on $(\chi^{-1} \circ e^{-1} \circ \lambda^{-1} \circ e)[\mathbb{S}^{n-1}(d)]$.

(3) $|x| = c, \chi(x) \in \mathbb{S}^{n-1}$.

In this case we have from (ii) that $(e \circ \chi)(x) \notin B^n$, and hence $(\lambda \circ e \circ \chi)(x) = (e \circ \chi)(x)$. Thus

$$e \circ \mu \circ e^{-1} \circ \lambda \circ e \circ \chi = e \circ \mu \circ e^{-1} \circ e \circ \chi = e \circ \mu \circ \chi = e(x),$$

and we conclude that e_0 is well defined on $\mathbb{S}^{n-1}(c)$.

This examination reveals that e_0 is well defined and continuous. The Jordan-Brouwer separation theorem and invariance of domain shows that e_0 is an embedding. Then, since all the mappings used in the construction are quasiconformal, e_0 is quasiconformal.

We record two properties of e_0:

(1) If $x \in \bar{A}(c, 1)$, then $e_0(x) = e(x)$.

(2) If $|x| = \frac{1}{2}$, then $e_0(x) = \frac{a}{b}\, e_0(2cx)$.

Basically what we have done is to alter e in the annulus $A(1/2, c)$ to get e_0, which essentially repeats on $\mathbb{S}^{n-1}(\frac{1}{2})$ the action of e on $\mathbb{S}^{n-1}(c)$. The point is that we are now able to exploit this repetitive feature of e_0 which will allow us to extend it to the entire ball and which will present us with the embedding $\hat{e}$ proving the theorem.

We recall that the setup from the start has $N = \bar{A}(1/2, 1)$. Then for $x \in N$, we set $\hat{e}(x) = e_0(x)$, and also we set $\hat{e}(0) = 0$.

For $x \in A(0, 1/2)$ there is a unique nonnegative integer k for which $(2c)^k x \in N$, and we then define

$$\hat{e}(x) = \left(\frac{a}{b}\right)^{-k} e_0[(2c)^k x].$$

By property (ii) of e_0, $\hat{e}$ is well defined and continuous except possibly at 0; however continuity at 0 follows from the fact that $a < b$, so $(a/b)^{-k} \to 0$ as $k \to \infty$.

That $\hat{e}$ is an embedding follows just as that for e_0 above, and it remains to verify the quasiconformality of $\hat{e}$ when e is quasiconformal.

The embedding e_0 is quasiconformal (say K-quasiconformal) since each of the auxiliary functions is as well. On each open annulus

$$A\left(\frac{1}{2^{k+1}c^k}, \frac{1}{2^k c^{k-1}}\right)$$

the mapping $\hat{e}$ is obtained composing $\hat{e}$ on N with two conformal scalings and is thus K-quasiconformal on each such annulus. The exceptional set,

$$\bigcup_{k=0}^{\infty} \mathbb{S}^{n-1}\left(\frac{1}{2^{k+1}c^k}\right) \cup \{0\},$$

is of finite $(n-1)$-Lebesgue measure and hence is removable by Theorem 6.4.22. Thus $\hat{e}$ is K-quasiconformal on B^n. This final observation completes the proof of the higher-dimensional quasiconformal version of the Schoenflies theorem. □

We make the remark that all the constructions used in the proof of the theorem are canonical in the sense that if we give the various function spaces the compact-open topologies, then the real numbers and embeddings constructed depend continuously on the given data. While these will not be important to us, they are important in some applications.

Thus the result above includes what is known as a canonical (quasiconformal) Schoenflies theorem.

7.4.4. Some consequences. We now state and prove a few corollaries of Theorems 7.4.2 and 7.4.5.

COROLLARY 7.4.6. *A Jordan domain D in $\mathbb{R}^n$ is quasiconformally equivalent to the unit ball B^n if and only if ∂D is locally quasiconformally flat in D.*

PROOF. Assume first that D is quasiconformally equivalent to B^n and a Jordan domain. Then there is $f : B^n \to D$, and as f extends to a map $\hat{f} : \bar{B}^n \to \bar{D}$ by Corollary 6.5.12, the map $\hat{f}|\partial D$ is locally quasiconformally flat in D.

Conversely, by Theorem 7.4.2, ∂D is quasiconformally flat in D if it is locally flat. Thus there is $h : \bar{A}(1/2, 1) \hookrightarrow \bar{D}$ with h quasiconformal on $A(1/2, 1)$—the inverse of the embedding provided by the flattening, perhaps modified by a radial map. Then Theorem 7.4.5 provides an extension $\hat{h} : B^n \to D$ with $\hat{h} = h$ on, say, $A(3/4, 1)$ and quasiconformal on B^n. Corollary 6.5.12 implies that h extends homeomorphically to the boundary. □

COROLLARY 7.4.7. *Suppose that D is a domain in $\mathbb{R}^n$ whose boundary is homeomorphic to the $(n-1)$-ball B^{n-1} by a homeomorphism $h : \partial D \to \bar{B}^{n-1}$ so that for each point x of ∂D, h extends to a quasiconformal embedding of a neighbourhood (in $\mathbb{R}^n$) of x in $\mathbb{R}^n$. Then h extends to a quasiconformal homeomorphism of $\mathbb{R}^n$. In particular, D is quasiconformally equivalent to B^n.*

PROOF. By Theorem 7.4.4 the map h extends to a quasiconformal embedding $\hat{h} : U \to \mathbb{R}^n$ where U is some neighbourhood of ∂D in $\mathbb{R}^n$. We can find a thin ellipsoid E with $\bar{B}^{n-1} \subset E \subset \hat{h}(U)$. E is the image of the sphere under a linear quasiconformal homeomorphism of $\mathbb{R}^n$ so the exterior of $\hat{h}^{-1}(E)$ in $\hat{\mathbb{R}}^n$ has a quasiconformally flat boundary. Corollary 7.4.6 allows us to extend $\hat{h}|\partial\hat{h}^{-1}(E)$ to a quasiconformal mapping of this exterior onto $\mathbb{R}^n \setminus E$. The result easily follows. □

The reader is left to prove the next result which follows directly from the above and the removability theorems.

COROLLARY 7.4.8. *Let $e : \mathbb{S}^{n-1} \to \hat{\mathbb{R}}^n$ be an embedding. By the Jordan-Brouwer separation theorem $\hat{\mathbb{R}}^n \setminus e(\mathbb{S}^{n-1})$ consists of two components, each being a Jordan domain. Suppose that e^{-1} is locally quasiconformally flat in each of these Jordan domains. Then e extends to a quasiconformal homeomorphism of $\hat{\mathbb{R}}^n$ and $e(\mathbb{S}^{n-1})$ is a quasisphere.*

Another corollary that is worth noting is the following local to global result.

COROLLARY 7.4.9. *Let $f : B^n \to \mathbb{R}^n$ be a K-quasiconformal mapping. Then for each r, $0 < r < 1$, there is a K_r-quasiconformal mapping $f_r : \mathbb{R}^n \to \mathbb{R}^n$ such that $f_r|B^n(0, r) = f$ and K_r depends only on K, r and the dimension n.*

PROOF. Let $D_r = f(B^n(0, r))$. Then obviously $\partial D_r = f(\mathbb{S}^{n-1}(r))$ is quasiconformally bicollared. The result then follows from the Schoenflies theorem. □

COROLLARY 7.4.10. *Let $f : B^n \to \mathbb{R}^n$ be a K-quasiconformal mapping. Then for each r, $0 < r < 1$, $S_r = f[\mathbb{S}(r)]$ is a K_r-quasisphere.*

It is obvious that in complete generality we must have $K_r \to \infty$ as $r \to 1$. In two dimensions it is possible to get quite good estimates here by using the theory of holomorphic motions, [**15**]. For instance one can prove that if $f : B^2 \to \mathbb{R}^2$ is a

K-quasiconformal mapping, then for each r, $0 < r < 1$, $S_r = f[\mathbb{S}(r)]$ is a $K\,\frac{1+r}{1-r}$-quasicircle. However in higher dimensions explicit estimates are very hard to come by.

7.4.4.1. *Proof for Theorem* 7.4.2. We already considered the "only if" part. For the rest, we need to check that the hypotheses imply that the boundary mapping is locally quasiconformally flat, therefore quasiconformally flat by Lemma 7.4.3, and the result follows from Theorem 7.4.5.

7.4.4.2. *A warning!* There is one thing we must tidy up lest there be confusion. This is the subtle difference between locally flat and and locally collared. The former was used extensively in this section and the later in our section on removability and boundary extension.

Let D be a Jordan domain. Recall that ∂D is locally flat if there is a homeomorphism $h : \mathbb{S}^{n-1} \to \partial D$ such that for each $x \in \partial D$ there is a relatively open neighbourhood U of x in ∂D and the map $h|h^{-1}(U) \to U$ has an extension $\hat{h} : \bar{V} \subset \bar{B}^n \to \bar{D}$, $\hat{h}|V$ quasiconformal and $\hat{h}|V \cap \partial D = h$. Thus *the same homeomorphism admits local extensions.* Of course these local extensions may not agree except on the boundary. The point of Theorem 7.4.2 is that they can be chosen to do so. The definition for locally collared does not require this. For each $x \in \partial D$ there is an open set V containing x and a homeomorphism $h_x : V \cap \bar{D} \to B^n_+$ quasiconformal on $V \cap D$. Thus different points can have different homeomorphisms.

Certainly locally flat implies locally collared, but the converse is not obviously true. In fact this is a difficult question as a moment's thought should convince the reader. Various techniques, such as engulfing, have been developed in geometric topology to address it. There are quasiconformal versions of these sorts of results coming from geometric topology; see [**102, 158**] as a place to start as well as the references therein.

There are a couple of situations where we can get a fairly complete result, though it relies on very sophisticated extension results we have yet to prove but will do so in the next chapter.

THEOREM 7.4.11. *Let $h : \mathbb{S}^{n-1} \to \mathbb{R}^n$ be quasisymmetric and suppose $h(\mathbb{S}^{n-1})$ is locally quasisymmetrically collared in D, the bounded component of $\mathbb{R}^n \setminus h(\mathbb{S}^{n-1})$. Then h is quasiconformally flat and extends to a quasiconformal mapping of B^n.*

PROOF. We work in $\hat{\mathbb{R}}^n$ and replace $\mathbb{S}^{n-1}$ with $\hat{\mathbb{R}}^{n-1} \hookrightarrow \hat{\mathbb{R}}^n$. In light of what we have already proven, we only need to show that h is quasiconformally locally flat. Let $x \in \partial D$; we may assume $h^{-1}(x) \neq \infty$. By hypothesis there is a neighbourhood U of x in $\mathbb{R}^n$ and a quasisymmetric mapping $f : U \cap \bar{D} \to B^n_+$. Now the map

$$f \circ h : h^{-1}(U \cap \partial D) \subset \mathbb{R}^n \to f(U \cap \partial D) = B^{n-1} \subset \mathbb{R}^n$$

is quasisymmetric—the composition of quasisymmetric mappings is quasisymmetric. Thus $f \circ h$ is quasiconformal as a mapping between domains of $\hat{\mathbb{R}}^{n-1}$. The extension theorem, Theorem 8.3.4, gives us an open set V in $\mathbb{R}^n$,

$$V \cap \mathbb{R}^{n-1} = h^{-1}(U \cap \partial D), \qquad V \subset U,$$

and a homeomorphism

$$\widehat{f \circ h} : V \cap \bar{\mathbb{H}}^n \to \bar{\mathbb{H}}^n,$$

which is quasiconformal on $V \cap \mathbb{H}^n$ and equal to $f \circ h$ on $V \cap \mathbb{R}^{n-1}$. The local extension of h that we seek is then given by

$$f^{-1} \circ \widehat{f \circ h},$$

which agrees with h on $h^{-1}(U \cap \partial D)$. The result follows. □

We have had to use quasisymmetry since there is no good notion of quasiconformality for a map $\mathbb{S}^{n-1} \hookrightarrow \mathbb{R}^n$ except this. Quasisymmetry certainly implies quasiconformality, but the converse is not always true. What is true however is that a quasiconformal map is locally quasisymmetric—indeed we proved this in Theorem 6.6.8. This, together with Theorem 7.4.11, has the immediate consequence below which we leave the reader to prove.

THEOREM 7.4.12. *Let $h : \mathbb{S}^{n-1} \to \mathbb{R}^n$ be quasisymmetric and suppose $h(\mathbb{S}^{n-1})$ is locally quasiconformally bi-collared. Then $h(\mathbb{S}^{n-1})$ is a quasisphere. Conversely, every quasisphere is the quasisymmetric image of $\mathbb{S}^{n-1}$ and is bi-collared.*

7.5. Väisälä's theorem on cylindrical domains

In this section we aim to give a precise characterisation of those domains $D = \Omega \times \mathbb{R}$, where $\Omega \subset \mathbb{R}^2$ is a planar domain, which admit a quasiconformal homeomorphism $f : B^3 \to D$. When $\Omega = B^2$ we have seen above that there exists such a mapping, and any such mapping has the bound on its outer distortion as given at (7.2), and we will not seek better lower bounds.

Thus we seek to determine when there is a quasiconformal mapping $f : B^2 \times \mathbb{R} \to \Omega \times \mathbb{R}$ and will give a proof for the aforementioned impressive result of J. Väisälä, [**164**].

We are already in a reasonable situation to motivate a potential theorem. Suppose that $D = \Omega \times \mathbb{R}$ is a quasiball in $\mathbb{R}^3$ where Ω is a planar domain. If Ω has an outward directed cusp, then $D = \Omega \times \mathbb{R}$ would have an outward directed ridge—an obstruction to D being a quasiball which was established in Theorem 7.3.4. An example of a domain with inward directed cusps is the unit disk with the nonnegative real axis deleted, $\Omega = \mathbb{D} \setminus [0, 1)$. It is not difficult to see that there is a quasiconformal mapping $\Omega \times \mathbb{R} \mapsto \mathbb{D}_+ \times \mathbb{R}$ by opening up the cusp $(r, \theta, z) \mapsto (r, \theta/2, z)$ and that $\mathbb{D}_+ \times \mathbb{R}$ is a quasiball. Thus we may allow inward cusps but should not allow outward cusps. Next, we might reasonably expect that any arc in $\partial\Omega$ should be rectifiable from our discussion of Rickman's rugs, in particular Lemma 7.3.6 and its consequences. We might therefore hope that a necessary and sufficient condition for $\Omega \times \mathbb{R}$ to be a quasiball is that Ω should have locally rectifiable boundary and admit no external cusps. This is basically correct.

We have the following theorem.

THEOREM 7.5.1. *Let Ω be a simply connected proper subdomain of $\mathbb{R}^2$. Then there is a quasiconformal homeomorphism $f : B^3 \to \Omega \times \mathbb{R}$ if and only if Ω is finitely connected on the boundary and satisfies an internal chord-arc condition.*

As suspected, this condition implies that the boundary of the planar domain Ω is rectifiable. Väisälä also gives a strong uniqueness result by showing that if Ω is bounded, then $K_O(f) \leq q_o$ (see (7.2)), and equality is only possible when Ω is a round disk. For unbounded domains the corresponding lower bound is trivially one, which is attained when Ω is a half-plane.

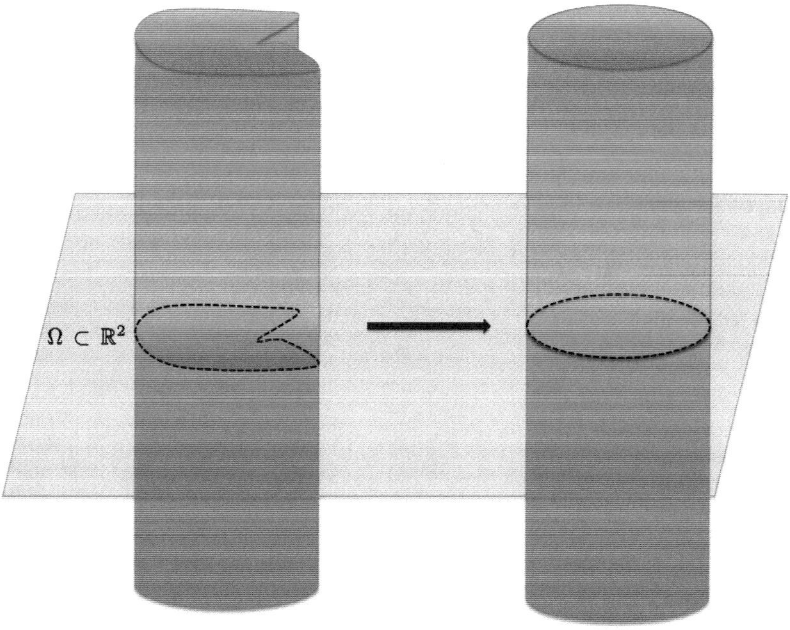

Mapping a cylindical domain $\Omega \times \mathbb{R}$ to a cyclinder $B^2 \times \mathbb{R}$.

We owe the reader a few definitions for the theorem to make sense. First, we have already discussed the condition "finitely connected on the boundary" in the introduction to subsection 6.5.3 concerning the version of Carathéodory's boundary extension for quasiconformal mappings between domains. It is straightforward to see that if Ω is finitely connected on the boundary, then so is $D = \Omega \times \mathbb{R}$.

7.5.1. Internal metrics. A key idea in the proof of Väisälä's theorem—and indeed a concept used extensively in the theory of quasiconformal mappings—is that we use quasiconformality in the usual sense to prove the quasisymmetry of a mapping in metrics adapted to the geometry of the domains in question. Here are some of the metrics in question.

Let $D \subset \mathbb{R}^n$ be a domain. For $a, b \in D$ we set

$$\begin{aligned} \delta_D(a,b) &= \inf\{\operatorname{diam}(\alpha) : \alpha \text{ is an arc joining } a \text{ to } b \text{ in } D\}, \\ \lambda_D(a,b) &= \inf\{\ell(\alpha) : \alpha \text{ is an arc joining } a \text{ to } b \text{ in } D\}. \end{aligned}$$

The metrics δ_D and λ_D generate the usual topology of D, and we have the inequalities

$$d \leq \delta_D \leq \lambda_D.$$

It is an easy exercise to see that if $E \subset D$ is a connected set, then the diameters are equal:

$$\delta_D(E) = \operatorname{diam}(E). \tag{7.25}$$

Next we use these metrics to discuss the internal chord-arc condition for planar domains.

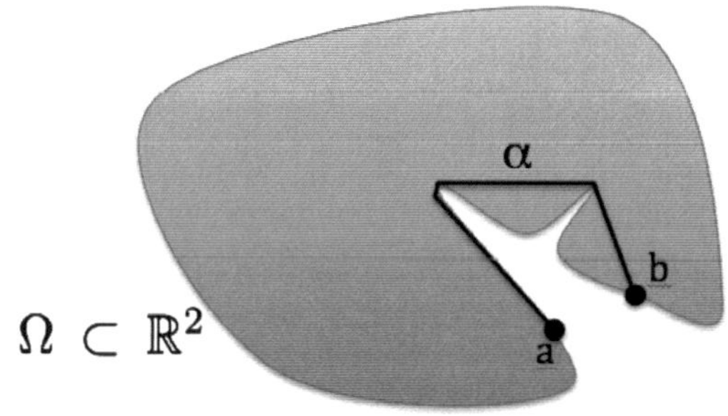

$$\begin{aligned}\delta_\Omega(a,b) &= \text{the diameter of } \alpha = \min\left\{2r : \alpha \subset \bar{B}^2(z,r), z \in \mathbb{R}^2\right\},\\ \lambda_\Omega(a,b) &= \text{the length of } \alpha.\end{aligned}$$

7.5.2. Chord-arc conditions. Let (X, d) be an arbitrary metric space and $\hat{X} = X \cup \{\infty\}$ be its one-point compactification. Let $\Gamma \subset \hat{X}$ be a locally rectifiable Jordan curve. If $a, b \in \Gamma \setminus \{\infty\}$, we let $\sigma(a, b)$ denote the length of the shorter component of $\Gamma \setminus \{a, b\}$. If there is a constant $c \geq 1$ such that

$$\sigma(a,b) \leq c \operatorname{dist}(a,b)$$

for all $a, b \in \Gamma \setminus \{\infty\}$, we say that Γ is a c-chord-arc curve. It is not difficult to prove that Γ is a c-chord-arc curve if and only if Γ is an L-bilipschitz image of the unit circle $\mathbb{S}$ or the extended real line $\hat{\mathbb{R}}$ with a quantifiable relationship between c and L.

If Ω is a simply connected proper subdomain of $\mathbb{R}^2$ which is finitely connected on the boundary, then standard results from planar point set topology (see for instance G.T. Whyburn's book [**168**]) tell us that the prime end boundary $\Gamma = \partial^*\Omega$ of Ω can be viewed as a Jordan curve, possibly passing through ∞, in $\hat{\mathbb{R}}^2$. We discuss this in more detail below.

Suppose this boundary curve, call it Γ, is locally rectifiable in the Euclidean metric and, as above, for $a, b \in \Gamma \setminus \{\infty\}$ we let $\sigma_\Gamma(a, b)$ denote the length of the shorter component of $\Gamma \setminus \{a, b\}$. Next, for $a, b \in \Gamma \setminus \{\infty\}$ we let

$$\delta_\Omega(a,b) = \inf\{\operatorname{diam}(\gamma) : \gamma \subset \Omega \text{ is a path connecting } a \text{ and } b\}. \tag{7.26}$$

The metric δ_Ω induces the usual topology of the domain Ω, and it is clear that $|a - b| \leq \delta_\Omega(a, b)$; however in general there is no control of $\delta_\Omega(a, b)$ by $|a - b|$. This is the point to the following definition: a planar domain Ω bounded by a locally rectifiable Jordan curve $\Gamma = \partial\Omega$ is said to be an *inner chord-arc* domain if there is a constant $c \geq 1$ such that for all $a, b \in \Gamma$ we have

$$\sigma(a,b) \leq c\, \delta_\Omega(a,b). \tag{7.27}$$

The internal chord-arc condition basically allows internally directed cusps, but not externally directed cusps that we were hoping for.

7.5.3. Prime ends. Suppose that a domain D of $\mathbb{R}^n$ is finitely connected on the boundary. An *endcut* of D is a path

$$\alpha : [a, b) \to D$$

such that $\alpha(t) \to z \in \partial D$ as $t \to b$. To denote this correspondence we write $z = h(\alpha)$. A subendcut of α is a restriction of α to a subinterval $[c, b)$. If U is a neighbourhood of $h(\alpha)$, there is a unique component $A(U, a)$ of $U \cap D$ containing a subendcut of α. Two endcuts α and β are equivalent, written $\alpha \sim \beta$, if $h(\alpha) = h(\beta)$ and if $A(U, \alpha) = A(U, \beta)$ for every neighbourhood U of $h(\alpha)$.

The equivalence class $[\alpha]$ of α is a *prime end* of D, and the collection of all prime ends of D is called the *prime end boundary* of D and is denoted $\partial^* D$. We write

$$D^* = D \cup \partial^* D.$$

There is a natural *impression map*

$$i_D : D^* \to \bar{D}, \tag{7.28}$$

defined by $i_D([\alpha]) = h(\alpha)$ for $[\alpha] \in \partial^* D$ and by $i_D = identity$ on D. If D is locally connected at a point $z \in \partial D$, then $(i_D)^{-1}(z)$ consists of a single point which, following a standard convention, we identify with z. In particular, if ∂D is homeomorphic to $\mathbb{S}^{n-1}$, we can identify $\partial^* D$ and ∂D.

There is a technical generalisation of our results on the boundary extension in Section 6.5.3 that we need here but will not prove. This can be found in [**162**, 17.10, 17.14]. The result we require is actually a refinement of our Theorem 6.5.8, and there are enough ideas about it so far that the reader will have a good chance of proving this himself.

In particular, if $f : B^n \to D$ is quasiconformal, then Väisälä has shown that f admits a continuous extension $f : \bar{B}^n \to \bar{D}$ such that every point-inverse $f^{-1}(y)$ is totally disconnected.

If α is an endcut of B^n, then using the continuity of f we see $f(\alpha)$ is an endcut of D. We now show that $h(\alpha) = h(\beta)$ if and only if $[\alpha] \sim [\beta]$. If $h(\alpha) = h(\beta) = b \in \partial D$ and if U is a neighborhood of $h(\alpha) = h(\beta) = b \in \partial D$, then there is an $r > 0$ such that $f[B^n \cap B^n(b, r)]$ is contained in a component of $U \cap D$ and contains subendcuts of both $f(\alpha)$ and $f(\beta)$ which means that $f(\alpha) \sim f(\beta)$. Conversely, if $f(\alpha) \sim f(\beta)$ and if, for the purposes of contradiction, we have $h(\alpha) \neq h(\beta)$, then $h[f(\alpha)] = h[f(\beta)] = y$. Since $f^{-1}(y)$ is totally disconnected, there is a compact set $F \subset \bar{B}^n \setminus f^{-1}(y)$ separating $h(\alpha)$ and $h(\beta)$ in B^n. We may assume that

$$\alpha \cap F = \beta \cap F = \emptyset.$$

Choose a connected neighborhood U of y such that $U \cap f(F) = \emptyset$. Since $f(F \cap D)$ separates $f(\alpha)$ and $f(\beta)$ in D, the neighbourhood $A(U, f(\alpha)) \neq A(U, f(\beta))$, which is a contradiction.

If β is an endcut of D, then $\mathbb{S}^{n-1} \cap \overline{f^{-1}(\beta)}$ is a connected set in $f^{-1}[h(\beta)]$, and hence a point. Thus $f^{-1}(\beta)$ is an endcut of B^n. The next lemma follows.

LEMMA 7.5.2. *If $f : B^n \to D$ is quasiconformal, then f has a unique bijective extension $f^* : \bar{B}^n \to D^*$ satisfying $f^*([\alpha]) = [f(\alpha)]$. This also implies that $i_D f^* = f$.*

If $\alpha : [a, b) \to D$ is an endcut of D, we say that α joins $\alpha(a)$ and $[\alpha] = u \in \partial^* D$. Similarly, an open path α in D joins elements $u, v \in \partial^* D$ if α has subpaths representing u and v. We can then extend the definition of the internal distance $\delta_D(a, b)$ to all $a, b \in Q = D^* \setminus (i_D)^{-1}(\infty)$. It is then easy to see that δ_D is a metric of Q.

LEMMA 7.5.3. *With the notation above, δ_D is consistent with the topology of Q so that f^* defines a homeomorphism.*

PROOF. Let $\epsilon > 0$, $u \in Q \cap \partial^* D$, and $b = (f^*)^{-1}(u)$. Since f is continuous, there is $U \subset B^n \cap B^n(b,r)$ such that $f(U) \subset B^n(i_D(u), \epsilon/2)$. Then $\delta_D(f(x), u) < \epsilon$ for all $x \in U$. Hence f^* is continuous at b in the metric δ_D. Next, with the same U as above, we choose a compact set $F \subset B^n \setminus f^{-1}[i_D(u)]$ separating b and $\mathbb{S}(b,r) \cap B^n$ in B^n. Then $\operatorname{dist}(f(F), i_D(u)) = q > 0$. Since $\delta_D(y,u) < q$ implies $(f^*)^{-1}(y) \in U$, we see that $(f^*)^{-1}$ is continuous at u in δ_D. □

7.5.4. Chord-arcs and prime ends. Later on we will need a two-dimensional result relating the internal chord-arc conditions and the rectifiability of the boundary—also expressed as a chord-arc condition. This is the content of the next lemma.

LEMMA 7.5.4. *Let Ω be a simply connected proper subdomain of $\mathbb{R}^2$ which is finitely connected on the boundary. Then the following are equivalent:*

(i) *Ω satisfies a c-internal chord-arc condition,*
(ii) *$\partial^*\Omega$ is a c-chord-arc in the metric δ_Ω.*

PROOF. Suppose that the impression $i_\Omega(u)$ of $u \in \partial^*\Omega$ is ∞ for at most one u, which we will write as $u = \infty$. It now suffices to show that if $\alpha \subset \partial^*\Omega \setminus \{\infty\}$ is a compact arc, then its length $\ell(\alpha)$ in the metric δ_Ω is equal to that of $\ell(\alpha)$.

Thus, let $u, v \in \partial^*\Omega \setminus \{\infty\}$. If there is a path β which joins u and v in Ω, then

$$|i_\Omega(u) - i_\Omega(v)| \leq \operatorname{diam}(\beta)$$

and hence $|i_\Omega(u) - i_\Omega(v)| \leq \delta_\Omega(u,v)$. This of course implies that $\ell(\alpha) \leq \ell_\delta(\alpha)$, the δ length of α.

To obtain the converse we begin by choosing an $\epsilon > 0$. If β is an arc in $\partial^*\Omega \setminus \{\infty\}$ with endpoints u and v, then there is a path β' joining u to v in the set $i_\Omega(\beta) + B^n(\epsilon)$. Since we have $\operatorname{diam}(\beta') \leq \operatorname{diam}[i_\Omega(\beta)] + 2\epsilon \leq \ell(\beta) + 2\epsilon$ we see that $\delta_\Omega(u,v) \leq \ell(\beta)$. If the points $u_0, u_1, \ldots, u_k$ divide up α into subarcs $\alpha_1, \alpha_2, \ldots, \alpha_k$, we obtain from the above estimate that

$$\sum_{j=1}^{k} \delta_\Omega(u_j, u_{j-1}) \leq \sum_{j=1}^{k} \ell(\alpha_j) = \ell(\alpha)$$

so that $\ell_\delta(\alpha)$ is indeed no more than $\ell(\alpha)$, and this completes the proof. □

The existence of a quasiconformal mapping $f : B^2 \times \mathbb{R} \to \Omega \times \mathbb{R}$ would easily follow if there were a locally L-bilipschitz homeomorphism $\varphi : B^2 \to \Omega$; if Ω were unbounded, we would look for such a mapping $\varphi : \mathbb{H}^2 \to \Omega$. This is because under these circumstances the map

$$f : B^2 \times \mathbb{R} \to \Omega \times \mathbb{R}, \quad f : (z,t) \mapsto (\varphi(z), t)$$

is also locally L-bilipschitz and therefore quasiconformal. We are expecting that Ω is quite regular and so a bilipschitz parameterisation may well be expected.

Therefore a key step in Väisälä's proof will be to establish the following theorem.

THEOREM 7.5.5. *Let Ω be a planar domain which is finitely connected on the boundary and which satisfies an internal chord-arc condition with constant c. Then there is a number $L = L(c) \geq 1$ such that if Ω is bounded, then there is a locally L-bilipschitz homeomorphism $\varphi : \Omega \to B^2$.*

We will want a similar result if Ω is unbounded. This result is a consequence of Theorem 7.5.15 below.

We will also want a rough converse to this result—if $\Omega \times \mathbb{R}$ is a quasiball, we seek a bilipschitz parameterisation of Ω. This would of course quickly imply that Ω is internal chord-arc. There is no obvious path to this result, but it is natural to expect that one should promote the quality of the mapping to a ball from quasiconformality (and hence local quasisymmetry) to quasisymmetry as a first step. This promotion will require further regularity properties of Ω—these will be encapsulated in the notion of a John domain.

7.5.5. John domains. John domains were first considered by F. John [**89**, p. 402]; the term is due to O. Martio and J. Sarvas [**112**], as these domains were an integral part of their 1979 study of uniform domains and the injectivity properties of functions. There are numerous characterisations of John domains, however, we follow Väisälä and use the characterisation via cigars.

An arc in $\mathbb{R}^n$ is the homeomorphic image of $[-1, 1]$. Given an arc $\alpha \subset \mathbb{R}^n$ with endpoints a, b, and if $x \in \alpha$, then we define $\alpha_a(x)$ and $\alpha_b(x)$ as those components of $\alpha \setminus \{x\}$ which contain a and b respectively. We then define the function

$$d_\alpha(x) = \min\{\operatorname{diam}[\alpha_a(x)], \operatorname{diam}[\alpha_b(x)]\}. \tag{7.29}$$

For $c \geq 1$ the open set

$$\operatorname{cigar}(\alpha, c) = \bigcup_{x \in \alpha} B^n[x, d_\alpha(x)/c] \tag{7.30}$$

is called a (diameter) c-cigar joining a and b. A domain $\Omega \subset \mathbb{R}^n$ is a c-John domain if each pair of points in Ω can be joined by a c-cigar lying in Ω.

We leave the verification of the following elementary, but useful, lemma to the reader.

LEMMA 7.5.6. *Let D be a John domain in $\mathbb{R}^n$. Then D is finitely connected on the boundary.*

As indicated above, we are going to need a slightly refined version of the notion of quasisymmetry in Euclidean spaces which is adapted to the internal geometry of a domain.

Let X and Y be metric spaces with distance written as $|a - b|$, let $\eta : [0, \infty) \to [0, \infty)$ be a homeomorphism, and let $f : X \to Y$ be an embedding.

If $|a - x| \leq t|b - x|$ implies $|f(a) - f(x)| \leq \eta(t)|f(b) - f(x)|$ for all $a, b, x \in X$ and $t > 0$, then we say f is η-quasisymmetric.

If $H \geq 1$ and if $|a - x| \leq |b - x|$ implies $|f(a) - f(x)| \leq H|f(b) - f(x)|$, then we say f is weakly H-quasisymmetric. Of course an η-quasisymmetric map is weakly H-quasisymmetric with $H = \eta(1)$.

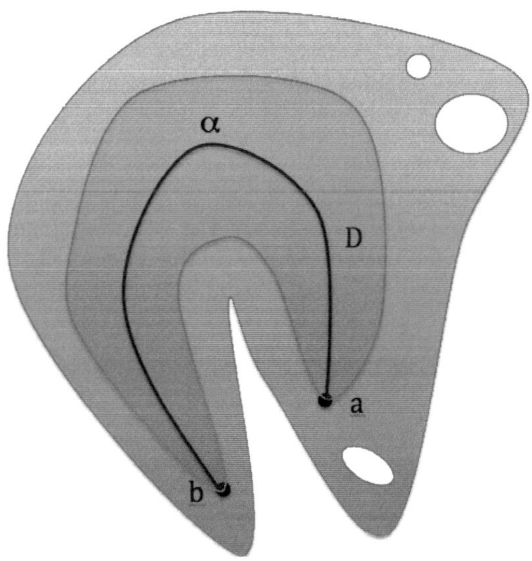

A John domain with a "cigar" connecting a to b along a central curve α—the width of the cigar at x is proportional to the diameter of the shorter of the two subarcs of $\alpha \setminus \{x\}$.

Of course these definitions are consistent with our earlier use of quasisymmetry in the Euclidean setting and also our discussion around Rickman's rug.

What is interesting here is that weak-quasisymmetry implies quasisymmetry for fairly general sorts of metric spaces, and we discuss this now.

A metric space X is *k-homogeneously totally bounded* if there is an increasing function $k : [1/2, \infty) \to [1, \infty)$ with the following property: For each $a \geq \frac{1}{2}$, every closed ball $\overline{B^n(x, r)} \subset X$ can be covered with sets $A_1, A_2, \ldots, A_s$ with $s \leq k(a)$ and for $j = 1, 2, \ldots, s$, $\operatorname{diam}(A_j) < r/a$.

Clearly if $t > 0$ and if X is a bounded and k-homogeneously totally bounded set whose points have mutual distances at least t, then the cardinality of X is less than $k(\operatorname{diam}(A)/t)$. The following theorem [**164**, Theorem 2.9] builds on an earlier well-known work of P. Tukia and J. Väisälä, [**158**].

THEOREM 7.5.7. *Suppose that X and Y are k-homogeneously totally bounded metric spaces and that X is also pathwise connected. Then every weakly H-quasisymmetric mapping $f : X \to Y$ is η-quasisymmetric with η depending only on H and the function k.*

PROOF. Let $a, b, x \in X$ be three distinct points with $|a - x| = t|b - x|$. We have to find an estimate of the form

$$|f(a) - f(x)| \leq \eta(t)|f(b) - f(x)|, \tag{7.31}$$

where $\eta(t) \to 0$ as $t \to 0$. We know that (7.31) is valid for $t = 1$ with $\eta(t) = H$. Thus we need to define η for all other t and therefore suppose initially that $t > 1$.

Put $r = |b - x|$ and let α be an arc from x to a. Inductively define points $a_0, a_1, \ldots, a_s$ of α such that $a_0 = x$; then a_{j+1} is the last point of α in $\overline{B^n(a_j, r)}$

and finally a_s is the first of these points outside $B^n(x, |x-a|)$. By construction $|a_i - a_j| \geq r$ for all $0 \leq i < j < s$. Since X is k-homogeneously totally bounded, we have

$$s \leq k(|x-a|/r) = k(t).$$

Since f is weakly H-quasisymmetric, we obtain

$$|f(a_1) - f(x)| \leq H|f(b) - f(x)|,$$

and so by induction we find

$$|f(a_{j+1}) - f(a_j)| \leq H|f(a_j) - f(a_{j-1})| \leq H^{j+1}|f(b) - f(x)|$$

for all $0 \leq i < j < s$.

This implies that

$$|f(a_s) - f(x)| \leq sH^s|f(b) - f(x)|.$$

Since $|a - x| \leq |a_s - x|$, we obtain (7.31) with $\eta(t) = sH^{s+1}$ with $s = k(t)$.

Next assume that $t < 1$. Set $r = |x - b|$ and for $j \geq 0$ we choose points $b_j \in \partial B^n(x, 3^{-j}r)$, with $b_0 = b$. Let s be the smallest integer with $3^{-s}r \leq |x-a|$. Then

$$s \geq \frac{\log(1/t)}{\log 3} = s_0(t).$$

Now if $0 \leq i < j < s$, we have $2|x - b_j| \leq |b_i - b_j|$, which implies that $|a - b_j| \leq |b_i - b_j|$. Hence

$$|f(a) - f(b_j)| \leq H|f(b_i) - f(b_j)|, \quad |f(x) - f(b_j)| \leq H|f(b_i) - f(b_j)|, \tag{7.32}$$

and thus

$$|f(a) - f(x)| \leq 2H|f(b_i) - f(b_j)|. \tag{7.33}$$

On the other hand, $|b_j - x| \leq |b - x|$ implies that the points

$$f(b_0), f(b_1), \ldots, f(b_{s-1})$$

lie in the ball $B^n(x, H|f(b) - f(x)|)$. Since Y is k-homogeneously totally bounded, we get

$$s \leq k\left(2H^2 \frac{|f(b) - f(x)|}{|f(a) - f(x)|}\right).$$

Finally as $s_0(t) \to \infty$ as $t \to 0$, this and (7.33) together give (7.31) with some $\eta(t)$ which converges to 0 as $t \to 0$. This is what we wanted to prove. □

7.5.6. Internal metrics and quasisymmetry. The following lemma allows us to connect weak-quasisymmetry and quasisymmetry in the internal metrics of a domain in the presence of some geometric control.

If α is an arc in $\mathbb{R}^n$ with endpoints a, b, and if $c \geq 1$, the set

$$car_d(\alpha, c) = \bigcup \{B(x, \operatorname{diam}[\alpha_a(x)]/c) : x \in \alpha\} \tag{7.34}$$

(α_a is defined in (7.29) above) is called a (diameter) c-carrot with vertex a and joining a to b. The possibility that $b = \infty$ is allowed with the obvious minor modifications.

A subset A of a domain Ω has the c-carrot property in Ω with center $x_0 \in \Omega \cup \infty$ if each $x \in A$ can be joined to x_0 by a c-carrot in Ω.

LEMMA 7.5.8. *Suppose that $\Omega \subset \mathbb{R}^n$ is a domain, that $A \subset \Omega$ has the c-carrot property in Ω with center x_0, and that e is another metric of A with $\delta_\Omega \le e \le d$. Then the metric space (A, e) is k-homogeneously totally bounded with a function k depending only on the constant c and the dimension n.*

PROOF. Let $x \in A$ and $r > 0$ and consider the closed ball $B_e(x, r)$. If $x_1, x_2, \dots, x_s \in B_e(x, t)$ with $e(x_i, x_j) \ge r/2$ for $i \ne j$, then it will suffice to find an upper bound on s depending only on c and n. For each j there are arcs α_j joining x_0 to x_j so that $car_d(\alpha_j, c) \subset \Omega$. As $\delta_\Omega(x_i, x_j) \ge r/2$, $d(\alpha_j) < r/4$ for at most one j, so we may assume that $d(\alpha_j) \ge r/4$ for all j. We then choose $y_j \in \alpha_j$ such that the subarcs $\beta_j = \alpha_j[x_j, y_j]$ have $d(\beta_j) = r/8$. Then the balls $B_j = B^n(y_j, r/8c)$ are contained in Ω and we want to see that these balls are disjoint. If B_i meets B_j for some $i \ne j$, the set $\gamma = \beta_i \cup [y_i, y_j] \cup \beta_j$ joins x_i to x_j in Ω so $\delta_\Omega(x_i, x_j) \le \operatorname{diam}(\gamma)$. Since $\delta_\Omega(x_i, x_j) \le r/2$ and since

$$\operatorname{diam}(\gamma) \le \operatorname{diam}(\beta_i) + \operatorname{diam}(\beta_j) + |y_i - y_j| < r/8 + r/8 + r/4c < r/2,$$

this gives a contradiction. It follows that $|y_i - y_j| \ge r/4c$ for $i \ne j$. On the other hand,

$$|y_i - x| \le |y_i - x_i| + |x_i - x| \le \operatorname{diam}(\beta_i) + e(x_i, x) \le 9r/8.$$

Since $(\mathbb{R}^n, d)$ itself is homogeneously totally bounded, this gives the result. □

Just for convenience we now recall from the extremality of the Teichmüller ring (see subsection 5.4.1) the following formulation. If C_0 and C_1 are disjoint continua in $\hat{\mathbb{R}}^n$, if $t > 0$, and if

$$\operatorname{dist}(C_0, C_1) \le t \min\{\operatorname{diam}(C_0), \operatorname{diam}(C_1)\}, \tag{7.35}$$

then the path family $\Gamma = \Delta(C_0, C_1; \mathbb{R}^n)$ of all paths joining C_0 and C_1 in $\mathbb{R}^n$ satisfies the modulus estimate

$$M(\Gamma) \ge \Psi_n(t) \tag{7.36}$$

where the function $\Psi_n : (0, \infty) \to (0, \infty)$ is decreasing and onto. Following Väisälä, we say that a pair of disjoint continua C_0, C_1 in a domain D of $\mathbb{R}^n$ is in *t-standard position* in D if

$$\delta_D(C_0, C_1) \le t \min\{\operatorname{diam}(C_0), \operatorname{diam}(C_1)\}. \tag{7.37}$$

Then, given $\phi : (0, \infty) \to (0, \infty)$, a decreasing homeomorphism, a domain D of $\mathbb{R}^n$ is called *ϕ broad* if for each $t > 0$ and each t-standard position pair (C_0, C_1) in D, the path family $\Gamma = \Delta(C_0, C_1; D)$ satisfies the inequality

$$M(\Gamma) \ge \phi(t). \tag{7.38}$$

Heinonen and Koskela [**76**] used the term *ϕ-Loewner* to describe this property.

In the proof of Väisälä's mapping theorem we only need these notions and structures in the case when D is the half-space and δ_D is the usual metric. However, there are potentially other applications, and so we follow the more general path laid out by him.

LEMMA 7.5.9. *Let $D \subset \mathbb{R}^n$ be a ϕ-broad domain and let e be a metric in D with $d \le e \le \delta_D$. Then (D, e) is k-homogeneously totally bounded with the function k depending only on ϕ and the dimension n.*

PROOF. Consider the closed ball $B_e(x,r)$ and points $x_1, x_2, \ldots, x_s \in \overline{B_e(x,r)}$ with $|x_i - x_j| > r/2$ for $i \neq j$; as before, our aim again is to bound s appropriately. Choose a positive number $q = q(\phi, n) < \frac{1}{16}$ such that

$$2\omega_{n-1}\left(\log \frac{1-4q}{8q}\right)^{1-n} \leq \phi(1). \tag{7.39}$$

Join x_1 to x by an arc $\alpha_i \subset D$. Since $\alpha_i \cup \alpha_j$ joins x_i and x_j in D, we have

$$r/2 \leq e(x_i, x_j) \leq \delta_D(x_i, x_j) \leq \operatorname{diam}(\alpha_i) + \operatorname{diam}(\alpha_j)$$

for $i \neq j$ so that $\operatorname{diam}(\alpha_i) < r/4$ for at most one i, and we may assume $\operatorname{diam}(\alpha_i) \geq r/4$ for all i.

Choose subarcs α_i^1 and α_i^2 of α_i such that x_i is an endpoint of α_i^1 and

$$\operatorname{diam}(\alpha_i^1) = \operatorname{diam}(\alpha_i^2) = \delta_D(\alpha_i^1, \alpha_i^2) = qr.$$

This is possible as $q < \frac{1}{16}$. Then (α_i^1, α_i^2) is a 1-standard position pair in D, and hence

$$M(\Gamma_i) \geq \phi(1), \quad \Gamma_i = \Delta(\alpha_i^1, \alpha_i^2; D).$$

Next setting $B_i = B_e(x_i, r/4)$, $a_i = m(B_i)$, and $\Gamma^* = \{\gamma \in \Gamma_i \cap B_i\}$, we have

$$M(\Gamma_i^*) \leq a_i (qr)^{-n}.$$

We next want to estimate the modulus $M(\Gamma_i \setminus \Gamma_i^*)$. If $\gamma \in \Gamma_i \setminus \Gamma_i^*$, there is $y \in \gamma$ with $e(y, x_i) \geq r/4$. Since $\alpha_i^1 \cup \gamma$ joins x_i and y_i in D, we have

$$\delta_D(y, x_i) \leq \operatorname{diam}(\alpha_i^1) + \operatorname{diam}(\gamma) = qr + \operatorname{diam}(\gamma).$$

Since $e \leq \delta_D$, this yields $\operatorname{diam}(\gamma) \geq r/4 - qr$. Hence γ meets the complement of the ball $B^n(x_i, r/8 - qr/2)$. On the other hand, $\operatorname{diam}(\alpha_i^1) = qr$ implies that $\alpha_i^1 \subset B^n(x_i, qr)$. Hence γ also meets $\overline{B^n(x_i, qr)}$, and we obtain the estimate

$$M(\Gamma_i \setminus \Gamma_i^*) \leq \omega_{n-1}\left(\log \frac{r/8 - qr/2}{qr}\right)^{1-n} \leq \phi(1)/2$$

by (7.39). Consequently,

$$\phi(1) \leq M(\Gamma_i) \leq M(\Gamma_i^*) + M(\Gamma_i \setminus \Gamma_i^*) \leq a_i (qr)^{-n} + \phi(1)/2$$

and $a_i \geq q^n r^n \phi(1)/2$.

Since $e(x_i, x_j) \geq r/2$, the balls B_i are disjoint and are contained in $B_e(x, 5r/4) \subset B^n(x, 2r)$, and hence

$$|B^n| 2^n r^n \geq \sum_{i=1}^{s} a_i \geq s q^n r^r \phi(1)/2,$$

which, when rearranged, gives us the desired bound on s. □

In what follows we will need the following lemma whose proof, due to J. Sarvas, can be found in [**164**, Theorem 2.4].

LEMMA 7.5.10. *Suppose that D is open in $\mathbb{R}^n$, $n \geq 2$, and that $f : D \to \mathbb{R}^n$ is K-quasiconformal. Suppose also that $x_0 \in D$, $a > 1$, $r > 0$ such that $B^n(x_0, ar) \subset D$. Then $f|B^n(x_0, r)$ is η-quasisymmetric where η depends only on n, K, and a.*

PROOF. By our previous work, in particular by Theorem 7.5.7, it suffices to show that f is weakly H-quasisymmetric with $H = H(n, K, a)$.

By using a preliminary similarity transformation we may assume $r = 1$. Let $h : B^n(x_0, 4) \to B^n(x_0, a)$ be a K_1-quasiconformal radial stretching with $h|B^n(x_0, 1) =$ *identity* and $K_1 = K_1(n, a)$. Then $g = f \circ h : B^n(x_0, 4) \to \mathbb{R}^n$ is a $K \times K_1$-quasiconformal homeomorphism and $g = f$ on $B^n(x_0, 1)$.

Now let $a, b, x \in B^n(x_0, 1)$ with $0 < |a - x| < |b - x|$. The ratio

$$|f(a) - f(x)|/|f(b) - f(x)| \leq L/\ell,$$

where

$$L = L(x, f, |a - x|) = \sup\{|f(x - f(y)| : |x - y| = |x - a|\}, \tag{7.40}$$

$$\ell = \ell(x, f, |a - x|) = \inf\{|f(x - f(y)| : |x - y| = |x - a|\}. \tag{7.41}$$

Observe that $|a - x| < 2$ and thus $B^n(x, |a - x|) \subset B^n(x_0, 3)$. We now choose $z \in \partial B^n(x, |x - a|)$ such that $|g(z) - g(x)| = L$. Let F be the ray from $g(x)$ through $g(z)$, and let $J \subset F$ be a segment joining $g(z)$ to $g[\partial B^n(x, 3)]$ in $g[B^n(x, 3)]$. Then $g^{-1}(J)$ is a continuum joining the spheres $\mathbb{S}^{n-1}(x, |x - a|)$ and $\mathbb{S}^{n-1}(x, 3)$. Set $A = B^n(f(x), 1)$. Then $g^{-1}(A)$ is a continuum joining x and $\mathbb{S}^{n-1}(x, |x - a|)$.

Let Γ be the family of all paths joining $g^{-1}(A)$ and $g^{-1}(J)$ in $B^n(x, 3)$. Then $M(\Gamma) > c_n$. On the other hand,

$$M[g(\Gamma)] \leq \omega_{n-1}\Big(\log \frac{L}{\ell}\Big)^{1-n}.$$

Since g is KK_1-quasiconformal, $M(\Gamma) < KK_1 M(g(\Gamma))$, and so we have the bound $L/\ell < e^t$ with $t = (KK_1\omega_{n-1}/c_n)^{1/(n-1)}$ which establishes the lemma. □

The next theorem is a key step. It shows that under certain circumstances quasiconformal mappings will be quasisymmetric in the internal metrics of a domain if one has some knowledge of the geometry of the image.

THEOREM 7.5.11. *Suppose that $f : D \to D'$ is a K-quasiconformal mapping between subdomains of $\mathbb{R}^n$, suppose D is ϕ-broad, and suppose also that $A \subset D$ is a pathwise connected set and that $f(A)$ has the c_1-carrot property in D' with center $y_0 \in \overline{D'}$.*

- *If $y_0 \neq \infty$, we assume that there is $c_2 > 0$ such that*
$$\operatorname{diam}(A) < c_2 \operatorname{dist}[f^{-1}(y_0), \partial D].$$
- *If $f(y_0) = \infty$, assume f extends continuously to a map $D \cup \{\infty\} \to D' \cup \{\infty\}$.*

Then $f|A$ is η-quasisymmetric in the metrics δ_D and $\delta_{D'}$, with η depending only on c_1, c_2, K, ϕ, and n.

PROOF. From Lemmas 7.5.8 and 7.5.9 we see that there is a function k depending only on the data such that the spaces $(f(A), \delta_{D'})$ and (A, δ_D) are k-homogeneously totally bounded. Therefore, by Theorem 7.5.7 it suffices to show that $f|A$ is weakly H-quasisymmetric, with H depending only on the data. We now do this.

Let $a, b, x \in A$ be distinct points with $\delta_D(a, x) \leq \delta_D(b, x) = r$. Set $a' = f(a)$, $b' = f(b)$, $x' = f(x)$,

$$u = \delta_{D'}(a', x'), \quad \text{and} \quad v = \delta_{D'}(b', x').$$

In view of Theorem 6.6.7 may assume that $D \neq \mathbb{R}^n \neq D'$. We set $\tau = \operatorname{dist}(x, \partial D)$, $\tau' = \operatorname{dist}(x', \partial D')$. We will subsequently choose constants $M_j \geq 1$, $0 \leq q_j < 1$ depending only on the data.

The proof breaks down into five cases, the first two of which are auxiliary cases. If $y_0 = \infty$ the third case is completely general.

Case 1: $\tau \geq 2r$. Then a and b lie in the ball $B = B^n(x, 3r/2)$, and $\delta_D(a,x) = |a-x|$, $\delta_D(b,x) = |b-x| = r$. Lemma 7.5.10 shows that $f|B$ is η-quasisymmetric depending only on the data, and weak-quasisymmetry holds with

$$\frac{u}{v} = \frac{\delta_{D'}(a',x')}{\delta_{D'}(b',x')} \leq 2\eta(1)^2.$$

Case 2: $B(x',u) \subset D'$. Now $u = |a'-x'|$. We may assume that $v/3 < u$ and hence $v = |b'-x'|$. Let R' be the ring $B^n(x',u) \setminus \bar{B}^n(x',v)$. The two components of the complement of $R = f^{-1}(R')$ are $C_0 = f^{-1}[B^n(x',v)]$ and $C_1 = \mathbb{R}^n \setminus f^{-1}[B^n(x',u)]$.

The continuum C_0 is bounded and contains x and b, while C_1 is unbounded and contains a and $\mathbb{R}^n \setminus D$.

If the segment $[x,b]$ meets $\mathbb{R}^n \setminus D$, then $\operatorname{dist}(C_0, C_1) \leq |b-x| \leq \operatorname{diam}(C_0)$. This is also true if $[x,b] \subset D$, as then

$$\operatorname{dist}(C_0, C_1) \leq |a-x| \leq \delta_D(a,x) \leq \delta_D(b,x) = |b-x| \leq \operatorname{diam}(C_0).$$

Let Γ_R be the path family joining the boundary components of the ring R. The Teichmüller estimate (7.35) provides a lower bound $M(\Gamma_R) \geq q_1$ for a constant q_1. Hence

$$q_1 \leq KM(\Gamma_{R'}) = K\omega_{n-1}\Big(\log\frac{u}{v}\Big)^{1-n},$$

which gives the bound

$$\frac{u}{v} = \frac{\delta_{D'}(a',x')}{\delta_{D'}(b',x')} \leq \exp\left[\frac{K\omega_{n-1}}{q_1}\right]^{1/(n-1)}.$$

Case 3: $\delta_D(X, X_0) \geq 2r$, $x_0 = f^{-1}(y_0)$. If $y_0 = \infty$, then $x_0 = \infty$. We join x' to b' by an arc $\alpha_0 \subset D'$ with $\operatorname{diam}(\alpha_0) < 2v$ and then join a' to y_0 by a carrot $car_d(E,c) \subset D'$ with central curve E.

For $y \in E \setminus \{a', y_0\}$ we set $\sigma(y) = \operatorname{diam}(E[a',y])$, the diameter of that part of E connecting a' and y. Then $B^n[y, \sigma(y)/c_1] \subset D'$. We examine two possibilities and for ease of notation write $\Theta = \Theta_n^K$ for the distortion function defined in (6.48).

First, suppose there is $y \in E$ with $|y-x'| < \sigma(y)/2c_1$. As $\tau' > \sigma(y)/2c_1$, if we put $t_0 = \Theta^{-1}(1/2)$ and if $|b'-x'| \leq t_0\tau'$, then $|b-x| \leq \tau/2$ which implies $r \leq \tau/2$, and this is Case 1 again. If $|b'-x'| \geq t_0\tau'$, then $[x',y] \cup E[y,a']$ joins x' to a' in D'. Then

$$u \leq |x'-y| + \sigma(y) \leq \sigma(y)/2c_1 + \sigma(y) \leq 2\sigma(y),$$

and as $v \geq |b'-x'| \geq t_0\tau' > t_0\sigma(y)/2c_1$ we obtain

$$\frac{u}{v} \leq 4c_1 t_0.$$

Secondly, if $|y-x'| \leq \sigma(y)/2c_1$ for all $y \in E$, then $C_0 \cap E = \emptyset$. We therefore consider the path families $\Gamma' = \Delta(C_0, E; D')$ and $\Gamma = f^{-1}(\Gamma')$. By the equality of diameters as seen in (7.25),

$$\begin{aligned}\operatorname{diam}[f^{-1}(C_0)] &= \delta_D(f^{-1}C_0) \geq \delta_D(x,b) = r,\\ \operatorname{diam}[f^{-1}(E)] &= \delta_D[f^{-1}(E)] \geq \delta_D(x_0,a)\\ &\geq \delta_D(x_0,x) - \delta_D(a,x) \geq 2r,\\ \delta_D[f^{-1}(C_0), f^{-1}(E)] &\leq \delta_D(a,x) \leq r.\end{aligned}$$

Thus the pair $(f^{-1}(C_0), f^{-1}(E))$ is in 1-standard position within D, and as D is ϕ-broad we have $M(\Gamma) \geq \phi(1)$. We want to bound the number

$$\delta_0 = \inf_{\gamma\in\Gamma'} \left\{ \frac{\operatorname{diam}(\gamma)}{\operatorname{diam}(C_0)} \right\}$$

which may we assume is at least 2 and in which case each $\gamma \in \Gamma'$ meets the spheres $\mathbb{S}^{n-1}(x', \operatorname{diam}[C_0])$ and $\mathbb{S}^{n-1}(x', \delta_0 \operatorname{diam}(C_0)/2)$; this implies

$$M(\Gamma') \leq \omega_{n-1}\big(\log \delta_0/2\big)^{1-n}.$$

Since f is K-quasiconformal, $M(\Gamma') \geq M(\Gamma)/K \geq \phi(1)/K$, and this bounds δ_0 by a number M say. Finally, as $\operatorname{diam}(C_0) < 2v$, there is $\gamma \in \Gamma'$, with $\operatorname{diam}(\gamma) \leq 2Mv + 2v$. If $y \in \gamma \cap E$, then

$$\sigma(y)/2c_1 \leq |y - x'| \leq \operatorname{diam}(\gamma) + \operatorname{diam}(C_0) < 2Mv + 2v$$

so that

$$u \leq \operatorname{diam}(C_0) + \operatorname{diam}(\gamma) + \sigma(y) < 2v + 2Mv + 2c_1(M+1)v,$$

and this last estimate gives

$$\frac{u}{v} \leq \frac{1}{2(1 + M + c_1(M+1))},$$

which completes the proof of Case 3.

In Cases 4 and 5 we assume that $y_0 \neq \infty$ so that $y_0 \in D'$, $x_0 \in D$ and that $\operatorname{diam}(A) \leq c_2 \operatorname{dist}(x_0, \partial D)$.

After scaling and translating we may further assume that $\operatorname{dist}(y_0, \partial D') = 1$. For every $y \in f(A)$ there is a carrot $car_d(E,c)$ joining y to y_0. Then

$$B^n[y_0, \operatorname{diam}(E)/c_1] \subset D',$$

which by our normalisation puts $\operatorname{diam}(E)/c_3 \leq 1$, and thus $\delta_{D'}(y, y_0) \leq c_1$.

Consequently, we will always have that

$$u \leq 2c_1 \tag{7.42}$$

and we will seek to bound v from below.

Case 4. $|x' - y_0| < \frac{1}{2}$. Put $r_0 = \ell(x', f^{-1}, 1/2)$, where ℓ is defined above at (7.41). We then have $B^n(x, r_0) \subset D$.

If $r \leq r_0$, then $\delta_D(a,x) \leq r$ implies $a \in B^n(x, r_0)$ and hence $a' \in B^n(x', 1/2) \subset D$. Then $u = |a' - x'| \leq 1/2$, and we find Case 2 again.

Choosing $a_0, b_0 \in \partial B^n(x, r_0)$ with

$$|f(a_0) - x'| = L(x, f, r_0), \qquad |f(b_0) - x'| = \ell(x, f, r_0)$$

we have Case 2 for the triple (x, a_0, b_0). Hence $1/2 = L(x, f, r_0) \leq H_1 \ell(x, f, r_0)$ with H_1 depending only on the data. If $r > r_0$, then

$$v \geq \ell(x, f, r) > \ell(x, f, r_0) \geq 1/2H_1.$$

This together with (7.42) gives

$$\frac{u}{v} \leq \frac{c_1}{H_1}.$$

Case 5. $\delta_D(x, x_0) \leq 2r$ *and* $|x' - y_0| \leq \frac{1}{2}$. We may assume that $v \leq \frac{1}{8}$, for otherwise we are done. Join x' and b' by an arc $C_0 \in D'$ with $\operatorname{diam}(C_0) < 2v$. We consider the path families

$$\Gamma' = \Delta(C_0, \bar{B}^n(y_0, 1/4); D')$$

and $\Gamma = f^{-1}(\Gamma')$. Since $C_0 \subset B^n(x', 2v)$ and $\operatorname{dist}[x', \bar{B}^n(y_0, 1/4)] \geq 1/4$, we have

$$M(\Gamma') \leq \omega_{n-1}\Big(\log \frac{1}{8v}\Big)^{1-n}. \tag{7.43}$$

In view of (7.25),

$$\begin{aligned} \operatorname{diam}(f^{-1}(C_0)) &= \delta_D(f^{-1}(C_0)) \geq \delta_D(b, x) = r, \\ d(f^{-1}(C_0)) &\geq \ell(y_0, f^{-1}, 1/4) \geq \Theta^{-1}(1/4)\operatorname{dist}(x_0, \partial D) \\ &\geq \Theta^{-1}(1/4)\delta_D(A)/c_2 \geq \Theta^{-1}(1/4)\delta_D(b, x)/c_2 \\ &= r/M_4, \\ \delta_D(f^{-1}(C_0), f^{-1}(\overline{B^n(y_0, 1/4)})) &\leq \delta_D(x, x_0) \leq 2r, \end{aligned}$$

the pair $(f^{-1}(C_0), f^{-1}(\bar{B}^n[y_0, 1/4]))$ is in $2M_4$-standard position in D which is ϕ-broad, and so we obtain $M(\Gamma) \geq \phi(2M_4)$. Since $M(\Gamma) \leq KM(\Gamma')$, this and (7.43) yield $v \geq q_3$, a constant depending only on the data. This finishes Case 5, and with it the proof of the theorem is complete. □

7.5.7. Proof for Väisälä's mapping theorem. We begin the proof of the mapping theorem by introducing a class of mappings—essentially the locally bilipschitz mappings.

A homeomorphism $f : D \to D'$ between domains of $\mathbb{R}^n$ is of L-*bounded length distortion*, abbreviated L-BLD, if

$$\frac{1}{L}\,\ell(\alpha) \leq \ell[f(\alpha)] \leq L\,\ell(\alpha) \tag{7.44}$$

for every path α in D. Here, as usual, $\ell(\alpha)$ denotes the length of the path α. This condition is equivalent to assuming that each point in D has a neighbourhood in which f is L-bilipschitz.

What is going to be useful here is that these maps behave well with respect to products whereas quasiconformal mappings do not. We have the following elementary lemma for which we leave the reader to provide a proof.

LEMMA 7.5.12. *Let* $f : D_1 \subset \mathbb{R}^n \to \mathbb{R}^n$ *and* $g : D_2 \subset \mathbb{R}^m \to \mathbb{R}^m$ *be* L_1 *and* L_2*-BLD respectively. Then*

$$f \times g : D_1 \times D_2 \to \mathbb{R}^{n+m}$$

is L*-BLD with* $L = \max\{L_1, L_2\}$.

Clearly an L-BLD homeomorphism is K-quasiconformal with $K = L^n$.

The mapping theorem that we want to prove, Theorem 7.5.1, is an obvious consequence of the following more general theorem.

THEOREM 7.5.13. *Let G be a simply connected proper subdomain of $\mathbb{R}^2$. Then the following conditions are equivalent:*

1. *There is a K-quasiconformal mapping $B^3 \to \Omega \times \mathbb{R}$.*
2. *Ω is finitely connected on the boundary and satisfies a c-internal chord-arc condition.*
3. *There is an L-BLD homeomorphism $\Omega \to \mathbb{D}$ (the unit disk) if Ω is bounded or $\Omega \to \mathbb{H}^2$ if Ω is unbounded.*
4. *There is an L-BLD homeomorphism $\Omega \times \mathbb{R} \to \mathbb{D} \times \mathbb{R}$ or $\mathbb{H}^2 \times \mathbb{R}$ as above.*

Here the constants K, c, and L depend only on each other.

The implications (3) $\Rightarrow$ (4) and (4) $\Rightarrow$ (1) are simple. The implication (1) $\Rightarrow$ (2) is where the bulk of the work is. The implication (2) $\Rightarrow$ (3) is really part of the theory of quasiconformal mappings in the plane, and so we now only discuss the problem lightly.

(2) $\Rightarrow$ (3). For the less general case of bilipschitz mappings, a geometric characterisation of those domains Ω in $\mathbb{R}^n$ which are bilipschitz equivalent to the unit disk was given in the early eighties by P. Tukia, [**153, 154**], D.S. Jerison and C.E. Kenig [**88**], and T.G. Latfullin [**96**]. They all prove various versions of the following theorem.

THEOREM 7.5.14. *A bounded planar domain Ω is bilipschitz homeomorphic to the disk $\mathbb{D}$ if and only if the boundary of Ω is a rectifiable Jordan curve satisfying the internal chord-arc condition: There is a constant $c > 1$ such that*

$$\delta_\Omega(x, y) \leq c|x - y|. \tag{7.45}$$

The result we want, generalising the above by considering locally bilipschitz or BLD-mappings, was proved in 1987 by J.Väisälä [**163**, Theorems 3.4 and 3.8].

THEOREM 7.5.15. *A simply connected planar domain Ω is BLD-homeomorphic to $\mathbb{H}^2$ if and only if Ω is unbounded, finitely connected on the boundary, and satisfies the internal chord-arc condition.*

A simply connected planar domain Ω is BLD-homeomorphic to $\mathbb{D}$ if and only if Ω is bounded, finitely connected on the boundary, and satisfies the internal chord-arc condition.

We will sketch a proof of this result when Ω is unbounded; the bounded case follows by reworking this proof in a fairly straightforward way and is left to the reader. The "only if" part of the proof is clear, so we discuss the converse.

Choose a conformal map $f_1 : \mathbb{H}^2 \to \Omega$. The hypotheses allow a homeomorphic extension $f_1 : \overline{\mathbb{H}} \to \Omega^*$ (Ω together with its prime ends) and we can assume $f_1(\infty) = \infty$. Since $\partial^*\Omega$ is locally rectifiable from the internal chord-arc condition, there is a homeomorphism $g : \mathbb{R} \to \partial^*\Omega \setminus \{\infty\}$ such that for all $x, y \in \mathbb{R}$,

$$\sigma_\Omega[g(x), g(y)] = |x - y|,$$

where $\sigma_\Omega(a, b)$ is the length of the boundary of Ω between finite points a, b. We can choose g such that

$$s = g^{-1} \circ f_1 : \mathbb{R} \to \mathbb{R}$$

is increasing. Next it can be shown using modulus estimates and the internal chord-arc condition that $s : \mathbb{R} \to \mathbb{R}$ is quasisymmetric (this is quite close to Ahlfors' proof relating the bounded turning condition and quasicircles we alluded to earlier).

We may therefore extend s by the Beurling-Ahlfors construction in (8.1) to a quasiconformal homeomorphism $f_2 : \mathbb{H}^2 \to \mathbb{H}^2$ which is locally bilipschitz in the hyperbolic metric. Finally, $f = f_1 \circ f_2^{-1} : \mathbb{H}^2 \to \Omega$ can be shown to be the desired BLD-homeomorphism. The hard part here is to show, again using modulus estimates and the internal chord-arc property, that there is a uniform constant such that if $z = x + iy \in \mathbb{H}^2$, then

$$\frac{y}{M} \leq \operatorname{dist}[f(z), \partial\Omega] \leq M\, y.$$

Since the hyperbolic metric of Ω and the quasi-hyperbolic metric $|dz|/\operatorname{dist}(z, \partial\Omega)$ are comparable in the simply connected domain Ω, we would therefore obtain from this the required local bilipschitz estimate.

We now return to the proof of Theorem 7.5.13.

(1) $\Rightarrow$ (2). We may replace B^3 by its Möbius equivalent $\mathbb{H}^3$ so as to assume that there is a K-quasiconformal map

$$f : \mathbb{H}^3 \to \Omega \times \mathbb{R} = D.$$

We first show that Ω is a c_1-John domain. Here and later, we let $c_1, c_2 \dots$ and $q_1, q_2, \dots$ denote constants depending only on K with $c_j \geq 1$ and $0 < q_j < 1$.

Let $x_0 \in \mathbb{R}^2$, let $r > 0$, and suppose that $B^2(x_0, r) \setminus \Omega$ has two components E_1 and E_2 both meeting $B^2(x_0, r/c)$. A standard characterisation of John domains shows us that it suffices to find an upper bound $c \leq c_2$ depending only on K.

Now $E_1 \times \{0\}$ and $E_2 \times \{0\}$ lie in different components of $B^3[(x_0, 0), r] \setminus \Omega$. Thus Gehring and Väisälä's distortion estimate that we established in (7.13) gives the upper bound $c \leq e^{MK}$, where $M = \sqrt{\Psi(1/2)/4\pi}$ is a universal constant.

So Ω is a c_1-John domain. It follows that Ω is finitely connected on the boundary by Lemma 7.5.6.

We separately consider the cases where Ω is bounded or unbounded.

Ω is bounded. We extend f to a homeomorphism $f* : \overline{\mathbb{H}^3} \to \overline{\Omega}$ and put $D^* = (\Omega^* \times \mathbb{R}) \cup \{-\infty, +\infty\}$. Performing an auxiliary Möbius transformation of $\mathbb{R}^3$ we may assume that $f*(0) = -\infty$ and $f*(\infty) = +\infty$. We have already proved that Ω is a John domain, so we can assume that Ω has the c_1-carrot property with center $x_0 \in \Omega$. We normalize so that $\operatorname{dist}(x_0, \partial\Omega) = 1$. Then

$$\delta_\Omega(x, x_0) < c_1 \tag{7.46}$$

for all $x \in \Omega$.

For $r > 0$, set $S^+(r) = \mathbb{H}^3 \cap \mathbb{S}^2(r)$. Let $r_0 \leq r_1$ be the endpoints of the projection of $f[S^+(r)]$ into the x_3-axis—this may be an interval or a point.

A straightforward distortion estimate gives the uniform bound

$$r_1 - r_0 < c_3; \tag{7.47}$$

in fact $c_3 = \sqrt{Km(\Omega)/\Psi(1)}$ will work, following [**54**, Lemma 8.1].

Next, we choose a pair of spheres $\mathbb{S}^2(r)$ and $\mathbb{S}^2(r')$ so that the projection of $f(S^+(r))$ into the x_3-axis lies in $(-\infty, 0]$ and the projection of $f(S^+(r'))$ into the x_3-axis lies in $[1, \infty)$. Again, by normalising via a similarity transformation, we may assume $r = 1$.

We now want to apply Theorem 7.5.11 to the mapping $f : \mathbb{H}^3 \to D$ with $A = \mathbb{H}^3 \cap (B^3(r') \setminus \bar{B}^3(r))$, and in order to apply this theorem we must show that $f(A)$ has the carrot property.

Choose $z_0 \in \mathbb{H}^3 \cap S^2$ such that $y_0 = f(z_0)$ is of the form (x_0, t_0), and hence by construction $r_0 \le t_0 \le 0$. We now show that the image $f(A)$ has the carrot property in D with center y_0, where the constant c_4 can be chosen to be $c_4 = 1 + c_1 + 2c_3$.

Assume that $y_1 = (x_1, t_1) \in f(A) \subset \Omega \times [r_0, r_1']$. Join x_1 to x_2 by a 2-dimensional carrot $car_d(E_0, c) \subset \Omega$ and put

$$E_1 = E_0 \times \{t_1\}, \quad E_2 = \{x_0\} \times [t_0, t_1], \quad \text{and } E - E_1 \cup E_2.$$

Then E is an arc joining y_0 to y_1. If $y \in E$ and if $\delta(y) = \operatorname{diam}(E[y_1, y])$, we should verify that $B^3[y, \delta(y)/c_4] \subset D$.

If $y \in E_1$, we can write $y = (x, t_1)$ with $B^2(x, \delta(y)/c_1) \subset \Omega$. Hence

$$B^3(y, \delta(y)/c_4) \subset B^3(y, \delta(y)/c_1) \subset \Omega \times \mathbb{R} = D.$$

If $y \in E_2$, we write $y = (x_0, t)$, and (7.47) implies that

$$\delta(y) \le \delta_\Omega(E) \le \delta_\Omega(E_1) + r' - r_0 \le \delta_\Omega(E_1) + 1 + 2c_3.$$

Since $\operatorname{dist}(x_0, \partial\Omega) = 1$, $\delta_\Omega(E_0) \le c_1$ and hence $\delta(y) \le 1 + c_1 + 2c_3 = c_4$. Thus

$$B^2(y, \delta(y)/c_4) \subset B^3(y, 1) \subset B^2(x_0, 1) \times \mathbb{R} \subset \Omega.$$

It follows that $f(A)$ has the desired property in Ω.

We still need to provide an upper bound for $\operatorname{diam}(A)/\operatorname{dist}(z_0, \partial\mathbb{H}^3)$ so we put $s = \operatorname{dist}(z_0, \partial\mathbb{H}^3)$ and observe that $\operatorname{diam}(A) = 2r'$ so we are reduced to showing that $r' \le c_5 s$, and we will do this with modulus estimates as follows.

Let Γ be the family of paths joining the upper hemispheres $S^+(1)$ and $S^+(r')$ in A. Then $M(\Gamma) = 2\pi \log^{-1} r'$. Since $B^n(x_0, 1) \subset \Omega \subset B^n(x_0, c_1)$, the usual modulus estimates with (7.47) give

$$\frac{\pi}{(1+2c_3)^2} \le M[f(\Gamma)] \le \pi c_1^2.$$

Since f is K-quasiconformal, we obtain

$$1 + q_1 \le r' \le c_6 \tag{7.48}$$

so that we are reduced to showing $s \ge q_2$.

We may assume that $s \le q_1$ (in (7.48) above); otherwise we are done. Let C_0 be the vertical segment of length s joining z_0 to $\partial\mathbb{H}^3$, let $C_1 = S^+(r')$ be the upper half-sphere, and let $\Gamma_1 = \Delta(C_0, C_1; \mathbb{H}^3)$. Then

$$M(\Gamma_1) \le 4\pi \log^{-2}\left(\frac{q_1}{s}\right). \tag{7.49}$$

Set $r'' = 1 - q_1$. In more or less the same way as the arguments above using path families, we can get the estimate $r_0'' \ge -c_7$, a constant depending on the data. Then $f(C_0)$ lies between the planes $\{x_3 = -c_7\}$ and $\{x_3 = 0\}$. Moreover, $f(C_1)$ lies between the planes $\{x_3 = 1\}$ and $\{x_3 = r_1' < l + c_3\}$.

Let Z be the cylinder $B^2(x_0, 1) \times [-c_7, 1 + c_3] \subset D$. Then there are continua $C_0' \subset Z \cap f(C_0)$ and $C_1' \subset Z \cap f(C_1)$ both with diameters at least $1/2$. Then comparing curve families in Z and $\mathbb{R}^3$ we easily see that

$$M(f(\Gamma)) \ge M(\Delta(C_0', C_1'; Z)) \ge M(\Delta(C_0', C_1'; \mathbb{R}^3))/c_8$$

for some universal constant c_8. This gives $M(f(\Gamma)) \ge q_3$, and since $M[f(\Gamma)] \le KM(\Gamma_1)$ we achieve our desired estimate of a lower bound for s.

We have now verified all hypotheses of Theorem 7.5.11, and so it follows that $f|A$ is η_1-quasisymmetric with respect to the Euclidean metric of A and the metric

δ_Ω of $f(A)$. Then $(f^*)^{-1} : \overline{f(A)} \to \mathbb{H}^3$ is η_2-quasisymmetric in the metric δ_Ω which is 2-bilipschitz equivalent to the product metric of $\Omega^* \times \mathbb{R}$.

This means the restriction $f_1 : \partial^*\Omega \times [0,1] \to \mathbb{R}^2$ of $(f^*)^{-1}$ is η_3-quasisymmetric in this product metric. It follows that the Jordan curve $\partial^*\Omega$ is a c_9-chord-arc in δ_Ω.

Finally, by Lemma 7.5.4 this means that Ω is a c_9-internal chord-arc.

Ω is unbounded. We again extend f to a homeomorphism and normalise it as above, but this time we use $A = \mathbb{H}^3$ in Theorem 7.5.11.

If $y = (x,t) \in \Omega$, as Ω is a c_1-John domain, there is a 2- dimensional carrot $car_d(E, 2c_1)$ joining x to ∞ in Ω. Then $car_d(E \times \{t\}, 2c_1)$ joins y to ∞ in Ω. Hence Ω has the $2c_1$-carrot property in Ω. We then apply Theorem 7.5.11 and conclude that f is quasisymmetric in δ_Ω and the remainder of the proof follows as above.

7.6. Quasiconformal homogeneity

We close this chapter with a brief discussion of quasiconformal homogeneity. This concept was first introduced by F.W. Gehring and B.P. Palka in their paper [**52**] in an initial attempt to characterize quasiballs. It is motivated by the following observation on conformally homogeneous domains basically proved by B. Kimel'fel'd [**91**].

THEOREM 7.6.1. *Let D be a domain in $\hat{\mathbb{R}}^n$, $n \geq 3$. Suppose that the conformal automorphisms of D act transitively on D—meaning that given $x, y \in D$ there is a conformal self-mapping $\phi : D \to D$ such that $\phi(x) = y$. Then ∂D has 0, 1, or 2 components. Further, either $D = \hat{\mathbb{R}}^n$, $D = \mathbb{R}^n$, there is a k, $0 \leq k \leq n-2$ such that D is conformally equivalent to $\hat{\mathbb{R}}^n \setminus \mathbb{S}^k$, or in the case $k = n-1$, D is conformally equivalent to B^n.*

PROOF. We leave the reader to deal with the case of $n = 2$ which, while quite different in character, is elementary. In dimension $n \geq 3$ the Liouville theorem asserts that each $\phi : D \to D$ is actually the restriction of a Möbius transformation of $\hat{\mathbb{R}}^n$. Thus the conformal automorphisms of D form a group. This group will be Lie subgroup of SO$(1,n)$. These were completely classified by L. Greenberg in 1962, [**60**], and the resulting list of possible orbits of a point is presented. It is easy to see that each example listed is in fact conformally homogeneous. □

This proof, relying as it does on the classification, is a bit unsatisfying. However, even in the case where we know D to be topologically equivalent to a ball, it seems not so easy to show that D is a round ball. Notice though that every Möbius transformation of D extends to the boundary of D (indeed $\hat{\mathbb{R}}^n$) and it is not difficult to deduce that ∂D is the closure of an orbit, and so it is a real analytic submanifold by results of D. Montgomery and L. Zippin [**119**]. Once this is known, the fact that D is a ball or half-space is straightforward as we will see.

Now Theorem 7.6.1 above suggests a possible route to identifying quasiballs. Perhaps it is enough that they be homogeneous with regard to some restricted family of quasiconformal transformations? The first thing to note is that it is easy to prove that any domain is quasiconformally homogeneous—at least in one sense.

THEOREM 7.6.2. *Let D be a domain and x', y' be points of D. Then there is a quasiconformal mapping $g : D \to D$ such that $g(x') = y'$.*

PROOF. First we recall that for any $y \in B^n$, there is a mapping $h_y : B^n \to B^n$, $h_y|\mathbb{S}^{n-1} = \mathit{identity}$, and $h_y(0) = y$, with h being K-quasiconformal. This map—with obvious modifications—is described in (6.62) in the case of the half-space $\mathbb{H}^n$, and it has the distortion $K^{1/(n-1)} \leq \exp \rho_{\mathbb{H}}(0, y) < \infty$. Then any two points x', y' of D lie in a smoothly bounded topological ball $B^* = g(B^n)$ with g a diffeomorphism—for instance take a smooth regular neighbourhood of a smooth arc in D connecting x' to y'. We put $x = g^{-1}(x')$ and $y = g^{-1}(y')$. The map we seek is

$$g \circ h_y \circ h_x^{-1} \circ g^{-1} : g(B^n) \to g(B^n) \subset D,$$

and we trace out that $x' \mapsto x \mapsto 0 \mapsto y \mapsto y'$ and also that this map is the identity on the boundary of $g(B^n)$. It therefore extends by the identity to be defined throughout D. The resulting homeomorphism is easily seen to be quasiconformal. □

Thus the characterisation of quasiballs we seek could only be true with additional hypotheses. The key point to the proof of Theorem 7.6.1 is that the collection of conformal self-mappings of a domain forms a Lie group and is therefore locally compact. The family of quasiconformal self-mappings of a domain does form a group, but it is not locally compact and is therefore not a Lie group. Locally compact groups of quasiconformal mappings are in fact Lie groups, [**106**], but there seems no chance of a classification theorem for them except possibly in low dimensions—however one could explore this further. Here we prefer to consider the following generalisation which retains as much compactness as we could expect. A domain D is said to be *K-quasiconformally homogeneous* if for each pair of points x, y of D, there is a K-quasiconformal mapping $f : D \to D$ with $f(x) = y$.

The next theorem suggests that we are on the right track. It shows that the topological structure of the boundary is similar to what we might expect.

THEOREM 7.6.3. *Let D be a K-quasiconformally homogeneous domain in $\hat{\mathbb{R}}^n$. Then ∂D has 0, 1, 2, or infinitely many components. In the case of two components, each of them is a point.*

PROOF. We may suppose that $0 \in D$ and that ∂D has finitely many components. Let $\{x_n\}_{n=1}^{\infty} \subset D$ be a sequence such that $x_n \to x \in \partial D$. For each n, let f_n be a K-quasiconformal self-mapping of D for which $f_n(0) = x_n$ is guaranteed to us by the definition of K-quasiconformal homogeneity. The kernel convergence Theorem 6.6.25 assures us that the sequence f_n converges locally uniformly in D to a constant $x \in \partial D$ and moreover, examining that theorem in our case, shows that a constant limit is a possibility only if D^c has at most two points or if D^c is a continuum, since we have excluded the possibility of infinitely many components. □

This leaves two cases of interest: that where ∂D is a continuum and that where ∂D is infinitely connected. We address the former case and leave the reader to explore the latter to find out for themselves that very little of interest can be said.

What we shall now show is that if ∂D is a continuum and if this boundary is "nice" at a single point, then we can identify D up to a quasiconformal mapping. Therefore, let D be a domain in $\mathbb{R}^n$ and $x_0 \in \partial D$. We say that ∂D admits an m-tangent hyperplane at x_0 if there is a sequence of real numbers $\lambda_i \to \infty$ and orthogonal transformations O_i so that $\lambda_i O_i(\partial D - x_0)$ converges to $\mathbb{R}^m$ where we view $\mathbb{R}^m \subset \mathbb{R}^n$ via the inclusion. We ask that this convergence hold in the spherical Hausdorff metric,

$$d_{\mathcal{H},\sigma}(E,F) = \max\big\{\inf_{x\in E} \operatorname{dist}_\sigma(x,F), \inf_{y\in F} \operatorname{dist}_\sigma(y,E)\big\}.$$

This sort of convergence certainly implies that the kernel of the sequence of domains $\varphi_i(D)$ is $\mathbb{R}^n \setminus \mathbb{R}^m$, or if $m = n-1$, the kernel of the sequence is $\mathbb{H}^n$, when we set $\varphi_i(x) = \lambda_i O_i(x - x_0)$.

The proof of the next theorem has many clear adaptions and extensions. We establish the most obvious of these.

THEOREM 7.6.4. *Let D be a K-quasiconformally homogeneous domain with an m-tangent at $x_0 \in \partial D$. Then D is quasiconformally equivalent to $\mathbb{R}^n \setminus \mathbb{R}^m$, $m < n-1$, or when $m = n-1$, D is quasiconformally equivalent to $\mathbb{H}^n$, and hence a quasiball.*

PROOF. We may suppose that $e_n \in D$ and that $0 \in \partial D$ admits an m-tangent with a sequence φ_i defined as above. If $\varphi^{-1}(e_n) \notin D$ except for finitely many i, then ∂D does not admit an m-tangent, so passing to a subsequence if necessary, we may assume $\varphi^{-1}(e_n) \in D$ for all i. Then, using the assumption that D is K-quasiconformally homogeneous, let $f_i : D \to D$ be K-quasiconformal with $f(e_n) = \varphi_i(e_n)$. Let $h_i(x) = (\varphi_i \circ f_i)(x)$. Then h_i is K-quasiconformal, $h_i(e_n) = e_n$, $e_n \in Ker(\{\varphi_i(D)\}) = \mathbb{R}^n \setminus \mathbb{R}^m$, and the kernel convergence Theorem 6.6.25 assures us that h_i converges locally uniformly in D to a quasiconformal mapping onto the kernel $\mathbb{R}^n \setminus \mathbb{R}^m$. □

COROLLARY 7.6.5. *If D is K-quasiconformally homogeneous and if ∂D is quasiconformally collared at some point $x_0 \in \partial D$, then D is a quasiball.*

PROOF. If ∂D is collared at x_0, then after a global quasiconformal change of coordinates (as per Corollary 7.4.9) D becomes a domain which admits an $(n-1)$-tangent at (the image of) x_0. From this the result follows. □

Finally, and unfortunately, we note that P. Tukia has observed in [**156**] (and we reported in (7.3.9) above) that the complementary domains of Rickman's rug are K-quasiconformally homogeneous (with respect to a quasiconformal group in fact) and are not quasiballs. Also J. McKemie [**115**] has shown that it is possible to modify Tukia's examples so as to achieve K as close to 1 as we would like, but of course not equal to 1.

Nevertheless the theory of quasiconformally homogeneous spaces has life and there are some quite surprising results about it. For instance it is known that there is a constant $K_n > 1$ such that if M^n is a K-quasiconformally homogeneous hyperbolic n-manifold, then $K \geq K_n$ [**20**]. Surprisingly this is still unknown in two dimensions, although there are many partial results which the reader can find in the survey [**21**]. Another aspect of quasiconformal homogeneity which has attracted the attention of researchers concerns the quasiconformal homogeneity of various sorts of continua. Here one studies continua $X \subset \mathbb{R}^n$ which have the property that there is $K < \infty$ such that for each $x, y \in X$ we can find a K-quasiconformal mapping $f : \mathbb{R}^n \to \mathbb{R}^n$ with $f(X) = X$ and $f(x) = y$. See for instance the recent paper [**131**] and the references therein.

CHAPTER 8

The Tukia-Väisälä Extension Theorem

In this chapter we return to a question that raised itself earlier in this book.

Given a quasiconformal self-mapping f of $\mathbb{R}^n$—when $n = 1$ and for convenience we use "quasiconformal" as a synonym for "quasisymmetric"—we asked: "Does there exist a quasiconformal mapping F of $\mathbb{H}^{n+1}$ onto itself whose homeomorphic extension F^* to $\bar{\mathbb{H}}^{n+1}$ coincides with f in $\mathbb{R}^n$?"

The case $n = 1$ of this problem was settled by Lars Ahlfors and Arne Beurling in a classic paper from 1956, [**11**]. Surprisingly the solution involved an explicit integral formula for a mapping F with the desired properties.

In the situation where f is increasing on $\mathbb{R}$, the *Beurling-Ahlfors extension* of f takes the form

$$F(x,y) = (u(x,y), v(x,y)), \tag{8.1}$$

with

$$u(x,y) = \frac{1}{2}\int_0^1 \big[f(x+ty) + f(x-ty)\big]\,dt$$

and

$$v(x,y) = \frac{1}{2}\int_0^1 \big[f(x+ty) - f(x-ty)\big]\,dt\,.$$

As a matter of fact, the Beurling-Ahlfors extension F of a quasisymmetric map f is a quasiconformal diffeomorphism of $\mathbb{H}^2$ and can be modified so as to show that any quasisymmetric homeomorphism of the real line actually admits a real-analytic quasiconformal extension.

In 1964 Ahlfors provided an affirmative response to the extension question in dimension two, [**6**].

His answer consisted in first giving an integral representation for a quasiconformal extension to $\mathbb{H}^3$ of any K-quasiconformal self-mapping of $\mathbb{R}^2$ with K suitably close to 1. He then appealed to the fact that an arbitrary quasiconformal mapping of $\mathbb{R}^2$ to itself can be realized as the finite composition of $(1+\varepsilon)$-quasiconformal mappings of $\mathbb{R}^2$ for any preassigned $\varepsilon > 0$. It is still unknown whether a factorization of this sort is possible for quasiconformal mappings of $\mathbb{R}^n$ for dimensions $n \geq 3$.

By combining renormalization arguments with some piecewise linear topology, L. Carleson dealt with the case $n = 3$ in 1974, at the same time contributing new proofs for the lower-dimensional cases [**28**]. Carleson's treatment of the three-dimensional situation hinged on a piecewise linear approximation result of E. Moise [**116**] that does not carry over to dimension four.

It took another 10 years before the general case of the problem was solved by P. Tukia and J. Väisälä [**158**]. They adopted and refined Carleson's renormalization technique, but they replaced the elements of piecewise linear topology in his approach with the more flexible Lipschitz methods that D. Sullivan had been developing over the previous years in order to solve other outstanding problems in the area. These methods are carefully developed and explained in [**157**].

The presentation of the Tukia-Väisälä extension theorem that follows is a blend of material from their 1982 paper [**158**] with simplifications suggested in their 1984 sequel [**159**]. The proof we give is complete apart from Sullivan's approximation theorem. This theorem follows the lines of the "furling" technique from geometric topology developed by R. Edwards and R. Kirby [**35**] and Kirby and L. Siebenmann [**92**]. In that case a key idea involves using that fact that the n-torus minus a point $\mathbb{S} \times \mathbb{S} \times \cdots \times \mathbb{S} \setminus \{x\}$ immerses in $\mathbb{R}^n$. Sullivan mimics that construction but uses hyperbolic geometry. For this he needs the existence of a compact hyperbolic n-manifold which immerses in $\mathbb{R}^n$ once a point is deleted. The technical term is parallelizable in the complement of a point. The existence of these manifolds is quite nontrivial—joint work of Sullivan and J.P. Serre—and obviously well outside the scope of this book, and so we leave the issue well alone (as did Tukia and Väisälä).

8.1. Lipschitz embeddings

As we have just suggested, the keystone of the Tukia-Väisälä scheme for extending a quasiconformal mapping from one dimension to the next is Dennis Sullivan's work on Lipschitz embeddings [**150, 157**].

Unfortunately, and as noted above, the limited scope and confines of this book do not afford us the luxury of including an in-depth account of Sullivan's ideas. We must be content to summarize and accept those of his results that are pertinent to the matter at hand, and to proceed from there.

Let X and X' be metric spaces whose distance functions we denote by d and d', respectively. A function $f : X \to X'$ is termed *bilipschitz* provided there is a constant $\lambda \geq 1$ such that

$$\lambda^{-1} d(p,q) \leq d'\big[f(p), f(q)\big] \leq \lambda d(p,q)$$

for all points p and q of X. If we wish to be explicit about λ, we shall refer to f as *λ-bilipschitz*.

We say that f is *locally λ-bilipschitz* when each point p of X has an open neighbourhood $U = U_p$ in X with the property that the restriction of f to U is λ-bilipschitz.

Assume now that $f : X \to X'$ is an embedding, meaning that f is a homeomorphism of X onto $f(X)$. We call f a *Lipschitz embedding*—or, to use the language favoured by those who regularly work with such mappings, a *LIP-embedding*—if each point p of X has an open neighbourhood $U = U_p$ in X such that the restriction of f to U is bilipschitz but not necessarily λ-bilipschitz with fixed λ.

We stress that a LIP-embedding is not required in this usage of the term to satisfy a global Lipschitz condition on X—the concept is purely a local one. A *LIP-homeomorphism* from X onto X' is a surjective LIP-embedding from X to X'.

Suppose that $f : X \to X'$ and $g : X' \to X''$ are LIP-embeddings. Then $g \circ f$ and f^{-1}, which is defined in $f(X)$, are also LIP-embeddings. If A is a relatively compact subset of X, then the restriction of f to A is locally λ-bilipschitz for some λ that is determined by f and A.

Almost exclusively we shall be concerned with Euclidean LIP-embeddings $f : U \to \mathbb{R}^n$, where U is an open set in $\mathbb{R}^n$, and hyperbolic LIP-embeddings $f : U \to \mathbb{H}^n$, where U is an open set in $\mathbb{H}^n$. We register in the form of lemmas some preparatory remarks about such mappings. The first is an immediate consequence of the following fact from (3.41): if E is a compact set in $\mathbb{H}^n$, then there are constants $a > 0$ and $b > 0$, which depend only on E, such that

$$a|x-y| \le d_{\mathbb{H}}(x,y) \le b|x-y| \tag{8.2}$$

whenever x and y are points of E.

We remind the reader that, by the notational conventions we have used, the symbol $\bar{A}$ for a subset A of $\mathbb{R}^n$ means the closure of A in $\hat{\mathbb{R}}^n$. Thus $\bar{A}$ is always a compact set.

LEMMA 8.1.1. *Let U be a nonempty open set in $\mathbb{H}^n$ such that $\bar{U}$ lies in $\mathbb{H}^n$ and let $f : U \to E$ be an embedding in which E is a compact subset of $\mathbb{H}^n$. Then f is (locally) λ-bilipschitz for some λ with respect to the Euclidean metric if and only if f is (locally) λ'-bilipschitz for some λ' with respect to the hyperbolic metric. In fact, if λ is known, then an appropriate λ' is determined by λ, U, and E, and vice versa.*

In particular, an embedding $f : V \to \mathbb{H}^n$, where V is an open set in $\mathbb{H}^n$, is a Euclidean LIP*-embedding if and only if it is a hyperbolic LIP-embedding.*

The second lemma brings to the hyperbolic context a fact we were already aware of in the Euclidean setting.

LEMMA 8.1.2. *Suppose that D is a subdomain of $\mathbb{H}^n$ and that f is an embedding of D into $\mathbb{H}^n$ which is locally λ-bilipschitz with respect to the hyperbolic metric. Then f is a λ^{2n-2}-quasiconformal mapping of D onto $D' = f(D)$.*

PROOF. It suffices to check that $f|B$ is λ^{2n-2}-quasiconformal whenever B is an open ball such that $\bar{B}$ is contained in D and $f|B$ is λ-bilipschitz in the hyperbolic metric. Fix a ball B of this type. Lemma 8.1.1 implies that $f|B$ is bilipschitz with respect to the Euclidean metric and hence quasiconformal by Lemma 6.2.3.

For any point x of B at which f is differentiable with $J_f(x) \ne 0$, we compute

$$\begin{aligned} \|f'(x)\| &= \limsup_{h\to 0} \frac{|f(x+h)-f(x)|}{|h|} \\ &= \limsup_{h\to 0} \left(\frac{|f(x+h)-f(x)|}{d_{\mathbb{H}}[f(x+h),f(x)]} \cdot \frac{d_{\mathbb{H}}[f(x+h),f(x)]}{d_{\mathbb{H}}(x+h,x)} \cdot \frac{d_{\mathbb{H}}(x+h,x)}{|h|} \right) \\ &\le \frac{\lambda f_n(x)}{x_n}. \end{aligned}$$

Similarly, for any such x we obtain

$$\ell[f'(x)] \ge \frac{f_n(x)}{\lambda x_n},$$

whence

$$H_f(x) = H\big[f'(x)\big] = \frac{\|f'(x)\|}{\ell[f'(x)]} \le \lambda^2 .$$

Since the last inequality is true for almost every point x of B, Corollary 6.4.19 yields $K(f|B) \le \lambda^{2n-2}$, as desired. □

Finally, we point out one situation in which a locally λ-bilipschitz mapping is automatically λ-bilipschitz on a global scale.

LEMMA 8.1.3. *Suppose that f is a homeomorphism of a hyperbolically convex subdomain D of $\mathbb{H}^n$ onto a domain D' of the same type. If f is locally λ-bilipschitz with respect to the hyperbolic metric, then f is λ-bilipschitz in that metric.*

PROOF. Let x and y be distinct points of D, and let A be the hyperbolic geodesic segment joining x and y. We choose points $x = z_0, z_1, \ldots, z_{r-1}, z_r = y$ in sequence along A in such a way that f is hyperbolically λ-lipschitz on each of the hyperbolic segments into which these points partition A. Then

$$d_{\mathbb{H}}\big[f(x), f(y)\big] \le \sum_{j=1}^{r} d_{\mathbb{H}}\big[f(x_{j-1}), f(x_j)\big] \le \lambda \sum_{j=1}^{r} d_{\mathbb{H}}(x_{j-1}, x_j) = \lambda d_{\mathbb{H}}(x, y) .$$

We conclude that f is hyperbolically λ-lipschitz in D. As the same is true of f^{-1} in D', f is a λ-bilipschitz mapping with respect to the hyperbolic metric. □

Lemma 8.1.3 has an obvious Euclidean analogue.

8.1.1. Sullivan's theorem. Sullivan's theory of LIP-embeddings contains much more information than is actually required to deal with the extension problem for quasiconformal mappings.

Following Tukia and Väisälä's presentation of this material, we state—and subsequently treat as a given—a quantitative version of the particular Sullivan result that gives precisely the information we need. A few words concerning notation: if S is a set and $\varphi : S \to \mathbb{R}^n$, then for any nonempty subset A of S we set $\|\varphi\|_A = \sup\{|\varphi(x)| : x \in A\}$; we denote the identity mapping of $\mathbb{R}^n$ by *id*.

THEOREM 8.1.4. *Suppose that U is an open set in $\mathbb{R}^n$, C is a compact subset of U, V is a neighbourhood of C in U, and $\varepsilon > 0$. Then there is a $\delta > 0$, which depends only on U, C, V, and ε, such that for each* LIP*-embedding $f : V \to \mathbb{R}^n$ satisfying $\|f - \text{identity}\|_V \le \delta$, there exists a* LIP*-homeomorphism $f^* : \mathbb{R}^n \to \mathbb{R}^n$ endowed with the following properties:*

(i) $\|f^* - \text{identity}\|_{\mathbb{R}^n} \le \varepsilon$;
(ii) $f^* = f$ *on* C;
(iii) $f^* = \text{identity}$ *in* $\mathbb{R}^n \setminus V$.

Furthermore, if f is locally λ-bilipschitz, then f^* can be chosen so that it is λ^*-bilipschitz for some λ^* which is solely a function of λ and n.

Theorem 8.1.4 and its proof are found in [**157**], which contains a wealth of material related to the quasiconformal extension problem and also provides a reasonably self-contained exposition of the requisite Sullivan theory.

We shall say that an embedding $g : U \to \mathbb{R}^n$, U being an open set in $\mathbb{R}^n$, is *uniformly approximable by* LIP*-embeddings* if for each $\varepsilon > 0$ there exists a LIP-embedding $h : U \to \mathbb{R}^n$ for which $\|h - g\|_U < \varepsilon$.

We apply Theorem 8.1.4 to derive a rather technical—but, as things turn out, crucial—approximation and smoothing result.

LEMMA 8.1.5. *Let U, U_0, V and W be open sets in $\mathbb{H}^n$ that exhibit the following relationships:*

$$\bar{U} \subset \mathbb{H}^n\,, \quad W \subset V \subset U\,, \quad U \cap \bar{W} \subset V\,, \quad \emptyset \neq \bar{U}_0 \subset U\,.$$

Assume that $\mathcal{G}$ is a family of embeddings $g : U \to \mathbb{H}^n$ which are uniformly approximable by LIP*-embeddings. Assume, further, that $\mathcal{G}$ is compact in the topology of Euclidean uniform convergence on U and that*

$$E = \overline{\{g(x) : g \in \mathcal{G},\ x \in U\}} \subset \mathbb{H}^n.$$

To each $\varepsilon > 0$ there corresponds a $\delta > 0$, where δ depends on n, ε, $\mathcal{G}$, and the four sets listed above, for which the following statement is true.

If $g \in \mathcal{G}$ and $h : V \to \mathbb{H}^n$ is a LIP*-embedding satisfying $\|h - g\|_V \leq \delta$, then there exists a* LIP*-embedding $\hat{h} : U \to \mathbb{H}^n$ such that $\|\hat{h} - g\|_{U_0} \leq \varepsilon$ and such that $\hat{h} = h$ in $U_0 \cap W$.*

Moreover, if h is locally λ-bilipschitz, then $\hat{h}$ can be chosen so that its restriction to U_0 is locally $\hat{\lambda}$-bilipschitz for some $\hat{\lambda}$ determined by λ, n, ε, $\mathcal{G}$ and the sets U, U_0, V, and W.

PROOF. Fix open sets A and B in U for which

$$\bar{U}_0 \subset A \subset \bar{A} \subset B \subset \bar{B} \subset U\,.$$

This can be done in a manner which makes A and B completely dependent on U and U_0. For instance, $A = \{x : \operatorname{dist}(x, U_0) < r\}$ and $B = \{x : \operatorname{dist}(x, U_0) < 2r\}$, where $r = \operatorname{dist}(U_0, \partial U)/4)$.

Since the family $\mathcal{G}$ is compact and since the function $g \mapsto \operatorname{dist}[g(\bar{A}), \partial g(B)]$ is positive and continuous on $\mathcal{G}$, we can be certain that

$$m = \min\Big\{d\big[g(\bar{A}), \partial g(B)\big] : g \in \mathcal{G}\Big\} > 0\,. \tag{8.3}$$

Let $\varepsilon > 0$ be given. The Arzelá-Ascoli theorem ensures that $\mathcal{G}$ is equicontinuous at each point of U—hence, uniformly equicontinuous on the compact set $\bar{B}$. In particular, we can choose η in $(0, d(\bar{A}, \partial B))$ such that

$$|g(x) - g(y)| \leq \varepsilon/4 \tag{8.4}$$

whenever g belongs to $\mathcal{G}$, x is a point of A, and y satisfies $|y - x| \leq \eta$.

The choice of η is controlled completely by ε, $\mathcal{G}$, U, and U_0. Next we apply Theorem 8.1.4 to the open set $A \cap V$, its compact subset $C = \bar{U}_0 \cap \bar{W}$, the neighbourhood $A \cap V$ of C, and the number $\eta > 0$ to obtain $\zeta > 0$ for which the following assertion is valid: if $f : A \cap V \to \mathbb{R}^n$ is a LIP-embedding with $\|f - \text{identity}\|_{A\cap V} \leq \zeta$, then there exists a LIP-homeomorphism f^* of $\mathbb{R}^n$ onto itself such that

$$\|f^* - \text{identity}\|_{\mathbb{R}^n} \leq \eta\,, \tag{8.5}$$

$$f^* = f \quad \text{in} \quad \bar{U}_0 \cap \bar{W} \tag{8.6}$$

and

$$f^* = \text{identity} \quad \text{in} \quad \mathbb{R}^n \setminus (A \cap V)\,. \tag{8.7}$$

In fact (8.7) carries with it the implication that any such f^* has $f^*(A) = A$ and $f^*(U) = U$.

The ζ arising here is determined by η, U, U_0, V, and W. Thus, ζ is dependent on ε, $\mathcal{G}$, and the sets appearing in the statement of the lemma. If, in addition, f is known to be locally λ-bilipschitz, then we may assume that f^* is λ^*-bilipschitz for some λ^* which is a function of λ and n.

Finally, we choose δ in $(0, m/4)$ so that the following condition is met:

$$|g^{-1}(w) - g^{-1}(z)| \leq \zeta \tag{8.8}$$

whenever g is in $\mathcal{G}$, z is in $g(A)$, and $|w-z| < 4\delta$. Note that by (8.3) any such w lies in $g(B)$, and if this were not possible, we could find sequences $\langle g_\nu \rangle$ in $\mathcal{G}$, $\langle x_\nu \rangle$ in A, and $\langle y_\nu \rangle$ in B for which $|y_\nu - x_\nu| > \zeta$, yet for which $|g_\nu(y_\nu) - g_\nu(x_\nu)| < d/(4\nu+1)$. Passing to subsequences, we could assume that $g_\nu \to g$ uniformly on U, where g is a member of $\mathcal{G}$, while $x_\nu \to x$ and $y_\nu \to y$, with x and y being points of $\bar{B}$ with $|x-y| \geq \zeta$. Thus $x \neq y$, whereas

$$g(x) = \lim_{\nu\to\infty} g_\nu(x_\nu) = \lim_{\nu\to\infty} g_\nu(y_\nu) = g(y)\ ,$$

contradicting the fact that g is an embedding. We can therefore choose δ with the stated property.

We claim that this δ, which depends only on the parameters specified, satisfies the requirements of the lemma.

The compactness of $\mathcal{G}$ entitles us to select a finite set of mappings $g_1, g_2, \ldots, g_p$ from $\mathcal{G}$ with the feature that

$$\min_{1\leq j\leq p} \|g - g_j\|_U < \min\{\varepsilon/8,\ \delta/2\} \tag{8.9}$$

whenever g belongs to $\mathcal{G}$. If we assume, as we may, that the number p of mappings involved here is the minimal one for which relation (8.9) can hold, then p and the mappings $g_1, g_2, \ldots, g_p$ are controlled by $\mathcal{G}$, ε, and δ—hence, by the proper data. By hypothesis we are free to fix for $j = 1, 2, \ldots, p$ a LIP-embedding $\tilde{g}_j : U \to \mathbb{R}^n$ such that

$$\|\tilde{g}_j - g_j\|_U < \min\{\varepsilon/8,\ \delta/2,\ \operatorname{dist}(E, \partial\mathbb{H}^n)\}\,. \tag{8.10}$$

Thus $\tilde{g}_j(U)$ lies in $\mathbb{H}^n$ and, as B is a relatively compact subset of U, we can fix a $\tilde{\lambda}$ for which all of the mappings $\tilde{g}_j$ become locally $\tilde{\lambda}$-bilipschitz when restricted to B.

Now consider a member g of $\mathcal{G}$ and a LIP-embedding $h : V \to \mathbb{H}^n$ that satisfies the condition $\|h - g\|_V \leq \delta$. In view of (8.9) and (8.10) we can pick a j such that $\tilde{g} = \tilde{g}_j$ has

$$\|\tilde{g} - g\|_U < \min\{\varepsilon/4,\ \delta\}\,. \tag{8.11}$$

Moreover,

$$|\tilde{g}^{-1}(w) - \tilde{g}^{-1}(z)| \leq \zeta \tag{8.12}$$

whenever z is a point of $\tilde{g}(A)$ and $|w - z| \leq 2\delta$. To see this, observe that by (8.11)

$$d\big[\tilde{g}(\bar{A}), \partial\tilde{g}(B)\big] > d\big[g(\bar{A}), \partial g(B)\big] - 2\delta \geq m - 2\delta > 4\delta - 2\delta = 2\delta\,.$$

In particular, $\tilde{g}(B)$ includes any point w within a distance 2δ of $\tilde{g}(A)$. Let $z = \tilde{g}(x)$ with x from A, and let w satisfy $|w - z| \leq 2\delta$. Then $w = \tilde{g}(y)$ for a point y of B, and

$$|g(y) - g(x)| \leq |g(y) - \tilde{g}(y)| + |\tilde{g}(y) - \tilde{g}(x)| + |\tilde{g}(x) - g(x)| < 4\delta\,.$$

By the choice of δ,

$$|\tilde{g}^{-1}(w) - \tilde{g}^{-1}(z)| = |y - x| = |g^{-1}[g(y)] - g^{-1}[g(x)]| \leq \zeta\,.$$

On the basis of the information that

$$\|\tilde{g} - h\|_{V \cap A} \leq \|\tilde{g} - g\|_U + \|g - h\|_V < 2\delta\,,$$

we conclude that $h(A \cap V)$ is contained in $\tilde{g}(B)$. It follows that $f = \tilde{g}^{-1} \circ h \mid A \cap V$ is a well-defined LIP-embedding and that, because of (8.12),

$$|f(x) - x| = |\tilde{g}^{-1}[h(x)] - \tilde{g}^{-1}[\tilde{g}(x)]| \leq \zeta$$

for every x in $A \cap V$; i.e., $\|f - \text{identity}\|_{A \cap V} \leq \zeta$.

We can thus select a LIP-homeomorphism $f^* : \mathbb{R}^n \to \mathbb{R}^n$ for which (8.5), (8.6), and (8.7) hold. If h happens to be locally λ-bilipschitz, then f is locally $(\tilde{\lambda}\lambda)$-bilipschitz, so we may take f^* to be λ^*-bilipschitz for some λ^* dependent only on λ, $\tilde{\lambda}$, and n. As noted earlier, $f^*(U) = U$ and $f^*(A) = A$. We infer that $\hat{h} = \tilde{g} \circ f^*$ is a well-defined LIP-embedding of U into $\mathbb{H}^n$ which, in the case of a locally λ-bilipschitz h, is locally $\hat{\lambda}$-bilipschitz in A for $\hat{\lambda} = \tilde{\lambda}\lambda^*$, a number depending only on the allowable parameters. For x in $U_0 \cap W$ we conclude on the grounds of (8.6) and the definition of f that

$$\hat{h}(x) = \tilde{g}[f^*(x)] = \tilde{g}[f(x)] = h(x)\,.$$

In other words, $\hat{h} = h$ in $U_0 \cap W$.

To complete the proof we must only verify that $\|\hat{h} - g\|_{U_0} \leq \varepsilon$. The inequalities (8.5) and (8.4) imply that

$$|g[f^*(x)] - g(x)| \leq \varepsilon/4$$

for every x in A, meaning that $\|g \circ f^* - g\|_A \leq \varepsilon/4$. From this fact, from the knowledge that $f^*(A) = A$, and from (8.11) we learn that

$$\begin{aligned}
\|\hat{h} - g\|_A &\leq \|\hat{h} - \tilde{g}\|_A + \|\tilde{g} - g\|_A = \|\tilde{g} \circ f^* - \tilde{g}\|_A + \|\tilde{g} - g\|_A \\
&\leq \|\tilde{g} \circ f^* - g \circ f^*\|_A + \|g \circ f^* - g\|_A + 2\|\tilde{g} - g\|_A \\
&= \|\tilde{g} - g\|_A + \|g \circ f^* - g\|_A + 2\|\tilde{g} - g\|_A \leq \varepsilon\,.
\end{aligned}$$

Since U_0 is contained in A, $\|\hat{h} - g\|_{U_0} \leq \varepsilon$. □

As we intend to apply Lemma 8.1.5 in a hyperbolic setting, a reformulation appropriate to that context will prove convenient. We use the following notation: if S is a set and if φ and ψ are functions from S into $\mathbb{H}^n$, then for any nonempty subset A of S we define

$$d_{\mathbb{H}}(\varphi, \psi; A) = \sup_{x \in A} d_{\mathbb{H}}\big[\varphi(x), \psi(x)\big]\,.$$

In light of (8.2) it is clear that if $\varphi(A)$ and $\psi(A)$ are subsets of a compact set E in $\mathbb{H}^n$, then

$$a\|\varphi - \psi\|_A \leq d_{\mathbb{H}}(\varphi, \psi; A) \leq b\|\varphi - \psi\|_A \tag{8.13}$$

for constants $a > 0$ and $b > 0$ that depend only on E. The result we seek is an immediate consequence of Lemma 8.1.5, (8.13), and Lemma 8.1.1.

LEMMA 8.1.6. *Let U, U_0, V, W and $\mathcal{G}$ be as in Lemma 8.1.5. To each $\varepsilon > 0$ there corresponds a $\delta > 0$, depending on ε and the items just listed, for which the following statement holds:*

If $g \in \mathcal{G}$ and $h : V \to \mathbb{H}^n$ is a LIP*-embedding satisfying $d_{\mathbb{H}}(h, g; V) \leq \delta$, then there exists a* LIP*-embedding $\hat{h} : U \to \mathbb{H}^n$ such that $d_{\mathbb{H}}(\hat{h}, g; U_0) \leq \varepsilon$ and such that $\hat{h} = h$ in $U_0 \cap W$. Moreover, if h is locally λ-bilipschitz with respect to the hyperbolic metric, then $\hat{h}$ can be chosen so that its restriction to U_0 is locally $\hat{\lambda}$-bilipschitz relative to the hyperbolic metric, where $\hat{\lambda}$ is determined by λ, n, ε, $\mathcal{G}$, and the sets U, U_0, V, and W.*

8.2. Preliminaries

We now fix the dimension $n \geq 1$ and keep it fixed for the rest of this chapter. In this section we put together a number of preliminary results.

8.2.1. The extension F_f. Given a domain D in $\mathbb{R}^n$ and a number $r \geq 1$ we write $D^*(r)$ for the subset of $\mathbb{R}^{n+1}_+ = \mathbb{R}^n \times [0, \infty)$ defined as follows:

$$D^*(r) = \{(x, t) : x \in D\ ,\ 0 \leq t < r^{-1}\text{dist}(x, \partial D)\}\,.$$

When $D = \mathbb{R}^n$ we have $\text{dist}(x, \partial D) = \infty$ for every x in D, so $D^*(r) = \mathbb{R}^{n+1}_+$.

The set $D^*(r)$ is connected and is relatively open in $\mathbb{R}^{n+1}_+$. Clearly $D^*(s)$ is contained in $D^*(r)$ when $r \leq s$. For the rest of this chapter the symbol D^* will serve as an abbreviation for $D^*(1)$. Notice that if (x, t) is a point of $D^*(r)$, then the ball $\bar{B}^n(x, rt)$ is contained in D.

In particular the following observation helps clarify the geometric situation:

$$\bar{B}^n(x, t) \subset D \Longleftrightarrow (x, t) \in D^*.$$

Observe, too, that if φ is a member of $\text{Sim}(\mathbb{H}^{n+1})$, the group of similarity transformations of $\mathbb{R}^{n+1}$ which preserve $\mathbb{H}^{n+1}$ (hence, leave $\mathbb{R}^n$ and $\mathbb{R}^{n+1}_+$ invariant as well), then $\varphi(D)^* = \varphi(D^*)$.

Let $f : D \to \mathbb{R}^n$ be an embedding, where once again D is a domain in $\mathbb{R}^n$. We define a continuous function $\tau_f : D^* \to [0, \infty)$ through the rule of correspondence

$$\tau_f(x, t) = \max_{|h|=t} |f(x + h) - f(x)|\,.$$

Thus $\tau_f(x, t) = L_f(x, t) > 0$ when $t > 0$, whereas $\tau_f(x, 0) = 0$. With the aid of τ_f we can create an embedding F_f of D^* into $\mathbb{R}^{n+1}_+$:

$$F_f(x, t) = \big(f(x), \tau_f(x, t)\big)\,.$$

Certainly $F_f(x, t) \neq F_f(y, s)$ when $x \neq y$. On the other hand, if (x, t) and (x, s) lie in D^* and if $0 \leq s < t$, then $f[\bar{B}^n(x, s)]$ is a compact subset of $f[B^n(x, t)]$, which implies that $\tau_f(x, s) < \tau_f(x, t)$ and, as a result, that $F_f(x, t) \neq F_f(x, s)$. This demonstrates that the function F_f is injective. The continuity of F_f is evident, while the fact that F_f is an open mapping of D^* into $\mathbb{R}^{n+1}_+$ is readily verified. Hence F_f is an embedding.

Treating $\mathbb{R}^n$ in the standard way as a subset of $\mathbb{R}^{n+1}$ and remarking that $F_f(x, 0) = (f(x), 0)$ for x in D, we regard F_f as an extension of f from D to D^*.

The correspondence $f \mapsto F_f$ is continuous in a sense that the next lemma makes precise.

LEMMA 8.2.1. *Suppose that D is a domain in $\mathbb{R}^n$, that f and g are embeddings of D into $\mathbb{R}^n$, and that A is a domain whose closure lies in D. Then*

$$\|F_g - F_f\|_{A^*} \le 3\|g - f\|_A . \tag{8.14}$$

In particular, if $\langle f_\nu \rangle$ is a sequence of embeddings of D into $\mathbb{R}^n$ that converges in the Euclidean locally uniform sense to an embedding f of D into $\mathbb{R}^n$, then $\langle F_{f_\nu} \rangle$ is locally uniformly convergent in D^* to F_f.

PROOF. Let (x, t) be a point of A^*. If $|y - x| = t$, then y belongs to A and

$$\begin{aligned} |g(y) - g(x)| &\le |g(y) - f(y)| + |f(y) - f(x)| + |f(x) - g(x)| \\ &\le \|g - f\|_A + \tau_f(x, t) + \|g - f\|_A , \end{aligned}$$

from which we infer that

$$\tau_g(x, t) \le \tau_f(x, t) + 2\|g - f\|_A .$$

By similar reasoning

$$\tau_f(x, t) \le \tau_g(x, t) + 2\|g - f\|_A ,$$

so

$$|\tau_g(x, t) - \tau_f(x, t)| \le 2\|g - f\|_A .$$

It follows that

$$\begin{aligned} |F_g(x, t) - F_f(x, t)| &\le |g(x) - f(x)| + |\tau_g(x, t) - \tau_f(x, t)| \\ &\le \|g - f\|_A + 2\|g - f\|_A = 3\|g - f\|_A \end{aligned}$$

for each element (x, t) of A^*, whence

$$\|F_g - F_f\|_{A^*} \le 3\|g - f\|_A .$$

The convergence statement is a direct consequence of the estimate (8.14) and the observation that each compact subset of D^* lies in A^* for some domain A whose closure is contained in D. □

The following lemma points out the naturality of the correspondence $f \mapsto F_f$ with respect to the action of the similarity group, hardly unexpected when one reflects for an instant on the definition of F_f.

LEMMA 8.2.2. *Let D be a domain in $\mathbb{R}^n$, and let f be an embedding of D into $\mathbb{R}^n$. Then*

$$F_{\psi \circ f \circ \varphi} = \psi \circ F_f \circ \varphi \tag{8.15}$$

whenever φ and ψ are similarity transformations of $\mathbb{R}^{n+1}$ that preserve $\mathbb{R}^{n+1}_+$.

PROOF. The domains of the functions appearing on the opposite sides of (8.15) are the same—namely, the set $A = \varphi^{-1}(D)^* = \varphi^{-1}(D^*)$.

Let λ and μ be the dilation factors of φ and ψ, respectively. Given a point (x, t) of A, we set $y = \varphi(x)$ and $z = f(y)$. Recalling Theorem 3.3.13, we write

$$\varphi(x, t) = (\varphi(x), \lambda t) \quad \text{and} \quad \psi(z, s) = (\psi(z), \mu s).$$

For w in $\varphi^{-1}(D)$ we see that $|w - x| = t$ if and only if $|\varphi(w) - y| = \lambda t$. Accordingly,

$$\tau_{\psi \circ f \circ \varphi}(x, t) = \mu \tau_f(y, \lambda t) ,$$

which leads to

$$F_{\psi \circ f \circ \varphi}(x, t) = \big(\psi(z), \mu \tau_f(y, \lambda t)\big) .$$

As for the other side of (8.15), we have

$$\psi \circ F_f \circ \varphi(x,t) = \psi \circ F_f(y, \lambda t) = \psi\big(z, \tau_f(y,\lambda t)\big) = \big(\psi(z), \mu\tau_f(y,\lambda t)\big) ,$$

which is what we wanted to show. □

8.2.2. The families $\mathcal{G}_K$ and $\mathcal{G}_K^*$. For any point z of $\mathbb{H}^{n+1}$ we let α_z denote the similarity transformation of $\mathbb{R}^{n+1}$ given by the formula

$$\alpha_z(x) = z + z_{n+1}(x - e_{n+1}) .$$

Put differently, α_z is the unique similarity that maps e_{n+1} to z and has the structure $x \mapsto \lambda x + b$ with $\lambda > 0$ and b in $\mathbb{R}^n$.

Suppose that D is a subdomain of $\mathbb{R}^n$. With each interior point z of D^* and each embedding $f : D \to \mathbb{R}^n$, we associate a similarity transformation $\beta_{z,f}$ belonging to $\sigma_{\mathbb{S}}(H^{n+1})$—namely,

$$\beta_{z,f} = \alpha_w^{-1}, \qquad w = F_f(z) .$$

To be more explicit,

$$\beta_{z,f}(x) = e_{n+1} + w_{n+1}^{-1}(x - w) . \tag{8.16}$$

We next establish some notation that will be in use for the remainder of this chapter. First, we set

$$r_0 = 3 + \sqrt{n}, \qquad B_0 = B^n(r_0), \qquad D_0 = \operatorname{int}(B_0^*) = B_0^* \cap \mathbb{H}^{n+1} .$$

The choice of r_0 ensures that the cube

$$Q = [-1,1] \times \cdots \times [-1,1] \times [0,2] \subset B_0^* \subset \mathbb{R}^{n+1}.$$

A small observation will prove its worth in upcoming constructions.

LEMMA 8.2.3. *If D is a subdomain of $\mathbb{R}^n$ and z is an interior point of $D^*(r_0)$, then $\alpha_z(D_0)$ is contained in D^*.*

PROOF. Write $z = (c,t)$, where c is in D and $0 < tr_0 < d(c, \partial D)$. Then

$$\alpha_z(x) = (c_1 + tx_1, \ldots, c_n + tx_n, tx_{n+1}) .$$

Consider a point x of D_0, say $x = (b,s)$ in which b lies in B_0 and $0 \le s < r_0 - |b|$. Notice that

$$|tb| < \frac{d(c,\partial D)}{r_0} r_0 = d(c, \partial D) ,$$

whence $c + tb$ belongs to D. Moreover,

$$d(c+tb, \partial D) \ge d(c, \partial D) - t|b| > tr_0 + t(s - r_0) = ts .$$

Therefore

$$\alpha_z(x) = (c + tb, ts) \in D^* ,$$

as maintained. □

When D is a domain in $\mathbb{R}^n$ and $1 \le K < \infty$ the notation $\mathcal{Q}_K(D)$ denotes the family of all K-quasiconformal embeddings of D into $\mathbb{R}^n$. Note that if $n = 1$ we take "K-quasiconformal" to be synonymous with "K-quasisymmetric".

Corresponding to each domain D in $\mathbb{R}^n$, each interior point z of $D^*(r_0)$, and each embedding f of D into $\mathbb{R}^n$, we introduce embeddings f_z of B_0 into $\mathbb{R}^n$ and $F_{z,f}$ of D_0 into $\mathbb{H}^{n+1}$ by setting

$$f_z = \beta_{z,f} \circ f \circ \alpha_z|_{B_0}, \qquad F_{z,f} = \beta_{z,f} \circ F_f \circ \alpha_z|_{D_0} .$$

Recalling Lemma 8.2.2, we note that $F_{z,f}$ is just the restriction to D_0 of F_g, where $g = f_z$. We then define for K in $[1, \infty)$ the families of mappings

$$\mathcal{G}_K = \{f_z : D \text{ is a domain in } \mathbb{R}^n \ , \ z \in \operatorname{int}[D^*(r_0)] \ , \ f \in \mathcal{Q}_K(D)\}$$

and

$$\begin{aligned} \mathcal{G}_K^* &= \{F_{z,f} : D \text{ is a domain in } \mathbb{R}^n \ , \ z \in \operatorname{int}[D^*(r_0)] \ , \ f \in \mathcal{Q}_K(D)\} \\ &= \{F_g|_{D_0} : g \in \mathcal{G}_K\} . \end{aligned}$$

One can view $\mathcal{G}_K$ as the family of "germs" of K-quasiconformal mappings in $\mathbb{R}^n$, since each member f of $\mathcal{Q}_K(D)$ sees its essential geometric structure near any point of D captured, modulo renormalization, in some member of $\mathcal{G}_K$.

Our next order of business is to demonstrate that the families $\mathcal{G}_K$ and $\mathcal{G}_K^*$ are relatively compact in the topology of (Euclidean) locally uniform convergence. As we proceed with the verification of this fact, two technical lemmas will come in handy.

LEMMA 8.2.4. *There exists a constant $H = H(n, K)$ for which the following statement is true: if D is a domain in $\mathbb{R}^n$, if f is a member of the family $\mathcal{Q}_K(D)$, if x belongs to D, and if $t > 0$ is such that the ball $B^n(x, r_0 t)$ lies in D, then*

$$\frac{L_f(x,t)}{\ell_f(x,t)} \le H . \tag{8.17}$$

PROOF. We assume that $n \ge 2$, the assertion being trivial when $n = 1$. Because of the way in which the ratio $L_f(x,t)/\ell_f(x,t)$ behaves when f is composed with similarity transformations, it suffices to consider the situation for $D = B_0$, $x = 0$, $t = 1$ and to produce a constant H such that (8.17) holds for every member of the family $\mathcal{F} = \{f \in \mathcal{Q}_K(B_0) : f(0) = 0, \ \ell_f(0,1) = |f(e_1)| = 1\}$. Since $q[f(0), \infty] = 2$ and $q[f(e_1), \infty] = \sqrt{2}$ for each f in $\mathcal{F}$, Theorem 6.6.12 (ii) informs us that $\mathcal{F}$ is a normal family in the chordal metric. The condition $f(0) = 0$, in conjunction with Theorem 6.6.26, shows that $\mathcal{F}$ is also normal with respect to the Euclidean metric.

We define

$$H = \sup_{f \in \mathcal{F}} L_f(0,1) .$$

To complete the proof we must verify that $H < \infty$. Let $\langle f_\nu \rangle$ be a sequence from $\mathcal{F}$ such that $L_{f_\nu}(0,1) \to H$. By passing to a subsequence we may assume that $f_\nu \to f$ in a Euclidean locally uniform fashion in B_0, where $f : B_0 \to \mathbb{R}^n$ is continuous. In particular, $f_\nu \to f$ uniformly on $\mathbb{S}^{n-1}$, so

$$H = \lim_{\nu \to \infty} L_{f_\nu}(0,1) = L_f(0,1) < \infty .$$

As defined, the constant H depends on n, K, and the number $r_0 = 3 + \sqrt{n}$, hence, entirely on n and K. □

The second technical result we require reads as follows.

LEMMA 8.2.5. *If g belongs to the family $\mathcal{G}_K$, then $g(0) = 0$ and*

$$\frac{1}{H} \le |g(e_1)| \le 1 , \tag{8.18}$$

in which H is the constant from Lemma 8.2.4.

PROOF. Write $g = f_z|B_0$, where f comes from $\mathcal{Q}_K(D)$ for some subdomain D of $\mathbb{R}^n$ and z from the interior of $D^*(r_0)$, and let $w = F_f(z)$. If we express z in the form $z = (x,t)$ with x in D and $0 < r_0 t < \operatorname{dist}(x, \partial D)$, then $w = (f(x), \tau_f(x,t))$. By using (8.16) and in the last step abbreviating $(y,0)$ to y, we obtain

$$\begin{aligned} g(0) &= \beta_{z,f} \circ f \circ \alpha_z(0) = \beta_{z,f}\big[\big(f(x),0\big)\big] \\ &= e_{n+1} + w_{n+1}^{-1}\big[\big(f(x),0\big) - w\big] = w_{n+1}^{-1}\big[f(x) - f(x)\big] = 0\,. \end{aligned}$$

Similarly,

$$g(e_1) = w_{n+1}^{-1}\big[f(x+te_1) - f(x)\big] = \tau_f(x,t)^{-1}\big[f(x+te_1) - f(x)\big]\,.$$

Now $B^n(x, r_0 t)$ is contained in D, so we can assert on the strength of Lemma 8.2.4 that

$$\frac{1}{H} \le \frac{\ell_f(x,t)}{L_f(x,t)} \le \frac{|f(x+te_1) - f(x)|}{\tau_f(x,t)} \le 1\,,$$

leading to

$$\frac{1}{H} \le |g(e_1)| \le 1$$

and thereby confirming (8.18). □

Armed with Lemma 8.2.5, we can establish the relative compactness of $\mathcal{G}_K$.

THEOREM 8.2.6. *The family $\bar{\mathcal{G}}_K$, the closure of $\mathcal{G}_K$ with respect to the topology of Euclidean locally uniform convergence in B_0, is a compact family of K-quasiconformal embeddings of B_0 into $\mathbb{R}^n$.*

PROOF. We treat the case $n \ge 2$ and leave the case $n = 1$ as an exercise for the reader. From Lemma 8.2.5 we extract the information that

$$q\big[g(0), \infty\big] = 2, \qquad q\big[g(e_1), \infty\big] \ge \sqrt{2}$$

whenever g is a member of $\mathcal{G}_K$. Once again invoking Corollary 6.6.12 (ii), we deduce that $\mathcal{G}_K$ is a normal family relative to the chordal metric. As $g(0) = 0$ holds for each g in $\mathcal{G}_K$, Theorem 6.6.26 confirms that the same is true for the Euclidean metric.

The upshot of the foregoing statements is that, as far as sequences from $\mathcal{G}_K$ are concerned, chordal and Euclidean locally uniform convergence in B_0 are equivalent notions.

For a mapping $\bar{g} : B_0 \to \mathbb{R}^n$ to qualify as a member of $\bar{\mathcal{G}}_K$ we require that there be a sequence $\langle g_\nu \rangle$ in $\mathcal{G}_K$ such that $g_\nu \to \bar{g}$ in the Euclidean—hence, also, chordal—locally uniform sense in B_0. In particular, $\bar{g}(0) = 0$. Lemma 8.2.5 implies that $|g(e_1)| \ge \frac{1}{H} > 0$ for any such function, so $\bar{g}$ is definitely nonconstant in B_0. By Theorem 6.6.23, each $\bar{g}$ in $\bar{\mathcal{G}}_K$ must be a K-quasiconformal mapping of B_0 into $\mathbb{R}^n$.

Let $\langle \bar{g}_\nu \rangle$ be an arbitrary sequence from $\bar{\mathcal{G}}_K$. We can choose for each ν a member g_ν of $\mathcal{G}_K$ with the property that $|\bar{g}_\nu(x) - g_\nu(x)| \le \nu^{-1}$ whenever $|x| \le r_0 - \nu^{-1}$. The normality of $\mathcal{G}_K$ guarantees the existence of a subsequence $\langle g_{\nu_k} \rangle$ of $\langle g_\nu \rangle$ such that $g_{\nu_k} \to \bar{g}$ locally uniformly in B_0. By definition $\bar{g}$ belongs to $\bar{\mathcal{G}}_K$ and by construction $\bar{g}_{\nu_k} \to \bar{g}$ locally uniformly in B_0. This shows $\bar{\mathcal{G}}_K$ to be a compact family of mappings relative to the topology of locally uniform convergence in B_0. □

It is now an easy task to verify that the family $\mathcal{G}_K^*$ is itself relatively compact.

THEOREM 8.2.7. *The family $\bar{\mathcal{G}}_K^*$, the closure of $\mathcal{G}_K^*$ with respect to the topology of Euclidean locally uniform convergence in D_0, is a compact family of embeddings of D_0 into $\mathbb{H}^{n+1}$.*

Each member of $\bar{\mathcal{G}}_K^*$ is the restriction to D_0 of F_g for some $g \in \bar{\mathcal{G}}_K$.

PROOF. Let $\langle \bar{G}_\nu \rangle$ be a sequence from $\bar{\mathcal{G}}_K^*$. We must produce a subsequence $\langle \bar{G}_{\nu_k} \rangle$ such that $\bar{G}_{\nu_k} \to \bar{G}$ locally uniformly in D_0, where $\bar{G}$ belongs to $\bar{\mathcal{G}}_K^*$.

By the definition of $\bar{\mathcal{G}}_K^*$ we can choose a sequence $\langle G_\nu \rangle$ from $\mathcal{G}_K^*$ such that $\|\bar{G}_\nu - G_\nu\|_A \to 0$ as $\nu \to \infty$ for each compact set A in D_0. We can then write $G_\nu = F_{g_\nu}|D_0$, where g_ν is a mapping from $\mathcal{G}_K$.

By Theorem 8.2.6, there is a subsequence $\langle g_{\nu_k} \rangle$ of $\langle g_\nu \rangle$ such that $g_{\nu_k} \to \bar{g}$, a member of $\bar{\mathcal{G}}_K$, locally uniformly in B_0. Appealing to Lemma 8.2.1 we conclude that $G_{\nu_k} \to \bar{G} = F_{\bar{g}}|D_0$ locally uniformly in D_0, which places $\bar{G}$ in $\bar{\mathcal{G}}_K^*$. Moreover, from the choice of G_ν it is clear that $\bar{G}_\nu \to \bar{G}$ locally uniformly in D_0. Finally, since each $\bar{G}$ in $\bar{\mathcal{G}}_K^*$ can be obtained in the above way, each has the structure suggested by the theorem. □

Consider a mapping f from the family $\mathcal{Q}_K(D)$. The mapping F_f is not the extension of f we seek: unfortunately F_f is not generally quasiconformal in $V \cap \mathbb{H}^{n+1}$ for any open neighbourhood V of D in $\mathbb{R}^{n+1}$.

Instead, F_f plays an auxiliary role of providing us with something defined $\mathbb{H}^{n+1}$ to work with. In particular we will seek to approximate F_f in the hyperbolic metric by a quasiconformal mapping. Notice that any quasiconformal mapping a bounded hyperbolic distance from F_f agrees with f on $\mathbb{R}^{n+1}$. The other nice property F_f has is that at large scales it behaves like a mapping which is bilipschitz in the hyperbolic metric.

The next lemma is typical of the way in which we use these properties of F_f: roughly speaking, it asserts that a mapping which is sufficiently close to F_f on a set A and is locally injective in A must be globally injective on A.

LEMMA 8.2.8. *Let M be a number satisfying*

$$0 < M < d_{\mathbb{H}}(e_{n+1}, \partial D_0 \cap \mathbb{H}^{n+1}).$$

There exists a constant $\rho > 0$, which depends only on n, K, and M, with the following property: if D is a domain in $\mathbb{R}^n$, if $f \in \mathcal{Q}_K(D)$, if A is a nonempty set in the interior of D^*, and if $h : A \to \mathbb{H}^{n+1}$ is a function such that $d_{\mathbb{H}}(h, F_f : A) \leq \rho$ and such that $h(z) \neq h(z')$ whenever z and z' are points of A with $0 < d_{\mathbb{H}}(z, z') \leq M$, then h is injective.

PROOF. The sphere $S = \mathbb{S}_{\mathbb{H}^{n+1}}(e_{n+1}, M)$ is a compact subset of D_0. Therefore, the set $\bar{\mathcal{G}}_K^* \times S$ is compact in the product topology, while the function $(g, x) \mapsto d_{\mathbb{H}}[g(x), g(e_{n+1})]$ is continuous and positive on $\bar{\mathcal{G}}_K^* \times S$. It follows that

$$\rho = \frac{1}{3} \min\Big\{ d_{\mathbb{H}}\big[g(x), g(e_{n+1})\big] : x \in S \ , \ g \in \bar{\mathcal{G}}_K^* \Big\} > 0 \, .$$

Now let f and h be as in the statement of the theorem, and let z and z' be distinct points of A. We claim that $h(z) \neq h(z')$. This is true by hypothesis if $d_{\mathbb{H}}(z, z') \leq M$, so we suppose that $d_{\mathbb{H}}(z, z') > M$. Because F_f is an embedding, it is evident that

$$d_{\mathbb{H}}\big[F_f(z), F_f(z')\big] \geq d_{\mathbb{H}}\big[F_f(z), F_f(S')\big] \ ,$$

where $S' = S_{\mathbb{H}}^{n-1}(z, M) = \alpha_z(S)$. Let $g = \beta_{z,f} \circ F_f \circ \alpha_z | D_0$, a member of $\mathcal{G}_K^*$. As $\beta_{z,f}$ is a hyperbolic isometry we deduce that

$$\begin{aligned} d_{\mathbb{H}}\big[F_f(z), F_f(S')\big] &= d_{\mathbb{H}}\big[\beta_{z,f} \circ F_f \circ \alpha_z(e_{n+1}) \ , \ \beta_{z,f} \circ F_f \circ \alpha_z(S)\big] \\ &= d_{\mathbb{H}}\big[g(e_{n+1}), g(S)\big] \geq 3\rho \, . \end{aligned}$$

According to our assumptions the function h satisfies the condition $d_{\mathbb{H}}(h, F_f; A) \leq \rho$.

The triangle inequality thus tells us that

$$\begin{aligned} d_{\mathbb{H}}\big[h(z), h(z')\big] &\geq d_{\mathbb{H}}\big[F_f(z), F_f(z')\big] - d_{\mathbb{H}}\big[h(z), F_f(z)\big] - d_{\mathbb{H}}\big[h(z'), F_f(z')\big] \\ &\geq 3\rho - \rho - \rho = \rho > 0 \ , \end{aligned}$$

whence $h(z) \neq h(z')$. □

We close the present section by identifying the last purely topological ingredient in the proof of the extension theorem. For $n \neq 3$ the result we shall merely quote follows directly from standard, albeit nontrivial, approximation theorems in piecewise linear topology; see [**158**]. When $n = 3$ a special argument is called for: one that rests on Lemma 8.2.1 and the easy-to-verify fact, true in all dimensions, that $F_f : B_0^* \to \mathbb{R}_+^{n+1}$ is a LIP-embedding whenever $f : B_0 \to \mathbb{R}^n$ is.

LEMMA 8.2.9. *If D is a domain in $\mathbb{H}^{n+1}$ such that $\bar{D}$ lies in D_0 and if g is a member of $\bar{\mathcal{G}}_K^*$, then $g|D$ is uniformly approximable by* LIP-*embeddings.*

Since the set $g(\bar{D})$ arising in Lemma 8.2.9 is a compact subset of $\mathbb{H}^{n+1}$, it is clear that by applying this result we can substitute uniform approximation in the hyperbolic metric for uniform convergence in the Euclidean metric. Thus, for each $\varepsilon > 0$ there is a LIP-embedding $h : D \to \mathbb{H}^{n+1}$ for which $d_{\mathbb{H}}(g, h; D) < \varepsilon$.

8.2.3. The Whitney decomposition of $\mathbb{H}^{n+1}$. Let Q be a closed cube in $\mathbb{R}^{n+1}$ whose edges are parallel to the coordinate axes. We use z_Q to designate the center of Q and $2\lambda_Q$ to indicate its edge-length.

Thus we can represent Q in the manner $Q = z_Q + \lambda_Q I^{n+1}$, where $I = [-1, 1]$. For $t > 0$ we denote

$$Q(t) = z_Q + t\lambda_Q(-1, 1)^{n+1}.$$

Thus $Q(t)$ is the open cube concentric with Q but t times it in size.

The symbol $\mathcal{K}$ will stand for the collection of all closed cubes in $\mathbb{R}^{n+1}$ with unit edge-length whose vertices lie in the set $\mathbb{Z}^n \times \{0, 1\}$. In other words, $\mathcal{K}$ consists of all translates of the cube $[0, 1]^{n+1}$ by vectors from the lattice $\mathbb{Z}^n$ in $\mathbb{R}^n$.

Distinct members of $\mathcal{K}$ are nonoverlapping, and the union of all the cubes in $\mathcal{K}$ is $\mathbb{R}^n \times [0, 1]$. We denote by $\mathcal{W}$ the collection of all cubes of the type $2^\nu e_{n+1} + 2^\nu Q$, where Q comes from $\mathcal{K}$ and ν is an integer:

$$\mathcal{W} = \{2^\nu e_{n+1} + 2^\nu Q : Q \in \mathcal{K}, \nu \in \mathbb{Z}\}.$$

As with $\mathcal{K}$, distinct cubes in $\mathcal{W}$ do not overlap; the union of all the cubes in $\mathcal{W}$ is $\mathbb{H}^{n+1}$. We shall refer to $\mathcal{W}$ as the *Whitney decomposition* of $\mathbb{H}^{n+1}$.

Notice that the members of $\mathcal{W}$ with fixed ν provide a decomposition $\mathcal{K}_\nu$ of $\mathbb{R}^n \times [2^\nu, 2^{\nu+1}]$ which, except for scale, mimics the decomposition of $\mathbb{R}^n \times [0, 1]$ given by $\mathcal{K}$.

Each cube in this "level ν" of $\mathcal{W}$ has below it 2^n cubes from "level $\nu - 1$". Observe, too, that any two cubes in $\mathcal{W}$ are congruent in the hyperbolic geometry of

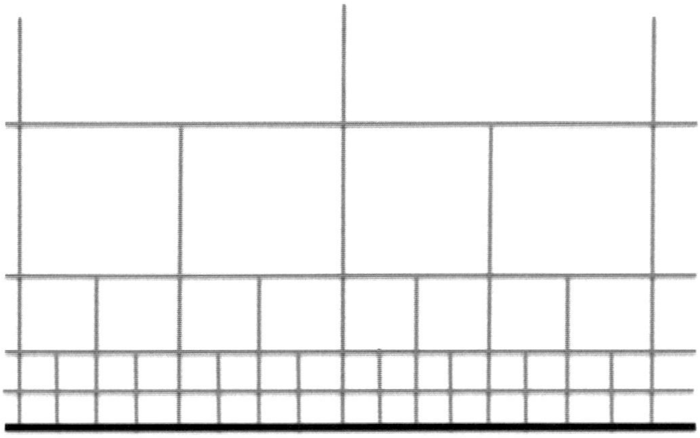

The Whitney decomposition of $\mathbb{H}^2$.

$\mathbb{H}^{n+1}$. For ease of reference we report several elementary facts about $\mathcal{W}$ in a lemma. In the lemma—and from this point on in the chapter—we take $N = 2^{n+1}$. The first two assertions of the lemma are readily confirmed by straightforward geometric arguments.

LEMMA 8.2.10. *The Whitney decomposition $\mathcal{W}$ of $\mathbb{H}^{n+1}$ enjoys the following properties:*

(i) *if Q belongs to $\mathcal{W}$, then $(z_Q)_{n+1} = 3\lambda_Q$;*
(ii) *if Q and Q' are disjoint members of $\mathcal{W}$, then $Q(3/2)$ and $Q'(3/2)$ are also disjoint;*
(iii) *there exist nonempty, pairwise disjoint subcollections $\mathcal{W}_1, \mathcal{W}_2, \dots, \mathcal{W}_N$ of $\mathcal{W}$ such that $\mathcal{W} = \mathcal{W}_1 \cup \mathcal{W}_2 \cup \cdots \cup \mathcal{W}_N$ and such that the cubes in each $\mathcal{W}_j$ are pairwise disjoint.*

PROOF OF (iii). Let $\mathcal{L}$ denote the collection of closed unit cubes in $\mathbb{R}^n$ that comprises all translates of the cube $J^n = [0,1]^n$ by vectors from the lattice $\mathbb{Z}^n$. It is a trivial observation that we can write

$$\mathcal{L} = \mathcal{L}_1 \cup \mathcal{L}_2 \cup \cdots \cup \mathcal{L}_M ,$$

where $M = 2^n$, where $\mathcal{L}_1, \mathcal{L}_2, \dots, \mathcal{L}_M$ are nonempty, pairwise disjoint subcollections of $\mathcal{L}$, and where the cubes in each $\mathcal{L}_j$ are pairwise disjoint.

The simplest way to arrange this is to let each $\mathcal{L}_j$ consist of Q_j, one of the 2^n cubes of $\mathcal{L}$ whose union is $2J^n$, together with all translates of Q_j by vectors from the lattice $2\mathbb{Z}^n$. This decomposition of $\mathcal{L}$ leads immediately to a corresponding decomposition for the collection $\mathcal{K}$; namely, $\mathcal{K} = \mathcal{K}_1 \cup \mathcal{K}_2 \cup \cdots \cup \mathcal{K}_M$, where

$$\mathcal{K}_j = \{Q \times [0,1] : Q \in \mathcal{L}_j\} .$$

By translating and scaling we can transport the above decomposition of $\mathcal{K}$ to each level of $\mathcal{W}$, thereby obtaining a decomposition of $\mathcal{K}_\nu$ into subcollections $\mathcal{K}_{\nu,1}$, $\mathcal{K}_{\nu,2}, \dots, \mathcal{K}_{\nu,M}$ with features analogous to those of $\mathcal{K}_1, \mathcal{K}_2, \dots, \mathcal{K}_M$. Let

$$\mathcal{W}_j^0 = \{Q : Q \in \mathcal{K}_{\nu,j}\,,\ \nu \text{ odd}\}, \qquad \mathcal{W}_j^e = \{Q : Q \in \mathcal{K}_{\nu,j}\,,\ \nu \text{ even}\},$$

and for $j = 1, 2, \dots, N$ take $\mathcal{W}_j = \mathcal{W}_j^0$ if j is odd and $\mathcal{W}_j = \mathcal{W}_{j/2}^e$ if j is even. Then the subcollections $\mathcal{W}_1, \mathcal{W}_2, \dots, \mathcal{W}_N$ of $\mathcal{W}$ have the stated properties. □

Some additional abbreviations and notation will later turn out to be helpful. First, we abbreviate

$$\alpha_{z_Q} = \alpha_Q, \qquad \beta_{z_Q,f} = \beta_{Q,f}.$$

Next, we set

$$Q_0 = e_{n+1} + (1/3)I^{n+1}$$

and remark that $Q_0(3)$, which is the interior of $Q = e_{n+1} + I^{n+1}$, lies in D_0. From Lemma 8.2.10(i) it follows easily that

$$Q_0 = \alpha_Q^{-1}(Q)$$

for every Q in $\mathcal{W}$, while part (ii) of the same lemma ensures that

$$Q_0(3/2) \cap \alpha_Q^{-1}\big[Q'(3/2)\big] = \emptyset \tag{8.19}$$

whenever Q and Q' are disjoint members of $\mathcal{W}$. One consequence of (8.19) is the following lemma.

LEMMA 8.2.11. *For t in $(0, 3/2]$, the collection $\mathcal{C}_t$ of (nonempty) open $(n+1)$-intervals $Q_0(t) \cap \alpha_Q^{-1}[Q'(1+2^{-j})]$, generated as Q and Q' vary over $\mathcal{W}$ and j ranges over $\{1, 2, \ldots, N+2\}$, is a finite collection, one that depends entirely upon n and t.*

Indeed, it is perfectly clear that $\mathcal{C}_t$ can be arrived at by fixing Q and letting Q' vary over the finitely many members of $\mathcal{W}$ that intersect Q.

Recalling that $r_0 = 3 + \sqrt{n}$, we introduce numbers r_1 and r_2:

$$r_1 = 2r_0 + \sqrt{n} = 6 + 3\sqrt{n}, \qquad r_2 = (3r_1 + \sqrt{n}\,)/9 = 9 + 5\sqrt{n}\,.$$

Then $1 < r_0 < r_1 < r_2$, so

$$D^*(r_2) \subset D^*(r_1) \subset D^*(r_0) \subset D^*$$

whenever D is a domain in $\mathbb{R}^n$. For any such domain D, we define

$$\mathcal{W}(D) = \big\{Q \in \mathcal{W} : z_Q \in D^*(r_1)\big\}$$

and make two observations.

LEMMA 8.2.12. *Let D be a domain in $\mathbb{R}^n$. Then $\overline{Q(3)}$ is contained in $D^*(r_0)$ whenever Q belongs to $\mathcal{W}(D)$. Moreover, the collection $\mathcal{W}(D)$ covers the interior of $D^*(r_2)$.*

PROOF. First let Q be a member of $\mathcal{W}(D)$ and let z be a point of $\overline{Q(3)}$. Write $z_Q = (x, t)$, where x is in D and $0 < r_1 t < \operatorname{dist}(x, \partial D)$, and $z = (y, s)$, where y is in $\mathbb{R}^n$ and $s \geq 0$. Since $z - z_Q$ lies in the set $3\lambda_Q I^{n+1}$, we see that $y - x$ is in $3\lambda_Q I^n$. In view of Lemma 8.2.10(i),

$$|y - x| \leq 3\lambda_Q\sqrt{n} = t\sqrt{n} < r_1 t < d(x, \partial D)\ ,$$

so y belongs to D. Furthermore,

$$\begin{aligned} \operatorname{dist}(y, \partial D) &\geq \operatorname{dist}(x, \partial D) - |y - x| \geq \operatorname{dist}(x, \partial D) - t\sqrt{n} \\ &\geq \Big(1 - \frac{\sqrt{n}}{r_1}\Big)\operatorname{dist}(x, \partial D) = \frac{2r_0 \operatorname{dist}(x, \partial D)}{r_1}\ , \end{aligned}$$

which in conjunction with the inequality $|s - t| \leq 3\lambda_Q = t$ implies that

$$s \le 2t < \frac{2\mathrm{dist}(x, \partial D)}{r_1} < \frac{\mathrm{dist}(y, \partial D)}{r_0}$$

and thus places z in $D^*(r_0)$.

Next, consider an interior point z of the set $D^*(r_2)$. Again write $z = (y, s)$, where now y belongs to D and $0 < r_2 s < \mathrm{dist}(y, \partial D)$. Choose a cube Q from $\mathcal{W}$ such that z lies in Q and, as earlier, let $z_Q = (x, t)$. We show that z_Q is a point of $D^*(r_1)$, which makes Q a member of $\mathcal{W}(D)$. Since z belongs to Q it is definitely the case that $t \le 3s/2$. In the present situation, $x - y$ is a point of $\lambda_Q I^n$, implying that

$$|x - y| \le \lambda_Q \sqrt{n} = \frac{t\sqrt{n}}{3} \le \frac{s}{2}\sqrt{n} < r_2 s < \mathrm{dist}(y, \partial D)\,.$$

Accordingly, x lies in D and

$$\begin{aligned}
\mathrm{dist}(x, \partial D) &\ge \mathrm{dist}(y, \partial D) - |x - y| \ge \mathrm{dist}(y, \partial D) - \frac{s\sqrt{n}}{2} \\
&\ge \Big(1 - \frac{\sqrt{n}}{2r_2}\Big)\mathrm{dist}(y, \partial D) = \frac{3r_1 \mathrm{dist}(y, D)}{2r_2} \\
&\ge \frac{3r_1 s}{2} \ge r_1 t\,.
\end{aligned}$$

We conclude that z_Q is in $D^*(r_1)$, as asserted. □

8.3. The Tukia-Väisälä extension theorem

Finally we come to the proof of the extension theorem.

8.3.1. The construction. To begin we fix, once and for all, the subcollections $\mathcal{W}_1, \mathcal{W}_2, \ldots, \mathcal{W}_N$ of the Whitney decomposition $\mathcal{W}$ of $\mathbb{H}^{n+1}$ as specified in Lemma 8.2.12 (iii) and the proof thereof.

Let D be a subdomain of $\mathbb{R}^n$. We write $\mathcal{W}_j(D) = \mathcal{W}_j \cap \mathcal{W}(D)$. Our choice of $\mathcal{W}_j$ makes certain that $\mathcal{W}_j(D)$ is a nonempty collection of pairwise disjoint cubes. For $j = 1, 2, \ldots, N$ we define

$$\mathcal{W}_j^*(D) = \bigcup_{k=1}^{j} \mathcal{W}_k(D)\,.$$

Therefore we have

$$\mathcal{W}_1(D) = \mathcal{W}_1^*(D) \subset \mathcal{W}_2^*(D) \subset \cdots \subset \mathcal{W}_N^*(D) = \mathcal{W}(D)\,.$$

Next we set

$$V_0(D) = \bigcup_{Q \in \mathcal{W}_1(D)} Q(3/2), \qquad W_0(D) = \bigcup_{Q \in \mathcal{W}_1(D)} Q(5/4),$$

and for $j = 1, 2, \ldots, N$ define sets $V_j(D)$ and $W_j(D)$ by

$$V_j(D) = \bigcup_{Q \in \mathcal{W}_j^*(D)} Q(1 + 2^{-j-1}), \qquad W_j(D) = \bigcup_{Q \in \mathcal{W}_j^*(D)} Q(1 + 2^{-j-2})\,.$$

Finally, for $j = 1, 2, \ldots, N$, for Q in $\mathcal{W}_j(D)$, and for t in $(0, 3/2]$, we let

$$V_Q(D, t) = Q_0(t) \cap \alpha_Q^{-1}\big[V_{j-1}(D)\big] \quad , \quad W_Q(D, t) = Q_0(t) \cap \alpha_Q^{-1}\big[W_{j-1}(D)\big]\,.$$

Thus we make the following remark.

LEMMA 8.3.1. *For fixed t in $(0, 3/2]$ the number of different sets $V_Q(D,t)$ (respectively, $W_Q(D,t)$), generated as D varies over all subdomains of $\mathbb{R}^n$ and Q over all cubes in $\mathcal{W}(D)$, is finite and depends only on n and t.*

PROOF. Any set $V_Q(D,t)$ is the union of sets of the type $Q_0(t) \cap \alpha_Q^{-1}[Q'(1+2^{-j})]$, where Q and Q' come from $\mathcal{W}$ and where $1 \le j \le N+1$; i.e., $V_Q(D,t)$ is either empty or is the union of sets from the finite collection $\mathcal{C}_t$. This fact dictates that at most finitely many different sets $V_Q(D,t)$ arise. The same reasoning obviously applies to sets of the kind $W_Q(D,t)$. □

The next result is where Sullivan's theorem is used in the solution of the extension problem.

LEMMA 8.3.2. *Let $\varepsilon > 0$ and $K \ge 1$ be given. There exists a δ in $(0, \varepsilon]$, where δ depends entirely on n, K, and ε, for which the following statement holds:*

- *if D is a domain in $\mathbb{R}^n$,*
- *if Q is a cube from $\mathcal{W}(D)$ such that $V = V_Q(D, 3/2)$ is nonempty,*
- *if g is a member of the family $\bar{\mathcal{G}}_K^*$, and*
- *if $h : V \to \mathbb{H}^{n+1}$ is a* LIP*-embedding for which $d_{\mathbb{H}}(h, g; V) \le \delta$,*

then there exists a LIP*-embedding*

$$\hat{h} : Q_0(3/2) \to \mathbb{H}^{n+1}$$

such that $d_{\mathbb{H}}[\hat{h}, g; Q_0(4/3)] \le \varepsilon$ and such that $\hat{h} = h$ in $W = W_Q(D, 4/3)$.

Furthermore, if h is locally λ-bilipschitz with respect to the hyperbolic metric, then $\hat{h}$ can be chosen so that its restriction to $Q_0(4/3)$ is locally $\hat{\lambda}$-bilipschitz in the hyperbolic metric, where $\hat{\lambda}$ is a function of λ, n, K, and ε.

PROOF. Fix V and W as indicated. We wish to apply Lemma 8.1.6 with $U = Q_0(3/2)$, $U_0 = Q_0(4/3)$, V, W, and $\mathcal{G} = \{g|U : g \in \bar{\mathcal{G}}_K^*\}$. Theorem 8.2.7 tells us that $\mathcal{G}$ is compact in the topology of Euclidean uniform convergence on U. In the present situation $E = \{g(x) : g \in \bar{\mathcal{G}}_K^*,\ x \in \bar{U}\}$, which is a compact set because $\bar{\mathcal{G}}_K^* \times \bar{U}$ is compact in the product topology and the function $(g, x) \mapsto g(x)$ is continuous on $\bar{\mathcal{G}}_K^* \times \bar{U}$. The appeal to Lemma 8.1.6 is therefore justified.

Although its wording is not this explicit, Lemma 8.1.6 guarantees the existence of a function

$$\Delta_n^{V,W} : [1, \infty) \times (0, \infty) \to (0, \infty),$$

which we may assume satisfies $\Delta_n^{V,W}(s,t) \le t$, so that $\delta = \Delta_n^{V,W}(K, \varepsilon)$ fulfills the requirements of the present lemma for the particular pair of sets (V, W).

At the same time it ensures the existence of a function

$$\Lambda_n^{V,W} : [1, \infty) \times [1, \infty) \times (0, \infty) \to [1, \infty).$$

In this instance we may suppose that $\Lambda_n^{V,W}(s,t,u) \ge s$, with the property that if the given embedding h is locally λ-bilipschitz relative to the hyperbolic metric, then $\hat{h}$ can be chosen so that its restriction to $Q_0(4/3)$ is locally $\hat{\lambda}$-bilipschitz with respect to the hyperbolic metric for $\hat{\lambda} = \Lambda_n^{V,W}(\lambda, K, \varepsilon)$.

Lemma 8.3.1 points out that the number of pairs (V, W) that can turn up in the context of Lemma 8.3.2 is finite. Letting Δ_n denote the minimum of the functions $\Delta_n^{V,W}$ for all pairs (V, W) that actually occur, we see that $\delta = \Delta_n(K, \varepsilon)$ is a number with property stated in the lemma.

Similarly, if Λ_n denotes the maximum of the functions $\Lambda_n^{V,W}$, then $\hat{\lambda} = \Lambda_n(\lambda, K, \varepsilon)$ meets the final stipulation of the lemma. □

In an attempt to simplify the statement of the lemma that gives the details of the extension procedure, we establish ahead of time certain parameters that enter into either the result or its proof.

Thus, given $\varepsilon > 0$ and $K \geq 1$, we do the following:

(i) We fix $\rho = \rho(n, K)$ corresponding to $M = d_{\mathbb{H}}[Q_0(5/4), \partial Q_0(4/3)]$ as in Lemma 8.2.8.

(ii) Letting $\delta_N = \min\{\varepsilon, \rho\}$, we define $\delta_{N-1}, \delta_{N-2}, \ldots, \delta_2, \delta_1$ by

$$\delta_j = \Delta_n(K, \delta_{j+1}),$$

where Δ_n is the function that appeared in the proof of Lemma 8.3.2. Then

$$\delta_N \geq \delta_{N-1} \geq \cdots \geq \delta_1 = \delta.$$

(iii) Exploiting the compactness of $\bar{\mathcal{G}}_K^*$ (Theorem 8.2.7) and invoking (8.13) for the compact set $E = \{g(x) : g \in \bar{\mathcal{G}}_K^*,\, x \in \overline{Q_0(3/2)}\,\}$, we select a finite set of mappings $g_1, g_2, \ldots, g_p$ from $\bar{\mathcal{G}}_K^*$—the choice will vary, of course, with K and ε—so that

$$\min_{1 \leq \nu \leq p} d_{\mathbb{H}}\big[g_\nu, g; Q_0(3/2)\big] < \delta/2 \tag{8.20}$$

holds for every g in $\bar{\mathcal{G}}_K^*$.

(iv) For each domain D in $\mathbb{R}^n$, each cube Q in $\mathcal{W}(D)$, and each mapping f from $\mathcal{Q}_K(D)$ we use (6.19) to fix an integer $\nu(Q, f)$ in $\{1, 2, \ldots, p\}$ for which the inequality

$$d_{\mathbb{H}}\big[g_{\nu(Q,f)}, \beta_{Q,f} \circ F_f \circ \alpha_Q; Q_0(3/2)\big] < \delta/2 \tag{8.21}$$

is valid.

(v) We apply Lemma 8.2.9 and choose for $\nu = 1, 2, \ldots, p$ a LIP-embedding $h_\nu : Q_0(3/2) \to \mathbb{H}^{n+1}$ with the property that

$$d_{\mathbb{H}}\big[g_\nu, h_\nu; Q_0(3/2)\big] < \delta/2\,. \tag{8.22}$$

(vi) Lemma 8.1.1 enables us to fix a number $\lambda_1 \geq 1$ so that the restriction of each of the embeddings $h_1, h_2, \ldots, h_p$ to $Q_0(5/4)$ is locally λ_1-bilipschitz with respect to the hyperbolic metric. We then define $\lambda_2, \lambda_3, \ldots, \lambda_N$ by $\lambda_j = \Lambda_n(\lambda_{j-1}, K, \delta_{j-1})$, where Λ_n is the other function that surfaced in the proof of Lemma 8.3.2. Here $\lambda_1 \leq \lambda_2 \leq \cdots \leq \lambda_N$.

The stage is now set for the construction.

LEMMA 8.3.3. *Let $\varepsilon > 0$ and $K \geq 1$ be given, and let the preconditions specified by* (i)–(vi) *above hold. If D is a domain in $\mathbb{R}^n$ and f is a member of the family $\mathcal{Q}_K(D)$, then there exists for $j = 1, 2, \ldots, N$ an embedding $F_f^j : V_j(D) \to \mathbb{H}^{n+1}$ such that F_f^j is locally λ_j-bilipschitz with respect to the hyperbolic metric and such that $d_{\mathbb{H}}[F_f^j, F_f : V_j(D)] \leq \delta_j$.*

PROOF. The proof proceeds by finite induction. Let Q be a cube from $\mathcal{W}_1^*(D)$. From (8.20) and (8.21) we infer that

$$d_{\mathbb{H}}\big[h_{\nu(Q,f)}, \beta_{Q,f} \circ F_f \circ \alpha_Q; Q_0(3/2)\big] \leq \delta\,.$$

Equivalently, since α_Q and $\beta_{Q,f}$ are hyperbolic isometries,

$$d_{\mathbb{H}}\big[\beta_{Q,f}^{-1} \circ h_{\nu(Q,f)} \circ \alpha_Q^{-1}, F_f; Q(3/2)\big] \le \delta\,. \tag{8.23}$$

We define $F_f^1 : V_1(D) \to \mathbb{H}^{n+1}$ by requiring that, for each cube Q belonging to $\mathcal{W}_1^*(D)$, F_f^1 coincide in $Q(5/4)$ with the mapping $\beta_{Q,f}^{-1} \circ h_{\nu(Q,f)} \circ \alpha_Q$.

Because $Q(5/4)$ and $Q'(5/4)$ are disjoint whenever Q and Q' are disjoint cubes from $\mathcal{W}$ by Lemma 8.2.10, F_f^1 is a well-defined function that satisfies the condition

$$d_{\mathbb{H}}\big[F_f^1, F_f; V_1(D)\big] \le \delta_1\,. \tag{8.24}$$

Certainly F_f^1 is one-to-one in $Q(5/4)$ for each cube Q in $\mathcal{W}_1^*(D)$. Indeed, we infer from (vi) and the definition of F_f^1 that the restriction of F_f^1 to $Q(5/4)$ is an embedding which is locally λ_1-bilipschitz with respect to the hyperbolic metric.

Let z and z' be distinct points of $V_1(D)$ for which $d_{\mathbb{H}}(z, z') \le M$. For the reason that

$$M = d_{\mathbb{H}}\big[Q(5/4), \partial Q(4/3)\big] < d_{\mathbb{H}}\big[Q(5/4), Q'(5/4)\big]$$

whenever Q and Q' are different cubes from $\mathcal{W}$, the choice of M dictates that both z and z' be in $Q(5/4)$ for some Q from $\mathcal{W}_1^*(D)$, implying that $F_f^1(z) \ne F_f^1(z')$.

In conjunction with (8.24) and the assumption that $\delta \le \rho$, this fact tells us that F_f^1 is an injection as per Lemma 8.2.8. Consequently, F_f^1 is an embedding of $V_1(D)$ into $\mathbb{H}^{n+1}$ which is locally λ_1-bilipschitz in the hyperbolic metric.

Assume now that $2 \le j \le N$ and that an embedding $F_f^{j-1} : V_{j-1}(D) \to \mathbb{H}^{n+1}$ has been constructed which is locally λ_{j-1}-bilipschitz with respect to the hyperbolic metric and which satisfies the condition

$$d_{\mathbb{H}}\big[F_f^{j-1}, F_f; V_{j-1}(D)\big] \le \delta_{j-1}\,. \tag{8.25}$$

We use F_f^{j-1} in the construction of an embedding $F_f^j : V_j(D) \to \mathbb{H}^{n+1}$ with the desired characteristics.

Noting that

$$V_j(D) = W_{j-1}(D) \cup \bigcup_{Q \in \mathcal{W}_j(D)} Q(1 + 2^{-j-1})\,,$$

we start by setting $F_f^j(z) = F_f^{j-1}(z)$ for z in $W_{j-1}(D)$. Then F_f^j embeds $W_{j-1}(D)$ into $\mathbb{H}^{n+1}$ in a locally λ_{j-1}-bilipschitz fashion with respect to the hyperbolic metric, and the inequality

$$d_{\mathbb{H}}\big[F_f^j, F_f; W_{j-1}(D)\big] \le d\big[F_f^{j-1}, F_f; V_{j-1}(D)\big] \le \delta_{j-1} \le \delta \tag{8.26}$$

holds.

Next, if Q is a cube from $\mathcal{W}_j(D)$ that is disjoint from every cube in $\mathcal{W}_{j-1}(D)$ (by Lemma 8.2.10, this means that $Q(1 + 2^{-j-1})$ does not meet $W_{j-1}(D)$), then we can revert to the procedure used to construct F_f^1 and demand that F_f^j coincide in $Q(1 + 2^{-j-1})$ with $\beta_{Q,f}^{-1} \circ h_{\nu(Q,f)} \circ \alpha_{Q,f}$. In this case the restriction of F_f^j to $Q(1 + 2^{-j-1})$ is locally λ_1-bilipschitz relative to the hyperbolic metric. Moreover, (8.23) applies here and informs us that

$$d_{\mathbb{H}}\big[F_f^j, F_f; Q(1 + 2^{-j-1})\big] \le \delta \le \delta_j \tag{8.27}$$

for any cube Q of the variety under discussion.

Finally, we consider cubes Q from $\mathcal{W}_j(D)$ that have nonempty intersection with one or more cubes from $\mathcal{W}^*_{j-1}(D)$. This guarantees that $V_Q = V_Q(D, 3/2)$ is not empty, so for each cube of the type in question we obtain an embedding $h_Q : V_Q \to \mathbb{H}^{n+1}$ by setting

$$h_Q = \beta_{Q,f} \circ F_f^{j-1} \circ \alpha_Q | V_Q\,.$$

The embedding h_Q is locally λ_{j-1}-bilipschitz with respect to the hyperbolic metric. By (8.25),

$$d_{\mathbb{H}}(h_Q, \beta_{Q,f} \circ F_f \circ \alpha_Q; V_Q) \leq d_{\mathbb{H}}\big[F_f^{j-1}, F_f; V_{j-1}(D)\big] \leq \delta_{j-1}\,.$$

According to Lemma 8.3.2 we can produce for each Q in this category a LIP-embedding $\hat{h}_Q : Q_0(3/2) \to \mathbb{H}^{n+1}$ such that $\hat{h}_Q = h_Q$ in $W_Q = W_Q(D, 4/3) = Q_0(4/3) \cap \alpha_Q^{-1}[W_{j-1}(D)]$ and such that

$$d_{\mathbb{H}}\big[\hat{h}_Q, \beta_{Q,f} \circ F_f \circ \alpha_Q; Q_0(4/3)\big] \leq \delta_j\,. \tag{8.28}$$

Here we recall that $\delta_{j-1} = \Delta_n(K, \delta_j)$.

Furthermore, we may assume that the restriction of $\hat{h}_Q$ to $Q_0(4/3)$ is locally $\hat{\lambda}$-bilipschitz in the hyperbolic metric, where $\hat{\lambda} = \Lambda_n(\lambda_{j-1}, K, \delta_{j-1}) = \lambda_j$. For cubes Q of the present class we define F_f^j in $Q(1 + 2^{-j-1})$ by insisting that it agree in this set with the mapping $\beta_{Q,f}^{-1} \circ \hat{h}_Q \circ \alpha_Q^{-1}$. Since $\hat{h}_Q = h_Q$ in W_Q, this definition forces $F_f^j(z) = F_f^{j-1}(z)$ to hold at any point z of $Q(1 + 2^{-j-1}) \cap W_{j-1}(D)$, so F_f^j is well defined in $Q(1 + 2^{-j-1})$.

Note that if Q and Q' are different cubes from $\mathcal{W}_j(D)$, then $Q(3/2)$ and $Q'(3/2)$ are disjoint. Thus no problems arise concerning the well-definedness of F_f^j in $Q(1 + 2^{-j-1}) \cap Q'(1 + 2^{-j-1})$. The restriction of F_f^j to $Q(1 + 2^{-j-1})$ is locally λ_j-bilipschitz with respect to the hyperbolic metric, and by (8.28)

$$d_{\mathbb{H}}\big[F_f^j, F_f; Q(1 + 2^{-j-1})\big] = d_{\mathbb{H}}\big[\hat{h}_Q, \beta_{Q,f} \circ F_f \circ \alpha_Q; Q_0(1 + 2^{-j-1})\big] \leq \delta_j\,. \tag{8.29}$$

We conclude that F_f^j is a well-defined mapping of $V_j(D)$ into $\mathbb{H}^{n+1}$, that F_f^j is locally λ_j-bilipschitz with respect to the hyperbolic metric as $\lambda_1 \leq \lambda_{j-1} \leq \lambda_j$, and, because of (8.26), (8.27), and (8.29), that

$$d_{\mathbb{H}}\big[F_f^j, F_f; V_j(D)\big] \leq \delta_j\,. \tag{8.30}$$

To complete the induction step—and, with it, the proof of the lemma—we need only confirm that F_f^j is injective.

In view of (8.30), the assumption that $\delta_j \leq \delta_N \leq \rho$, and Lemma 8.2.8, it suffices to check that $F_f^j(z) \neq F_f^j(z')$ whenever z and z' are points of $V_j(D)$ for which $0 < d_{\mathbb{H}}(z, z') \leq M$.

Fix z and z' fitting this description. If z and z' lie in $W_{j-1}(D)$, then $F_f^j(z) = F_f^{j-1}(z) \neq F_f^{j-1}(z') = F_f^j(z')$ by the induction hypothesis. Assume, therefore, that one of the two points z or z' is in $Q(1 + 2^{-j-1})$, where Q is a cube from $\mathcal{W}_j(D)$. For definiteness, say z has this property. Since z then belongs to $Q(5/4)$ and since

$$d_{\mathbb{H}}(z, z') \leq M = d_{\mathbb{H}}\big[Q(5/4), \partial Q(4/3)\big] < d_{\mathbb{H}}\big[Q(1 + 2^{-j-1}), Q'(1 + 2^{-j-1})\big]$$

for any Q' in $\mathcal{W}_j(D)$ other than Q, either z and z' are both points of $Q(1+2^{-j-1})$, in which case it is evident that $F_f^j(z) \neq F_f^j(z')$, or z' is a point of $W_{j-1}(D) \cap Q(4/3)$.

In the latter event $w = \alpha_Q^{-1}(z)$ and $w' = \alpha_Q^{-1}(z')$ are distinct points of W_Q, a set in which $h_Q = \hat{h}_Q$. Accordingly,

$$F_f^j(z') = F_f^{j-1}(z') = \beta_{Q,f}^{-1} \circ h_Q(w') = \beta_{Q,f}^{-1} \circ \hat{h}_Q(w') \neq \beta_{Q,f}^{-1} \circ \hat{h}_Q(w) = F_f^f(z).$$

We infer that F_f^j is an injection, which finishes the proof. □

8.3.2. The extension theorem. The effort that went into Lemma 8.3.3 has an immediate payoff, as we are about to find out. For a domain D in $\mathbb{R}^n$ we shall use $\hat{D}$ as an abbreviation for the interior of $D^*(r_2)$. The extension theorem can now be formulated as follows:

THEOREM 8.3.4. *Suppose that n is a positive integer, that $1 \le K < \infty$, and that $\varepsilon > 0$. Then there exists a constant $\lambda \ge 1$, determined entirely by n, K, and ε, for which the following assertion is true: each K-quasiconformal embedding $f : D \to \mathbb{R}^n$, where D is a domain in $\mathbb{R}^n$, admits an extension to an embedding $F^* : D^*(r_2) \to \mathbb{R}_+^{n+1}$ whose restriction F to the domain $\hat{D}$ is a locally λ-bilipschitz mapping with respect to the hyperbolic metric, and hence a λ^{2n}-quasiconformal mapping of $\hat{D}$ onto a domain in $\mathbb{H}^{n+1}$, which satisfies the condition that $d_{\mathbb{H}}(F, F_f; \hat{D}) \le \varepsilon$.*

PROOF. Fix a domain D in $\mathbb{R}^n$ and a K-quasiconformal embedding $f : D \to \mathbb{R}^n$. Let δ_N, $\lambda = \lambda_N$ and $F_f^N : V_N(D) \to \mathbb{H}^{n+1}$ be as in Lemma 8.3.3. It follows from Lemma 8.2.12 that $V_N(D)$ contains the domain $\hat{D}$. As a result, $F = F_f^N | \hat{D}$ provides an embedding of $\hat{D}$ into $\mathbb{H}^{n+1}$ which is locally λ-bilipschitz in the hyperbolic metric—Lemma 8.1.2 certifies that F is a λ^{2n}-quasiconformal mapping—and obeys the condition $d_{\mathbb{H}}(F, F_f; \hat{D}) \le \delta_N \le \varepsilon$. This inequality makes it plain that

$$\lim_{z \to x} F(z) = \lim_{z \to x} F_f(z) = f(x)$$

for every x in D, so the function $F^* : D^*(r_2) \to \mathbb{R}_+^{n+1}$ given by $F^*(z) = F(z)$ if z belongs to $\hat{D}$ and $F^*(x) = f(x)$ if x is in D provides an embedding with the desired properties. □

There is a special case of Theorem 8.3.4 that deserves a statement of its own; see also Theorem 6.5.20.

THEOREM 8.3.5. *Suppose that n is a positive integer, that $1 \le K < \infty$, and that $\varepsilon > 0$.*

Then there is a constant λ, which depends only on n, K, and ε, such that each K-quasiconformal self-mapping f of $\mathbb{R}^n$ admits an extension to a homeomorphism F^* of $\mathbb{R}_+^{n+1}$ whose restriction F to $\mathbb{H}^{n+1}$ is a λ-bilipschitz homeomorphism of $\mathbb{H}^{n+1}$ with respect to the hyperbolic metric, and hence is a λ^{2n}-quasiconformal self-mapping of $\mathbb{H}^{n+1}$, that satisfies the condition

$$d_{\mathbb{H}}(F, F_f; \mathbb{H}^{n+1}) \le \varepsilon.$$

PROOF. We apply Theorem 8.3.4 in the case $D = \mathbb{R}^n$. Then $D^*(r_2) = \mathbb{R}_+^{n+1}$ and $\hat{D} = \mathbb{H}^{n+1}$. Let f be a K-quasiconformal mapping of $\mathbb{R}^n$ onto itself, and let F^* be an extension of f that meets the specifications of Theorem 8.3.4. The mapping F^* can be extended by reflection to a λ^{2n}-quasiconformal embedding $\hat{F}$ of $\mathbb{R}^{n+1}$ into itself. By the remark after Theorem 6.4.25, $\hat{F}$ maps $\mathbb{R}^{n+1}$ onto itself. It follows that $F^*(\mathbb{R}^{n+1}_+) = \mathbb{R}^{n+1}_+$ and $F(\mathbb{H}^{n+1}) = \mathbb{H}^{n+1}$. Lemma 8.1.3 accounts for the fact that F is globally λ-bilipschitz in $\mathbb{H}^{n+1}$ with respect to the hyperbolic metric. □

A refinement of Theorem 8.3.5 also bears mentioning.

COROLLARY 8.3.6. *Suppose that $n \geq 2$ and that $1 \leq K < \infty$. Then there is a constant λ, which is a function of n and K, such that each K-quasiconformal self-mapping f of $\hat{\mathbb{R}}^n$ admits an extension to a homeomorphism F^* of $\bar{\mathbb{H}}^{n+1}$ whose restriction F to $\mathbb{H}^{n+1}$ is a λ-bilipschitz homeomorphism of $\mathbb{H}^{n+1}$ with respect to the hyperbolic metric and is thus a λ^{2n}-quasiconformal self-mapping of $\mathbb{H}^{n+1}$.*

PROOF. Let λ be a constant that has the feature described in Theorem 8.3.5 for $\varepsilon = 1$. Given a K-quasiconformal mapping f of $\hat{\mathbb{R}}^n$ onto itself, we select a mapping φ from Möb($\mathbb{H}^{n+1}$) that transforms $f(\infty)$ to ∞. We apply Theorem 8.3.5 to the function $g = \varphi \circ f|\mathbb{R}^n$ and arrive at an extension G^* of g to a homeomorphism of $\mathbb{R}^{n+1}_+$ whose restriction to $\mathbb{H}^{n+1}$ is a λ-bilipschitz mapping when viewed from the perspective of the hyperbolic metric.

Moreover, by defining $G^*(\infty) = \infty$ we turn G^* into a homeomorphism of $\bar{\mathbb{H}}^{n+1}$ as the extension $\hat{G}$ of G^* obtained via reflection in $\mathbb{R}^n$ is a quasiconformal mapping of $\mathbb{R}^{n+1}$ onto itself, so $\hat{G}(x) \to \infty$ as $x \to \infty$.

Since φ is both a 1-quasiconformal mapping and a hyperbolic isometry, $F^* = \varphi^{-1} \circ G^*$ yields an extension of f with the stated properties. □

Corollary 8.3.6 has an analogue on the sphere $\mathbb{S}^n$. Indeed, when $n \geq 2$ there is a natural definition of quasiconformality for a homeomorphism f of $\mathbb{S}^n$: f is K-quasiconformal if and only if $\pi^{-1} \circ f \circ \pi$ is K-quasiconformal, where $\pi : \hat{\mathbb{R}}^n \to \mathbb{S}^n$ is the stereographic projection.

For example, if F is a K-quasiconformal mapping of B^{n+1} onto itself and F^* is the homeomorphic extension of F to $\overline{B}^{n+1}$, then $f = F^*|\mathbb{S}^n$ is K-quasiconformal. More generally, if $0 < r < 1$ and F^* is an embedding of $\overline{B}^{n+1} \setminus \overline{B}^{n+1}(r)$ into $\hat{\mathbb{R}}^{n+1}$ such that $F^*(\mathbb{S}^n) = \mathbb{S}^n$ and such that F^* is K-quasiconformal in $B^{n+1} \setminus \overline{B}^{n+1}(r)$, then $f = F^*|\mathbb{S}^n$ is K-quasiconformal.

To see this, let Φ be the Möbius transformation of $\hat{\mathbb{R}}^{n+1}$ that maps $\mathbb{H}^{n+1}$ to B^{n+1} and coincides with π on $\hat{\mathbb{R}}^n$. Theorem 6.5.20 and the remark subsequent to it, applied to $\Phi^{-1} \circ F^* \circ \Phi$, show that $\pi^{-1} \circ f \circ \pi$ is K-quasiconformal. Here we note that the proof of Theorem 6.5.20 and the remark alluded to are actually valid in the more general situation indicated above.

The counterpart of Corollary 8.3.6 for $\mathbb{S}^n$ provides a converse to this observation.

COROLLARY 8.3.7. *Suppose that $n \geq 2$ and that $1 \leq K < \infty$. There is a constant λ, which is a function of n and K, such that each K-quasiconformal self-mapping of $\mathbb{S}^n$ admits an extension to a homeomorphism F^* of $\overline{B}^{n+1}$ whose restriction to B^{n+1} is a λ-bilipschitz homeomorphism of B^{n+1} with respect to the hyperbolic metric and is thus a λ^{2n}-quasiconformal self-mapping of B^{n+1}.*

PROOF. Let Φ be the Möbius transformation of $\hat{\mathbb{R}}^{n+1}$ that entered into the discussion immediately preceding the statement of the corollary. We have noted that Φ is an isometry of $\mathbb{H}^{n+1}$ onto B^{n+1} when these spaces are given their hyperbolic metrics. If f is a K-quasiconformal self-mapping of $\mathbb{S}^n$, then Corollary 8.3.7 informs us that $g = \pi^{-1} \circ f \circ \pi$ has an extension to a homeomorphism G^* of $\overline{\mathbb{H}}n + 1$

whose restriction to $\mathbb{H}^{n+1}$ is hyperbolically λ-bilipschitz, where λ depends only on n and K.

Clearly $F^* = \Phi \circ G^* \circ \Phi^{-1}$ is a homeomorphic extension of f to $\overline{B}^{n+1}$ that in B^{n+1} is λ-bilipschitz relative to the hyperbolic metric. □

Corollaries 8.3.6 and 8.3.7 are by no means the end of the story as far as quasiconformal extensions from $\hat{\mathbb{R}}^n$ to $\mathbb{H}^{n+1}$ or from $\mathbb{S}^n$ to B^{n+1} are concerned. In fact, these results have triggered a hunt for more direct methods of constructing quasiconformal extensions and more explicit ways of representing them. Moreover, it is sometimes required to find extensions that obey certain side conditions. As they stand, Corollaries 8.3.6 and 8.3.7 do not have such flexibility. What one ultimately seeks are extension operators—integral operators, for instance, like the one yielding the Beurling-Ahlfors extension—that produce extensions when the boundary maps are fed into them.

In the case of B^n ($n = 2$ is allowed here) one looks for an operator $\mathbf{E}$ that assigns to each homeomorphism f of $\mathbb{S}^{n-1}$ a homeomorphic extension $F^* = \mathbf{E}(f)$ to $\overline{B}^n$ with the following properties:

(i) $\mathbf{E}(\text{identity}) = \text{identity}$.

(ii) If f arises as the boundary correspondence of some quasiconformal self-mapping of B^n (when $n \geq 3$ this is equivalent to saying that f is a quasiconformal homeomorphism of $\mathbb{S}^{n-1}$), then $F = F^*|B^n$ is itself a quasiconformal mapping and $K(F)$ depends completely on data derived from f.

(iii) $\mathbf{E}$ is continuous with respect to the topology of uniform convergence on $\mathbb{S}^{n-1}$ and some reasonable topology for $C(\overline{B}^n, \mathbb{R}^n)$—for instance the topology of Euclidean uniform convergence on $\overline{B}^n$ or the topology of Euclidean locally uniform convergence in B^n.

The above are the minimal requirements for $\mathbf{E}$. There are numerous other special features which it would be highly desirable for $\mathbf{E}$ to incorporate. We continue our list with several of these.

(iv) F is always a (real analytic?) diffeomorphism of B^n.

(v) If f is a homeomorphism of $\mathbb{S}^{n-1}$ induced by a quasiconformal self-mapping of B^n, then F is λ-bilipschitz in the hyperbolic metric, where λ depends only on data associated with f.

(vi) $\mathbf{E}$ is natural in relation to the action of $\text{Möb}(B^n)$; that is, with the obvious meaning

$$\mathbf{E}(\psi \circ f \circ \varphi) = \psi \circ \mathbf{E}(f) \circ \varphi$$

whenever φ and ψ are Möbius transformations of $\hat{\mathbb{R}}^n$ that preserve B^n—hence, also leave $\mathbb{S}^{n-1}$ invariant Thus, in conjunction with (i), (vi) implies that $\mathbf{E}(\varphi) = \varphi$ for every φ from $\text{Möb}(B^n)$.

(vii) $\mathbf{E}(g \circ f) = \mathbf{E}(g) \circ \mathbf{E}(f)$ for all homeomorphisms f and g of $\mathbb{S}^{n-1}$ or, at least, for any such homeomorphisms that admit quasiconformal extensions to B^n.

In the case $n = 2$ an operator $\mathbf{E}$ with properties (i)–(vi) was discovered by Douady and Earle [**34**]. Although the Douady-Earle operator does have direct analogues in all dimensions, only in the plane is it certain to produce homeomorphic extensions. The existence of an operator enjoying all seven of these properties remains an open question even when $n = 2$.

CHAPTER 9

The Mostow Rigidity Theorem and Discrete Möbius Groups

9.1. Introduction and statement of the theorem

Mostow rigidity is one of the seminal applications of the theory of higher-dimensional quasiconformal mappings and one of the most important results in geometry and topology discriminating, as it does, between the theory of Riemann surfaces and their deformation spaces—called Teichmüller spaces—in two dimensions, and their higher-dimensional counterparts which admit no nontrivial deformations. The theorem essentially states that the geometry of a finite volume hyperbolic manifold of dimension $n \geq 3$ is determined by the fundamental group and hence by the topology. The theorem was proven for closed manifolds by G. Mostow in 1968 [**125**] and extended to finite volume manifolds by G. Prasad in 1973 [**136**].

The applications of many of the basic ideas in the proof of Mostow rigidity have grown significantly in recent years in connection with M. Gromov's theory of hyperbolic groups [**62**] and now encompass the quasi-isometric classification of groups and the study of quasisymmetric mappings between the abstract boundaries of groups—both key ideas in Mostow's original proof. As to wider applications, Mostow rigidity was used by William Thurston to prove the uniqueness of circle packing representations of triangulated planar graphs.

In this chapter we therefore aim to present a relatively comprehensive discussion of the theorem and its proof from which the reader should be able to approach these more general topics in geometric group theory and elsewhere with concrete examples at hand.

George (Dan) Mostow was born in 1923 and was awarded the Leroy P. Steele Prize for Seminal Contribution to Research in 1993 for his book *Strong Rigidity of Locally Symmetric Spaces* from 1973. The rigidity phenomenon for lattices in Lie groups that he discovered and explored is known as Mostow rigidity. A. Weil, A. Borel and others had earlier proved a number of closely related theorems (see [**166, 167**] and the memorial written by A. Borel [**24**]). These implied that cocompact—meaning $\mathbb{H}^n/\Gamma$ is compact—discrete groups of isometries of hyperbolic space of dimension at $n \geq 3$ have no nontrivial continuous deformations.

This was part of a more general theory which sought to study those deformations of a discrete subgroup, with compact quotient, of a connected semi-simple Lie group which has no connected compact normal subgroup. These things are discussed fairly comprehensively in Raghunathan's book [**137**].

Although there are now other proofs of Mostow rigidity which avoid much of the quasiconformal analysis, such as that presented in the lecture notes of M. Gromov and P. Pansu [**63**] from 1991 (and earlier work) and proofs using the theory of harmonic mappings initiated by J. Jost, Sui, and S-T. Yau, and others from the 1980s (the survey of R. Spatzier [**148**] is worth looking at here for these things), these proofs are influenced by Mostow's in important ways.

The proof given here follows and simplifies some of the ideas of S. Agard [**1**] as refined by P. Tukia [**155**] in avoiding the use of various ergodic theory results (the most commonly used of which is the lemma due to Mautner as refined by L.V. Ahlfors). In doing this we will also manage to avoid a lot of hyperbolic geometry necessary in proving the ergodicity of the action of a lattice on its boundary (perhaps not a good thing), but the proof becomes much more transparent as a result and we will pick up many of the key ideas in the theory of discrete groups anyway. We will therefore give a complete and elementary proof of this classical version of Mostow rigidity theorem:

THEOREM 9.1.1. *Let M^n and N^n be diffeomorphic closed hyperbolic manifolds of dimension $n \geq 3$. Then M^n and N^n are isometric.*

Of course Theorem 9.1.1 has significant extensions and in its most general form is part of the Mostow-Margulis super-rigidity theory concerning lattices in semi-simple Lie groups. The proof we give will actually prove the following version of this theorem which replaces the hypothesis that M^n and N^n are closed and diffeomorphic with the weaker assumption that M^n and N^n are closed and homotopy equivalent.

THEOREM 9.1.2. *Let M^n and N^n be closed hyperbolic manifolds of dimension $n \geq 3$. If $\pi_1(M^n) \cong \pi_1(N^n)$, then M^n and N^n are isometric. In particular M^n and N^n have the same volume.*

Perhaps the most well-known version of this theorem is the next result we state, which is stronger in the sense that the hypothesis that the manifolds concerned are closed is replaced by a hypothesis regarding their volume. We will go some way toward proving this more general statement, but at the end of the day there are complications inherent in moving from the case of closed manifolds to finite volume manifolds that we cannot hope to cover here in any reasonable way. We will however make some brief comments about this transition.

THEOREM 9.1.3. *Let M^n and N^n be finite volume hyperbolic manifolds of dimension $n \geq 3$. If $\pi_1(M^n) \cong \pi_1(N^n)$, then M^n and N^n are isometric. In particular M^n and N^n have the same volume.*

This theorem says that in dimension greater than two an algebraic isomorphism between the fundamental groups of two finite volume hyperbolic manifolds of the same dimension is enough to imply that the manifolds are isometric.

This last theorem is obviously false in two dimensions. For instance the two Riemann surfaces $F_1 = \mathbb{C} \setminus \{0, -1, 1\}$ and $F_2 = \mathbb{C} \setminus \{0, -1, 2\}$ (both diffeomorphic to each other) admit hyperbolic metrics, and by the Signature Formula—found in A. Beardon's introductory book [**16**, Theorem 10.4.3]—of A. Hurwitz and L. Siegel each F_i has finite area 4π.

Any isometry $F_1 \to F_2$ will certainly be conformal and therefore extend over the isolated singularities to a conformal map of $\hat{\mathbb{C}}$ which must yield a linear fractional transformation of the Riemann sphere to itself. This linear fractional transformation must map $\{-1, 0, 1, \infty\}$ to $\{-1, 0, 2, \infty\}$. This is a contradiction, as the cross ratios of these two sets of four points is different—yet linear fractional transformations preserve cross ratios, (3.3.1).

This contradiction implies that F_1 and F_2 are not isometric, nor indeed are they conformally equivalent. It is the theory of Teichmüller and the associated moduli spaces which shows that for a surface of genus g with n punctures, there is a $3g - 3 + n$-complex-dimensional space of conformally distinct (and therefore not isometric) surfaces which are smoothly and quasiconformally equivalent. Note that these are not the same thing for noncompact surfaces.

In the case of the four-times punctured sphere which we have just discussed above, there is a one-complex-dimensional moduli space of conformally inequivalent surfaces; the complex cross ratio locally parameterizes this space. This is a striking contrast to Theorem 9.1.1.

Remarkably the proof we give for Theorem 9.1.1 boils down to a question of the regularity of the boundary values of a quasiconformal self-homeomorphism of B^n, for, as we shall see, this is the only place that the hypothesis on dimension is used.

In three dimensions, the induced boundary map of the two sphere will be quasiconformal and differentiable almost everywhere. In two dimensions the induced boundary map of the circle will, as we have observed, be a quasisymmetric mapping that may be singular. This fact will at least allow the possibility of deformation spaces, and of course such deformation spaces not only exist, but have an incredibly interesting and useful theory of their own, far too vast to expose here.

The hypothesis regarding finite volume is necessary. William Thurston [**152**], his co-authors, and many others have shown that there is a very rich and interesting theory of deformation spaces of hyperbolic structures on infinite volume manifolds, particularly in dimension three.

We also note here that the results of Grigori Perelman and Richard Hamilton which establish Thurston's geometrisation conjecture show that in three dimensions the most prevalent geometry a manifold could be expected to admit is hyperbolic. Mostow's rigidity theorem then shows the fundamental group to be the primary topological invariant in three dimensions.

We will prove Theorem 9.1.1 but we will omit some details for the proof of Theorem 9.1.2. This is because some of the necessary additional tools are quite technical and would need a significant digression into the geometry of Kleinian groups and the structure of the ends of noncompact hyperbolic spaces.

There are a number of steps in the proof of Mostow's rigidity theorem, and many of them are completely general in the sense that they are valid in all dimensions and are actually just aspects of the theory of covering spaces. Some require a little knowledge of the geometry of Möbius groups which we shall develop. Thus we discuss hyperbolic manifolds, covering spaces, and Möbius groups, and then quasi-isometries and the limit set of a Möbius group before establishing the rigidity theorem.

9.2. Hyperbolic manifolds, covering spaces and Möbius groups

We recall from our discussion in Section 3.4 that $\mathbb{H}^n$ is the n-dimensional hyperbolic space identified as the upper half-space

$$\{x = (x_1, x_2, \ldots, x_n) \in \mathbb{R}^n : x_n > 0\} \tag{9.1}$$

endowed with the Riemannian metric

$$ds_{hyp} = \frac{|dx|}{x_n}. \tag{9.2}$$

The boundary of $\mathbb{H}^n$ is therefore naturally identified with $\partial\mathbb{H} = \hat{\mathbb{R}}^{n-1}$, and we have already called this the *sphere at infinity.*

9.2.1. Hyperbolic manifolds. A hyperbolic n-manifold M^n is a manifold locally modelled on hyperbolic space. Thus there is a cover of M^n by charts $\{(U_i, \varphi_i)\}_{i\geq 0}$, $U_i \subset M^n$, $M^n \subset \bigcup_{i\geq 0} U_i$ with $\varphi : U_i \to \mathbb{H}^n$ homeomorphisms such that for each i, j the map

$$\varphi_i \circ \varphi_j^{-1} : \varphi_j(U_i \cap U_j) \to \varphi_i(U_i \cap U_j) \subset \mathbb{H}^n \tag{9.3}$$

is a hyperbolic isometry between these open subsets of hyperbolic n-space.

In Theorem 3.4.4 we identified the full isometry group of hyperbolic space as $\text{Möb}(\mathbb{H}^n)$ and exhibited that the Poincaré extension 3.3.2 and (3.27) showed a one-to-one correspondence between $\text{Möb}(\mathbb{H}^n)$ and $\text{Möb}^+(n-1)$, the group of Möbius transformations of $\hat{\mathbb{R}}^{n-1} = \partial\mathbb{H}^n$.

As a local hyperbolic isometry, the map $\varphi_i \circ \varphi_j^{-1}$ is actually the restriction of a hyperbolic isometry of $\mathbb{H}^n$. There are various ways to see this—it is in fact a general property of isometry groups of all reasonable metrics. Perhaps the easiest way in our case is to use the ball model, let $x_0 \in \varphi_j(U_i \cap U_j)$, and put $y_0 = (\varphi_i \circ \varphi_j^{-1})(x_0)$. Now choose isometries g and h with $h(0) = x_0$ and $g(y_0) = 0$. Then $g \circ \varphi_i \circ \varphi_j^{-1} \circ h$ fixes the origin, and is an isometry in a neighbourhood of that point. Thus it preserves distances to 0 and maps the standard orthogonal basis to another orthogonal basis. It quickly follows that this map is actually the restriction of an orthogonal transformation, say O. Thus $\varphi_i \circ \varphi_j^{-1} = g^{-1} \circ O \circ h^{-1}$ defines a global extension. Another way—at least in $n \geq 3$—is to realise that the map $\varphi_i \circ \varphi_j^{-1}$ will certainly be conformal and, therefore by the Liouville theorem, the restriction of a Möbius transformation of $\hat{\mathbb{R}}^n$. This transformation must be a hyperbolic isometry.

In any case, it follows that the map defined in (9.3) is the restriction to $\varphi_j(U_i \cap U_j)$ of an isometry of $\mathbb{H}^n$.

For a hyperbolic manifold, an *atlas* is a maximal (ordered by inclusion) collection of charts (U_i, φ_i) for which (9.3) holds.

We have already seen that the Möbius group acts transitively on $\mathbb{H}^n$ and that for each $x \in \mathbb{H}^n$ the point stabilizers

$$\Gamma_x = \{\gamma \in \text{Möb}(\mathbb{H}^n) : \gamma(x) = x\} \cong \text{O}(n-1).$$

In what follows we shall restrict our attention to the orientation-preserving isometries of hyperbolic n-space; we continue to denote this space as $\text{Isom}^+(\mathbb{H}^n)$.

An alternative definition of a hyperbolic manifold is that it is a manifold M^n which admits a Riemannian metric of constant negative sectional curvature. We do not wish to develop all the necessary differential geometry and the theory of covering spaces to define all the terms here, especially as this is so well done in Joseph Wolf's excellent book *Spaces of Constant Curvature* [**169**]. We simply wish to record the following 1891 theorem of W. Killing [**90**] as refined in 1926 by H. Hopf [**81**], but see [**169**, Corollary 2.4.10] for a more modern account.

THEOREM 9.2.1. *Let M^n be a Riemannian manifold of dimension $n \geq 2$ and let K be a real number. Then M^n is complete, connected, and of constant curvature K if and only if it is isometric to a quotient*

- $\mathbb{S}^n/\Gamma$ *with* $\Gamma \subset \mathrm{O}(n)$, *the orthogonal group, if* $K > 0$,
- $\mathbb{R}^n/\Gamma$ *with* $\Gamma \subset \mathbb{E}(n)$, *the group of Euclidean isometries, if* $K = 0$,
- $\mathbb{H}^n/\Gamma$ *with* $\Gamma \subset \mathrm{Isom}(\mathbb{H}^n)$, *the group of hyperbolic isometries, if* $K < 0$,

where the group Γ *acts freely and properly discontinuously.*

There are a few things that need a little explanation here.

First the quotient space X/Γ is defined as $X/\sim$, where the equivalence relation is defined by $x \sim y$ if and only if there is $\gamma \in \Gamma$ with $\gamma(x) = y$.

Next, the group Γ acts *freely* if the only element with a fixed point is the identity. Note that in the second two cases the group action naturally extends to the boundary ($\{\infty\}$ for $\mathbb{R}^n$ and $\hat{\mathbb{R}}^{n-1}$ for $\mathbb{H}^n$), where the transformation must have fixed points.

Finally a group acts *properly discontinuously* on X if for each compact $F \subset X$ we have

$$\#\{\gamma \in \Gamma : \gamma(F) \cap F \neq \emptyset\} < \infty.$$

Thus the translates of any compact set can return back to that set only finitely often.

As each of the spaces $\mathbb{S}^n, \mathbb{R}^n$, and $\mathbb{H}^n$ is simply connected, we see that $\Gamma \cong \pi_1(M^n)$, the *fundamental group*. More on fundamental groups and covering spaces can be found in almost any book on algebraic topology, for instance [**61**].

That the action is free and properly discontinuous implies that for each $x \in X$ there is $r > 0$, so

$$\gamma(B^n(x,r)) \cap B^n(x,r) = \emptyset \quad \text{if } \gamma \in \Gamma \setminus \{identity\}.$$

Such a small round ball $B^n(x,r)$ will then project isometrically into the quotient space, and this implies that the quotient map $\Pi : \mathbb{H}^n \to M^n$ is locally an isometry, and so in particular, it is conformal. This also provides charts which give the hyperbolic manifold structure on the quotient.

9.2.2. Discrete groups. A subgroup Γ of $\mathrm{Isom}^+(\mathbb{H}^n)$ is said to be *discrete* if the identity is isolated in Γ in the topology of local uniform convergence in $\mathbb{H}^n$. This means that if $\{\gamma_i\}_{i\geq 0}$ is a sequence in Γ and if $\gamma_i \to identity$ locally uniformly in $\mathbb{H}^n$ as $i \to \infty$, then $\gamma_i = identity$ for all i sufficiently large.

The strong convergence properties of sequences of Möbius transformations, such as those exhibited in Theorem 3.6.7, and the equivalence of the various topologies on the Möbius group, either coming from their action on $\hat{\mathbb{R}}^n$ or from their representation as matrix groups, shows that almost any weak discreteness property will imply discreteness in the stronger topology, say uniform convergence in the spherical metric.

In fact Theorem 3.6.7 quickly shows us that a properly discontinuous group is discrete. For groups of isometries the converse is not difficult.

THEOREM 9.2.2. *A subgroup* Γ *of* $\mathrm{Isom}^+(\mathbb{H}^n)$ *is properly discontinuous if and only if it is discrete.*

PROOF. As we have observed above the only real issue here is to show that the discrete subgroup Γ of $\mathrm{Isom}^+(\mathbb{H}^n)$ is properly discontinuous on $\mathbb{H}^n$. If $F \subset \mathbb{H}^n$ is compact and $\{\gamma_i\}_{i\geq 0}$ is a distinct sequence of elements of Γ with $\gamma_i(F) \cap F \neq \emptyset$, then all $x_0 \in F$ have the property that

$$\rho_{\mathbb{H}}[x_0, \gamma_i(x_0)] \leq 2\mathrm{diam}_H(F).$$

Theorem 3.6.7 now asserts that the sequence $\{\gamma_i\}_{i\geq 0}$ contains a subsequence $\{\gamma_{i_k}\}_{k\geq 0}$ which converges to γ locally uniformly in $\mathbb{H}^n$. Of course γ must be a Möbius transformation in $\mathrm{Isom}^+(\mathbb{H}^n)$. Then the sequence $\gamma_{i_{k+1}} \circ \gamma_{i_k}^{-1}$ converges locally uniformly in $\mathbb{H}^n$ to the identity by Lemma 3.6.8.

If Γ is discrete, this is not possible unless $\gamma_{i_{k+1}} \circ \gamma_{i_k}^{-1} = \mathit{identity}$ for all sufficiently large k. This means that $\gamma_{i_{k+1}} = \gamma_{i_k}$ for sufficiently large k and contradicts the choice of the initial sequence consisting of distinct elements. We deduce that it cannot be possible to find the γ_i and that Γ is therefore properly discontinuous, as we required. □

Two tesselations of hyperbolic 2-space.

A discrete group of hyperbolic isometries gives rise to a tessellation of hyperbolic space, the orbit of a fundamental domain (a domain constructed so as to contain exactly one point from each orbit equivalence class). Two examples are illustrated above, one where the fundamental domain is compact and one where it is not. On the left the orbit of a point is R-dense for some finite R (see Theorem 9.5.1 below) while on the right it is not; an orbit may miss the "horoballs" illustrated. In both cases the fundamental domain has finite volume—area in this case. We note in passing, and just for general interest, that Poincaŕe gave a straightforward method to construct a fundamental domain. Let Γ be a discrete group of hyperbolic isometries, none of which fix the point $0 \in B^n$. For $\gamma \in \Gamma$ define the half-space $H_\gamma = \{x \in B^n : \rho_{\mathbb{H}}(0,x) \leq \rho_{\mathbb{H}}(\gamma(0),x)\}$. Then put

$$\mathcal{D} = \bigcap_{\gamma \in \Gamma} H_\gamma. \tag{9.4}$$

Then an elementary argument based on discreteness shows that $\mathcal{D}$ is a nonempty convex set containing 0 and contains at least one point from every orbit, and that the interior of $\mathcal{D}$ contains exactly one point from any orbit. Thus $\Gamma(\mathcal{D}) = B^n$ defines a tessellation by a convex body. In the two cases above, this is a hyperbolic triangle.

It will be useful to have these examples in mind when we discuss normal limit points and so forth below.

Having considered the notion of properly discontinuous as it pertains to Möbius groups, we next consider the conclusion that the covering space action is free in Theorem 9.2.1.

It is immediate that any element $\gamma \in \Gamma$, a discrete group of Möbius transformations of B^n, which fixes a point $x \in B^n$ must have finite order. Simply put $F = \{x\}$ in the definition of proper discontinuity to see that $\#\{\gamma \in \Gamma : \gamma(x) = x\} < \infty$, and so in particular if $\gamma(x) = x$, then $\{\gamma^n : n \in \mathbb{Z}\}$ can have only finitely many distinct elements.

Such an element is called a torsion or elliptic element of Γ. The *order* of such an element is

$$\operatorname{ord}(\gamma) = \min_{n \geq 1}\{n : \gamma^n = \mathit{identity}\}.$$

Elements of finite order are conjugate into the orthogonal group and have a non-trivial fixed point set in $\mathbb{H}^n$. To see this, let γ have finite order m, choose $x_0 \in B^n$ and consider the finite set of points $X = \{\gamma^k(x_0) : k \in \mathbb{Z}\}$. There is a unique smallest hyperbolic ball—and so a round ball $B = B^n(y_0, r)$—which contains this finite set. We have used this fact before; uniqueness is easily deduced from the hyperbolic cosine law and some elementary trigonometry. Then $\gamma(B)$ is a hyperbolic ball of the same radius containing $\gamma(X) = X$, so $\gamma(B) = B$, and hence $\gamma(y_0) = y_0$ by uniqueness. This shows that the fixed point set of γ in B^n is nonempty. Next conjugate the group $\Gamma' = \langle\gamma\rangle$ by an isometry α for which $\alpha(y_0) = 0$. Then

$$\Gamma'^{\alpha} = \alpha \circ \Gamma' \circ \alpha^{-1}$$

is an isomorphic group which fixes the origin and the $(n-1)$-sphere boundary of B^n. Thus Γ'^{α} must be a cyclic subgroup of the orthogonal group.

A discrete group Γ is *torsion free* if it contains no elliptic elements. This discussion has shown us that the following lemma is true. It implies that a hyperbolic manifold can found as the quotient space (sometimes the term *orbit space* is used) of a torsion free discrete group of Möbius transformations acting on $\mathbb{H}^n$, or equivalently on B^n.

LEMMA 9.2.3. *Let Γ be a discrete subgroup of the isometry group* $\operatorname{Isom}^+(\mathbb{H}^n)$. *Then Γ acts freely on $\mathbb{H}^n$ if and only if Γ is torsion free. In this case the orbit space $M = \mathbb{H}^n/\Gamma$ is a hyperbolic manifold.*

The *volume* of the hyperbolic manifold $M = \mathbb{H}^n/\Gamma$ is defined to be the hyperbolic volume of a fundamental domain for Γ. For instance,

$$\mathrm{vol}_{\mathbb{H}}(\mathbb{H}^n/\Gamma) := \mathrm{vol}_{\mathbb{H}}(\mathcal{D}) = \int_{\mathcal{D}} \frac{1}{x_n^n}\, dx, \tag{9.5}$$

where $\mathcal{D}$ is the fundamental domain defined at (9.4). In the ball model of hyperbolic space this integral is

$$\mathrm{vol}_{\mathbb{H}}(B^n/\Gamma) = \mathrm{vol}_{\mathbb{H}}(\mathcal{D}) = \int_{\mathcal{D}} \frac{dx}{(1-|x|^2)^n},$$

where of course $\mathcal{D}$ has changed.

Notice that if $\mathcal{D}$ is compact (in which case we call Γ *cocompact* in $\mathbb{H}^n$), then x_n is bounded above and below by a number greater than 0, and so the integral at (9.5) is finite. In order for a noncompact convex fundamental domain to have finite volume, it must have cusps of certain type at the boundary. We do not have the space to go into this, but there are quite strong restrictions on a noncompact convex body in $\mathbb{H}^n$ to have finite hyperbolic volume.

If Γ is a discrete subgroup of the isometry group $\mathrm{Isom}^+(\mathbb{H}^n)$, then Γ is often called the *orbifold* fundamental group of $\mathbb{H}^n/\Gamma$. In the case where the group is torsion free, Γ is algebraically isomorphic to the usual fundamental group as we have noted.

Also, as noted earlier, the restriction to torsion free subgroups, necessary to guarantee a free action, implies that the quotient space is a manifold. However the theory of discrete groups of Möbius transformations is often enriched by studying these more general groups without the restrictive hypothesis of being torsion free, and in much of what follows we will simply consider discrete groups of Möbius transformations of $\mathbb{H}^n$ so as to have application in greater generality. Of course it is a consequence of a well-known result of Atle Selberg, together with the fact that $\mathrm{Isom}^+(\mathbb{H}^n)$ can be represented as a matrix group, that a finitely generated Möbius group has a torsion free subgroup of finite index, [**144**], so the geometric actions of these two groups, the one with torsion and the finite index subgroup without, are quite closely related. In fact we will see below that they are quasi-isometric groups.

9.3. Quasiconformal manifolds and quasiconformal mappings

So now we have a reasonable description of the objects to which Mostow rigidity theorem relates, namely hyperbolic manifolds. We will need to consider the mappings between them. However at this point we take a digression to define what it means for a map between Riemannian (or more general) manifolds to be quasiconformal since there is another striking result of Dennis Sullivan sitting in the background here—one which we will not be able to prove but to which the reader should be made aware.

First we need to define what it means for a manifold to admit a quasiconformal structure and then what it means for a map between quasiconformal manifolds to be quasiconformal.

Let M^n be a topological manifold. We say M^n admits a quasiconformal structure if M^n admits an atlas $\{U_i, \varphi_i\}_{i\in I}$ such that for each $i, j \in I$ the transition charts

$$\varphi_i \circ \varphi_j^{-1} : \varphi_j(U_i \cap U_j) \to \mathbb{R}^n$$

are quasiconformal.

If M^n and N^n are topological manifolds which admit a quasiconformal structure, with $\{U_i^M, \varphi_i^M\}_{i\in I}$ an atlas of M^n and $\{U_j^N, \varphi_j^N\}_{j\in J}$ an atlas of N^n, then a homeomorphism $f : M^n \to N^n$ is quasiconformal if for each $i \in I$ and $j \in J$ the mapping between the open subset of $\mathbb{R}^n$ defined by

$$\varphi_j^N \circ f \circ (\varphi_i^M)^{-1} : \varphi_i^M(U_i^M) \to \varphi_j^N(U_j^N) \tag{9.6}$$

is quasiconformal (when it is in fact defined).

This local definition does not allow us to define what it means for a manifold to be K-quasiconformal or what it means for a map between these quasiconformal manifolds to be K-quasiconformal. This is easily remedied for particular manifolds, but the terminology becomes particularly clumsy and the number K depends on the structures.

Since locally a diffeomorphism between open subsets of $\mathbb{R}^n$ is quasiconformal, we see that smooth manifolds are quasiconformal. Evidently hyperbolic manifolds are 1-quasiconformal, as indeed are all the space forms of Theorem 9.2.1. But 1-quasiconformal manifolds (or more precisely conformal manifolds) form a more general class. For instance $\mathbb{S}^{n-1} \times \mathbb{S}^1$ admits a conformal structure given as the quotient of the simply connected space $\mathbb{R}^n \setminus \{0\}$ by the properly discontinous free group of Möbius transformations $\langle x \mapsto \lambda^n x : n \in \mathbb{Z}\rangle$ and $\lambda \neq 1$. This is not a space form since it is a product and therefore has some sectional curvatures equal to 0, while others will be $+1$. Notice that for different λ few of these manifolds are conformally equivalent, yet all of them are quasiconformally homeomorphic.

We now state Sullivan's theorem [**150**].

THEOREM 9.3.1. *Every topological manifold of dimension n, $n \neq 4$, admits a unique quasiconformal structure.*

Uniqueness means the following thing. If $\{U_i, \varphi_i\}_{i\in I}$ and $\{V_i, \psi_i\}_{i\in I}$ are both quasiconformal atlas' for M^n, then there is a quasiconformal map

$$f : (M^n, \{(U_i, \varphi_i)\}) \to (M^n, \{(V_i, \psi_i)\}).$$

Note that this map might not be the identity! The situation in dimension 4 is discussed in the paper of Simon Donaldson and Sullivan [**33**] and is not yet fully clarified.

Actually, with Sullivan's approximation result already in hand in Theorem 8.1.4, the proof for the existence of quasiconformal structures given by Theorem 9.3.1 follows using a well laid out path in geometric topology when studying structures of various types, for instance, smooth, piecewise linear, Lipschitz, and so forth. However, this path lies in a different direction to ours and we leave the matter there.

Hyperbolic manifolds admit an easier characterisation of quasiconformal maps between them which we shall now explore.

9.3.1. Maps between hyperbolic manifolds. Let $M^n = \mathbb{H}^n/\Gamma_M$ and $N^n = \mathbb{H}^n/\Gamma_N$ be hyperbolic manifolds, and let $\Pi_M : \mathbb{H}^n \to M^n$ and $\Pi_N : \mathbb{H}^n \to N^n$ be the conformal covering projection maps. A mapping $f : M^n \to N^n$ is K-*quasiconformal* if the locally defined homeomorphism

$$\Pi_N^{-1} \circ f \circ \Pi_M$$

between subsets of $\mathbb{H}^n$ is locally K-quasiconformal.

Since $\mathbb{H}^n$ is simply connected it is the universal covering space of both M^n and N^n, and the theory of covering spaces shows us that this locally defined map is a well-defined quasiconformal mapping of $\mathbb{H}^n$; see for instance [**169**, Section 1.8]. We therefore have the following theorem.

THEOREM 9.3.2. *Let M^n and N^n be hyperbolic manifolds and let $f : M^n \to N^n$ be a homeomorphism. Then there is a homeomorphism $\tilde{f} : \mathbb{H}^n \to \mathbb{H}^n$ such that the following diagram commutes:*

$$\begin{array}{ccc} \mathbb{H}^n & \xrightarrow{\tilde{f}} & \mathbb{H}^n \\ \Pi_M \downarrow & \circ & \downarrow \Pi_N \\ M^n & \xrightarrow{f} & N^n \end{array}$$

That is to say,

$$(f \circ \Pi_M)(x) = (\Pi_N \circ \tilde{f})(x) \qquad \textit{for each } x \in \mathbb{H}^n \ . \tag{9.7}$$

The map f is K-quasiconformal if and only if the map $\tilde{f}$ is K-quasiconformal.

Another way of looking at equation (9.7) is the following. Any continuous map $g : M^n \to N^n$ induces a homomorphism $g_* : \pi_1(M^n) \to \pi_1(N^n)$ of the fundamental groups. Thus we identify a homomorphism $\phi = g^* : \Gamma_M \to \Gamma_N$ between discrete Möbius groups. Then g lifts to the universal cover and equation (9.7) tells us that

$$\tilde{g} \circ \gamma = \phi(\gamma) \circ \tilde{g}$$

or equivalently $\tilde{g} \circ \Gamma_M = \Gamma_N \circ \tilde{g}$, and if $\tilde{g}$ is a homeomorphism, $\tilde{g} \circ \Gamma_M \circ \tilde{g}^{-1} = \Gamma_N$.

At this point there is no restriction on the dimension and we have arrived at a fairly general setup. We now want to discuss a generalisation of the notion of quasiconformality in the hyperbolic setting, although it initially seems as though we are looking at a generalisation of the notion of a bilipschitz map; this impression is dispelled in Theorem 9.4.3.

9.4. Quasi-isometries

A continuous map $f : \mathbb{H}^n \to \mathbb{H}^n$ is called a *quasi-isometry* if there are constants $L \geq 1$ and $D \geq 0$ such that for all $x, y \in \mathbb{H}^n$,

$$\frac{1}{L}\,\rho_{\mathbb{H}}(x,y) - D \leq \rho_{\mathbb{H}}\big[f(x), f(y)\big] \leq L\,\rho_{\mathbb{H}}(x,y) + D, \tag{9.8}$$

and if for each $w \in \mathbb{H}^n$,

$$\mathrm{dist}_{\mathbb{H}}[w, f(\mathbb{H}^n)] \leq D. \tag{9.9}$$

These two conditions can be used to define a quasi-isometry between general metric spaces; for instance, in various generalisations of the case at hand these metric spaces might be the boundaries of groups.

The condition (9.8) shows that a quasi-isometry is like a bilipschitz mapping at large scales, and the additional constant D allows the mapping to be noninjective locally—or exhibit a different modulus of continuity from bilipschitz mappings, for instance if it were locally Hölder continuous—but it must be injective at large scales. The next lemma is obvious.

LEMMA 9.4.1. *Let $f : \mathbb{H}^n \to \mathbb{H}^n$ be an (L, D)-quasi-isometry. If $\rho_{\mathbb{H}}(x, y) > LD$, then $f(x) \neq f(y)$.*

9.4.1. Quasiconformal mappings of $\mathbb{H}^n$ are quasi-isometries. The condition (9.9) implies that f should be nearly a surjection. In particular as $w \to \partial\mathbb{H}^n$, $w \in \mathbb{H}^n$, we must have the Euclidean distance $\mathrm{dist}(y, f(\mathbb{H}^n)) \to 0$ and we might expect that f extends to the boundary. Let us first prove that quasiconformal maps are quasi-isometries, a result first established in by F.W Gehring and B. Osgood, [**53**]. Before doing this we make an elementary geometric observation in the next lemma.

LEMMA 9.4.2. *Let $x, y \in \mathbb{H}^n$ with*

$$\frac{|x-y|}{\mathrm{dist}(x, \partial\mathbb{H}^n)} < \frac{3}{4}.$$

Then

$$\frac{1}{2}\,\rho_{\mathbb{H}}(x,y) \le \frac{|x-y|}{\mathrm{dist}(x,\partial\mathbb{H}^n)} \le 2\,\rho_{\mathbb{H}}(x,y). \tag{9.10}$$

PROOF. Since x, y and the hyperbolic geodesic between them all lie in some 2-dimension hyperbolic subspace, it is enough to prove the result in $\mathbb{H}^2$. We can use a similarity preserving $\mathbb{H}^2$ so as to assume $x = i$; note that the quantity $|x-y|\mathrm{dist}^{-1}(x, \partial\mathbb{H}^n)$ is preserved by such similarities. If $|i - z| = t < 1$, then

$$\log(1+t) \le \rho_{\mathbb{H}}(i,z) \le \log\frac{1}{1-t}.$$

Then $-\log(1-t) \le 2t$ if $t \le 0.79$, and $\log(1+t) > \frac{1}{2}t$ when $t < 1$. □

We now have the quasi-isometry estimate of the following theorem. A number $L = L(n, K)$ appears in the proof, but presumably it is far from best possible.

THEOREM 9.4.3. *Let $f : \mathbb{H}^n \to \mathbb{H}^n$ be K-quasiconformal. Then f is an $(L,1)$-quasi-isometry in the hyperbolic metric where L depends only on the dimension n and on K.*

PROOF. We only need to prove the right-hand inequality of (9.8) since we can consider f^{-1} which also satisfies the hypotheses to obtain the left-hand one. First recall Theorem 6.6.4 with $D = \mathbb{H}^n$ to obtain the inequality

$$\frac{|f(y)-f(x)|}{\mathrm{dist}[f(x), \partial\mathbb{H}^n]} \le \Theta_K^n\left(\frac{|y-x|}{\mathrm{dist}(x,\partial\mathbb{H}^n)}\right) \tag{9.11}$$

whenever x belongs to D and $|y - x| < \mathrm{dist}(x, \partial\mathbb{H}^n)$. We want to integrate this inequality—the relationship between (9.11) and the hyperbolic metric is clear. Let α be the hyperbolic geodesic between x and y.

We partition α into k-subarcs α_i, $i = 1, 2, \ldots, k$, each with endpoints x_{i-1} and x_i, and $x_0 = x$, such that

$$\frac{|x_i - x_{i-1}|}{\mathrm{dist}(x_i, \partial\mathbb{H}^n)} = a < \frac{3}{4}.$$

Then from (9.11)

$$\frac{|f(x_i) - f(x_{i-1})|}{\mathrm{dist}[f(x_i), \partial\mathbb{H}^n]} \le \Theta_K^n(a).$$

We now assume that $\Theta_K^n(a) < \frac{3}{4}$ as well, and sum over k:

$$\begin{aligned}
\rho_{\mathbb{H}}(f(x), f(y)) &\leq \sum_k \rho_{\mathbb{H}}(f(x_i), f(x_{i-1})) \leq 2\sum_k \frac{|f(x_i) - f(x_{i-1})|}{\operatorname{dist}[f(x_i), \partial \mathbb{H}^n]} \\
&\leq 2k\Theta_K^n(a) = 2k\left[\frac{\Theta_K^n(a)}{a}\right] a = \frac{2\Theta_K^n(a)}{a} \sum_k \frac{|x_i - x_{i-1}|}{\operatorname{dist}(x_i, \partial\mathbb{H}^n)} \\
&\leq \frac{4\Theta_K^n(a)}{a} \sum_k \rho_{\mathbb{H}}(x_i, x_{i-1}) = \frac{4\Theta_K^n(a)}{a} \rho_{\mathbb{H}}(x, y). \qquad (9.12)
\end{aligned}$$

Our earlier estimates on the distortion function Θ_K^n assure us that $4\Theta_K^n(a)/a \to \infty$ as $a \to 0$.

Thus one might seek to balance the choice of a and the constant D in the definition of (L, D)-quasi-isometry.

If we choose to put $D = 1$, then there is nothing to prove unless $\rho_{\mathbb{H}}[f(x), f(y)] > 1$. By what we have already proved in (9.12) and if we make sure we satisfy the constraints on a,

$$\rho_{\mathbb{H}}(x, y) \geq \frac{a}{4\Theta_K^n(a)} = \delta.$$

Any hyperbolic geodesic of hyperbolic length at least δ can be partitioned into an integral number of subarcs of length b, where $\delta/2 \leq b \leq \delta$. Therefore (with our a priori bound on a) we have an integral number of subarcs $\alpha[x_i, x_{i-1}]$ with $|x_i - x_{i-1}|\operatorname{dist}^{-1}(x_i, \partial\mathbb{H}^n) = b$ with $b \in [\delta/4, \delta]$. We now put $b = \delta/4$, instead of a, into our estimate in (9.12). To satisfy our constraints we put $a = (\Theta_K^n)^{-1}(3/4) < \frac{3}{4}$, $\delta = \frac{1}{3}(\Theta_K^n)^{-1}(3/4) = \frac{a}{3}$. Then we achieve

$$L = \frac{48\Theta_K^n(a/12)}{a}, \qquad a = (\Theta_K^n)^{-1}(3/4)$$

and the quasi-isometry estimate

$$\rho_{\mathbb{H}}[f(x), f(y)] \leq L\, \rho_{\mathbb{H}}(x, y) + 1. \qquad (9.13)$$

The near surjectivity requirement (9.9) is redundant in our situation, and thus the proof is complete. □

What is missing from this result, and what is clear from compactness, is that we should have $L \to 1$ as $K \to 1$. The proof we offered cannot achieve this, no matter how careful we are with various choices, because of the bad behavior of the distortion function as $K \to 1$. We recall that

$$\Theta_K^n(t) \approx \lambda_n^{2-K^{1/(1-n)}} t^{K^{1/(1-n)}},$$

so our L grows like $R^{K^{1/(n-1)}}$, for a constant $R = R(n) > 1$.

Actually, in the proof of the above Theorem 9.4.3 we have relied on a distortion result which is valid for all domains. Minor modifications subsequently lead to the conclusion that quasiconformal mappings are quasi-isometries in the quasi-hyperbolic metric. This is actually one of the results of the previously cited [**53**]. We leave the reader to figure out the details here—part of the solution will be to establish the existence of quasi-hyperbolic geodesics; see [**52**].

Next, we would like the converse. This cannot be true in general since quasi-isometries do not need to be continuous or injective and quasiconformal mappings do. The answer lies in consideration of the boundary values, and the theorem we want was basically established in the 1963/64 result of V.A. Efremovich and E.S. Tihomirova [**36**]. Here we present a refined and updated version of that result.

THEOREM 9.4.4. *Let $f : \mathbb{H}^n \to \mathbb{H}^n$ be an (L, D)-quasi-isometry. Then f is a bounded hyperbolic distance from a K-quasiconformal mapping $F : \mathbb{H}^n \to \mathbb{H}^n$. The mapping F extends continuously to the boundary $\partial\mathbb{H}^n = \mathbb{R}^{n-1}$, and this boundary mapping $F_0 : \mathbb{R}^{n-1} \to \mathbb{R}^{n-1}$ is also K-quasiconformal when $n \geq 3$ and quasisymmetric when $n = 2$. Here K depends only on the dimension n and the numbers L and D.*

Moreover if f is compatible with a Möbius group in the sense that there are Möbius groups Γ and $\tilde{\Gamma}$ acting on $\bar{\mathbb{H}}^n$ so that $f \circ \Gamma = \tilde{\Gamma} \circ f$, then we also have

$$F_0 \circ \Gamma = \tilde{\Gamma} \circ F_0.$$

In order to prove this result we will need to develop some additional machinery, in particular the Morse lemma on quasigeodesics. Before we do this, we present some motivation and explain how quasi-isometries arise in the Mostow rigidity theorem. The proof will be completed in 9.6.2.

9.5. Groups as geometric objects

One of the important applications of the theory of quasi-isometries arises when we start treating groups as geometric objects. This was a study initiated by Micha Gromov in his seminal paper [**62**]. A finitely presented group

$$G = \langle a_1, a_2, \ldots, a_n : W_1 = W_2 = \ldots W_N = 1 \rangle \tag{9.14}$$

has a natural metric—namely the reduced word length (on this choice of generators) or equivalently the path metric in the Cayley graph—which again depends on the choice of generators.

Without going into too much detail we can briefly explain this, as it is necessary in what follows.

If $w \in G$ we have $w = \prod_{k=1}^{\ell} a_{i_k}^{m_k}$, $i_k \in \{1, \ldots, n\}$, and $m_k \in \mathbb{Z}$, $m_k \neq 0$, and we may also assume that w is a maximally reduced word, that is, reduced by the relations $W_i = 1$ at (9.14). Then we set

$$d_G(id_G, w) = \sum_k |m_k|,$$

where id_G is the identity element of G, and extend the metric by the rule

$$d_G(w, \tilde{w}) = d(id_G, \tilde{w}w^{-1}).$$

We leave it as an exercise to show that this is a well-defined metric—the *word metric*— and that with a different choice of a finite generating set the identity map from G to itself is a quasi-isometry. Notice also that with this definition the group G acts by right multiplication ($g(w) = wg$) on the metric space (G, d_G), and this action is by isometries: if $w, \tilde{w}$ and g are elements of G, then

$$d_G[g(w), g(\tilde{w})] = d_G(wg, \tilde{w}g) = d(id_G, \tilde{w}gg^{-1}w^{-1}) = d(id_G, \tilde{w}w^{-1}) = d_G(w, \tilde{w}).$$

What we want to show next is that the groups involved in the rigidity theorem are actually quasi-isometric to hyperbolic space itself when given this word metric.

The proof we give for the next theorem follows the ideas of E. Ghys and P. de La Harpe [**57, 58**]. A subgroup Γ of $\mathrm{Isom}^+(\mathbb{H}^n)$ is said to have *R-dense orbit* if there is $R > 0$ and $x_0 \in \mathbb{H}^n$ such that for every $x \in \mathbb{H}^n$,

$$\mathrm{dist}_{\mathbb{H}}(x, \{\gamma(x_0) : \gamma \in \Gamma\}) \leq R.$$

In particular, if there is a compact set $\bar{D}$ which tessellates $\mathbb{H}^n$ under the action of the group, so $\bigcup_{\gamma\in\Gamma} \gamma(\bar{D}) = \mathbb{H}^n$, then Γ has an R-dense orbit, with $R = \mathrm{diam}_{\mathbb{H}}(D)$. Of course, the converse is true as well.

THEOREM 9.5.1. *Let Γ be a discrete Möbius group acting on hyperbolic space $\mathbb{H}^n$ with R-dense orbit. Then the function $f : \Gamma \to \mathbb{H}$ defined by*

$$f(\gamma) = \gamma(\mathbf{j}), \qquad \mathbf{j} = (0, 0, \ldots, 0, 1) \in \mathbb{H}^n,$$

is a quasi-isometry between Γ with the word metric and $\mathbb{H}^n$ with the hyperbolic metric.

In particular this means that there are constants L and D so that

$$\frac{1}{L}\,\rho_{\mathbb{H}}(x, y) - D \leq d_\Gamma\big[f(\gamma), f(\tilde{\gamma})\big] \leq L\,\rho_{\mathbb{H}}(x, y) + D. \tag{9.15}$$

PROOF. We may assume that it is the orbit of $\mathbf{j}$ that is R-dense. Let $B_R = B_\rho(\mathbf{j}, R)$—the hyperbolic ball about $\mathbf{j}$ of radius R. As Γ is discrete and therefore discontinuous on $\mathbb{H}^n$, there is an $r > 0$ so that with $B_r = B_\rho(\mathbf{j}, r)$ we have $\gamma(B_r) \cap B_r = \emptyset$.

Let $\eta_1, \eta_2, \ldots, \eta_n$ be a list of those finitely many words in $\Gamma \setminus \{id_\Gamma\}$ for which $\gamma_i(B_R) \cap B_R \neq \emptyset$. We claim that $\{\eta_1, \eta_2, \ldots, \eta_n\}$ is a generating set of Γ and, then using this generating set to get the metric d_Γ, we have that for each $\gamma \in \Gamma$

$$d_\Gamma(id_\Gamma, \gamma) \leq \rho_{\mathbb{H}}[x_0, \gamma(x_0)]/r + 1. \tag{9.16}$$

Let $\gamma \in \Gamma$ and α be the hyperbolic geodesic from $\mathbf{j}$ to $\gamma(\mathbf{j})$ and choose points x_i, $i = 0, \ldots, m-1$, along α with $x_0 = \mathbf{j}$, $x_m = \gamma(x_0)$, $\rho_{\mathbb{H}}(x_i, x_{i+1}) = r$, $i = 0, \ldots, n-1$, and $0 < \rho_{\mathbb{H}}(x_{m-1}, x_m) \leq r$.

Since $\Gamma(\mathbf{j})$ is R-dense, for each $j = 0, 1, \ldots, m$ we have $x_j \in \gamma_j(B_R)$, although this choice may not be unique.

Notice that for $j = 0, \ldots, m-1$, $\gamma_j(B_R) \cap \gamma_{j+1}(B_R) \neq \emptyset$ as the distance between the centers of these balls is $r \leq R$. This implies $B_R \cap \gamma_j^{-1}\gamma_{j+1}(B_R) \neq \emptyset$, and so $\gamma_j^{-1}\gamma_{j+1}$ must lie among our chosen $\eta_1, \eta_2, \ldots, \eta_n$.

Thus, there is some k with $\gamma_{j+1} = \eta_k \gamma_j$. Inductively we see $\gamma = \eta_{i_1} \circ \cdots \circ \eta_{i_\ell}$, and so the η form a generating set and we evidently have (9.16) as $d_\Gamma(id_\Gamma, \gamma) \leq m \leq \rho(\mathbf{j}, \gamma(\mathbf{j}))/r + 1$. By definition $d_\Gamma(\gamma, \eta\gamma) = d_\Gamma(id_\Gamma, \eta)$, and so we also have

$$d_\Gamma(\gamma, \eta\gamma) = d_\Gamma(id_\Gamma, \eta) \leq m \leq \rho(\mathbf{j}, \eta(\mathbf{j}))/r + 1 = \rho(\gamma(\mathbf{j}), \gamma \circ \eta(\mathbf{j}))/r + 1.$$

This is the right-hand side of the definition of quasi-isometry. Next, since for every $\eta \in \{\eta_1, \eta_2, \ldots, \eta_n\}$ we have $\rho(\mathbf{j}, \eta(\mathbf{j})) \leq 2R$, it is simply the triangle inequality which yields

$$\begin{aligned} \rho(f(\gamma), f(\tilde{\gamma}(\mathbf{j}))) &= \rho(\gamma(\mathbf{j}), \tilde{\gamma}(\mathbf{j})) = \rho(\mathbf{j}, \gamma^{-1}\tilde{\gamma}(\mathbf{j})) \\ &\leq 2R d_\Gamma(id_\Gamma, \gamma^{-1}\tilde{\gamma}) = 2R d_\Gamma(\gamma, \tilde{\gamma}), \end{aligned}$$

and this completes the proof. □

Notice that the proof has shown that if there is an R dense orbit, the group Γ is finitely generated. As another consequence we have the following corollary whose proof amounts to unwinding the quasi-isometry condition (9.15) and noting that if Γ is torsion free, then $\gamma(\mathbf{j}) = \tilde{\gamma}(\mathbf{j})$ implies $\gamma = \tilde{\gamma}$. Actually for an arbitrary finitely generated discrete group of Möbius transformations Selberg's lemma shows there is only a finite indeterminacy here (the maximal order of a point stabiliser), and so the result remains true without the assumption that Γ is torsion free, but we will not need this.

COROLLARY 9.5.2. *Let Γ be a torsion free discrete Möbius group acting on $\mathbb{H}^n$ with R dense orbit $\Gamma(\mathbf{j})$. Then there is $P : \mathbb{H}^n \to \Gamma(\mathbf{j})$, an equivariant nearest point retraction, so that the function $P^* : \mathbb{H}^n \to \Gamma$ defined by*

$$P^*(x) = \gamma,$$

where $\gamma(\mathbf{j}) = P(x)$, is a quasi-isometry.

PROOF. The nearest point retraction is not well defined at every point. Since the group is discrete the orbit $\Gamma(\mathbf{j})$ is discrete as well. Then it is clear that the nearest point retraction is a well-defined and continuous function (locally constant in fact) on an open dense set. Let $x, y \in \mathbb{H}^n$ and suppose $\gamma_x(\mathbf{j})$ and $\gamma_y(\mathbf{j})$ are respective closest orbit points. Then, by the triangle inequality we have both

$$\begin{aligned} \rho_{\mathbb{H}}[\gamma_x(\mathbf{j}), \gamma_y(\mathbf{j})] &\leq \rho_{\mathbb{H}}(x,y) + \rho_{\mathbb{H}}[x, \gamma_x(\mathbf{j})] + \rho_{\mathbb{H}}[y, \gamma_y(\mathbf{j})] \leq \rho_{\mathbb{H}}(x,y) + 2R \quad \text{and} \\ \rho_{\mathbb{H}}[\gamma_x(\mathbf{j}), \gamma_y(\mathbf{j})] &\geq \rho_{\mathbb{H}}(x,y) - \rho_{\mathbb{H}}[x, \gamma_x(\mathbf{j})] - \rho_{\mathbb{H}}[y, \gamma_y(\mathbf{j})] \geq \rho_{\mathbb{H}}(x,y) - 2R. \end{aligned}$$

Thus however a nearest point is chosen and we will find ourselves with a $(1, 2R)$-quasi-isometry. However, there is a slight technical problem with equivariance when we try to assign values to points where the retract is not well defined. Clearly if the retract is well defined at x, then it is well defined at $\gamma(x)$ since $\gamma[\Gamma(\mathbf{j})] = \Gamma(\mathbf{j})$.

To ensure equivariance we partition $\mathbb{H}^n$ into orbit equivalence classes, pick an x from each class and choose the nearest point $\gamma_x(\mathbf{j})$ to x, or make a random choice if such a point is not unique. Then define $P[\gamma(x)] = \gamma[\gamma_x(\mathbf{j})]$ on the equivalence class. Since Γ acts by isometry, it is not difficult to see that our construction does not change the nearest point retraction when that is well defined—there is a unique closest point. By construction this map is now well defined and equivariant. □

9.5.1. Isomorphisms and quasi-isometries. Clearly an algebraic isomorphism between abstract groups induces an isometry in the word metrics (of course with respect to the image/preimage of a generating set). We have therefore arrived at the following theorem which is the first step to relating an isomorphism of Möbius groups and geometric mappings of $\mathbb{H}^n$.

THEOREM 9.5.3. *Let Γ and $\tilde{\Gamma}$ be torsion discrete Möbius groups acting on $\mathbb{H}^n$, each with an R dense orbit. Suppose there is an isomorphism $\varphi : \Gamma \to \tilde{\Gamma}$. Then there is a quasi-isometry $f : \mathbb{H}^n \to \mathbb{H}^n$ which commutes with the group actions in the sense that $f \circ \Gamma = \tilde{\Gamma} \circ f$, or more precisely,*

$$f \circ \gamma = \varphi(\gamma) \circ f, \qquad \text{for every } \gamma \in \Gamma. \tag{9.17}$$

PROOF. We may suppose that in both instances the orbit of $\mathbf{j}$ is R-dense. Let P_Γ and $P_{\tilde{\Gamma}}$ denote the equivariant nearest point retractions as per Corollary 9.5.2. Then the map of hyperbolic space defined by

$$x \mapsto P_\Gamma(x) = \gamma(\mathbf{j}) \mapsto \gamma \in \Gamma \mapsto \varphi(\gamma) \in \tilde{\Gamma} \mapsto [\varphi(\gamma)](\mathbf{j}) \in \mathbb{H}^n$$

is the composition of quasi-isometries and so is a quasi-isometry. If $\gamma_0 \in \Gamma$, then

$$\begin{aligned}\gamma_0(x) &\mapsto P_\Gamma(\gamma_0(x)) = \gamma_0[P_\Gamma(x)] = \gamma_0[\gamma(\mathbf{j})] = (\gamma_0 \circ \gamma)(\mathbf{j}) \\ &\mapsto \gamma_0 \circ \gamma \mapsto \varphi(\gamma_0 \circ \gamma) \mapsto \varphi(\gamma_0 \circ \gamma)(\mathbf{j}) = \varphi(\gamma_0)[\varphi(\gamma)(\mathbf{j})].\end{aligned}$$

This shows us the map commutes with the group actions. □

We must now connect quasi-isometries of $\mathbb{H}^n$ with the theory of quasiconformal mappings and establish Theorem 9.4.4. This is done through the study of the boundary values of quasi-isometric mappings. The easiest way to get to the boundary is to study what happens to geodesic lines under quasi-isometic mappings, and this is the content of the next section.

9.5.2. Quasigeodesics and the Morse Lemma. The image $\tilde{\alpha} = f(\alpha)$ of a hyperbolic geodesic α under a quasi-isometry f of $\mathbb{H}^n$ is called a hyperbolic *quasigeodesic*. We will henceforth drop the term hyperbolic. If the quasi-isometry constants associated with the quasi-isometry f are (L, D), as per (9.8), then we refer to $\tilde{\alpha}$ as an (L, D)-quasigeodesic.

Negative curvature forces some differences between quasigeodesics in Euclidean space (under Euclidean quasi-isometries) and hyperbolic space. For instance, as an exercise the reader is encouraged to prove that the logarithmic spiral in $\mathbb{R}^2$ is a quasigeodesic in $\mathbb{R}^2$. This spiral deviates by any arbitrary amount from a straight line.

The next lemma shows that this cannot happen in hyperbolic space. We say that a function $f : \mathbb{H}^n \to \mathbb{H}^n$ has the *fellow traveller* property with constant $A \geq 0$ if whenever α is a hyperbolic geodesic, then there is a hyperbolic geodesic α_f such that

$$\rho_{\mathbb{H}}[f(x), \alpha_f] \leq A, \quad \rho_{\mathbb{H}}[f(\alpha), y] \leq A \qquad \text{for all } x \in \alpha \text{ and } y \in \alpha_f. \tag{9.18}$$

Note that there is no continuity assumption in the definition of fellow traveller functions. The key observation here is that under these assumptions on a fellow travelling function, the only accumulation points of the image $f(\alpha)$ on the boundary $\partial\mathbb{H}^n$ are the endpoints of α_f; moreover the image $f(\alpha)$ lies in a hyperbolic cylindrical neighbourhood of α_f. This gives us a way of defining the boundary values of a mapping with the fellow traveller property which will be injective if it is well defined. The next theorem is most often called the Morse–Mostow lemma, or simply the Morse lemma.

THEOREM 9.5.4. *Given $L \geq 1$ and $D \geq 0$ there exists a constant $A = A(L, D) > 0$ such that any (L, D)-quasi-isometry has the A-fellow traveller property.*

PROOF. Let α be the (L, D)-quasigeodesic. Then there is a hyperbolic geodesic β and an (L, D)-quasi-isometry $h : \mathbb{H}^n \to \mathbb{H}^n$ with $h(\beta) = \alpha$.

We need a little argument to circumvent the fact that h may not be continuous or have a locally defined inverse. Choose $x_0^\beta \in \beta$ and let x_i^β, $i \in \mathbb{Z}$, be ordered points along β at hyperbolic distance 1 from each other; so $\rho_{\mathbb{H}}(x_i^\beta, x_{i+N}^\beta) = |N|$. Let $x_i = h(x_i^\beta)$ and $B_i = B_\rho(x_i, L/2 + D)$. Since h is an (L, D)-quasi-isometry, if we denote by $[x_i, x_{i+1}]$ the hyperbolic segment, then

$$\alpha \subset h\Big[\bigcup_{i\in\mathbb{Z}} B_\rho(x_i^\beta, 1/2)\Big] \subset \bigcup_{i\in\mathbb{Z}} h[B_\rho(x_i^\beta, 1/2)] \subset \bigcup_{i\in\mathbb{Z}} B_\rho(x_i, L/2 + D) = \bigcup_{i\in\mathbb{Z}} B_i\,.$$

Now the balls B_i form a connected chain. Put

$$A = 2(L + D) \geq 1.$$

Let M be a large integer and let $\tilde{\alpha}_M$ be the hyperbolic geodesic which connects x_{-M} and x_M. We want to know the length of the largest chain of balls which lie outside the hyperbolic semi-cylinder $\mathcal{C}_A = \{x \in \mathbb{H} : \rho_{\mathbb{H}}(x, \tilde{\alpha}_M) \leq A\}$. Suppose it is N and label the sequence of centers $x_k \ldots x_{k+N}$. Now

$$\rho_{\mathbb{H}}(x_k, x_{k+N}) \geq \frac{1}{L}\rho_{\mathbb{H}}(x_k^\beta, x_{k+N}^\beta) - D = \frac{N}{L} - D. \tag{9.19}$$

There is another path from x_k to x_{k+n} that we should consider. Namely we traverse $x_k \to x_{k-1} \to x' \to x'' \to x_{k-N+1} \to x_{k+N}$, where x' is the closest point of $\tilde{\alpha}_M$ to x_{k-1} and x'' is the closest point of $\tilde{\alpha}_M$ to x_{k+N+1}. By construction $\rho_{\mathbb{H}}(x', x_{k-1}) \leq A + L/2 + D$ and $\rho_{\mathbb{H}}(x'', x_{k+N+1}) \leq A + L/2 + D$ since the balls B_{k-1} and B_{k+N+1} both come within distance A of $\tilde{\alpha}_M$. To estimate the distance from x' to x'' we note that the segment between them is covered by the projections of the balls B_{k+i}.

We need the following elementary lemma of hyperbolic trigonometry.

LEMMA 9.5.5. *Let α be a hyperbolic geodesic and $B = B_\rho(x, r)$ a ball with* $\operatorname{dist}_{\mathbb{H}}(B, \alpha) = R$. *Let δ be the length of the nearest point projection of B on α. Then*

$$\sinh(\delta/2) = \sinh(r)/\cosh(R + r).$$

In particular

$$\delta \leq 2e^{-R}. \tag{9.20}$$

PROOF. We only need to consider the Lambert quadrilateral in the hyperbolic plane spanned by α and x and which is formed by the shortest geodesic segment connecting x to α of length $R + r$, the endpoint of this geodesic to an endpoint x' of the projection of B on α, the geodesic segment perpendicular to α starting at x' and tangent to B at x'', and the radius of B from x to x''. The relevant formulas of hyperbolic trigonometry can be found in [**16**, Theorem 7.17.1].

The estimate in (9.20) follows directly. The reader should not be alarmed that there is no term involving r in (9.20); it is "built in" by the assumption $\rho_{\mathbb{H}}(x, \alpha) \geq r + R$. □

Returning to our proof of Theorem 9.5.4 we see that the diameter of the projection of B_i is at most $2e^{-A-L/2-D}$. The triangle inequality now yields

$$\rho_{\mathbb{H}}(x_k, x_{k+N}) \leq 2A + 2(L/2 + d) + 2(N + 2)e^{-A}. \tag{9.21}$$

Combining the equations (9.19) and (9.21) we have

$$\begin{aligned} \frac{N}{L} - D &\leq 2A + 2(L/2 + D) + 2(N + 2)e^{-A}, \\ N(1 - 2e^{-2(L+D)}) &\leq 4(L + D)L + 3LD + L^2 + 2 = 7LD + 5L^2 + 2, \end{aligned}$$

and with a bit of simplification this gives $N \leq 8(L + D)^2$. Thus we have obtained a bound on the length of any chain of balls which lies outside the A neighbourhood of $\tilde{\alpha}_M$. It follows that the distance of any point in such a chain from $\tilde{\alpha}_M$ is at most

$$A + (L + D)(N + 2)/2 = 2(L + D) + (4(L + D)^2 + 1)(L/2 + D) \leq 4(L + D)^3,$$

the factor $(N + 2)/2$ since the chain must go out and come back.

At this point we have established that *independently of* M the hyperbolic distance from the approximating geodesic $\tilde{\alpha}_M$ and α is at most $4(L+D)^3$.

Now that we have the local result, we can get the global result by an application of the Arzela-Ascoli theorem, that is, the nested sequence of subarcs $\{\tilde{\alpha}_M\}_{M\geq 0}$, exhausting α with associated geodesic segments $\tilde{\alpha_n}$ with $\rho_{\mathbb{H}}(\alpha_n, \alpha) < A$. We complete $\tilde{\alpha}_n$ to geodesic lines and note that $\rho_{\mathbb{H}}(x_0, \tilde{\alpha}_n) \leq A$. The space of such lines is compact (viewed as subarcs of circles in $\mathbb{H}^n$), so we can pass to a subsequence to find a limit geodesic line $\tilde{\alpha}$. The convergence of these lines is easily seen to be locally uniform in $\mathbb{H}^n$ in the hyperbolic metric. Now if $x \in \alpha$, then $\rho_{\mathbb{H}}(x, \tilde{\alpha}_n) \leq A$ for all sufficiently large n, so $\rho_{\mathbb{H}}(x, \tilde{\alpha}) \leq A$.

Uniqueness simply comes from the fact that distinct geodesics in hyperbolic space cannot stay a bounded distance from each other. □

The reason we established the Morse lemma lies the following section, which is of independent interest.

9.6. The boundary values are quasiconformal

One of the interesting features of quasi-isometries of negatively curved spaces is that they induce homeomorphisms of the boundary at infinity (when appropriately defined) and appear to have much more regularity there than on the interior. In fact this underpins much of the Mostow rigidity theory for lattices in semisimple lie groups. Here we give the first recognised example of this, due to Mostow of course, although our version has various refinements as the extension theorem for quasiconformal maps was unknown in 1968! Theorem 9.6.1 also covers a large part of the Efremovich–Tihomirova Theorem referred to earlier.

THEOREM 9.6.1. *Let* $f : \mathbb{H}^n \to \mathbb{H}^n$ *be an* (L, D)*-quasi-isometry. Then there is a* K*-quasiconformal mapping* $F : \mathbb{H}^n \to \mathbb{H}^n$*, with* $K = K(L, D)$*, and a constant* $M = M(L, D)$ *such that*

$$\rho_{\mathbb{H}}[f(x), F(x)] < M. \tag{9.22}$$

Therefore f *admits a homeomorphic* K*-quasiconformal extension to* $\partial\mathbb{H}^n$*.*

PROOF. If such a quasiconformal mapping as F exists, then F extends to a homeomorphism $\hat{F} : \bar{\mathbb{H}}^n \to \bar{\mathbb{H}}^n$ and the boundary map $\hat{F}|\mathbb{R}^{n-1} \to \mathbb{R}^{n-1}$ is quasiconformal, Theorem 6.5.21. Then (9.22) implies that f extends to the boundary—the induced boundary map being $\hat{F}|\mathbb{R}^{n-1}$, which is quasiconformal. Thus the second part of the theorem follows from the first. However we cannot follow this approach; it is imperative that we construct F from the boundary values of f which we must first show to exist.

Extending f *to* ∂B^n. Theorem 9.5.4 shows that f has the fellow traveller property with constant $a(L, D)$. It is clear that if $\phi \in \mathrm{Isom}^+(\mathbb{H}^n)$, then both $\phi \circ f$ and $f \circ \phi$ are both (L, D)-quasi-isometries and both have the fellow traveller property with the same constant as f as well. We can therefore replace f by a function (again denoted by f) such that $f(0) = 0$. We wish to extend f to the boundary $\mathbb{S}^{n-1}$. We do this as follows. For $\zeta \in \mathbb{S}^{n-1}$ let $[0, \zeta]$ denote the geodesic ray emanating from 0 and terminating at ζ. This is an infinite subarc of a unique geodesic α. By hypothesis $f(\alpha)$ fellow travels a unique geodesic α_f and we set $f_0(\zeta) = \zeta_f$, where z_f is the endpoint of that ray of α_f a bounded distance from $f([\mathbf{j}, \zeta])$—there is clearly only one such endpoint. This gives us a well-defined extension $f_0 : \mathbb{S}^{n-1} \to \mathbb{S}^{n-1}$.

The boundary map f_0 is injective. Suppose $\zeta_0, \zeta_1 \in \partial\mathbb{H}^n$ and $f_0(\zeta_0) = f_0(\zeta_1)$. Let α be the geodesic between ζ_0 and ζ_1. Then there is α_f, a geodesic fellow travelling $f(\alpha)$. In particular α_f is a bounded distance from $f(\alpha)$. However, by assumption $f(\alpha)$ meets $\mathbb{S}^{n-1}$ at a single point and $f(\alpha) \not\subset \mathbb{S}^{n-1}$. It therefore cannot be a bounded distance from any geodesic. This contradiction ensures that f_0 is injective.

9.6.0.1. *The boundary map f_0 is continuous.* We prove a uniform continuity estimate which will be useful for what is to come. Our first observation is an elementary exercise in hyperbolic geometry.

Suppose that ζ_1 and ζ_2 are points of $\mathbb{S}^{n-1} = \partial B^n$ and that $|\zeta_1 - \zeta_2| = \epsilon$. Let α be the hyperbolic geodesic joining ζ_1 to ζ_2. Then

$$\operatorname{dist}_{\mathbb{H}}(0, \alpha) = \operatorname{arccosh}\left(\frac{2}{\epsilon}\right). \tag{9.23}$$

To see this, first note that the angle θ at the vertex 0 of the hyperbolic triangle formed by ζ_1, ζ_2 and 0 is $\sin(\theta/2) = \epsilon/2$.

If we drop a perpendicular from 0 to α, we form a right angled hyperbolic triangle, and the second cosine law of hyperbolic trigonometry, [**16**], tells us that

$$\cosh \operatorname{dist}_{\mathbb{H}}(0, \alpha) = \frac{1}{\sin(\theta/2)}.$$

This establishes (9.23) above.

Next, suppose that $|\zeta_1 - \zeta_2| \le \delta$ and α is the hyperbolic geodesic joining them. Let α_f be the geodesic joining $f(\zeta_1)$ to $f(\zeta_2)$. If $|f(\zeta_1) - f(\zeta_2)| \ge \epsilon$, then by (9.23) we have the two inequalitities

$$\begin{aligned} \operatorname{dist}_{\mathbb{H}}(0, \alpha_f) &\le \operatorname{arccosh}\left(\frac{2}{\epsilon}\right), \\ \operatorname{dist}_{\mathbb{H}}(0, \alpha) &\ge \operatorname{arccosh}\left(\frac{2}{\delta}\right). \end{aligned}$$

Let $x \in \alpha$ be a point for which $\rho_{\mathbb{H}}[0, f(x)]$ is minimal. Then, since f is an (L, D)-quasi-isometry which fixes the point 0, and since f has the $A(L, D)$ fellow traveller property, we may calculate that

$$\begin{aligned} \operatorname{arccosh}\left(\frac{2}{\epsilon}\right) &\ge \operatorname{dist}_{\mathbb{H}}(0, \alpha_f) \\ &\ge \operatorname{dist}_{\mathbb{H}}(0, f(\alpha)) - A(L, D) = \rho_{\mathbb{H}}[0, f(x)] - A(L, D) \\ &\ge \frac{1}{L}\rho(x, 0) - A(L, D) - D \ge \frac{1}{L}\operatorname{dist}_{\mathbb{H}}(0, \alpha) - A(L, D) - D \\ &\ge \frac{1}{L}\operatorname{arccosh}\left(\frac{2}{\delta}\right) - A(L, D) - D. \end{aligned} \tag{9.24}$$

Since the right-hand side here tends to ∞ as $\delta \to 0$ and since the dependence between ϵ and δ depends only on the constants L and D, we find that (9.24) provides a uniform modulus of continuity. As a continuous injection between compact Hausdorff spaces, f_0 is a homeomorphism.

Note further that there is a uniform continuity estimate for the inverse of the boundary map f_0. Namely, with the same reasoning as above, if $|\zeta_1 - \zeta_2| = \delta$, then

$\mathrm{dist}_{\mathbb{H}}(0,\alpha) = \operatorname{arccosh}\left(\frac{2}{\delta}\right)$ and

$$\begin{aligned}\operatorname{arccosh}\Big(\frac{2}{|f(\zeta_1)-f(\zeta_2)|}\Big) &= \mathrm{dist}_{\mathbb{H}}(0,\alpha_f)\\ &\leq L\,\mathrm{dist}_{\mathbb{H}}(0,\alpha)+D+A(L,D)\\ &= L\operatorname{arccosh}\left(\frac{2}{\delta}\right)+D+A(L,D).\\ &= L\operatorname{arccosh}\left(\frac{2}{|\zeta_1-\zeta_2|}\right)+D+A(L,D).\end{aligned}$$

Then, as $|f(\zeta_1)-f(\zeta_2)|\to 0$, we must have $|\zeta_1-\zeta_2|\to 0$ with a uniform estimate depending only on the constants L and D.

Thus if L and D are fixed, and $\{f_{0,k}\}_{k=1}^{\infty}$ is a sequence of homeomorphisms, each of which is the boundary value of an (L,D)-quasi-isometry f_k for which $f_k(0)=0$, then the sequence is precompact by the Arzela-Ascoli Theorem 3.6.4 as the sequences $\{f_{0,k}\}_{k=1}^{\infty}$ and $\{(f_{0,k})^{-1}\}_{k=1}^{\infty}$ are equicontinuous. As a consequence any limit of such a sequence is a homeomorphism.

9.6.0.2. *The boundary map f_0 is quasiconformal.* Let $\mathcal{F}$ be the family of boundary values of (L,D)-quasi-isometries of the hyperbolic metric of B^n. We have just shown that the members of $\mathcal{F}$ are homeomorphisms. It is clear that $\mathcal{F}$ is Möbius invariant—if $f\in\mathcal{F}$ and ϕ and ψ are Möbius transformations of the ball, then $\phi\circ f\circ\psi\in\mathcal{F}$.

Now we want to show that the family $\mathcal{F}$ is precompact under normalisation, in particular the three-point normalisation discussed in Beurling's Compactness Theorem, 6.6.31. We have just shown above in the equicontinuity estimates (9.24) and (9.25) that the subfamily $\mathcal{F}_0=\{f_0\in\mathcal{F}: f_0=f|\partial\mathbb{H}, f(0)=0\}$ is precompact. Therefore the result we seek will follow if we can show that the set of normalised elements of $\mathcal{F}$ is a compact perturbation by Möbius transformations of $\mathcal{F}_0$.

We therefore collect the normalised elements of $\mathcal{F}$ as

$$\mathcal{F}_*=\{f\in\mathcal{F}: f(\pm e_1)=e_1, f(e_n)=e_n\}.$$

Given $f\in\mathcal{F}_*$ we first seek to bound $\rho_{\mathbb{H}}[0,f(0)]$. Let α_1 join $\pm e_1$, α_2 join e_1 to e_n, and α_3 join $-e_1$ to e_n. Notice that $0\in\alpha_1$ while

$$\mathrm{dist}_{\mathbb{H}}(0,\alpha_i)=\operatorname{arccosh}\left(\frac{1}{\sqrt{2}}\right),\qquad i=2,3.$$

Since f fixes the endpoints of these three geodesics, the fellow traveller of $f(\alpha_i)$ is α_i. Then

$$\begin{aligned}\mathrm{dist}_{\mathbb{H}}[f(0),\alpha_i] &\leq \mathrm{dist}_{\mathbb{H}}[f(0),f(\alpha_i)]+\mathrm{dist}_{\mathbb{H}}[\alpha_i,f(\alpha_i)]\\ &\leq L\,\mathrm{dist}_{\mathbb{H}}(0,\alpha_i)+D+A(L,D)\\ &\leq L\operatorname{arccosh}\left(1/\sqrt{2}\right)+A(L,D)+D-M.\end{aligned}$$

Now M is an absolute constant depending only on L and D and not on f.

It is an elementary exercise in hyperbolic geometry to see that the set of points in B^n whose hyperbolic distance from the three geodesics α_i, $i=1,2,3$, is at most M is a compact hyperbolically convex region in B^n—this is all we need, but the reader can make an explicit calculation along the following lines: symmetry considerations reduce the problem to two dimensions. Then convexity identifies the extremal point x_0 as the unique point at hyperbolic distance M from both α_1 and α_2 (or its symmetric point equidistant from α_1 and α_3). The bound is then $\rho_{\mathbb{H}}(0, x_0)$. One can move the situation to $\mathbb{H}^2$ with $0 \mapsto \mathbf{j}$ and consider the hyperbolic triangle formed from the lines $\ell_1 = \{\Im m(z) = 0\}$, $\ell_2 = \{\Im m(z) = 1\}$, and the geodesic segment β of length $M > \mathrm{dist}_{\mathbb{H}}(\mathbf{j}, \ell_2)$ perpendicular to ℓ_1 (meeting ℓ_2 at the image of x_0). Then β meets ℓ_1 at a point, say it, $t > 1$. The triangle inequality shows that the distance we seek to bound is less than $\rho_{\mathbb{H}}(\mathbf{j}, it) + \rho_{\mathbb{H}}(it, x_0) = \log t + M$. Then t can be found from the computation of the hyperbolic length,

$$\ell(\beta) = \frac{1}{t} \int_{\arccos(1/t)}^{\pi/2} \frac{d\theta}{\sin(\theta)} = \frac{1}{t} \log \left[\cot \left[\frac{\mathrm{arcsech}(t)}{2} \right] \right] = M.$$

The family of all Möbius transformations, or hyperbolic isometries, of B^n moving the origin $0 \in B^n$ no more than a fixed finite distance is compact, a direct consequence of Theorem 3.6.7 and Lemma 3.6.9. We have therefore shown that $\mathcal{F}_* \subset \mathcal{C} \circ \mathcal{F}_0$, with $\mathcal{C}$ a compact set of Möbius transformations. Thus $\mathcal{F}_*$ is precompact, and every limit is a homeomorphism—as $\mathcal{F}_0$ has this latter property.

We have now shown that $\mathcal{F}$ is a precompact modulo two-point normalization and that $\mathcal{F}$ is stable with respect to similarity transformations. Beurling's Compactness Theorem 6.6.31 implies that the family $\mathcal{F}$ consists of K-quasiconformal homeomorphisms of $\mathbb{S}^{n-1}$, for some finite K. Thus K depends only on L, D, and the dimension n.

We have now proved that the boundary values f_0 of an (L, D) quasi-isometry f are K-quasiconformal with K depending only on L and D. The proof of Theorem 9.6.1 will be complete once we show that any two quasi-isometries with the same boundary values are a bounded distance apart in the hyperbolic metric—this is Theorem 9.6.3 below. This follows as f_0 admits a quasiconformal extension to $\mathbb{H}^n$ which is bilipschitz in the hyperbolic metric by the extension result of Corollary 8.3.7. □

9.6.1. Quasi-isometries and boundary values. Here we give the appropriate version of Corollary 6.7.3 where we showed that quasiconformal mappings of $\mathbb{H}^n$ with the same boundary values are a bounded hyperbolic distance apart. First we compare quasi-isometries with the identity of $\mathbb{S}^{n-1}$ as boundary values.

LEMMA 9.6.2. *If $f : \bar{\mathbb{H}}^n \to \bar{\mathbb{H}}^n$ is an (L, D)-quasi-isometry on $\mathbb{H}^n$ and $f\big|\partial\mathbb{H}^n =$ identity, then for all $x \in \mathbb{H}$,*

$$\rho_{\mathbb{H}}[x, f(x)] \leq A(L, D),$$

where $A(L, D)$ is the fellow traveller constant.

PROOF. Let $x \in \mathbb{H}^n$ and choose a Möbius transformation φ so that $\varphi(0) = x$. Then $\rho_{\mathbb{H}}[x, f(x)] = \rho_{\mathbb{H}}[\varphi(0), f(\varphi(0))] = \rho_{\mathbb{H}}[0, (\varphi^{-1} \circ f \circ \varphi)(0)]$. Now $\varphi^{-1} \circ f \circ \varphi$ is an (L, D)-quasi-isometry and has the $A(L, D)$ fellow traveller property. It is also the identify on the boundary, and hence $\alpha_f = \alpha$ for every geodesic line α. Hence $\rho_{\mathbb{H}}(\alpha, f(0)) \leq A(L, D)$ for every geodesic line passing through 0, since $f(0) \in f(\alpha)$, which fellow travels with α. Hence $\rho_{\mathbb{H}}(0, f(0)) \leq A(L, D)$. □

THEOREM 9.6.3. *There are numbers $a = a(L, D, n)$ and $K = K(L, D, n)$ with the following property. Let $f : \mathbb{H}^n \to \mathbb{H}^n$ be an (L, D)-quasi-isometry; then there is a K-quasiconformal mapping $F : \mathbb{H}^n \to \mathbb{H}^n$ for which*

$$\rho_{\mathbb{H}}[f(x), F(x)] \leq a. \tag{9.25}$$

PROOF. Let $f : \mathbb{H}^n \to \mathbb{H}^n$ be an (L, D)-quasi-isometry. Then f admits a boundary value mapping $f_0 : \partial\mathbb{H}^n \to \partial\mathbb{H}^n$ which is $K_1 = K_1(L, D, n)$-quasiconformal by what we have already proved in Theorem 9.6.1. The mapping f_0 admits an extension $F : \bar{\mathbb{H}}^n \to \bar{\mathbb{H}}^n$ which is M-bilipschitz in the hyperbolic metric, Corollary 8.3.7, and $M = M(K_1(L, D, n))$, and so depends only on L, D, and n. So F is M^2 quasiconformal and $F^{-1} \circ f$ is an (ML, MD)-quasi-isometry which is the identity on the boundary. Note that F is quasiconformal and therefore has an inverse, whereas f may not. Finally, Lemma 9.6.2 shows that Theorem 9.6.3 holds with $a(L, D, n) = A(ML, MD)$. □

THEOREM 9.6.4. *Let $f, g : \mathbb{H}^n \to \mathbb{H}^n$ be a pair of (L, D)-quasi-isometries with the same boundary values. Then*

$$\rho_{\mathbb{H}}[f(x), g(x)] \leq 2a, \tag{9.26}$$

where $a = a(L, D, n)$ is found in Theorem 9.6.3.

PROOF. Let these boundary values define the map f_0 and let $x \in \mathbb{H}^n$. Then Theorem 9.6.3 provides us with a quasiconformal mapping for which $\rho_{\mathbb{H}}[f(x), F(x)] \leq a$ and $\rho_{\mathbb{H}}[g(x), F(x)] \leq a$. The result follows from the triangle inequality. □

9.6.2. Proof of the Efremovich–Tihomirova Theorem 9.4.4. In view of Theorem 9.6.1 the only thing left to prove of Theorem 9.4.4 is the compatibility property with respect to Möbius transformations. This is evident from the uniqueness of the boundary values. If ϕ and ψ are Möbius transformations of $\mathbb{H}^n$ and if f is an (L, D)-quasi-isometry, then $\phi \circ f \circ \psi$ is an (L, D)-quasi-isometry. If f has quasiconformal boundary values f_0 with a K-quasiconformal extension F and $\rho_{\mathbb{H}}[f(y), F(y)] \leq a$ for all $y \in \mathbb{H}^n$, then $\phi \circ F \circ \psi$ is K-quasiconformal, has the same boundary values as $\phi \circ f \circ \psi$, and

$$\begin{aligned} \rho_{\mathbb{H}}[(\phi \circ F \circ \psi)(x), (\phi \circ f \circ \psi_(x)] &= \rho_{\mathbb{H}}[(F \circ \psi)(x), (f \circ \psi_(x)] \\ &= \rho_{\mathbb{H}}(F[\psi(x)], f[\psi(x)]) \leq a. \end{aligned}$$

9.7. The limit set of a Möbius group

A discrete group Γ of Möbius transformations of $\mathbb{H}^n$ acts naturally on the boundary $\hat{\mathbb{R}}^{n-1} = \partial\mathbb{H}^n$. This action on the boundary is also by a discrete group of Möbius transformations of course—typically called a *Kleinian group*. We often refer to $\partial\mathbb{H}^n$ as being "at infinity"; we have already introduced the sphere at infinity, and sometimes the term "ideal boundary" is also used in the literature. The group $\Gamma|\hat{\mathbb{R}}^{n-1}$ is a discrete subgroup of $GM(n-1)$. The converse is also true. A discrete subgroup Γ of $GM(n-1)$ has an extension to a discrete group of Möbius transformations acting on $\mathbb{H}^n$. This is an obvious consequence of the Poincaré extension operator which commutes with the group operation of composition and the compactness results of Theorem 3.6.7 and Lemma 3.6.9.

Let Γ be a discrete subgroup of isometries of $\mathbb{H}^n$. The *orbit* of a point $x \in \mathbb{H}^n$ under the group Γ is the set

$$O_\Gamma(x) = \{\gamma(x) : \gamma \in \Gamma\}.$$

The *limit set* of Γ, denoted $\Lambda(\Gamma)$, is the set of all accumulation points of the orbit of the point $\mathbf{j} = (0, \ldots, 0, 1) \in \mathbb{H}^n$:

$$O_\Gamma(\mathbf{j}) = \{\gamma(\mathbf{j}) : \gamma \in \Gamma\}.$$

THEOREM 9.7.1. *The limit set $\Lambda = \Lambda(\Gamma)$ of a discrete group of Möbius transformations of $\mathbb{H}^n$ is a closed Γ-invariant subset of $\partial\mathbb{H}^n$. If $x \in \mathbb{H}^n$, then*

$$\Lambda = \overline{O_\Gamma(x)} \setminus O_\Gamma(x).$$

PROOF. The limit set consists of all accumulation points and is closed by definition. For $\tilde{\gamma} \in \Gamma$

$$\tilde{\gamma}(O_\Gamma(\mathbf{j})) = \{(\tilde{\gamma} \circ \gamma)(\mathbf{j}) : \gamma \in \Gamma\} = \{\gamma(\mathbf{j}) : \gamma \in \tilde{\gamma}^{-1}\Gamma = \Gamma\} = O_\Gamma(\mathbf{j}).$$

Thus $O_\Gamma(\mathbf{j})$ is Γ invariant and so the set of accumulation points is also invariant.

In general the limit set of a Kleinian group can be a very complicated object. For instance the Apollonian gasket illustrated below as a subset of $\mathbb{C}$ is the limit set of a Kleinian group.

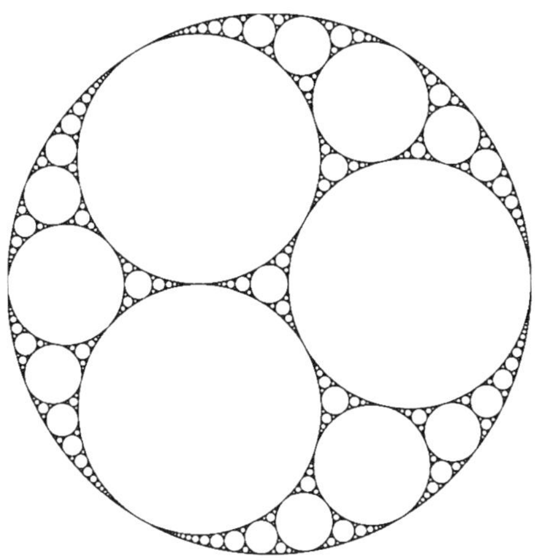

The limit set of a Kleinian group of $\hat{\mathbb{R}}^2 = \partial\mathbb{H}^3$.

Unfortunately, the limit sets for the groups of interest for the Mostow rigidity theorem are simply the spheres at infinity themselves, as we will shortly see. However these objects are intrinsically interesting, and a good place to start to see a thorough discussion—and lots of very nice pictures—is the book *Indra's Pearls* by D. Mumford, C. Series, and D. Wright, [**126**]. These sets are also interesting from the analytical perspective of quasisymmetric mappings, conformal-dimensional, and other things as well. The reader could look at the work of M. Bonk, B. Kleiner and S. Merenkov (see [**22, 23**] as a place to start).

We return to our discussion of limit sets and will next show that $\Lambda = \Lambda(\Gamma) \subset \partial\mathbb{H}^n$. Suppose

$$\gamma_j(\mathbf{j}) \to x_0 \in \mathbb{H}^n.$$

Since we also have $\gamma_j(\partial\mathbb{H}^n) = \partial\mathbb{H}^n$ and $\partial\mathbb{H}^n$ is compact in $\hat{\mathbb{R}}^{n+1}$, Theorem 3.6.7 asserts that the sequence $\{\gamma_j\}_{j=1}^{\infty}$ contains a subsequence $\{\gamma_{j_k}\}_{k=1}^{\infty}$ converging uniformly in $\hat{\mathbb{R}}^{n+1}$ to a Möbius transformation and Lemma 3.6.8 implies that $\{\gamma_{j_k}^{-1}\}_{k=1}^{\infty}$ converges uniformly to its inverse. Then the sequence $\{\gamma_{j_k}^{-1} \circ \gamma_{j_{k+1}}\}_{k=1}^{\infty}$ of elements of Γ converges uniformly in $\hat{\mathbb{R}}^{n+1}$ to the identity element of Γ. This contradicts the assumption that Γ is discrete.

Next, let $x \in \mathbb{H}^n$. Then the hyperbolic distance $\rho_{\mathbb{H}}(x, \mathbf{j}) = a < \infty$ and since each $\gamma \in \Gamma$ is a hyperbolic isometry we have $a = \rho_{\mathbb{H}}(\gamma(x), \gamma(\mathbf{j}))$. However if $\gamma_j(\mathbf{j}) \to \partial\mathbb{H}^n$, then comparing the hyperbolic and spherical metrics of $\mathbb{H}^n \subset \hat{\mathbb{R}}^{n+1}$ gives $\sigma(\gamma_j(\mathbf{j}), \gamma_j(x)) \to 0$. Thus $O_\Gamma(x)$ and $O(\mathbf{j})$ have the same accumulation points on $\partial\mathbb{H}^n$ and the result follows. □

It is not quite true that the limit set is the set of points of accumulation of an orbit of any point of $\mathbb{H}^n$. To see this consider the discrete group $\Gamma = \langle x \mapsto 2x \rangle$ of isometries of $\mathbb{H}^n$ whose limit set consists of $\{0, \infty\}$. Note that the orbit of the point $0 \in \partial\mathbb{H}^n$ does not accumulate at ∞. Fortunately the cases where this sort of thing happens are classified. This classification itself is a beautiful thing and we will discuss it in a moment, but first we recall that the isometries of hyperbolic space fall into three types. We already saw this in 3.5.2 but recall it here. A nontrivial isometry of $\mathbb{H}^n$ extends to a Möbius transformation of $\hat{\mathbb{R}}^{n-1}$:

- *elliptic* if it fixes some point of $\mathbb{H}^n$.
- *parabolic* if it fixes a unique point at infinity (that is, a point of $\hat{\mathbb{R}}^{n-1} = \partial\mathbb{H}^n$).
- *loxodromic* if it fixes two points at infinity.

If the limit set of a discrete group of hyperbolic isometries contains more than three points, then it is true that the orbit of any point $x_0 \in \partial\mathbb{H}^n$ accumulates on the limit set and then the limit set is perfect (that is, closed and uncountable). This will be an easy consequence of our subsequent discussion, but we will not pursue the matter. Such groups are called *nonelementary* and are the groups of Möbius transformations which will primarily concern us. One then seeks to classify the *elementary groups*. Evidently elementary groups have a limit set which consists of $0, 1$, or 2 points.

9.7.1. Classification of discrete elementary Möbius groups. Let Γ be a discrete elementary group of Möbius transformations of $\mathbb{H}^n$. Then $\Lambda(\Gamma)$ consists of $0, 1$, or 2 points, and we have the following:

- $\#\Lambda(\Gamma) = 0$; then Γ consists only of elliptic elements and is isomorphic to a subgroup of the orthogonal group $\mathrm{O}(n)$ (the group of isometries of the $n-1$-sphere, a space of constant positive sectional curvature).
- $\#\Lambda(\Gamma) = 1$; then Γ consists only of parabolic and elliptic elements and is isomorphic to a subgroup of the Euclidean group (the group of isometries of $\mathbb{R}^{n-1}$, a space of identically zero sectional curvature).

- $\#\Lambda(\Gamma) = 2$; then Γ consists only of loxodromic and elliptic elements and is isomorphic to a subgroup of the similarity group $\mathrm{GS}(n-1)$ (the group of isometries of $\mathbb{S}^{n-2} \times \mathbb{R} \equiv \mathbb{R}^{n-1} \setminus \{0\}$ with the metric $ds = |dx|/|x|$).

Let us take a moment to see how this classification arises. Do note that there is a close connection between Stallings' classification of the ends of groups [**149**] and this classification of the limit set.

$\Lambda(\Gamma) = \emptyset$. In order for $\Lambda(\Gamma) = \emptyset$ we can have no accumulation points for the orbit $O = O(\mathbf{j})$ whose closure $\bar{O}$ is therefore a compact subset of $\mathbb{H}^n$. It is an elementary exercise in hyperbolic trigonometry to see that $\bar{O}$ lies in a unique closed hyperbolic ball of smallest radius. As Γ acts by isometries, this uniqueness forces each $\gamma \in \Gamma$ to fix the center of this smallest ball, call it w_0.

Let $\phi : \mathbb{H}^n \to B^n$ be a Möbius transformation with $\phi(w_0) = 0$. Then $\tilde{\Gamma} = \phi \circ \Gamma \circ \phi^{-1}$ is a discrete group of Möbius transformations of the unit ball fixing the origin. Thus $\tilde{\Gamma}$ is a subgroup of the orthogonal group and the remainder of the claim follows.

$\Lambda(\Gamma) = \{x_0\}$. Since the limit set is Γ invariant we must have $\gamma(x_0) = x_0$ for all $\gamma \in \Gamma$. Choose a Möbius transformation ϕ with $\phi(x_0) = \infty$ and replace Γ with $\phi \circ \Gamma \circ \phi^{-1}$ so we can assume that each $\gamma \in \Gamma$ fixes $\infty \in \partial\mathbb{H}^n = \hat{\mathbb{R}}^{n-1}$. There can be no loxodromic elements in Γ because each such element has two fixed points, both of which are clearly limit points. Thus, from the description of the orientation-preserving Möbius transformation of $\mathbb{R}^n$ in Section 2.4 every $\gamma \in \Gamma$ is of the form

$$\gamma(x) = O_\gamma x + a_\gamma, \qquad O_\gamma \in \mathrm{O}(n),\ a_\gamma \in \mathbb{R}^{n-1}.$$

If all the a_γ are equal to 0, then Γ is a discrete subgroup of $\mathrm{O}(n)$ and so finite. This implies $\Lambda(\Gamma) = \emptyset$. Thus some $a_\gamma \neq 0$ and Γ contains a parabolic element. It is also now evident that after our normalisation every element is a Euclidean isometry.

$\Lambda(\Gamma) = \{x_0, y_0\}$ *and* $x_0 \neq y_0$. We choose ϕ so that $\phi(x_0) = 0$ and $\phi(y_0) = \infty$ and then replace Γ by the group $\phi \circ \Gamma \circ \phi^{-1}$ so that now each element $\gamma \in \Gamma$ fixes $\{0, \infty\}$. As the limit set is invariant each element $\gamma \in \Gamma$ fixes or interchanges the points 0 and ∞, $\gamma^2 = \gamma \circ \gamma$ fixes both 0 and ∞, and hence there are no parabolic elements and every element is a similarity. The claim follows. It is just a little more work to see that there is a finite index infinite cyclic subgroup, but we will not digress.

9.7.2. Normal limit points. We now come to the definition of a normal limit point. We use this terminology for two reasons. First normal limit points are generic (in a precisely defined way) in the limit set of a discrete Möbius group, and second there are some convergence properties associated with these points.

Suppose x_0 is a limit point of a discrete Möbius group Γ. Then x_0 is called a *normal limit point* if there is a Möbius transformation Φ and a sequence $\{\mu_j\}_{j=1}^{\infty}$ with $\mu_j \to \infty$ and $\{x_j\}_{j=1}^{\infty}$ with $x_j \to 0$ so that upon defining $\varphi_j(x) = \mu_j(x - x_j)$ we have

$$(\varphi_j \circ \gamma_j) \to \Phi$$

uniformly in $\hat{\mathbb{R}}^{n-1}$.

The following lemma is an elementary consequence of the definition.

LEMMA 9.7.2. *The set of normal limit points of* Γ *is invariant.*

There are two other notions which correspond closely to the normal limit point which we shall define below. They both have the virtue that they are defined geometrically, and in particular points of approximation do not need the Poincaré extension and thus are intrinsic to the action of the Möbius group action on the Riemann sphere $\hat{\mathbb{R}}^{n-1}$. This has important topological consequences for the generalisations of discrete Möbius groups, both uniformly quasiconformal groups and convergence groups; see [**51**]. We will quickly prove that both these types of points are normal. First the definitions:

Points of approximation. Let Γ be a discrete Möbius group. A point $x_0 \in L(\Gamma)$ is called a *point of approximation* if there is a sequence $\{\gamma_j\}_{j=1}^{\infty} \subset \Gamma$ and a $y_0 \in \hat{\mathbb{R}}$ such that

(1) $\gamma_j(x) \to x_0$ locally uniformly in $\hat{\mathbb{R}} \setminus \{y_0\}$,
(2) $\gamma_j(y_0) \not\to x_0$.

This term was introduced by G. Hedlund [**72**] before being used extensively by A. Beardon and B. Maskit in their study [**17**] which will be quite relevant to us. Clearly if x_0 is a loxodromic fixed point, then it is a point of approximation. The converse is not true, but points of approximation behave like loxodromic fixed points. The next lemma is elementary.

LEMMA 9.7.3. *If x_0 is a point of approximation for Γ and φ is a Möbius transformation, then $\varphi(x_0)$ is a point of approximation for $\varphi \circ \Gamma \circ \varphi^{-1}$.*

PROOF. From the definition of a point of approximation, we have a sequence $\{\gamma_j\}_{j=1}^{\infty} \subset \Gamma$ and $y_0 \in \hat{\mathbb{R}}$ such that $\gamma_j(x) \to x_0$ locally uniformly in $\hat{\mathbb{R}} \setminus \{y_0\}$ and $\gamma_j(y_0) \not\to x_0$. It is easy to check using Lemma 3.6.8 that the sequence $\{\varphi \circ \gamma_j \circ \varphi^{-1}\}_{j=1}^{\infty} \subset \varphi \circ \Gamma \circ \varphi^{-1}$ has

(1) $\varphi \circ \gamma_j \circ \varphi^{-1} \to \varphi(x_0)$ locally uniformly in $\hat{\mathbb{R}} \setminus \{\varphi(y_0)\}$,
(2) $(\varphi \circ \gamma_j \circ \varphi^{-1})[\varphi(y_0)] \not\to \varphi(x_0)$.

Thus $\varphi(x_0)$ is a point of approximation. □

Next,

LEMMA 9.7.4. *Suppose x_0 is a point of approximation. Then there is a sequence $\{\mu_j\}_{j=1}^{\infty}$ with $\mu_j \to \infty$ and $\{x_j\}_{j=1}^{\infty}$ with $x_j \to 0$ so that upon defining $\varphi_j(x) = \mu_j(x - x_j)$ we have*

$$\varphi_j \circ \gamma_j(x) \to \Phi(x)$$

uniformly in $\hat{\mathbb{R}}$ for a Möbius transformation Φ. Thus x_0 is a normal limit point.

PROOF. With an appropriate choice of Möbius transformation together with Lemma 9.7.3 we may assume that $x_0 = 0$ and $y_0 = \infty$. Thus $\gamma_j(x) \to 0$ locally uniformly in $\mathbb{R}^n$ and $\gamma_j(\infty) \to z_0 \neq 0$. Let $x_j = \gamma_j(0)$ and $y_j = \gamma_j((1, 0, \ldots, 0))$. Then both $x_j, y_j \to 0$ and hence $\mu_j = 1/|y_j - x_j| \to \infty$. Set

$$\varphi_j(x) = \frac{x - x_j}{|y_j - x_j|}.$$

Since

$$\varphi[\gamma_j(0)] = 0, \quad \big|\varphi(\gamma_j[(1, 0, \ldots, 0)])\big| = 1, \quad \text{and} \quad \varphi[\gamma_j(\infty)] \to \infty,$$

then Lemma 3.6.8 asserts that $\varphi_j \circ \gamma_j$ converges uniformly in $\hat{\mathbb{R}}$ to a Möbius transformation Φ. □

If $\gamma : \mathbb{H}^n \to \mathbb{H}^n$ is a loxodromic Möbius transformation, then the hyperbolic line joining the two fixed points of γ is called the *axis* of γ and is denoted $ax(\gamma)$. It is the unique semicircle containing both the fixed points of γ which is orthogonal to $\mathbb{R}^{n-1}$, and so centered on $\mathbb{R}^{n-1}$. Notice that if $x \in \mathbb{H}^n$, then

$$\gamma^n(x) = (\gamma \circ \gamma \circ \cdots \circ \gamma)(x)$$

converges to the attractive fixed point of γ, while

$$\mathrm{dist}_{\mathbb{H}}(\gamma^n(x), ax(\gamma)) = \mathrm{dist}_{\mathbb{H}}(x, ax(\gamma)).$$

Conical or radial limit points will also have this sort of geometric convergence.

Conical limit points. Let Γ be a discrete Möbius group. A point $x_0 \in \Lambda(\Gamma)$ in the limit set of Γ is called a *conical limit point* if there is an infinite distinct sequence $\{\gamma_j\}_{j=1}^{\infty} \subset \Gamma$ so that for each $x \in \mathbb{H}^n$ there is $M_x < \infty$:

$$\mathrm{dist}_{\mathbb{H}}[\gamma_j(x), \alpha] < M_x, \tag{9.27}$$

where α is the hyperbolic ray with starting point $\mathbf{j} \in \mathbb{H}^n$ and ending on the point $x_0 \in \partial\mathbb{H}^n$.

Notice that if β is any other hyperbolic line ending at x_0, then $\mathrm{dist}_{\mathbb{H}}(\alpha, \beta) = 0$, and so in (9.27) we could replace α by any geodesic ray ending at x_0. In particular we could have $\beta = \{(x_0, t) : 0 < t < 1\}$ and then the set

$$\mathcal{C}_M(x_0) = \{x \in \mathbb{H}^n : \rho_{\mathbb{H}}(x, \beta) < M\},$$

which is a Euclidean cone with base at x_0. We arrive at the equivalent definition of a conical limit point.

LEMMA 9.7.5. *A point $x_0 \in \Lambda(\Gamma)$ is a conical limit point if and only if for some M the cone $\mathcal{C}_M(x_0)$ meets the orbit $O_\Gamma(\mathbf{j})$ infinitely often. Then for any $x \in \mathbb{H}^n$ there is an $M_x < \infty$ so that $O_\Gamma(x)$ meets $\mathcal{C}_{M_x}(x_0)$ infinitely often.*

PROOF. The only thing to do is to choose $M_x = \rho(\mathbf{j}, x) + M$, and the result follows from the triangle inequality. □

Again it is elementary to establish that the conical limit points are invariant: if x_0 is a conical limit point and φ is a Möbius transformation, then $\varphi(x_0)$ is a conical limit point of the discrete Möbius group $\varphi \circ \Gamma \circ \varphi^{-1}$.

We need an estimate on the solid angle at x_0 of the cone $\mathcal{C}_M(x_0)$. To get this we consider the geodesic from the intersection of the upper semi-circle centered at x_0 and of radius 1. Calculating the hyperbolic length from $(x_0, 1) \in \mathbb{H}^n$ along this semicircle gives

$$\int_0^\theta \frac{d\phi}{\cos(\phi)} = M,$$

so

$$\theta = 2\arctan\left[\tanh\left(\frac{M}{2}\right)\right]. \tag{9.28}$$

LEMMA 9.7.6. *Suppose x_0 is a conical limit point. Then there is a sequence $\{\mu_j\}_{j=1}^{\infty}$ with $\mu_j \to \infty$ and $\{x_j\}_{j=1}^{\infty}$ with $x_j \to 0$ and a Möbius transformation Φ, so that upon defining $\varphi_j(x) = \mu_j(x - x_j)$ we have*

$$\varphi_j \circ \gamma_j \to \Phi$$

uniformly in $\hat{\mathbb{R}}^{n-1}$. Thus x_0 is a normal limit point.

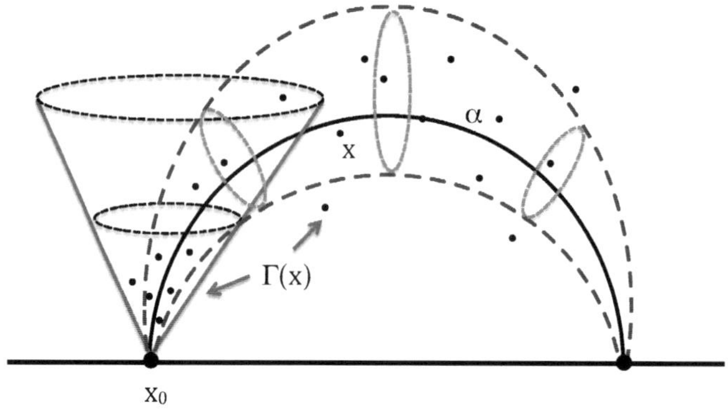

A conical limit point.

Proof. We choose a sequence $\{g_j\}_{j=1}^{\infty}$ so that $g_j[(x_0,1)]$ accumulates to x_0 in the cone $\mathcal{C}_M(x_0)$. Set $x_j = x_0$ and $\mu_j = 1/|g_j((x_0,1)) - x_0|$. Then from (9.28) and the fact that scalar multiplication preserves cones based at the origin, we have the nth coordinate at least $\cos\left(2\arctan\left[\tanh\left(\frac{M}{2}\right)\right]\right)$ and at most 1,

$$\operatorname{sech}(M) \leq \varphi_j\big(g_j[(x_0,1)]\big)_n \leq 1, \quad \text{and} \quad \big|\varphi_j\big(g_j[(x_0,1)]\big)\big| = 1,$$

and also the sequence of Möbius transformations $\varphi_j \circ g_j$ leaves the boundary $\partial\mathbb{H}^n$ invariant. Therefore this sequence forms a normal family and our convergence theorems assert that some subsequence converges to a Möbius transformation uniformly on $\hat{\mathbb{R}}^{n-1}$ and indeed $\hat{\mathbb{R}}^n$. $\square$

We wish to circumvent a discussion of the fundamental domains for the action of a Möbius group on $\mathbb{H}^n$, but we need to be able to prove that most limit points are normal. We do this in the following way. We recall our earlier discussion concerning dense orbits.

A discrete Möbius group Γ acting on $\mathbb{H}^n$ has an M-dense orbit if there is some point x so that for every $y \in \mathbb{H}^n$

$$\operatorname{dist}_{\mathbb{H}}[y, O_\Gamma(x)] \leq M. \tag{9.29}$$

As noted earlier, if Γ has an M-dense orbit, say of x, then the orbit of $\mathbf{j} = (0,\ldots,0,1) \in \mathbb{H}^n$ is $M' = M + \rho_{\mathbb{H}}(x,\mathbf{j})$ dense. We want to prove the following theorem.

Theorem 9.7.7. *Suppose Γ has an M-dense orbit. Then $\Lambda(\Gamma) = \partial\mathbb{H}^n$ and every point of $\partial\mathbb{H}^n$ is a conical and therefore a normal limit point.*

Proof. Replacing M by a slightly larger constant if necessary, we can assume the orbit of $\mathbf{j}$ is M-dense. Now for each $k = 1,2,\ldots$ and by the definition of an M-dense orbit there is $\gamma_k \in \Gamma$ so that

$$\rho_{\mathbb{H}}\left[(x_0, e^{-kM}), \gamma_k(\mathbf{j})\right] \leq M. \tag{9.30}$$

This not only puts $\gamma_k(\mathbf{j}) \in \mathcal{C}_M(x_0)$, but also since

$$\rho_{\mathbb{H}}\left[(x_0, e^{-kM}), (x_0, e^{-(k+2)M})\right] = \int_{e^{-(j+2)M}}^{e^{-jM}} \frac{ds}{s} = 2M \tag{9.31}$$

we see that as $k \to \infty$ we must have distinct orbit points accumulating at x_0 since clearly $(x_0, e^{-kM}) \to (x_0, 0) \in \partial\mathbb{H}^n$. □

9.8. Mappings compatible with a Möbius group

earlier, in our discussion of groups as geometric objects 9.5 we saw that any discrete Möbius groups of $\mathbb{H}^n$ with M-dense orbit gave us a quasi-isometry between the group with the word metric and hyperbolic space $\mathbb{H}^n$. Further, we also saw that an isomorphism $\varphi : \Gamma \to \tilde{\Gamma}$ between two groups with M-dense orbits induced a quasi-isometry of hyperbolic space ϕ which was compatible in the sense that

$$(\phi \circ \gamma)(x) = [\varphi(\gamma) \circ \phi](x), \qquad x \in \mathbb{H}^n.$$

This is the content of the two Theorems 9.4.4 and 9.5.3. For simplicity we typically write $\varphi(\gamma) = \tilde{\gamma}$, so the above equation becomes

$$\phi \circ \gamma = \tilde{\gamma} \circ \phi \qquad x \in \mathbb{H}^n. \tag{9.32}$$

We begin to explore the implications of the functional equation (9.32) in the next theorem.

THEOREM 9.8.1. *Let $n \geq 2$ and let γ and $\tilde{\gamma}$ be Möbius transformations with γ loxodromic. Suppose that the map $f : \hat{\mathbb{R}}^{n-1} \to \hat{\mathbb{R}}^{n-1}$ solves the functional equation*

$$f \circ \gamma = \tilde{\gamma} \circ f \tag{9.33}$$

and that f is differentiable at the attractive fixed point of γ with nonvanishing Jacobian. Then there are Möbius transformations ϕ and ψ so that $\phi \circ f \circ \psi$ is linear and $\tilde{\gamma}$ is loxodromic with the same multiplier as γ.

PROOF. First, an elementary induction shows that for each $k \in \mathbb{Z}$

$$f \circ \gamma^k = \tilde{\gamma}^k \circ f. \tag{9.34}$$

Suppose that x_0 and y_0 are the attracting and repelling fixed points of γ. The continuity of f tells us that for each $x \in \hat{\mathbb{R}}^{n-1} \setminus \{y_0\}$ we have $f \circ \gamma^k(x) \to f(x_0)$ as $k \nearrow \infty$ and for each $y \in \hat{\mathbb{R}}^{n-1} \setminus \{y_0\}$ we have $f \circ \gamma^k(y) \to f(y_0)$ as $k \searrow -\infty$. It follows that $\tilde{\gamma}$ is either parabolic or loxodromic and that $f(x_0)$ is a fixed point of $\tilde{\gamma}$. Choose Möbius transformations φ_1, φ_2 so that $\varphi_1(0) = x_0$, $\varphi_1(\infty) = y_0$ and $\varphi_2(f(x_0)) = 0$. Then

$$\varphi_2 \circ f \circ \varphi_1 \circ \varphi_1^{-1} \circ \gamma \circ \varphi_1 = \varphi_2 \circ \tilde{\gamma} \circ \varphi_2^{-1} \circ \varphi_2 \circ f \circ \varphi_1. \tag{9.35}$$

Set $\alpha = \varphi_1^{-1} \circ \gamma \circ \varphi_1$ and $\tilde{\alpha} = \varphi_2^{-1} \circ \tilde{\gamma} \circ \varphi_2$. Then α is a loxodromic Möbius transformation with fixed points 0 (attractive) and ∞ (repulsive), and hence

$$\alpha(x) = \lambda O x,$$

where $0 < \lambda < 1$ is a scalar and O lies in the orthogonal group $O(n)$. Writing $h = \varphi_2 \circ f \circ \varphi_1$ we now have the functional equation

$$h \circ \alpha = \tilde{\alpha} \circ h. \tag{9.36}$$

Notice that h is differentiable at 0 with $h(0) = 0$ and as the φ_i are smooth we may calculate that

$$\lambda Dh(0)O = D\tilde{\alpha}(0)Dh(0), \tag{9.37}$$

and after computing the determinant of both sides we also see $|D\tilde{\alpha}(0)| = \lambda < 1$, which shows that both $\tilde{\alpha}$ and $\tilde{\gamma}$ are loxodromic with the same multiplier as γ.

Reviewing (9.35) we see that we could now choose φ_2 so that the repelling fixed point of $\tilde{\gamma}$ is mapped to ∞. Then $\tilde{\alpha}(x) = \lambda U x$ with $U \in \mathrm{O}(n)$. This now shows us that (9.36) reduces to

$$h(\lambda^k O^k x) = \lambda^k U^k h(x) \tag{9.38}$$

for all $x \in \mathbb{R}^n$. In particular we have the formula

$$h(x) = U^{-k}\lambda^{-k} h(\lambda^k O^k x). \tag{9.39}$$

Since h is differentiable at 0 we may write $h(\zeta) = h(0) + Dh(0)\zeta + \epsilon(\zeta)|\zeta|$ where $\epsilon(\zeta) \to 0$ as $\zeta \to 0$. In particular, for any x we have

$$h(\lambda^k O^k x) = \lambda^k Dh(0) O^k x + \lambda^k \epsilon(\lambda^k O^k x)$$

as soon as k is sufficiently large, depending on x. From (9.39) this gives

$$h(x) = \lim_{k\to\infty} U^{-k} Dh(0) O^k x \tag{9.40}$$

as well as an assurance that the limit exists. Further, since the orthogonal group is compact, we can pass to subsequences k_j so $O^{k_j} \to \tilde{O}$ and $U^{k_j} \to \tilde{U}$, whereupon we have the formula

$$h(x) = \tilde{U}^{-1} Dh(0) \tilde{O} x \tag{9.41}$$

revealing h to be linear. Since $h = \varphi_2 \circ f \circ \varphi_1$ we see the proof is complete once we set $\phi = \varphi_2^{-1}$ and $\psi = \varphi_1$. □

The situation described in the theorem is entirely possible for an arbitrary linear mapping $A \in \mathrm{GL}(n, \mathbb{R})$. We would like to infer that A is conformal and so we need to make some additional hypotheses.

THEOREM 9.8.2. *Let γ and $\tilde{\gamma}$ be Möbius transformations with γ loxodromic and suppose the fixed point set of γ does not contain 0 or ∞. If the linear map $A : \hat{\mathbb{R}}^{n-1} \to \hat{\mathbb{R}}^{n-1}$ solves the functional equation*

$$A\gamma(x) = \tilde{\gamma}(Ax), \tag{9.42}$$

then A is conformal. That is, there is a $\lambda \in \mathbb{R}$ and an $O \in \mathrm{O}(n)$ such that $Ax = \lambda O x$.

PROOF. Clearly $\tilde{\gamma} = (A \circ \gamma \circ A^{-1})(x)$ is loxodromic with attractive fixed point $\tilde{x_0} = Ax_0$, repulsive fixed point $\tilde{y_0} = Ay_0$, and the same multiplier as γ. Let

$$\phi(x) = \frac{x - x_0}{|x - x_0|^2} - \frac{y_0 - x_0}{|y_0 - x_0|^2}, \quad \psi(x) = \frac{x - Ax_0}{|x - Ax_0|^2} - \frac{Ay_0 - Ax_0}{|Ay_0 - Ax_0|^2}. \tag{9.43}$$

Then ϕ maps x_0 to ∞ and y_0 to 0 and ψ maps Ax_0 to ∞ and Ay_0 to 0. The functional relationship (9.42) can be rewritten as

$$A \circ \phi^{-1} \circ \phi \circ \gamma \circ \phi^{-1} \circ \phi = \psi^{-1} \circ \psi \circ \tilde{\gamma} \circ \psi^{-1} \circ \psi \circ A, \tag{9.44}$$

which we rewrite as

$$[\psi \circ A \circ \phi^{-1}] \circ \phi \circ \gamma \circ \phi^{-1} = \psi \circ \tilde{\gamma} \circ \psi^{-1} \circ [\psi \circ A \circ \phi^{-1}]. \tag{9.45}$$

If $\eta = \phi \circ \gamma \circ \phi^{-1}$ and $\tilde{\eta} = \psi \circ \tilde{\gamma} \circ \psi^{-1}$ and if we set

$$B = \psi \circ A \circ \phi^{-1}, \tag{9.46}$$

then we arrive at the functional equation

$$B\eta = \tilde{\eta} B$$

with loxodromic η and $\tilde{\eta}$ with fixed points $\{0,\infty\}$. Following the argument in (9.40) we see that B is linear. Writing (9.46) as $\psi \circ A = B\phi$ we have

$$\frac{Ax - Ax_0}{|Ax - Ax_0|^2} - \frac{Ay_0 - Ax_0}{|Ay_0 - Ax_0|^2} = \frac{Bx - Bx_0}{|x - x_0|^2} - \frac{By_0 - Bx_0}{|y_0 - x_0|^2}. \tag{9.47}$$

We let $x \to \infty$ to see that

$$\frac{Ay_0 - Ax_0}{|Ay_0 - Ax_0|^2} = \frac{By_0 - Bx_0}{|y_0 - x_0|^2},$$

and hence with $y = x - x_0$ (9.47) yields

$$\frac{Ay}{|Ay|^2} = \frac{By}{|y|^2} \tag{9.48}$$

for all $y \in \mathbb{R}^n$. If $u, v \in \mathbb{R}^n$, then as B is linear,

$$0 = Bu + Bv - B^n(u+v) = Au\frac{|u|^2}{|Ay|^2} + Av\frac{|v|^2}{|Av|^2} - A(u+v)\frac{|u^2+v^2|}{|A(u+v)|^2}$$

giving the equality

$$Au\Big(\frac{|u|^2}{|Au|^2} - \frac{|u^2+v^2|}{|A(u+v)|^2}\Big) = -Av\Big(\frac{|v|^2}{|Av|^2} - \frac{|u^2+v^2|}{|A(u+v)|^2}\Big).$$

As soon as u and v are linearly independent we see that the scalar terms on each side must vanish. That is to say we must have

$$\frac{|u|^2}{|Au|^2} = \frac{|u^2+v^2|}{|A(u+v)|^2} \qquad \text{for all } u, v \in \mathbb{R}^n.$$

We fix some $u \neq 0$ and let v vary to see that there is a positive μ so that for all $x \in \mathbb{R}^n$

$$|Ax| = \mu|x|.$$

This implies that A is a scalar multiple of an orthogonal transformation as, for instance, $|A^tA\zeta| = \mu^2$ for all $|\zeta| = 1$. Hence all the eigenvalues of A^tA are equal to μ. □

We want to use this result to establish the linearisation technique in the proof of Theorem 9.8.1. This is the next easy lemma.

LEMMA 9.8.3. *Suppose that $f : \mathbb{R}^n \to \mathbb{R}^n$ is differentiable at 0 with nonvanishing Jacobian, that $\phi_j(x) = \mu_j(x - x_j)$ with $x_j \to 0$ and $\mu_j \to \infty$, and that $\psi_j(y) = \eta_j y + y_j$ with $\eta_j \to 0$ and $y_j \to 0$. Suppose the limit*

$$\lim_{j\to\infty} \phi_j \circ f \circ \psi_j(x) = g(x)$$

exists and converges uniformly on $\hat{\mathbb{R}}^{n-1}$ to a mapping g. Then g is affine.

PROOF. We can surely assume that $f(0) = 0$. We can write $f(x) = Df(0)x + \epsilon(x)|x|$, where $\epsilon(x) \to 0$ as $x \to 0$. Then, for any $x \in \mathbb{R}^n$

$$\begin{aligned}\phi_j \circ f \circ \psi_j(x) &= \mu_j[f(\eta_j x + y_j) - x_j] \\ &= \mu_j\eta_j Df(0)x + \mu_j\eta_j Df(0)y_j + \mu_j\eta_j\epsilon(\eta_j x + y_j)|\eta_j x + y_j| - \mu_j x_j.\end{aligned}$$

At $x = 0$ we see

$$g(0) = \lim_{j\to\infty} \mu_j\eta_j Df(0)y_j + \mu_j\eta_j\epsilon(y_j)|y_j| - \mu_j x_j$$

so that

$$g(x) - g(0) = \lim_{j\to\infty} \mu_j \eta_j \Big(Df(0)x + \epsilon(\eta_j x + y_j)|\eta_j x + y_j| - \epsilon(y_j)|y_j| \Big). \tag{9.49}$$

For $x \neq 0$, since $\eta_j \to 0$ and $y_j \to 0$, we see that as $j \to \infty$,

$$\frac{|\epsilon(\eta_j x + y_j)|\eta_j x + y_j| - \epsilon(y_j)|y_j||}{|Df(0)x|} \to 0. \tag{9.50}$$

Thus the limit in (9.49) can exist only if $\mu_j \eta_j \to \lambda < \infty$ and consequently

$$g(x) - g(0) = \lambda Df(0)x, \tag{9.51}$$

and if g is nonconstant, $\lambda \neq 0$ and g is affine. □

Here is the main theorem of this section.

THEOREM 9.8.4. *Let x_0 be a point of approximation for the sequence of Möbius transformations $\{\gamma_j\}_{j=1}^{\infty}$. Let the map $f : \hat{\mathbb{R}}^{n-1} \to \hat{\mathbb{R}}^{n-1}$ solve the functional equations*

$$f \circ \gamma_j = \tilde{\gamma_j} \circ f, \tag{9.52}$$

where each $\tilde{\gamma_j}$ is a Möbius transformation. If f is differentiable at x_0 with nonvanishing Jacobian, then f is a Möbius transformation.

PROOF. We replace f by $\psi \circ f \circ \phi$, γ_j by $\phi^{-1} \circ \gamma_j \circ \phi$, and $\tilde{\gamma_j}$ by $\psi \circ \tilde{\gamma_j} \circ \psi^{-1}$ where $\phi(x_0) = 0$ and $\psi(f(x_0)) = 0$. Thus we may assume that $f(0) = 0$ and that 0 is a point of approximation for the sequence γ_j at which f is differentiable with nonvanishing Jacobian. From Lemma 9.7.4 we can choose Möbius φ_j and $\tilde{\varphi_j}$ so that

$$\varphi_j \circ \gamma_j \to \Phi, \quad \tilde{\varphi_j} \circ \tilde{\gamma_j} \to \tilde{\Phi}$$

for Möbius transformations Φ and $\tilde{\Phi}$. Then

$$\begin{aligned} f \circ \varphi_j^{-1} \circ \varphi_j \circ \gamma_j &= \tilde{\varphi}_j^{-1} \circ \tilde{\varphi_j} \circ \tilde{\gamma_j} \circ f, \\ \tilde{\varphi_j} \circ f \circ \varphi_j^{-1} &= (\tilde{\varphi_j} \circ \tilde{\gamma_j}) \circ f \circ (\varphi_j \circ \gamma_j)^{-1}. \end{aligned} \tag{9.53}$$

Then since

$$\tilde{\varphi_j} \circ \tilde{\gamma_j} \circ f \circ (\varphi_j \circ \gamma_j)^{-1} \to \tilde{\Phi} \circ f \circ \Phi$$

uniformly in $\hat{\mathbb{R}}^{n-1}$, Lemma 9.8.3 applied to the sequence $\{\tilde{\varphi_j} \circ f \circ \varphi_j^{-1}\}$ and (9.53) implies that $\tilde{\Phi} \circ f \circ \Phi$ is an affine map. Finally, Theorem 9.8.2 implies that this affine map is in fact conformal. □

9.9. The proof of Mostow's theorem

All the pieces are now in place to give a proof of the various forms of Mostow rigidity. We prove the following theorem before discussing what is missing for the more general result and how it implies the results we discussed in the introduction to this chapter.

THEOREM 9.9.1. *Let $n \geq 3$ and let M and N be closed hyperbolic n-manifolds with isomorphic fundamental groups. Then M and N are isometric.*

PROOF. First, as M and N are hyperbolic manifolds there are discrete torsion free groups Γ and $\tilde{\Gamma}$ of hyperbolic isometries so that

$$M = \mathbb{H}^n/\Gamma, \qquad N = \mathbb{H}^n/\tilde{\Gamma}.$$

We relate the key points for our subsequent discussion.

(i) **Both Γ and $\tilde{\Gamma}$ have R-dense orbits.**
Since M and N are closed, their respective diameters are bounded. Let

$$R = \max\{\mathrm{diam}_{\mathbb{H}}(M), \mathrm{diam}_{\mathbb{H}}(N)\} < \infty.$$

This means the image of the point $\mathbf{j} \in \mathbb{H}^n$ under the projection $\Pi_M : \mathbb{H}^n \mapsto \mathbb{H}^n/\Gamma = M$ is a distance at most R from any other point of M, and of course the same is true for N and the projection Π_N. Thus $B^n_{\mathbb{H}}(\mathbf{j}, R)$ contains a point from every orbit equivalence class of Γ, and this is the same as saying that the orbit of $\mathbf{j}$ is R-dense under Γ. The same is true for the orbit of $\mathbf{j}$ under $\tilde{\Gamma}$.

(ii) **Every limit point of Γ and $\tilde{\Gamma}$ is a normal limit point.**
This is a consequence of Theorem 9.7.7.

(iii) **There is a quasiconformal mapping which commutes with the group actions.** That is, we assert the existence of a quasiconformal mapping $f : \hat{\mathbb{R}}^n \to \hat{\mathbb{R}}^n$ such that

$$f \circ \Gamma = \tilde{\Gamma} \circ f. \tag{9.54}$$

To see this, first note that the groups Γ and $\tilde{\Gamma}$ are the fundamental groups, or more precisely, $\pi_1(M) \cong \Gamma$ and $\pi_1(N) \cong \tilde{\Gamma}$. Therefore our hypotheses assert the existence of an algebraic isomorphism $\varphi : \Gamma \to \tilde{\Gamma}$. Then Theorem 9.5.3, together with the R-density we have already established, provides us with a quasi-isometry $F : \mathbb{H}^n \to \mathbb{H}^n$ which commutes with the group actions. Theorem 9.6.1 shows that the uniquely defined boundary values $f : \partial\mathbb{H}^n \to \partial\mathbb{H}^n$ of the quasi-isometry F are quasiconformal, and also commutes with the group actions as indicated in (9.54).

(iv) **The quasiconformal mapping of (iii) is a Möbius transformation.** As the dimension $n \geq 3$, the quasiconformal mapping $f : \hat{\mathbb{R}}^{n-1} \to \hat{\mathbb{R}}^{n-1}$ is differentiable almost everywhere with a nonvanishing Jacobian determinant, Theorem 6.4.13. Then Theorem 9.8.4 implies that $f : \hat{\mathbb{R}}^{n-1} \to \hat{\mathbb{R}}^{n-1}$ is a Möbius transformation.

(v) **The manifolds M and N are isometric.** We are now in the situation that f in (9.54) above is in fact a Möbius transformation, call it $\phi : \hat{\mathbb{R}}^{n-1} \to \hat{\mathbb{R}}^{n-1}$, so $\phi \circ \Gamma = \tilde{\Gamma} \circ \phi$. The Poincaré extension $\Phi : \bar{\mathbb{H}}^n \to \bar{\mathbb{H}}^n$ of ϕ is a hyperbolic isometry, and since Möbius transformations are uniquely defined by their boundary values the functional equation also holds in $\mathbb{H}^n$. That is,

$$(\Phi \circ \Gamma)(x) = (\tilde{\Gamma} \circ \Phi)(x), \qquad \text{for all } x \in \mathbb{H}^n. \tag{9.55}$$

Now equation (9.55) can be used to establish that the following map ψ defined by

$$\psi = \Pi_N \circ \Phi \circ (\Pi_M)^{-1} : M \to N$$

is an isometry.

Let $x' \in M$ and let $x \in \mathbb{H}^n$ such that $\Pi_M(x) = x'$. Then $(\Pi_M)^{-1}(x') = \Gamma(x)$. Thus

$$(\Pi_N \circ \Phi \circ (\Pi_M)^{-1})(x') = (\Pi_N \circ \Phi)(\Gamma(x)) = \Pi_N(\tilde{\Gamma}[\Phi(x)]) = y'$$

so $\psi(x') = y'$ for a unique well defined $y' \in N$. Since Π_M and Π_N are local isometries (as Γ and $\tilde{\Gamma}$ are discrete and torsion free) ψ is a well-defined homeomorphism. Further, simply by the definition of the metrics on M and N that we have for $x'_1, x'_2 \in M$ with $\Pi_M(x_i) = x'_i$, $i = 1, 2$ (and using similar notation for N), we may use the fact that Φ is a hyperbolic isometry to see that

$$\begin{aligned}
\rho_M(x'_1, x'_2) &= \operatorname{dist}_{\mathbb{H}}[\Gamma(x_1), \Gamma(x_2)] = \operatorname{dist}_{\mathbb{H}}\big(\Phi[\Gamma(x_1)], \Phi[\Gamma(x_2)]\big) \\
&= \operatorname{dist}_{\mathbb{H}}\big(\tilde{\Gamma}[\Phi(x_1)], \tilde{\Gamma}[\Phi(x_2)]\big) \\
&= \rho_N\big(\Pi_N\tilde{\Gamma}[\Phi(x_1)], \Pi_N\tilde{\Gamma}[\Phi(x_2)]\big) \\
&= \rho_N\big[\psi(x'_1), \psi(x'_2)\big].
\end{aligned}$$

This shows that ψ is an isometry. □

We note that the only point where the dimension $n \geq 3$ is used is in part (iv) above. We also note that all that is really required above is that almost every limit point is normal and that there is a quasiconformal map which commutes with the fundamental group actions Γ of M and $\tilde{\Gamma}$ of N.

Now Theorem 9.1.1 is an obvious consequence of Theorem 9.9.1 as a diffeomorphism induces an isomorphism of fundamental groups. In fact a number of steps in the proof above can be simplified under this assumption. A diffeomorphism φ between closed manifolds is quasiconformal. Since the projections are local isometries (and locally conformal) the lift of φ to $F : \mathbb{H}^n \to \mathbb{H}^n$, commuting with the group actions, is quasiconformal and therefore has quasiconformal boundary values $f : \partial\mathbb{H}^n \to \partial\mathbb{H}^n$. The proof follows in the same manner.

However, in order to address these issues we have to delve further into the geometric theory of discrete groups of hyperbolic isometries and their end structure which we cannot do here. However we make the following observations. If M has finite volume and $M = \mathbb{H}^n/\Gamma$, then Γ is "geometrically finite" and Beardon and Maskit [**17**] have shown that every limit point of Γ is either a parabolic fixed point or a normal limit point—a limit point cannot be both by the way, a consequence of discreteness. Since Γ is discrete, the set of parabolic fixed points is countable. Since $\mathbb{H}^n/\Gamma$ has finite volume it is easy to see that $\Lambda(\Gamma) = \partial\mathbb{H}^n$ and therefore almost every point is a normal limit point. Obtaining the quasi-isometry $M \to N$ whose lift will provide us with the desired quasiconformal conjugacy is a little more problematic. General facts from topology provide a continuous map commuting with the group actions. Then a close examination of the structure of the fundamental domains and their infinite ends is necessary to promote this map to a quasi-isometry. We leave the details well alone! We do however note a corollary of what we have proven together with what we have noted above.

COROLLARY 9.9.2. *Let M and N be finite volume hyperbolic n-manifolds, $n \geq 3$, and suppose there is a quasi-isometry (perhaps even quasiconformal) mapping $f : M \to N$. Then M and N are isometric.*

PROOF. We can lift $f : M \to N$ to a map $F : \mathbb{H}^n \to \mathbb{H}^n$, and as f is assumed to be a quasi-isometry, it easily follows that F is as well. The proof now follows as for Theorem 9.9.1, knowing that almost every limit point is a normal limit point by what was stated above. □

Finally we note that the assumption that M and N are manifolds is not necessary. If Γ and $\tilde{\Gamma}$ are simply *lattices*—meaning that $\mathbb{H}^n/\Gamma$ and $\mathbb{H}^n/\tilde{\Gamma}$ both have finite volume—and if $\Gamma \cong \tilde{\Gamma}$, then there is a hyperbolic isometry $\Phi : \mathbb{H}^n \to \mathbb{H}^n$ such that $\Phi \circ \Gamma \circ \Phi^{-1} = \tilde{\Gamma}$. The proof of this fact is more or less the same as that given above once we know a little about the torsion structure of the groups—there is a bound (depending on Γ) on the order of any elliptic element in Γ and therefore also of $\tilde{\Gamma}$. Otherwise, one can deduce this more general result from Selberg's lemma asserting the existence of a torsion free subgroup of finite index and the Mostow rigidity theorem mentioned above in the finite volume case.

Basic Notation

Here we have collected together some of the standard notation used throughout the text.

- $\mathbb{C}$, the complex plane
- $\hat{\mathbb{C}} = \mathbb{C} \cup \{\infty\}$, the Riemann sphere
- $\hat{\mathbb{R}}^n = \mathbb{R}^n \cup \{\infty\}$, the Riemann n-sphere or Möbius space
- $B^n(a,r) = \{x \in \mathbb{R}^n : |x-a| \leq r\}$
- $B^n = B^n(0,1)$, the open unit ball
- $\bar{B}^n(a,r) = \{x \in \mathbb{C} : |x-a| \leq r\}$, the closed ball about a of radius r
- $\bar{B}^n$, the closed unit ball
- $\operatorname{diam}(E)$, the diameter of the set $E \subset \mathbb{C}$
- $|E|$, the Lebesgue measure of the set E
- $\operatorname{dist}(E,F)$, the distance between the sets E and F,

$$\operatorname{dist}(E,F) = \inf_{x \in E, w \in F} |x-w|$$

- $\mathcal{H}^s(E)$, the s-dimensional Hausdorff measure of a set E
- $\dim_{\mathcal{H}}(E)$, the Hausdorff dimension of a set E
- $\mathcal{M}^s(E)$, the s-dimensional content of a set E
- $\chi_F(x)$, the characteristic function of the set F,

$$\chi_F(x) = \begin{cases} 1, & x \in F, \\ 0, & x \notin F \end{cases}$$

- $\chi_{\text{identity}(0,R)}$, the characteristic function of the disk identity$(0,R)$
- $\mathrm{GL}(n,\mathbb{R})$, the general linear group, that is, the space of invertible $n \times n$ matrices with real entries
- $\mathrm{SL}(n,\mathbb{R})$, those matrices $A \in \mathrm{GL}(n,\mathbb{R})$ with determinant equal to 1, $\det(A) = 1$
- $\mathrm{SO}(n,\mathbb{R})$, the orthogonal matrices in $\mathrm{SL}(n,\mathbb{R})$
- $|A|$, the operator norm of $A \in \mathrm{GL}(n,\mathbb{R})$,

$$|A| = \max_{|\zeta|=1} |A\zeta|$$

- $\|A\|$, the Hilbert-Schmidt norm of $A \in \mathrm{GL}(n,\mathbb{R})$,

$$\|A\| = \left(\sum_{i,j=1}^{n} a_{i\,j}^2 \right)^{1/2}$$

- Df, the differential matrix of the function $f(x) = (f^1(x), f^2(x), \ldots, f^n(x))$,

$$Df(x) = \begin{bmatrix} \frac{\partial f^1}{\partial x_1}(x) & \frac{\partial f^1}{\partial x_2}(x) & \ldots & \frac{\partial f^1}{\partial x_n}(x) \\ \frac{\partial f^2}{\partial x_1}(x) & \frac{\partial f^2}{\partial x_2}(x) & \ldots & \frac{\partial f^2}{\partial x_n}(x) \\ \vdots & \vdots & \ddots & \vdots \\ \frac{\partial f^n}{\partial x_1}(x) & \frac{\partial f^n}{\partial x_2}(x) & \ldots & \frac{\partial f^n}{\partial x_n}(x) \end{bmatrix}$$

- $D^t f = (Df)^t$, the transpose differential matrix
- $J(x, f) = \det[Df(x)]$, the Jacobian determinant
- $f|E$, the function f restricted to the set E
- $L_f(x)$, the maximal derivative of a function f
- $|dx|$, ds, line-elements in integrating with respect to arc length
- $\text{supp}(f)$, the support of the function f, $\text{supp}(f) = \overline{\{x : f(x) \neq 0\}}$
- $C(\Omega)$, the space of continuous real-valued functions defined on an open set Ω
- $C_0(\Omega)$, those functions in $C(\Omega)$ whose support is compactly contained in Ω
- $C^\infty(\Omega)$, the space of infinitely differentiable real-valued functions defined on an open set Ω
- $C_0^\infty(\Omega)$, those functions in $C^\infty(\Omega)$ whose support is compactly contained in Ω
- $C^{1,\alpha}(\Omega)$, those functions in $C^1(\Omega)$ whose first derivatives satisfy a Hölder estimate with exponent α
- $L^p(\Omega)$, the Banach space ($p \geq 1$) of functions f with $|f|^p$ integrable in Ω
- $L^p_{loc}(\Omega)$, the Banach space ($p \geq 1$) of functions f with $|f|^p$ locally integrable in Ω, that is, integrable in each compact subset of Ω
- $L^\infty(\Omega)$, the Banach space of essentially bounded measurable functions
- $W^{k,p}(\Omega, \mathbb{V})$, $1 \leq p \leq \infty$, $k \in \mathbb{N}$, the *Sobolev space* of all distributions $f \in D'(\Omega, \mathbb{V})$ whose derivatives up to the kth order are represented by functions in $L^p(\Omega, \mathbb{V})$ and equipped with the norm

$$\|f\|_{k,p} = \left(\sum_{|\alpha+\beta| \leq k} \int_\Omega \left| \frac{\partial^{\alpha+\beta} f(\zeta)}{\partial x^\alpha \, \partial \bar{x}^\beta} \right|^p \right)^{1/p}$$

for $p < \infty$ and

$$\|f\|_{k,\infty} = \text{essup}_{|\alpha+\beta| \leq k} \left| \frac{\partial^{\alpha+\beta} f(\zeta)}{\partial x^\alpha \, \partial \bar{x}^\beta} \right| ;$$

when $\mathbb{V} = \mathbb{C}$, we denote these spaces by $W^{k,p}(\Omega)$. The space $C^\infty(\Omega, \mathbb{V})$ is a dense subspace of $W^{k,p}(\Omega, \mathbb{V})$ for all k and p, $1 \leq p < \infty$

- $W^{k,p}_{loc}(\Omega)$, the space of functions f with $f \in W^{k,p}(\Omega')$ for every relatively compact subdomain with $\overline{\Omega'} \subset \Omega$
- $\mathbb{W}^{1,p}(\Omega)$, the Banach space (modulo constants) of functions whose gradient lies in $L^p(\Omega)$v

Bibliography

[1] S. Agard, *A geometric proof of Mostow's rigidity theorem for groups of divergence type*, Acta Math., **151**, (1983), 231–252.

[2] L.V. Ahlfors, *Zur theorie der Überlagerungsflächen*, Acta Math., **65**, (1935), 157–194.

[3] L.V. Ahlfors, *Lectures on quasiconformal mappings*, Van Nostrand, Princeton 1966; Reprinted by Wadsworth Inc. Belmont, 1987.

[4] L.V. Ahlfors, *Commentary on: "Zur theorie der Überlagerungsflächen" (1935)*, Fields Medallists' Lectures, 8–9, World Sci. Ser. 20th Century Math., **5**, World Sci. Publishing, River Edge, NJ, 1997.

[5] L.V. Ahlfors, *On quasiconformal mappings*, J. d'Analyse Math., **3**, (1953/54), 1–58. (correction loc. cit., pp. 207–208).

[6] L.V. Ahlfors, *Extension of quasiconformal mappings from two to three dimensions*, Proc. Nat. Acad. Sci. U.S.A., **51**, (1964), 768–771.

[7] L.V. Ahlfors, *Quasiconformal reflections*, Acta Math., **109**, (1963), 291–301.

[8] L.V. Ahlfors, *Conformal invariants. Topics in geometric function theory.* Reprint of the 1973 original. With a foreword by Peter Duren, F. W. Gehring and Brad Osgood. AMS Chelsea Publishing, Providence, RI, 2010.

[9] L.V. Ahlfors, *Möbius transformations in several dimensions*, Ordway Professorship Lectures in Mathematics. University of Minnesota, School of Mathematics, Minneapolis, Minn., 1981. ii+150 pp.

[10] L.V. Ahlfors and A. Beurling, *Conformal invariants and function theoretic null sets*, Acta Math., **83**, (1950), 101–129.

[11] L.V. Ahlfors and A. Beurling, *The boundary correspondence under quasiconformal mappings*, Acta Math., **96**, (1956), 125–142.

[12] G.D. Anderson, M.K. Vamanamurthy, and M. Vuorinen, *Conformal invariants, inequalities, and quasiconformal maps*, Canadian Mathematical Society Series of Monographs and Advanced Texts. A Wiley-Interscience Publication. John Wiley & Sons, Inc., New York, 1997. xxviii+505 pp.

[13] C. Andreian Cazacu, *Foundations of quasiconformal mappings*, Handbook of complex analysis: Geometric Function Theory, Vol. 2, 687–753, Elsevier, Amsterdam, 2005.

[14] M. Arsenovič, V. Manojlović, and R. Näkki, *Boundary modulus of continuity and quasiconformal mappings*, Ann. Acad. Sci. Fenn. Math., **37**, (2012), 107–118.

[15] K. Astala, T. Iwaniec and G. Martin, *Elliptic partial differential equations and quasiconformal mappings in the plane*, Princeton Mathematical Series, **48**, Princeton University Press, Princeton, NJ, 2009. xviii+677 pp.

[16] A.F. Beardon, *The geometry of discrete groups*, Graduate Texts in Mathematics, **91**, Springer-Verlag, New York, 1983. xii+337 pp.

[17] A.F. Beardon and B. Maskit, *Limit points of Kleinian groups and finite sided fundamental polyhedra*, Acta Math., **132**, (1974), 1–12.

[18] B.A. Bhayo and M. Vuorinen, *On Mori's theorem for quasiconformal mappings in n-space*, Trans. Amer. Math. Soc., **363**, (2011), no. 11, 5703–5719.

[19] B. Bojarski and T. Iwaniec, *Another approach to Liouville Theorem*, Math. Nachr., **107**, (1982), 253–262.

[20] P. Bonfert-Taylor, R. Canary, G.J. Martin, and E.C. Taylor, *Quasiconformal homogeneity of hyperbolic manifolds*, Math. Ann., **331**, (2005), 281–295.

[21] P. Bonfert-Taylor, R. Canary, and E.C. Taylor, *Quasiconformal homogeneity after Gehring and Palka*, Comput. Methods Funct. Theory, **14**, (2014), 417–430.

[22] M. Bonk, B. Kleiner and S. Merenkov, *Rigidity of sets*, Amer. J. Math., **131**, (2009), 409–443.

[23] M. Bonk, *Uniformization of Sierpiński carpets in the plane*, Invent. Math., **186**, (2011), 559–665.

[24] A. Borel, *André Weil and algebraic topology*, André Weil (1906–1998). Gaz. Math., **80**, (1999), 63–74.

[25] M. Brown, *A proof of the generalised Schoenflies theorem*, Bull. Amer. Math. Soc., **66**, (1960), 74–76.

[26] D. Bump, *Lie groups*, second edition. Graduate Texts in Mathematics, **225**, Springer, New York, 2013.

[27] D.E. Callender, *Hölder continuity of n-dimensional quasiconformal mappings*, Pacific J. Math., **10**, (1960), 499–515.

[28] L. Carleson, *The extension problem for quasiconformal mappings*, Contributions to analysis (a collection of papers dedicated to Lipman Bers), pp. 39–47. Academic Press, New York, 1974.

[29] J.A. Clarkson, *Uniformly convex spaces*, Trans. Amer. Math. Soc., **40**, (1936), 396–414.

[30] A. Connes, D. Sullivan, and N. Teleman, *Quasiconformal mappings, operators on Hilbert space, and local formulae for characteristic classes*, Topology, **33**, (1994), 663–681.

[31] A. Connes, D. Sullivan, and N. Teleman, *Formules locales pour les classes de Pontrjagin topologiques*, C. R. Acad. Sci. Paris Sr. I Math., **317**, (1993), 521–526.

[32] E. de Giorgi, *Sulla differenziabilità e l'analiticità delle estremali degli integrali multipli regolari*, Mem. Accad. Sci. Torino. Cl. Sci. Fis. Mat. Nat., **3**, (1957), 25–43.

[33] S.K. Donaldson and D. P. Sullivan, *Quasiconformal 4-manifolds*, Acta Math., **163**, (1989), 181–252.

[34] A. Douady and C.J. Earle, *Conformally natural extension of homeomorphisms of the circle*, Acta Math., **157**, (1986), 23–48.

[35] R.D. Edwards and R. C. Kirby, *Deformations of spaces of imbeddings*, Ann. Math., **93**, (1971), 63–88.

[36] V.A. Efremovich and E.S. Tihomirova, *Equimorphisms of hyperbolic spaces*, Izv. Akad. Nauk SSSR Ser. Mat., **28**, (1964), 1139–1144.

[37] H. Federer, *Geometric measure theory*, Die Grundlehren der mathematischen Wissenschaften, **153** Springer-Verlag New York Inc., New York, 1969, xiv+676 pp.

[38] R. Finn and J. Serrin, *On the Hölder continuity of quasi-conformal and elliptic mappings*, Trans. Amer. Math. Soc., **89**, (1958), 1–15.

[39] R.H. Fox and E. Artin, *Some wild cells and spheres in three-dimensional space*, Annals of Math., **49**, (1948), 979–990.

[40] K.O. Friedrichs, *On Clarkson's inequalities*, Communications on Pure and Applied Math., **23**, (1970), 603–607.

[41] V.N. Dubinin, *Condenser capacities and symmetrization in geometric function theory*, translated from the Russian by N.G. Kruzhilin, Springer, Basel, 2014.

[42] B. Fuglede, *Extremal length and functional completion*, Acta Math., **98**, (1957), 171–219.

[43] D. Gabai, *Convergence groups are Fuchsian groups*, Ann. of Math., **136**, (1992), 447–510.

[44] D.B. Gauld and M.K. Vamanamurthy, *Quasiconformal extensions of mappings in n-space*, Ann. Acad. Sci. Fenn. Ser. A I Math., **3**, (1977), 229–246.

[45] D.B. Gauld and M.K. Vamanamurthy, *A special case of Schönflies' theorem for quasiconformal mappings in n-space*, Ann. Acad. Sci. Fenn. Ser. A I Math., **3**, (1977), 311–316.

[46] F.W. Gehring, *Rings and quasiconformal mappings in space*, Trans. Amer. Math. Soc., **103**, (1962), 353–393.

[47] F.W. Gehring, *The L^p-integrability of the partial derivatives of a quasiconformal mapping*, Acta Math., **130**, (1973), 265–277.

[48] F.W. Gehring, *Quasiconformal mappings in Euclidean spaces*, Handbook of complex analysis: Geometric Function Theory, Vol 2, 1–29, Elselvier, Amsterdam, 2005.

[49] F.W. Gehring and T. Iwaniec, *The limit of mappings with finite distortion*, Ann. Acad. Sci. Fenn. Math., **24**, (1999), 253–264.

[50] F.W. Gehring and O. Lehto, *On the total differentiability of functions of a complex variable*, Ann. Acad. Sci. Fenn. A I, **272**, (1959), 9pp.

[51] F.W. Gehring and G. J. Martin, *Discrete quasiconformal groups*, **I**, Proc. London Math. Soc., **55**, (1987), 331–358.

[52] F.W. Gehring and B.P. Palka, *Quasiconformally homogeneous domains*, J. Analyse Math., **30**, (1976), 172–199.

[53] F.W. Gehring and B. Osgood, *Uniform domains and the quasi-hyperbolic metric*, Journal d'Analyse Mathématique, **36**, (1979), 50–74.

[54] F.W. Gehring and J. Väisälä, *The coefficients of quasiconformality of domains in space*, Acta Math., **114**, (1965), 1–70.

[55] M. Giaquinta, *Multiple integrals in the calculus of variations and non-linear elliptic systems*, Annals of Math. Stud., **105**, Princeton University Press, 1983.

[56] M. Giaquinta and E. Giusti, *On the regularity of the minima of variational integrals*, Acta Math., **148**, (1982), 31–46.

[57] E. Ghys and P. de La Harpe, *Infinite groups as geometric objects (after Gromov)*, Ergodic theory, symbolic dynamics, and hyperbolic spaces (Trieste, 1989), 299–314, Oxford Sci. Publ., Oxford Univ. Press, New York, 1991.

[58] E. Ghys and P. de La Harpe, *Quasi-isométries et quasi-géodésiques*, Sur les groupes hyperboliques d'après Mikhael Gromov (Bern, 1988), 79–102, Progr. Math., 83, Birkhäuser Boston, Boston, MA, 1990.

[59] J. D. Gray and S. A. Morris, *When is a Function that Satisfies the Cauchy-Riemann Equations Analytic?* American Math. Monthly, **85**, (1978), 246 -256

[60] L. Greenberg, *Discrete subgroups of the Lorentz group*, Math. Scand., **10**, (1962), 85–107.

[61] M.J. Greenberg, J.R. Harper, *Algebraic topology. A first course*, Mathematics Lecture Note Series, **58**, Benjamin/Cummings Publishing Co., Inc., Advanced Book Program, Reading, Mass., 1981. xi+311 pp.

[62] M. Gromov, *Hyperbolic groups*, Essays in group theory, 75–263, Math. Sci. Res. Inst. Publ., **8**, Springer, New York, 1987.

[63] M. Gromov and P. Pansu, *Rigidity of lattices: an introduction*, Geometric topology: recent developments (Montecatini Terme, 1990), 39–137, Lecture Notes in Math., **1504**, Springer, Berlin, 1991.

[64] W. Gross, *Über das Fläshenmass von Punktmengen*, Monat. Math. Physik, **29**, (1918), 145–176.

[65] W. Gross, *Über das lineare Mass von Punktmengen*, Monat. Math. Physik, **29**, (1918), 177–193.

[66] H. Grötzsch, *Über die Verzerrung bei schlichten nichtconformen Abbildungen und übereine damit zusammenhängende Erweiterung des Picardschen Satzes*, Ber. Verh. Sächs. Akad. Wiss. Leipzig, **80**, (1928), 503–507.

[67] P. Gruber, *Convex Geometry*, Grundlehren der Mathematischen Wissenschaften, Springer Verlag, 2007.

[68] W.K. Hayman, *Symmetrization in the theory of functions*, Tech. Rep. no. 11, Navy Contract N6-ori-106 Task Order 5, Stanford University, Calif., 1950. 38 pp.

[69] W.K. Hayman, *Multivalent functions*, second edition. Cambridge Tracts in Mathematics, **110**, Cambridge University Press, Cambridge, 1994. xii+263 pp

[70] G. H. Hardy, J. E. Littlewood, and G. Pólya, *Inequalities*, Cambridge University Press, 1952.

[71] P. Hartman *On isometries and on a theorem of Liouville*, Math. Z., **69**, (1958), 202–210.

[72] G.A. Hedlund, *Fuchsian groups and transitive horocycles* , Duke Math. J., **2**, (1936), 530–542.

[73] J. Heinonen, *Lectures on Lipschitz Analysis*, Report., University of Jyväskylä Department of Mathematics and Statistics, **100**, University of Jyväskylä, Jyväskylä, 2005.

[74] J. Heinonen, T. Kilpeläinen, and O. Martio, *Nonlinear Potential Theory of Degenerate Elliptic Equations*, Oxford Math. Monographs, 1993.

[75] J. Heinonen and P. Koskela, *Definitions of quasiconformality*, Invent. Math., **120**, (1995), 61–79.

[76] J. Heinonen and P. Koskela, *Quasiconformal maps in metric spaces with controlled geometry*, Acta Math., **181**, (1998), 1–61.

[77] J. Hesse, *p-extremal length and p-measurable curve families*, Proc. Amer. Math. Soc., **53**, (1975), 356–360.

[78] J. Hesse, *A p-extremal length and p-capacity equality*, Ark. Mat., **13**, (1975), 131–144.

[79] E. Hopf, *A remark on linear elliptic differential equations of second order*, Proc. Amer. Math. Soc., **3**, (1952), 791–793.

[80] E. Hopf and N. Wiener, *Über eine Klasse singulärer Integralgleichungen*, Sitzungber. Akad. Wiss. Berlin, (1931), pp. 696–706
[81] H. Hopf, *Zum Clifford-Kleinschen Raumproblem*, Math. Annalen, **95**, (1926), 313–339.
[82] T. Iwaniec, *p-Harmonic tensors and quasiregular mappings*, Ann. of Math., **136**, (1992), 589–624.
[83] T. Iwaniec, *The failure of lower semicontinuity for the linear dilatation*, Bull. London Math. Soc., **30**, (1998), 55–61.
[84] T. Iwaniec and G. Martin, *Geometric function theory and non-linear analysis*, Oxford Mathematical Monographs. The Clarendon Press, Oxford University Press, New York, 2001. xvi+552. ISBN: 0-19-850929-4
[85] T. Iwaniec and G. Martin, *Quasiregular mappings in even dimensions*, Acta Math., **170**, (1993), 29–81.
[86] T. Iwaniec and J. Onninen, *n-harmonic mappings between annuli: the art of integrating free Lagrangians*, Mem. Amer. Math. Soc., **218**, (2012), no. 1023, viii+105 pp. ISBN: 978-0-8218-5357-3
[87] T. Iwaniec and J. Onninen, *An invitation to n-harmonic hyperelasticity*, Pure Appl. Math. Q., **7**, (2011), Special Issue: In honor of Frederick W. Gehring, Part 2, 319–343.
[88] D.S. Jerison and C.E. Kenig, *Hardy spaces, A_∞ and singular integrals on chord-arc domains*, Math. Scand., **50**, (1982), 221–247.
[89] F. John, *Rotation and strain*, Comm. Pure Appl. Math., **14**, (1961), 391–413.
[90] W. Killing, *Über die Clifford-Klein'schen Raumformen*, Math. Annalen, **39**, (1891), 257–278.
[91] B.N. Kimel'fe'ld, *Homogeneous regions on the conformal sphere*, Mat. Zametki, **8**, (1970), 321–328.
[92] R.C. Kirby and L.C. Siebenmann, *Foundational essays on topological manifolds, smoothings, and triangulations.* With notes by John Milnor and Michael Atiyah. Annals of Mathematics Studies, **88**. Princeton University Press, Princeton, N.J.; University of Tokyo Press, Tokyo, 1977.
[93] R. Kühnau, *Bibliography of geometric function theory*, Handbook of complex analysis: Geometric Function Theory, Vol. 2, 809–828, Elsevier, Amsterdam, 2005.
[94] T. Kuusalo, *Quasiconformal mappings without boundary extensions*, Ann. Acad. Sci. Fenn. Ser. A I Math., **10**, (1985), 331– 338.
[95] S. Lang, *Fundamentals of Differential Geometry*, Graduate Texts in Mathematics, **191**, Springer-Verlag, New York, 1999.
[96] T.G. Latfullin, *Geometric characterisation of the quasi-isometric image of a half plane*, Theory of mappings, its generalizations and applications, 116–126, "Naukova Dumka", Kiev, 1982.
[97] O. Lehto and K. Virtanen, *Quasiconformal mappings in the plane*, Springer–Verlag, 1973.
[98] J. Lindenstrauss and D. Preiss, *On Fréchet differentiability of Lipschitz maps between Banach spaces*, Annals of Math., **157**, (2003), 257–288.
[99] J. Liouville, *Extension au cas de trois dimensions de la question du tracé géographique*, Note IV in *Application de l'Anallyse à la Géometrie*, Bachelier, Paris, 1850, 609–616.
[100] J. Liouville, *Théorèm sur l'équation* $dx^2 + dy^2 + dz^2 = \lambda(d\alpha^2 + d\beta^2 + d\gamma^2)$, J. Math. Pures Appl., **1**, (15), (1850), 103.
[101] J. Lützen, *Joseph Liouville 1809–1882: master of pure and applied mathematics*, Studies in the History of Mathematics and Physical Sciences, **15**, Springer-Verlag, New York, 1990. xx+884 pp.
[102] J. Luukkainen, *Lipschitz and quasiconformal approximation of homeomorphism pairs*, Topology Appl., **109**, (2001), 1–40.
[103] J. Malý, *A simple proof of the Stepanov theorem on differentiability almost everywhere*, Exposition. Math., **17**, (1999), 59–61.
[104] G.J. Martin, *Infinite group actions on spheres*, Revista Mat. Iberoamericana, **4**, No. 3–4 (1988), 407–451.
[105] G.J. Martin, *Algebraic convergence of discrete isometry groups of negative curvature*, Pacific J. Math., **160**, (1992), 109–127.
[106] G.J. Martin, *The Hilbert-Smith conjecture for quasiconformal actions*, Electron. Res. Announc. Amer. Math. Soc., **5**, (1999), 66–70.
[107] G.J. Martin and R. Skora, *Group actions on $\mathbb{S}^2$*, Amer. J. Math., **111**, (1989), 387–402.

[108] P. Mattila, *Geometry of sets and measures in Euclidean spaces. Fractals and rectifiability*, Cambridge Studies in Advanced Mathematics, **44**, Cambridge University Press, Cambridge, 1995. xii+343 pp.

[109] O. Martio, S. Rickman, and J. Väisälä, *Definitions for Quasiregular mappings*, Ann. Acad. Sci. Fenn. Ser. A.I., **448**, (1969), 1-40.

[110] O. Martio, S. Rickman, and J. Väisälä, *Distortion and Singularities of Quasiregular Mappings*, Ann. Acad. Sci. Fenn. Ser. A I Math., **465**, (1970), 1–13.

[111] O. Martio, S. Rickman, and J. Väisälä, *Topological and metric properties of quasiregular mappings*, Ann. Acad. Sci. Fenn. Ser. A I Math., **488**, (1971), 1–31.

[112] O. Martio and J. Sarvas, *Injectivity theorems in plane and space*, Ann. Acad. Sci. Fenn. Ser. A I Math., **4**, (1979), 383–401.

[113] M. Mateljevicá, *Isoperimetric Inequality, F. Gehring's Problem on Linked Curves and Capacity*, Filomat, **29**, (3), (2015), 629–650.

[114] V. Maz'ya, *Sobolev spaces with applications to elliptic partial differential equations*, Second, revised and augmented edition, Grundlehren der Mathematischen Wissenschaften **342**, Springer, Heidelberg, 2011. xxviii+866 pp.

[115] M. J. McKemie, *Quasiconformal groups with small dilatation*, Ann. Acad. Sci. Fenn. Ser. A I Math., **12**, (1987), 95–118.

[116] E. E. Moise, *Geometric topology in dimensions 2 and 3*, Graduate Texts in Mathematics, **47**, Springer-Verlag, New York-Heidelberg, 1977. x+262 pp.

[117] G. Monge *Application de l'Anallyse à la Géometrie*, Bachelier, Paris, 1850.

[118] P. Montel, *Sur les suites infinies de fonctions*, Gauthier-Villars, Paris, 1907.

[119] D. Montgomery and L. Zippin, *Topological transformation groups*, Reprint of the 1955 original. Robert E. Krieger Publishing Co., Huntington, N.Y., 1974. xi+289 pp.

[120] A. Mori *An absolute constant in the theory of quasiconformal mappings*, J. Math. Soc. Japan, **8**, (1956), 156–166.

[121] A. Mori *On quasiconformality and pseudo-analyticity*, Trans. Amer. Math. Soc., **84**, (1957), 56–77.

[122] C.B. Morrey, *Multiple integrals in the calculus of variations*, Springer–Verlag, 1966.

[123] J. Moser, *On Harnack's theorem for elliptic differential equations*, Communications on Pure and Applied Mathematics, **14**, (1961), 577–591.

[124] J. Moser, *A rapidly convergent iteration method and non-linear partial differential equations. I & II*, Ann. Scuola Norm. Sup. Pisa, **20**, (1966), 265–315, Ann. Scuola Norm. Sup. Pisa, **20**, (1966), 499–535.

[125] G.D. Mostow, *Quasi-conformal mappings in n-space and the rigidity of hyperbolic space forms*, Inst. Hautes Études Sci. Publ. Math., **34**, (1968), 53–104.

[126] D. Mumford, C. Series and D. Wright, *Indra's pearls*, Cambridge University Press, 2002.

[127] J.R. Munkres, *Topology: a first course*, Prentice-Hall, Inc., Englewood Cliffs, N.J., 1975. xvi+413 pp.

[128] R. Näkki, *Cluster sets and quasiconformal mappings*, Complex Var. Elliptic Equ., **55**, (2010), 31–47.

[129] R. Näkki and O. Martio, *Boundary accessibility of a domain quasiconformally equivalent to a ball*, Bull. London Math. Soc., **36**, (2004), 115–118.

[130] R. Näkki and B.P. Palka, *Lipschitz conditions and quasiconformal mappings*, Indiana Univ. Math. J., **31**, (1982), 377–401.

[131] R. Näkki and B.P. Palka, *Swiss cheese, dendrites, and quasiconformal homogeneity*, Comput. Methods Funct. Theory, **14**, (2014), 525–539.

[132] G. Narasimhan, *Complex analysis in one variable*, Birkhäuser Boston, Inc., Boston, MA, 1985. xvi+266 pp.

[133] R. Nevanlinna *On differentiable mappings*, Analytic functions, pp. 3–9, Princeton Univ. Press, Princeton, N.J., 1960.

[134] R. Nevanlinna *Die konformen Selbstabbildungen des euklidischen Raumes*, Rev. Fac. Sci. Univ. Istanbul. Sér. A., **19**, (1954), 133–139.

[135] M. Ohtsuka, *Extremal length and precise functions*. With a preface by Fumi-Yuki Maeda. GAKUTO International Series. Mathematical Sciences and Applications, **19**. Gakkotosho Co., Ltd., Tokyo, 2003.

[136] G. Prasad, *Strong rigidity of Q-rank* 1 *lattices*, Invent. Math., **21**, (1973), 255–286.

[137] M.S. Raghunathan, *Discrete subgroups of Lie groups*, Ergebnisse der Mathematik und ihrer Grenzgebiete, **68**, Springer-Verlag, New York-Heidelberg, 1972. ix+227 pp

[138] J. G. Ratcliffe, *Foundations of hyperbolic manifolds*, Graduate Texts in Mathematics, **149**, Springer-Verlag, New York, 1994. xii+747 pp.

[139] R. Remmert and L. Kay, *Classical Topics in Complex Function Theory*, Springer-Verlag, 1998.

[140] Yu. G. Reshetnyak, *Space mappings with bounded distortion*, Sibirsk. Mat. Zh., **8**, (1967), 629–659.

[141] Yu. G. Reshetnyak, *Liouville's Theorem on conformal mappings under minimal regularity assumptions*, Sibirsk. Mat. Zh., **8**, (1967), 835–840.

[142] W. Rudin, *Real and Complex Analysis*, (third ed.), McGraw-Hill, 1986.

[143] J. Sarvas, *Symmetrization of condensers in n-space*, Ann. Acad. Sci. Fenn. Ser., A I, **522**, (1972), 44 pp.

[144] A. Selberg, *On discontinuous groups in higher-dimensional symmetric spaces*, Contributions to function theory (Internat. Colloq. Function Theory, Bombay, 1960), pp. 147–164. Tata Institute of Fundamental Research, Bombay.

[145] J. L. Schiff, *Normal families*, Universitext, Springer-Verlag, New York, 1993.

[146] V.A. Shlyk, *K-capacity and the Rado problem for mappings with bounded distortion*, Sibirsk. Mat. Zh., **31**, (1990), 179–186, 222; translation in Siberian Math. J., **31**, (1990), 152–158.

[147] V.A. Shlyk, *On the equality between p-capacity and p-modulus*, Sibirsk. Mat. Zh., **34**, (1993), 216–221; translation in Siberian Math. J., **34**, (1993), 1196–1200.

[148] R.J. Spatzier, *Harmonic Analysis in Rigidity Theory*, in Petersen, Karl E.; Salama, Ibrahim A., Ergodic Theory and its Connection with Harmonic Analysis, Proceedings of the 1993 Alexandria Conference, Cambridge University Press, (1995), 153–205,

[149] Stallings, *Group theory and three-dimensional manifolds*, A James K. Whittemore Lecture in Mathematics given at Yale University, 1969, Yale Mathematical Monographs, **4**, Yale University Press, New Haven, Conn.-London, 1971. v+65 pp.

[150] D. Sullivan, *Hyperbolic geometry and homeomorphisms*, Geometric topology (proceedings Georgia Topology Conf., Athens, Ga., 1977) edited by J. C. Cantrell, Academic Press, New York, N. Y.-London, 1979, 543–555.

[151] C.H. Taubes, *Differential geometry. Bundles, connections, metrics and curvature*, Oxford Graduate Texts in Mathematics, **23**, Oxford University Press, Oxford, 2011. xiv+298 pp

[152] W.P. Thurston *Three-dimensional geometry and topology*, Vol. 1. Edited by Silvio Levy. Princeton Mathematical Series, **35**, Princeton University Press, Princeton, NJ, 1997. x+311 pp.

[153] P. Tukia, *Extension of quasisymmetric and Lipschitz embeddings of the real line into the plane*, Ann. Acad. Scie. Fenn. Ser. A. I. Math., **6**, (1981), 89–94.

[154] P. Tukia, *The planar Schönfliess theorem for Lipschitz maps*, Ann. Acad. Sci. Fenn. Ser. A I Math., **5**, 1980, 49–72.

[155] P. Tukia, *On isomorphisms of geometrically finite Möbius groups*, Inst. Hautes Études Sci. Publ. Math., **61**, (1985), 171–214.

[156] P. Tukia, *A quasiconformal group not isomorphic to a Möbius group*, Ann. Acad. Sci. Fenn. Ser. A I Math., **6**, (1981), 149–160.

[157] P. Tukia and J. Väisälä, *Lipschitz and quasiconformal approximation and extension*, Ann. Acad. Sci. Fenn. Ser. A I Math., **6**, (1981), 303–342.

[158] P. Tukia and J. Väisälä, *Quasiconformal extension from dimension n to n + 1*, Ann. of Math., **115**, (1982), 331–348.

[159] P. Tukia and J. Väisälä, *Bi-Lipschitz extensions of maps having quasiconformal extensions*, Math. Ann., **269**, (1984), 561–572.

[160] J. Väisälä, *Lectures on n-dimensional quasiconformal mappings*, Lecture Notes in Math., **229**, Springer-Verlag, 1971.

[161] J. Väisälä, *On quasiconformal mappings of a ball*, Ann. Acad. Sci. Fenn. Ser. A I, **304**, (1961), 7 pp.

[162] J. Väisälä, *On quasiconformal mappings in space*, Ann. Acad. Sci. Fenn. Ser. A I, **298**, (1961), 36 pp.

[163] J. Väisälä, *Homeomorphisms of bounded length distortion*, Ann. Acad. Sci. Fenn. Ser. A I Math., **12**, (1987), 303–312.

[164] J. Väisälä, *Quasiconformal maps of cylindrical domains*, Acta Math., **162**, (1989), 201–225.

[165] M. Vuorinen, *Conformal geometry and quasiregular mappings*, Lecture Notes in Mathematics, **1319**, Springer-Verlag, Berlin, 1988. xx+209 pp.

[166] A. Weil, *On discrete subgroups of Lie groups*, Annals of Math., **72**, (1960), 369–384.

[167] A. Weil, *On discrete subgroups of Lie groups. II*, Annals of Math., **75**, (1962), 578–602.

[168] G.T. Whyburn, *Analytic Topology*, American Mathematical Society Colloquium Publications 28, American Mathematical Society, Providence, R.I., 1942.

[169] J. Wolf, *Spaces of Constant Curvature*, Third edition. Publish or Perish, Inc., Boston, Mass., 1974. xv+408 pp.

[170] W.P. Ziemer, *Extremal length and p-capacity*, Michigan Math. J., **16**, (1969), 43–51.

[171] V.A. Zorich, *The theorem of M.A. Lavrent'ev on quasiconformal mappings in space*, Mat. Sb., **74**, (1967), 417–433.

[172] V.A. Zorich, *Homeomorphity of space quasiconformal mappings* (Russian) Dokl. Acad. Nauk. SSSR, **176**, (1967), 31–34; (English) Sov. Math. Dokl. **8**, (1967), 1039–1042.

Index